Springer Handbooks of Computational Statistics

Series Editors

James E. Gentle
Wolfgang K. Härdle
Yuichi Mori

For further volumes:
http://www.springer.com/series/7286

Jin-Chuan Duan • Wolfgang Karl Härdle
James E. Gentle
Editors

Handbook of Computational Finance

Springer

Editors
Jin-Chuan Duan
National University of Singapore
Risk Management Institute
21 Heng Mui Keng Terrace, Level 4
119613 Singapore
Singapore
bizdjc@nus.edu.sg

Prof. James E. Gentle
George Mason University
Department of Computational
and Data Sciences
University Drive 4400
22030 Fairfax Virginia
USA
jgentle@gmu.edu

Prof.Dr. Wolfgang Karl Härdle
Ladislaus von Bortkiewicz Chair of Statistics
C.A.S.E. Centre for Applied Statistics and Economics
School of Business and Economics
Humboldt-Universität zu Berlin
Unter den Linden 6
10099 Berlin
Germany
haerdle@wiwi.hu-berlin.de

ISBN 978-3-642-17253-3 e-ISBN 978-3-642-17254-0
DOI 10.1007/978-3-642-17254-0
Springer Heidelberg Dordrecht London New York

Library of Congress Control Number: 2011937712

Cover design: WMXDesign GmbH

Printed on acid-free paper

Springer is part of Springer Science+Business Media (www.springer.com)

Contents

Part I Introduction

1 **Computational Finance: An Introduction** 3
Jin-Chuan Duan, James E. Gentle, and Wolfgang Karl Härdle

Part II Asset Pricing Models

2 **Modeling Asset Prices** 15
James E. Gentle and Wolfgang Karl Härdle

3 **Diffusion Models of Asset Prices** 35
Jérôme Detemple and Marcel Rindisbacher

4 **Jump-Diffusion Models Driven by Lévy Processes** 61
José E. Figueroa-López

5 **Multivariate Time Series Models for Asset Prices** 89
Christian M. Hafner and Hans Manner

6 **Option Data and Modeling BSM Implied Volatility** 117
Matthias R. Fengler

7 **Interest Rate Derivatives Pricing with Volatility Smile** 143
Haitao Li

8 **Volatility Investing with Variance Swaps** 203
Wolfgang Karl Härdle and Elena Silyakova

Part III Statistical Inference in Financial Models

9 **Evaluation of Asset Pricing Models Using Two-Pass Cross-Sectional Regressions** 223
Raymond Kan and Cesare Robotti

10 Parametric Estimation of Risk Neutral Density Functions 253
Maria Grith and Volker Krätschmer

11 Nonparametric Estimation of Risk-Neutral Densities 277
Maria Grith, Wolfgang Karl Härdle, and Melanie Schienle

12 Value at Risk Estimation 307
Ying Chen and Jun Lu

13 Volatility Estimation Based on High-Frequency Data 335
Christian Pigorsch, Uta Pigorsch, and Ivaylo Popov

14 Identifying Jumps in Asset Prices 371
Johan Bjursell and James E. Gentle

15 Simulation-Based Estimation Methods for Financial Time Series Models 401
Jun Yu

Part IV Computational Methods

16 Filtering Methods 439
Andras Fulop

17 Fitting High-Dimensional Copulae to Data 469
Ostap Okhrin

18 Numerical Methods for Nonlinear PDEs in Finance 503
Peter A. Forsyth and Kenneth R. Vetzal

19 Numerical Solution of Stochastic Differential Equations in Finance 529
Timothy Sauer

20 Lattice Approach and Implied Trees 551
Rüdiger U. Seydel

21 Efficient Options Pricing Using the Fast Fourier Transform 579
Yue Kuen Kwok, Kwai Sun Leung, and Hoi Ying Wong

22 Dynamic Programming and Hedging Strategies in Discrete Time 605
Shih-Feng Huang and Meihui Guo

23 Approximation of Dynamic Programs 633
Michèle Breton and Javier de Frutos

24 Computational Issues in Stress Testing 651
Ludger Overbeck

25 Portfolio Optimization 675
Jérôme Detemple and Marcel Rindisbacher

26 Low-Discrepancy Simulation 703
Harald Niederreiter

27 Introduction to Support Vector Machines and Their Applications in Bankruptcy Prognosis 731
Yuh-Jye Lee, Yi-Ren Yeh, and Hsing-Kuo Pao

Part V Software Tools

28 MATLAB® as a Tool in Computational Finance 765
James E. Gentle and Angel Martinez

29 R as a Tool in Computational Finance 781
John P. Nolan

Contributors

Johan Bjursell George Mason University, Fairfax, VA 22030, USA, cbjursel@gmu.edu

Michèle Breton GERAD, HEC Montréal, 3000, chemin de la Côte-Sainte-Catherine, Montréal, QC, Canada H3T 2A7, michele.breton@hec.ca

Ying Chen Department of Statistics and Applied Probability, National University of Singapore, Singapore 117546, Singapore, stacheny@nus.edu.sg

Jérôme Detemple Boston University School of Management, Boston University, 595 Commonwealth Avenue, Boston, MA 02215, USA, detemple@bu.edu

Jin-Chuan Duan National University of Singapore, Singapore 117546, Singapore, bizdjc@nus.edu.sg

Matthias R. Fengler University of St. Gallen, Gallen, Switzerland, matthias.fengler@unisg.ch

José E. Figueroa-López Department of Statistics, Purdue University, West Lafayette, IN 47907-2066, USA, figueroa@stat.purdue.edu

Peter A. Forsyth Cheriton School of Computer Science, University of Waterloo, Waterloo, ON, Canada, paforsyt@uwaterloo.ca

Javier de Frutos GERAD and Dpto de Matemática Aplicada, Universidad de Valladolid, Valladolid, Palencia, Soria and Segovia, Spain, frutos@mac.uva.es

Andras Fulop ESSEC Business School, Paris, France, fulop@essec.fr

James E. Gentle George Mason University, Fairfax, VA 22030, USA, jgentle@gmu.edu

Maria Grith Ladislaus von Bortkiewicz Chair of Statistics, Humboldt-Universität zu Berlin, Spandauer Straße 1, 10178 Berlin, Germany, gritmari@wiwi.hu-berlin.de

Meihui Guo Department of Applied Mathematics, National Sun Yat-sen University, Kaohsiung, Taiwan, guomh@math.nsysu.edu.tw

Christian M. Hafner Institut de statistique and CORE, Université catholique de Louvain, Voie du Roman Pays 20, B-1348 Louvain-la-Neuve, Belgium, christian.hafner@uclouvain.be

Wolfgang Karl Härdle Ladislaus von Bortkiewicz Chair of Statistics and CASE - Center for Applied Statistics and Economics, Humboldt-Universität zu Berlin, Spandauer Straße 1, 10178 Berlin, Germany
and
Graduate Institute of Statistics, CDA - Centre for Complex Data Analysis, National Central University, No. 300, Jhongda Rd., Jhongli City, Taoyuan County 32001, Taiwan, (R.O.C.), haerdle@wiwi.hu-berlin.de

Shih-Feng Huang, Department of Applied Mathematics, National University of Kaohsiung, Kaohsiung, Taiwan, huangsf@nuk.edu.tw

Raymond Kan Joseph L. Rotman School of Management, University of Toronto, 105 St. George Street, Toronto, Ontario, Canada M5S 3E6, kan@chass.utoronto.ca

Volker Krätschmer Weierstrass Institute of Applied Analysis and Stochastics, Mohrenstrasse 39, D-10117 Berlin, Germany, kraetsch@wias-berlin.de

Yue Kuen Kwok Hong Kong University of Science and Technology, Clear Water Bay, NT, Hong Kong, maykwok@ust.hk

Yuh-Jye Lee Department of Computer Science and Information Engineering, National Taiwan University of Science and Technology, Taipei, Taiwan 10607, yuh-jye@mail.ntust.edu.tw

Kwai Sun Leung The Chinese University of Hong Kong, Sha Tin, NT, Hong Kong, ksleung@se.cuhk.edu.hk

Haitao Li Professor of Finance, Stephen M. Ross School of Business, University of Michigan, Ann Arbor, MI 48109, htli@umich.edu

Jun Lu Department of Statistics and Applied Probability, National University of Singapore, Singapore 117546, Singapore, g0700712@nus.edu.sg

Hans Manner Department of Social and Economic Statistics, University of Cologne, manner@statistik.uni-koeln.de

Angel Martinez Strayer University, Fredericksburg, VA 22406-1094, USA

Harald Niederreiter RICAM, Austrian Academy of Sciences, Altenbergerstr. 69, A-4040 Linz, Vienna, Austria, ghnied@gmail.com

John P. Nolan Department of Mathematics and Statistics, American University, Washington, DC 20016-8050, USA, jpnolan@american.edu

Ostap Okhrin Ladislaus von Bortkiewitcz Chair of Statistics, C.A.S.E. - Center of Applied Statistics and Economics, Humboldt-Universität zu Berlin, D-10178 Berlin, Germany, ostap.okhrin@wiwi.hu-berlin.de

Ludger Overbeck Institute of Mathematics, University of Giessen, 35390 Gießen, Hessen, Germany, ludger.overbeck@math.uni-giessen.de

Hsing-Kuo Pao Department of Computer Science and Information Engineering, National Taiwan University of Science and Technology, Taipei, Taiwan 10607, pao@mail.ntust.edu.tw

Christian Pigorsch Department of Economics, University of Bonn, Adenauerallee 24-42, D-53113 Bonn, Germany, christian.pigorsch@uni-bonn.de

Uta Pigorsch Department of Economics, University of Mannheim, L7, 3-5, D-68131 Mannheim, Baden-Württemberg, Germany, uta.pigorsch@vwl.uni-mannheim.de

Ivaylo Popov Business School of the University of Mannheim, L5, 5, D-68131 Mannheim, Baden-Württemberg, Germany, ipopov@mail.uni-mannheim.de

Marcel Rindisbacher Boston University School of Management, Boston University, 595 Commonwealth Avenue, Boston, MA 02215, USA, rindisbm@bu.edu

Cesare Robotti Research Department, Federal Reserve Bank of Atlanta, 1000 Peachtree St. N.E., Atlanta, GA 30309, USA, cesare.robotti@atl.frb.org

Timothy Sauer Department of Mathematics, George Mason University, Fairfax, VA 22030, USA, tsauer@gmu.edu

Melanie Schienle Chair of Econometrics and CASE - Center for Applied Statistics and Economics, Humboldt-Universität zu Berlin, Spandauer Straße 1, 10178 Berlin, Germany, melanie.schienle@wiwi.hu-berlin.de

Rüdiger U. Seydel Mathematisches Institut der Universität zu Köln, Weyertal 86, D-50931 Köln, Germany, seydel@math.uni-koeln.de, www.compfin.de

Elena Silyakova Ladislaus von Bortkiewicz Chair of Statistics and CASE-Center for Applied Statistics and Economics, Humboldt-Universität zu Berlin, Spandauer Straße 1, 10178 Berlin, Germany, silyakoe@cms.hu-berlin.de

Kenneth R. Vetzal School of Accounting and Finance, University of Waterloo, Waterloo, ON, Canada, kvetzal@uwaterloo.ca

Hoi Ying Wong The Chinese University of Hong Kong, Sha Tin, NT, Hong Kong, hywong@cuhk.edu.hk

Jun Yu School of Economics, Lee Kong Chian School of Economics and Sim Kee Boon Institute for Financial Economics, Singapore Management University, 90 Stamford Road, Singapore 178903, yujun@smu.edu.sg

Yi-Ren Yeh Research Center for Information Technology Innovation, Academia Sinica, Taipei 11529, Taiwan, yryeh@citi.sinica.edu.tw

Part I
Introduction

Chapter 1
Computational Finance: An Introduction

Jin-Chuan Duan, James E. Gentle, and Wolfgang Karl Härdle

1.1 Computational Statistics, Finance, and Computational Finance

This book is the fourth volume of the *Handbook of Computational Statistics*. As with the other handbooks in the series, it is a collection of articles on specific aspects of the broad field, written by experts in those areas. The purpose is to provide a survey and summary on each topic, ranging from basic background material through the current frontiers of research. The development of the field of computational statistics has been rather fragmented. We hope that the articles in this handbook series can provide a more unified framework for the field.

The methods of computational statistics have pervaded most areas of application, particularly such data-rich fields as finance. The tools of computational statistics include efficient computational algorithms, graphical methods, simulation, and resampling. These tools allow processing of massive amounts of data and simulation of complex data-generating processes, leading to better understanding of those processes.

J.-C. Duan (✉)
Department of Mathematics, National University of Singapore, Singapore 119077
e-mail: bizdjc@nus.edu.sg

J.E. Gentle
Department of Computer Science, George Mason University, VA, USA
e-mail: jgentle@gmu.edu

W.K. Härdle
Ladislaus von Bortkiewicz Chair of Statistics and CASE - Center for Applied Statistics and Economics, Humboldt-Universität zu Berlin, Spandauer Straße 1, 10178 Berlin, Germany
and
Graduate Institute of Statistics, CDA - Centre for Complex Data Analysis, National Central University, No. 300, Jhongda Rd., Jhongli City, Taoyuan County 32001, Taiwan, (R.O.C.)
e-mail: haerdle@wiwi.hu-berlin.de

J.-C. Duan et al. (eds.), *Handbook of Computational Finance*, Springer Handbooks of Computational Statistics, DOI 10.1007/978-3-642-17254-0_1,

1.1.1 Finance

The field of finance is concerned with asset prices, how they vary over time, and risk associated with that variation.

Anything that is openly traded has a market price that may be more or less than some "fair" price. For shares of corporate stock, the fair price is likely to be some complicated function of intrinsic (or "book") current value of identifiable assets owned by the company, expected rate of growth, future dividends, and other factors. Some of these factors that affect the price can be measured at the time of a stock transaction, or at least within a relatively narrow time window that includes the time of the transaction. Most factors, however, relate to future expectations and to subjective issues, such as current management and corporate policies, that are believed to affect future financial performance of the corporation.

There are many motivating factors for the study of financial data. Investors, speculators, and operators seek an advantage over others in the trading of financial assets. Academics often find a different motive for studying financial data just because of the challenges of developing models for price movements. Finally, government regulators and others are motivated by an interest in maintaining a fair and orderly market. With more and more retirees depending on equity investments for their livelihood, it becomes very important to understand and control the risk in portfolios of corporate stock.

Study of characteristics of a corporation and their relationship to the current and future financial status of the corporation is a major topic in the field of finance. A general objective is to measure "fair price", "value", or "worth" of corporate stock. The price, either the market price or the fair price, varies over time.

In the subfield of finance known variously as "mathematical finance", "financial engineering", or "computational finance", a major objective is to develop and study models of the *movement* of the *market price* of basic financial assets. A second important objective in computational finance is to develop useful models of the *fair price* of derivative assets; that is, financial assets that convey rights and/or obligations to execute prespecified transactions in some basic underlying assets. The fair price of such a derivative asset depends on the expected movements of the market price of the underlying asset.

1.1.2 Models for Movement of Prices of Basic Assets

We first consider the question of how market prices change over time.

As a first simple approximation, we assume discrete time, $t_0, t_1, t_2, \ldots$ or $t, t+1, t+2, \ldots$. In more realistic models, we assume continuous time, but in applications of course, we have to revert to some type of discreteness. We also generally study aggregated prices or index prices The prices of individual securities, even if they follow similar models, behave in a way peculiar to the security. There are more security-specific extraordinary events that affect the price of a given security, than there are extraordinary events that affect the overall market.

A stochastic model of the price of a stock (or index) may view the price as a random variable that depends on previous prices and some characteristic parameters of the particular stock. For example, in discrete time:

$$S_{t+1} = f(S_t, \mu, \sigma), \tag{1.1}$$

where t indexes time, μ and σ are parameters, and f is some function that contains a random component. The randomness in f may be assumed to reflect all variation in the price that is not accounted for in the model.

In the absence of exogenous forces, the movement of stock prices is similar to a random walk with steps that are neither independent nor identically distributed. A simple random walk could take the prices negative. Also, it seems intuitive that the random walk should have a mean step size that is proportional to the magnitude of the price. The proportional rate of change, $(S_{t+1}-S_t)/S_{t+1}$, therefore, is more interesting than the prices themselves, and is more amenable to fitting to a probability model.

Two different types of models of price movements are obvious choices. One type of model is based on a stochastic diffusion process and another uses an autoregressive moving average process. Each approach has a wide range of variations. The most difficult choice is the probability distribution for the random components in the model. The vexing questions relate to the tails of the distributions and the nature of the dependencies of random elements. Assumptions of a single probability model or of independence rarely can be supported by observational data.

1.1.3 Pricing Models for Derivative Assets

There are many kinds of derivative assets that are based on the price of some underlying asset, which may be a commodity such as an extractive or agricultural product, a financial security, or a financial index. Although these derivative assets are often used by speculators looking for fast gains, an important role of derivatives is to add a measure of order to financial markets by allowing for hedging and spreading of risk.

A derivative can be either a right or an obligation either to buy an asset, to sell an asset, or otherwise to settle an obligation at some future time or within some fixed time period.

The fair price of a derivative asset depends on the movement of the market price of the underlying asset or index that the right or obligation of the settlement is based on.

The value of a derivative is difficult to assess not only because of the uncertainty of the price of the underlying, but also because of the range options the holder of the derivative may have in closing out the derivative.

There are various approaches to setting a fair price for derivatives. These approaches generally are based on some financial principle, such as an assumption that prices of various securities, even in different markets, have rational relationships to each other, that is, an assumption that arbitrage is not possible. Because the value of a derivative asset depends strongly on the passage of time, some fixed time-value measure, that is, a risk-free return, must be assumed. Following these kinds of assumptions, a hedged portfolio that includes the derivative asset can be constructed under an additional assumption of a complete market. The balance required by the hedge portfolio yields the relative values of the assets.

1.1.4 Statistical Inference in Financial Models

Statistical inference means the use of observational data to make decisions about the process that yields the data. Statistical inference involves development of models of processes that exhibit some type of randomness. These models generally consist of a systematic component, possibly with various parameters, and a stochastic component.

Prior to statistical inference about a model, exploratory methods of data analysis are employed in order to establish some general aspects of the data-generating process. Following the exploratory data analysis, the development of a model generally begins with some assumptions about the nature of the systematic and stochastic components and their relationships with each other.

Formal statistical inference usually begins either with estimation or testing of some parameters in the systematic component of the model. This is followed by inference about the stochastic component and comparison of the model residuals with observational data. The assessment of the adequacy of the model by study of the residuals is one of the most important types of statistical inference and is the basis for the feedback loops that are a vital component of model building.

Simulation methods are used to study the models. Because the stochastic components of the models are so important, to use the models for prediction often requires Monte Carlo simulation.

1.1.5 Computational Methods for Financial Models

Many problems in finance, especially those involving pricing of financial assets, cannot be formulated into simple models. The numerical methods for dealing with such models are generally computationally-intensive.

The relevant areas of numerical analysis for financial model include most of the standard methods: optimization, filtering, solution of differential equations, and simulation. Computations in linear algebra are generally basic to most of these more specific numerical methods.

1.2 The Organization and Contents of This Handbook

The purpose of this handbook is to provide a survey of the important concepts and methods of computational finance. A glance at the table of contents reveals a wide range of articles written by experts in various subfields. The articles are expository, taking the reader from the basic concepts to the current research trends.

1.2.1 Organization

After this introductory part, this handbook is divided into four parts: "Pricing Models", "Statistical Inference in Financial Models", "Computational Methods", and "Software Tools". The chapters in each part generally range from more basic topics to more specialized topics, but in many cases there is may be no obvious sequence of topics. There often considerable interrelationships of a chapter in one part with chapters in other parts of this handbook.

1.2.2 Asset Pricing Models (Part II)

The second part begins with an article by Gentle and Härdle that surveys the general approaches to modeling asset prices. The next three chapters address specific approaches. First, Detemple and Rindisbacher consider general diffusion models, and then Figueroa-López discusses diffusion models with a superimposed jump component, which also allows for stochastic volatility and clustering of volatility. Next, Hafner and Manner discuss multivariate time series models, such as GARCH and linear factor models, that allow for stochastic volatility.

The next two chapters in Part II address pricing of derivatives. Fengler reviews the basic Black-Scholes-Merton (BSM) option pricing formula for stock options, and then discusses the concept of implied volatility, which derives from an inverse of the formula using observed prices of options. Especially since 1987, it has been observed that a plot of implied volatility versus moneyness exhibits a convex shape, or "smile". The volatility smile, or volatility surface when a term structure dimension is introduced, has been a major impetus for the development of option pricing models. For other derivatives, the term structure of implied volatility or its relationship to moneyness has not been as thoroughly investigated. Li explores the "smile" of implied volatility in the context of interest rate derivatives.

Financial markets can be built on anything that varies. If volatility varies, then it can be monetized. In the final chapter of Part II, Härdle and Silyakova discuss the market in variance swaps.

1.2.3 Statistical Inference in Financial Models (Part III)

While Part II addressed the descriptive properties of financial models, the chapters in Part III consider issues of statistical inference, estimation and testing, with these models. The first chapter in this section, by Kan, develops criteria for evaluating the correspondence of asset pricing models to actual observed prices, and the discusses statistical methods of comparing one model with another. The next two chapters consider the general problem of estimation of the probability density of asset option prices, both under the assumption that the prices conform to the risk-neutral valuation principle. The first of these chapters, by Kraetschmer and Grith, uses parametric models, and the other chapter, by Härdle, Grith, and Schienle, uses nonparametric and semiparametric models.

A topic that is receiving a great deal of attention currently is value at risk (VaR). Chen and Lu provide a survey of recent developments in estimation of VaR and discuss the robustness and accuracy of the methods, comparing them using simulated data and backtesting with real data.

An important parameter in any financial model is the volatility, whether it is assumed to be constant or stochastic. In either case, data-based estimates of its magnitude or of its distribution are necessary if the model is to be used. (As noted above, the model can be inverted to provide an "implied volatility", but this is not of much value for the primary purpose for which the model was developed.) The basic statistic for estimation of the volatility is "realized volatility", which is just the standard deviation of a sample of returns. The sample is actually a sequence, and hence cannot be considered a random sample. Furthermore, the sampling interval has a very large effect on the estimator. While certain statistical properties of the realized volatility require ever-increasing frequencies, other effects ("noise") become confounded with the volatility at high frequencies. Christian Pigorsch, Uta Pigorsch, and Popov address the general problem of estimation of the volatility using realized volatility at various frequencies.

Bjursell and Gentle discuss the problem of identifying jumps in a jump-diffusion model. Their focus is energy futures, particularly in brief periods that include the release of official build statistics.

Several of the chapters in Parts II and III use simulation to illustrate the points being discussed. Simulation is also one of the most useful tools for statistical inference in financial models. In the final chapter of Part III, Yu discusses various simulation-based methods for use with financial time series models.

1.2.4 Computational Methods (Part IV)

Many financial models require extensive computations for their analysis. Efficient numerical methods have thus become an important aspect of computational finance.

As we indicated above, statistical models generally consist of systematic components ("signals") and random components ("noise"), and a primary aspect of statistical analysis is to identify the effects of these components. An important method of doing this is filtering. In the first chapter of Part IV, Fulop describes filtering techniques in the setting of a hidden dynamic Markov process, which underlies many financial models.

The stochastic components of financial models are often assumed to have some simple parametric form, and so fitting the probability model to empirical data merely involves the estimation of parameters of the model. Use of nonparametric models often results in greater fidelity of the model to observational reality. The greatest problem in fitting probability models to empirical data, however, occurs when multiple variables are to be modeled. The simple assumption of independence of the variables often leads to gross underestimation of risk. Simple variances and covariances do not adequately capture the relationships. An effective method of modeling the relationships of the variables is by use of copulae. These are not as simple to fit to data as are variances and covariances, especially if the number of variables is large. Okhrin discusses the use of copulae and numerical methods for fitting high-dimensional copulae to data.

The next two chapters in Part IV discuss numerical methods of solution of partial differential equations (PDEs) in finance. Forsyth and Vetzal describe numerical solution of nonlinear deterministic PDEs, and Sauer discusses numerical methods for stochastic partial differential equations (SDEs). Both of these chapters are focused on the financial applications of PDEs.

One of the most important problems in computational finance is the development of accurate and practical methods for pricing derivatives. Seydel discusses lattice or tree-based methods, and Kwok, Leung, and Wong discuss the use of discrete Fourier transforms implemented by the fast Fourier transform (FFT) of course.

Some of the earliest studies in computational finance led to the development of dynamic programming. This continues to be an important tool in computational finance. Huang and Guo discuss its use in hedging strategies, and Breton and Frutos describe the use of approximation of dynamic programs for derivative pricing.

An important concern about any model or inference procedure is the robustness to unusual situations. A model that serves very well in "normal" times may be completely inappropriate in other regimes. Evaluation of models involves "stress testing"; that is, assessment of the validity of the model in unusual situations, such as bubble markets or extended bear markets. Overbeck describes methods of stress testing for risk management.

One of the earliest problems to which modern computational methods were addressed is that of selection of an optimal portfolio, given certain characteristics of the available securities and restrictions on the overall risk. Rindisbacher and Detemple discuss portfolio optimization in the context of modern pricing models.

As Yu discussed in Part III, simulation-based methods have widespread applications in financial models. The efficiency of these computationally-intensive methods can be greatly increased by the use of better methods of covering the sample space. Rather than simulating randomness of sampling, it is more efficient to proceed through the sample space deterministically in a way that guarantees a certain uniformity of coverage. In the next chapter of Part IV, Niederreiter describes the concepts of low discrepancy simulation, and discusses how quasi Monte Carlo methods can be much more efficient than ordinary Monte Carlo.

An important tool in computational finance is statistical learning; that is, the identification of rules for classification of features of interest. There are various approaches to statistical learning, and in the last chapter of Part IV, Lee, Yeh, and Pao discuss support vector machines, which is one of the most useful of the methods of classification.

1.2.5 Software Tools (Part V)

Financial modeling and analysis require good software tools. In Part V Gentle and Martinez briefly discuss the various types of software available for financial applications and then proceed to discuss one specific software package, Matlab. This flexible and powerful package is widely used not only for financial analyses but for a range of scientific applications. Another software package, which is open source and freely distributed, is Nolan discusses R, and gives several examples of its use in computational finance.

1.3 The Computational Statistics Handbook Series

The first handbook in the series, published in 2004, was on concepts and fundamentals. It had thirty expository chapters, written by experts in various subfields of computational statistics. The chapters, which were organized into parts on statistical computing, statistical methodology, and applications (including financial applications), covered a wide range of topics and took the reader from the basic concepts to the current research trends. As mentioned above, there are several chapters in this more fundamental handbook, such as those in the part on statistical computing, that provide more background on the topics of this handbook on computational finance.

The handbook on concepts and fundamentals set the stage for future handbooks that will go more deeply into the various subfields of computational statistics. These handbooks will each be organized around either a specific class of theory and methods, or else around a specific area of application. Two subsequent handbooks on specific topics in computational statistics have appeared, one on visualization and one on partial least squares.

The current handbooks in the Springer Handbooks of Computational Statistics, published by Springer in Berlin, Heidelberg, and New York are the following.

- *Handbook of Computational Statistics. Concepts and Methods*, edited by James E. Gentle, Wolfgang Härdle, and Yuichi Mori (2004).
- *Handbook of Data Visualization*, edited by Chun-houh Chen, Wolfgang Härdle, and Antony Unwin (2008).
- *Handbook of Partial Least Squares. Concepts, Methods and Applications in Marketing and Related Fields*, edited by Vincenco Esposito Vinzi, Wynne W. Chin, Jörg Henseler, Huiwen Wang (2009).

Part II
Asset Pricing Models

Chapter 2
Modeling Asset Prices

James E. Gentle and Wolfgang Karl Härdle

Abstract As an asset is traded, its varying prices trace out an interesting time series. The price, at least in a general way, reflects some underlying value of the asset. For most basic assets, realistic models of value must involve many variables relating not only to the individual asset, but also to the asset class, the industrial sector(s) of the asset, and both the local economy and the general global economic conditions. Rather than attempting to model the value, we will confine our interest to modeling the price. The underlying assumption is that the price at which an asset trades is a "fair market price" that reflects the actual value of the asset.

Our initial interest is in models of the price of a basic asset, that is, not the price of a derivative asset. Usually instead of the price itself, we consider the relative change in price, that is, the rate of return, over some interval of time.

The purpose of asset pricing models is not for prediction of future prices; rather the purpose is to provide a description of the stochastic behavior of prices. Models of price changes have a number of uses, including, for investors, optimal construction of portfolios of assets and, for market regulators, maintaining a fair and orderly market. A major motivation for developing models of price changes of given assets is to use those models to develop models of fair value of derivative assets that depend on the given assets.

J.E. Gentle (✉)
Department of Computational and Data Sciences, George Mason University, Fairfax, VA, USA
e-mail: jgentle@gmu.edu

W.K. Härdle
Ladislaus von Bortkiewicz Chair of Statistics and CASE - Center for Applied Statistics and Economics, Humboldt-Universität zu Berlin, Spandauer Straße 1, 10178 Berlin, Germany
and
Graduate Institute of Statistics, CDA - Centre for Complex Data Analysis, National Central University, No. 300, Jhongda Rd., Jhongli City, Taoyuan County 32001, Taiwan, (R.O.C.)
e-mail: haerdle@wiwi.hu-berlin.de

J.-C. Duan et al. (eds.), *Handbook of Computational Finance*, Springer Handbooks of Computational Statistics, DOI 10.1007/978-3-642-17254-0_2,

The rate of return has a strong stochastic component, and in this chapter, we describe various stochastic models of the rate of return. We also briefly discuss statistical inference in these models, and applications of these models for pricing derivative assets. Our presentation is quite general. We refer to readily-available literature, some in the present volume, for details on the analysis and applications of the models.

The models we consider in this chapter are for the prices of a single asset, although, of course, that asset may be a portfolio of individual assets. Pricing models of more than one asset must take into account the correlations among their prices. Multivariate pricing models are discussed by Hafner and Manner (2010, this volume).

In most models of asset prices such as those we discuss in Sects. 2.2–2.4, the basic observable components are the prices themselves, and the stochastic components of interest are the changes in asset prices. Such models assume rational and independent traders. Models of asset prices depend on principles of general economic theory such as equilibrium and arbitrage.

Another approach to modeling asset prices is based on modeling the stochastic aspects in terms of behavior of the traders who collectively determine the asset prices. This agent-based approach allows incorporation of human behavior in the model and so instead of relying solely on classical economic theory, the results of behaviorial economics can be included in the model. In the agent-based approach, which we briefly discuss in Sect. 2.6, the actions of the agents include a random component and their actions determine the prices.

In discussing models, it is always worthwhile to recall the dictum, generally attributed to George Box, "All models are wrong, but some are useful." The usefulness of models of asset prices is not because of the opportunity for financial gain, but rather for determining fair prices, for better understanding of market dynamics, and possibly for regulatory policy development.

2.1 Characteristics of Asset Price Data

Asset prices are directly observable and are readily available from the various markets in which trading occurs. Instead of the prices themselves, however, we are often more interested in various derived data and statistical summaries of the derived data. The most common types of derived data are a first-order measure of change in the asset prices in time, and a second-order measure of the variation of the changes.

The scaled change in the asset price is called the rate of return, which in its simplest form is just the price difference between two time points divided by the price at the first time point, but more often is the difference in the logarithm of the price at the first time point and that at the second time point. The length of the time period of course must be noted. Rates of return are often scaled in some simple way to correspond to an annual rate. In the following, when we refer to "rate of return," we will generally mean the *log-return*, that is, the difference in the logarithms. This derived measure is one of the basic quantities we seek to model.

The log-return depends on the length of the time interval, and so we may speak of "weekly" log-returns, "daily" returns, and so on. As the time interval becomes very short, say of the order of a few minutes, the behavior of the returns changes in a significant way. We will briefly comment on that high-frequency property in Sect. 2.2.7 below.

One of the most important quantities in financial studies is some measure of the variability of the log-returns. The standard deviation of the log-return is called the *volatility*.

A standard deviation is not directly observable, so an important issue in financial modeling is what derived measures of observable data can be used in place of the standard deviation. The sample standard deviation of measured log-returns over some number of time intervals, of course, is an obvious choice. This measure is called *statistical volatility* or *realized volatility*.

Before attempting to develop a model of an empirical process, we should examine data from the process. Any reasonable model must correspond at least to the grossest aspects of the process. In the case of asset prices, there may be various types of empirical processes. We will just focus on one particular index of the price of a set of assets, the S&P 500 Index.

We will examine some empirical data for the S&P 500. First we compute the log-rate for the S&P 500 from January 1, 1990, to December 31, 2005. A histogram for this 15 year period is shown in Fig. 2.1.

With a first glance at the histogram, one may think that the log-returns have a distribution similar to a Gaussian. This belief, however, does not receive affirmation by the q–q plot in Fig. 2.2.

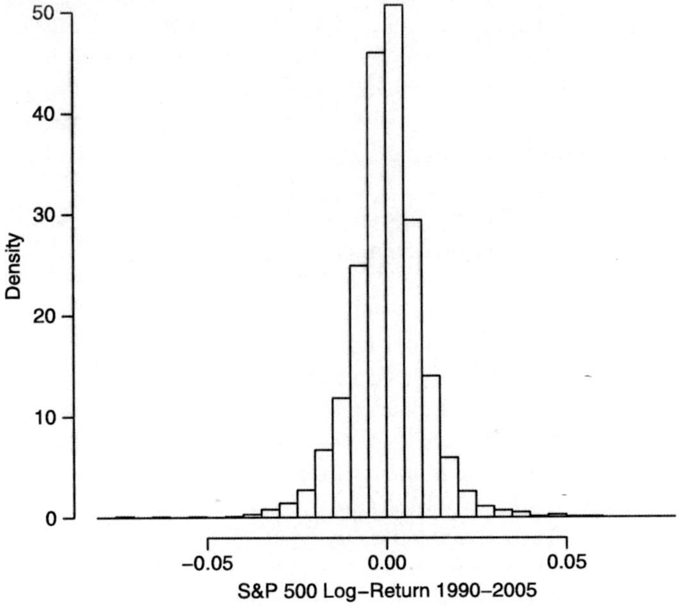

Fig. 2.1 Histogram of log-rates of return 1990–2005

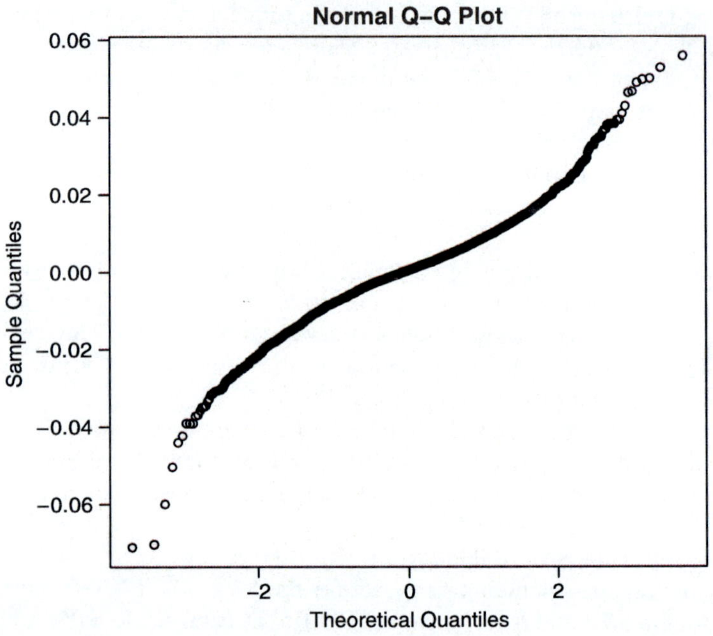

Fig. 2.2 Normal q–q plot of log-rates of return 1990–2005

Some may argue, however, that data models based on a normal distribution are often robust, and can accommodate a wide range of distributions that are more-or-less symmetric and unimodal.

One who is somewhat familiar with the performance of the US stock market will recognize that we have been somewhat selective in our choice of time period for examining the log-return of the S&P 500. Let us now look at the period from January 1, 1987, to September 30, 2009. The belief – or hope – that a normal distribution is an adequate model of the stochastic component is quickly dispelled by looking at the q–q plot in Fig. 2.3.

Figure 2.3 indicates that the log-rates of the S&P 500 form a distribution with very heavy tails. We had only seen a milder indication of this in Figs. 2.1 and 2.2 of the histogram and q–q plots for the 1990–2005 period.

The previous graphs have shown only the static properties of the log-return over fixed periods. It is instructive to consider a simple time series plot of the rates of log-returns of the S&P 500 over the same multi-year period, as shown in Fig. 2.4.

Even a cursory glance at the data in Fig. 2.4 indicates the modeling challenges that it presents. We see the few data points with very large absolute values relative to the other data. A visual assessment of the range of the values in the time series gives us a rough measure of the volatility, at least in a relative sense. Figure 2.4 indicates that the volatility varies over time and that it seems to be relatively high for some periods and relatively low for other periods. The extremely large values of the log-returns seem to occur in close time-proximity to each other.

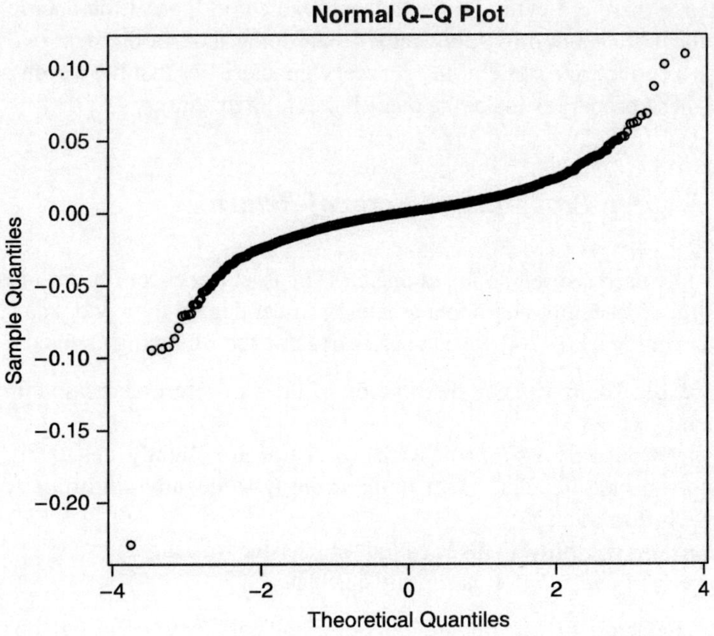

Fig. 2.3 Normal q–q plot of log-rates of return 1987–2009

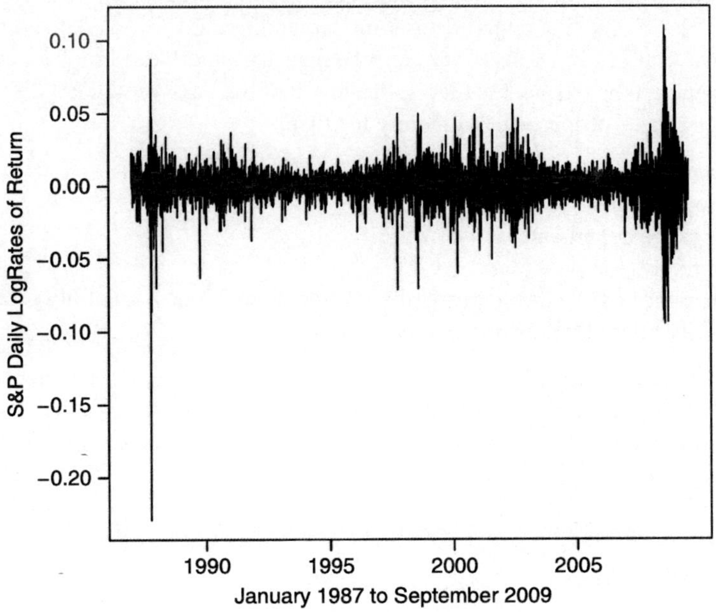

Fig. 2.4 Rates of return

Of course there are many more ways that we could look at the data in order to develop ideas for modeling it, but rather than doing that, in the next two sections we will just summarize some of the general characteristics that have been observed. Many of these properties make the data challenging to analyze.

2.1.1 Stylized Properties of Rates of Return

We have only used a single index of one class of asset prices for illustrations, but the general properties tend to hold to a greater or lesser degree for a wide range of asset classes. From Figs. 2.1–2.4, we can easily observe the following characteristics.

- *Heavy tails.* The frequency distribution of rates of return decrease more slowly than $\exp(-x^2)$.
- *Asymmetry in rates of return.* Rates of return are slightly negatively skewed. (Possibly because traders react more strongly to negative information than to positive information.)
- *Nonconstant volatility.* (This is called "stochastic volatility.")
- *Clustering of volatility.* (It is serially correlated.)

These characteristics are apparent in our graphical illustrations, but the detection of other properties requires computations of various statistics. There are some characteristics that we could observe by using two other kinds of similar plots. In one approach, we compare rates of return at different frequencies, and in the other, we study lagged data. Lagged data is just an additional form of derived measure, much like rate of return itself is a derived measure, and like rate of return it may also depend on the frequency; that is, the length of the lag. We will not display plots illustrating these properties, but merely list them.

- Asymmetry in lagged correlations.
- Aggregational normality.
- Long range dependence.
- Seasonality.
- Dependence of stochastic properties on frequency. Coarse volatility predicts fine volatility better than the other way around.

These stylized properties have been observed through analysis of financial data of various classes over many years. Some of the most interesting of these properties depend on how the volatility changes. We will now note some more properties of the volatility itself.

2.1.2 Volatility

A standard deviation is defined in terms of a probability model, so defining volatility as the standard deviation of the log-return implies a probability model for the log-return. It is this probability model that is central to more general models of asset prices.

Our preliminary graphical analyses showed that there is a problem with a simple interpretation of volatility; it is not constant in time. In some cases, it is clear that news events, that is, shocks to financial markets, cause an increase in volatility. In fact, it appears that both "positive" news and "negative" news lead to higher levels of volatility, but negative news tends to increase future volatility more than positive news does. It also appears that there are two distinct components to the effect of news on volatility, one with a rapid decay and one with a slow decay.

Another aspect of volatility, as we mentioned above, it that it is not directly observable, as is the price of an asset or even the change in price of an asset.

The point of this discussion is that the concept of volatility, despite its simple definition, is neither easy to model nor to measure.

Volatility, however, is one of the most important characteristics of financial data, and any useful model of changes in asset prices must include a component representing volatility. Increased volatility, however it is measured, has the practical effect of increasing the risk premium on financial assets.

2.2 The Basic Models

Asset prices and their rates of change are stochastic processes. We will represent the general form of the stochastic process modeling the asset prices as $\{X_t : t \in \mathcal{I}\}$, for some (well-ordered) index set $\mathcal{I}$. We assume a general probability space $(\Omega, \mathcal{F}, P)$. The specific form of the stochastic process is determined by the nature of $\mathcal{I}$ and $(\Omega, \mathcal{F}, P)$, and by the stochastic relations between X_t and X_s for $t, s \in \mathcal{I}$ and $s < t$; that is, relations between X_t and the sequence $\{X_s : s \in \mathcal{I}, s < t\}$.

In this section we consider various forms of models of asset prices and of changes in asset prices. We begin with an abstract description. The purpose of this approach is to emphasize that the models used in conventional financial analyses are just particular choices that are made to simplify the analysis.

As we discuss pricing models from simple to more complex, we should bear in mind the empirical properties discussed in Sect. 2.1.1 of the processes we are attempting to model. We will consider various formulations of models to capture various properties, but in the end we see that the models do not fully capture all of those stylized properties.

2.2.1 *Systematic Factors and Random Perturbations*

Many mathematical models of interesting processes take the form of a systematic component that involves various measurable factors, plus a random component that represents unobservable or non-quantifiable factors and/or truly "random" factors:

$$Y = f(y_s) + E. \tag{2.1}$$

(Here we are using different notation so as to focus on the abstract model.) The function f may take on a variety of forms. In preliminary models, it almost always linear. As a model is refined, it may assume more complicated forms. The input y_s may represent directly observable variables or it may represent derived variables such as rates. As models are built or evolve, in addition to changes in the function form of f, the factors included in the input y_s may change. In preliminary models, y_s may include a large number of factors that are of potential interest, and as part of the model-building process, some of these factors are removed from the model. Alternatively, in preliminary models, y_s may include only one or two factors that are believed to be important, and as part of the model-building process, other factors are added the model.

In many models, the random component E is the most important term in the model. A mathematical model may be very precise in the description of E, for example, the model may state that $E \sim \mathrm{N}(0, \sigma^2)$, or the model may be looser, stating only, for example, that the expectation of E is 0, and that in set of E's, they are exchangeable.

Before we can build models of stochastic processes in time $\{X_t : t \in \mathcal{I}\}$, we must address the nature of the index set $\mathcal{I}$.

2.2.2 *Indexing Time*

There are essentially two types of index sets. A "discrete time" index set is countable, and, hence, can be taken as the set of integers. A "continuous time" index can be taken as an interval in the reals. These two ways of treating time lead to two general classes of models.

For discrete time, the models evolve from moving average and autoregressive models. The continuous time models are diffusion processes, possibly in combination with a Poisson process. Although discrete time and continuous time may appear to yield completely different kinds of models, there are various limiting equivalent forms.

For either discrete or continuous time, there are various possibilities for choice of the probability space. A standard approach, of course, is to use a normal distribution, at least as a first approximation, as a model for the stochastic component. The important consideration is the nature of the conditional distribution of X_t given $\{X_s : s \in \mathcal{I}, s < t\}$.

In this chapter we will review the types of models that have been used for changes in asset prices over time. We first describe these briefly, and then indicate some of the ways in which the models are inadequate. Several other papers in this Handbook are concerned with various modifications of these models.

2.2.3 Discrete Time Series Models

Discrete time series models describe the behavior of a stochastic process in terms of a functional relationship of the general form

$$X_t = f(X_{t-1}, \ldots, X_{t-p}, \epsilon_t, \epsilon_{t-1}, \ldots, \epsilon_{t-q}). \tag{2.2}$$

In models of this form, the ϵ_i are generally assumed to be random variables, and so if their effects are additive, this is of the same form as model (2.1). More specific assumptions about their distribution allow various methods of statistical inference to be used in making refinements to the model. In most models of this form, the function f is linear. We will briefly describe various forms of the model (2.2). These models are the subject of the well-established field of *time series analysis in the time domain*. We begin with a few definitions.

A *white noise* process $\{\epsilon_t\}$ is one in which for each t, $\epsilon_t \sim \mathrm{N}(0, 1)$, that is, it has a Gaussian or normal distribution with mean 0 and variance 1, and for $s \neq t$, $\mathrm{Cov}(\epsilon_s, \epsilon_t) = 0$; that is, ϵ_s and ϵ_t are independent (because of normality).

The most useful forms of the function f in (2.2) are linear. A particularly simple form yields a *linear process*. We say $\{X_t\}$ is a linear process if it has the form

$$X_t = \mu + \sum_{i=-\infty}^{\infty} a_i \epsilon_{t-i}, \tag{2.3}$$

where $\sum_{i=-\infty}^{\infty} a_i < \infty$ and $\{\epsilon_t\}$ is a white noise process.

One of the most important properties of a stochastic process is *stationarity*, which refers to a distributional measure remaining constant in time. The mean of the linear process is stationary: $\mathrm{E}(X_t) = \mu$. The linear process is also *covariance stationary* since

$$\mathrm{Cov}(X_t, X_{t+h}) = \sum_{i=-\infty}^{\infty} \sum_{j=-\infty}^{\infty} a_i a_j \mathrm{I}_{\{(i,j)|i+j=h\}}(i, j) = \sum_{i=-\infty}^{\infty} a_i a_{i-h}$$

and $\mathrm{V}(\epsilon_t) = 1$. Note that covariance stationary means that the covariance between X_s and X_t depends only on $|t - s|$.

In general, we say that a process is *weakly stationary* (or just *stationary*) if it is mean and covariance stationary.

If the linear model involves only the ϵ_i, that is,

$$X_t = \beta_1 \epsilon_{t-1} + \cdots + \beta_q \epsilon_{t-q} + \epsilon_t, \tag{2.4}$$

it is called a moving average model with q terms. We refer to this model as MA(q). Assuming $\{\epsilon_t\}$ is a white noise process, the MA(q) model is a linear process, and the normality of the stochastic components allows use of relatively simple statistical analyses. For example, we can use maximum likelihood methods, which require specification of probability density functions, and these are particularly straightforward when the stochastic components are normal.

If the linear model involves only the X_{t-j} and ϵ_t, that is,

$$X_t = \alpha_0 + \alpha_1 X_{t-1} + \cdots + \alpha_p X_{t-p} + \epsilon_t, \tag{2.5}$$

it is called an autoregressive model with p terms. We refer to this model as AR(p). Again, specific assumptions about the distributions of

$$\ldots, \epsilon_{t-2}, \epsilon_{t-1}, \epsilon_t, \epsilon_{t+1}, \epsilon_{t+2}, \ldots$$

allow various methods for statistical inference about their distribution and about the parameters α_j.

Combining the MA(q) model of (2.4) with the AR(p) model of (2.5), we have the autoregressive moving average model of order p and q, that is, ARMA(p, q),

$$X_t = \alpha_0 + \alpha_1 X_{t-1} + \cdots + \alpha_p X_{t-p} + \beta_1 \epsilon_{t-1} + \cdots + \beta_q \epsilon_{t-q} + \epsilon_t. \tag{2.6}$$

Assumptions about the relative values of the β_j and α_k imply certain interesting properties of the time series.

The usefulness of ARMA models can be greatly extended by applying it to differences of the time series. If the X's in (2.6) are replaced by dth-order differences, the "integrated" model in the same form as (2.6) is called an ARIMA(p, d, q) model. The differences allow the model to accommodate seasonal effects.

The simple AR, MA, ARMA, and ARIMA models we have just described can be applied to a time series of prices or to a series of returns. The nature of the series and the assumptions about the stochastic component determine the kind of analyses. For example, given the price process $\{X_y\}$, an AR(1) model of returns $Y_t = (X_t - X_{t-1})/X_{t-1}$ from (2.5) would have the form of a pure noise,

$$Y_t = \delta_t. \tag{2.7}$$

The random variable δ_t does not have the same distribution as that of ϵ_t. In fact, if $\{\epsilon_t\}$ is a white noise, then δ_t is a Cauchy process, which has infinite moments of all orders. Clearly, the specific assumptions about the distributions of $\{\epsilon_t\}$ determine the methods for statistical inference about their distribution and about the parameters in the model.

2.2.4 *Continuous Time Diffusion Models*

Differential equations are effective models of continuous change of quantities over time. Such models are widely used for expressing diffusion of a substance or of energy over a physical space. At a macro level the laws governing diffusion are deterministic. Furthermore, substances and energy can be treated as ensembles over a physical space, and so the diffusion model represents an average density. Thus, such a model contains no stochastic component.

Empirical evidence clearly indicates that a deterministic differential equation could not effectively model price movements of an asset such as a stock.

The first step must be to introduce a stochastic component into the differential equation, and the simplest way to this is for the differential to be from a Brownian motion. This is what Bachelier proposed in 1900 (see, e.g., Steele (2001)). In Bachelier's stochastic differential equation, the Brownian motion represented the change in price. This model is

$$\mathrm{d}X_t = \mu X_t \mathrm{d}t + \sigma X_t \mathrm{d}W_t, \tag{2.8}$$

where W_t is a Brownian motion. Clearly, $\mathrm{d}W_t$ could represent some other type of stochastic differential, although the existence of a stochastic differential with appropriate properties would need to be established. (Wiener established this for Brownian motion. See, again for example, Steele (2001).)

Samuelson (1965) modified the model (2.8) to one he called geometric Brownian motion:

$$\frac{\mathrm{d}X_t}{X_t} = \mu \mathrm{d}t + \sigma \mathrm{d}W_t. \tag{2.9}$$

This is a model for the *rate of change* of asset prices. Note that this is similar to forming (2.7) from (2.5), and then changing the assumptions about the distribution of the random component so that the random variable in the derived equation has a simple distribution.

The geometric Brownian motion model (2.9) has been widely used in financial analysis. In the context of a riskless portfolio of an asset and an option on the asset, the geometric Brownian motion model leads to the Black-Scholes-Merton differential equation for the fair price P of an option:

$$\frac{\partial P_t}{\partial t} + rX_t \frac{\partial P_t}{\partial X_t} + \frac{1}{2}\sigma^2 X_t^2 \frac{\partial^2 P_t}{\partial X_t^2} = rP, \tag{2.10}$$

where r is a risk-free interest rate.

Detemple and Rindisbacher (2010, this volume) provide a more extensive discussion of diffusion models. We will briefly consider some modifications of the basic diffusion models in Sect. 2.4.

2.2.5 *Accounting for Jumps*

Looking at the data in Fig. 2.4, we notice a few extremely large returns, both positive and negative. These outliers are called "jumps." Figures 2.2 and 2.3 indicate that the presence of these outliers is inconsistent with the assumption that the underlying random variables in either model (2.6) or model (2.9) have Gaussian distributions. (In model (2.6) the random variable is ϵ, and in model (2.9) it is $\mathrm{d}W_t$.)

In standard statistical analyses, there are two simple ways of accounting for outliers. One way is to use an "outlier-generating distribution" or "jump process," that is, a heavy-tailed distribution, such as stable distribution other than the Gaussian. Figueroa-López (2010, this volume) describes the use of Lévy processes in diffusion models. Other discussions of models with non-Gaussian random components are in Jondeau et al. (2007) and Rachev et al. (2005).

Another method of accounting for jumps is to use a mixture of distributions. Even mixtures of Gaussian distributions result in outlier-generating distributions. Instead of using simple mixtures of Gaussians, however, a more common approach is to use a mixture of a continuous distribution, such as a Gaussian, and a discrete Poisson process, possibly associated with an effect with a random magnitude. Bjursell and Gentle (2010, this volume) and Cont and Tankov (2004) describe the use of mixtures that include Poisson processes. We will briefly consider jump-diffusion models in Sect. 2.4.2.

Either of these modifications to the models results in more difficult data analyses.

2.2.6 *Accounting for Stochastic Volatility*

The ARMA model of (2.6) incorporates the volatility of the stochastic process in the standard deviation of the random variables ϵ, and the diffusion model of (2.9) incorporates the volatility in the standard deviation of the random variables $\sigma \mathrm{d}W_t$. An assumption of either model is that this standard deviation is constant; hence, a serious deficiency of either of the two basic models (2.6) and (2.9) is that the model does not account for the stochastic volatility that is apparent in Fig. 2.4.

To be realistic, either type of model must be modified to allow for the volatility to be nonconstant. Further, as we note from Fig. 2.4, the modification must include a serial correlation of the volatility.

2.2.7 *Market Microstructure*

Pricing data represent the value exchanged in a specific trade. The price at which a specific transaction occurs should be exactly the same as the price (within the minimum unit of money) of the same transaction at the same time. It turns out, for a variety of reasons, that this is not the case. *Tick* data, that is, data on each transaction (also called "high-frequency data") exhibit characteristics that are different from price data collected less frequently, say at the close of each trading day.

Some stylized properties of tick data include intraday periodicity; nonsynchronicity, that is, a sequence of prices over a short time interval do not form an equally-spaced time series; price clustering; and negative lag-1 autocorrelations.

These properties constitute what is called "market microstructure." See Lai and Xing (2008) for more discussion of microstructure.

Bjursell and Gentle (2010, this volume) discuss the use of microstructure noise to test for jumps superimposed on a diffusion model.

2.3 GARCH-Type Models

The AR, MA, ARMA, and ARIMA models described in Sect. 2.2.3 assume a constant variance. There are various ways of modifying the model to make the variance change over time.

For a model of the form (2.7), we first introduce a scale on the random component:

$$Y_t = \sigma_t \delta_t. \tag{2.11}$$

Then, following the empirical observation that the standard deviation of a process is proportional to the magnitude (that is, the coefficient of variation is relatively constant), we may assume a model for the variance of the form

$$\sigma_t^2 = \alpha_0 + \alpha_1 Y_{t-1}^2. \tag{2.12}$$

The variance is conditional on the value of Y_{t-1}^2, and so this kind of model is called an ARCH (autoregressive conditionally heteroscedastic) model; specifically the model of (2.11) and (2.12) is called an ARCH(1) model (recall that it originated as an AR(1) model).

The ARCH models can be generalized further by modeling the variance as an AR process; that is, (2.12) may become, for example,

$$\sigma_t^2 = \alpha_0 + \alpha_1 Y_{t-1}^2 + \beta_1 \sigma_{t-1}^2. \tag{2.13}$$

Such models are called GARCH (generalized autoregressive conditionally heteroscedastic) models; specifically, the model of (2.11) and (2.13) is a GARCH(1,1) model, because both components are lag 1 processes.

Notice that the simple ARCH(1) model of (2.11) and (2.12) could be reformulated by squaring both sides of (2.11), then subtracting (2.12) and then rearrange terms to obtain

$$Y_t^2 = \alpha_0 + \alpha_1 Y_{t-1}^2 + \gamma_t, \tag{2.14}$$

in which, if δ_t is a $\mathrm{N}(0, 1)$ random variable, then γ_t is a scaled and shifted chi-squared random variable with one degree of freedom.

The purpose of this re-expression is only to show that the ARCH(1) model is related to an AR(1) model with a change of distribution of the random component. The ARCH and GARCH models, while they do incorporate stochastic volatility, if the underlying distribution of the stochastic component is normal, the models will not display the heavy-tailed and asymmetric returns that are observed empirically.

Many variations of GARCH models have been studied; see, for example, Christoffersen et al. (2010, this volume) and Gouriéroux (1997). Most of these variations are still based on an underlying normal distribution, however.

2.3.1 *GARCH with Jumps*

As we mentioned previously, jumps can be modeled either through an outlier-generating distribution or by superimposition of a jump process. The most common way of incorporating jumps in a discrete time series model is by use of a heavy-tailed distribution, such as stable distribution other than the Gaussian. This, of course, presents problems in the statistical analysis of data using such models.

2.3.2 *Inference on the Parameters*

Statistical inference on autoregressive moving average models is usually based on the likelihood. Given a distribution for the random components in any such model, it is usually rather simple to formulate the associated likelihood. The likelihood rarely can be maximized analytically, but there are efficient numerical methods. These methods are usually two-stage optimizations, and are similar to methods originally used in the ARIMA models of Box and Jenkins. Gouriéroux (1997) describes maximum likelihood methods for various GARCH models.

Just fitting the parameters, of course, is only one part of the problem of statistical inference. Various assumptions about the distributions of the stochastic components require different methods for statistical inference such as tests and confidence regions. Even if the underlying distribution is not assumed to be normal, most inference methods end up using approximate normal distributions.

2.4 Diffusion Models

The basic geometric Brownian motion diffusion model (2.9),

$$\frac{\mathrm{d}X_t}{X_t} = \mu \mathrm{d}t + \sigma \mathrm{d}W_t,$$

misses most of the salient empirical properties of Sect. 2.1.1.

Brownian motion is a rather complex process, and given our understanding of it – and our lack of understanding of a similar process not based on Gaussianity – we would seek to build modifications onto the Brownian motion, rather than to replace the Gaussian distribution with some other distribution that is either heavy-tailed or asymmetric. (Recall our elliptical reference above to the *existence* of Brownian motion.)

There are several possible modifications of the Brownian motion. We will formulate two modifications below that address stochastic volatility and jumps. Before doing so, however, we mention a simple modification that allows for long range dependencies in a model of the form (2.9). In this modification, instead of the Brownian motion W_t, we use a fractional Brownian motion, W_t^H, where $0 < H < 1$ is the Hurst index. (An index of 0.5 is ordinary Brownian motion.) The essential characteristic of a fractional Brownian motion,

$$\mathrm{Cov}(W_t^H, W_s^H) = \frac{1}{2}\left(|t|^{2H} + |s|^{2H} - |s-t|^{2H}\right),$$

allows for the modified model (2.9) to exhibit long range dependencies, which, as we remarked without elaboration in Sect. 2.1.1, is an empirical property of rates of return. Fractional Brownian motion is in spirit related to the reformulation of the ARCH(1) model of (2.11) and (2.12) as the AR(1) model (2.14).

2.4.1 Coupled Diffusion Models

The modification of an AR model that yields a GARCH model is merely to apply to a function of the volatility the same basic time series model that is used for returns. This way of handling stochastic volatility in the case of diffusion models would result in coupled diffusion models in which a secondary diffusion model is applied to a function of the volatility:

$$\frac{\mathrm{d}X_t}{X_t} = \mu \mathrm{d}t + \sigma_t \mathrm{d}(W_1)_t \tag{2.15}$$

$$\mathrm{d}\sigma_t^2 = \alpha(\mu_{\sigma_t^2} - \sigma_t^2)\mathrm{d}t + \beta(\sigma_t^2)^\gamma \mathrm{d}(W_2)_t, \tag{2.16}$$

where α, $\mu_{\sigma_t^2}$, β, and γ are constants and $(W_1)_t$ and $(W_2)_t$ are Brownian motions.

Equations (2.15) and (2.16) are sometimes called the Hull and White model (although that term is usually used for a different model used for interest rate derivatives). For the special case of $\gamma = 0.5$, it is also called the Heston model.

There are many variations on models of this form. Notice that this model does not tie the magnitude of the volatility to the magnitude of the return, as the simple ARCH model did. This could be remedied by an incorporation of X into (2.16). An important consideration is the relationship between the two Brownian motions $(W_1)_t$ and $(W_2)_t$. The simplest assumption is that they are independent. An alternative, but still very simple assumption, is that $(W_2)_t$ is a linear combination of $(W_1)_t$ and an independent Brownian motion.

While the coupled diffusion model do incorporate stochastic volatility, just as with the ARCH and GARCH models, because the underlying distribution of the stochastic component is normal, the models will not display the heavy-tailed and asymmetric returns that are observed empirically.

2.4.2 Diffusion with Jumps

A modification of any of the models that we have discussed above that can display both heavy-tailed and asymmetric returns is to superimpose a Poisson process onto the model. Starting with the simple geometric Brownian motion diffusion model (2.9), we write

$$\frac{\mathrm{d}X_t}{X_t} = \mu \mathrm{d}t + \sigma \mathrm{d}W_t + \kappa_t \mathrm{d}q_t, \tag{2.17}$$

where W_t is the standard Wiener process; q_t is a counting process with intensity λ_t, that is, $P(\mathrm{d}q_t = 1) = \lambda_t \mathrm{d}t$; and κ_t is the size of the price jump at time t if a jump occurred. If X_{t-} denotes the price immediately prior to the jump at time t, then $\kappa_t = X_t - X_{t-}$.

2.4.3 Inference on the Parameters

If restrictive assumptions are made about the constancy of parameters and independence of the events in the process, there are fairly simple statistical estimators for most of the parameters in the single-equation models. Parameters in coupled equations can often be estimated using two-stage likelihood methods. The parameters in a model such as (2.17) are difficult to estimate because we do not know which of the two processes is operating. One approach to the fitting the parameters in a model with a superimposed process is to set an arbitrary threshold for the return, and to assume the Poisson process generates any realization greater than that threshold.

For models with time-varying parameters, analysis generally depends on the use of Monte Carlo methods.

2.5 How Simple Can a Realistic Model Be?

At this point, we must ask how simple can a pricing model be and still capture all of the empirical properties that we have observed. Clearly, the basic models of Sect. 2.2 fail drastically.

The first modification to the simple ARMA or geometric Brownian motion model is usually to address the stochastic volatility. An approach in either case is to couple the basic process with a similar process for the volatility. So long as the underlying stochastic components are Gaussian, two-stage maximum likelihood methods can be used in the analysis.

The issue of heavy tails and asymmetric distributions could perhaps be addressed by replacing the Gaussian processes with some asymmetric heavy-tailed process, perhaps a stable process. The loss of the simplicity of the normal distribution, however, is a very steep price to pay. An alternative approach is to superimpose a Poisson jump process, as in model (2.17). Such a model has a stochastic volatility (due to the firing of the Poisson process), but it is not the slowly-varying volatility that we observe. Hence, the jump process needs to be superimposed on a model that already accounts for stochastic volatility, such as a GARCH model or a coupled diffusion model.

It is clear that the amount of a jump, κ_t, is not constant. A simple modification would be to take κ_t as an independent random variable. Its distribution would seem to be heavy-tailed and to have a negative mean. Empirically (see Fig. 2.4) a negative (positive) jump tends to be followed immediately by a positive (negative) jump, This may suggest that jumps be modeled as paired events instead of trying to accommodate these positive and negative values in the distribution of κ_t.

A further glance at Fig. 2.4 indicates two additional considerations (assuming a somewhat arbitrary visual identification of jumps): jumps do not follow a time-homogeneous Poisson process, and jumps and (ordinary) volatility are not independent. This means that λ_t (the Poisson intensity) must be stochastic and it must depend on q_s, for $s < t$. Also, σ_t must depend on q_s, for $s < t$. Furthermore, σ_t and λ_t must be correlated.

Rather than suggesting a comprehensive and realistic model, in this section, we have just discussed some of the relevant considerations. We seek a realistic model that accounts for the peculiar properties of the rate-of-return process, but we must realistically limit the degrees of freedom in the model.

2.6 Agent-Based Models

The pricing models discussed in Sects. 2.2–2.5 are developed from a macro perspective on the prices themselves. This perspective excludes aspects of the market that results from irrational human behavior, where "irrational" is defined subjectively and usually means that the market participants are attempting to optimize a simple objective function. In a rational approach to modeling market behavior, what individual traders are doing has no affect on the decision of a trader to buy or sell; that is the market does not have "momentum." There is an instantaneous adjustment of prices to some "fair market value." No matter how attractive a rational approach to financial modeling is, its attractive simplicity cannot make it so. Market participants do not act independently of each other. Traders do not have share the same processed data. Traders do not identify the same objective function. Traders do not all share a similar model of the market. The proportion of traders who behave in a certain way, that is, who do share a similar model varies in time.

The ultimate dependence of prices on the beliefs and actions of individual traders suggests another approach to financial modeling. This approach begins with models of behavior of the market participants. In this kind of approach to scientific modeling, called "agent-based," the actions of a set of individual "agents" are governed by control parameters that can depend on the actions of other agents.

We will not pursue this approach here. LeBaron (2006) provides a survey of the micro perspective modeling incorporated in an agent-based approach.

2.7 Applications of Pricing Models

We must emphasize again that the role of pricing models is not to predict prices. Pricing models provide a description of stochastic behavior, and for that reason they have important applications in a number of areas, such as in the regulation of financial markets, in management of risk, and in pricing of derivative assets.

Options pricing is probably the highest profile application of asset pricing models. This application soared to prominence in the early 1970s when Black and Scholes used the differential (2.10) derived from the geometric Brownian motion model (2.9) to develop exact formulas for fair prices of European puts and calls.

As we have pointed out, the simple geometric Brownian motion model does not correspond very well with empirical data. Although prices yielded by the Black-Scholes options pricing formulas were useful for traders, they quickly noticed that the prices set by the market differed from the Black-Scholes prices in systematic ways. If the market price is inserted as the price in a Black-Scholes formula, any other single variable in the formula can be solved for. The time to expiry, the current market price of the asset, and the strike price are all directly observable, so the only variable in the model that might be considered questionable is the volatility. An interesting fact emerged; if the formula is applied to options on the same underlying asset and at the same time to expiry but at different strike prices, the value of the volatility that satisfies the formula is not constant, but rather a convex function of the strike price. This was called the "volatility smile." Likewise, if the same strike price but different times to expiry are entered into the formula, the volatility exhibits systematic curvature. Fengler (2010, this volume) provides more details on this kind of result from the Black-Scholes formula.

Although we have taken the definition of "volatility" simply to be "standard deviation of rates of returns," we have already indicated in Sect. 2.1.2 the difficulties in assigning a value to volatility. The value of volatility implied by the inverted use of the Black-Scholes formula with observed prices of derivatives therefore has intrinsic interest. Volatility defined by inversion of a pricing formula is called "implied volatility," and so volatility defined as originally in terms of a standard deviation is now often called "statistical volatility." The inverted use of pricing models together with observed prices of derivatives to define a type of asset price volatility is probably more common now than use of the pricing models for their earlier purpose of determining fair prices for derivatives.

There are now markets in implied volatility of various market indexes, and this kind of market provides another tool for hedging investment risks. The most widely traded such implied volatility index is the VIX, which follows the implied volatility of the S&P 500. Traded implied volatility indexes use rather complicated asset pricing models; none currently use the simple Black-Scholes formula.

The simpler models such as ARMA/ARIMA or geometric Brownian motion can often be analyzed by well-established statistical methods. The most impressive result of such an analysis is probably the Black-Scholes formulas. For more realistic models, the analysis is often by Monte-Carlo methods. In the case of stochastic models, the Monte Carlo methods are often coupled with numerical solutions to the stochastic differential equations; see, for example, Sauer (2010, this volume).

Realistic asset pricing models generally present analysis problems that can feasibly be addressed only by Monte Carlo methods. See Yu (2010, this volume) or Glasserman (2004) for more detailed discussion of Monte Carlo methods in the application of asset pricing models.

References

Bjursell, J., & Gentle, J. E. (2010). Identifying jumps in asset prices. In J.-C. Duan, J. E. Gentle, & W. Härdle (Eds.), *Handbook of computational finance*. Berlin: Springer.

Christoffersen, P., Jacobs, K., & Ornthanalai, C. (2010). GARCH Option pricing: Theory and evidence. In J.-C. Duan, J. E. Gentle, & W. Härdle (Eds.), *Handbook of computational finance*. Berlin: Springer.

Cont, R., & Tankov, P. (2004). *Financial modelling with jump processes*. Boca Raton: Chapman & Hall.

Detemple, J., & Rindisbacher, M. (2010). Diffusion models of asset prices. In J.-C. Duan, J. E. Gentle, & W. Härdle (Eds.), *Handbook of computational finance*. Berlin: Springer.

Fengler, M. (2010). Option data and modelling BSM implied volatility. In J.-C. Duan, J. E. Gentle, & W. Härdle (Eds.), *Handbook of computational finance*. Berlin: Springer.

Figueroa-López, J. E. (2010). Jump-diffusion models driven by Lévy processes. In J.-C. Duan, J. E. Gentle, & W. Härdle (Eds.), *Handbook of computational finance*. Berlin: Springer.

Glasserman, P. (2004). *Monte Carlo methods in financial engineering*. New York: Springer.

Gouriéroux, C. (1997). *ARCH models and financial applications*. New York: Springer.

Hafner, C. M., & Manner, H. (2010). Multivariate time series models for asset prices. In J.-C. Duan, J. E. Gentle, & W. Härdle (Eds.), *Handbook of computational finance*. Berlin: Springer.

Jondeau, E., Poon, S.-H., & Rockinger, M. (2007). *Financial modeling under non-gaussian distributions*. London: Springer.

Lai, T.-L., & Xing, H. (2008). *Statistical models and methods for financial markets*. New York: Springer.

LeBaron, B. (2006). Agent-based computational finance. In L. Tesfatsion & K. L. Judd (Eds.), *Handbook of computational economics* (Vol. 2, pp. 1187–1232). Amsterdam: North-Holland.

Rachev, S. T., Menn, C., & Fabozzi, F. J. (2005). *Fat-tailed and skewed asset return distributions*. Hoboken: Wiley.

Samuelson, P. A. (1965). Proof that properly anticipated prices fluctuate randomly. *Industrial Management Review*, *6*, 41–49.

Sauer, T. (2010). Numerical solution of stochastic differential equations in finance. In J.-C. Duan, J. E. Gentle, & W. Härdle (Eds.), *Handbook of computational finance*. Berlin: Springer.

Steele, J. M. (2001). *Stochastic calculus and financial applications*. New York: Springer.

Yu, J. (2010). Simulation-based estimation methods for financial time series models. In J.-C. Duan, J. E. Gentle, & W. Härdle (Eds.), *Handbook of computational finance*. Berlin: Springer.

Chapter 3
Diffusion Models of Asset Prices

Jérôme Detemple and Marcel Rindisbacher

Abstract This paper reviews the literature on asset pricing in diffusion models. The first part is devoted to equilibrium models based on the representative agent paradigm. Explicit formulas for fundamental equilibrium quantities such as the state price density, the interest rate and the market price of risk are presented. Valuation formulas for stocks and contingent claims are also provided. Numerical implementation of the model is carried out in a setting with constant relative risk aversion. The second part of the paper focuses on multiagent models with complete financial markets. Characterizations of equilibrium are reviewed and numerical algorithms for computation are proposed.

3.1 Introduction

This paper provides a review of asset pricing models cast in general equilibrium settings with diffusive uncertainty structure. It identifies and discusses the basic relations that tie consumption to equilibrium quantities such as state price densities, interest rates and market prices of risk. It also reviews the structure of asset risk premia and return volatilities. Both single and multiple agent economies are considered. Numerical algorithms for implementation are provided. Numerical illustrations are given in the context of simple examples with constant relative risk aversion.

The topic of asset pricing has a long history, dating back centuries, and has been the subject of an enormous literature. This review does not seek to give a comprehensive presentation. Rather, it will focus on a limited, but essential, part of the modern literature, that dealing with complete or effectively complete

J. Detemple (✉) · M. Rindisbacher
Boston University School of Management, Boston University, 595 Commonwealth Avenue, Boston, MA 02215, USA
e-mail: detemple@bu.edu; rindisbm@bu.edu

J.-C. Duan et al. (eds.), *Handbook of Computational Finance*, Springer Handbooks of Computational Statistics, DOI 10.1007/978-3-642-17254-0_3,

financial markets. Moreover, the presentation will concentrate on continuous time pure exchange economies in which primitives follow diffusion processes. These choices are driven by the desire to present the most basic relations tying endogenous and exogenous variables and to provide clear and analytically tractable formulas for equilibrium quantities. Complete market models, which are easy to handle, are particulary useful for these purposes. Continuous time facilitates tractability and permits the derivation of transparent and implementable formulas for a variety of endogenous variables. Particularly illuminating examples include formulas for interest rates, market prices of risk and asset return volatilities.

Classic papers in the continuous time diffusion tradition include Merton (1973), Breeden (1979), Cox et al. (1985) and Huang (1987). In these settings, stock prices and state variables (such as dividends) are typically modelled as joint diffusion processes. In standard continuous time pure exchange economies, all equity-financed firms operate technologies that produce flows of a perishable consumption good. Equity shares (stocks) represent claims to this perishable output. Stock dividends are thus naturally modelled as flows of production per unit time. Stock prices, which are determined in equilibrium, reflect the structure of dividend processes. General processes can be used to model the evolution of dividends over time (see, for instance, Duffie and Zame (1989); Karatzas et al. (1990); Detemple and Zapatero (1991); Back (1991)). For computational tractability, it is nevertheless convenient to assume that they follow more specialized Markovian or diffusion processes. The models reviewed in this paper focus on the diffusion setting.

Section 3.2 presents the basic economic model with a representative agent and a single perishable consumption good. The financial market structure and the agent's choices, preferences and decision problem are described. Equilibrium is defined. Equilibrium prices and returns are presented in Sect. 3.3. The structure of the interest rate, market prices of risk and stock return volatilities are examined and discussed. Section 3.4 describes the computational method and provides an illustrative example. Models with heterogeneous agents are reviewed in Sect. 3.5. Computational algorithms are described. A numerical illustration is presented in the context of an economy with two agents. Concluding remarks are formulated last.

3.2 The Single Agent Model

A continuous time asset pricing model is developed in the context of a pure exchange economy with a single perishable consumption good, a representative agent and complete markets. The good serves as the numeraire. The economy has a finite horizon $[0, T]$. The uncertainty is carried by a d-dimensional Brownian motion process W. Brownian increments represent economic shocks. There are k state variables Y. Aggregate consumption C, dividends D and state variables Y follow diffusion processes.

3.2.1 The Financial Market

The financial market has d_s risky assets (stocks) and 1 locally riskless asset (a money market account). Stocks are claims to dividends, which are paid in units of the consumption good. The vector of dividends $D = (D_1, \ldots, D_{d_s})$ evolves according to

$$dD_t = I_t^D \left[\gamma\left(t, D_t, Y_t\right) dt + \lambda\left(t, D_t, Y_t\right) dW_t\right] \tag{3.1}$$

$$dY_t = \mu^Y\left(t, Y_t\right) dt + \sigma^Y\left(t, Y_t\right) dW_t, \tag{3.2}$$

where I_t^D is a diagonal matrix with vector of dividends D_t on its diagonal, $\gamma(t, D_t, Y_t)$ is the d_s-dimensional vector of expected dividend growth rates and $\lambda(t, D_t, Y_t)$ is the $d_s \times d$ volatility matrix of dividend growth rates. Likewise, $\mu^Y(t, Y_t)$ is the k-dimensional vector of expected changes in the state variables and $\sigma^Y(t, Y_t)$ their $k \times d$ matrix of volatility coefficients. Coefficients of the stochastic differential equations (3.1)–(3.2) are assumed to satisfy standard conditions for the existence of a unique strong solution (D, Y). Stocks are in unit supply (the number of shares is normalized to one). The aggregate dividend (aggregate consumption) is $C \equiv \mathbf{1}'D$.

Stock prices are determined in a competitive equilibrium. Equilibrium prices are assumed to have a representation

$$dS_t + D_t dt = I_t^S \left[\mu\left(t, D_t, Y_t\right) dt + \sigma\left(t, D_t, Y_t\right) dW_t\right], \tag{3.3}$$

where $\mu(t, D_t, Y_t)$ is the d_s-dimensional expected return and $\sigma(t, D_t, Y_t)$ the $d_s \times d$ matrix of return volatilities. The coefficients $\mu(t, D_t, Y_t)$ and $\sigma(t, D_t, Y_t)$ are endogenous.

The locally riskless asset is a money market account which pays interest at some rate $r(t, D_t, Y_t)$ per unit time. There is no exogenous supply of this asset (the money market account is an inside asset in zero net supply). The interest rate, representing the return on the asset, is also endogenously determined in equilibrium.

To simplify notation the arguments of drift, volatility and other functions are sometimes omitted. For example r_t will be used to denote $r(t, D_t, Y_t)$, μ_t to denote $\mu(t, D_t, Y_t)$, etc.

The following assumptions are made

Assumption 1. *Candidate equilibrium prices processes satisfy the following conditions*

(i) $\int_0^T |r_v|\, dv < \infty, \quad \mathbf{P}-a.s.$
(ii) $\int_0^T \left(\sum_i |\mu_{iv}| + \sum_{i,j} \left|\left[\sigma_v \sigma_v'\right]_{i,j}\right|\right) dv < \infty, \quad \mathbf{P}-a.s.$

Assumption 1 is a set of restrictions on the space of candidate equilibrium prices processes. These assumptions are weak. Condition (i) ensures that the discount factor at the riskfree rate $b_t \equiv \exp\left(-\int_0^t r_v dv\right)$ is strictly positive for all $t \in [0, T]$.

Condition (ii) ensures that the cumulative expected returns and return variances exist. This condition is sufficient for the existence of the total return process in (3.3).

3.2.2 Consumption, Portfolios and Wealth

The economy has a representative agent endowed with 1 share of each stock. The endowment of the money market account is null. The standing agent consumes and allocates wealth among the different assets available. Let X_t be the wealth at date t. Consumption is c_t and π_t is the $d \times 1$ vector of wealth proportions invested in the risky assets (thus $1 - \pi_t' \mathbf{1}$ is the proportion invested in the riskless asset). Consumption satisfies the physical nonnegativity constraint $c \geq 0$. No sign restrictions are placed on the proportions π invested in the various assets: long as well as short positions are permitted. The evolution of wealth is governed by the stochastic differential equation

$$dX_t = (X_t r_t - c_t)\, dt + X_t \pi_t' \left[(\mu_t - r_t \mathbf{1})\, dt + \sigma_t dW_t\right] \tag{3.4}$$

subject to some initial condition $X_0 = x \equiv \mathbf{1}' S$. For this evolutionary equation to make sense the following integrability condition is imposed on the policy (c, π)

$$\int_0^T \left(|c_t| + \left| X_t \pi_t' (\mu_t - r_t \mathbf{1}) \right| + \left| X_t \pi_t' \sigma_t \sigma_t' \pi_t X_t \right| \right) dt < \infty, \quad (\mathbf{P} - a.s.). \tag{3.5}$$

Under (3.5) the stochastic integral on the right hand side of (3.4) is well defined. Condition (3.5) is a joint restriction on consumption-portfolio policies and candidate equilibrium price processes.

3.2.3 Preferences

Preferences are assumed to have the time-separable von Neumann-Morgenstern representation. The felicity provided by a consumption plan (c) is

$$\mathcal{U}(c) \equiv \mathbf{E}\left[\int_0^T u(c_v, v) dv \right], \tag{3.6}$$

where the utility function $u : [A, \infty) \times [0, T] \to \mathbb{R}$ is strictly increasing, strictly concave and differentiable over its domain. The consumption lower bound is assumed to be nonnegative, $A \geq 0$. The limiting conditions $\lim_{c \to A} u'(c, t) = \infty$ and $\lim_{c \to \infty} u'(c, t) = 0$ are also assumed to hold, for all $t \in [0, T]$. If $[A, \infty)$ is a proper subset of $\mathbb{R}_+$ (i.e. $A > 0$) the function u is extended to $\mathbb{R}_+ \times [0, T]$ by setting $u(c, t) = -\infty$ for $c \in \mathbb{R}_+ \backslash [A, \infty)$ and for all $t \in [0, T]$.

This class of utility functions includes the HARA specification

$$u(c,t) = \frac{1}{1-R}(c-A)^{1-R},$$

where $R > 0$ and $A \geq 0$. If $A > 0$ the function has the required properties over the subset $[A, \infty) \subset \mathbb{R}_+$. The function is then extended by setting $u(c,t) = -\infty$ for $0 \leq c < A$. This particular HARA specification corresponds to a model with subsistence consumption A.

Under these assumptions the inverse $I : \mathbb{R}_+ \times [0,T] \rightarrow [A, \infty)$ of the marginal utility function $u'(c_t, t)$ with respect to its first argument exists and is unique. It is also strictly decreasing with limiting values $\lim_{y \to 0} I(y,t) = \infty$ and $\lim_{y \to \infty} I(y,t) = A$.

3.2.4 The Consumption-Portfolio Choice Problem

The consumer-investor seeks to maximize expected utility

$$\max_{(c,\pi)} \mathcal{U}(c) \equiv \mathbf{E}\left[\int_0^T u(c_v, v)\, dv\right] \tag{3.7}$$

subject to the constraints

$$dX_t = (r_t X_t - c_t)\, dt + X_t \pi_t' \left[(\mu_t - r_t)\, dt + \sigma_t dW_t\right]; \; X_0 = x_i \tag{3.8}$$

$$c_t \geq 0, X_t \geq 0 \tag{3.9}$$

for all $t \in [0,T]$, and the integrability condition (3.5). The first constraint, (25.6), describes the evolution of wealth given a consumption-portfolio policy (c, π). The quantity x represents initial resources, given by the value of endowments $x = \mathbf{1}' S_0$. The next one (25.7) has two parts. The first captures the physical restriction that consumption cannot be negative. The second is a non-default condition requiring that wealth can never become negative.

A policy (c, π) is said to be *admissible*, written $(c, \pi) \in \mathcal{A}$, if and only if it satisfies (25.6) and (25.7). A policy (c^*, π^*) is *optimal*, written $(c^*, \pi^*) \in \mathcal{A}^*$, if and only if it cannot be dominated, i.e., $\mathcal{U}(c^*) \geq \mathcal{U}(c)$ for all $(c, \pi) \in \mathcal{A}$.

3.2.5 Equilibrium

A *competitive equilibrium* is a collection of stochastic processes $(c, \pi, S_0, r, \theta, \sigma)$ such that:

1. *Individual rationality:* $(c, \pi) \in \mathcal{A}^*$, where (S_0, r, μ, σ) is taken as given.
2. *Market clearing:* (a) commodity market: $c = C \equiv \mathbf{1}' D$, (b) equity market: $X\pi = S$ and (c) money market: $\pi' \mathbf{1} = 1$.

This notion of equilibrium involves rational expectations. The representative agent correctly forecasts the evolution of asset returns when making individual decisions (condition 1). In equilibrium, forecasted return processes and market clearing return processes coincide (condition 2).

For later developments it is also useful to note that clearing of the commodity market implies clearing of the equity and money markets.

3.3 Equilibrium

The optimal consumption demand is characterized in Sect. 3.3.1. Formulas for the equilibrium state price density and the values of securities are given in Sect. 3.3.2.

3.3.1 Optimal Consumption Policy

The competitive equilibrium pins down the returns associated with intertemporal transfers of funds. The interest rate r measures the instantaneous return on the money market account (the locally riskless asset). The market price of risk θ_j captures the expected instantaneous return on a claim with unit exposure to the Brownian motion risk W_j. Let $\theta \equiv (\theta_1, \ldots, \theta_d)'$ be the d-dimensional vector of market prices of risk. Both r and θ are endogenous in equilibrium and reflect the economic structure and conditions.

The state price density (SPD) associated with a pair of candidate equilibrium processes (r, θ) is

$$\xi_v \equiv \exp\left(-\int_0^v \left(r_s + \frac{1}{2}\theta_s'\theta_s\right) ds - \int_0^v \theta_s' dW_s\right), \tag{3.10}$$

where $v \in [0, T]$. The SPD ξ_v is the stochastic discount factor that matters for the valuation at date 0 of a cash flow received at $v \geq 0$. The conditional state price density (CSPD) $\xi_{t,v} \equiv \xi_v/\xi_t$ determines the value at t of a cash flow at v. The SPD (CSPD) also represents the cost at 0 (at t) of a state-contingent dollar received at time v.

Assumption 2. *Candidate market prices of risk satisfy $\int_0^T \theta_v'\theta_v dv < \infty$ $(P - a.s.)$.*

Under Assumption 2 the stochastic integral $\int_0^v \theta_s' dW_s$ exists and is a local martingale. In combination with Assumption 1(i), it implies that the growth rate of the state price density is well defined.

The following characterization of the consumption demand (optimal consumption) is derived in Pliska (1986), Karatzas et al. (1987) and Cox and Huang (1989).

Theorem 1. *Consider the dynamic consumption-portfolio problem (25.5)–(25.7) and suppose that Assumptions 1 and 2 hold. Also assume that the minimum wealth condition*

$$x \geq A\mathbf{E}\left[\int_0^T \xi_v dv\right] \tag{3.11}$$

is satisfied. A consumption plan c^* is optimal if and only if it satisfies the first order conditions

$$u'\left(c_v^*, v\right) = y\xi_v \tag{3.12}$$

$$\mathbf{E}\left[\int_0^T \xi_v I(y^*\xi_v, v)dv\right] = x \tag{3.13}$$

for some constant $y > 0$.

Condition (3.12) shows that optimal consumption is set so as to equate the marginal utility of consumption to its marginal cost. Given the utility assumptions in Sect. 3.2.3, the unique solution is $c_v^* = I(y^*\xi_v, v)$ where $I(\cdot, v)$ is the inverse marginal utility function. Condition (3.13) is the static budget constraint. It ensures that initial resources are exhausted at the optimum.

The first order condition (3.12) can be rewritten in terms of the value function associated with the agent's optimization problem, more specifically its derivative with respect to wealth. The resulting equation corresponds to the optimality condition derived by Merton (1971), based on dynamic programming principles.

3.3.2 Equilibrium State Price Density

At equilibrium, the demand for the consumption good c equals its supply $C = \mathbf{1}'D$ where

$$\frac{dC_t}{C_t} = \frac{\mathbf{1}'dD_t}{\mathbf{1}'D_t} \equiv \mu_t^C dt + \sigma_t^C dW_t. \tag{3.14}$$

Substituting into the first order condition (3.12) leads to the following closed form solution for the equilibrium state price density and its components.

Theorem 2. *Consider the representative agent economy (u) and suppose that the aggregate dividend suffices to finance the subsistence consumption level (i.e., (3.11) holds). The equilibrium state price density is given by*

$$\xi_t = \frac{u_c\left(C_t, t\right)}{u_c\left(C_0, 0\right)} \tag{3.15}$$

for $t < T$ where $u_c\left(\cdot, t\right)$ is the derivative of the instantaneous utility function with respect to consumption. The equilibrium interest rate and market price of risk are given by

$$r_t = \beta_t + R_t\mu_t^C - \frac{1}{2}R_t P_t \sigma_t^C \left(\sigma_t^C\right)' \tag{3.16}$$

$$\theta_t = R_t \sigma_t^C, \tag{3.17}$$

where

$$\beta_t \equiv -\frac{u_{ct}(C_t,t)}{u_c(C_t,t)}, \quad R_t \equiv -\frac{u_{cc}(C_t,t)\,C_t}{u_c(C_t,t)}, \quad P_t \equiv -\frac{u_{ccc}(C_t,t)\,C_t}{u_{cc}(C_t,t)}. \tag{3.18}$$

and $u_{ct}(\cdot,t)$, $u_{cc}(\cdot,t)$, $u_{ccc}(\cdot,t)$ *are the first and second derivatives of the marginal utility function with respect to time and consumption. The coefficient* β_t *is the representative agent's subjective discount rate,* R_t *is a measure of relative risk aversion and* P_t *a measure of relative prudence.*

Equation (3.15) in Theorem 2 shows that the equilibrium SPD is the marginal rate of substitution between aggregate consumption at dates t and 0. An application of Ito's lemma to this expression leads to the equilibrium formulas for the interest rate and the market price of risk. The relation (25.64) between the instantaneous interest rate and the moments of the consumption (or production) growth rate was discovered by Breeden (1986). Formula (3.17) for the market price of risk leads to the well known CCAPM (consumption CAPM). This relation, between asset risk premia and the consumption growth rate, was identified and discussed by Breeden (1979). It builds on the standard intertemporal capital asset pricing model derived by Merton (1973).

The following characterization of stock prices is a direct consequence of Theorem 2.

Theorem 3. *Equilibrium stock prices are given by the present value formula*

$$S_{jt} = \mathbf{E}_t\left[\int_t^T \xi_{t,v} D_{jv} dv\right] = D_{jt}\mathbf{E}_t\left[\int_t^T \xi_{t,v} D_{jt,v} dv\right], \tag{3.19}$$

where $D_{jt,v} \equiv D_{jv}/D_{jt}$, *for* $j = 1, \ldots, d_s$. *Asset return volatilities are*

$$\begin{aligned} S_{jt}\sigma_{jt} &= \mathbf{E}_t\left[\int_t^T \xi_{t,v}\mathcal{D}_t D_{jv} dv\right] + \mathbf{E}_t\left[\int_t^T \xi_{t,v}\left(R_t\frac{\mathcal{D}_t C_t}{C_t} - R_v\frac{\mathcal{D}_t C_v}{C_v}\right) D_{jv} dv\right] \\ &= \lambda_j(t, D_t, Y_t)\,S_{jt} + \sigma_t^C \mathbf{E}_t\left[\int_t^T \xi_{t,v}(R_t - R_v)\,D_{jv} dv\right] \\ &\quad + D_{jt}\left(\mathbf{E}_t\left[\int_t^T \xi_{t,v}\left(\mathcal{D}_t D_{jt,v} - R_v(\mathcal{D}_t \log C_{t,v})\,D_{jt,v}\right) dv\right]\right), \end{aligned} \tag{3.20}$$

where $\mathcal{D}_t D_{jv}$, $\mathcal{D}_t Y_s$ *solves the stochastic differential equations*

$$\begin{aligned} d\mathcal{D}_t D_{jv} &= \left[\gamma_j(v, D_v, Y_v)\,dv + \lambda_j(v, D_v, Y_v)\,dW_v\right]\mathcal{D}_t D_{jv} && (3.21) \\ &\quad D_{jv}\left[\partial_D \gamma_j(v, D_v, Y_v)\,dv + (dW_v)'\,\partial_D \lambda_j(v, D_v, Y_v)'\right]\mathcal{D}_t D_v \\ &\quad + D_{jv}\left[\partial_Y \gamma_j(v, D_v, Y_v)\,dv + (dW_v)'\,\partial_Y \lambda_j(v, D_v, Y_v)'\right]\mathcal{D}_t Y_v && (3.22) \end{aligned}$$

$$d\mathcal{D}_t Y_s = \left[\partial\mu^Y(s, Y_s)ds + \sum_{j=1}^{d} \partial_Y \sigma_j^Y(s, Y_s)dW_s^j\right]\mathcal{D}_t Y_s \tag{3.23}$$

with initial conditions $\mathcal{D}_t D_{jt} = D_{jt}\lambda_j(t, D_t, Y_t)$ *and* $\mathcal{D}_t Y_t = \sigma^Y(t, Y_t)$. *The Malliavin derivative of aggregate consumption is* $\mathcal{D}_t C_v = \sum_j \mathcal{D}_t D_{jv}$. *In these expressions* $\partial_D f(D, Y)$ *and* $\partial_Y f(D, Y)$ *stand for the gradients of a function* f *with respect to the vectors* D *and* Y. *If* f *is a* $k\times 1$ *vector the differential* $\partial_Y f(Y)$ *is a* $k\times k$ *matrix. Expected stock returns satisfy the CCAPM:* $\mu_{jv} - r_v = \sigma_{jv}\theta_v = R_v\sigma_{jv}\sigma_v^C$.

The present value formula (3.19) is standard. Expression (3.20) shows that stock volatilities are made up of two components. The first one (first line of (3.20)) consists of static terms associated with the instantaneous volatilities of the stock's dividend and the aggregate consumption growth rates. The second one (second line of (3.20)) is associated with intertemporal components due to fluctuations in the coefficients of the dividend and consumption growth rates. Formula (3.20) is a special case of the volatility expression in Detemple and Zapatero (1991), which also accounts for habit formation effects and assumes a more general uncertainty structure (Ito processes). A variation of the formula is presented in Detemple and Serrat (2003) for the case of wealth constraints, in a model with constant coefficients. Recent contributions, such as Berrada (2006) and Gallmeyer and Hollifield (2008), have also analyzed versions of the formula for certain types of economies with heterogeneous beliefs.

The CCAPM and the present value formula in Theorem 3 are the basis for a wide literature discussing properties of asset pricing models and in particular their ability to explain empirical regularities. The volatility formula sheds light on some of the debates. For instance, the original argument for the volatility puzzle identified in Grossman and Shiller (1981) relies on the assumption of constant relative risk aversion (CRRA) and constant moments of the dividend growth rate. As (3.20) shows, stock return and dividend growth rate volatilities are the same under these assumptions. In this context, low volatility of a dividend growth rate implies low volatility of the stock's return, in contradiction with the empirical evidence. Follow-up literature, e.g. Hansen and Singleton (1983) and Grossman et al. (1987), highlighted the high levels of relative risk aversion estimates implied by this simple economic model (with CRRA). This theme is central to the equity premium puzzle popularized by the analysis in Mehra and Prescott (1985).

Likewise, contingent claim values satisfy the standard present value formulas

Theorem 4. *Consider a contingent claim* (f, F, τ) *where* f *is an adapted stochastic process representing intermediate cash flows,* F *is an* $\mathcal{F}_\tau$*-measurable terminal payment and* $\tau < T$ *is a stopping time representing the random maturity date of the claim. The equilibrium value of the claim is*

$$V_t = \mathbf{E}_t\left[\int_t^\tau \xi_{t,v} f_v dv + \xi_{t,v} F_\tau\right]. \tag{3.24}$$

In the case of a pure discount bond $(0, 1, v)$ *with maturity date* $v < T$, *the equilibrium price becomes* $B_t^v = E_t[\xi_{t,v}]$. *Similarly, the European call option* $(0, (S_{jv} - K)^+, v)$ *written on stock* j *and maturing at* $v < T$ *is worth* $C_t^e = E_t\left[\xi_{t,v}(S_{jv} - K)^+\right]$.

Valuation formulas for contingent claims, such as (3.24), can be found in numerous places. A derivation in the context of a production model and a PDE characterization of the price can be found in Cox et al. (1985).

3.4 Computation

Equilibrium asset prices and contingent claims are characterized in Theorems 3 and 4 as conditional expectations of their discounted payoffs where discounting is performed using the equilibrium SPD in Theorem 2. For general diffusion processes, these expectations cannot be simplified and expressed in explicit form as known functions of the relevant state variables. Numerical computations are therefore necessary for implementation.

In principle, various numerical approaches can be employed to carry out computations. Lattice methods (such as PDE schemes or finite dimensional trees) and Monte Carlo methods are two possible approaches. In many practical implementations, however, the relevant state space is large. Lattice methods, which suffer from the curse of dimensionality, quickly become infeasible as the dimensionality of the problem increases. In these instances, Monte Carlo simulation remains the only tractable approach.

This section describes an implementation based on Monte Carlo simulation. Section 25.4 presents the computational algorithm. An illustrative example is presented in Sect. 25.5.

3.4.1 A Monte Carlo Method

Numerical implementation is carried out using Monte Carlo simulation. The following basic scheme can be employed for the valuation of stocks (taken as an example) at date 0:

1. Select a discretization with $K+1$ points of the time interval $[0,T]$: $\{t_k : k=0,\ldots,K\}$. Let $h \equiv t_{k+1} - t_k$ be the common size of the partition (equidistant partition).
2. Along this time discretization simulate M trajectories, $m=1,\ldots,M$, of the Brownian motion $\underline{W}$ and construct the corresponding trajectories for the pair (D,Y). This can be done using various discretization schemes for stochastic differential equations (see Kloeden and Platen (1999)). For the Euler scheme

$$D^m_{t_{k+1}} = D^m_{t_k} + I^{D^m}_{t_k}\left[\gamma\left(t_k, D^m_{t_k}, Y^m_{t_k}\right)h + \lambda\left(t_k, D^m_{t_k}, Y^m_{t_k}\right)\Delta W^m_{t_k}\right] \quad (3.25)$$

$$Y^m_{t_{k+1}} = Y^m_{t_k} + \mu^Y\left(t_k, Y^m_{t_k}\right)h + \sigma^Y\left(t_k, Y^m_{t_k}\right)\Delta W^m_{t_k} \quad (3.26)$$

for $k=0,\ldots,K-1$ and $m=1,\ldots,M$, where $\Delta W_{t_k}^{m} \equiv W_{t_{k+1}}^{m} - W_{t_k}^{m}$. Initial values are $D_{t_0}^{m} = D_0$ and $Y_{t_0}^{m} = Y_0$ for all m.

3. Construct the corresponding trajectories of the equilibrium SPD ξ using (3.15). Construct the weighted dividend processes ξD and their cumulative value $P_{jT} \equiv \int_0^T \xi_v D_{jv} dv$ using

$$P_{jt_{k+1}}^{m} = P_{jt_k}^{m} + \xi_{t_k}^{m} D_{jt_k}^{m} h, \quad \text{for } k = 0,\ldots,K-1 \text{ and } m = 1,\ldots,M.$$

Initial values are $P_{jt_0}^{m} = 0$ for $m = 1,\ldots,M$.

4. Calculate the stock prices by taking the Monte Carlo average of the cumulative discounted payoff over the set of simulated trajectories

$$S_{j0} = \frac{1}{M}\sum_{m=1}^{M} P_{jt_K}^{m} = \frac{1}{M}\sum_{m=1}^{M}\left(\sum_{k=0}^{K-1} \xi_{t_k}^{m} D_{jt_k}^{m}\right) h.$$

The same algorithm applies in order to price securities at an arbitrary date $t \in [0,T)$. In that case the simulation estimates the evolution of the conditional state price density $\xi_{t,v} = u_c(C_v, v)/u_c(C_t, t)$ and the associated discounted dividends over the subinterval $[t, T)$.

The volatility process in Theorem 3, which is also expressed as a conditional expectation, can be estimated in the same manner. Step 2, in this instance, constructs the trajectories of the vector diffusion process $(D, Y, \mathcal{D}D, \mathcal{D}Y)$ according to (3.25) and (3.26) and

$$\begin{aligned}
&\left(\mathcal{D}_{t_0} D_{jt_{k+1}}\right)^{m} \\
&= \left(\mathcal{D}_{t_0} D_{jt_k}\right)^{m} + \left[\gamma_j\left(t_k, D_{t_k}^{m}, Y_{t_k}^{m}\right) h + \lambda_j\left(t_k, D_{t_k}^{m}, Y_{t_k}^{m}\right) \Delta W_{t_k}^{m}\right] \left(\mathcal{D}_{t_0} D_{jt_k}\right)^{m} \\
&\quad + D_{jt_k}^{m}\left[\partial_D \gamma_j\left(t_k, D_{t_k}^{m}, Y_{t_k}^{m}\right) h + \left(\Delta W_{t_k}^{m}\right)' \partial_D \lambda_j\left(t_k, D_{t_k}^{m}, Y_{t_k}^{m}\right)'\right] \left(\mathcal{D}_{t_0} D_{t_k}\right)^{m} \\
&\quad + D_{jt_k}^{m}\left[\partial_Y \gamma_j\left(t_k, D_{t_k}^{m}, Y_{t_k}^{m}\right) h + \left(\Delta W_{t_k}^{m}\right)' \partial_Y \lambda_j\left(t_k, D_{t_k}^{m}, Y_{t_k}^{m}\right)'\right] \left(\mathcal{D}_{t_0} Y_{t_k}\right)^{m}
\end{aligned} \tag{3.27}$$

$$\begin{aligned}
&\left(\mathcal{D}_{t_0} Y_{t_{k+1}}\right)^{m} \\
&= \left(\mathcal{D}_{t_0} Y_{t_k}\right)^{m} + \left[\partial_Y \mu^Y\left(t_k, Y_{t_k}^{m}\right) h + \sum_{j=1}^{d} \partial_Y \sigma_j^Y\left(t_k, Y_{t_k}^{m}\right) \Delta W_{t_k}^{j,m}\right] \left(\mathcal{D}_{t_0} Y_{t_k}\right)^{m}
\end{aligned}$$

for $k=0,\ldots,K-1$ and $m=1,\ldots,M$, where $\Delta W_{t_k}^{j,m} \equiv W_{t_{k+1}}^{j,m} - W_{t_k}^{j,m}$. Initial conditions are $\mathcal{D}_{t_0} D_{jt_0}^{m} = D_{jt_0} \lambda_j\left(t_0, D_{t_0}, Y_{t_0}\right)$ *and* $\mathcal{D}_{t_0} Y_{t_0}^{m} = \sigma^Y\left(t_0, Y_{t_0}\right)$ for all $m=1,\ldots,M$. In some cases, a dimensionality reduction can be achieved by expressing these Malliavin derivatives in terms of the related tangent processes (see Detemple et al., 2008).

3.4.2 Numerical Example

The example is based on the model with CRRA and a single source of uncertainty (hence a single stock) capturing the risk factor underlying the stock market. In this setting aggregate consumption is the dividend paid by the asset, $C = D$. Equilibrium is summarized in the following corollary.

Corollary 1. *Consider the economy of Theorem 2 and suppose that the agent has constant relative risk aversion $u(c,t) = \exp(-\beta t) c^{1-R}/(1-R)$ where R is the relative risk aversion coefficient and β is a constant subjective discount rate. Also suppose $d = d_s = 1$ (hence $C = D$). The equilibrium state price density and its components are*

$$\xi_t = \exp(-\beta t)\left(\frac{C_t}{C_0}\right)^{-R} \tag{3.28}$$

$$r_t = \beta + R\mu_t^C - \frac{1}{2}R(1+R)\sigma_t^C\left(\sigma_t^C\right)' \tag{3.29}$$

$$\theta_t = R\sigma_t^C. \tag{3.30}$$

The equity premium and the volatility of the market return are

$$\mu_t - r_t = R\sigma_t\sigma_t^C \tag{3.31}$$

$$\begin{aligned}\sigma_t &= \sigma_t^C + (1-R)\frac{\mathbf{E}_t\left[\int_t^T \xi_{t,v}\mathcal{D}_t C_{t,v}dv\right]}{\mathbf{E}_t\left[\int_t^T \xi_{t,v}C_{t,v}dv\right]}\\ &= R\sigma_t^C + (1-R)\frac{\mathbf{E}_t\left[\int_t^T \xi_{t,v}\mathcal{D}_t C_v dv\right]}{\mathbf{E}_t\left[\int_t^T \xi_{t,v}C_v dv\right]}.\end{aligned} \tag{3.32}$$

In these expressions $\mu_t^C = \gamma(t, D_t, Y_t)$ and $\sigma_t^C = \lambda(t, C_t, Y_t)$.

For illustration purposes assume that aggregate consumption (dividend) follows the nonlinear mean reverting process

$$dC_t = C_t\left[\kappa\left(\overline{C} - C_t\right)dt + \lambda dW_t\right],$$

where κ, λ are constants. This process has linear speed of mean reversion (the drift has therefore a quadratic component) and proportional volatility. The Malliavin derivative process solves

$$d\mathcal{D}_t C_s = \mathcal{D}_t C_s\left[\kappa\left(\overline{C} - 2C_s\right)ds + \lambda dW_s\right]; \quad \mathcal{D}_t C_t = C_t\lambda$$

so that

$$\mathcal{D}_t C_s = C_t \lambda \exp\left(\int_t^s \left(\kappa\left(\overline{C} - 2C_v\right) - \frac{1}{2}\lambda^2\right) dv + \lambda\left(W_s - W_t\right)\right)$$
$$= C_s \lambda \exp\left(-\kappa \int_t^s C_v dv\right).$$

It follows that $\mathcal{D}_t C_s > 0$ for all $s \in [t, T]$ and that the second component in the stock volatility (on the second line of (3.32)) is positive (resp. negative) for $R < 1$ (resp. $R > 1$).

Figure 3.1 displays the market return volatility as a function of relative risk aversion and the volatility of the consumption growth rate. For the parameter values selected the market return volatility is increasing in both variables. It exceeds (resp. falls below) the consumption growth rate volatility $\lambda = 3\%$ when risk aversion exceeds 1 (resp. falls below 1). Maximum market volatility is $\sigma_0 = 4.5\%$ over the range of values examined.

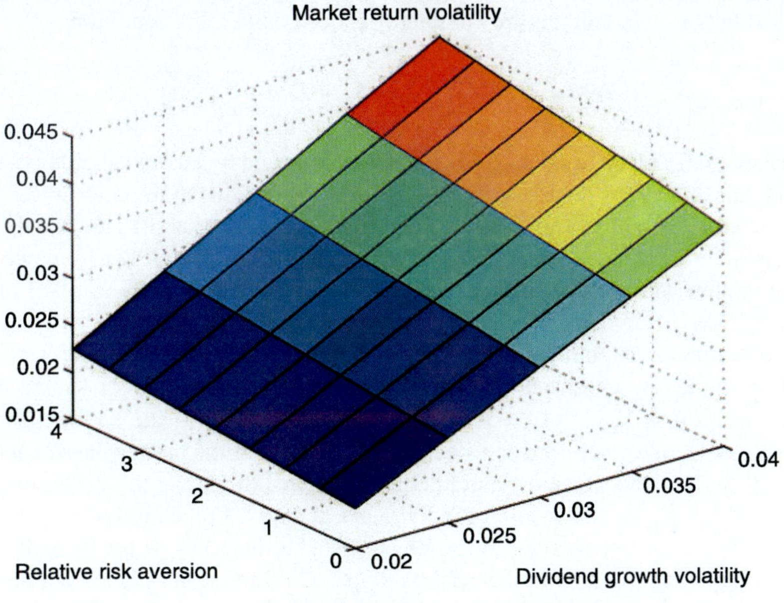

Fig. 3.1 This figure shows the market volatility as a function of relative risk aversion and aggregate consumption (dividend) growth rate volatility. Parameter values are $C_0 = 10^9$, $\kappa = 0.001 \times 10^{-9}$, $\overline{C} = 10^9$, $\beta = 0.01$, $T = 100$. Risk aversion varies from 0 to 4. Consumption volatility from 2 to 4%. The number of trajectories and time discretization points are $M = 1{,}000$, $N = 100$

3.5 Multiagent Models

Section 3.5.1 outlines the structure of the canonical complete market model with multiple agents. Section 3.5.2 describes the structure of consumption and portfolio demands. Equilibrium characterizations are in Sect. 3.5.3. Algorithms for numerical computation and illustrations are presented in Sect. 3.5.4.

3.5.1 Model Structure

The uncertainty structure and the financial market are the same as in Sect. 25.2.1. But it is now assumed that there are as many risky stocks as Brownian motions ($d_s = d$). In addition to Assumption 1, it is also postulated that asset prices are such that the financial market is complete in equilibrium.

Assumption 3. *Candidate equilibrium price processes are such that* σ^{-1} *exists.*

Assumption 3 is a non-degeneracy condition which ensures that all the risks are hedgeable (i.e., that markets are complete). Indeed, under this condition

$$dW_t = \sigma_t^{-1}\left(\left(I_t^S\right)^{-1}(dS_t + D_t dt) - \mu_t dt\right)$$

which indicates that Brownian shocks can be duplicated by taking suitable positions in the stocks and the riskless asset. In addition, under this assumption the market price of risk is uniquely implied by the returns on traded assets and equals $\theta \equiv \sigma^{-1}(\mu - r)$. It represents the risk premium per unit risk. The SPD associated with this complete market structure is given by (25.2) evaluated at the implied market price of risk.

The economy's population is comprised of $i = 1, \ldots, N$ diverse investors. Each individual is endowed with a number $n_i \equiv (n_{i,1} \ldots, n_{i,d_s})$ of shares of the stocks, such that $\sum_{i=1}^N n_{i,j} = 1$ for all $j = 1, \ldots, d_s$ (aggregate endowments are normalized to one share for each stock). Endowments of the money market account are null. Individuals consume and allocate their wealth among the different assets available. Let X_{it} be the wealth of individual i at date t. Consumption is c_{it} and π_{it} is the $d \times 1$ vector of wealth proportions invested in the risky assets (thus $1 - \pi_{it}'\mathbf{1}$ is the proportion invested in the riskless asset). Consumption satisfies the physical nonnegativity constraint $c_i \geq 0$. No sign restrictions are placed on the proportions π_i invested in the various assets: long as well as short positions are permitted. The evolution of individual wealth is governed by the stochastic differential equation

$$dX_{it} = (X_{it} r_t - c_{it})\, dt + X_{it}\pi_{it}'\left[(\mu_t - r_t\mathbf{1})\, dt + \sigma_t dW_t\right] \tag{3.33}$$

subject to the initial condition $X_{i0} = x_i = n_i S_0$. For this evolutionary equation to make sense the following integrability condition is imposed on the policy (c_i, π_i)

$$\int_0^T \left(|c_{it}| + \left|X_{it}\pi'_{it}(\mu_t - r_t\mathbf{1})\right| + \left|X_{it}\pi'_{it}\sigma_t\sigma'_t\pi_{it}X_{it}\right|\right) dt < \infty, \quad (P-a.s.). \tag{3.34}$$

Under (3.34) the stochastic integral on the right hand side of (3.33) is well defined. Condition (3.34) is a joint restriction on consumption-portfolio policies and candidate equilibrium price processes.

Preferences for each individual are assumed to have the time-separable von Neumann-Morgenstern representation. The felicity provided by a consumption plan (c_i) is

$$\mathcal{U}(c_i) \equiv \mathbf{E}\left[\int_0^T u_i(c_{iv}, v)dv\right], \tag{3.35}$$

where the utility function $u_i : [A_i, \infty) \times [0,T] \to \mathbb{R}$ is strictly increasing, strictly concave and differentiable over its domain. The consumption bound is assumed to be nonnegative, $A_i \geq 0$. The limiting conditions $\lim_{c\to A_i} u'_i(c,t)$ and $\lim_{c\to\infty} u'_i(c,t) = 0$ are also assumed to hold, for all $t \in [0,T]$. If $[A_i, \infty)$ is a proper subset of $\mathbb{R}_+$ (i.e. $A_i > 0$) the function u_i is extended to $\mathbb{R}_+ \times [0,T]$ by setting $u_i(c,t) = -\infty$ for $c \in \mathbb{R}_+ \backslash [A_i, \infty)$ and for all $t \in [0,T]$.

Each consumer-investor seeks to maximize expected utility

$$\max_{(c_i,\pi_i)} \mathcal{U}(c_i) \tag{3.36}$$

subject to the constraints

$$dX_{it} = (r_t X_{it} - c_{it})\,dt + X_{it}\pi'_{it}\left[(\mu_t - r_t)\,dt + \sigma_t dW_t\right]; \; X_{i0} = x_i \tag{3.37}$$

$$c_{it} \geq 0, X_{it} \geq 0 \tag{3.38}$$

for all $t \in [0,T]$, and the integrability condition (3.34). Initial resources are $x_i = n_i S_0$.

A policy (c_i, π_i) is *admissible* for agent i, written $(c_i, \pi_i) \in \mathcal{A}_i$, if and only if it satisfies (3.37) and (3.38). A policy $\left(c_i^*, \pi_i^*\right)$ is *optimal* for i, written $\left(c_i^*, \pi_i^*\right) \in \mathcal{A}_i^*$, if and only if it cannot be dominated, i.e., $\mathcal{U}\left(c_i^*\right) \geq \mathcal{U}(c_i)$ for all $(c_i, \pi_i) \in \mathcal{A}_i$.

A *competitive equilibrium* is a collection of stochastic processes

$$\{(c_i, \pi_i) : i = 1, \ldots N, (S_0, r, \theta, \sigma)\}$$

such that:

1. *Individual rationality:* $(c_i, \pi_i) \in \mathcal{A}_i^*$, for $i = 1, \ldots, N$ where (S_0, r, θ, σ) is taken as given.
2. *Market clearing:* (i) commodity market: $\sum_{i=1}^N c_i = C$, (ii) equity market: $\sum_{i=1}^N X_i \pi_i = S$ and (iii) money market: $\sum_{i=1}^N X_i\left(1 - \pi_i'\mathbf{1}\right) = 0$.

3.5.2 Consumption and Portfolio Demands

Optimal consumption and portfolio demands are obtained by using the methods in Pliska (1986), Karatzas et al. (1987), Cox and Huang (1989), Ocone and Karatzas (1991) and Detemple et al. (2003).

Theorem 5. *Consider the dynamic consumption-portfolio problem (25.5)–(25.7) and suppose that the minimum wealth condition holds*

$$x_i \geq A_i \mathbf{E}\left[\int_0^T \xi_v dv\right]. \tag{3.39}$$

Optimal consumption is $c_{iv}^* = I_i\left(y_i^* \xi_{iv}, v\right)$ *where* y_i^* *solves*

$$\mathbf{E}\left[\int_0^T \xi_v I_i(y_i^* \xi_v, v) dv\right] = x_i. \tag{3.40}$$

Intermediate wealth satisfies

$$X_{it}^* = \mathbf{E}_t\left[\int_t^T \xi_{t,v} c_{iv}^* dv\right]. \tag{3.41}$$

Define the absolute risk tolerance function

$$\Gamma_{iu}(c, v) = -\frac{u_i'(c, v)}{u_i''(c, v)} \tag{3.42}$$

and let $\Gamma_{iv}^* \equiv \Gamma_{iu}(c_{iv}^*, v)$ *be the risk tolerance function evaluated at optimal consumption at date v. Define the random variable* $I_v^i \equiv I_i\left(y_i^* \xi_v, v\right)$*. The optimal portfolio has the decomposition* $\pi_{it}^* = \pi_{it}^m + \pi_{it}^r + \pi_{it}^\theta$ *with*

$$\pi_{it}^m = \mathbf{E}_t\left[\int_t^T \xi_{t,v} \Gamma_{iv}^* dv\right] (\sigma_t')^{-1} \theta_t \tag{3.43}$$

$$\pi_{it}^r = -(\sigma_t')^{-1} \mathbf{E}_t\left[\int_t^T \xi_{t,v}\left(c_{iv}^* - \Gamma_{iv}^*\right) H_{t,v}^r dv\right] \tag{3.44}$$

$$\pi_{it}^\theta = -(\sigma_t')^{-1} \mathbf{E}_t\left[\int_t^T \xi_{t,v}\left(c_{iv}^* - \Gamma_{iv}^*\right) H_{t,v}^\theta dv\right], \tag{3.45}$$

where the random variables $H_{t,v}^r$, $H_{t,v}^\theta$ *are*

$$(H_{t,v}^r)' = \int_t^v \partial r(Y_s, s) \mathcal{D}_t Y_s ds \tag{3.46}$$

$$(H^{\theta}_{t,v})' = \int_t^v \theta_s' \partial\theta(Y_s, s) \mathcal{D}_t Y_s ds + \int_t^v dW_s' \partial\theta(Y_s, s) \mathcal{D}_t Y_s \tag{3.47}$$

and $\mathcal{D}_t Y_s$ satisfies the stochastic differential equation (3.23).

The consumption demand is obtained as in the single agent model, as the inverse marginal utility evaluated at the (normalized) SPD. The portfolio financing the consumption demand has three terms, described in (3.43)–(3.45). The first component (3.43) is a mean-variance term, motivated by the desire to diversify. The next two components are dynamic hedging terms, designed to hedge fluctuations in the opportunity set: (3.44) is an interest rate hedge and (3.45) is a market price of risk hedge.

3.5.3 Equilibrium Prices and Allocations

Aggregating over individual consumption demands yields the aggregate demand function

$$\sum_{i=1}^{N} I_i \left(y_i^* \xi_t, t \right).$$

The clearing condition in the commodity market leads to the following standard characterization of the equilibrium state price density.

Theorem 6. *Consider the multiagent economy with population $\{(u_i, n_i) : i = 1, \ldots, N\}$ and suppose that individual wealth finances the subsistence consumption for each agent (i.e., (3.39) holds for each $i = 1, \ldots, N$). The equilibrium state price density is given by*

$$\xi_t = \frac{f(C_t, t; z)}{f(C_0, 0; z)}, \tag{3.48}$$

where $f(C_t, t; z)$ is the unique solution of the nonlinear equation

$$I_1 \left(f(C_t, t; z), t \right) + \sum_{i=2}^{N} I_i \left(z_i f(C_t, t; z), t \right) = C_t \tag{3.49}$$

and the $N - 1$ dimensional vector of relative Lagrange multipliers $z \equiv (z_2, \ldots, z_N)$ solves the system of nonlinear equations

$$x_i = \mathbf{E} \left[\int_0^T \frac{f(C_v, v; z)}{f(C_0, 0; z)} I_i \left(z_i f(C_v, v; z), v \right) dv \right] \tag{3.50}$$

for $i = 2, \ldots, N$. In (3.50) $x_i = n_i S_0$ where S_0 is given by the present value formula in Theorem 3 (with ξ as in (3.48)).

Questions pertaining to the existence and uniqueness of equilibria related to those described in Theorem 6 are addressed in Karatzas et al. (1990) and Karatzas et al. (1991) (see also Karatzas and Shreve (1998), Sect. 4.6).

Equation (3.48) expresses the state price density in terms of a function $f(C_t, t; z)$ of aggregate consumption C_t and of the vector of Lagrange multipliers z, which satisfies the market clearing condition (3.49) at each point in time $t \in [0, T]$. The function $f(C_t, t; z)$ is the unnormalized state price density. It corresponds to the marginal utility of the aggregator. Lagrange multipliers are also endogenous: they satisfy the vector of static budget constraints (3.50) which are parameterized by the unknown stochastic process $\{f(C_t, t; z) : t \in [0, T]\}$.

An application of Ito's lemma now gives the equilibrium interest rate and market price of risk.

Theorem 7. *Consider the multiagent economy with population $\{(u_i, n_i) : i = 1, \ldots, N\}$ and suppose that individual wealth finances the subsistence consumption of each agent. The equilibrium interest rate and market price of risk are given by*

$$r_t = \beta_t^a + R_t^a \mu_t^C - \frac{1}{2} R_t^a P_t^a \sigma_t^C \left(\sigma_t^C\right)' \tag{3.51}$$

$$\theta_t = R_t^a \sigma_t^C, \tag{3.52}$$

where

$$\beta_t^a \equiv \frac{\sum_{i=1}^N I_i'(z_i f(C_t, t; z)) \beta_{it}}{\sum_{i=1}^N I_i'(z_i f(C_t, t; z)) z_i f(C_t, t; z)} \tag{3.53}$$

$$R_t^a \equiv \frac{C_t}{\sum_{i=1}^N I_i'(z_i f(C_t, t; z)) z_i f(C_t, t; z)} \tag{3.54}$$

$$P_t^a \equiv \frac{1}{R_t^a} \frac{\sum_{i=1}^N I_i''(z_i f(C_t, t; z)) z_i f'(C_t, t; z)}{\sum_{i=1}^N I_i'(z_i f(C_t, t; z)) z_i f(C_t, t; z)}. \tag{3.55}$$

The coefficient β_t^a is the aggregate discount rate, R_t^a is a measure of aggregate relative risk aversion and P_t^a is a measure of aggregate prudence.

Stock prices satisfy the standard present value formula based on the equilibrium SPD.

Theorem 8. *Equilibrium stock prices are given by the present value formula (3.19). Asset return volatilities are*

$$S_{jt}\sigma_{jt} = \mathbf{E}_t\left[\int_t^T \xi_{t,v} \mathcal{D}_t D_{jv} dv\right] + \mathbf{E}_t\left[\int_t^T (\mathcal{D}_t \xi_{t,v}) D_{jv} dv\right], \tag{3.56}$$

where $\mathcal{D}_t \xi_{t,v}$ is calculated on the basis of the equilibrium SPD formula (3.48).

The volatility formula in Theorem 8 concentrates on the contributions of future dividends and of the CSPD. A more refined version, identifying static and dynamic components, can be derived by calculating the Malliavin derivatives $\mathcal{D}_t D_{jv}, \mathcal{D}_t \xi_{t,v}$.

Equilibrium portfolios can also be characterized as

Theorem 9. *Equilibrium portfolios are* $\pi_{it}^* = \pi_{it}^m + \pi_{it}^r + \pi_{it}^\theta$ *with*

$$\pi_{it}^m = \mathbf{E}_t \left[\int_t^T \xi_{t,v} \Gamma_{iv}^* dv \right] (\sigma_t')^{-1} \theta_t \tag{3.57}$$

$$\pi_{it}^r = -(\sigma_t')^{-1} \mathbf{E}_t \left[\int_t^T \xi_{t,v} \left(c_{iv}^* - \Gamma_{iv}^* \right) H_{t,v}^r dv \right] \tag{3.58}$$

$$\pi_{it}^\theta = -(\sigma_t')^{-1} \mathbf{E}_t \left[\int_t^T \xi_{t,v} \left(c_{iv}^* - \Gamma_{iv}^* \right) H_{t,v}^\theta dv \right], \tag{3.59}$$

where $H_{t,v}^r, H_{t,v}^\theta$ *are*

$$(H_{t,v}^r)' = \int_t^v \mathcal{D}_t r_s ds \quad \text{and} \quad (H_{t,v}^\theta)' = \int_t^v \theta_s' \mathcal{D}_t \theta_s ds + \int_t^v dW_s' \mathcal{D}_t \theta_s \tag{3.60}$$

and the Malliavin derivatives $\mathcal{D}_t r_s, \mathcal{D}_t \theta_s$ *are calculated on the basis of the formulas in Theorem 7.*

3.5.4 Computation

This section first presents two algorithms that can be used to compute equilibrium for general economic settings. It then specializes to a class of economies with two agents. A numerical illustration is presented in this context.

3.5.4.1 General Economies

On the basis of the result in Theorem 6, numerical computation of the state price density can be performed by using an iteration-simulation algorithm of the following type:

1. Select a discretization with $K+1$ points of the time interval: $\{t_k : k = 0, \ldots, K\}$.
2. Along this discretization simulate M trajectories of W and (D, Y). For the Euler scheme this gives the approximations (3.25) and (3.26).
3. Fix a vector of multipliers $z^{(0)}$:

   (a) At each time t_k and for each trajectory m calculate the solution of (3.49) using a zero finding procedure. If a Newton-Raphson scheme is used, the iterative algorithm is

$$f^{(n+1,m)}\left(C_{t_k}^m, t_k; z^{(0)}\right) = f^{(n,m)}\left(C_{t_k}^m, t_k; z^{(0)}\right) + \left(\sum_{i=1}^{N} I_i'\left(z_i^{(0)} f^{(n,m)}\left(C_{t_k}^m, t_k; z^{(0)}\right)\right) z_i^{(0)}\right)^{-1} \times \left(C_{t_k}^m - \sum_{i=1}^{N} I_i\left(z_i^{(0)} f^{(n,m)}\left(C_{t_k}^m, t_k; z^{(0)}\right)\right)\right)$$

for $n = 0, \ldots,$ where $f^{(0,m)}(C_{t_k}^m, t_k; z^{(0)})$ is a selected initial condition. Stop when a selected convergence criterion is satisfied. This produces an approximate stochastic process

$$\left\{ f_k^{(n^0,m)} \equiv f^{(n^0,m)}\left(C_k^m, t_k; z^{(0)}\right) : k = 0, \ldots, K-1 \right\},$$

where n^0 is the stopping point.

(b) Feed $\left\{ f_k^{(n^0,m)} : k = 0, \ldots, K-1 \right\}$ into (3.50) and calculate the updated vector of multipliers $z^{(1)}$ which solves (3.50). In this calculation, expectations are estimated by Monte Carlo averaging over the trajectories of the (Euler) approximate process $\left\{ f_k^{(n^0,m)} : k = 0, \ldots, K-1 \right\}$. A Newton-Raphson scheme can again be used to calculate the fixed point. This Newton-Raphson Monte Carlo Euler scheme is

$$z_i^{(j+1)} = z_i^{(j)} + \left(\sum_{m=1}^{M}\sum_{k=0}^{K-1}\left(f_k^{(n^0,m)}\right)^2 I_i'\left(z_i^{(j)} f_k^{(n^0,m)}\right) h\right)^{-1} \times \sum_{m=1}^{M}\sum_{k=0}^{K-1} f_k^{(n^0,m)}\left(n_i C_k^m - I_i\left(z_i^{(j)} f_k^{(n^0,m)}\right)\right) h$$

for $j = 0, \ldots$ and $i = 2, \ldots, N$. Stop when a selected convergence criterion is satisfied.

4. Repeat step 3 until some desired convergence criterion is satisfied.

The overall procedure is a two-stage Newton-Raphson Monte Carlo Euler scheme. This procedure is computationally intensive, as it is exponential in the number of stages.

An alternative is a one stage Newton-Raphson Monte Carlo Euler scheme. Choose starting values $z_i^{(0)}$ and $f^{(0,m)}\left(C^m; z^{(0)}\right)$ and iterate to obtain

$$\left\{\left(f^{(n+1,m)}, z_i^{(n+1)}\right) : m = 1, \ldots, M \text{ and } i = 2, \ldots, N\right\}$$

from $\left\{\left(f^{(n,m)}, z_i^{(n)}\right) : m = 1, \ldots, M \text{ and } i = 2, \ldots, N\right\}$ as follows

$$\begin{bmatrix} f_0^{(n+1,m)} \\ \vdots \\ f_{K-1}^{(n+1,m)} \\ z_2^{(n+1)} \\ \vdots \\ z_N^{(n+1)} \end{bmatrix} = \begin{bmatrix} f_0^{(n,m)} \\ \vdots \\ f_{K-1}^{(n,m)} \\ z_2^{(n)} \\ \vdots \\ z_N^{(n)} \end{bmatrix} + \left(H^{(n,m)}\right)^{-1} L^{(n,m)},$$

where

$$L^{(n,m)} \equiv \begin{bmatrix} C_0^m - \sum_{i=1}^N I_i\left(z_i^{(n)} f_0^{(n,m)}\right) \\ \vdots \\ C_{K-1}^m - \sum_{i=1}^N I_i\left(z_i^{(n)} f_{K-1}^{(n,m)}\right) \\ \sum_{m=1}^M \sum_{k=0}^{K-1} f_k^{(n,m)}\left(n_2 C_k^m - I_2\left(z_2^{(n)} f_k^{(n,m)}\right)\right) h \\ \vdots \\ \sum_{m=1}^M \sum_{k=0}^{K-1} f_k^{(n,m)}\left(n_N C_k^m - I_N\left(z_N^{(n)} f_k^{(n,m)}\right)\right) h \end{bmatrix}$$

and

$$H^{(n,m)} \equiv \begin{bmatrix} H_{11}^{(n,m)} & H_{12}^{(n,m)} \\ H_{21}^{(n,m)} & H_{22}^{(n,m)} \end{bmatrix}$$

with blocks

$$H_{11}^{(n,m)} \equiv diag\left[\sum_{i=1}^N I_i'\left(z_i^{(n)} f_k^{(n,m)}\right) z_i^{(n)}\right]$$

$$H_{12}^{(n,m)} \equiv \left[I_i'\left(z_i^{(n)} f_k^{(n,m)}\right) f_k^{(n,m)}\right]_{\substack{k=0,\ldots,K-1 \\ i=2,\ldots,N}}$$

$$H_{21}^{(n,m)} \equiv H_{211}^{(n,m)} + H_{212}^{(n,m)}$$

$$H_{211}^{(n,m)} \equiv \left[\sum_{m=1}^M \left(-n_i C_k^m + I_i\left(z_i^{(n)} f_k^{(n,m)}\right)\right) h\right]_{\substack{i=2,\ldots,N \\ k=0,\ldots,K-1}}$$

$$H_{212}^{(n,m)} \equiv \left[\sum_{m=1}^M \left(f_k^{(n,m)} I_i'\left(z_i^{(n)} f_k^{(n,m)}\right) z_i^{(n)}\right) h\right]_{\substack{i=2,\ldots,N \\ k=0,\ldots,K-1}}$$

$$H_{22}^{(n,m)} \equiv diag\left[\sum_{m=1}^{M}\sum_{k=0}^{K-1}\left(f_k^{(n,m)}\right)^2 I_i'\left(z_i^{(n)} f_k^{(n,m)}\right) h\right].$$

For a vector $V \equiv [V_1, \ldots, V_p]'$, the notation $diag[V_i]$ indicates the diagonal matrix with vector elements V_i on the diagonal, and $[V_{ij}]_{\substack{i=1,\ldots,p \\ j=1,\ldots,q}}$ is the matrix with rows $i = 1, \ldots, p$ and columns $j = 1, \ldots, q$. The iteration continues until a selected tolerance threshold is attained (Step 4).

The one-stage Newton-Raphson Monte Carlo Euler scheme updates the vector of Lagrange multipliers at each iteration. It therefore cuts down on the number of fixed point computations required to approximate the equilibrium SPD. But the number of equations involved at each iteration increases.

3.5.4.2 A Class of Economies with Two Types of Agents

Certain economies, in which agents' risk attitudes are related, are more amenable to computations. Consider for instance economies populated by two types of agents, both with constant relative risk aversion, where $R_2 = 2R_1$. Economies in that class have been examined by Dumas (1989) and Wang (1996).

Under these restrictions, the market clearing condition becomes

$$\left(\frac{f(C_t, t; z_2)}{\exp(-\beta t)}\right)^{-1/R_1} + \left(z_2 \frac{f(C_t, t; z_2)}{\exp(-\beta t)}\right)^{-1/R_2} = C_t$$

leading to the quadratic equation $G^2 + z_2^{-1/R_2} G - C_t = 0$, where

$$G \equiv \left(\frac{f(C_t, t; z_2)}{\exp(-\beta t)}\right)^{-1/R_2}.$$

A characterization of equilibrium is provided next

Corollary 2. *Consider the multiagent economy with population* $\{(u_i, n_i) : i = 1, 2\}$ *where* $u_i(c, t) = \exp(-\beta t) c^{1-R_i} / (1 - R_i)$ *with constant relative risk aversion* R_i *and constant subjective discount rate* β*. Assume furthermore that* $R_2 = 2R_1$*. The equilibrium state price density is given by (3.48) where*

$$f(C_t, t; z_2) = \exp(-\beta t)\left(\frac{-z_2^{-1/R_2} + \sqrt{z_2^{-2/R_2} + 4C_t}}{2}\right)^{-R_2}. \tag{3.61}$$

and the transformed multiplier $\varphi \equiv z_2^{-1/R_2}$ *solves the nonlinear equation*

$$x_2 = \mathbf{E}\left[\int_0^T \exp(-\beta v)\left(\frac{-\varphi+\sqrt{\varphi^2+4C_v}}{-\varphi+\sqrt{\varphi^2+4C_0}}\right)^{-R_2} \varphi\left(\frac{-\varphi+\sqrt{\varphi^2+4C_v}}{2}\right)dv\right] \tag{3.62}$$

with

$$x_2 = n_2\mathbf{E}\left[\int_0^T \exp(-\beta v)\left(\frac{-\varphi+\sqrt{\varphi^2+4C_v}}{-\varphi+\sqrt{\varphi^2+4C_0}}\right)^{-R_2} C_v dv\right]. \tag{3.63}$$

When $R_2 = 2R_1$ the equilibrium state price density is known up to the constant $\varphi \equiv z_2^{-1/R_2}$, that satisfies (3.62) and (3.63). The nonlinear equation for φ can also be written as $\varphi = G(\varphi)$ with

$$G(\varphi) \equiv 2n_2 \frac{\mathbf{E}\left[\int_0^T \exp(-\beta v)\left(-\varphi+\sqrt{\varphi^2+4C_v}\right)^{-R_2} C_v dv\right]}{\mathbf{E}\left[\int_0^T \exp(-\beta v)\left(-\varphi+\sqrt{\varphi^2+4C_v}\right)^{1-R_2} dv\right]}.$$

Simple derivations show that $G(0) > 0$ and $\lim_{\varphi\to\infty} G(\varphi)/\varphi < 1$. This ensures the existence of a fixed point. For $R_2 < 1$, it can be shown that $G'(\varphi) < 1$ at any arbitrary fixed point, which guarantees uniqueness (see also Karatzas and Shreve (1998), Theorems 6.1 and 6.4, for results in a related model).

Computation of equilibrium reduces to the resolution of this nonlinear equation. This is a one-dimensional zero finding problem, which can be solved by standard methods (Newton-Raphson scheme, bisection method, secant method, etc.,...).

To complete the description of equilibrium, return components are given next

Corollary 3. *Consider the economy of Corollary 2 and suppose* $d = d_s = 1$ *(hence* $C = D$*). The equilibrium state price density and its components are*

$$\xi_t = \frac{f(C_t, t; z_2)}{f(C_0, 0; z_2)} = \exp(-\beta t)\frac{\left(-\varphi+\sqrt{\varphi^2+4C_t}\right)^{-R_2}}{\left(-\varphi+\sqrt{\varphi^2+4C_0}\right)^{-R_2}} \tag{3.64}$$

$$r_t = \beta + R_t^a \mu_t^C - \frac{1}{2}R_t^a P_t^a \left(\sigma_t^C\right)^2 \tag{3.65}$$

$$\theta_t = R_t^a \sigma_t^C, \tag{3.66}$$

where the aggregate relative risk aversion and prudence coefficients are

$$R_t^a = 2R_2 \left(\frac{(\varphi^2 + 4C_t)^{-1/2}}{-\varphi + \sqrt{\varphi^2 + 4C_t}} \right) C_t \tag{3.67}$$

$$P_t^a = \left(\frac{2}{\varphi^2 + 4C_t} \right) \left(1 + (1 - R_2) \frac{\sqrt{\varphi^2 + 4C_t}}{-\varphi + \sqrt{\varphi^2 + 4C_t}} \right) C_t. \tag{3.68}$$

The equity premium and the volatility of the market return are given by

$$\mu_t - r_t = R_t^a \sigma_t \sigma_t^C \tag{3.69}$$

$$\sigma_t = \sigma_t^C + \frac{\mathbf{E}_t \left[\int_t^T \xi_{t,v} \left(\mathcal{D}_t C_{t,v} + \frac{\mathcal{D}_t \xi_{t,v}}{\xi_{t,v}} C_{t,v} \right) dv \right]}{\mathbf{E}_t \left[\int_t^T \xi_{t,v} C_{t,v} dv \right]}, \tag{3.70}$$

where

$$\mathcal{D}_t C_{t,v} = \frac{\mathcal{D}_t C_v}{C_t} - C_{t,v} \frac{\mathcal{D}_t C_t}{C_t} \tag{3.71}$$

$$\frac{\mathcal{D}_t \xi_{t,v}}{\xi_{t,v}} = -\left(R_v^a \frac{\mathcal{D}_t C_v}{C_v} - R_t^a \frac{\mathcal{D}_t C_t}{C_t} \right) \tag{3.72}$$

In these expressions $\mu_t^C = \gamma(t, D_t, Y_t)$ *and* $\sigma_t^C = \lambda(t, C_t, Y_t)$.

3.5.4.3 Numerical Example

Consider the two-agent economy of Sect. 3.5.4.2 and suppose that aggregate consumption follows the nonlinear process described in Sect. 25.5. Numerical computation of the transformed multiplier is based on (a variation of) the iterative scheme $\varphi^{(n+1)} = G(\varphi^{(n)})$.

Figure 3.2 displays the stock return volatility when relative risk aversion R_2 and consumption growth rate volatility vary. The return volatility behavior mimicks the patterns found in the single agent model, but with a milder impact of risk aversion. In this example aggregate risk aversion is lower than the risk aversion of agent 2, which helps to explain the smaller slope.

3.6 Conclusions

Equilibrium considerations have long been central in the asset pricing literature. Analyses based on complete market models have derived expressions for interest rates, risk premia and return volatilities that highlight the relation to fundamentals, such as consumption. Numerical analysis permits a new level of understanding of these relations and the complex phenomena that affect the behavior of security prices.

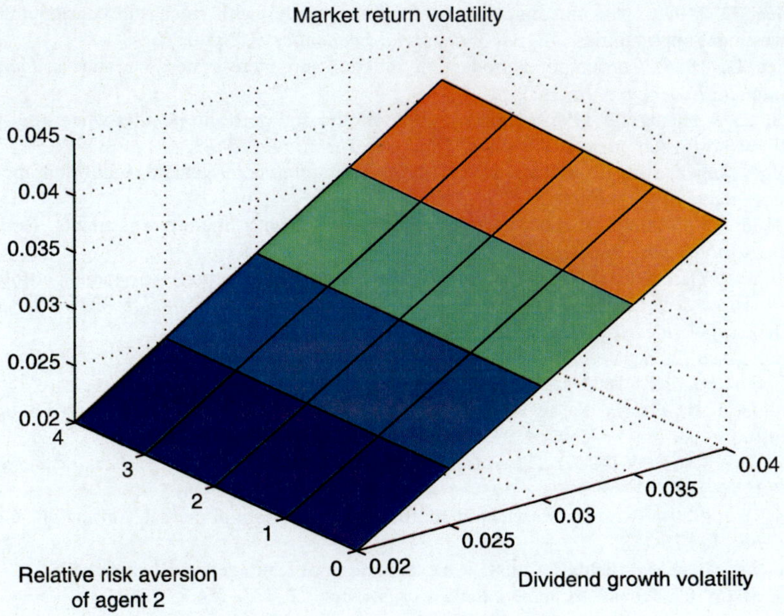

Fig. 3.2 Market volatility as a function of relative risk aversion of agent 2 and aggregate consumption (dividend) growth rate volatility. Parameter values are $C_0 = 10^9$, $\kappa = 0.001 \times 10^{-9}$, $\overline{C} = 10^9$, $\beta = 0.01$, $n_2 = 0.5$, $T = 100$. Risk aversion R_2 varies from 0 to 4. Consumption volatility from 2 to 4%. The number of trajectories and time discretization points are $M = 1,000$ $N = 100$

While complete market models are useful in helping us develop a basic understanding of equilibrium phenomena, they are clearly limited in their ability to capture certain features of the economic environment. For instance, it is fairly clear that risk factors can not all be hedged. Sudden economic events, emerging factors, individual-specific risks and restrictions on trading are a few examples of relevant elements that are set aside by complete market analyses. These aspects have been at the center of research efforts during the past two decades. Yet, progress has been slow due to the complexity of the issues and the lack of tractability of the models seeking to incorporate these elements. Further efforts are undoubtedly required to get a fuller grasp of the effects at play. Advances in methodology and in numerical analysis are likely to prove instrumental for gaining new insights about asset prices in these constrained environments.

References

Back, K. (1991). Asset pricing for general processes. *Journal of Mathematical Economics*, *20*, 371–395.

Berrada, T. (2006). Incomplete information, heterogeneity and asset pricing. *Journal of Financial Econometrics*, *4*, 136–160.

Breeden, D. (1979). An intertemporal asset pricing model with stochastic consumption and investment opportunities. *Journal of Financial Economics, 7*, 265–296.
Breeden, D. (1986). Consumption, production, inflation and interest rates: A synthesis. *Journal of Financial Economics, 16*, 3–39.
Cox, J. C., & Huang, Cf. (1989). Optimal consumption and portfolio policies when asset prices follow a diffusion process. *Journal of Economic Theory, 49*, 33–83.
Cox, J. C., Ingersoll, J. E., & Ross, S. A. (1985). An intertemporal general equilibrium model of asset prices. *Econometrica, 53*, 363–384.
Detemple, J., & Serrat, A. (2003). Dynamic equilibrium with liquidity constraints. *Review of Financial Studies, 16*, 597–629.
Detemple, J., Garcia, R., & Rindisbacher, M. (2008). Simulation methods for optimal portfolios. In J. R. Birge & V. Linetsky (Eds.), *Handbooks in operations research and management science, Financial engineering* (Vol. 15, pp. 867–923). Amsterdam: Elsevier.
Detemple, J. B., & Zapatero, F. (1991). Asset prices in an exchange economy with habit formation. *Econometrica, 59*, 1633–1657.
Detemple, J. B., Garcia, R., & Rindisbacher, M. (2003). A Monte-Carlo method for optimal portfolios. *Journal of Finance, 58*, 401–446.
Duffie, D., & Zame, W. (1989). The consumption-based capital asset pricing model. *Econometrica, 57*, 1279–1297.
Dumas, B. (1989). Two-person dynamic equilibrium in the capital market. *Review of Financial Studies, 2*, 157–188.
Gallmeyer, M., & Hollifield, B. (2008). An examination of heterogeneous beliefs with a short-sale constraint in a dynamic economy. *Review of Finance, 12*, 323–364.
Grossman, S. J., & Shiller, R. J. (1981). The determinants of the variability of stock market prices. *American Economic Review, 71*, 222–227.
Grossman, S. J., Melino, A., & Shiller, R. J. (1987). Estimating the continuous-time consumption-based asset-pricing model. *Journal of Business and Economic Statistics, 5*, 315–328.
Hansen, L. P., & Singleton, K. J. (1983). Stochastic consumption, risk aversion, and the temporal behavior of asset returns. *Journal of Political Economy, 91*, 249–268.
Huang, Cf. (1987). An intertemporal general equilibrium asset pricing model: The case of diffusion information. *Econometrica, 55*, 117–142.
Karatzas, I., & Shreve, S. E. (1998). *Methods of mathematical finance*. Berlin: Springer.
Karatzas, I., Lehoczky, J. P., & Shreve, S. E. (1987). Optimal portfolio and consumption decisions for a "small investor" on a finite horizon. *SIAM Journal of Control and Optimization, 25*, 1557–1586.
Karatzas, I., Lehoczky, J. P., & Shreve, S. E. (1990). Existence, uniqueness of multi-agent equilibrium in a stochastic, dynamic consumption/investment model. *Mathematics of Operations Research, 15*, 80–128.
Karatzas, I., Lakner, P., Lehoczky, J. P., & Shreve, S. E. (1991). Dynamic equilibrium in a simplified stochastic economy with heterogeneous agents. In *Stochastic analysis: liber amicorum for Moshe Zakai* (pp. 245–272). New York: Academic.
Kloeden, P. E., & Platen, E. (1999). *Numerical solution of stochastic differential equations* (3rd ed.). Berlin: Springer.
Mehra, R., & Prescott, E. C. (1985). The equity premium: A puzzle. *Journal of Monetary Economics, 15*, 145–61.
Merton, R. C. (1971). Optimum consumption and portfolio rules in a continuous time model. *Journal of Economic Theory, 3*, 373–413.
Merton, R. C. (1973). An intertemporal capital asset pricing model. *Econometrica, 41*, 867–87.
Ocone, D., & Karatzas, I. (1991). A generalized Clark representation formula, with application to optimal portfolios. *Stochastics and Stochastics Reports, 34*, 187–220.
Pliska, S. (1986). A stochastic calculus model of continuous trading: Optimal portfolios. *Mathematics of Operations Research, 11*, 371–382.
Wang, J. (1996). The term structure of interest rates in a pure exchange economy with heterogeneous investors. *Journal of Financial Economics, 41*, 75–110.

Chapter 4
Jump-Diffusion Models Driven by Lévy Processes

José E. Figueroa-López

Abstract During the past and this decade, a new generation of continuous-time financial models has been intensively investigated in a quest to incorporate the so-called *stylized empirical features of asset prices* like fat-tails, high kurtosis, volatility clustering, and leverage. Modeling driven by "memoryless homogeneous" jump processes (Lévy processes) constitutes one of the most viable directions in this enterprise. The basic principle is to replace the underlying Brownian motion of the Black-Scholes model with a type of jump-diffusion process. In this chapter, the basic results and tools behind jump-diffusion models driven by Lévy processes are covered, providing an accessible overview, coupled with their financial applications and relevance. The material is drawn upon recent monographs (cf. Cont and Tankov (2004). *Financial modelling with Jump Processes*. Chapman & Hall.; Sato (1999). *Lévy Processes and Infinitely Divisible Distributions*. Cambridge University Press.) and papers in the field.

4.1 An Overview of Financial Models with Jumps

The seminal Black-Scholes model Black and Scholes (1973) provides a framework to price options based on the fundamental concepts of hedging and absence of arbitrage. One of the key assumptions of the Black-Scholes model is that the stock price process $t \to S_t$ is given by a *geometric Brownian motion* (GBM), originally proposed by Samuelson (1965). Concretely, the time-t price of the stock is postulated to be given by

$$S_t = S_0 e^{\sigma W_t + \mu t}, \tag{4.1}$$

J.E. Figueroa-López (✉)
Department of Statistics, Purdue University, West Lafayette, IN 47907-2066, USA
e-mail: figueroa@stat.purdue.edu

J.-C. Duan et al. (eds.), *Handbook of Computational Finance*, Springer Handbooks of Computational Statistics, DOI 10.1007/978-3-642-17254-0_4,

where $\{W_t\}_{t\geq 0}$ is a standard Brownian motion. This model is plausible since Brownian motion is the model of choice to describe the evolution of a random measurement whose value is the result of a large-number of small shocks occurring through time with high-frequency. This is indeed the situation with the log return process $X_t = \log(S_t/S_0)$ of a stock, whose value at a given time t (not "very" small) is the superposition of a high number of small movements driven by a large number of agents posting bid and ask prices almost at "all times".

The limitations of the GBM were well-recognized almost from its inception. For instance, it well known that the time series of log returns, say $\log\{S_\Delta/S_0\}, \ldots, \log\{S_{k\Delta}/S_{(k-1)\Delta}\}$, exhibit *leptokurtic* distributions (i.e. fat tails with high kurtosis distributions), which are inconsistent with the Gaussian distribution postulated by the GBM. As expected the discrepancy from the Gaussian distribution is more marked when Δ is small (say a day and smaller). Also, the volatility, as measured for instance by the square root of the realized variance of log returns, exhibits *clustering* and *leverage* effects, which contradict the random-walk property of a GBM. Specifically, when plotting the time series of log returns against time, there are periods of high variability followed by low variability periods suggesting that high volatility events "cluster" in time. Leverage refers to a tendency towards a volatility growth after a sharp drop in prices, suggesting that volatility is negatively correlated with returns. These and other *stylized statistical features* of asset returns are widely known in the financial community (see e.g. Cont 2001 and Barndorff-Nielsen and Shephard (2007) for more information). In the risk-neutral world, it is also well known that the Black-Scholes implied volatilities of call and put options are not flat neither with respect to the strike nor to the maturity, as it should be under the Black-Scholes model. Rather implied volatilities exhibit smile or smirk curve shapes.

In a quest to incorporate the stylized properties of asset prices, many models have been proposed during the last and this decade, most of them derived from natural variations of the Black-Scholes model. The basic idea is to replace the Brownian motion W in (4.1), with another related process such as a Lévy process, a Wiener integral $\int_0^t \sigma_s \mathrm{d}W_s$, or a combination of both, leading to a "*jump-diffusion model*" or a *semimartingale* model. The simplest jump-diffusion model is of the form

$$S_t := S_0 e^{\sigma W_t + \mu t + Z_t}, \tag{4.2}$$

where $Z := \{Z_t\}_{t\geq 0}$ is a "pure-jump" Lévy process. Equivalently, (4.2) can be written as

$$S_t := S_0 e^{X_t}, \tag{4.3}$$

where X_t is a general Lévy process. Even this simple extension of the GBM, called *geometric Lévy model* or *exponential Lévy model*, is able to incorporate several stylized features of asset prices such as heavy tails, high-kurtosis, and asymmetry of log returns. There are other reasons in support of incorporating *jumps* in the dynamics of the stock prices. On one hand, certain event-driven information often produces "sudden" and "sharp" price changes at discrete unpredictable times.

Second, in fact stock prices are made up of discrete trades occurring through time at a very high frequency. Hence, processes exhibiting infinitely many jumps in any finite time horizon $[0, T]$ are arguably better approximations to such high-activity stochastic processes.

Merton (1976), following Press (1967), proposed one of the earliest models of the form (4.2), taking a compound Poisson process Z with normally distributed jumps (see Sect. 4.2.1). However, earlier Mandelbrot (1963) had already proposed a pure-jump model driven by a stable Lévy process Z. Merton's model is considered to exhibit light tails as all exponential moments of the densities of $\log(S_t/S_0)$ are finite, while Mandelbrot's model exhibit very heavy tails with not even finite second moments. It was during the last decade that models exhibiting appropriate tail behavior were proposed. Among the better known models are the *variance Gamma model* of Carr et al. (1998), the *CGMY model* of Carr et al. (2002), and the *generalized hyperbolic motion* of Barndorff-Nielsen (1998); Barndorff-Nielsen and Shephard (2001) and Eberlein and Keller (1995); Eberlein (2001). We refer to Kyprianou et al. 2005, Chapter 1 and Cont and Tankov 2004, Chapter 4 for more extensive reviews and references of the different types of geometric Lévy models in finance.

The geometric Lévy model (4.2) cannot incorporate volatility clustering and leverage effects due to the fact that log returns will be independent identically distributed. To cope with this shortcoming, two general classes of models driven by Lévy processes have been proposed. The first approach, due to Barndorff-Nielsen and Shephard (see e.g. Barndorff-Nielsen and Shephard 2001 and references therein), proposes a stochastic volatility model of the form

$$S_t := S_0 e^{\int_0^t b_u \mathrm{d}u + \int_0^t \sigma_u \mathrm{d}W_u}, \tag{4.4}$$

where σ is a stationary non-Gaussian Ornstein–Uhlenbeck process

$$\sigma_t^2 = \sigma_0^2 + \int_0^t \alpha \sigma_s^2 \mathrm{d}s + Z_{\alpha t},$$

driven by a subordinator Z (i.e. a non-decreasing Lévy process) (see Shephard 2005 and Andersen and Benzoni 2007 for two recent surveys on these and other related models). The second approach, proposed by Carr et al. (2003); Carr and Wu (2004), introduces stochastic volatility via a random clock as follows:

$$S_t = S_0 e^{Z_{\tau(t)}}, \quad \text{with} \quad \tau(t) := \int_0^t r(u)\mathrm{d}u. \tag{4.5}$$

The process τ plays the role of a "business" clock which could reflect non-synchronous trading effects or a "cumulative measure of economic activity". Roughly speaking, the rate process r controls the volatility of the process; for instance, in time periods where r is "high", the "business time" τ runs faster resulting in more frequent jump times. Hence, positive *mean-revering diffusion processes* $\{r(t)\}_{t\geq 0}$ are plausible choices to incorporate volatility clustering.

To account for the leverage phenomenon, different combinations of the previous models have been considered leading to semimartingale models driven by Wiener and Poisson random measures. A very general model in this direction assumes that the log return process $X_t := \log(S_t/S_0)$ is given as follows (c.f. Jacod 2006; Todorov 2008):

$$X_t = X_0 + \int_0^t b_s \mathrm{d}s + \int_0^t \sigma_s \mathrm{d}W_s + \int_0^t \int_{|x|\leq 1} \delta(s,x)\bar{M}(\mathrm{d}s,\mathrm{d}x)$$

$$+ \int_0^t \int_{|x|>1} \delta(s,x) M(\mathrm{d}s,\mathrm{d}x)$$

$$\sigma_t = \sigma_0 + \int_0^t \tilde{b}_s \mathrm{d}s + \int_0^t \tilde{\sigma}_s \mathrm{d}W_s + \int_0^t \int_{|x|\leq 1} \tilde{\delta}(s,x)\bar{M}(\mathrm{d}s,\mathrm{d}x)$$

$$+ \int_0^t \int_{|x|>1} \tilde{\delta}(s,x) M(\mathrm{d}s,\mathrm{d}x),$$

where W is a $d-$dimensional Wiener process, M is the *jump measure* of an independent Lévy process Z, defined by

$$M(B) := \#\{(t,\Delta Z_t) \in B : t > 0 \text{ such that } \Delta Z_t \neq 0\},$$

and $\bar{M}(\mathrm{d}t,\mathrm{d}x) := M(\mathrm{d}t,\mathrm{d}x) - \nu(\mathrm{d}x)dt$ is the compensate Poisson random measure of Z, where ν is the Lévy measure of Z. The integrands (b, σ, etc.) are random processes themselves, which could even depend on X and σ leading to a system of stochastic differential equations. One of the most active research fields in this very general setting is that of statistical inference methods based on high-frequency (intraday) financial data. Some of the researched problems include the prediction of the integrated volatility process $\int_0^t \sigma_s^2 \mathrm{d}s$ or of the Poisson integrals $\int_0^t \int_{\mathbb{R}\setminus\{0\}} g(x)M(\mathrm{d}x,\mathrm{d}s)$ based on realized variations of the process (see e.g. Jacod 2006, 2007; Mancini 2009; Woerner 2003, 2006; Podolskij 2006; Barndorff-Nielsen and Shephard 2006), testing for jumps (Barndorff-Nielsen and Shephard 2006; Podolskij 2006; Ait-Sahalia and Jacod 2006), and the estimation in the presence of "microstructure" noise (Aït-Sahalia et al. 2005; Podolskij and Vetter 2009 2009).

In this work, the basic methods and tools behind jump-diffusion models driven by Lévy processes are covered. The chapter will provide an accessible overview of the probabilistic concepts and results related to Lévy processes, coupled whenever is possible with their financial application and relevance. Some of the topics include: construction and characterization of Lévy processes and Poisson random measures, statistical estimation based on high- and low-frequency observations, density transformation and risk-neutral change of measures, arbitrage-free option pricing and integro-partial differential equations. The material is drawn upon recent monographs (c.f. Cont and Tankov 2004; Sato 1999) and recent papers in the field.

4.2 Distributional Properties and Statistical Estimation of Lévy Processes

4.2.1 Definition and Fundamental Examples

A Lévy process is a probabilistic model for an unpredictable measurement X_t that evolves in time t, in such a way that the change of the measurement in disjoint time intervals of equal duration, say $X_{s+\Delta} - X_s$ and $X_{t+\Delta} - X_t$ with $s + \Delta \leq t$, are independent from one another but with identical distribution. For instance, if S_t represents the time-t price of an asset and X_t is the *log return during* $[0, t]$, defined by

$$X_t = \log\left(S_t/S_0\right),$$

then the previous property will imply that daily or weekly log returns will be independent from one another with common distribution. Formally, a Lévy process is defined as follows:

Definition 1. A **Lévy process** $X = \{X_t\}_{t \geq 1}$ is a $\mathbb{R}^d$-valued stochastic process (collection of random vectors in $\mathbb{R}^d$ indexed by time) defined on a probability space $(\Omega, \mathcal{F}, \mathbb{P})$ such that:

(i) $X_0 = 0$.
(ii) X has **independent increments**: $X_{t_1} - X_{t_0}, \ldots, X_{t_n} - X_{t_{n-1}}$ are independent for any $0 \leq t_0 < \ldots < t_n$.
(iii) X has **stationary increments**: the distribution of $X_{t+\Delta} - X_t$ is the same as X_Δ, for all $t, \Delta \geq 0$.
(iv) its paths are right-continuous with left-limits (rcll).
(v) it has **no fixed jump-times**; that is, $\mathbb{P}(\Delta X_t \neq 0) = 0$, for any time t.

The last property can be replaced by asking that X is continuous in probability, namely, $X_s \xrightarrow{\mathbb{P}} X_t$, as $s \to t$, for any t. Also, if X satisfies all the other properties except (iv), then there exists a rcll version of the process (see e.g. Sato 1999).

There are three fundamental examples of Lévy processes that deserve some attention: Brownian motion, Poisson process, and compound Poisson process.

Definition 2. A (standard) Brownian motion W is a real-valued process such that (i) $W_0 = 0$, (ii) it has independent increments, (iii) $W_t - W_s$ has normal distribution with mean 0 and variance $t - s$, for any $s < t$, and (iv) it has continuous paths.

It turns out that the only real Lévy processes with continuous paths are of the form $X_t = \sigma W_t + bt$, for constants $\sigma > 0$ and b.

A Poisson process is another fundamental type of Lévy process that is often used as building blocks of other processes.

Definition 3. A Poisson process N is an integer-valued process such that (i) $N_0 = 0$, (ii) it has independent increments, (iii) $N_t - N_s$ has Poisson distribution with parameter $\lambda(t-s)$, for any $s < t$, and (iv) its paths are rcll. The parameter λ is called the intensity of the process.

The Poisson process is frequently used as a model to count events of certain type (say, car accidents) occurring randomly through time. Concretely, suppose that $T_1 < T_2 < \ldots$ represent random occurrence times of a certain event and let N_t be the number of events occurring by time t:

$$N_t = \sum_{i=1}^{\infty} \mathbf{1}_{\{T_i \leq t\}}. \tag{4.6}$$

Then, if the events occur independently from one another, homogeneously in time, and with an intensity of λ events per unit time, $\{N_t\}_{t \geq 0}$ given by (4.6) will be approximately a Poisson process with intensity λ. This fact is a consequence of the Binomial approximation to the Poisson distribution (see, e.g., Feller 1968 for this heuristic construction of a Poisson process). It turns out that any Poisson process can be written in the form (4.6) with $\{T_i\}_{i \geq 1}$ (called arrival times) such that the waiting times

$$\tau_i := T_i - T_{i-1},$$

are independent exponential r.v.'s with common mean $1/\lambda$ (so, the bigger the λ, the smaller the expected waiting time between arrivals and the higher the intensity of arrivals).

To introduce the last fundamental example, the compound Poisson process, we recall the concept of probability distribution. Given a random vector J in $\mathbb{R}^d$ defined on some probability space $(\Omega, \mathbb{P})$, the distribution of J is the mapping ρ defined on sets $A \subset \mathbb{R}^d$ as follows:

$$\rho(A) := \mathbb{P}(J \in A).$$

Thus, $\rho(A)$ measures the probability that the random vector J belongs to the set A. A *compound Poisson process* with jump distribution ρ and jump intensity λ is a process of the form

$$Z_t := \sum_{i=1}^{N_t} J_i,$$

where $\{J_i\}_{i \geq 1}$ are independent with common distribution ρ and N is a Poisson process with intensity λ that is independent of $\{J_i\}_i$. When $d = 1$, one can say that the compound Poisson process $Z := \{Z_t\}_{t \geq 0}$ is like a Poisson process with random jump sizes independent from one another. A compound Poisson process is the only Lévy process that has piece-wise constant paths with finitely-many jumps in any time interval $[0, T]$. Note that the distribution of the compound Poisson process Z

is characterized by the finite measure:

$$\nu(A) := \lambda\rho(A), \quad A \subset \mathbb{R}^d,$$

called the *Lévy measure* of Z. Furthermore, for any finite measure ν, one can associate a compound Poisson process Z with Lévy measure ν (namely, the compound Poisson process with intensity of jumps $\lambda := \nu(\mathbb{R}^d)$ and jump distribution $\rho(dx) := \nu(dx)/\nu(\mathbb{R}^d)$).

For future reference, it is useful to note that the characteristic function of Z_t is given by

$$\mathbb{E}e^{i\langle u,Z_t\rangle} = \exp\left\{t\int_{\mathbb{R}^d}\left(e^{i\langle u,x\rangle}-1\right)\nu(\mathrm{d}x)\right\} \tag{4.7}$$

Also, if $\mathbb{E}|J_i| = \int |x|\rho(\mathrm{d}x) < \infty$, then $\mathbb{E}Z_t = t\int x\rho(\mathrm{d}x)$ and the so-called *compensated compound Poisson process* $\bar{Z}_t := Z_t - \mathbb{E}Z_t$ has characteristic function

$$\mathbb{E}e^{i\langle u,\bar{Z}_t\rangle} = \exp\left\{t\int_{\mathbb{R}^d}\left(e^{i\langle u,x\rangle}-1-i\langle u,x\rangle\right)\nu(\mathrm{d}x)\right\}. \tag{4.8}$$

One of the most fundamental results establishes that any Lévy process can be approximated arbitrarily close by the superposition of a Brownian motion with drift, $\sigma W_t + bt$, and an independent compound Poisson process Z. The reminder $R_t := X_t - (\sigma W_t + bt + Z_t)$ is a pure-jump Lévy process with jump sizes smaller than say an $\varepsilon > 0$, which can be taken arbitrarily small. The previous fundamental fact is a consequence of the Lévy-Itô decomposition that we review in Sect. 4.3.2.

4.2.2 *Infinitely Divisible Distributions and the Lévy–Khintchine Formula*

The marginal distributions of a Lévy process X are *infinitely-divisible*. A random variable ξ is said to be *infinitely divisible* if for each $n \geq 2$, one can construct n i.i.d. r.v's . $\xi_{n,1}, \ldots, \xi_{n,n}$ such that

$$\xi \stackrel{\mathfrak{D}}{=} \xi_{n,1} + \ldots + \xi_{n,n}.$$

That X_t is infinitely divisible is clear since

$$X_t = \sum_{k=0}^{n-1}\left(X_{(k+1)t/n} - X_{kt/n}\right),$$

and $\{X_{(k+1)t/n} - X_{kt/n}\}_{k=0}^{n-1}$ are i.i.d. The class of infinitely divisible distributions is closely related to limits in distribution of an array of row-wise i.i.d. r.v.'s:

Theorem 1 (Kallenberg 1997). *ξ is infinitely divisible iff for each n there exists i.i.d. random variables $\{\xi_{n,k}\}_{k=1}^{k_n}$ such that*

$$\sum_{k=1}^{k_n} \xi_{n,k} \xrightarrow{\mathfrak{D}} \xi, \quad \text{as} \quad n \to \infty.$$

In term of the characteristic function $\varphi_\xi(u) := \mathbb{E}e^{i\langle u,\xi\rangle}$, ξ is infinitely divisible if and only if $\varphi_\xi(u) \neq 0$, for all u, and its distinguished n^{th}-root $\varphi_\xi(u)^{1/n}$ is the characteristic function of some other variable for each n (see Lemma 7.6 in Sato 1999). This property of the characteristic function turns out to be sufficient to determine its form in terms of three "parameters" (A, b, ν), called the *Lévy triplet* of ξ, as defined below.

Theorem 2 (Lévy-Khintchine formula). *ξ is infinitely divisible iff*

$$\mathbb{E}e^{i\langle u,\xi\rangle} = \exp\left\{ i\,\langle b, u\rangle - \frac{1}{2}\langle u, Au\rangle + \int \left(e^{i\langle u,x\rangle} - 1 - i\,\langle u, x\rangle\, \mathbf{1}_{|x|\leq 1} \right) \nu(\mathrm{d}x) \right\}, \tag{4.9}$$

for some symmetric nonnegative-definite matrix A, a vector $b \in \mathbb{R}^d$, and a measure ν (called the Lévy measure) on $\mathbb{R}_0^d := \mathbb{R}^d \backslash \{0\}$ such that

$$\int_{\mathbb{R}_0^d} (|x|^2 \wedge 1)\nu(\mathrm{d}x) < \infty. \tag{4.10}$$

Moreover, all triplets (A, b, ν) with the stated properties may occur.

The following remarks are important:

Remark 1. The previous result implies that the time-t marginal distribution of a Lévy process $\{X_t\}_{t\geq 0}$ is identified with a Lévy triplet (A_t, b_t, ν_t). Given that X has stationary and independent increments, it follows that $\mathbb{E}e^{i\langle u,X_t\rangle} = \left\{\mathbb{E}e^{i\langle u,X_1\rangle}\right\}^t$, for any rational t and by the right-continuity of X, for any real t. Thus, if (A, b, ν) is the Lévy triplet of X_1, then $(A_t, b_t, \nu_t) = t(A, b, \nu)$ and

$$\varphi_{X_t}(u) := \mathbb{E}e^{i\langle u,X_t\rangle} = e^{t\psi(u)}, \quad \text{where} \tag{4.11}$$

$$\psi(u) := i\,\langle b, u\rangle - \frac{1}{2}\langle u, Au\rangle + \int \left(e^{i\langle u,x\rangle} - 1 - i\,\langle u, x\rangle\, \mathbf{1}_{|x|\leq 1} \right) \nu(\mathrm{d}x). \tag{4.12}$$

The triple (A, b, ν) is called the Lévy or characteristic triplet of the Lévy process X.

Remark 2. The exponent (4.12) is called the *Lévy exponent* of the Lévy process $\{X_t\}_{t\geq 0}$. We can see that its first term is the Lévy exponent of the Lévy process bt. The second term is the Lévy exponent of the Lévy process ΣW_t, where $W = (W^1, \ldots, W^d)^T$ are $d-$independent Wiener processes and Σ is a $d \times d$ lower triangular matrix in the Cholesky decomposition $A = \Sigma\Sigma^T$. The last term in the

Lévy exponent can be decomposed into two terms:

$$\psi^{cp}(u) = \int_{|x|>1} \left(e^{i\langle u,x\rangle} - 1\right) \nu(\mathrm{d}x),$$

$$\psi^{lccp}(u) = \int_{|x|\leq 1} \left(e^{i\langle u,x\rangle} - 1 - i\,\langle u, x\rangle\right) \nu(\mathrm{d}x).$$

The first term above is the Lévy exponent of a compound Poisson process X^{cp} with Lévy measure $\nu_1(\mathrm{d}x) := \mathbf{1}_{|x|>1}\nu(\mathrm{d}x)$ (see (4.7)). The exponent ψ^{lccp} corresponds to the limit in distribution of *compensated compound Poisson processes*. Concretely, suppose that $X^{(\varepsilon)}$ is a compound Poisson process with Lévy measure $\nu_\varepsilon(dx) := \mathbf{1}_{\varepsilon<|x|\leq 1}\nu(dx)$, then the process $X_t^{(\varepsilon)} - \mathbb{E}X_t^{(\varepsilon)}$ converges in distribution to a process with characteristic function $\exp\{t\psi^{lccp}\}$ (see (4.8)). Lévy-Khintchine formula implies that, in distribution, X is the superposition of four independent Lévy processes as follows:

$$X_t \stackrel{\mathfrak{D}}{=} \underbrace{bt}_{\text{Drift}} + \underbrace{\Sigma\, W_t}_{\text{Brownian part}} + \underbrace{X_t^{cp}}_{\text{Cmpnd. Poisson}} + \underbrace{\lim_{\varepsilon\searrow 0}\left(X_t^{(\varepsilon)} - \mathbb{E}X_t^{(\varepsilon)}\right)}_{\text{Limit of cmpstd cmpnd Poisson}}, \tag{4.13}$$

where equality is in the sense of finite-dimensional distributions. The condition (4.10) on ν guarantees that the X^{cp} is indeed well defined and the compensated compound Poisson converges in distribution.

In the rest of this section, we go over some other fundamental distributional properties of the Lévy process and their applications.

4.2.3 Short-Term Distributional Behavior

The characteristic function (4.11) of X determines uniquely the Lévy triple (A, b, ν). For instance, the uniqueness of the matrix A is a consequence of the following result:

$$\lim_{h\to 0} h \cdot \log \varphi_{X_t}\left(h^{-1/2}u\right) = -\frac{t}{2}\langle u, Au\rangle\,; \tag{4.14}$$

see pp. 40 in Sato (1999). In term of the process X, (4.14) implies that

$$\left\{\frac{1}{\sqrt{h}} X_{ht}\right\}_{t\geq 0} \stackrel{\mathfrak{D}}{\longrightarrow} \{\Sigma\, W_t\}_{t\geq 0}, \quad \text{as } h \to 0. \tag{4.15}$$

where $W = (W^1, \ldots, W^d)^T$ are d–independent Wiener processes and Σ is a lower triangular matrix such that $A = \Sigma\Sigma^T$.

From a statistical point of view, (4.15) means that, when $\Sigma \neq 0$, the short-term increments $\{X_{(k+1)h} - X_{kh}\}_{k=1}^{n}$, properly scaled, behave like the increments of a Wiener process. In the context of the exponential Lévy model (4.34), the result (4.15) will imply that the log returns of the stock, properly scaled, are normally distributed when the Brownian component of the Lévy process X is non-zero. This property is not consistent with the empirical heavy tails of high-frequency financial returns. Recently, Rosiński (2007) proposes a pure-jump class of Lévy processes, called *tempered stable (TS) Lévy processes*, such that

$$\left\{ \frac{1}{h^{1/\alpha}} X_{ht} \right\}_{t \geq 0} \stackrel{\mathfrak{D}}{\longrightarrow} \{Z_t\}_{t \geq 0}, \quad \text{as } h \to 0, \tag{4.16}$$

where Z is a stable process with index $\alpha < 2$.

4.2.4 Moments and Short-Term Moment Asymptotics

Let $g : \mathbb{R}^d \to \mathbb{R}_+$ be a nonnegative locally bounded function and X be a Lévy process with Lévy triplet (A, b, ν). The expected value $\mathbb{E}g(\xi)$ is called the g-moment of a random variable ξ. Let us now consider submultiplicative or subadditive moment functions g. Recall that a nonnegative locally bounded function g is submultiplicative (resp. subadditive) if there exists a constant $K > 0$ such that $g(x + y) \leq Kg(x)g(y)$ (resp. $g(x + y) \leq K(g(x) + g(y))$), for all x, y. Examples of this kind of functions are $g(x_1, \ldots, x_d) = |x_j|^p$, for $p \geq 1$, and $g(x_1, \ldots, x_d) = \exp\{|x_j|^\beta\}$, for $\beta \in (0, 1]$. In the case of a compound Poisson process, it is easy to check that

$$\mathbb{E}g(X_t) < \infty, \text{ for any } t > 0 \text{ if and only if } \int_{|x|>1} g(x)\nu(\mathrm{d}x) < \infty.$$

The previous fact holds for general Lévy processes (see Kruglov 1970 and (Sato, 1999, Theorem 25.3)). In particular, $X(t) := (X_1(t), \ldots, X_d(t)) := X_t$ has finite mean if and only if $\int_{\{|x|>1\}} |x|\nu(\mathrm{d}x) < \infty$. In that case, by differentiation of the characteristic function, it follows that

$$\mathbb{E}X_j(t) = t \left(\int_{\{|x|>1\}} x_j \nu(\mathrm{d}x) + b_j \right),$$

Similarly, $\mathbb{E}|X(t)|^2 < \infty$ if and only if $\int_{\{|x|>1\}} |x|^2 \nu(\mathrm{d}x) < \infty$, in which case,

$$\mathrm{Cov}\left(X_j(t), X_k(t)\right) = t \left(A_{jk} + \int x_j x_k \nu(\mathrm{d}x) \right).$$

The two above equations show the connection between the the Lévy triplet (A, b, ν), and the mean and covariance of the process. Note that the variance rate

$\mathrm{Var}(X_j(t))/t$ remains constant over time. It can also be shown that the kurtosis is inversely proportional to time t. In the risk-neutral world, these properties are not empirically supported under the exponential Lévy model (4.2), which rather support a model where both measurements increase with time t (see e.g. Carr et al. 2003 and references therein).

The Lévy measure ν controls the short-term ergodic behavior of X. Namely, for any bounded continuous function $\varphi : \mathbb{R}^d \to \mathbb{R}$ vanishing on a neighborhood of the origin, it holds that

$$\lim_{t\to 0} \frac{1}{t} \mathbb{E}\varphi(X_t) = \int \varphi(x)\nu(\mathrm{d}x); \tag{4.17}$$

cf. (Sato, 1999, Corollary 8.9). For a real Lévy processes X with Lévy triplet (σ^2, b, ν), (4.17) can be extended to incorporate unbounded functions and different behaviors at the origin. Suppose that $\varphi:\mathbb{R} \to \mathbb{R}$ is $\nu-$continuous such that $|\varphi| \leq g$ for a subadditive or submultiplicative function $g:\mathbb{R} \to \mathbb{R}_+$. Furthermore, fixing $I:=\{r \geq 0: \int (|x|^r \wedge 1)\, \nu(dx) < \infty\}$, assume that φ exhibits any of the following behaviors as $x \to 0$:

(a) i. $\varphi(x) = o(|x|^2)$.
   ii. $\varphi(x) = O(|x|^r)$, for some $r \in I \cap (1,2)$ and $\sigma = 0$.
   iii. $\varphi(x) = o(|x|)$, $1 \in I$ and $\sigma = 0$.
   iv. $\varphi(x) = sO(|x|^r)$, for some $r \in I \cap (0,1)$, $\sigma = 0$, and $\bar{b} := b - \int_{|x|\leq 1} x\nu(\mathrm{d}x) = 0$.

(b) $\varphi(x) \sim x^2$.

(c) $\varphi(x) \sim |x|$ and $\sigma = 0$.

Building on results in Woerner (2003) and Jacod (2007), Figueroa-López (2008) proves that

$$\lim_{t\to 0} \frac{1}{t} \mathbb{E}\varphi(X_t) := \begin{cases} \int \varphi(x)\nu(\mathrm{d}x), & \text{if (a) holds,} \\ \sigma^2 + \int \varphi(x)\nu(\mathrm{d}x), & \text{if (b) holds,} \\ |\bar{b}| + \int \varphi(x)\nu(\mathrm{d}x), & \text{if (c) holds.} \end{cases} \tag{4.18}$$

Woerner (2003) and also Figueroa-López (2004) used the previous short-term ergodic property to show the consistency of the statistics

$$\hat{\beta}^{\pi}(\varphi) := \frac{1}{t_n} \sum_{k=1}^{n} \varphi\left(X_{t_k} - X_{t_{k-1}}\right), \tag{4.19}$$

towards the integral parameter $\beta(\varphi) := \int \varphi(x)\nu(\mathrm{d}x)$, when $t_n \to \infty$ and $\max\{t_k - t_{k-1}\} \to 0$, for test functions φ as in (a). When $\nu(\mathrm{d}x) = s(x)\mathrm{d}x$, Figueroa-López (2004) applied the estimators (4.19) to analyze the asymptotic properties of nonparametric *sieve-type* estimators $\hat{s}$ for s. The problem of model selection was analyzed further in Figueroa-López and Houdré (2006); Figueroa-López (2009), where it was proved that sieve estimators $\widetilde{s}_T$ can match the rate of convergence of the minimax risk of estimators $\hat{s}$. Concretely, it turns out that

$$\limsup_{T\to\infty} \frac{\mathbb{E}\|s - \widetilde{s}_T\|^2}{\inf_{\hat{s}} \sup_{s\in\Theta} \mathbb{E}\|s - \hat{s}\|^2} < \infty,$$

where $[0, T]$ is the time horizon over which we observe the process X, Θ is certain class of smooth functions, and the infimum in the denominator is over all estimators $\hat{s}$ which are based on whole trajectory $\{X_t\}_{t\leq T}$. The optimal rate of the estimator $\widetilde{s}_T$ is attained by choosing appropriately the dimension of the sieve and the sampling frequency in function of T and the smoothness of the class of functions Θ.

4.2.5 Extraction of the Lévy Measure

The Lévy measure ν can be inferred from the characteristic function $\varphi_{X_t}(u)$ of the Lévy process (see, e.g., Sato 1999, pp. 40–41). Concretely, by first recovering $\langle u, Au\rangle$ from (4.14), one can obtain

$$\Psi(u) := \log \varphi_{X_1}(u) + \frac{1}{2}\langle u, Au\rangle .$$

Then, it turns out that

$$\int_{[-1,1]^d} (\Psi(u) - \Psi(u+w))\, \mathrm{d}w = \int_{\mathbb{R}^d} e^{i\langle z,x\rangle}\tilde{\nu}(\mathrm{d}x), \qquad (4.20)$$

where $\tilde{\nu}$ is the finite measure

$$\tilde{\nu}(\mathrm{d}x) := 2^d \left(1 - \prod_{j=1}^{d} \frac{\sin x_j}{x_j}\right) \nu(\mathrm{d}x).$$

Hence, ν can be recovered from the inverse Fourier transform of the left-hand side of (4.20).

The above method can be applied to devise non-parametric estimation of the Lévy measure by replacing the Fourier transform φ_{X_1} by its empirical version:

$$\hat{\varphi}_{X_1}(u) := \frac{1}{n}\sum_{k=1}^{n} \exp\{i\langle u, X_k - X_{k-1}\rangle\} .$$

given discrete observations $X_1, \ldots, X_n$ of the process. Recently, similar nonparametric methods have been proposed in the literature to estimate the Lévy density $s(x) = \nu(\mathrm{d}x)/\mathrm{d}x$ of a real Lévy process X (c.f. Neumann and Reiss 2007; Comte and Genon-Catalot 2008; Gugushvili 2008). For instance, based on the increments $X_1 - X_0, \ldots, X_n - X_{(n-1)}$, Neumann and Reiss (2007) consider a nonparametric estimator for s that minimizes the distance between the "population" characteristic function $\varphi_{X_1}(\cdot; s)$ and the empirical characteristic function $\hat{\varphi}_{X_1}(\cdot)$. By appropriately defining the distance metric, Neumann and Reiss (2008) showed the consistency of the proposed estimators. Another approach, followed for instance by Watteel and Kulperger (2003) and Comte and Genon-Catalot (2008), relies on an "explicit" formula for the Lévy density s in terms of the derivatives of the characteristic function φ_{X_1}. For instance, under certain regularity conditions,

$$\mathcal{F}(xs(x))(\cdot) = -i \frac{\varphi'_{X_1}(\cdot)}{\varphi_{X_1}(\cdot)},$$

where $\mathcal{F}(f)(u) = \int e^{iux} f(x)dx$ denotes the Fourier transform of a function f. Hence, an estimator for s can be built by replacing ψ by a smooth version of the empirical estimate $\hat{\varphi}_{X_1}$ and applying inverse Fourier transform $\mathcal{F}^{-1}$.

4.3 Path Decomposition of Lévy Processes

In this part, we show that the construction in (4.13) holds true a.s. (not only in distribution) and draw some important consequences. The fundamental tool for this result is a probabilistic characterization of the random points $\{(t, \Delta X_t) : t \text{ s.t. } \Delta X_t \neq 0\}$ as a *Poisson point process* on the semi-plane $\mathbb{R}_+ \times \mathbb{R}\backslash\{0\}$. Due to this fact, we first review the properties of Poisson random measures, which are also important building blocks of financial models.

4.3.1 Poisson Random Measures and Point Processes

Definition 4. Let S be a Borel subset of $\mathbb{R}^d$, let $\mathcal{S}$ be the set of Borel subsets of S, and let m be a $\sigma-$finite measure on $\mathcal{S}$. A collection $\{M(B) : B \in \mathcal{S}\}$ of $\bar{\mathbb{Z}}_+$-valued random variables defined on a probability space $(\Omega, \mathcal{F}, \mathbb{P})$ is called a **Poisson random measure** (PRM) (or process) on S with mean measure m if:

(1) For every $B \in \mathcal{S}$, $M(B)$ is a Poisson random variable with mean $m(B)$.
(2) If $B_1, \ldots, B_n \in \mathcal{S}$ are disjoint, then $M(B_1), \ldots, M(B_1)$ are independent.
(3) For every sample outcome $\omega \in \Omega$, $M(\cdot; \omega)$ is a measure on $\mathcal{S}$.

Above, we used some basic terminology of real analysis. For all practical purposes, Borel sets of $\mathbb{R}^d$ are those subsets that can be constructed from basic operations (complements, countable unions, and intersections) of elementary sets of the form $(a_1, b_1] \times \ldots (a_d, b_d]$. A measure m is a mapping from $\mathcal{S}$ to $[0, \infty]$ such that

$$m(\emptyset) = 0, \quad \text{and} \quad m\left(\bigcup_{i=1}^{\infty} B_i\right) = \sum_{i=1}^{\infty} m(B_i),$$

for any mutually disjoints Borel sets $\{B_i\}_{i \geq 1}$. A measure is said to be σ-finite if there exists mutually disjoints $\{B_i\}_{i \geq 1}$ such that $\mathbb{R}^d = \bigcup_{i=1}^{\infty} B_i$ and $m(B_i) < \infty$, for any i.

It can be proved that (a.s.), a Poisson random measure $M(\cdot;\omega)$ is an atomic measure; that is, there exist countably many (random) points $\{\mathbf{x}_i\}_i \subset S$ (called atoms) such that

$$M(B) = \#\{i : \mathbf{x}_i \in B\} = \sum_{i=1}^{\infty} \delta_{\mathbf{x}_i}(B). \tag{4.21}$$

Similarly, if a sequence of finitely many or countably many random points $\{\mathbf{x}_i\}_i$ is such that the measure (4.21) satisfies (1)–(3) above, then we say that $\{\mathbf{x}_i\}_i$ is a **Poisson point process** on S with mean measure m. The following is a common procedure to construct a realization of a Poisson random measure or point process:

1. Suppose that $B_1, B_2, \ldots$ is a partition of S such that $m(B_j) < \infty$
2. Generate $n_j \sim Poiss(m(B_j))$
3. Independently, generate n_j-points, say $\{\mathbf{x}_i^j\}_{i=1}^{n_j}$, according to the distribution $m(\cdot)/m(B_j)$
4. Define $M(B) = \#\{(i,j) : \mathbf{x}_i^j \in B\}$

4.3.1.1 Transformation of Poisson Random Measures

Among the most useful properties of PRM is that certain transformations of a Poisson point process are still a Poisson point process. The following is the simplest version:

Proposition 1. *Suppose that $T : S \to S' \subset \mathbb{R}^{d'}$ is a one-to-one measurable function. Then, the random measure associated with the transformed points $\mathbf{x}_i' := T(\mathbf{x}_i)$, namely $M'(\cdot) = \sum_{i=1}^{\infty} \delta_{\mathbf{x}_i'}(\cdot)$, is also a Poisson random measure with mean measure $m'(B) := m(\{\mathbf{x} : T(\mathbf{x}) \in B\})$.*

The following result shows that a *marked Poisson point process* is still a Poisson point process. Suppose that we associate a $\mathbb{R}^{d'}$-valued score $\mathbf{x}_i'$ to each point $\mathbf{x}_i$ of M. The scores are assigned independently from one another. The distribution of the scores can actually depend on the point $\mathbf{x}_i$. Concretely, let $\sigma(\mathbf{x}, d\mathbf{x}')$ be a probability measure on $S' \subset \mathbb{R}^{d'}$, for each $\mathbf{x} \in S$ (hence, $\sigma(\mathbf{x}, S') = 1$). For each i, generate a r.v. $\mathbf{x}_i'$ according $\sigma(\mathbf{x}_i, d\mathbf{x}')$ (*independently from any other variable*). Consider the so-called *marked Poisson process*

$$M'(\cdot) = \sum_{i=1}^{\infty} \delta_{(\mathbf{x}_i, \mathbf{x}_i')}(\cdot).$$

Proposition 2. *M' is a Poisson random measure on $\mathcal{S} \times \mathcal{S}'$ with mean measure*

$$m'(\mathrm{d}\mathbf{x}, \mathrm{d}\mathbf{x}') = \sigma(\mathbf{x}, \mathrm{d}\mathbf{x}')m(\mathrm{d}\mathbf{x}).$$

As an example, consider the following experiment. We classify the points of the Poisson process M into k different types. The probability that the point $\mathbf{x}_i$ is of type j is $p_j(\mathbf{x}_i)$ (necessarily $p_j(\cdot) \in [0,1]$), independently from any other classification. Let $\{\mathbf{y}_i^j\}$ be the points of $\{\mathbf{x}_i\}$ of type j and let M^j be the counting measure associated with $\{\mathbf{y}_i^j\}$:

$$M^j := \sum \delta_{\{\mathbf{y}_i^j\}}$$

We say that the process M^1 is constructed from M by *thinning*.

Proposition 3. *$M^1, \ldots, M^k$ are independent Poisson random measures with respective mean measures $m_1(\mathrm{d}\mathbf{x}) := p_1(\mathbf{x})m(\mathrm{d}\mathbf{x}), \ldots, m_k(\mathrm{d}\mathbf{x}) := p_k(\mathbf{x})m(\mathrm{d}\mathbf{x})$.*

Example 1. Suppose that we want to simulate a Poisson point process on the unit circle $S := \{(x, y){:}x^2 + y^2 \leq 1\}$ with mean measure:

$$m'(B) = \iint_{B \cap S} \sqrt{x^2 + y^2}\mathrm{d}x\mathrm{d}y.$$

A method to do this is based on the previous thinning method. Suppose that we generate a "homogeneous" Poisson point process M on the square $R := \{(x, y){:}|x| \leq 1, |y| \leq 1\}$ with an intensity of $\lambda = 8$ points per unit area. That is, the mean measure of M is

$$m(B) = \frac{1}{4}\iint_B \lambda\mathrm{d}x\mathrm{d}y.$$

Let $\{(x_i, y_i)\}_i$ denote the atoms of the Poisson random measure M. Now, consider the following thinning process. We classify the point (x_i, y_i) of type 1 with probability $p(x_i, y_i) := \frac{1}{2}\sqrt{x_i^2 + y_i^2}$ and of type 2 with probability $1 - p(x_i, y_i)$. Suppose that $\{(x_i^1, y_i^1)\}_i$ are the point of type 1. Then, this process is a Poisson point process with mean measure m'.

4.3.1.2 Integration with Respect to a Poisson Random Measure

Let M be a Poisson random measure as Definition 4. Since $M(\cdot;\omega)$ is an atomic random measure for each ω, say $M(\cdot;\omega) = \sum_{i=1}^{\infty} \delta_{\mathbf{x}_i(\omega)}(\cdot)$, one can define the integral

$$M(f) := \int_S f(\mathbf{x}) M(d\mathbf{x}) = \sum_{i=1}^{\infty} f(\mathbf{x}_i),$$

for any measurable nonnegative deterministic function f. This is a $\bar{\mathbb{R}}_+ = \mathbb{R} \cup \{\infty\}$-valued r.v. such that

$$\mathbb{E}\left[e^{-\int f(\mathbf{x})M(d\mathbf{x})}\right] = \exp\left\{-\int \left(1 - e^{-f(\mathbf{x})}\right) m(d\mathbf{x})\right\},$$

$$\times\, \mathbb{E}\left[\int f(\mathbf{x}) M(d\mathbf{x})\right] = \int f(\mathbf{x}) m(d\mathbf{x});$$

see Kallenberg 1997, Lemma 10.2. Also, if $B \in \mathcal{S}$ is such that $m(B) < \infty$, then

$$\int_B f(\mathbf{x}) M(d\mathbf{x}) := \sum_{i:\mathbf{x}_i \in B} f(\mathbf{x}_i),$$

is a well-defined $\mathbb{R}^d$–valued r.v. for any measurable function $f : \mathcal{S} \to \mathbb{R}^d$. Its characteristic function is given by

$$\mathbb{E}\left[e^{\mathrm{i}\langle \int_B f(\mathbf{x})M(d\mathbf{x}), \mathbf{u}\rangle}\right] = \exp\left\{\int_B \left(e^{\mathrm{i}\langle f(\mathbf{x}), \mathbf{u}\rangle} - 1\right) m(d\mathbf{x})\right\}.$$

Furthermore, if $B_1, \ldots, B_m$ are disjoints sets in $\mathcal{S}$ with finite measure, then

$$\int_{B_1} f(\mathbf{x}) M(d\mathbf{x}), \ldots, \int_{B_m} f(\mathbf{x}) M(d\mathbf{x}).$$

are independent (see (Sato, 1999, Proposition 19.5)).

In the general case, determining conditions for the integral $\int_S f(\mathbf{x})M(d\mathbf{x})$ to be well-defined requires some care. Let us assume that m is a radon measure (that is, $m(K) < \infty$, for any compact $K \subset S$). Then, $\int_S f(\mathbf{x})M(d\mathbf{x}) = \sum_{i=1}^{\infty} f(\mathbf{x}_i)$ is well-defined for any bounded function $f : S \to \mathbb{R}$ of compact support. We say that the integral $\int_S f(\mathbf{x})M(d\mathbf{x})$ exists if

$$\int_S f_n(\mathbf{x}) M(d\mathbf{x}) \xrightarrow{\mathbb{P}} X, \quad \text{as} \quad n \to \infty,$$

for a random variable X and any sequence f_n of bounded functions with compact support such that $|f_n| \leq |f|$ and $f_n \to f$. In that case, the so-called *Poisson integral* $\int_S f(\mathbf{x})M(d\mathbf{x})$ is defined to be that common limit X. We define in a similar way the so-called *compensated Poisson integral of* f, denoted by $\int_S f(\mathbf{x})(M - m)(d\mathbf{x})$. The following theorem gives conditions for the existence of the Poisson integrals (see (Kallenberg, 1997, Theorem 10.15)):

Proposition 4. *Let M be a Poisson random measure as in Definition 4. Then,*

(a) $M(f) = \int_S f(\mathbf{x})M(\mathrm{d}\mathbf{x})$ *exists iff* $\int_S (|f(\mathbf{x})| \wedge 1)m(\mathrm{d}\mathbf{x}) < \infty$.
(b) $(M-m)(f) := \int_S f(\mathbf{x})(M-m)(\mathrm{d}\mathbf{x})$ *exists iff* $\int_S (|f(\mathbf{x})|^2 \wedge |f(\mathbf{x})|)m(\mathrm{d}\mathbf{x}) < \infty$.

4.3.2 The Lévy-Itô Decomposition

The following result, called the Lévy-Itô decomposition, is fundamental for the theory of Lévy processes. It says that any Lévy process X is the superposition of a constant drift bt, a Brownian component ΣW_t, a compound Poisson process X_t^{cp}, and the limit of compensated Poisson processes. As stated below, it characterizes not only Lévy processes but also processes with independent increments (called *additive processes*).

Theorem 3. *[13.4, Kallenberg] Let $\{X_t\}_{t\geq 0}$ be a rcll process in $\mathbb{R}^d$ with $X(0) = 0$. Then, X has independent increments without fixed jumps times if and only if, there exists $\Omega_0 \in \mathcal{F}$ with $\mathbb{P}(\Omega_0) = 1$ such that for any $\omega \in \Omega_0$,*

$$X_t(\omega) = b_t(\omega) + G_t(\omega) + \int_0^t \int_{\{|x|>1\}} x\; M(\omega; \mathrm{d}s, \mathrm{d}x) + \int_0^t \int_{\{|x|\leq 1\}} x\; (M - m)(\omega; \mathrm{d}s, \mathrm{d}x), \tag{4.22}$$

for any $t \geq 0$, and for a continuous function b with $b_0 = 0$, a continuous centered Gaussian process G with independent increments and $G_0 = 0$, and an independent Poisson random measure M on $[0, \infty) \times \mathbb{R}_0^d$ with mean measure m satisfying

$$\int_0^t \int_{\mathbb{R}_0^d} (|x|^2 \wedge 1)\; m(\mathrm{d}s, \mathrm{d}x) < \infty, \quad \forall t > 0. \tag{4.23}$$

The representation is almost surely unique, and all functions b, processes G, and measures m with the stated properties may occur.

Note that the above theorem states that the *jump random measure* M_X of X, defined by

$$M_X((s,t] \times B) := \sum_{u \in (s,t]: \Delta X_u \neq 0} \mathbf{1}\{\Delta X_u \in B\},$$

is almost surely a Poisson process with mean measure $m(dt, dx)$. In the case of a Lévy process (that is, we also assume that X has stationary increments), the previous theorem implies that M_X is a Poisson random measure in $\mathbb{R}_+ \times \mathbb{R}_0$ with mean measure $m(dt, dx) = \nu(dx)dt$, for a measure ν satisfying (4.10). In that case, the representation (4.22) takes the following form:

$$X_t = bt + \Sigma W_t + \int_0^t \int_{\{|x|>1\}} x\, M(\mathrm{d}s, \mathrm{d}x) + \int_0^t \int_{\{|x|\le 1\}} x\, (M - m)(\mathrm{d}s, \mathrm{d}x), \tag{4.24}$$

where W is a d-dimensional Wiener process. The third term is a compound Poisson process with intensity of jumps $\nu(|x| > 1)$ and jump distribution $\mathbf{1}_{\{|x|>1\}}\nu(\mathrm{d}x)/\nu(|x| > 1)$. Similarly, the last term can be understood as the limit of compensated Poisson processes as follows:

$$\int_0^t \int_{\{|x|\le 1\}} x(M - m)(\mathrm{d}s, \mathrm{d}x) = \lim_{\varepsilon \downarrow 0} \int_0^t \int_{\{\varepsilon < x| \le 1\}} x\, (M - m)(\mathrm{d}s, \mathrm{d}x). \tag{4.25}$$

Furthermore, the convergence in (4.25) is uniform on any bounded interval of t (c.f. [19.2, Sato]).

4.3.3 Some Sample Path Properties

One application of the Lévy-Itô decomposition (4.24) is to determine conditions for certain path behavior of the process. The following are some cases of interest (see Sect. 19 in Sato 1999 for these and other path properties):

1. *Path-continuity*: The only continuous Lévy processes are of the form $bt + \sigma W_t$.
2. *Finite-variation*: A necessary and sufficient condition for X to have a.s. paths of bounded variation is that $\sigma = 0$ and

$$\int_{\{|x|\le 1\}} |x|\nu(\mathrm{d}x) < \infty.$$

   Note that in that case one can write

$$X_t = b_0 t + \int_0^t \int x\, M(\mathrm{d}s, \mathrm{d}x),$$

   where $b_0 := b - \int_{|x|\le 1} x\nu(\mathrm{d}x)$, called the *drift* of the Lévy process, is such that

$$\mathbb{P}\left(\lim_{t\to 0} \frac{1}{t} X_t = b_0 \right) = 1.$$

   A process of finite-variation can be written as the difference of two non-decreasing processes. In the above representation, this processes will be $b_0 t + \int_0^t \int_{x>0} x\, M(\mathrm{d}s, \mathrm{d}x)$, and $\int_0^t \int_{x<0} x\, M(\mathrm{d}s, \mathrm{d}x)$ when $b_0 > 0$.
3. A non-decreasing Lévy process is called a *subordinator*. Necessary and sufficient conditions for X to be a subordinator are that $b_0 > 0, \sigma = 0$, and $\nu((-\infty, 0)) = 0$.

4.4 Simulation of Lévy Processes

4.4.1 *Approximation by Skeletons and Compound Poisson Processes*

Accurate path simulation of a pure jump Lévy processes $X = \{X(t)\}_{t\in[0,1]}$, regardless of the relatively simple statistical structure of their increments, present some challenges when dealing with *infinite jump activity* (namely, processes with infinite Lévy measure). One of the most popular simulation schemes is based on the generation of *discrete skeletons*. Namely, the discrete skeleton of X based on equally spaced observations is defined by

$$\widetilde{X}_t = \sum_{k=1}^{\infty} X_{\frac{k-1}{n}} \mathbf{1}_{[\frac{k-1}{n} \le t < \frac{k}{n}]} = \sum_{k=1}^{\infty} \Delta_k \mathbf{1}_{\{t \ge \frac{k}{n}\}},$$

where $\Delta_k = X_{k/n} - X_{(k-1)/n}$ are i.i.d. with common distribution $\mathcal{L}(X_{1/n})$. Popular classes where this method is applicable are Gamma, variance Gamma, Stable, and Normal inverse Gaussian processes (see (Cont and Tankov, 2004, Section 6.2)). Lamentably, the previous scheme has limited applications since in most cases a r.v. with distribution $\mathcal{L}(X_{1/n})$ is not easy to generate.

A second approach is to approximate the Lévy process by a finite-jump activity Lévy processes. That is, suppose that X is a pure-jump Lévy process, then, in light of the Lévy-Itô decomposition (4.24), the process

$$X_t^{0,\varepsilon} \equiv t\left(b - \int_{\{|x|\ge\varepsilon\}} x\,\nu(\mathrm{d}x)\right) + \sum_{s\le t} \Delta X_s \mathbf{1}_{\{|\Delta X_s|\ge\varepsilon\}} \tag{4.26}$$

converges uniformly on any bounded interval to X a.s. (as usual $\Delta X_t \equiv X_t - X_t-$). The process $\sum_{s\le t} \Delta X_s \mathbf{1}_{\{|\Delta X_s|\ge\varepsilon\}}$ can be simulated using a *compound Poisson process* of the form $\sum_{i=1}^{N_t^\varepsilon} J_i^\varepsilon$, where N_t^ε is a homogeneous Poisson process with intensity $\nu(|x| \ge \varepsilon)$ and $\{J_i^\varepsilon\}_{i=1}^\infty$ are i.i.d with common distribution $\nu_\varepsilon(\mathrm{d}x) \equiv \mathbf{1}_{\{|x|\ge\varepsilon\}}\nu(\mathrm{d}x)/\nu(|x| \ge \varepsilon)$. Clearly, such a scheme is unsatisfactory because all jumps smaller than ε are totally ignored. An alternative method of simulation approximates the small jumps with a Wiener motion.

4.4.2 *Approximation of the Small Jumps of a Lévy Processes*

Consider a Lévy process with Lévy triple (σ^2, b, ν). Define the following processes:

$$X_t^{1,\varepsilon} := b_\varepsilon t + \sigma W_t + \int_0^t \int_{|x|\ge\varepsilon} x M(\mathrm{d}x, \mathrm{d}s),$$

where $b_\varepsilon = b - \int_{\varepsilon<|x|\leq 1} x\nu(\mathrm{d}x)$ and M is the jump-measure of X (a posterior a Poisson measure on $\mathbb{R}_+ \times \mathbb{R}_0^d$ with mean measure $\nu(\mathrm{d}x)\mathrm{d}t$). Consider the following pure jump Lévy process

$$X_t^\varepsilon := X_t - X_t^{1,\varepsilon} = \int_0^t \int_{\{|x|<\varepsilon\}} x\left\{M(\mathrm{d}x, \mathrm{d}s) - \nu(\mathrm{d}x)\mathrm{d}s\right\}.$$

Also, consider the jump-diffusion model

$$X_t^{2,\varepsilon} := b_\varepsilon t + (\sigma^2 + \sigma^2(\varepsilon))^{1/2} W_t + \int_0^t \int_{|x|\geq\varepsilon} x N(\mathrm{d}x, \mathrm{d}s),$$

where $\sigma^2(\varepsilon) = \int_{|x|\leq\varepsilon} x^2 \nu(\mathrm{d}x)$. Asmussen and Rosiński (2001) establish the following approximation method:

Theorem 4. *Suppose that ν has no atoms in a neighborhood of the origin. Then:*

(a) $\{\sigma^{-1}(\varepsilon)X_t^\varepsilon\}_{t\geq 0}$ converges in distribution to a standard Brownian motion $\{B(t)\}_{t\geq 0}$ if and only if

$$\lim_{\varepsilon\to 0} \frac{\sigma(\varepsilon)}{\varepsilon} = \infty. \tag{4.27}$$

(b) Under (4.27), it holds that

$$\sup_{x\in\mathbb{R}} \left|\mathbb{P}(X_t \leq x) - \mathbb{P}(X_t^{2,\varepsilon} \leq x)\right| \leq c\, \frac{\int_{|x|\leq\varepsilon} x^3 \nu(\mathrm{d}x)}{\sigma^3(\varepsilon)} \leq c\, \frac{\varepsilon}{\sigma(\varepsilon)}.$$

The first part of the above theorem provides a way to approximate the small-jumps component of X properly scaled by a Wiener process. Condition (4.27) can be interpreted as an assumption requiring that the size of the jumps of $\sigma^{-1}(\varepsilon)X_\varepsilon$ are asymptotically vanishing. Part (b) suggests that the distribution of certain Lévy processes (with infinite jump activity) can be approximated closely by the combination of a Wiener process with drift and a compound Poisson process.

4.4.3 Simulations Based on Series Representations

Throughout, $X = \{X_t\}_{t\in[0,1]}$ is a Lévy process on $\mathbb{R}^d$ with Lévy measure ν and without Brownian component (which can be simulated separately). Let $M := M_X$ be the jump measure of the process X, which we assumed admits the following representation:

Condition 1. The following series representation holds:

$$M(\cdot) = \sum_{i=1}^\infty \delta_{(U_i, H(\Gamma_i, V_i))}(\cdot), \quad a.s. \tag{4.28}$$

for a homogeneous Poisson process $\{\Gamma_i\}_{i=1}^{\infty}$ on $\mathbb{R}_+$ with unit intensity rate, an independent random sample $\{U_i\}_{i=1}^{\infty}$ uniformly distributed on $(0, 1)$, and an independent random sample $\{V_i\}_{i=1}^{\infty}$ with common distribution F on a measurable space S, and a measurable function $H : (0, \infty) \times S \to \mathbb{R}^d$.

Remark 3. Representation (4.28) can be obtained (in law) if the Lévy measure has the decomposition

$$\nu(B) = \int_0^{\infty} \sigma(u; B)\mathrm{d}u, \tag{4.29}$$

where $\sigma(u; B) = \mathbb{P}[H(u, \mathbf{V}) \in B]$. It is not always easy to obtain (4.29). The following are typical methods: the inverse Lévy measure method, Bondensson's method, and Thinning method (see Rosiński 2001 for more details).

Define

$$A(s) = \int_0^s \int_S H(r, v) I\left(\|H(r, v)\| \leq 1\right) F(\mathrm{d}v)\mathrm{d}r. \tag{4.30}$$

Condition 2.

$$A(\Gamma_n) - A(n) \to 0, \quad a.s. \tag{4.31}$$

Lemma 1. *The limit in (4.31) holds true if any of the following conditions is satisfied:*

i. $b \equiv \lim_{s\to\infty} A(s)$ *exists in* $\mathbb{R}^d$*;*
ii. *the mapping* $r \to \|H(r, v)\|$ *is nonincreasing for each* $v \in S$.

Proposition 5. *If the conditions 1 and 2 are satisfies then, a.s.*

$$X_t = bt + \sum_{i=1}^{\infty} \left(H(\Gamma_i, V_i) I\left(U_i \leq t\right) - tc_i\right), \tag{4.32}$$

for all $t \in [0, 1]$, *where* $c_i \equiv A(i) - A(i - 1)$.

Remark 4. The series (4.32) simplifies further when $\int_{|x|\leq 1} |x|\nu(\mathrm{d}\mathbf{x}) < \infty$, namely, when X has paths of bounded variation. Concretely, a.s.

$$X_t = (b - a)\,t + \sum_{i=1}^{\infty} J_i \mathrm{I}\left(U_i \leq t\right), \tag{4.33}$$

where $a = \int_{|x|\leq 1} x\nu(\mathrm{d}x)$. The vector $b_0 \equiv b - a$ is the *drift* of the Lévy process.

4.5 Density Transformation of Lévy Processes

The following two results describe Girsanov-type theorems for Lévy processes. Concretely, the first result provides conditions for the existence of an equivalent probability measure under which X is still a Lévy process, while the second result provides the density process. These theorems have clear applications in mathematical finance as a device to define risk-neutral probability measures. The proofs can be found in Sect. 33 of Sato (1999). Girsanov-type theorems for more general processes can be found in Jacod and Shiryaev (2003) (see also Applebaum (2004) for a more accessible presentation).

Theorem 5. *Let $\{X_t\}_{t\leq T}$ be a real Lévy process with Lévy triple (σ^2, b, ν) under some probability measure $\mathbb{P}$. Then the following two statements are equivalent:*

(a) There exists a probability measure $\mathbb{Q} \sim \mathbb{P}$ such that $\{X_t\}_{t\leq T}$ is a Lévy process with triplet (σ'^2, b', ν') under $\mathbb{Q}$.
(b) All the following conditions hold.

  (i) $\nu'(\mathrm{d}x) = k(x)\nu(\mathrm{d}x)$, for some function $k : \mathbb{R} \to (0, \infty)$.
  (ii) $b' = b + \int x\mathbf{1}_{|x|<1}(k(x) - 1)\nu(\mathrm{d}x) + \sigma\eta$, for some $\eta \in \mathbb{R}$.
  (iii) $\sigma' = \sigma$.
  (iv) $\int \left(1 - \sqrt{k(x)}\right)^2 \nu(\mathrm{d}x) < \infty$.

Theorem 6. *Suppose that the equivalent conditions of the previous theorem are satisfied. Then, $\xi \equiv \frac{d\mathbb{Q}}{d\mathbb{P}}$, is given by the formula*

$$\xi \equiv \exp\left(\eta\sigma W_T - \frac{1}{2}\eta^2\sigma^2 T\right.$$
$$\left. + \lim_{\varepsilon \downarrow 0}\left(\int_0^T \int_{|x|>\varepsilon} \log k(x) M(\mathrm{d}s, \mathrm{d}x) - T\int_{|x|>\varepsilon}(k(x) - 1)\nu(\mathrm{d}x)\right)\right),$$

with $\mathbb{E}_{\mathbb{P}}[\xi] \equiv 1$. The convergence on the right-hand side of the formula above is uniform in t on any bounded interval.

4.6 Exponential Lévy Models

As it was explained in the introduction, the simplest extension of the GBM (4.1) is the Geometric or exponential Lévy model:

$$S_t = S_0 e^{X_t}, \tag{4.34}$$

where X is a general Lévy process with Lévy triplet (σ^2, b, ν) defined on a probability space $(\Omega, \mathbb{P})$. In this part, we will review the financial properties of this model. As in the Black-Scholes model for option pricing, we shall also assume the existence of a risk-free asset B with constant interest rate r. Concretely, B is given by any of the following two equivalent definitions:

$$\begin{aligned} \mathrm{d}B_t &= rB_t\mathrm{d}t \\ B_0 &= 1 \end{aligned}, \quad \text{or} \quad B_t = e^{rt}. \tag{4.35}$$

The following are relevant questions: (1) Is the market arbitrage-free?; (2) Is the market complete?; (3) Can the arbitrage-free prices of European simple claim $\mathcal{X} = \Phi(S_T)$ be computed in terms of a Black-Scholes PDE?.

4.6.1 Stochastic Integration and Self-Financing Trading Strategies

As in the classical Black-Scholes model, the key concept to define arbitrage opportunities is that of a self-financing trading strategy. Formally, this concept requires the development of a theory of stochastic integration with respect to Lévy processes and related processes such as (4.34). In other words, given a suitable trading strategy $\{\beta_t\}_{0\leq t\leq T}$, so that β_t represents the number of shares of the stock held at time t, we want to define the integral

$$G_t := \int_0^t \beta_u \mathrm{d}S_u, \tag{4.36}$$

which shall represent the net gain/loss in the stock at time t. Two different treatments of the general theory of stochastic integration with respect to semimartingales can be found in Jacod and Shiryaev (2003), and Protter (2004). More accessible presentations of the topic are given in, e.g., Applebaum (2004), and Cont and Tankov (2004). Our goal in this part is only to recall the general ideas behind (4.36) and the concept of self-financibility.

We first note that the process β should not only be adapted to the information process $\{\mathcal{F}_t\}_{0\leq t\leq T}$ generated by the stock price (i.e. β_t should depend only on the stock prices up to time t), but also should be *predictable*, which roughly speaking means that its value at any time t can be determined from the information available right before t. As usual, (4.36) can be defined for simple trading strategies in a natural manner and then, this definition can be extended to a certain class of processes β as the limits of stochastic integrals for simple trading strategies. Concretely, consider a "buy-and-hold" trading strategy of the form $\beta_t := \mathbf{1}_{\{\tau_1<t\leq\tau_2\}}$, for deterministic times $0 \leq \tau_1 < \tau_2 \leq T$. That is, β_t represents a strategy that buys one share of the stock "right-after" time τ_1 and holds it until time τ_2. Then, the net gain/loss process is $G_t = \int_0^t \beta_u \mathrm{d}S_u = S_{\tau_2\wedge t} - S_{\tau_1\wedge t}$. Combinations of buy and hold strategies can be defined similarly as

$$\beta_t := \xi_0 \mathbf{1}_{\{t=0\}} + \sum_{i=1}^{n} \xi_i \mathbf{1}_{\{\tau_{i-1}<t\leq\tau_i\}}, \tag{4.37}$$

where $0 = \tau_0 < \tau_1 < \ldots < \tau_n \leq T$ are deterministic trading times and the value of ξ_i is revealed at time τ_{i-1}, for $i = 1, \ldots, n$, while ξ_0 is deterministic. The net gain/loss of the strategy (4.37) at time t is then given by

$$G_t = \int_0^t \beta_u \mathrm{d}S_u = \xi_0 S_0 + \sum_{i=1}^{n} \xi_i \left(S_{\tau_i \wedge t} - S_{\tau_{i-1} \wedge t}\right).$$

The integral (4.36) can subsequently be defined for more general processes β that can be approximated by simple processes of the form (4.37). For instance, if β is an adapted process (thus, for any $t \geq 0$, the value of β_t is revealed at time t) having paths that are left-continuous with right limits (lcrl), then for any sequence $0 = \tau_0^n < \tau_1 < \ldots < \tau_n^n = T$ such that $\max_k \left(\tau_k^n - \tau_{k-1}^n\right) \to 0$, it holds that

$$\beta_0 S_0 + \sum_{i=1}^{n} \beta_{\tau_{i-1}} \left(S_{\tau_i \wedge t} - S_{\tau_{i-1} \wedge t}\right) \xrightarrow{\mathbb{P}} \int_0^t \beta_u \mathrm{d}S_u,$$

as $n \to \infty$, where the convergence is uniform in $[0, T]$. The times τ_i^n can be taken to be *stopping times*, which means that at any time t, one can decide whether the event $\tau_i^n \leq t$ occurs or not.

Once a trading strategy has been defined, one can easily define a *self-financing strategy* on the market (4.34–4.35), as a pair (α, β) of adapted processes such that the so-called value process $V_t := \alpha_t B_t + \beta_t S_t$, satisfies that

$$V_t = V_0 + \int_0^t \alpha_u B_u r \mathrm{d}u + \int_0^t \beta_u \mathrm{d}S_u,$$

or equivalently expressed in "differential form",

$$\mathrm{d}V_t = \alpha_t B_t r \mathrm{d}t + \beta_t \mathrm{d}S_t.$$

Intuitively, the change of the portfolio value $\mathrm{d}V_t$ during a small time interval $[t, t + \mathrm{d}t]$ is due only to the changes in the value of the primary assets in the portfolio and not due to the infusion or withdrawal of money into the portfolio.

4.6.2 Conditions for the Absence of Arbitrage

Let us recall that an arbitrage opportunity during a given time horizon $[0, T]$ is just a self-financing trading strategy (α, β) such that its value process $\{V_t\}_{0 \leq t \leq T}$ satisfies the following three conditions:

$$(i) \quad V_0 = 0, \quad (ii) \quad V_T \geq 0, \quad a.s. \quad (iii) \quad \mathbb{P}(V_T > 0) > 0.$$

According to the first fundamental theorem of finance, the market (4.34-4.35) is arbitrage-free if there exists an *equivalent martingale measure* (EMM) $\mathbb{Q}$; that is, if there exists a probability measure $\mathbb{Q}$ such that the following two conditions hold:

(a) $\mathbb{Q}(B) = 0$ if and only if $\mathbb{P}(B) = 0$.
(b) The discounted price process $S_t^* := B_t^{-1} S_t$, for $0 \leq t \leq T$, is a martingale under $\mathbb{Q}$.

In order to find conditions for the absence of arbitrage, let us recall that for any function k satisfying (iv) in Theorem 5, and any real $\eta \in \mathbb{R}$, it is possible to find a probability measure $\mathbb{Q}^{(\eta,k)}$ equivalent to $\mathbb{P}$ such that, under $\mathbb{Q}^{(\eta,k)}$, X is a Lévy process with Lévy triplet (σ^2, b', ν') given as in Theorem 5. Thus, under $\mathbb{Q}^{(\eta,k)}$, the discounted stock price S^* is also an exponential Lévy model

$$S_t^* = S_0 e^{X_t^*},$$

with X^* being a Lévy process with Lévy triplet $(\sigma^2, b' - r, \nu')$. It is not hard to find conditions for an exponential Lévy model to be a martingale (see, e.g., Theorem 8.20 in Cont and Tankov 2004). Concretely, S^* is a martingale under $\mathbb{Q}^{(\eta,k)}$ if and only if

$$b + \int x \mathbf{1}_{\{|x| \leq 1\}} (k(x) - 1) \nu(dx) + \sigma\eta - r + \frac{\sigma^2}{2} + \int_{\mathbb{R}_0} (e^x - 1 - x \mathbf{1}_{\{|x| \leq 1\}}) k(x) \nu(dx) = 0. \tag{4.38}$$

It is now clear that if $\nu \equiv 0$, there will exist a unique EMM of the form $\mathbb{Q}^{(\eta,k)}$, but if $\nu \not\equiv 0$, there will exist in general infinitely-many of such EMM. In particular, we conclude that the exponential Lévy market (4.34–4.35) is incomplete. One popular EMM for exponential Lévy models is the so-called Esscher transform, where $k(x) = e^{\theta x}$, and η is chosen to satisfy (4.38).

4.6.3 *Option Pricing and Integro-Partial Differential Equations*

As seen in the previous part, the exponential Lévy market is in general incomplete, and hence, options are not superfluous assets whose payoff can be perfectly replicated in an ideal frictionless market. The option prices are themselves subject to modeling. It is natural to adopt an EMM that preserve the Lévy structure of the log return process $X_t = \log(S_t / S_0)$ as in the previous section. From now on, we adopt exactly this option pricing model and assume that the time-t price of a European claim with maturity T and payoff $\mathcal{X}$ is given by

$$\Pi_t = \mathbb{E}_{\mathbb{Q}} \left\{ e^{-r(T-t)} \mathcal{X} \middle| S_u, u \leq t \right\},$$

where $\mathbb{Q}$ is an EMM such that

$$S_t = S_0 e^{rt + X_t^*},$$

with X^* being a Lévy process under $\mathbb{Q}$. Throughout, (σ^2, b^*, ν^*) denotes the Lévy triplet of X^* under $\mathbb{Q}$.

Note that in the case of a simple claim $\mathcal{X} = \Phi(S_T)$, there exists a function $C : \mathbb{R}_+ \times \mathbb{R}_+ \to \mathbb{R}_+$ such that $\Pi_t := C(t, S_t(\omega))$. Indeed, by the Markov property, one can easily see that

$$C(t,x) = e^{-r(T-t)}\mathbb{E}_Q\left[\Phi\left(xe^{r(T-t)+X^*_{T-t}}\right)\right]. \tag{4.39}$$

The following theorem shows that C satisfies an integro-partial differential equation (IPDE). The IPDE equation below is well-known in the literature (see e.g. Chan 1999 and Raible 2000) and its proof can be found in, e.g., (Cont and Tankov, 2004, Proposition 12.1).

Proposition 6. *Suppose the following conditions:*

1. $\int_{|x|\geq 1} e^{2x}\nu^*(\mathrm{d}x) < \infty$*;*
2. Either $\sigma > 0$ *or* $\liminf_{\varepsilon \searrow 0} \varepsilon^{-\beta} \int_{|x|\leq\varepsilon} |x|^2\nu^*(\mathrm{d}x) < \infty$.
3. $|\Phi(x) - \Phi(y)| \leq c|x-y|$*, for all* x, y *and some* $c > 0$.

Then, the function $C(t,x)$ *in (4.39) is continuous on* $[0,T]\times[0,\infty)$*,* $C^{1,2}$ *on* $(0,T)\times(0,\infty)$ *and verifies the integro-partial differential equation:*

$$\frac{\partial C(t,x)}{\partial t} + rx\frac{\partial C}{\partial x}(t,x) + \frac{1}{2}\sigma^2 x^2 \frac{\partial^2 C}{\partial x^2}(t,x) - rC(t,x)$$
$$+ \int_{\mathbb{R}_0}\left(C(t,xe^y) - C(t,x) - x(e^y-1)\frac{\partial C}{\partial x}(t,x)\right)\nu^*(\mathrm{d}y) = 0,$$

on $[0,T)\times(0,\infty)$ *with terminal condition* $C(T,x) = \Phi(x)$*, for all* x.

Acknowledgement Research partially supported by a grant from the US National Science Foundation (DMS-0906919). The author is indebted to the editors for their helpful suggestions that improve this chapter considerably.

References

Ait-Sahalia, Y., & Jacod, J. (2006). Testing for jumps in a discretely observed process. Technical report, Princeton University and Université de Paris VI,

Aït-Sahalia, Y., Mykland, P. A., & Zhang, L. (2005). How often to sample a continuous-time process in the presence of market microstructure noise. *The Review of Financial Studies, 18*(2), 347–350.

Andersen, T., & Benzoni, L. (2007). Stochastic Volatility. Technical report, Working paper, Northwestern University.

Applebaum, D. (2004). *Lévy processes and stochastic calculus*. Cambridge Univ. Press.

Asmussen, S., & Rosiński, J. (2001). Approximations of small jumps of Lévy processes with a view towards simulation. *Journal of Applied Probability, 38*(2), 482–493.

Barndorff-Nielsen, O. E. (1998). Processes of normal inverse Gaussian type. *Finance and Stochastics*, *2*(1), 41–68.

Barndorff-Nielsen, O. E., & Shephard, N. (2001). Modelling by Lévy processess for financial economics. *Lévy Processes. Theory and Applications, by O.E. Barndorff-Nielsen, T. Mikosch, and S.I. Resnick*, 283–318.

Barndorff-Nielsen, O. E., & Shephard, N. (2006). Econometrics of testing for jumps in financial economics using bipower variation. *Journal of Financial Econometrics*, *4*(1), 1–30.

Barndorff-Nielsen, O. E., & Shephard, N. (2007). Variation, jumps, Market Frictions and High Frequency Data in Fnancial Econometric. *In Advances in Economics and Econometrics. Theory and Applications, by R. Blundell, T. Persson, and W. Newey (Eds.).*

Black, F., & Scholes, M. (1973). The pricing of options and corporate liabilities. *Journal of Political Economy*, *81*(3), 637–659.

Carr, P., Geman, H., Madan, D., & Yor, M. (2002). The fine structure of asset returns: An empirical investigation. *Journal of Business*, *75*(2), 305–332.

Carr, P., Geman, H., Madan, D., & Yor, M. (2003). Stochastic volatility for Lévy processes. *Mathematical Finance*, *13*(3), 345–382.

Carr, P., Madan, D., & Chang, E. (1998). The variance gamma process and option pricing. *European Finance Review*, *2*(1), 79–105.

Carr, P., & Wu, L. (2004). Time-changed levy processes and option pricing. *Journal of Financial Economics*, *71*(1), 113–141.

Chan, T. (1999). Pricing contingent claims on stocks driven by Lévy processes. *The Annals of Applied Probability*, *9*(2), 504–528.

Comte, F., & Genon-Catalot, V. (2008). Nonparametric adaptive estimation for pure jump Lévy processes. *Working paper, arXiv:0806.3371.2008.*

Cont, R. (2001). Empirical properties of asset returns: stylized facts and statistical issues. *Quantitative Finance*, *1*(2), 223–236.

Cont, R., & Tankov, P. (2004). *Financial modelling with Jump Processes*. Chapman & Hall.

Eberlein, E. (2001). Application of Generalized Hyperbolic Lévy Motions to Finance. *Lévy Processes. Theory and Applications, by O.E. Barndorff-Nielsen, T. Mikosch, and S.I. Resnick*, 319–336.

Eberlein, E., & Keller, U. (1995). Hyperbolic distribution in finance. *Bernoulli*, *1*(3), 281–299.

Feller, W. (1968). *An Introduction to Probability Theory and Its Applications*, Vol. 1 (3rd Edn.), New York, Wiley. ISBN 0-471-25708-7 Section XIII.7.

Figueroa-López, J. E. (2004). *Nonparametric estimation of Lévy processes with a view towards mathematical finance*. PhD thesis, Georgia Institute of Technology, http://etd.gatech.edu. No. etd-04072004-122020.

Figueroa-López, J. E. (2008). Small-time moment asymptotics for Lévy processes. *Statistics and Probability Letters*, *78*, 3355–3365.

Figueroa-López, J. E. (2009). Nonparametric estimation for Lévy models based on discrete-sampling. *IMS Lecture Notes – Monograph Series. Optimality: The Third Erich L. Lehmann Symposium*, *57*, 117–146.

Figueroa-López, J. E., & Houdré, C. (2006). Risk bounds for the non-parametric estimation of Lévy processes. *IMS Lecture Notes – Monograph Series. High Dimensional Probability*, *51*, 96–116.

Gugushvili, S. (2008). Nonparametric estimation of the characteristic triplet of a discretely observed Lévy process. *Working paper, arXiv:0807.3469v1.2008.*

Jacod, J. (2006). Asymptotic property of realized power variations and related power variations and related functionals of semimartingales. *Preprint*.

Jacod, J. (2007). Asymptotic properties of power variations of Lévy processes. *ESAIM:P&S*, *11*, 173–196.

Jacod, J., & Shiryaev, A. N. (2003). *Limit Theorems for Stochastic Processes*. Springer.

Kallenberg, O. (1997). *Foundations of Modern Probability*. Berlin: Springer.

Kruglov, V. M. (1970). A note on infinitely divisible distributions. *Theory of probability and applications*, *15*, 319–324.

Kyprianou, A., Schoutens, W., & Wilmott, P. (2005). *Exotic option pricing and advanced Lévy models*. Wiley.
Mancini, C. (2009). Non parametric threshold estimation for models with stochastic diffusion coefficient and jumps. *Scandinavian Journal of Statistics*, *36*(2), 270–296.
Mandelbrot, B. (1963). The variation of certain speculative prices. *The Journal of Business*, *36*(4), 394–419.
Merton, R. C. (1976). Option pricing when underlying stock returns are discontinuous. *J. Financial Economics*, *3*(1-2), 125–144.
Neumann, M., & Reiss, M. (2007). Nonparametric estimation for Lévy processes from low-frequency observations. *Working paper, ArXiv:0709.2007v1.*
Podolskij, M. (2006). *New Theory on estimation of integrated volatility with applications*. PhD thesis, Ruhr-Universität Bochum, Available on the web.
Podolskij, M. & Vetter, M. (2009). Estimation of volatility functionals in the simultaneous presence of microstructure noise. *Bernoulli 15*(3), 634–658.
Podolskij, M., & Vetter, M. (2009). Bipower-type estimation in a noisy diffusion setting. *Stochastic processes and their applications 119*, 2803–2831.
Press, S. J. (1967). A compound event model for security prices. *The Journal of Business*, *40*, 317–335.
Protter, P. (2004). *Stochastic Integration and Differentil Equations*. (2nd ed.). Springer-Verlag, Berlin, Heidelberg, 2004.
Raible, S. (2000). *Lévy processes in Finance: Theory, Numerics, and Empirical Facts*. PhD thesis, Albert-Ludwigs-Universitat Freiburg.
Rosiński, J. (2001). Series representations of Lévy processes from the perspective of point processes. *In Lévy Processes-Theory and Applications*, 401–415.
Rosiński, J. (2007). Tempering stable processes. *Stochastic processes and their applications*, *117*(6), 677–707.
Samuelson, P. (1965). Rational theory of warrant pricing. *Industrial Management Review*, *6*(2), 13–32.
Sato, K. (1999). *Lévy Processes and Infinitely Divisible Distributions*. Cambridge University Press.
Shephard, N. (2005). General Introduction. *In N. Shephard (Ed.), Stochastic Volatility*, 451–479.
Todorov, V. (2008). Econometric analysis of jump-driven stochastic volatility models. *Forthcoming in Journal of Econometrics*, *16*, 12–21.
Watteel, R. N. & Kulperger, R. J. (2003). Nonparametric estimation of the canonical measure for innitely divisible distributions. *Journal of Statistical Computation and Simulation*, *73*(7), 525–542.
Woerner, J. (2003). Variational sums and power variation: a unifying approach to model selection and estimation in semimartingale models. *Statistics and Decisions*, *21*, 47–68.
Woerner, J. (2006). Power and multipower variations: inference for high frequency data. *In Stochastic Finance, A.N. Shiryaev, M. do Rosário Grosshino, P. Oliviera, M. Esquivel, Eds.*, 343–354.

Chapter 5
Multivariate Time Series Models for Asset Prices

Christian M. Hafner and Hans Manner

Abstract The modelling of multivariate financial time series has attracted an enormous interest recently, both from a theoretical and practical perspective. Focusing on factor type models that reduce the dimensionality and other models that are tractable in high dimensions, we review models for volatility, correlation and dependence, and show their application to quantities of interest such as value-at-risk or minimum-variance portfolio. In an application to a 69-dimensional asset price time series, we compare the performance of factor-based multivariate GARCH, stochastic volatility and dynamic copula models.

5.1 Introduction

In this chapter we review recent developments in time series analysis of financial assets. We will focus on the multivariate aspect since in most applications the dynamics of a broad variety of assets is relevant. In many situations in finance, the high dimensional characteristics of the data can lead to numerical problems in estimation algorithms. As a motivating example, we show that an application of a standard multivariate GARCH type model in high dimensions to determine the minimum variance portfolio (MVP) yields sub-optimal results due to biased parameter estimates. One possibility to avoid numerical problems is to impose more structure on the conditional covariance matrix of asset returns, for example a factor structure.

C.M. Hafner (✉)
Institut de statistique and CORE, Université catholique de Louvain, Voie du Roman Pays 20, 1348 Louvain-la-Neuve, Belgium
e-mail: christian.hafner@uclouvain.be

H. Manner
Department of Social and Economic Statistics, University of Cologne
e-mail: manner@statistik.uni-koeln.de

J.-C. Duan et al. (eds.), *Handbook of Computational Finance*, Springer Handbooks of Computational Statistics, DOI 10.1007/978-3-642-17254-0_5,

We first discuss recent advances in factor models, where factors can be observed as in the one-factor capital asset pricing model (CAPM) and the three-factor model of Fama and French (1993), or unobserved. The main idea of factor models is to capture common movements in asset prices while reducing the dimension substantially, allowing for flexible statistical modelling.

If factors exhibit specific dynamic features such as volatility clustering or fat tails, then these are typically inherited by the asset prices or returns. For example, fat tailed factor distributions may generate tail dependence and reduce the benefits of portfolio diversification. As for volatility clustering, the modelling of the volatility and the dependence between assets becomes essential for asset pricing models. We therefore review volatility models, again focusing on multivariate models. Since its introduction by Engle (1982) and Bollerslev (1986), the generalized autoregressive conditional heteroscedastic (GARCH) model has dominated the empirical finance literature and several reviews appeared, e.g. Bera and Higgings (1993) and Bauwens et al. (2006). We compare (multivariate) GARCH models to the alternative class of (multivariate) stochastic volatility (SV) models, where the volatility processes are driven by idiosyncratic noise terms. We consider properties and estimation of the alternative models.

With an increasing amount of intra-day data available, an alternative approach of volatility modelling using so-called realized volatility (RV) measures has become available. This approach goes back to an idea of Andersen and Bollerslev (1998). Rather than modelling volatility as an unobserved variable, RV tries to make volatility observable by taking sums of squared intra-day returns, which converges to the daily integrated volatility if the time interval between observations goes to zero. A similar approach is available to obtain realized covariances, taking sums of intra-day cross-products of returns. While this approach delivers more precise measures and predictions of daily volatility and correlations, it also uses another information set and is hence difficult to compare with standard GARCH or SV type models.

Correlation-based models are models of linear dependence, which are sufficient if the underlying distributions have an elliptical shape. However, one often finds empirically that there is an asymmetry in multivariate return distributions and that correlations change over time. In particular, clusters of large negative returns are much more frequent than clusters of large positive returns. In other words, there is lower tail dependence but no upper tail dependence. Copulas are a natural tool to model this effect and have the additional advantage of decoupling the models for the marginal distributions from those for the dependence. We review recent research on dynamic copula models and compare them to correlation-based models.

Finally, we consider approaches how to evaluate the quality of fitted models from a statistical and economic perspective. Two important criteria are, for example, the Value-at-Risk of portfolios and the portfolio selection problem.

5.2 The Investor Problem and Potential Complications

Since the seminal work of Markowitz (1959), portfolio selection has become one of the main areas of modern finance. Today, investment strategies based on mean-variance optimization are considered the benchmark. A first problem of the standard approach is that the obtained optimal portfolio weights depend on second moments (variances and covariances) of the underlying asset returns, which are notoriously time-varying. In other words, the optimal portfolio can only be considered optimal for a short period of time, after which a re-balancing becomes necessary. Another problem is that the formula for optimal portfolio weights depends on the inverse of the covariance matrix, and that in high dimensions the covariance matrix is typically ill-behaved. Hence, portfolio selection might lead to suboptimal results in high dimensions when the standard formulas are applied.

A somewhat related problem is the numerical complexity of standard multivariate volatility models, where the number of parameters may explode as the dimension increases, which leads to intractable estimation and inference of these models. Moreover, in those models where the number of parameters is constant (such as the DCC model of Engle (2002) see Sect. 5.4.2.1), there is no problem in terms of model complexity, but another problem occurs: as the dimension increases, parameter estimates are downward biased and variation in correlations is underestimated, see e.g. Engle et al. (2007). In the following, we illustrate this effect using data of the London stock exchange.

We use the estimated (time varying) covariance matrix for the DCC model to construct the MVP. For the estimated covariance matrix $\hat{H}_t$, the MVP weights are

$$w_t = \frac{\hat{H}_t^{-1}\iota}{\iota^\top \hat{H}_t^{-1}\iota}, \tag{5.1}$$

where ι is an $(N \times 1)$ vector of ones.

The measure of interest is then the variance of the MVP, which should be minimal across different models, and the variance of the standardized portfolio returns given by $r_{p,t} = w_t^\top r_t / \sqrt{w_t^\top \hat{H}_t w_t}$, which should be close to one.

To illustrate the potential problems that can occur when modelling large dimensional data sets we consider daily returns of 69 stocks that are part of the FTSE 100 index ranging from January 1995 until December 1996 we consider the problem of estimating conditional correlations and constructing the MVP between only the first two stocks in the data set. However, a model is fit to a larger data set and we look at the effect of including additional assets in the model.

Figure 5.1 shows the correlations between the first two assets of the sample estimated using the DCC Garch model by Engle in Engle (2002) as the number of assets in the sample K is increased. Surprisingly as the dimension of the data set increases the correlation dynamics are estimated with less precision and the conditional correlations become almost flat for K large as already noted in Engle

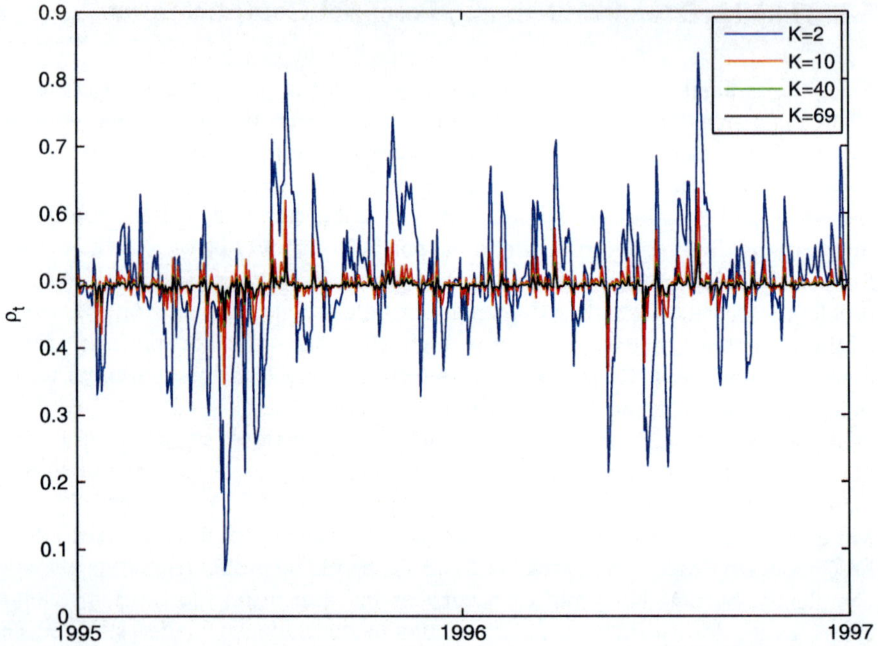

Fig. 5.1 Conditional correlations between two fixed assets for growing dimensions of the model

et al. (2007). Using the covariance matrix estimated using the same sample we constructed the MVP for the first two assets using (5.1). The number of assets is increased from 2 to 69 and the variance of the resulting portfolio is plotted in Fig. 5.2 as a function of K. The portfolio reaches the lowest variance for the model estimated using about ten assets thus implying that the additional information contained in the other series adds economic value. However, once K is increased further the variance grows again and the benefit of including more information in the data is outweighed by the numerical problems causing the flat estimates of the conditional correlations. As the dimension of the model grows further the problem is likely to become worse in addition to the computational complexity that makes estimating large dimensional models difficult.

5.3 Factor Models for Asset Prices

Let $r_t = (r_{1t}, \ldots, r_{Nt})^\top$ denote the vector of asset returns at time t, $t = 1, \ldots, T$. Factor models assume that there is a small number K, $K < N$ of factors f_{kt}, $k = 1 \ldots, K$, such that

$$r_t = a + Bf_t + \varepsilon_t, \tag{5.2}$$

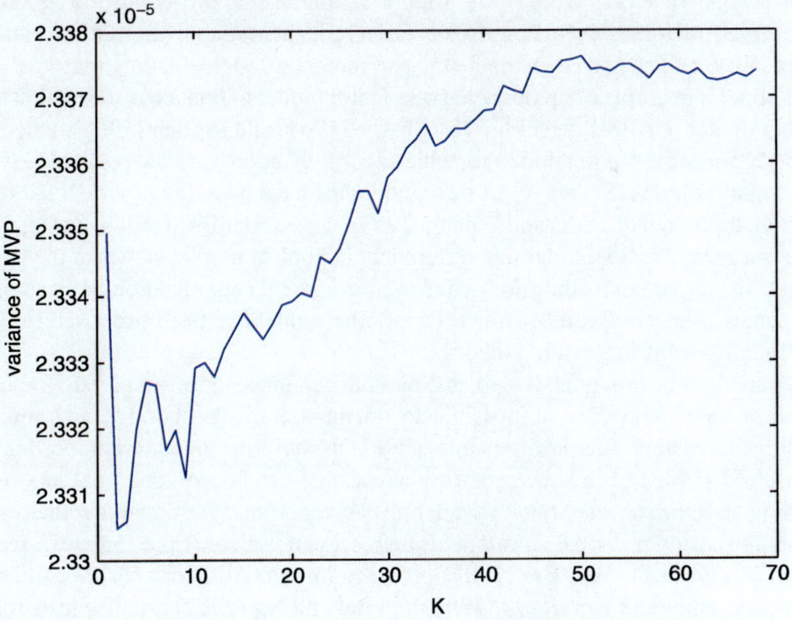

Fig. 5.2 Variance of the MVP of two fixed assets for growing dimensions of the model

where a is an $(N \times 1)$ vector, B an $(N \times K)$ loading matrix and ε_t a stochastic error term with mean zero and variance matrix Ω, uncorrelated with the factors. The idea of factor models is to separate common, non-diversifiable components from idiosyncratic, diversifiable ones. The idiosyncratic error terms are usually assumed to be uncorrelated so that Ω is diagonal, in which case one speaks of a *strict factor model*. If the factors are stationary with mean zero and variance matrix Σ, then returns are stationary with mean a and variance

$$H := \mathrm{Var}(r_t) = B\Sigma B^\top + \Omega. \tag{5.3}$$

Dynamic properties of the factors typically carry over to returns. For example, if factors are nonstationary with time-varying variance Σ_t, then returns will also be nonstationary with variance $H_t = B\Sigma_t B^\top + \Omega$. Another example is that of conditional heteroscedasticity, where factors can be stationary but conditioned on the information of lagged factors, the variance Σ_t is time-varying. Models for Σ_t and H_t will be discussed in the next section.

Note that factor models are identified only up to an invertible rotation of the factors and the loading matrix. To see this, let G be an invertible $(K \times K)$ matrix and write (5.2) equivalently as $r_t = a + BGG^{-1}f_t + \varepsilon_t$, then we have the same model but with factors $\tilde{f}_t = G^{-1}f_t$ and loading matrix $\tilde{B} = BG$. Thus, only the K-dimensional factor space can be identified, not the factors themselves.

Two types of factor models are usually distinguished: those with observed and unobserved factors. When factors are observed, then simple estimation methods such as OLS can be used to estimate the parameters a and the loading matrix B. The most popular example of an observed one-factor model in finance is the capital asset pricing model (CAPM), developed by Sharpe (1964) and Lintner (1965), where the single factor is the market portfolio, which is usually approximated by an observable broad market index. Several empirical anomalies have been found which led to the three-factor model of Fama and French (1993), where additional to the market factor there is a second factor explaining differences in book to market values of the stocks, and a third factor controlling for differences in market capitalization or sizes of the companies. A general multifactor asset pricing model has been proposed by Ross (1976) in his arbitrage pricing theory (APT).

When factors are unobserved, estimation becomes more involved. Imposing structure on Ω and Σ it is possible to do maximum likelihood estimation, but in high dimensions this is often infeasible. On the other hand, Chamberlain and Rothschild (1983) have shown that by allowing Ω to be non-diagonal and hence defining an *approximate factor model*, one can consistently estimate the factors (up to rotation) using principal components regression if both the time and cross-section dimension go to infinity. Bai (2003) provides inferential theory for this situation, whereas Connor and Korajczyk (1993) and Bai and Ng (2002) propose tests for the number of factors in an approximate factor model.

In order to render the factor model *dynamic*, several approaches have been suggested recently. A stationary dynamic factor model specifies the loading matrix B as a lag polynomial $B(L)$ where L is the lag operator and factors follow a stationary process, for example a vector autoregression. Forni et al. (2000) apply the dynamic principal components method by Brillinger (1981) to estimate the common component $B(L)f_t$ in the frequency domain. Forecasting using the dynamic factor model has been investigated e.g. by Stock and Watson (2002). A recent review of dynamic factor models is given by Breitung and Eickmeier (2006).

Rather than considering stationary processes, Motta et al. (2011) follow another approach where factors are stationary but the loading matrix B is a smooth function of time, and hence returns are non-stationary. Estimation is performed using localized principal components regression. To extend the idea of dynamic factor models to the nonstationary case, Eichler et al. (0000) let the lag polynomial $B(L)$ be a function of time and show asymptotic properties of the frequency domain estimator for the common components.

5.4 Volatility and Dependence Models

5.4.1 Univariate Volatility Models

In this section we review alternative univariate models for volatility: GARCH, stochastic volatility and realized volatility.

5.4.1.1 GARCH

The generalized autoregressive conditional heteroskedasticity (GARCH) model introduced by Engle (1982) and Bollerslev (1986) suggests the following specification for asset returns r_t,

$$\begin{aligned} r_t &= \mu_t + \varepsilon_t, \quad \varepsilon_t = \sigma_t \xi_t \\ \sigma_t^2 &= \omega + \alpha \varepsilon_{t-1}^2 + \beta \sigma_{t-1}^2, \end{aligned} \tag{5.4}$$

where $\xi_t \sim N(0,1)$ and μ_t is the mean, conditional on the information set at time $t-1$. For example, the CAPM mentioned in Sect. 5.3 implies that for the return on the market portfolio, $\mu_t = r_f + \lambda\sigma_t^2$, where r_f is the risk free interest rate, λ the market price of risk and σ_t^2 market volatility that could be explained by the GARCH model in (5.4). This is the so-called GARCH-in-mean or GARCH-M model of Engle et al. (1987).

For σ_t^2 in (5.4) to be a well defined variance, sufficient conditions for positivity are $\omega > 0$ and $\alpha \geq 0, \beta \geq 0$. Higher order models that include more lags of ε_t and σ_t^2 are possible but rarely used in practice. A more serious restriction of the standard GARCH model is that recent errors ε_t have a symmetric impact on volatility with respect to their sign. Empirically, one has often observed a *leverage effect*, meaning a higher impact of negative errors than positive ones. Many extensions of the standard GARCH model have been proposed, see e.g. Hentschel (1995) for a review of alternative specifications.

The GARCH(1,1) process in (5.4) is covariance stationary if and only if $\alpha + \beta < 1$, in which case the unconditional variance of ε_t is given by $\sigma^2 = \omega/(1-\alpha-\beta)$. In the GARCH-M case with $\mu_t = r_f + \lambda\sigma_t^2$, the unconditional first two moments of r_t are given by $E[r_t] = r_f + \lambda\sigma^2$ and $\text{Var}(r_t) = \lambda^2\text{Var}(\sigma_t^2) + \sigma^2$. Note that a positive autocorrelation of σ_t^2 induces a similar autocorrelation in returns in the GARCH-M model. This corresponds to empirical evidence of significant first order autocorrelations in daily or weekly stock returns, see e.g. Chap. 2 of Campbell et al. (1997). Straightforward calculations show that the τ-order autocorrelation of r_t is given by

$$\rho(\tau) = (\alpha+\beta)^\tau \frac{\lambda^2\text{Var}(\sigma_t^2)}{\lambda^2\text{Var}(\sigma_t^2) + \sigma^2}, \quad \tau \geq 1.$$

Compared with an AR(1) model with $\mu_t = \phi r_{t-1}$ for which $\rho(\tau) = \phi^\tau$, these autocorrelations could be matched for $\tau = 1$, but at higher orders the GARCH-M model would imply higher autocorrelation than the AR(1) model. Hafner and Herwartz (2000) compared the GARCH-M and AR(1) specifications and found that in most cases the AR(1) model, although without economic motivation, provides a better fit to the data. Obviously, if $\lambda = 0$, then r_t is white noise with $\rho(\tau) = 0$ for all $\tau \neq 0$. An effect of nonzero autocorrelation of returns does not violate the hypothesis of market efficiency, as the autocorrelation is explained by a time-varying risk premium, see e.g. Engle et al. (1987).

The GARCH model implies that returns y_t have a fat tailed distribution, which corresponds to empirical observations already found by Fama (1965) and Mandelbrot (1963). In particular, assuming $\xi_t \sim N(0,1)$ and finite fourth moments of r_t by the condition $\beta^2 + 2\alpha\beta + 3\alpha^2 < 1$, the GARCH(1,1) process in (5.4) has an unconditional kurtosis given by

$$\kappa = 3 + \frac{6\alpha^2}{1 - \beta^2 - 2\alpha\beta - 3\alpha^2},$$

where the second term is positive such that $\kappa > 3$. Thus, while the conditional distribution of r_t is Gaussian, the unconditional one is fat-tailed. Furthermore, there is volatility clustering in the sense that there are periods of high volatility and other periods of low volatility. This reflected by a positive autocorrelation of squared error terms.

Estimation of GARCH models is rather straightforward. Suppose one can separate the parameter ϕ that describes the conditional mean μ_t from the volatility parameter $\theta = (\omega, \alpha, \beta)'$. Assuming normality of ξ_t, one can write the log likelihood function for a sample of T observations up to an additive constant as

$$L(\phi, \theta) = -\frac{1}{2}\sum_{t=1}^{T}\left[\log \sigma_t^2(\theta) + \frac{\{y_t - \mu_t(\phi)\}^2}{\sigma_t^2(\theta)}\right]$$

which is maximized numerically w.r.t. ϕ and θ. Under weak regularity conditions, Bollerslev and Wooldridge (1992) show that $\sqrt{T}(\hat{\theta} - \theta) \to N(0, J^{-1})$ where J is the Fisher information matrix.

5.4.1.2 Stochastic Volatility

Stochastic volatility (SV) models offer a good alternative to capture time-varying variances of asset returns. They originated in different branches of the literature such as financial economics, option pricing and the modelling of financial markets in order to relax the constant variances assumption. For example, Hull and White (1987) allow volatility to follow a general diffusion in their option pricing model. Clark (1973) introduced a model where the information flow to the market is specified as a log-normal stochastic process, which results in a mixture of normal distributions for asset prices. Taylor (1986) accommodated the persistence in volatility and suggested the following autoregressive SV model, which is the most common formulation.

$$r_{it} = \mu_{it} + \exp(h_{it}/2)\xi_{it} \tag{5.5}$$

$$h_{it+1} = \delta_i + \gamma_i h_{it} + \sigma_{\eta_i} \eta_{it} \tag{5.6}$$

ξ_{it} and η_{it} are standard normal innovations and are potentially (negatively) correlated, which leads to a statistical leverage effect meaning that price drops lead to increases in future volatility. σ_{η_i} is assumed to be positive and for $|\gamma| < 1$ the returns r_{it} are strictly stationary. This basic specification is able to explain the fat-tailed return distributions and persistence in volatility well due to the flexibility introduced by the error term. In fact, the Gaussian SV model fit financial data considerably better than a Normal GARCH(1,1) model and it performs about as well as a GARCH model with Student-t innovations. Taylor (1994), Ghysels et al. (1996) and Andersen and Shephard (2009) are excellent reviews on SV models and some extensions. Estimation of SV models, which is reviewed in Broto and Ruiz (2004), is not trivial and probably the main reason why ARCH models are considered more often in empirical studies. Estimation can be done by many different techniques such as the method of moments (see Taylor (1986)), quasi maximum likelihood using the Kalman filter in Harvey et al. (1994), the simulated method of moments by Duffie and Singleton (1993), Gourieroux et al. (1993) and Gallant and Tauchen (1996), Markov Chain Monte Carlo (MCMC) estimation by Jacquier et al. (1994) and Kim et al. (1998), and simulation based maximum likelihood estimations using importance sampling (IS) by Danielsson (1994), Danielsson and Richard (1993) and Liesenfeld and Richard (2003). We recommend using either MCMC or IS methods for estimating the parameters and latent volatility process in a SV model, as these offer very efficient estimates and the considerable computational effort can be handled easily by modern computers.

5.4.1.3 Realized Volatility

With the availability of high-frequency data, by which we mean price data observed every 5 min or even more often, a new set of very powerful tools for volatility estimation and modelling has evolved, namely realized volatility and related concepts. The information contained in high-frequency data allows for improved estimation and forecasting of volatility compared to using only daily data. Furthermore, realized volatility measure relate closely to continuous time SV models and one only needs to assume that the return process is arbitrage free and has a finite instantaneous mean. This in turn implies that the price process is a semi-martingale that the returns can be decomposed into a predictable and integrable mean component and a local martingale. This includes the continuous time stochastic volatility diffusion

$$\mathrm{d}p_t = \mu_t \mathrm{d}t + \sigma_t \mathrm{d}W_t, \tag{5.7}$$

where W_t denotes Brownian motion and the volatility process σ_t is assumed to be stationary. Denote the continuously compounded h period return by $r_{t+h,h} \equiv p_{t+h} - p_t$, where one usually chooses $h = 1$ to be one trading day. Consider a sample of $1/\Delta$ observations per day. In practice Δ is often chosen to be 1/288 corresponding to 5-min returns, although this clearly depends on the data set. Sampling too frequently can lead to a bias due to microstructure noise in the data. Then realized variance for day t is defined as

$$RV = \sum_{j=1}^{h/\Delta} r_{t+j\Delta,\Delta}^2. \tag{5.8}$$

This is a consistent estimator of the quadratic variation and, if the price process does not exhibit any jumps, also of the integrated variance $\int_0^h \sigma_{t+s}^2 ds$. However, in the presence of jumps quadratic variation decomposes into integrated variance and the quadratic variation of the jump component. Barndorff-Nielsen and Shephard (2004b) propose a measure that consistently estimates the integrated variance even in the presence of jumps. This estimator, called bipower variation, is defined as

$$BPV = \frac{\pi}{2} \sum_{j=2}^{h/\Delta} |r_{t+j\Delta,\Delta}||r_{t+(j-1)\Delta,\Delta}|. \tag{5.9}$$

Thus it is possible to separate the continuous and the jump components of volatility by estimating both realized variance and bipower variation, and to identify the jumps by looking at the difference between the two.

Convergence in probability of RV was established by Andersen et al. (2003). Empirical properties of RV are documented in Andersen et al. (2001) and Andersen et al. (2001), such as approximate log-normality, high correlation across different RV series, and long memory properties of volatilities. Forecasting of volatility and the gains that can be made by using high frequency data are discussed in Andersen et al. (1999). Anderson and Vahid (2007) consider latent factor models for RV series and show that these can help forecasting volatilities. The asymptotic distribution of the RV measure and connections to SV models are provided in the notable contributions Barndorff-Nielsen and Shephard (2002a) and Barndorff-Nielsen and Shephard (2002b).

5.4.2 Multivariate Volatility Models

5.4.2.1 Multivariate GARCH Models

GARCH models have been vastly applied to multivariate problems in empirical finance. The typically large number of assets, however, caused problems in early years where models were too complex with too many parameters to estimate. For example, the BEKK model of Engle and Kroner (1995) specifies the conditional covariance matrix H_t as

$$H_t = C_0 C_0^\top + A\varepsilon_{t-1}\varepsilon_{t-1}^\top A^\top + BH_{t-1}B^\top, \tag{5.10}$$

where C_0, A and B are $N \times N$ parameter matrices and C_0 is upper triangular. The model (5.10) is the simplest version of a BEKK model, but higher order models are rarely used. An advantage of the classical BEKK model is its flexibility and generality while generating implicitly a positive definite H_t. However, the number of parameters to estimate is $O(N^2)$, which revealed to be infeasible in high dimensions.

In the following we will therefore concentrate on two model classes, factor GARCH and DCC models, that can be applied to hundreds or thousands of assets. Factor models can be shown to be restricted versions of the BEKK model in (5.10), while DCC type models form a separate, non-nested class of models. A broad overview of multivariate GARCH models has been given recently by Bauwens et al. (2006).

Suppose there are N asset returns, $r_{1t}, \ldots, r_{Nt}, t = 1, \ldots, T$. A model with K factors can be written as

$$r_{it} = b_{i1} f_{1t} + \ldots + b_{iK} f_{Kt} + \varepsilon_{it}, \quad i = 1, \ldots, N,$$

where ε_{it} is an idiosyncratic white noise sequence. In matrix notation this is just the model given in (5.2). If factors follow univariate GARCH processes with conditional variance σ_{it}^2 and are conditionally orthogonal, then the conditional variance of r_{it} can be written as

$$h_{it} = \sum_{k=1}^{K} b_{ik}^2 \sigma_{it}^2 + \omega_i,$$

where $\omega_i = \mathrm{Var}(\varepsilon_{it})$. Factors can be observed assets as in Engle et al. (1990) or latent and estimated using statistical techniques. For example, the Orthogonal GARCH model of Alexander (2001) uses principal components as factors and the eigenvalues of the sample covariance matrix to obtain the factor loadings, before estimating the univariate GARCH models of the factors. van der Weide (2002) generalizes the O-GARCH model to allow for multiplicities of eigenvalues while maintaining identifiability of the model.

A second class of models has attracted considerable interest recently, the class of dynamic conditional correlation (DCC) models introduced by Engle (2002) and Tse and Tsui (2002). In the standard DCC model of order (1,1), conditional variances h_{it} are estimated in a first step using e.g. univariate GARCH. Then, standardized residuals $e_{it} = (r_{it} - \mu_{it})/\sqrt{h_{it}}$ are obtained and the conditional correlation is given by

$$R_{ij,t} = \frac{Q_{ij,t}}{\sqrt{Q_{ii,t} Q_{jj,t}}},$$

where $Q_{ij,t}$ is the (i,j)-element of the matrix process Q_t,

$$Q_t = S(1 - \alpha - \beta) + \alpha e_{t-1} e_{t-1}^\top + \beta Q_{t-1} \qquad (5.11)$$

with S being the sample covariance matrix of e_{it}. In the special case of $\alpha = \beta = 0$, one obtains the constant conditional correlation (CCC) model of Bollerslev (1990).

Splitting the joint likelihood into conditional mean, variance and correlation parameters, the part of the likelihood corresponding to the correlation parameters can be written as

$$\log L(\alpha, \beta) = -\frac{1}{2}\sum_{t=1}^{T}(\log|R_t| + e_t^\top R_t^{-1} e_t) \tag{5.12}$$

An interesting feature of estimators that maximize (5.12) is that for increasing dimension N the α estimates appear to go to zero, as noted already by Engle and Sheppard (2001). Engle et al. (2007) argue that this may be due to the first stage estimation of the conditional variance parameters and the sample covariance matrix S. The parameters of the first stage can be viewed as nuisance parameters for the estimation of the second stage. The covariance targeting idea used in the specification of (5.11) depends on one of these nuisance parameters, S. The effect, clearly demonstrated in simulations by Engle et al. (2007) and Hafner and Franses (2009), is a negative bias for the α estimate, thus delivering very smooth correlation processes in high dimensions and eventually estimates that converge to the degenerate case of a CCC model. Engle et al. (2007) propose to use a so-called composed likelihood estimation, where the sum of quasi-likelihoods over subsets of assets is maximized. They show that this approach does not suffer from bias problems in high dimensions.

Another reason why maximization of (5.12) is not suitable in high dimensions is numerical instability due to almost singular matrices R_t and the problem of inverting this matrix at every t. The sample covariance matrix S is typically ill-conditioned, meaning that the ratio of its largest and smallest eigenvalue is huge. In this case, shrinkage methods as in Ledoit et al. (2003) could possibly be applied to S to improve the properties of the DCC estimates.

A limitation of the classical DCC model in (5.11) is that only two parameters, α and β, drive the dynamic structure of a whole covariance matrix, possibly of high dimension. This seems implausible if N is large, say 50 or higher. Hafner and Franses (2009) proposed to generalize the DCC model as

$$Q_t = S \odot (1 - \bar{\alpha}^2 - \bar{\beta}^2) + \alpha\alpha^\top \odot \varepsilon_{t-1}\varepsilon_{t-1}^\top + \beta\beta^\top \odot Q_{t-1},$$

where now α and β are $(N \times 1)$ vectors, $\odot$ is the Hadamard product, i.e. elementwise multiplication, and $\bar{\alpha} = (1/N)\sum_i \alpha_i$ and $\bar{\beta} = (1/N)\sum_i \beta_i$. This generalized version of the DCC model has the advantage of still guaranteeing a positive definite Q_t and R_t while being much more flexible in allowing some correlations to be very smooth and others to be erratic.

5.4.2.2 Multivariate Stochastic Volatility Models

The basic specification for a multivariate stochastic volatility model (MSV) introduced by Harvey et al. (1994) is given by

$$r_t = \mu_t + H_t^{1/2}\xi_t \tag{5.13}$$

$$H_t^{1/2} = \text{diag}\{\exp(h_{1t},\dots,\exp(h_{Nt}))$$

$$h_{it+1} = \delta_i + \gamma_i h_{it} + \eta_{it}, \text{ for } i = 1,\dots,N \tag{5.14}$$

$$\begin{pmatrix}\xi_t\\ \eta_t\end{pmatrix} \sim N\left[\begin{pmatrix}0\\0\end{pmatrix},\begin{pmatrix}P_\xi & 0\\ 0 & \Sigma_\eta\end{pmatrix}\right], \tag{5.15}$$

where $\mu_t = (\mu_{1t},\dots,\mu_{Nt})^\top$, $\xi_t = (\xi_{1t},\dots,\xi_{Nt})^\top$ and $\eta_t = (\eta_{1t},\dots,\eta_{Nt})^\top$. Σ_η is a positive-definite covariance matrix and P_ξ is a correlation matrix capturing the contemporaneous correlation between the return innovations. Of course, both correlations between the mean innovations and the volatility innovations can be restricted to be zero to reduce the number of parameters. If one only assumes that the off-diagonal elements of Σ_η are equal to zero this specification corresponds to the constant conditional correlation (CCC) GARCH model by Bollerslev (1990), since no volatility spillovers are possible.

This basic model has relatively few parameters to estimate $(2N + N^2)$, but Danielsson (1998) shows that it outperforms standard Vector-GARCH models that have a higher number of parameters. Nevertheless, a number of extensions of this model are possible. First, one can consider heavy tailed distributions for the innovations in the mean equation ξ_t in order to allow for higher excess kurtosis compared to the Gaussian SV model, although in most cases this seems to be unnecessary. Harvey et al. (1994) suggest using a multivariate t-distribution for that purpose.

A second simple and natural extension of the basic model can be achieved by introducing asymmetries into the model. One possibility is to replace (5.15) by

$$\begin{pmatrix}\xi_t\\ \eta_t\end{pmatrix} \sim N\left[\begin{pmatrix}0\\0\end{pmatrix},\begin{pmatrix}P_\xi & L\\ L & \Sigma_\eta\end{pmatrix}\right]$$

$$L = \text{diag}\{\lambda_1\sigma_{\eta,11},\dots,\lambda_N\sigma_{\eta,NN}\}, \tag{5.16}$$

where $\sigma_{\eta,ii}$ denotes the i'th diagonal element of Σ_η and λ_i is expected to be negative for $i = 1,\dots,N$. This specification allows for a statistical leverage effect. Asai et al. (2006) distinguish between leverage, denoting negative correlation between current returns and future volatility, and general asymmetries meaning negative returns have a different effect on volatility than positive ones. These asymmetric effects may be modeled as a threshold effect or by including past returns and their absolute values, in order to incorporate the magnitude of the past returns, in (5.14). The latter extension was suggested by Danielsson (1994) and is given by

$$h_{it+1} = \delta_i + \phi_{i1}y_{it} + \phi_{i2}|y_{it}| + \gamma_i h_{it} + \sigma_{\eta_i}\eta_{it}. \tag{5.17}$$

A potential drawback of the basic models and its extensions is that the number of parameters grows with N and it may become difficult to estimate the model with a high dimensional return vector. Factor structures in MSV models are a possibility to achieve a dimension reduction and make the estimation of high dimensional systems feasible. Furthermore, factor structures can help identify common features in asset returns and volatilities and thus relate naturally to the factor models described in Sect. 5.3. Diebold and Nerlove (1989) propose a multivariate ARCH model with latent factors that can be regarded as the first MSV model with a factor structure, although Harvey et al. (1994) are the first to propose the use of common factors in the SV literature. Two types of factor SV models exist: Additive factor models and multiplicative factor models. An additive K factor model is given by

$$\begin{aligned} r_t &= \mu_t + Df_t + e_t \\ f_{it} &= \exp(h_{it}/2)\xi_{it} \\ h_{it+1} &= \delta_i + \gamma_i h_{it} + \sigma_{\eta_i}\eta_{it}, \text{ for } i = 1,\ldots,K, \end{aligned} \tag{5.18}$$

with $e_t \sim N(0, \text{diag}(\sigma_1^2,\ldots,\sigma_N^2))$, $f_t = (f_{1t},\ldots,f_{Kt})^\top$, D is an $N \times K$ matrix of factor loadings and $K < N$. Identification is achieved by setting $D_{ii} = 1$ for all $i = 1,\ldots,N$ and $D_{ij} = 0$ for all $j < i$. As mentioned in Asai et al. (2006) a serious drawback of this specification is that homoscedastic portfolios can be constructed, which is unrealistic. Assuming a SV model for each element of e_t can solve this problem, although it does increase the number of parameters again. Furthermore, the covariance matrix of e_t is most likely not diagonal. A further advantage of the model is that it does not only accommodate time-varying volatility, but also time-varying correlations, which reflects the important stylized fact that correlations are not constant over time. A multiplicative factor model with K factors is given by

$$\begin{aligned} r_t &= \mu_t + \exp\left(\frac{wh_t}{2}\right)\xi_t \\ h_{it+1} &= \delta_i + \gamma_i h_{it} + \sigma_{\eta_i}\eta_{it}, \text{ for } i = 1,\ldots,K, \end{aligned} \tag{5.19}$$

where w is an $N \times K$ matrix of factor loadings that is of rank K and $h_t = (h_{1t},\ldots,h_{Kt})^\top$. This model is also called stochastic discount factor model.

Although factor MSV models allow for time-varying correlations these are driven by the dynamics in the volatility. Thus a further extension of the basic model is to let the correlation matrix P_ξ depend on time. For the bivariate case Yu and Meyer (2006) suggest the following specification for the correlation coefficient ρ_t.

$$\begin{aligned} \rho_t &= \frac{\exp(2\lambda_t) - 1}{\exp(2\lambda_t) + 1} \\ \lambda_{t+1} &= \delta_\rho + \gamma_\rho \lambda_t + \sigma_\rho z_t, \end{aligned} \tag{5.20}$$

where $z_t \sim N(0,1)$. A generalization to higher dimensions of this model is not straightforward. Yu and Meyer (2006) propose the following specification following the DCC specification of Engle (2002).

$$P_{\xi t} = \text{diag}(Q_t^{-1/2}) Q_t \text{diag}(Q_t^{-1/2}) \tag{5.21}$$
$$Q_{t+1} = (\iota\iota^\top - A - B) \odot S + B \odot Q_t + A \odot z_t z_t^\top,$$

where $z_t \sim N(0, I)$, ι is a vector of ones. An alternative to this is the model by Asai and McAleer (2009), which also uses the DCC specification, but the correlations are driven by a Wishart distribution.

Further specifications of MSV models along with a large number of references can be found in Asai et al. (2006), whereas Yu and Meyer (2006) compares the performance of a number of competing models. One main finding of this study is that models that allow for time-varying correlations clearly outperform constant correlation models.

Estimation can in principle be done using the same methods suggested for univariate models, although not each method may be applicable to every model. Still, simulated maximum likelihood estimation and MCMC estimation appear to be the most flexible and efficient estimation techniques available for MSV models.

5.4.2.3 Realized Covariance

The definition of realized volatility extends to the multivariate case in a straightforward fashion and thus the additional information contained in high frequency data can also be exploited when looking at covariance, correlation and simple regressions. Some references are Andersen et al. (2001) and Andersen et al. (2001) providing definitions, consistency results and empirical properties of the multivariate realized measures. Barndorff-Nielsen and Shephard (2004a) provide a distribution theory for realized covariation, correlation and regression, the authors discuss how to calculate confidence intervals in practice. A simulation study illustrates the good quality of their approximations in finite samples when Δ is small enough (about 1/288 works quite well). Let the h period return vector be $r_{t+h,h}$. Then realized covariance is defined as

$$RCOV = \sum_{j=1}^{h/\Delta} r_{t+j\Delta,\Delta} r_{t+j\Delta,\Delta}^\top. \tag{5.22}$$

The realized correlation between return on asset k, $r_{(k)t+h,h}$, and the return of asset l, $r_{(l)t+h,h}$, is calculated as

$$RCORR = \frac{\sum_{j=1}^{h/\Delta} r_{(k)t+j\Delta,\Delta} r_{(l)t+j\Delta,\Delta}}{\sqrt{\sum_{j=1}^{h/\Delta} r_{(k)t+j\Delta,\Delta}^2 \sum_{j=1}^{h/\Delta} r_{(l)t+j\Delta,\Delta}^2}}. \tag{5.23}$$

Finally, the regression slope when regressing variable l on variable k is given by

$$\hat{\beta}_{(lk),t} = \frac{\sum_{j=1}^{h/\Delta} r_{(k)t+j\Delta,\Delta} r_{(l)t+j\Delta,\Delta}}{\sum_{j=1}^{h/\Delta} r_{(k)t+j\Delta,\Delta}^2}. \tag{5.24}$$

All these quantities have been shown to follow a mixed normal limiting distribution. An application of the concept of realized regression is given in Andersen et al. (2006), where the authors compute the realized quarterly betas using daily data and discuss its properties.

5.4.2.4 Dynamic Copula Models

A very useful tool for specifying flexible multivariate versions of any class of distribution functions are copulas. A copula is, loosely speaking, that part of a multivariate distribution function that captures all the contemporaneous dependence. The most important results concerning copulas known as Sklar's theorem tells us that there always exists a copula such that any multivariate distribution function can be decomposed into the marginal distributions capturing the individual behavior of each series and a copula characterizing the dependence structure. This separation does not only allow for an easy and tractable specification of multivariate distributions, but also for a two-step estimation greatly reducing the computational effort. Thus any of the volatility models described above can be generalized to the multivariate case in a straightforward fashion by coupling the univariate models using copulas. Furthermore, dependence structures that go beyond linear correlation such as tail dependence and asymmetric dependencies, which is useful when markets or stocks show stronger correlation for negative than for positive returns, can be allowed for. Nelsen (2006) provides a mathematical introduction to the topic, whereas Joe (1997) treats the topic from a statistical viewpoint. Cherubini et al. (2004) and Franke et al. (2008) look at copulas and their applications for financial problems.

Consider the N-dimensional return vector $r_t = (r_{1t}, \ldots, r_{Nt})^\top$. Let F_i be the marginal distribution function of return i at let H be the joint distribution function of $\boldsymbol{r}_t$. Then by Sklar's theorem there exists a copula function C such that

$$H(r_{1t}, \ldots, r_{Nt}) = C\left\{F_1(r_{1t}), \ldots, F_N(r_{Nt})\right\}. \tag{5.25}$$

Additionally, if the marginals are continuous the copula is unique. Recalling that by the probability integral transform the variable $u_{it} = F_i(r_{it})$ follows a standard uniform distribution it becomes clear that a copula is simply a multivariate distribution function with $U(0,1)$ marginals.

A large number of examples of copula function and methods to simulate artificial data from them, which is extremely useful for the pricing of derivatives with multiple underlying assets, is discussed in the chapter "Copulae Modelling" in this handbook. However, here we focus our attention on the situation when the copula is allowed to vary over time, which accommodates the special case of time-varying correlations, a feature usually observed in financial data. Dynamic copulas can thus be used to construct extremely flexible multivariate volatility models that tend to fit the data better than models assuming a dependence structure that is fixed over time. In what follows we denote the time-varying parameter of a bivariate copula by θ_t.

Structural Breaks in Dependence

A formal test for the presence of a breakpoint in the dependence parameter of a copula was developed in Dias and Embrechts (2004). Denote $\boldsymbol{\eta}_t$'s the parameters of the marginal distributions, which are treated as nuisance parameters. Formally, the null hypothesis of no structural break in the copula parameter becomes

$$H_0 : \theta_1 = \theta_2 = ... = \theta_T \text{ and } \boldsymbol{\eta}_1 = \boldsymbol{\eta}_2 = ... = \boldsymbol{\eta}_T$$

whereas the alternative hypothesis of the presence of a single structural break is formulated as:

$$H_1 : \boldsymbol{\theta}_1 = ... = \boldsymbol{\theta}_k \neq \boldsymbol{\theta}_{k+1} = ... = \boldsymbol{\theta}_T \equiv \boldsymbol{\theta}_k^* \text{ and } \eta_1 = \eta_2 = ... = \eta_T.$$

In the case of a known break-point k, the test statistics can be derived as a generalized likelihood ratio test. Let $L_k(\theta, \eta)$, $L_k^*(\boldsymbol{\theta}, \eta)$ and $L_T(\boldsymbol{\theta}, \eta)$ be the log-likelihood functions corresponding to a copula based multivariate model using the first k observations, the observations from $k + 1$ to T and all observations, respectively. Then the likelihood ratio statistic can be written as

$$LR_k = 2[L_k(\hat{\boldsymbol{\theta}}_k, \hat{\boldsymbol{\eta}}_T) + L_k^*(\hat{\boldsymbol{\theta}}_k^*, \hat{\boldsymbol{\eta}}_T) - L_T(\hat{\boldsymbol{\theta}}_T, \hat{\boldsymbol{\eta}}_T)],$$

where a hat denotes the maximizer of the corresponding likelihood function. Note that $\hat{\boldsymbol{\theta}}_k$ and $\hat{\boldsymbol{\theta}}_k^*$ denote the estimates of $\boldsymbol{\theta}$ before and after the break, whereas $\hat{\boldsymbol{\theta}}_T$ and $\hat{\boldsymbol{\eta}}_T$ are the estimates of $\boldsymbol{\theta}$ and $\boldsymbol{\eta}$ using the full sample. In the case of an unknown break date k, a recursive procedure similar to the one proposed in Andrews (1993) can be applied. The test statistic is the supremum of the sequence of statistics for known k

$$Z_T = max_{1 \leq k < T} LR_k \tag{5.26}$$

and the asymptotic critical values of Andrews (1993) can be used. Manner and Candelon (2010) extended the procedure to additionally allow for a breakpoint in the unconditional variance of the individual series at a (possibly) different point in time and they discuss how to estimate the breakpoints in volatility and in dependence sequentially.

The Conditional Copula Model

Patton (2006a) showed that Sklar's theorem still holds for conditional distributions and suggested the following time varying specification for copulas. For the Gaussian copula correlation evolves, similarly to the DCC model, as

$$\rho_t = \Lambda\left\{\alpha + \beta_1 \cdot \rho_{t-1} + \beta_2 \cdot \frac{1}{p}\sum_{j=1}^{p} \Phi^{-1}(u_{1,t-j}) \cdot \Phi^{-1}(u_{2,t-j})\right\}, \tag{5.27}$$

where, $\Lambda(x) = \frac{1-e^{-x}}{1+e^{-x}}$ is the inverse Fisher transformation. The number of lags p is chosen to be 10, although this is a rather arbitrary choice that may be varied. For copulas different from the Gaussian the sum in (5.27) is replaced by $\sum_{j=1}^{p} |u_{1,t-j} - u_{2,t-j}|$ and Λ has to be replaced by a transformation appropriate to ensure the dependence parameter is in the domain of the copula of interest.

Adaptive Estimation of Time-Varying Copulas

In order to save some space we refer to the chapter "Copulae Modelling" in this handbook for a description of these techniques to estimate dynamic copulas introduced by Giacomini et al. (2009).

Stochastic Dynamic Copulas

While the model by Patton can be seen as the counterpart to a GARCH model, where correlations are a function of the past observation, in Hafner and Manner (2011) we propose to let the dependence parameter of a copula follow a transformation of a Gaussian stochastic process. That has, similar to stochastic volatility models, the advantage of being a bit more flexible than a DCC model or the specification by Patton at the cost of being more difficult to estimate. Furthermore, it is a natural approach for a multivariate extension of stochastic volatility models.

We assume that θ_t is driven by an unobserved stochastic process λ_t such that $\theta_t = \Psi(\lambda_t)$, where $\Psi : \mathbb{R} \to \Theta$ is an appropriate transformation to ensure that the copula parameter remains in its domain and whose functional form depends on the choice of copula. The underlying dependence parameter λ_t, which is unobserved, is assumed to follow a Gaussian autoregressive process of order one,

$$\lambda_t = \alpha + \beta\lambda_{t-1} + \nu\varepsilon_t, \tag{5.28}$$

where ε_t is an i.i.d. $N(0,1)$ innovation. Since λ_t is unobservable it must be integrated out of the likelihood function. Such a T dimensional integral cannot be solved analytically. However, λ_t can be integrated out by Monte Carlo integration using the efficient importance sampler of Liesenfeld and Richard (2003).

Local Likelihood Estimation of Dynamic Copulas

A model which allows θ_t to change over time in a non-parametric way is proposed in Hafner and Reznikova (2010). It is assumed that the copula parameter can be represented as a function $\theta(t/T)$ in rescaled time. If that function is sufficiently smooth then the bivariate return process is locally stationary. Estimation is done in two steps, where first GARCH models for the margins are estimated and in the second step the time-varying copula parameter is estimated by local maximum likelihood estimation. That means that the log-likelihood function is locally weighted by a kernel function. Additionally, a one step correction for the estimates of the GARCH parameters ensures semi-parametric efficiency of the estimator, which is shown to work well in simulations.

5.4.2.5 Assessing the Quality of the Models

For practical purposes it is important to have a way to distinguish among the many competing models. For testing a particular feature of a model such as the leverage effect one can often apply standard hypothesis tests such a t-tests or likelihood ratio tests. When competing models do not belong to the same model class and are non-nested this is usually not possible anymore. Here we do not only consider statistical criteria to assess how well a given model can describe that data, but we also look at some economic measures that compare the usefulness of competing models for certain investment decisions.

The simplest way to compare the in-sample fit of competing models is to look at the value of the log-likelihood function at the parameter estimates, which gives a good indication of how well the statistical model describes a given data set. Since not all models have the same number of parameters and since models with a larger number of parameters will most of the time fit the data better due to more flexibility, it is often recommendable to use some type of information criterion that penalizes a large number of parameters in a model. The two most commonly used information criteria are the Akaike information criterion given by

$$AIC = -2LL + 2p \tag{5.29}$$

and the Bayesian information criterion

$$BIC = -2LL + p\log(T), \tag{5.30}$$

where LL denotes the value of log-likelihood function, T is the sample size and p is the number of parameters in the model. The model with the smallest value for either AIC or BIC is then considered the best fitting one, where the BIC tends to favor more parsimonious models. However, even the best fitting model from a set of candidate models may not provide reasonable fit for the data, which is why distributional assumptions are often tested using specific goodness-of-fit tests such as the Jarque-Bera test for normality, the Kolmogorov-Smirnov test or the Anderson-Darling test. One may also want to test for i.i.d.'ness of the standardized residuals of the candidate model by testing for remaining autocorrelation and heteroscedasticity. Finally, one may be interested in comparing the out-of-sample performance of a number of models. We refer to Diebold et al. (1998) for possible procedures. When comparing the forecasting performance of volatility models realized volatility offers itself naturally as a measure for the (unobserved) variance of a series.

Although a good statistical fit of a model is a desirable feature of any model a practitioner may be more interested in the economic importance of using a certain model. A very simple, yet informative measure is the Value-at-Risk (VaR), which measures how much money a portfolio will loose at least with a given probability. For portfolio return y_t the VaR at quantile α is defined as $P[y_t < \text{VaR}_\alpha] = \alpha$. The VaR can be computed both in sample and out-of-sample and Engle and Manganelli (2004) suggest a test to assess the quality of a VaR estimate for both cases. A related measure is the expected shortfall (ES), which is the expected loss given that the portfolio return lies below a specific quantile, i.e. $\text{ES}_\alpha = E(y_t | y_t < \text{VaR}_\alpha)$. As portfolio managers are often interested to minimize the risk of their portfolio for a given target return models can be compared by their ability to construct the MVP as suggested by Chan et al. (1999). The MVP can be considered and the conditional mean can be ignored as it is agreed on that the mean of stock returns is notoriously difficult to forecast, especially for returns observed at a high frequency. A similar approach was taken in Fleming et al. (2001) to evaluate the economic values of using sophisticated volatility models for portfolio selection. Since portfolio manager often aim at reproducing a certain benchmark portfolio (Chan et al. (1999)) also suggest to compare models by their ability to minimize the tracking error volatility, which is the standard deviation of the difference between the portfolio's return and the benchmark return.

5.5 Data Illustration

In this section we want to illustrate some of the techniques mentioned above for modelling a multi-dimensional time series of asset prices. The data we consider are those 69 stocks from the FTSE 100 index that were included in that index over our whole sample period. We look at daily observations from the beginning of 1995 until the end of 2005 and calculate returns by taking the first difference of the natural logarithm. We multiply returns by 100 to ensure stability of the numerical procedures used for estimation. Modelling 69 assets is still less than the vast dimensions required for practical applicability, but it is already quite a large number for many multivariate time series models and much more than what is used in most studies. Fitting a 69 dimensional volatility model directly to the data is not possible for many of the models presented above, mainly because the number of parameters grows rapidly with the dimension of the problem and estimation becomes difficult or even impossible. We therefore impose a lower dimensional factor structure on the data in order to achieve a reduction of the dimension of the problem and fit different volatility models to the factors extracted by principal component analysis (PCA). The idiosyncratic components are assumed to be independent of each other and their time-varying volatilities are estimated by univariate GARCH and SV models. When estimating simple univariate GARCH or SV models to the factors this is very similar to the O-GARCH model of Alexander (2001), but we also consider multivariate GARCH and SV models to model the volatility of the factors jointly. Namely, we estimate DCC and BEKK GARCH models, and SV models with conditional correlations being described by the Patton and SCAR copula specification. For the last two cases conditional correlations can only be estimated for the case of two factors. Note that although the correlations between the factors extracted by PCA are unconditionally zero, conditional correlations may be different from zero and vary over time.

For the factor specification the covariance matrix for the full set of assets can be calculated using (5.3) in Sect. 5.3. For the number of factors we restrict our attention to a maximum of four factors. When estimating SV model the efficient importance sampler by Liesenfeld and Richard (2003) is used for estimation and for the time-varying volatility we consider the smoothed variance, i.e. an estimate of the volatility using the complete sample information. The volatilities of the first two factors estimated by GARCH and SV are shown in Fig. 5.3, whereas the conditional

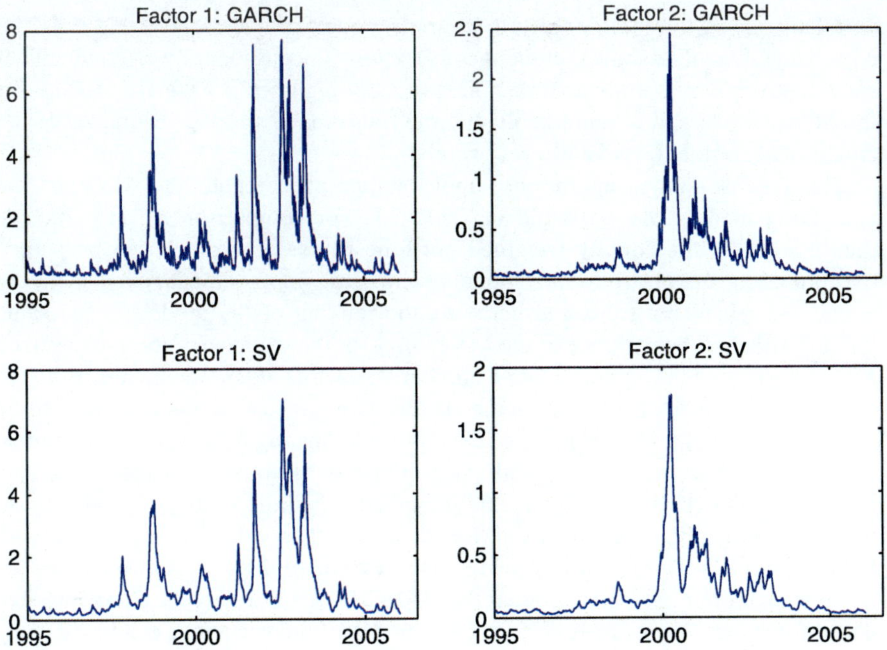

Fig. 5.3 Conditional volatilities of the first two factors

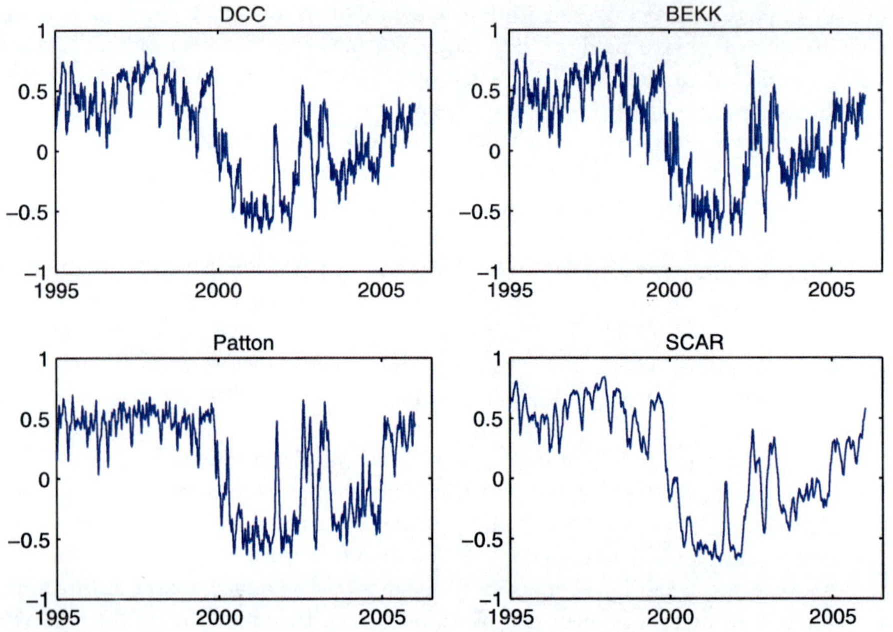

Fig. 5.4 Conditional correlations between the first two factors

correlation using the four competing bivariate models can be found in Fig. 5.4. The correlation dynamics show that the factors are only unconditionally orthogonal, but show a strong variation over time and extremely high persistence ($\beta = 0.99$ for the SCAR model). It is remarkable that the four models produce estimates of the conditional correlations that are very similar.

The results comparing the in-sample ability to compute the MVP of the competing models can be found in Table 5.1. For comparison we also include the variance of the equally weighted portfolio to see how much can be gained by optimizing the portfolio. All models yield clear improvements over using the equally weighted portfolio. Furthermore, the ranking of the models is the same looking either at the variance of the MVP, σ_{MVP}, or the variance of the standardized MVP, $\sigma_{MVP-std}$. The choice of the number of factors does not matter as much as one might expect. Still, two factor models give the best results and seem to be sufficient to estimate the volatility of the data set. Allowing for non-zero conditional correlations between the factors slightly improves the quality of the covariance matrix of the stock returns. Finally, the smoothed volatilities of the SV models seem to provide much better estimates of the covariance than volatilities estimated with GARCH models. This is not surprising, as the SV volatilities are estimated using the full information in the data, whereas the GARCH volatilities are based on one-step ahead forecasts. Hence, the two-factor SV model with correlations estimated using a SCAR specification provides the best fit for our data set based on the economic

Table 5.1 In-sample fit of competing volatility models

	Model	σ_{MVP}	$\sigma_{MVP-std}$	Model	σ_{MVP}	$\sigma_{MVP-std}$
Equally weighted		0.821				
	GARCH	0.307	2.369	SV	0.143	1.341
	O-GARCH	0.295	2.004	O-SV	0.135	1.100
1 Factor	DCC	0.292	1.965	SCAR	0.131	1.073
	BEKK	0.292	1.959	Patton	0.133	1.081
	O-GARCH	0.298	2.047	O-SV	0.137	1.129
2 Factors	DCC	0.297	2.006			
	BEKK	0.296	1.994			
	O-GARCH	0.302	2.041	O-SV	0.140	1.128
3 Factors	DCC	0.301	2.003			
4 Factors	BEKK	0.299	1.990			

Note: Variance of the MVPs (σ_{MVP}) and the standardized portfolio ($\sigma_{MVP-std}$) constructed using competing factor based multivariate volatility models for 69 stocks from the FTSE 100 during the period 1995–2005

criteria we have chosen. Nevertheless one has to keep in mind that the analysis we have done is entirely an in-sample comparison. Out-of-sample the models may be ranked quite differently. In particular, when considering one-step ahead forecasts the GARCH model is likely to perform quite well (in particular better than SV), due to the way it is designed. When considering multi-step ahead forecasts it is unclear which model will do better and this issue is worth investigating.

5.6 Outlook

In this chapter we have reviewed new developments in the dynamic modelling of financial asset returns. We have concentrated on the multivariate aspect, since the typical practical application concerns not only volatilities but also an adequate modelling of asset dependencies. We have paid attention to the use of factor structures in order to achieve some dimension reduction when modelling a large number of assets. Such factor structures combined with appropriate volatility models seem to provide a good fit to the data we examined, and not too much information is lost when computing the MVP, compared to modelling the full set of assets directly.

In future research the choice of the number of factors, a problem that has been discussed extensively in a theoretical way and for macroeconomic applications, needs to be analyzed concerning the model performance using economic criteria such as the construction of the MVP. Also the use of the class of locally stationary factor models by Motta et al. (2011) and Eichler et al. (0000) for financial applications needs to be considered. Furthermore, the modelling of vast dimensional data (i.e. over 100 assets) needs to be studied. Although some progress has been made for GARCH models, stochastic volatility models that are usable for such dimensions, and estimation techniques for them, need to be developed. Finally, time-varying copula models need to be extended to allow for dimensions larger than two in order to be relevant for realistic applications.

References

Alexander, C. (2001). Orthogonal GARCH. In *Mastering risk, financial times* (Vol 2, pp. 21–38). London: Prentice Hall.

Anderson, H. M., & Vahid, F. (2007). Forecasting the volatility of australian stock returns: do common factors help? *Journal of Business and Economic Statistics, 25*(1), 76–90.

Andersen, T. G., & Bollerslev, T. (1998). Answering the skeptics: Yes, standard volatility models do provide accurate forecasts. *International Economic Review, 39*, 885–905.

Andersen, T.G., Shephard, N. (2009): Stochastic volatility: origins and overview. In: T.G. Andersen, R.A. Davis, J.-P. Kreiss, T. Mikosch (Eds.) Handbook of Financial Time Series, p. 233–254, Springer Verlag: Berlin, Heidelberg New York.

Andersen, T. G., Bollerslev, T., & Lange, S. (1999). Forecasting financial market volatility: Sample frequency vis-a-vis forecast horizon. *Journal of Financial Economics, 61*, 43–76.

Andersen, T. G., Bollerslev, T., Diebold, F. X., & Ebends, H. (2001). The distribution of realized stock volatility. *Journal of Financial Economics, 61*, 43–76.

Andersen, T. G., Bollerslev, T., Diebold, F. X., & Labys, P. (2001). The distribution of realized exchange rate volatility. *Journal of the American Statistical Association, 96*, 42–55.

Andersen, T. G., Bollerslev, T., Diebold, F. X., & Labys, P. (2003). Modeling and forecasting realized volatility. *Econometrica, 71*(2), 579–625.

Andersen, T. G., Bollerslev, T., Diebold, F. X., & Wu, J. (2006). Realized beta: Persistence and predictability. In T. Fomby & D. Terrell (Eds.), *Advances in Econometrics: Econometric analysis of economic and financial time series in honor of R. F. Engle and C. W. J. Granger* (Vol. B, pp. 1–40).

Andrews, D. W. K. (1993). Tests for parameter instability and structural change with unknown change point. *Econometrica, 61*, 821–856.

Asai, M., & McAleer, M. (2009). The structure of dynamic correlations in multivariate stochastic volatility models. *Journal of Econometrics, 150*, 182–192.

Asai, M., McAleer, M., & Yu, J. (2006). Multivariate stochastic volatility: A review. *Econometric Reviews, 25*(2–3), 145–175.

Bai, J. (2003). Inferential theory for factor models of large dimensions. *Econometrica, 71*, 135–171.

Bai, J., & Ng, S. (2002). Determining the number of factors in approximate factor models. *Econometrica, 70*(1), 191–221.

Barndorff-Nielsen, O. E., & Shephard, N. (2002a). Econometric analysis of realized volatility and its use in estimating stochastic volatilty models. *Journal of the Royal Statistical Society, Ser. B, 64*, 253–280.

Barndorff-Nielsen, O. E., & Shephard, N. (2002b). Estimating quadratic variation using realized variance. *Journal of Applied Econometrics, 17*, 457–477.

Barndorff-Nielsen, O. E., & Shephard, N. (2004a). Econometric analyis of realized covariation: high frequency based covariance, regression, and correlation in financial economics. *Econometrica, 72*, 885–925.

Barndorff-Nielsen, O. E., & Shephard, N. (2004b). Power and bipower variation with stochastic volatility and jumps. *Journal of Financial Econometrics, 2*, 1–37.

Bauwens, L., Laurent, S., & Rombouts, J. (2006). Multivariate garch models: A survey. *Journal of Applied Econometrics, 7*, 79–109.

Bera, A., & Higgings, M. (1993). A survey of ARCH models: Properties, estimation and testing. *Journal of Economic Surveys, 7*, 305–366.

Bollerslev, T. (1986). Generalized autoregressive conditional heteroskedasticity. *Journal of Econometrics, 31*, 307–327.
Bollerslev, T. (1990). Modelling the coherence in short-run nominal exchange rates: A multivariate generalized ARCH approach. *Review of Economics and Statistics, 72*, 498–505.
Bollerslev, T., & Wooldridge, J. M. (1992). Quasi maximum likelihood estimation of dynamic models with time-varying covariances. *Econometric Reviews, 11*, 143–172.
Breitung, J., & Eickmeier, S. (2006). Dynamic factor models. In O. Hübler & J. Frohn (Eds.), *Modern econometric analysis*. Berlin: Springer.
Brillinger, D. R. (1981). *Time series: Data analysis and theory*. New York: Holt, Rinehart and Winston.
Broto, C., & Ruiz, E. (2004). Estimation methods for stochastic volatility models: a survey. *Journal of Economic Surveys, 18*(5), 613–649.
Campbell, J. Y., Lo, A. W., & MacKinlay, A. C. (1997). *The econometrics of financial markets*. Princeton: Princepton University Press.
Chamberlain, G., & Rothschild, M. (1983). Arbitrage, factor structure, and meanvariance analysis on large asset markets. *Econometrica, 51*, 1281–1304.
Chan, L. K. C., Karceski, J., & Lakonishok, J. (1999). On portfolio optimization: forcasting covariances and choosing the risk model. *The Review of Financial Studies, 12*(5), 937–974.
Cherubini, G., Luciano, E., & Vecchiato, W. (2004). *Copula methods in finance*. UK: Wiley.
Clark, P. K. (1973). A subordinate stochastic process model with finite variance for speculative prices. *Econometrica, 41*, 135–155.
Connor, G, & Korajczyk, R. A. (1993). A test for the number of factors in an approximate factor model. *Journal of Finance, 48*(4), 1263–1291.
Danielsson, J. (1994). Stochastic volatility in asset prices estimation with simulated maximum likelihood. *Journal of Econometrics, 64*(1–2), 375–400.
Danielsson, J. (1998). Multivariate stochastic volatility models: Estimation and comparison with VGARCH models. *Journal of Empirical Finance, 5*, 155–173.
Danielsson, J., & Richard, J. F. (1993). Accelerated Gaussian importance sampler with application to dynamic latent variable models. *Journal of Applied Econometrics, 8*, 153–174.
Dias, A., & Embrechts, P. (2004). Change-point analysis for dependence structures in finance and insurance. In G. Szegoe (Ed.), *Risk measures for the 21st century* (Chap. 16, pp. 321–335). New York: Wiley.
Diebold, F. X., & Nerlove, M. (1989). The dynamics of exchange rate volatility: a multivariate latent factor ARCH model. *Journal of Applied Econometrics, 4*, 1–21.
Diebold, F. X., Gunther, T. A., & Tay, A. S. (1998). Evaluating density forecasts. *International Economic Review, 39*, 863–883.
Duffie, D., & Singleton, K. J. (1993). Simulated moments estimation of markov models of asset prices. *Econometrica, 61*, 929–952.
Eichler, M., Motta, G., & von Sachs, R. Fitting dynamic factor models to non-stationary time series. *Journal of Econometrics 163*(1), 51–70.
Engle, R. F. (1982). Autoregressive conditional heteroskedasticity with estimates of the variance of U.K. inflation. *Econometrica, 50*, 987–1008.
Engle, R. F. (2002). Dynamic conditional correlation: A simple class of multivariate generalized autoregressive conditional heteroskedasticity models. *Journal of Business and Economic Statistics, 20*(3), 339–350.
Engle, R. F., & Kroner, K. F. (1995). Multivariate simultaneous generalized ARCH. *Econometric Theory, 11*, 122–150.
Engle, R. F., & Manganelli, S. (2004). CAViaR: Conditional autoregressive value at risk by regression quantiles. *Journal of Business and Economic Statistics, 22*(4), 367–381.
Engle, R. F., & Sheppard, K. (2001). Theoretical and empirical properties of dynamic conditional correlation multivariate GARCH. NBER working paper 8554, National Bureau of Economic Research.
Engle, R. F., Lilien, D. M., & Robins, R. P. (1987). Estimating time varying risk premia in the term structure: The ARCH-M model. *Econometrica, 55*, 391–407.

Engle, R. F., Ng, V. K., & Rothschild, M. (1990). Asset pricing with a factor ARCH covariance structure: Empirical estimates for treasury bills. *Journal of Econometrics*, *45*, 213–238.

Engle, R. F., Shephard, N., & Sheppard, K. (2007). Fitting and testing vast dimensional time-varying covariance models. NYU Working Paper No. FIN-07-046.

Fama, E. F. (1965). The behavior of stock market prices. *Journal of Business*, *38*, 34–105.

Fama, E. F., & French, K. R. (1993). Common risk factors in the returns on stocks and bonds. *Journal of Financial Economics*, *33*, 3–56.

Fleming, J., Kirby, C., & Ostdiek, B. (2001). The economic value of volatility timing. *The Journal of Finance*, *56*(1), 329–352.

Forni, M., Hallin, M., Lippi, M., & Reichlin, L. (2000). The generalized dynamic factor model: Identification and estimation. *The Review of Economics and Statistics*, *82*, 540–554.

Franke, J., Härdle, W., & Hafner, C. M. (2008). *Statistics of financial markets an introduction*. Berlin: Springer.

Gallant, A. R., & Tauchen, G. (1996). Which moments to match. *Econometric Theory*, *12*, 657–681.

Ghysels, E., Harvey, A. C., & Renault, E. (1996). Stochastic volatility. In G. Maddala & C. R. Rao (Eds.), *Handbook of statistics* (Vol. 14). Amsterdam: Elsevier.

Giacomini, E., Härdle, W., & Spokoiny, V. (2009). Inhomogeneous dependency modelling with time varying copulae. *Journal of Business and Economic Statistics*, *27*, 224–234.

Gourieroux, C., Monfort, A., & Renault, E. (1993). Indirect inference. *Journal of Applied Econometrics*, *8*, 85–118.

Hafner, C. M., & Franses, P. H. (2009). A generalized dynamic conditional correlation model: Simulation and application to many assets. *Econometric Reviews*, *28*, 612–631.

Hafner, C. M., & Herwartz, H. (2000). Testing linear autoregressive dynamics under heteroskedasticity. *The Econometrics Journal*, *3*, 177–197.

Hafner, C. M., & Manner, H. (2011). Dynamic stochastic copula models: Estimation, inference and applications. *Journal of Applied Econometrics*. doi: 10.1002/jae.1197.

Hafner, C. M., & Reznikova, O. (2010). Efficient estimation of a semiparametric dynamic copula model. *Computational Statistics and Data Analysis*, *54*, 2609–2627.

Harvey, A. C., Ruiz, E., & Shephard, N. (1994). Multivariate stochastic variance models. *Review of Economic Studies*, *61*, 247–264.

Hentschel, L. (1995). All in the family: Nesting symmetric and asymmetric garch models. *Journal of Financial Economics*, *39*, 71–104.

Hull, J., & White, A. (1987). The pricing of options with stochastic volatilities. *Journal of Finance*, *42*, 281–300.

Jacquier, E., Polson, N. G., & Rossi, P. E. (1994). Bayesian analysis of stochastic volatitliy models (with discussion). *Journal of Business and Economic Statistics*, *12*, 371–389.

Joe, H. (1997). *Multivariate models and dependence concepts*. London: Chapman & Hall.

Kim, S., Shephard, N., & Chib, S. (1998). Stochastic volatility: Likelihood inference and comparison with ARCH models. *Review of Economic Studies*, *65*, 361–393.

Ledoit, O, Santa-Clara, P, & Wolf, M. (2003). Flexible multivariate GARCH modeling with an application to international stock markets. *Review of Economics and Statistics*, *85*, 735–747.

Liesenfeld, R., & Richard, J. F. (2003). Univariate and multivariate volatility models: Estimation and diagnostics. *Journal of Empirical Finance*, *10*, 505–531.

Lintner, J. (1965). Security prices, risk and maximal gains from diversification. *Journal of Finance*, *20*, 587–615.

Mandelbrot, B. (1963). The variation of certain speculative prices. *Journal of Business*, *36*, 394–419.

Manner, H. & Candelon, B. (2010). *Testing for asset market linkages: A new approach based on time-varying copulas*. Pacific Economic Review **15**, 364–384. doi: 10.1111/j.1468-0106.2010.00508.x

Markowitz, H. (1959). *Portfolio selection: Efficient diversification of investments*. New York: Wiley.

Motta, G., Hafner, C., & von Sachs, R. (2011). Locally stationary factor models: Identification and nonparametric estimation. *Econometric Theory*, *27*(6).

Nelsen, R. B. (2006). *An introduction to copulas*. New York: Springer.

Patton, A. (2006a). Modelling asymmetric exchange rate dependence. *International Economic Review*, *47*(2), 527–556.

Ross, S. A. (1976). The arbitrage theory of capital asset pricing. *Journal of Economic Theory*, *13*, 341–360.

Sharpe, W. F. (1964). Capital asset prices: A theory of market equilibrium under conditions of risk. *Journal of Finance*, *19*, 425–442.

Stock, J. H., & Watson, M. W. (2002). Forecasting using principal components from a large number of predictors. *Journal of the American Statistical Association*, *97*, 1167–1179.

Taylor, S. J. (1986). *Modelling financial time series*. Chichester: Wiley.

Taylor, S. J. (1994). Modelling stochastic volatility: A review and comparative study. *Mathematical finance*, *4*, 183–204,.

Tse, Y. K., & Tsui, A. K. C. (2002). A multivariate GARCH model with time-varying correlations. *Journal of Business and Economic Statistics*, *20*(3), 351–362.

van der Weide, R. (2002). Go-garch: A multivariate generalized orthogonal GARCH model. *Journal of Applied Econometrics*, *17*, 549–564.

Yu, J., & Meyer, R. (2006). Multivariate stochastic volatility models: Bayesian estimation and model comparison. *Econometric Reviews*, *25*(2–3), 361–384.

Chapter 6
Option Data and Modeling BSM Implied Volatility

Matthias R. Fengler

Abstract The present handbook contributions introduces the notion of the Black-Scholes-Merton implied volatility surface and reviews its stylized facts. Static no-arbitrage conditions and recent theoretical results on the far expiry, short expiry and far strike asymptotics are surveyed. A discussion of the numerical aspects of computing implied volatility efficiently and accurately follows. We conclude by reviewing models of the implied volatility surface starting with parametric and non- and semiparametric approaches. The emphasis is on models observing financial no-arbitrage constraints.

6.1 Introduction

The discovery of an explicit solution for the valuation of European style call and put options based on the assumption a Geometric Brownian motion driving the underlying asset constitutes a landmark in the development of modern financial theory. First published in Black and Scholes (1973), but relying heavily on the notion of no-arbitrage in Merton (1973), this solution is nowadays known as the Black-Scholes-Merton (BSM) option pricing formula. In recognition of this achievement, Myron Scholes and Robert C. Merton were awarded the Nobel prize in economics in 1997 (Fischer Black had already died by this time).

Although it is widely acknowledged that the assumptions underlying the BSM model are far from realistic, the BSM formula still enjoys unrivalled popularity in financial practice. This is not so much because practitioners believe in the model as a good description of market behavior, but rather because it serves as a convenient mapping device from the space of option prices to a single real number called the

M.R. Fengler (✉)
University of St. Gallen, St. Gallen, Switzerland
e-mail: matthias.fengler@unisg.ch

J.-C. Duan et al. (eds.), *Handbook of Computational Finance*, Springer Handbooks of Computational Statistics, DOI 10.1007/978-3-642-17254-0_6,

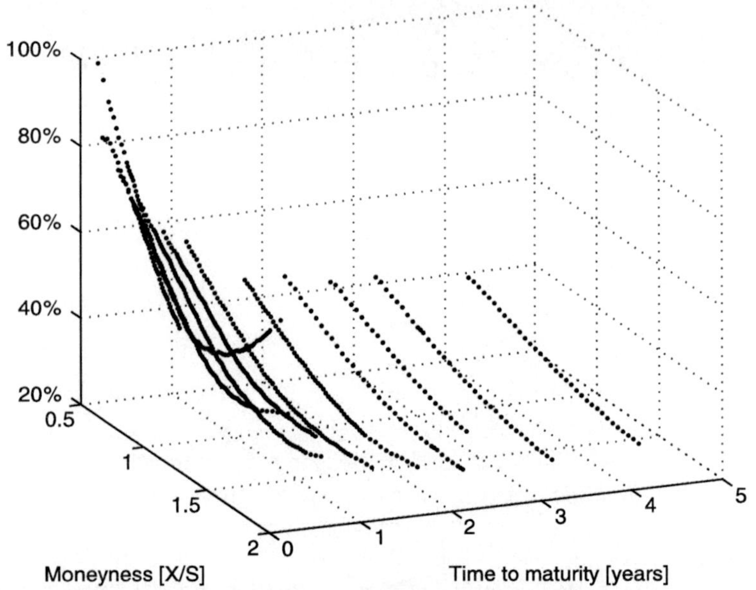

Fig. 6.1 IV surface of DAX index options from 28 October 2008, traded at the EUREX. IV given in percent across a spot moneyness metric, time to expiry in years

(BSM-)implied volatility. Indeed, the only unknown parameter involving the BSM formula is the volatility. Backed out of given option prices it allows for straight forward comparisons of the relative expensiveness of options across various strikes, expiries and underlying assets. In practice calls and puts are thus quoted in terms of implied volatility.

For illustration consider Fig. 6.1 displaying implied volatility (IV) as observed on 28 October 2008 and computed from options traded on the futures exchange EUREX, Frankfurt. IV is plotted against relative strikes and time to expiry. Due to institutional conventions, there is a very limited number of expiry dates, usually 1–3 months apart for short-dated options and 6–12 months apart for longer-dated ones, while the number of strikes for each expiry is more finely spaced. The function resulting for a fixed expiry is frequently called the 'IV smile' due to its U-shaped pattern. For a fixed (relative) strike across several expiries one speaks of the term structure of IV. Understandably, the non-flat surface, which also fluctuates from day to day, is in strong violation to the assumption of a Geometric Brownian motion underlying the BSM model.

Although IV observations are observed on this degenerate design, practitioners think of them as stemming from a smooth and well-behaved surface. This view is due to the following objectives in option portfolio management: (1) market makers quote options for strike-expiry pairs which are illiquid or not listed; (2) pricing engines, which are used to price exotic options and which are based on far more realistic assumptions than the BSM model, are calibrated against an observed IV surface; (3) the IV surface given by a listed market serves as the market of primary hedging instruments against volatility and gamma risk (second-order sensitivity with respect to the spot); (4) risk managers use stress scenarios defined on the IV surface to visualize and quantify the risk inherent to option portfolios.

Each of these applications requires suitably chosen interpolation and extrapolation techniques or a fully specified model of the IV surface. This suggests the following structure of this contribution: Section 6.2 introduces the BSM-implied volatility and in Sect. 6.3 we outline its stylized facts. No-arbitrage constraints on the IV surface are presented in Sect. 6.4. In Sect. 6.5, recent theoretical advances on the asymptotic behavior of IV are summarized. Approximation formulae and numerical methods to recover IV from quoted option prices are reviewed in Sect. 6.6. Parametric, semi- and nonparametric modeling techniques of IV are considered in Sect. 6.7.

6.2 The BSM Model and Implied Volatility

We consider an economy on the time interval $[0, T^*]$. Let $(\Omega, \mathcal{F}, \mathrm{P})$ be a probability space equipped with a filtration $(\mathcal{F}_t)_{0 \le t \le T^*}$ which is generated by a Brownian motion $(W_t)_{0 \le t \le T^*}$ defined on this space, see e.g. Steele (2000). A stock price $(S_t)_{0 \le t \le T^*}$, adapted to $(\mathcal{F}_t)_{0 \le t \le T^*}$ (paying no-dividends for simplicity) is modeled by the Geometric Brownian motion satisfying the stochastic differential equation

$$\frac{dS_t}{S_t} = \mu\, dt + \sigma\, dW_t, \tag{6.1}$$

where μ denotes the (constant) instantaneous drift and σ^2 measures the (constant) instantaneous variance of the return process of $(\log S_t)_{t \ge 0}$. We furthermore assume the existence of a riskless money market account paying interest r. A European style call is a contingent claim paying at some expiry date T, $0 < T \le T^*$, the amount $\psi_c(S_T) = (S_T - X)^+$, where $(\cdot)^+ \stackrel{\text{def}}{=} \max(\cdot, 0)$ and X is a fixed number, the exercise price. The payoff of a European style put is given by $\psi_p(S_T) = (X - S_T)^+$.

Under these assumptions, it can be shown that the option price $H(S_t, t)$ is a function in the space $\mathcal{C}^{2,1}\big(\mathbb{R}^+ \times (0, T)\big)$ satisfying the partial differential equation

$$0 = \frac{\partial H}{\partial t} + rS\frac{\partial H}{\partial S} + \frac{1}{2}\sigma^2 S^2 \frac{\partial^2 H}{\partial S^2} - rH \tag{6.2}$$

subject to $H(S_T, T) = \psi_i(S_T)$ with $i \in \{c, p\}$.

The celebrated BSM formula for calls solving (6.2) with boundary condition $\psi_c(S_T)$ is found to be

$$C_t^{BSM}(X, T) = S_t \Phi(d_1) - e^{-r(T-t)} X \Phi(d_2), \tag{6.3}$$

with

$$d_1 = \frac{\log(S_t/X) + (r + \frac{1}{2}\sigma^2)(T-t)}{\sigma\sqrt{T-t}}, \tag{6.4}$$

$$d_2 = d_1 - \sigma\sqrt{T-t}, \tag{6.5}$$

and where $\Phi(v) = \int_{-\infty}^{v} \varphi(u)\,du$ is the cdf of the standard normal distribution with pdf $\varphi(v) = \frac{1}{\sqrt{2\pi}} e^{-v^2/2}$ for $v \in \mathbb{R}$.

Given observed market prices $\widetilde{C}_t$, one defines – as first introduced by Latané and Rendelman (1976) – implied volatility as

$$\hat{\sigma} : C_t^{BSM}(X, T, \hat{\sigma}) - \widetilde{C}_t = 0. \tag{6.6}$$

Due to monotonicity of the BSM price in σ, there exists a unique solution $\hat{\sigma} \in \mathbb{R}^+$. Note that the definition in (6.6) is not confined to European options. It is also used for American options, which can be exercised at any time in $[0, T]$. In this case, as no explicit formulae for American style options exists, the option price is computed numerically, for instance by means of finite difference schemes (Randall and Tavella (2000)).

In the BSM model volatility is just a constant, whereas empirically, IV displays a pronounced curvature across strikes X and different expiry days T. This gives rise to the notion of an IV surface as the mapping

$$\widehat{\sigma} : (t, X, T) \to \widehat{\sigma}_t(X, T). \tag{6.7}$$

In Fig. 6.2, we plot the time series of 1 Y at-the-money IV of DAX index options (left axis, black line) together with DAX closing prices (right axis, gray line). An option is called at-the-money (ATM) when the exercise price is equal or close to the spot (or to the forward). The index options were traded at the EUREX, Frankfurt (Germany), from 2000 to 2008. As is visible IV is subject to considerable variations. Average DAX index IV was about 22% with significantly higher levels in times of market stress, such as after the World Trade Center attacks 2001, during the pronounced bear market 2002–2003 and the financial crisis end of 2008.

6.3 Stylized Facts of Implied Volatility

The IV surface displays a number of static and dynamic stylized facts which we demonstrate here using the present DAX index option data dating from 2000 to 2008. These facts can be observed for any equity index market. They similarly hold for stocks. Other asset classes may display different features, for instance, smiles may be more shallow, symmetric or even upward-sloping, but this does not

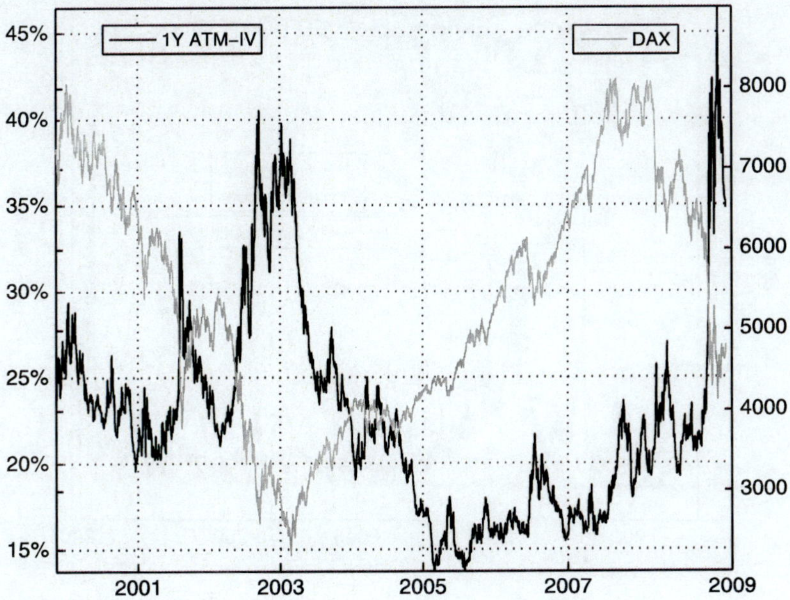

Fig. 6.2 Time series of 1Y ATM IV (*left axis, black line*) and DAX index closing prices (*right axis, gray line*) from 2000 to 2008

fundamentally change the smile phenomenon. A more complete synthesis can be found in Rebonato (2004).

Stylized facts of IV are as follows:

1. The smile is very pronounced for short expiries and becomes flattish for longer dated options. This fact was already visible in Fig. 6.1.
2. As noted by Rubinstein (1994) this has not always been the case. The strong asymmetry in the smile first appeared after the 1987 market turmoil.
3. For equity options, both index and stocks, the smile is negatively skewed. We define the 'skew' here by $\left.\frac{\partial\hat{\sigma}^2}{\partial m}\right|_{m=0}$, where m is log-forward moneyness as defined in Sect. 6.5. Figure 6.3 depicts the time series of the DAX index skew (left axis) for 1M and 1Y options. The skew is negative throughout and – particularly the short-term skew – increases during times of crisis. For instance, skews increase in the aftermath of the dot-com boom 2001–2003, or spike at 11 September 2001 and during the heights of the financial crisis 2008. As theory predicts, see Sect. 6.5, the 1Y IV skew has most of the time been flatter than the 1M IV skew.
4. Fluctuations of the short-term skew are much larger. Figure 6.4 gives the quantiles of the skew as a function of time to expiry. Similar patterns also apply to IV levels and returns.

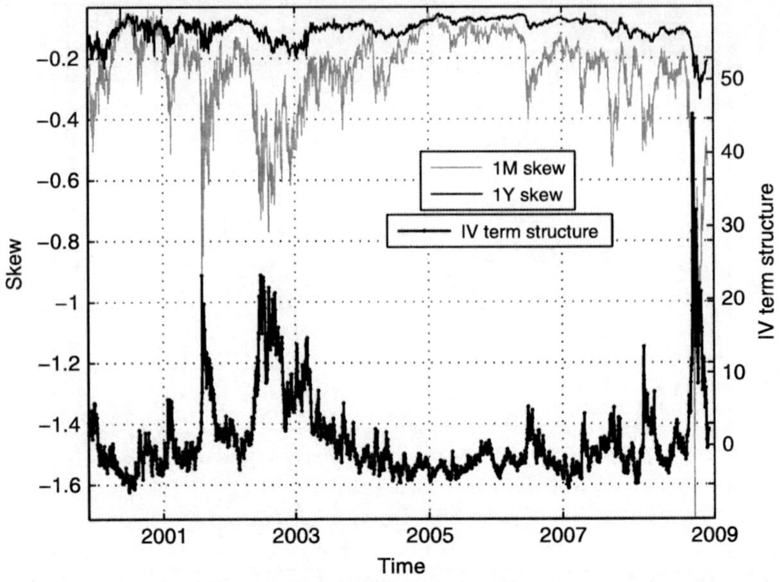

Fig. 6.3 Time series of 1M and 1Y IV skew (*left axis, gray line* and *black line* respectively) and time series of the IV term structure (*right axis, black dotted line*). Skew is defined as $\left.\frac{\partial\hat{\sigma}^2}{\partial m}\right|_{m=0}$, where m is log-forward moneyness. The derivative is approximated by a finite difference quotient. IV term structure is the difference between 1M ATM and 1Y ATM in terms of percentage points. Negative values indicate an upward sloping term structure

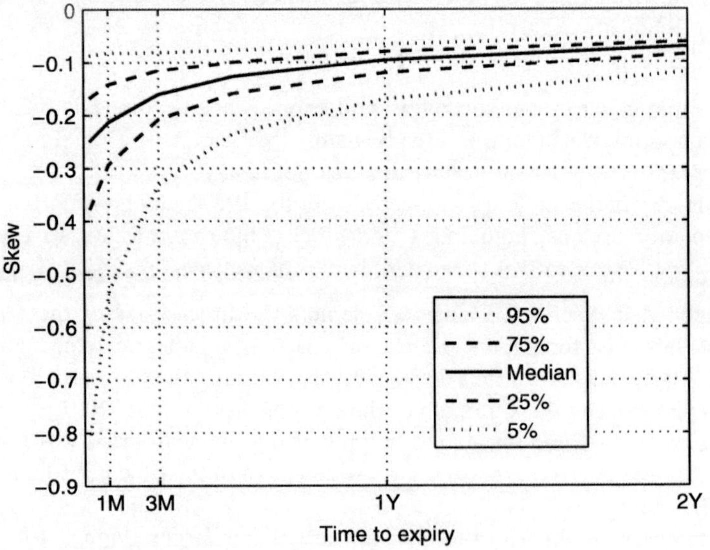

Fig. 6.4 Empirical quantiles of the ATM IV skew as a function of time to expiry. Skew is defined as $\left.\frac{\partial\hat{\sigma}^2}{\partial m}\right|_{m=0}$, where m is log-forward moneyness

5. The IV surface term structure is typically upward sloping (i.e. has increasing levels of IV for longer dated options) in calm times, while in times of crisis it is downward sloping with short dated options having higher levels of IV then longer dated ones. This is seen in Fig. 6.3 giving the difference of 1M ATM IV minus 1Y ATM in terms of percentage points on the right axis. A positive value therefore indicates a downward sloping term structure. During the financial crisis the term structure slope achieved unprecedented levels. Humped profiles can be observed as well.
6. Returns of the underlying asset and returns of IV are negatively correlated. For the present data set we find a correlation between 1M ATM IV and DAX returns of $\rho = -0.69$.
7. IV appears to be mean-reverting, see Fig. 6.2, but it is usually difficult to confirm mean reversion statistically, since IV data is often found to be nearly integrated, see Fengler et al. (2007) for a discussion.
8. Shocks cross the IV surface are highly correlated, as can be observed from the comovements of IV levels in Fig. 6.2 and the skew and the term structure in Fig. 6.3. In consequence IV surface dynamics can be decomposed into a small number of driving factors.

6.4 Arbitrage Bounds on the Implied Volatility Surface

Despite the rich empirical behavior, IV cannot simply assume any functional form. This is due to constraints imposed by no-arbitrage principles. For IV, these constraints are very involved, but are easily stated indirectly in the option price domain. From now on, we set $t = 0$ and suppress dependence on t for sake of clarity.

We state the bounds using a (European) call option; deriving the corresponding bounds for a put is straightforward. The IV function must be such that the call price is bounded by

$$\max\left(S - e^{-rT} X, 0\right) \le C(X,T) \le S. \tag{6.8}$$

Moreover, the call price must be a decreasing and convex function in X, i.e.

$$-e^{-rT} \le \frac{\partial C}{\partial X} \le 0 \quad \text{and} \quad \frac{\partial^2 C}{\partial X^2} \ge 0. \tag{6.9}$$

To preclude calendar arbitrage, prices of American calls for the same strikes must be nondecreasing across increasing expiries. This statement does not hold for European style calls because their theta can change sign. No-arbitrage implies, however, that there exists a monotonicity relationship along forward-moneyness corrected strikes (also in the presence of dividend yield), see Reiner (2000), Gatheral (2004), Kahalé (2004), Reiner (2004), Fengler (2009). Denote by $x = X/F^T$ forward-moneyness, where F^T is a forward with expiry T, and by $T_1 < T_2$ the expiry dates of two call options whose strike prices X_1 and X_2 are related by forward-moneyness, i.e. $x_1 = x_2$. Then

$$C(X_2, T_2) \geq C(X_1, T_1). \tag{6.10}$$

Most importantly, this results implies that total implied variance must be nondecreasing in forward-moneyness to preclude arbitrage. Defining total variance as $\nu^2(x,T) \stackrel{\text{def}}{=} \hat{\sigma}^2(x,T)\,T$, we have

$$\nu^2(x, T_2) > \nu^2(x, T_1). \tag{6.11}$$

Relationship (6.11) has the important consequence that one can visually check IV smiles for calendar arbitrage by plotting total variance across forward moneyness. If the lines intersect, (6.11) is violated.

6.5 Implied Volatility Surface Asymptotics

Many of the following results had the nature of conjectures and were generalized and rigorously derived only very recently. Understanding the behavior of IV for far expiries and far strikes is of utter importance for extrapolation problems often arising in practice.

Throughout this section set $r = 0$ and $t = 0$. This is without loss of generality since in the presence of nonzero interest rates and dividends yields, the option and underlying asset prices may be viewed as forward prices, see Britten-Jones and Neuberger (2000). Furthermore define log-(forward) moneyness by $m \stackrel{\text{def}}{=} \log x = \log(X/S)$ and total (implied) variance by $\nu^2 \stackrel{\text{def}}{=} \hat{\sigma}^2 T$. Let $S = (S_t)_{t \geq 0}$ be a nonnegative martingale with $S_0 > 0$ under a fixed risk-neutral measure.

6.5.1 Far Expiry Asymptotics

The results of this section can be found in more detail in Tehranchi (2010) whom we follow closely.

The first theorem shows that the IV surface flattens for infinitely large expiries.

Theorem 1 **(Rogers and Tehranchi (2009)).** *For any $M > 0$ we have*

$$\lim_{T\to\infty} \sup_{m_1, m_2 \in [-M, M]} |\hat{\sigma}(m_2, T) - \hat{\sigma}(m_1, T)| = 0.$$

Note that this result does not hinge upon the central limit theorem, mean-reversion of spot volatility etc., only the existence of the martingale measure. In particular, $\lim_{T\to\infty} \hat{\sigma}(m, T)$ does not need to exist for any m.

The rate of flattening of the IV skew can be made more precise by the following result. It shows that the flattening behavior of the IV surface as described in Sect. 6.3 is not an empirical artefact, but has a well-founded theoretical underpinning (for earlier, less general arguments see Hodges (1996), Carr and Wu (2003)).

Theorem 2 (Rogers and Tehranchi (2009)).

(i) For any $0 \le m_1 < m_2$ *we have*

$$\frac{\hat{\sigma}(m_2, T)^2 - \hat{\sigma}(m_1, T)^2}{m_2 - m_1} \le \frac{4}{T}.$$

(ii) For any $m_1 < m_2 \le 0$

$$\frac{\hat{\sigma}(m_2, T)^2 - \hat{\sigma}(m_1, T)^2}{m_2 - m_1} \ge -\frac{4}{T}.$$

(iii) If $S_t \xrightarrow{P} 0$ *as* $t \to \infty$*, for any* $M > 0$ *we have*

$$\limsup_{T\to\infty} \sup_{m_1, m_2 \in [-M, M], m_1 \neq m_2} T \left| \frac{\hat{\sigma}(m_2, T)^2 - \hat{\sigma}(m_1, T)^2}{m_2 - m_1} \right| \le 4.$$

As pointed out by Rogers and Tehranchi (2009), the inequality in *(iii)* is sharp in the sense that there exists a martingale $(S_t)_{t\ge0}$ with $S_t \xrightarrow{P} 0$ such that

$$T \frac{\partial}{\partial m} \hat{\sigma}(m, T)^2 \to -4.$$

as $T \to \infty$ uniformly for $m \in [-M, M]$. The condition $S_t \xrightarrow{P} 0$ as $t \to \infty$, is not strong. It holds for most financial models and is equivalent to the statement that

$$C(X, T) = \mathrm{E}[(S_T - X)^+] \to S_0$$

as $T \to \infty$ for some $X > 0$. The BSM formula (6.3) fulfills it trivially. Indeed one can show that if the stock price process does not converge to zero, then $\lim_{T\to\infty} \hat{\sigma}(m, T) = 0$, because $\nu^2 < \infty$.

Finally Tehranchi (2009) obtains the following representation formula for IV:

Theorem 3 (Tehranchi (2009)). *For any* $M > 0$ *we have*

$$\lim_{T\to\infty} \sup_{m\in[-M, M]} \left| \hat{\sigma}(m, T) - \sqrt{-\frac{8}{T} \log \mathrm{E}[S_T \wedge 1]} \right|$$

with $a \wedge b = \min(a, b)$*. Moreover there is the representation*

$$\hat{\sigma}_{\infty}^{2} = \lim_{T\to\infty} -\frac{8}{T}\log \mathrm{E}[S_T \wedge 1] \tag{6.12}$$

whenever this limit is finite.

A special case of this result was derived by Lewis (2000) in the context of the Heston (1993) model. For certain model classes, such as models based on Lévy processes, the last theorem allows a direct derivation of $\hat{\sigma}_{\infty}$.

The implication of these results for building an IV surface are far-reaching. The implied variance skew must be bounded by $\left|\frac{\partial v^2}{\partial m}\right| \leq 4$ and should decay at a rate of $1/T$ between expiries. Moreover, a constant far expiry extrapolation in $\hat{\sigma}(m, T_n)$ beyond the last extant expiry T_n is wrong, since the IV surface does not flatten in this case. A constant far expiry extrapolation in $v^2(m, T_n)$ beyond T_n is fine, but may not be a very lucky choice given the comments following Theorem 2 number (*iii*).

6.5.2 Short Expiry Asymptotics

Roper and Rutkowski (2009) consider the behavior of IV towards small times to expiry. They prove

Theorem 4 (Roper and Rutkowski (2009)). *If* $C(X,\epsilon) = (S - X)^+$ *for some* $\epsilon > 0$ *then*

$$\lim_{T\to 0+} \hat{\sigma}(X,T) = 0. \tag{6.13}$$

Otherwise

$$\lim_{T\to 0+} \hat{\sigma}(X,T) = \begin{cases} \lim_{T\to 0+} \frac{\sqrt{2\pi}C(X,T)}{S\sqrt{T}} & \text{if } S = X \\ \lim_{T\to 0+} \frac{|\log(S/X)|}{\sqrt{-2T\log\left[C(X,T)-(S-X)^+\right]}} & \text{if } S \neq X \end{cases}, \tag{6.14}$$

in the sense that the LHS is finite (infinite) whenever the RHS is finite (infinite).

The quintessence of this theorem is twofold. First, the asymptotic behavior of $\hat{\sigma}(X,T)$ as $T \to 0^+$ is markedly different for $S = X$ and $S \neq X$. Note that the ATM behavior of (6.14) is the well-established Brenner and Subrahmanyam (1988), Feinstein (1988) formula to be presented in Sect. 6.6.1. Second, convergent IV is not a behavior coming for granted. In particular no-arbitrage does not guarantee that a limit exists, see Roper and Rutkowski (2009) for a lucid example. However, the limit of time-scaled IV exists and is zero:

$$\lim_{T\to 0+} v(X,T) = \lim_{T\to 0+} \hat{\sigma}\sqrt{T} = 0. \tag{6.15}$$

6.5.3 Far Strike Asymptotics

Lee (2004) establishes the behavior of the IV surface as strikes tend to infinity. He finds a one-to-one correspondence between the large-strike tail and the number of moments of S_T, and the small-strike tail and the number of moments of S_T^{-1}. We retain the martingale assumption for $(S_t)_{t\geq 0}$ and $m \stackrel{\text{def}}{=} \log(X/S)$.

Theorem 5 **(Lee (2004)).** *Define*

$$\widetilde{p} = \sup\{p : \mathrm{E}\, S_T^{1+p} < \infty\} \quad \beta_R = \limsup_{m\to\infty} \frac{\nu^2}{|m|} = \limsup_{m\to\infty} \frac{\hat{\sigma}^2}{|m|/T}.$$

Then $\beta_R \in [0,2]$ *and*

$$\widetilde{p} = \frac{1}{2\beta_R} + \frac{\beta_R}{8} - \frac{1}{2},$$

with the understanding that $1/0 = \infty$. *Equivalently,*

$$\beta_R = 2 - 4(\sqrt{\widetilde{p}^2 + \widetilde{p}} - \widetilde{p}).$$

The next theorem considers the case $m \to -\infty$.

Theorem 6 **(Lee (2004)).** *Denote by*

$$\widetilde{q} = \sup\{q : \mathrm{E}\, S_T^{-q} < \infty\} \quad \beta_L = \limsup_{m\to-\infty} \frac{\nu^2}{|m|} = \limsup_{m\to-\infty} \frac{\hat{\sigma}^2}{|m|/T}.$$

Then $\beta_L \in [0,2]$ *and*

$$\widetilde{q} = \frac{1}{2\beta_L} + \frac{\beta_L}{8} - \frac{1}{2},$$

with $1/0 = \infty$, *or*

$$\beta_L = 2 - 4(\sqrt{\widetilde{q}^2 + \widetilde{q}} - \widetilde{q}).$$

Roger Lee's results have again vital implications for the extrapolation of the IV surface for far strikes. They show that linear or convex skews for far strikes are wrong by the $\mathcal{O}(|m|^{1/2})$ behavior. More precisely, the IV wings should not grow faster than $|m|^{1/2}$ and not grow slower than $|m|^{1/2}$, unless the underlying asset price is supposed to have moments of all orders. The elegant solution following from these results is to extrapolate ν^2 linearly in $|m|$ with an appropriately chosen $\beta_L, \beta_R \in [0,2]$.

6.6 Approximating and Computing Implied Volatility

6.6.1 Approximation Formulae

There is no closed-form, analytical solution to IV, even for European options. In situations when iterative procedures is not readily available, such as in the context of a spreadsheet, or when numerical approaches are not applicable, such as in real time applications, approximation formulae to IV are of high interest. Furthermore, they also serve as good inital values for the numerical schemes discussed in Sect. 6.6.2.

The most simple approximation to IV, which is due to Brenner and Subrahmanyam (1988) and Feinstein (1988), is given by

$$\hat{\sigma} \approx \sqrt{\frac{2\pi}{T}} \frac{C}{S}. \tag{6.16}$$

The rationale of this formula can be understood as follows. Define by $K \stackrel{\text{def}}{=} S = Xe^{-rT}$ the discounted ATM strike. The BSM formula then simplifies to

$$C = S\left(2\,\Phi(\sigma\sqrt{T}/2) - 1\right).$$

Solving for σ yields the semi-analytical formula

$$\sigma = \frac{2}{\sqrt{T}}\;\Phi^{-1}\left(\frac{C+S}{2S}\right), \tag{6.17}$$

where Φ^{-1} denotes the inverse function of the normal cdf. A first order Taylor expansion of (6.17) in the neighborhood of $\frac{1}{2}$ yields formula (6.16). In consequence, it is exact only, when the spot is equal to the discounted strike price.

A more accurate formula, which also holds for in-the-money (ITM) and out-of- the-money (OTM) options (calls are called OTM when $S \ll X$ and ITM when $S \gg X$), is based on a Taylor expansion of third order to Φ. It is due to Li (2005):

$$\hat{\sigma} \approx \begin{cases} 2z\sqrt{\frac{2}{T}} - \frac{1}{\sqrt{T}}\sqrt{8z^2 - \frac{6\alpha}{\sqrt{2}z}} & \text{if} \quad \rho \le 1.4 \\ \frac{1}{2\sqrt{T}}\left(\alpha + \sqrt{\alpha^2 - \frac{4(K-S)^2}{S(S+K)}}\right) & \text{if} \quad \rho > 1.4, \end{cases} \tag{6.18}$$

where $z = \cos\left[\frac{1}{3}\arccos\left(\frac{3\alpha}{\sqrt{32}}\right)\right]$, $\alpha = \frac{\sqrt{2\pi}}{S+K}(2C + K - S)$ and $\rho = |K - S|SC^{-2}$. The value of the threshold parameter ρ separating the first part, which is for nearly-ATM options, and the second part for deep ITM or OTM options, was found by Li (2005) based on numerical tests.

Other approximation formulae found in the literatur often lack a rigorous mathematical foundation. The possibly most prominent amongst these are those suggested by Corrado and Miller (1996) and Bharadia et al. (1996). The Corrado and Miller (1996) formula is given by

$$\hat{\sigma} \approx \frac{1}{\sqrt{T}} \frac{\sqrt{2\pi}}{S+K} \left[C - \frac{S-K}{2} + \sqrt{\left(C - \frac{S-K}{2}\right)^2 - \frac{(S-K)^2}{\pi}} \right]. \quad (6.19)$$

Its relative accuracy is explained by the fact that (6.19) is identical to the second formula in (6.18) after multiplying the second term under the square root by $\frac{1}{2}(1 + K/S)$, which is negligible in most cases, see Li (2005) for the details. Finally, the Bharadia et al. (1996) approximation is given by

$$\hat{\sigma} \approx \sqrt{\frac{2\pi}{T}} \frac{C - (S-K)/2}{S - (S-K)/2}. \quad (6.20)$$

Isengildina-Massa et al. (2007) investigate the accuracy of six approximation formulae. According to their criteria, Corrado and Miller (1996) is the best approximation followed by Li (2005) and Bharadia et al. (1996). This finding holds uniformly also for deviations to up to 1% around ATM (somewhat unfortunate, the authors do not consider a wider range) and up to maturities of 11 months. As a matter of fact, the approximation by Brenner and Subrahmanyam (1988) and Feinstein (1988) is of competing quality for ATM options only.

6.6.2 *Numerical Computation of Implied Volatility*

6.6.2.1 Newton–Raphson

The Newton–Raphson method, which will be the method of first choice in most cases, was suggested by Manaster and Koehler (1982). Denoting the observed market price by $\widetilde{C}$, the approach is described as

$$\sigma_{i+1} = \sigma_i - \left(C_i(\sigma_i) - \widetilde{C}\right) \Big/ \frac{\partial C}{\partial \sigma}(\sigma_i), \quad (6.21)$$

where $C_i(\sigma_i)$ is the option price and $\frac{\partial C}{\partial \sigma}(\sigma_i)$ is the option vega computed at σ_i. The algorithm is run until a tolerance criterion, such as $|\widetilde{C} - C_{i+1}| \leq \epsilon$, is achieved; IV is given by $\hat{\sigma} = \sigma_{i+1}$. The algorithm may fail, when the vega is close to zero, which regularly occurs for (short-dated) far ITM oder OTM options. The Newton–Raphson method has at least quadratic convergence, and combined with a good choice of the initial value, it achieves convergence within a very small number of steps. Originally, Manaster and Koehler (1982) suggested

$$\sigma_0 = \sqrt{\frac{2}{T}\,|\log(S/X) + rT|} \tag{6.22}$$

as initial value (setting $t = 0$). It is likely, however, that the approximation formulae discussed in Sect. 6.6.1 provide initial values closer to the solution.

6.6.2.2 Regula Falsi

The regula falsi is more robust than Newton–Raphson, but has linear convergence only. It is particularly useful when no closed-form expression for the vega is available, or when the price function is kinked as e.g. for American options with high probability of early exercise.

The regula falsi is initialized by two volatility estimates σ_L and σ_H with corresponding option prices $C_L(\sigma_L)$ and $C_H(\sigma_H)$ which need to include the solution. The iteration steps are:

1. Compute

$$\sigma_{i+1} = \sigma_L - \left(C_L(\sigma_L) - \widetilde{C}\right)\frac{\sigma_H - \sigma_L}{C_H(\sigma_H) - C_L(\sigma_L)}\;; \tag{6.23}$$

2. If $C_{i+1}(\sigma_{i+1})$ and $C_L(\sigma_L)$ have same sign, set $\sigma_L = \sigma_{i+1}$; if $C_{i+1}(\sigma_{i+1})$ and $C_H(\sigma_H)$ have same sign, set $\sigma_H = \sigma_{i+1}$. Repeat step 1.

The algorithm is run until $|\widetilde{C} - C_i| \leq \epsilon$, where ϵ the desired tolerance. Implied volatility is $\hat{\sigma} = \sigma_{i+1}$.

6.7 Models of Implied Volatility

6.7.1 Parametric Models of Implied Volatility

Since it is often very difficult to define a single parametric function for the entire surface (see Chap. 2 in Brockhaus et al. (2000) and Dumas et al. (1998) for suggestions in this directions), a typical approach is to estimate each smile independently by some nonlinear function. The IV surface is then reconstructed by interpolating total variances along forward moneyness as is apparent from Sect. 6.4. The standard method is linear interpolation. If derivatives of the IV surface with respect to time to expiry are needed, higher order polynomials for interpolation are necessary. Gatheral (2006) suggests the well-behaved cubic interpolation due to Stineman (1980). A monotonic cubic interpolation scheme can be found in Wolberg and Alfy (2002).

In practice a plethora of functional forms is used. The following selection of parametric approaches is driven by their respective popularity in three different asset classes (equity, fixed income, FX markets) and by the solid theoretical underpinnings they are derived from.

6.7.1.1 Gatheral's SVI Parametrization

The stochastic volatility inspired (SVI) parametrization for the smile was introduced by Gatheral (2004) and is motivated from the asymptotic extreme strikes behavior of a IV smile, which is generated by a Heston (1993) model. It is given in terms of log-forward moneyness $m = \log(X/F)$ as

$$\hat{\sigma}^2(m,T) = a + b\left(\rho(m-c) + \sqrt{(m-c)^2+\theta^2}\right), \tag{6.24}$$

where $a > 0$ determines the overall level of implied variance and $b \geq 0$ (predominantly) the angle between left and right asymptotes of extreme strikes; $|\rho| \leq 1$ rotates the smile around the vertex, and θ controls the smoothness of the vertex; c translates the graph.

The beauty of Gatheral's parametrization becomes apparent observing that implied variance behaves linear in the extreme left and right wing as prescribed by the moment formula due to Lee (2004), see Sect. 6.5.3. It is therefore straight forward to control the wings for no-arbitrage conditions. Indeed, comparing the slopes of the left and right wing asymptotes with Theorem 6, we find that

$$b(1+|\rho|) \leq \frac{2}{T},$$

to preclude arbitrage (asymptotically) in the wings. The SVI appears to fit a wide of range smile patterns, both empirical ones and those of many stochastic volatility and pure jump models, Gatheral (2004).

6.7.1.2 The SABR Parametrization

The SABR parametrization is a truncated expansion of the IV smile which is generated by the SABR model proposed by Hagan et al. (2002). SABR is an acronym for the 'stochastic $\alpha\beta\rho$ model', which is a two-factor stochastic volatility model with parameters α, the initial value of the stochastic volatility factor; $\beta \in [0,1]$, an exponent determining the dynamic relationship between the forward and the ATM volatility, where $\beta = 0$ gives rise to a 'stochastic normal' and $\beta = 1$ to a 'stochastic log-normal' behavior; $|\rho| \leq 1$, the correlation between the two Brownian motions; and $\theta > 0$, the volatility of volatility. The SABR approach is very popular in fixed income markets where each asset only has a single exercise date, such as swaption markets.

Denote by F the forward price, X is as usual the strike price. The SABR parametrization is a second order expansion given by

$$\hat{\sigma}(X,T) = \hat{\sigma}^0(X)\Big\{1 + \hat{\sigma}^1(X)\,T\Big\} + \mathcal{O}(T^2), \tag{6.25}$$

where the first term is

$$\hat{\sigma}^0(X) = \frac{\theta}{\chi(z)} \log \frac{F}{X} \tag{6.26}$$

with

$$z = \frac{\theta}{\alpha} \frac{F^{1-\beta} - X^{1-\beta}}{1-\beta}$$

and

$$\chi(z) = \log\left(\frac{\sqrt{1-2\rho z + z^2} + z - \rho}{1-\rho}\right);$$

the second term is

$$\hat{\sigma}^1(X) = \frac{(1-\beta)^2}{24}\frac{\alpha^2}{(FX)^{1-\beta}} + \frac{1}{4}\frac{\rho\beta\theta\alpha}{(FX)^{(1-\beta)/2}} + \frac{2-3\rho^2}{24}\theta^2. \tag{6.27}$$

Note that we display here the expansion in the corrected version as was pointed out by Obłój (2008); unlike the original fomula this version behaves consistently for $\beta \to 1$, as then $z(\beta) \to \frac{\theta}{\alpha}\log\frac{F}{X}$.

The formula (6.25) is involved, but explicit and can therefore be computed efficiently. For the ATM volatility, i.e. $F = X$, z and $\chi(z)$ disappear, and the first term in (6.25) collapses to $\hat{\sigma}^0(F) = \alpha F^{\beta-1}$.

As a fitting strategy, it is usually recommended to obtain β from a log-log plot of historical data of the ATM IV $\hat{\sigma}(F,F)$ against F and to exclude it from the subsequent optimizations. Parameter θ and ρ are inferred from a calibration to observed market IV; during that calibration α is found implicitly by solving for the (smallest) real root of the resulting cubic polynomial in α, given θ and ρ and the ATM IV $\hat{\sigma}(F,F)$:

$$\alpha^3\frac{(1-\beta)^2 T}{24\,F^{2-2\beta}} + \alpha^2\frac{\rho\beta\theta T}{4\,F^{(1-\beta)}} + \alpha\left(1 + \frac{2-3\rho^2}{24}\theta^2 T\right) - \hat{\sigma}(F,F)F^{1-\beta} = 0.$$

For further details on calibrations issues we refer to Hagan et al. (2002) and West (2005), where the latter has a specific focus on the challenges arising in illiquid markets. Alternatively, Mercurio and Pallavicini (2006) suggest a calibration procedure for all parameters (inluding β) from market data exploiting both swaption smiles and constant maturity swap spreads.

6.7.1.3 Vanna-Volga Method

In terms of input information, the vanna-volga (VV) approach is probably the most parsimonious amongst all constructive methods for building an IV surface, as it relies on as few as three input observations per expiry only. It is popular in FX markets. The VV method is based on the idea of constructing a replication portfolio that is locally risk-free up to second order in spot and volatility in a fictitious setting, where the smile is flat, but varies stochastically over time. Clearly, this setting is not only fictitious, but also theoretically inconsistent, as there is no model which generates a flat smile that fluctuates stochastically. It may however be justified by the market practice of using a BSM model with a regularly updated IV as input factor. The hedging costs incurred by the replication portfolio thus constructed are then added to the flat-smile BSM price.

To fix ideas, denote the option vega by $\frac{\partial C}{\partial \sigma}$, volga by $\frac{\partial^2 C}{\partial \sigma^2}$ and vanna by $\frac{\partial^2 C}{\partial \sigma \partial S}$. We are given three market observations of IV $\hat{\sigma}_i$ with associated strikes X_i, $i = 1, 2, 3$, with $X_1 < X_2 < X_3$, and same expiry dates $T_i = T$ for which the smile is to be constructed. In a first step, the VV method solves the following system of linear equations, for an arbitrary strike X and for some base volatility $\tilde{\sigma}$:

$$\begin{aligned}
\frac{\partial C^{BSM}}{\partial \sigma}(X, \tilde{\sigma}) &= \sum_{i=1}^{3} w_i(X) \frac{\partial C^{BSM}}{\partial \sigma}(X_i, \tilde{\sigma}) \\
\frac{\partial^2 C^{BSM}}{\partial \sigma^2}(X, \tilde{\sigma}) &= \sum_{i=1}^{3} w_i(X) \frac{\partial^2 C^{BSM}}{\partial \sigma^2}(X_i, \tilde{\sigma}) \\
\frac{\partial^2 C^{BSM}}{\partial \sigma \partial S}(X, \tilde{\sigma}) &= \sum_{i=1}^{3} w_i(X) \frac{\partial^2 C^{BSM}}{\partial \sigma \partial S}(X_i, \tilde{\sigma})
\end{aligned} \tag{6.28}$$

The system can be solved numerically or analytically for the weights $w_i(X)$, $i = 1, 2, 3$. In a second step, the VV price is computed by

$$C(X) = C^{BSM}(X, \tilde{\sigma}) + \sum_{i=1}^{3} w_i(X) \left[C^{BSM}(X_i, \hat{\sigma}_i) - C^{BSM}(X_i, \tilde{\sigma}) \right], \tag{6.29}$$

from which one obtains IV by inverting the BSM formula. These steps need to be solved for each X to construct the VV smile. For more details on the VV method, approximation formulae for the VV smile, and numerous practical insights we refer to the lucid description presented by Castagna and Mercurio (2007). As a typical choice for the base volatility, Castagna and Mercurio (2007) suggest $\tilde{\sigma} = \sigma_2$, where σ_2 would be an ATM IV, and σ_1 and σ_3 are 25Δ put and the 25Δ call IV, respectively. As noted there, the VV method is not arbitrage-free by construction, in particular convexity can not be guaranteed, but it appears to produce arbitage-free estimates of IV surfaces for usual market conditions.

6.7.2 *Non- and Semiparametric Models of Implied Volatility*

If potential arbitrage violations in the resulting estimate are of no particular concern, virtually any non- and semiparametric method can be applied to IV data. A specific choice can often be made from practical considerations. We therefore confine this section to pointing to the relevant examples in the literature.

Piecewise quadratic or cubic polynomials to fit single smiles was applied by Shimko (1993), Malz (1997), Ané and Geman (1999) and Hafner and Wallmeier (2001). Aït-Sahalia and Lo (1998), Rosenberg (2000), Cont and da Fonseca (2002) and Fengler et al. (2003) employ a Nadaraya-Watson smoother. Higher order local polynomial smoothing of the IV surface was suggested in Fengler (2005), when the aim is to recover the local volatility function via the Dupire formula, or by Härdle et al. (2010) for estimating the empirical pricing kernel. Least-squares kernel regression was suggested in Gouriéroux et al. (1994) and Fengler and Wang (2009). Audrino and Colangelo (2009) rely on IV surface estimates based on regression trees in a forecasting study. Model selection between fully parametric, semi- and nonparametric specifications is discussed in detail in Aït-Sahalia et al. (2001).

6.7.3 *Implied Volatility Modeling Under No-Arbitrage Constraints*

For certain applications, for instance for local volatility modeling, an arbitrage-free estimate of the IV surface is mandatory. Methods producing arbitrage-free estimates must respect the bounds presented in Sect. 6.4. They are surveyed in this section.

6.7.3.1 Call Price Interpolation

Interpolation techniques to recover a globally arbitrage-free call price function have been suggested by Kahalé (2004) and Wang et al. (2004). It is crucial for these algorithms to work that the data to be interpolated are arbitrage-free from the beginning. Consider the set of pairs of strikes and call prices $(X_i, C_i), i = 0, \ldots, n$. Then, applying to (6.9), the set does not admit arbitrage in strikes if the first divided differences associated with the data observe

$$-e^{-rT} < \frac{C_i - C_{i-1}}{X_i - X_{i-1}} < \frac{C_{i+1} - C_i}{X_{i+1} - X_i} < 0 \tag{6.30}$$

and if the price bounds (6.8) hold.

For interpolation Kahalé (2004) considers piecewise convex polynomials which are inspired from the BSM formula. More precisely, for a parameter vector $\Theta = (\theta_1, \theta_2, \theta_3, \theta_4)^\top$ with $\theta_1 > 0$, $\theta_2 > 0$ consider the function

$$c(X;\Theta) = \theta_1\,\Phi(d_1) - X\,\Phi(d_2) + \theta_3 X + \theta_4, \tag{6.31}$$

where $d_1 = \left[\log(\theta_1/X) + 0.5\,\theta_2^2\right]/\theta_2$ and $d_2 = d_1 - \theta_2$. Clearly, $c(X;\Theta)$ is convex in strikes $X > 0$, since it differs from the BSM formula by a linear term, only. It can be shown that on a segment $[X_i, X_{i+1}]$ and for C_i, C_{i+1} and given first order deratives C_i' and C_{i+1}' there exists a unique vector Θ interpolating the observed call prices.

Kahalé (2004) proceeds in showing that for a sequence (X_i, C_i, C_i') for $i = 0, \ldots, n+1$ with the (limit) conditions $X = 0$, $X_i < X_{i+1}$, $X_{n+1} = \infty$, $C_0 = S_0$, $C_{n+1} = 0$, $C_0' = -e^{-rT}$ and $C_{n+1}' = 0$ and

$$C_i' < \frac{C_{i+1} - C_i}{X_{i+1} - X_i} < C_{i+1}' \tag{6.32}$$

for $i = 1, \ldots, n$ there exists a unique $\mathcal{C}^1$ convex function $c(X)$ described by a series of vectors Θ_i for $i = 0, \ldots, n$ interpolating observed call prices. There are $4(n+1)$ parameters in Θ_i, which are matched by $4n$ equations in the interior segments $C_i = c(X_i;\Theta_i)$ and $C_i' = c'(X_i;\Theta_i)$ for $i = 1, \ldots, n$, and four additional equations by the four limit conditions in (X_0, C_0) and (X_{n+1}, C_{n+1}).

A $\mathcal{C}^2$ convex function is obtained in the following way: For $j = 1, \ldots, n$, replace the jth condition on the first order conditions by $\gamma_j = c'(X_j;\Theta_j)$ and $\gamma_j = c'(X_j;\Theta_{j-1})$, for some $\gamma_j \in]l_j, l_{j+1}[$ and $l_j = (C_j - C_{j-1})/(X_j - X_{j-1})$. Moreover add the condition $c''(X_j;\Theta_j) = c''(X_j;\Theta_{j-1})$. This way the number of parameters is still equal to the number of constraints.

Concluding, the Kahalé (2004) algorithm for a $\mathcal{C}^2$ call price function is as follows:

1. Put $C_0' = -e^{-rT}$, $C_{n+1}' = 0$ and $C_i' = (l_i + l_{i+1})/2$ for $i = 1, \ldots, n$, where $l_i = (C_i - C_{i-1})/(X_i - X_{i-1})$.
2. For each $j = 1, \ldots, n$ compute the $\mathcal{C}^1$ convex function with continuous second order derivative at X_j. Replace $C_j' = \gamma_j$.

Kahalé (2004) suggests to solve the algorithm using the Newton–Raphson method.

An alternative, cubic B-spline interpolation was suggested by Wang et al. (2004). For observed prices $(X_i, C_i), i = 0, \ldots, n$, $0 < a = X_0 < \ldots < X_n = b < \infty$ they consider the following minimization problem:

$$\begin{aligned} \min \quad & \|c''(X) - e^{-rT} h(X)\|_2^2 \\ \text{s.t.} \quad & c(X_i) = C_i, \qquad i = 0, \ldots, n, \\ & c''(X) \geq 0 \qquad X \in (0, \infty), \end{aligned} \tag{6.33}$$

where $\|\cdot\|_2$ is the (Lebesgue) L^2 norm on $[a, b]$, h some prior density (e.g., the log-normal density) and c the unknown option price function with absolutely continuous first and second order derivatives on $[a, b]$. By the Peano kernel theorem, the constraints $c(X_i) = C_i, i = 1, \ldots, n$ can be replaced by

$$\int_a^b B_i(X)\, c''(X)\, dX = d_i, \qquad i = 1, \ldots, n-2, \tag{6.34}$$

where B_i is a normalized linear B-spline with the support on $[X_i, X_{i+2}]$ and d_i the second divided differences associated with the data. Wang et al. (2004) show that this infinite-dimensional optimization problem has a unique solution for $c''(X)$ and how to cast it into a finite-dimensional smooth optimization problem. The resulting function for $c(X)$ is then a cubic B-spline. Finally they devise a generalized Newton method for solving the problem with superlinear convergence.

6.7.3.2 Call Price Smoothing by Natural Cubic Splines

For a sample of strikes and call prices, $\{(X_i, C_i)\}$, $X_i \in [a, b]$ for $i = 1, \ldots, n$, Fengler (2009) considers the curve estimate defined as minimizer $\widehat{g}$ of the penalized sum of squares

$$\sum_{i=1}^{n} \left\{C_i - g(X_i)\right\}^2 + \lambda \int_a^b \{g''(v)\}^2\, dv. \tag{6.35}$$

The minimizer $\widehat{g}$ is a natural cubic spline, and represents a globally arbitrage-free call price function. Smoothness is controlled by the parameter $\lambda > 0$. The algorithm suggested by Fengler (2009) observes the no-arbitrage constraints (6.8)–(6.10). For this purpose the natural cubic spline is converted into the value-second derivative representation suggested by Green and Silverman (1994). This allows to formulate a quadratic program solving (6.35). Put $g_i = g(u_i)$ and $\gamma_i = g''(u_i)$, for $i = 1, \ldots, n$, and define $g = (g_1, \ldots, g_n)^\top$ and $\gamma = (\gamma_2, \ldots, \gamma_{n-1})^\top$. By definition of a natural cubic spline, $\gamma_1 = \gamma_n = 0$. The natural spline is completely specified by the vectors g and γ, see Sect. 2.5 in Green and Silverman (1994) who also suggest the nonstandard notation of the entries in γ.

Sufficient and necessary conditions for g and γ to represent a valid cubic spline are formulated via the matrices Q and R. Let $h_i = u_{i+1} - u_i$ for $i = 1, \ldots, n-1$, and define the $n \times (n-2)$ matrix Q by its elements $q_{i,j}$, for $i = 1, \ldots, n$ and $j = 2, \ldots, n-1$, given by

$$q_{j-1,j} = h_{j-1}^{-1}, \qquad q_{j,j} = -h_{j-1}^{-1} - h_j^{-1}, \qquad \text{and } q_{j+1,j} = h_j^{-1},$$

for $j = 2, \ldots, n-1$, and $q_{i,j} = 0$ for $|i - j| \geq 2$, where the columns of Q are numbered in the same non-standard way as the vector γ.

The $(n-2) \times (n-2)$ matrix R is symmetric and defined by its elements $r_{i,j}$ for $i, j = 2, \ldots, n-1$, given by

$$\begin{aligned} r_{i,i} &= \tfrac{1}{3}(h_{i-1}+h_i) \text{ for } i = 2,\dots,n-1 \\ r_{i,i+1} = r_{i+1,i} &= \tfrac{1}{6}h_i \quad \text{ for } i = 2,\dots,n-2, \end{aligned} \tag{6.36}$$

and $r_{i,j} = 0$ for $|i-j| \geq 2$. R is strictly diagonal dominant, and thus strictly positive-definite.

Arbitrage-free smoothing of the call price surface can be cast into the following iterative quadratic minimization problem. Define a $(2n-2)$-vector $y = (y_1,\dots,y_n,0,\dots,0)^\top$, a $(2n-2)$-vector $\xi = (g^\top,\gamma^\top)^\top$ and the matrices, $A = (Q,-R^\top)$ and

$$B = \begin{pmatrix} I_n & 0 \\ 0 & \lambda R \end{pmatrix}, \tag{6.37}$$

where I_n is the unit matrix with size n. Then:

1. Estimate the IV surface by means of an initial estimate on a regular forward-moneyness grid $\mathcal{J} = [x_1, x_n] \times [T_1, T_m]$.
2. Iterate through the price surface from the last to the first expiry, and solve the following quadratic programs.
   For T_j, $j = m,\dots,1$, solve

$$\min_{\xi} \; -y^\top \xi + \frac{1}{2}\xi^\top B \xi \tag{6.38}$$

subject to

$$\begin{aligned} A^\top \xi &= 0 \\ \gamma_i &\geq 0 \\ \frac{g_2-g_1}{h_1} - \frac{h_1}{6}\gamma_2 &\geq -e^{-rT_j} \\ -\frac{g_n-g_{n-1}}{h_{n-1}} - \frac{h_{n-1}}{6}\gamma_{n-1} &\geq 0 \\ g_1 &\leq S_t \qquad \text{if } \; j = m \\ g_i^{(j)} &< g_i^{(j+1)} \qquad \text{if } \; j \in [m-1,1] \\ & \qquad\qquad \text{for } \; i = 1,\dots,n \quad (*) \\ g_1 &\geq S_t - e^{-rT_j}u_1 \\ g_n &\geq 0 \end{aligned} \tag{6.39}$$

where $\xi = (g^\top,\gamma^\top)^\top$. Note that we suppress the explicit dependence on j except in conditions $(*)$ to keep the notation more readable. Conditions $(*)$ implement (6.10); therefore $g_i^{(j)}$ and $g_i^{(j+1)}$ are related by forward-moneyness.

The resulting price surface is converted into IV. It can be beneficial obtain a first coarse estimate of the surface by gridding it on the estimation grid. This allows to more easily implement condition (6.10). The minimization problem can be solved by using the quadratic programming devices provided by standard statistical software packages. The reader is referred to Fengler (2009) for the computational details and the choice of the smoothing parameter λ. In contrast to the approach by Kahalé (2004), a potential drawback this approach suffers from is the fact that the call price function is approximated by cubic polynomials. This can turn out to be disadvantageous, since the pricing function is not in the domain of polynomials functions. It is remedied however by the choice of a sufficiently dense grid in the strike dimension in $\mathcal{J}$.

6.7.3.3 IV Smoothing Using Local Polynomials

As an alternative to smoothing in the call price domain Benko et al. (2007) suggest to directly smooth IV by means of constrained local quadratic polynomials. This implies minimization of the following (local) least squares criterion

$$\min_{\alpha_0,\alpha_1,\alpha_2} \sum_{i=1}^{n} \left\{\widetilde{\sigma}_i - \alpha_0 - \alpha_1(x_i - x) - \alpha_2(x_i - x)^2\right\}^2 \mathcal{K}_h(x - x_i), \tag{6.40}$$

where $\widetilde{\sigma}$ is observed IV. We denote by $\mathcal{K}_h(x - x_i) = h^{-1}\mathcal{K}\left(\frac{x-x_i}{h}\right)$ and by $\mathcal{K}$ a kernel function – typically a symmetric density function with compact support, e.g. $\mathcal{K}(u) = \frac{3}{4}(1 - u^2)\mathbf{1}(|u| \leq 1)$, the Epanechnikov kernel, where $\mathbf{1}(\mathcal{A})$ is the indicator function of some set $\mathcal{A}$. Finally, h is the bandwidth which governs the trade-off between bias and variance, see Härdle (1990) for the details on nonparametric regression. Since $\mathcal{K}_h$ is nonnegative within the (localization) window $[x-h, x+h]$, points outside of this interval do not have any influence on the estimator $\widehat{\sigma}(x)$.

No-arbitrage conditions in terms of IV are obtained by computing (6.9) for an IV adjusted BSM formula, see Brunner and Hafner (2003) among others. Expressed in forward moneyness $x = X/F$ this yields for the convexity condition

$$\begin{aligned}\frac{\partial^2 C^{BSM}}{\partial x^2} &= e^{-rT}\sqrt{T}\varphi(d_1)\\ &\times \left\{\frac{1}{x^2\widehat{\sigma}T} + \frac{2d_1}{x\widehat{\sigma}\sqrt{T}}\frac{\partial\widehat{\sigma}}{\partial x} + \frac{d_1 d_2}{\widehat{\sigma}}\left(\frac{\partial\widehat{\sigma}}{\partial x}\right)^2 + \frac{\partial^2\widehat{\sigma}}{\partial x^2}\right\}\end{aligned} \tag{6.41}$$

where d_1 and d_2 are defined as in (6.4) and (6.5).

The key property of local polynomial regression is that it yields simultaneously to the regression function its derivatives. More precisely, comparing (6.40) with the Taylor expansion of $\widehat{\sigma}$ shows that

$$\widehat{\sigma}(x_i) = \alpha_0,\ \widehat{\sigma}'(x_i) = \alpha_1,\ \widehat{\sigma}''(x_i) = 2\alpha_2. \tag{6.42}$$

Based on this fact Benko et al. (2007) suggest to miminize (6.40) subject to

$$e^{-rT}\sqrt{T}\varphi(d_1)\left\{\frac{1}{x^2\alpha_0 T}+\frac{2d_1\alpha_1}{x\alpha_0\sqrt{T}}+\frac{d_1 d_2}{\alpha_0}(\alpha_1)^2+2\alpha_2\right\}\geq 0, \quad (6.43)$$

with

$$d_1=\frac{\alpha_0^2 T/2-\log(x)}{\sigma\sqrt{T}},\ d_2=d_1-\alpha_0\sqrt{T}.$$

This leads to a nonlinear optimization problem in $\alpha_0, \alpha_1, \alpha_2$.

The case of the entire IV surface is more involved. Suppose the purpose is to estimate $\widehat{\sigma}(x,T)$ for a set of maturities $\{T_1,\ldots,T_L\}$. By (6.11), for a given value x, we need to ensure $\widehat{\upsilon}^2(x,T_l,)\leq\widehat{\upsilon}^2(x,T_{l'})$, for all $T_l<T_l'$. Denote by $\mathcal{K}_{h_x,h_T}$ $(x-x_i, T_l-T_i)$ a bivariate kernel function given by the product of the two univariate kernel functions $\mathcal{K}_{h_x}(x-x_i)$ and $\mathcal{K}_{h_T}(T-T_i)$. Extending (6.40) linearly into the time-to-maturiy dimension then leads to the following optimization problem:

$$\begin{aligned}\min_{\alpha(l)}\quad &\sum_{l=1}^{L}\sum_{i=1}^{n}\mathcal{K}_{h_x,h_T}(x-x_i,T_l-T_i)\Big\{\widetilde{\sigma}_i-\alpha_0(l)\\ &-\alpha_1(l)(x_i-x)-\alpha_2(l)(T_i-T)-\alpha_{1,1}(l)(x_i-x)^2\\ &-\alpha_{1,2}(l)(x_i-x)(T_i-T)\Big\}^2\end{aligned} \quad (6.44)$$

subject to

$$\sqrt{T_l}\varphi(d_1(l))\left\{\frac{1}{x^2\alpha_0(l)T_l}+\frac{2d_1(l)\alpha_1(l)}{x\alpha_0(l)\sqrt{T_l}}+\frac{d_1(l)d_2(l)}{a_0(l)}\alpha_1^2(l)+2\alpha_{1,1}(l)\right\}\geq 0,$$

$$d_1(l)=\frac{\alpha_0^2(l)T_l/2-\log(x)}{\alpha_0(l)\sqrt{T_l}},\quad d_2(l)=d_1(l)-a_0(l)\sqrt{T_l},\quad l=1,\ldots,L$$

$$2T_l\alpha_0(l)\alpha_2(l)+\alpha_0^2(l)>0\quad l=1,\ldots,L$$

$$\alpha_0^2(l)T_l<\alpha_0^2(l')T_l',\quad T_l<T_l'.$$

The last two conditions ensure that total implied variance is (locally) nondecreasing, since $\frac{\partial\upsilon^2}{\partial T}>0$ can be rewritten as $2T\alpha_0\alpha_2+\alpha_0^2>0$ for a given T, while the last conditions guarantee that total variance is increasing across the surface. From a computational view, problem (6.44) calculates for a given x the estimates for all given T_l in one step in order to warrant that $\widehat{\upsilon}$ is increasing in T.

The approach by Benko et al. (2007) yields an IV surface that respects the convexity conditions, but neglects the conditions on call spreads and the general price bounds. Therefore the surface may not be fully arbitrage-free. However, since convexity violations and calendar arbitrage are by far the most virulent instances of arbitrage in observed IV data occurring the surfaces will be acceptable in most cases.

References

Aït-Sahalia, Y., Bickel, P. J., & Stoker, T. M. (2001). Goodness-of-fit tests for regression using kernel methods. *Journal of Econometrics*, *105*, 363–412.

Aït-Sahalia, Y., & Lo, A. (1998). Nonparametric estimation of state-price densities implicit in financial asset prices. *Journal of Finance*, *53*, 499–548.

Ané, T., & Geman, H. (1999). Stochastic volatility and transaction time: An activity-based volatility estimator. *Journal of Risk*, *2*(1), 57–69.

Audrino, F., & Colangelo, D. (2009). Semi-parametric forecasts of the implied volatility surface using regression trees. *Statistics and Computing*, *20*(4), 421–434.

Benko, M., Fengler, M. R., Härdle, W. & Kopa, M. (2007). On extracting information implied in options. *Computational Statistics*, *22*(4), 543–553.

Bharadia, M. A., Christofides, N., & Salkin, G. R. (1996). Computing the Black-Scholes implied volatility – generalization of a simple formula. In P. P. Boyle, F. A. Longstaff, P. Ritchken, D. M. Chance & R. R. Trippi (Eds.), *Advances in futures and options research*, (Vol. 8, pp. 15–29.). London: JAI Press

Black, F., & Scholes, M. (1973). The pricing of options and corporate liabilities. *Journal of Political Economy*, *81*, 637–654.

Brenner, M., & Subrahmanyam, M. (1988). A simple formula to compute the implied standard deviation. *Financial Analysts Journal*, *44*(5), 80–83.

Britten-Jones, M., & Neuberger, A. J. (2000). Option prices, implied price processes, and stochastic volatility. *Journal of Finance*, *55*(2), 839–866.

Brockhaus, O., Farkas, M., Ferraris, A., Long, D., & Overhaus, M. (2000). *Equity derivatives and market risk models*. London: Risk Books.

Brunner, B., & Hafner, R. (2003). Arbitrage-free estimation of the risk-neutral density from the implied volatility smile. *Journal of Computational Finance*, *7*(1), 75–106.

Carr, P., & Wu, L. (2003). Finite moment log stable process and option pricing. *Journal of Finance*, *58*(2), 753–777.

Castagna, A., & Mercurio, F. (2007). Building implied volatility surfaces from the available market quotes: A unified approach. In I. Nelken (Ed.), *Volatility as an asset class* (pp. 3–59). London: Risk Books.

Cont, R., & da Fonseca, J. (2002). The dynamics of implied volatility surfaces. *Quantitative Finance*, *2*(1), 45–60.

Corrado, C. J., & Miller, T. W. (1996). A note on a simple, accurate formula to compute implied standard deviations. *Journal of Banking and Finance*, *20*, 595–603.

Dumas, B., Fleming, J., & Whaley, R. E. (1998). Implied volatility functions: Empirical tests, *Journal of Finance*, *53*(6), 2059–2106.

Feinstein, S. (1988). A source of unbiased implied volatility. *Technical Report 88–89*, Federal Reserve Bank of Atlanta.

Fengler, M. R. (2005). *Semiparametric modeling of implied volatility*, Lecture Notes in Finance. Berlin: Springer.

Fengler, M. R. (2009). Arbitrage-free smoothing of the implied volatility surface. *Quantitative Finance*, *9*(4), 417–428.

Fengler, M. R., & Wang, Q. (2009). Least squares kernel smoothing of the implied volatility smile. In W. Härdle, N. Hautsch & L. Overbeck (Eds.), *Applied Quantitative Finance* (2nd ed.). Berlin: Springer.

Fengler, M. R., Härdle, W., & Villa, C. (2003). The dynamics of implied volatilities: A common principle components approach. *Review of Derivatives Research, 6*, 179–202.

Fengler, M. R., Härdle, W., & Mammen, E. (2007). A semiparametric factor model for implied volatility surface dynamics. *Journal of Financial Econometrics, 5*(2), 189–218.

Gatheral, J. (2004). A parsimonious arbitrage-free implied volatility parameterization with application to the valuation of volatility derivatives, Presentation at the ICBI Global Derivatives and Risk Management, Madrid, España.

Gatheral, J. (2006). *The volatility surface: A practitioner's guide*, New Jersey: Wiley.

Gouriéroux, C., Monfort, A., & Tenreiro, C. (1994). Nonparametric diagnostics for structural models. *Document de travail 9405*. CREST, Paris.

Green, P. J., & Silverman, B. W. (1994). Nonparametric regression and generalized linear models. In: *Monographs on statistics and applied probability* (Vol. 58). London: Chapman and Hall.

Hafner, R., & Wallmeier, M. (2001). The dynamics of DAX implied volatilities. *International Quarterly Journal of Finance, 1*(1), 1–27.

Hagan, P., Kumar, D., Lesniewski, A., & Woodward, D. (2002). Managing smile risk. *Wilmott Magazine, 1*, 84–108.

Härdle, W. (1990). *Applied nonparametric regression*. Cambridge, UK: Cambridge University Press.

Härdle, W., Okhrin, O., & Wang, W. (2010). Uniform confidence bands for pricing kernels. *SFB 649 Discussion Paper 2010–03*. Berlin: Humboldt-Universität zu.

Heston, S. (1993). A closed-form solution for options with stochastic volatility with applications to bond and currency options. *Review of Financial Studies, 6*, 327–343.

Hodges, H. M. (1996). Arbitrage bounds on the implied volatility strike and term structures of European-style options. *Journal of Derivatives, 3*, 23–35.

Isengildina-Massa, O., Curtis, C., Bridges, W., & Nian, M. (2007). Accuracy of implied volatility approximations using "nearest-to-the-money" option premiums. *Technical report*, Southern Agricultural Economics Association.

Kahalé, N. (2004). An arbitrage-free interpolation of volatilities. *RISK, 17*(5), 102–106.

Latané, H. A., & Rendelman, J. (1976). Standard deviations of stock price ratios implied in option prices. *Journal of Finance, 31*, 369–381.

Lee, R. W. (2004). The moment formula for implied volatility at extreme strikes. *Mathematical Finance, 14*(3), 469–480.

Li, S. (2005). A new formula for computing implied volatility. *Applied Mathematics and Computation, 170*(1), 611–625.

Malz, A. M. (1997). Estimating the probability distribution of the future exchange rate from option prices. *Journal of Derivatives, 5*(2), 18–36.

Manaster, S., & Koehler, G. (1982). The calculation of implied variances from the black-and-scholes model: A note. *Journal of Finance, 37*, 227–230.

Mercurio, F., & Pallavicini, A. (2006). Smiling at convexity: Bridging swaption skews and CMS adjustments. *RISK, 19*(8), 64–69.

Merton, R. C. (1973). Theory of rational option pricing. *Bell Journal of Economics and Management Science, 4*, 141–183.

Obłój, J. (2008). Fine-tune your smile: Correction to Hagan et al., *Wilmott Magazine, 35*, 102–104.

Randall, C., & Tavella, D. (2000). *Pricing financial instruments: The finite difference method.* New York: Wiley.

Rebonato, R. (2004). *Volatility and Correlation*, 2nd edn, Wiley.

Reiner, E. (2000). Calendar spreads, characteristic functions, and variance interpolation. Mimeo.

Reiner, E. (2004). The characteristic curve approach to arbitrage-free time interpolation of volatility. Presentation at the ICBI Global Derivatives and Risk Management, Madrid, España.

Rogers, L. C. G., & Tehranchi, M. (2009). Can the implied volatility surface move by parallel shifts?. *Finance and Stochastics, 14*(2), 235–248.

Roper, M., & Rutkowski, M. (2009). On the relationship between the call price surface and the implied volatility surface close to expiry. *International Journal of Theoretical and Applied Finance, 12*(4), 427–441.

Rosenberg, J. (2000). Implied volatility functions: A reprise. *Journal of Derivatives*, *7*, 51–64.
Rubinstein, M. (1994). Implied binomial trees. *Journal of Finance*, *49*, 771–818.
Shimko, D. (1993). Bounds on probability. *RISK*, *6*(4), 33–37.
Steele, J. M. (2000). *Stochastic calculus and financial applications*. Berlin: Springer.
Stineman, R. W. (1980). A consistently well-behaved method of interpolation. *Creative Computing*, *6*(7), 54–57.
Tehranchi, M. (2009). Asymptotics of implied volatility far from maturity. *Journal of Applied Probability*, *46*(3), 629–650.
Tehranchi, M. (2010). Implied volatility: Long maturity behavior. In R. Cont (Ed.), *Encyclopedia of quantitative finance*. New York: Wiley.
Wang, Y., Yin., H., & Qi, L. (2004). No-arbitrage interpolation of the option price function and its reformulation. *Journal of Optimization Theory and Applications*, *120*(3), 627–649.
West, G. (2005). Calibration of the SABR model in illiquid markets. *Applied Mathematical Finance*, *12*(4), 371–385.
Wolberg, G., & Alfy, I. (2002). An energy-minimization framework for monotonic cubic spline interpolation. *Journal of Computational and Applied Mathematics*, *143*, 145–188.

Chapter 7
Interest Rate Derivatives Pricing with Volatility Smile

Haitao Li

Abstract The volatility "smile" or "skew" observed in the S&P 500 index options has been one of the main drivers for the development of new option pricing models since the seminal works of Black and Scholes (J Polit Econ 81:637–654, 1973) and Merton (Bell J Econ Manag Sci 4:141–183, 1973). The literature on interest rate derivatives, however, has mainly focused on at-the-money interest rate options. This paper advances the literature on interest rate derivatives in several aspects. First, we present systematic evidence on volatility smiles in interest rate caps over a wide range of moneyness and maturities. Second, we discuss the pricing and hedging of interest rate caps under dynamic term structure models (DTSMs). We show that even some of the most sophisticated DTSMs have serious difficulties in pricing and hedging caps and cap straddles, even though they capture bond yields well. Furthermore, at-the-money straddle hedging errors are highly correlated with cap-implied volatilities and can explain a large fraction of hedging errors of all caps and straddles across moneyness and maturities. These findings strongly suggest the existence of systematic unspanned factors related to stochastic volatility in interest rate derivatives markets. Third, we develop multifactor Heath–Jarrow–Morton (HJM) models with stochastic volatility and jumps to capture the smile in interest rate caps. We show that although a three-factor stochastic volatility model can price at-the-money caps well, significant negative jumps in interest rates are needed to capture the smile. Finally, we present nonparametric evidence on the economic determinants of the volatility smile. We show that the forward densities depend significantly on the slope and volatility of LIBOR rates and that mortgage refinance activities have strong impacts on the shape of the volatility smile. These results provide nonparametric evidence of unspanned stochastic volatility and suggest that the unspanned factors could be partly driven by activities in the mortgage markets.

H. Li (✉)
Professor of Finance, Stephen M. Ross School of Business, University of Michigan, Ann Arbor, MI 48109
e-mail: htli@umich.edu

J.-C. Duan et al. (eds.), *Handbook of Computational Finance*, Springer Handbooks of Computational Statistics, DOI 10.1007/978-3-642-17254-0_7,

7.1 Introduction

The extensive literature on multifactor dynamic term structure models (hereafter, DTSMs) of the last decade mainly focuses on explaining bond yields and swap rates (see Dai and Singleton 2003; Piazzesi 2009 for surveys of the literature). The pricing and hedging of over-the-counter interest rate derivatives such as caps and swaptions has attracted attention only recently. World-wide, caps and swaptions are among the most widely traded interest rate derivatives. According to the Bank for International Settlements, in recent years, their combined notional value exceeds 10 trillion dollars, which is many times larger than that of exchange-traded options. The accurate and efficient pricing and hedging of caps and swaptions is therefore of enormous practical importance. Moreover, because cap and swaption prices may contain information on term structure dynamics not contained in bond yields or swap rates (see Jagannathan et al. 2003 for a related discussion), Dai and Singleton (2003, p. 670) argue that there is an "enormous potential for new insights from using (interest rate) derivatives data in model estimations."

Since caps and swaptions are traded over-the-counter, the common data sources, such as Datastream, only supply at-the-money (ATM) option prices. As a result, the majority of the existing literature uses only ATM caps and swaptions, with almost no documentation of the relative pricing of caps with different strike prices. In contrast, the attempt to capture the volatility smile in equity option markets has been the driving force behind the development of the equity option pricing literature for the past few decades (for reviews of the equity option literature, see Duffie 2002; Campbell et al. 1997; Bakshi et al. 1997, and references therein). Analogously, studying caps and swaptions with different strike prices could provide new insights about existing term structure models that are not available from using only ATM options.

Using almost 4 years of daily interest rate caps price data, we provide a comprehensive documentation of volatility smiles in the caps market. We obtain daily prices of interest rate caps between August 1, 2000 and July 26, 2004 from SwapPX. Our data set is one of the most comprehensive ones available for caps written on dollar LIBOR rates. One advantage of our data is that we observe prices of caps over a wide range of strike prices and maturities. There are 15 different maturities ranging from 6 months to 10 years throughout the sample period, and for each maturity, there are 10 different strike prices. The data makes it possible to examine issues that have not been addressed in the literature.

The first question we study is the pricing and hedging of interest rate caps over different strike prices using one of the most popular classes of term structure models, the DTSMs. One main reason for the popularity of the DTSMs is their tractability. They provide closed-form solutions for the prices of not only zero-coupon bonds, but also of a wide range of interest rate derivatives (see, for example, Duffie et al. 2000; Chacko and Das 2002; Leippold and Wu 2002). The closed-form formulas significantly reduce the computational burden of implementing these models and simplify their applications in practice. However, almost all existing DTSMs assume that bonds and derivatives are driven by the same set of risk factors, which implies that derivatives are redundant and can be perfectly hedged using solely bonds. Interest rate caps and swaptions are derivatives written on Libor and swap rates. Therefore, according to DTSMs, their prices should be determined by the same set of risk factors that determine Libor and swap rates.

Li and Zhao (2006) study the pricing and hedging of caps over different strike prices using the quadratic term structure models (QTSMs) of Ahn et al. (2002) (hereafter, ADG). We choose the QTSMs over the affine term structure models (ATSMs) of Duffie and Kan (1996) in our analysis because of their superior performance in capturing the conditional volatility of bond yields, which is important for pricing derivatives. We find that the QTSMs have serious difficulties in hedging caps and cap straddles, even though they capture bond yields well. Furthermore, ATM straddle hedging errors are highly correlated with cap-implied volatilities and can explain a large fraction of hedging errors of all caps and straddles across moneyness and maturities. Our results strongly suggest the existence of systematic unspanned factors related to stochastic volatility in interest rate derivatives markets.

Li and Zhao (2006) contribute nicely to the literature on the "unspanned stochastic volatility" puzzle. Heidari and Wu (2003) show that while the three common term structure factors (i.e. the level, slope and curvature of the yield curve) can explain 99.5% of the variations of bond yields, they explain less than 60% of swaption implied volatilities. Similarly, Collin-Dufresne and Goldstein (2002) show that there is a very weak correlation between changes in swap rates and returns on ATM cap straddles: the R^2s of regressions of straddle returns on changes of swap rates are typically less than 20%. Furthermore, one principal component explains 80% of regression residuals of straddles with different maturities. As straddles are approximately delta neutral and mainly exposed to volatility risk, they refer to the factor that drives straddle returns but is not affected by the term structure factors as "unspanned stochastic volatility" (USV). Jagannathan et al. (2003) find that an affine three-factor model can fit the LIBOR swap curve rather well. However, they identify significant shortcomings when confronting the model with data on caps and swaptions, thus concluding that derivatives must be used when evaluating term structure models. Fan et al. (2003) (hereafter, FGR), however, challenge the findings of Heidari and Wu (2003) and Collin-Dufresne and Goldstein (2002), arguing that the linear regression approach used in these two studies could give misleading results of USV due to the highly nonlinear dependence of straddle returns on the underlying yield factors. Instead, FGR show that multifactor models with state variables linked solely to underlying LIBOR and swap rates can hedge swaptions and even swaption straddles very well. Our rigorous analysis of model-based hedging of caps and cap straddles based on QTSMs avoids the problems facing the linear regression approach of previous studies and helps resolve the controversy on USV.

Some recent studies also provide evidence in support of the existence of USV using bond data alone. They show the yield curve volatilities backed out from a cross-section of bond yields do not agree with the time-series filtered volatilities, via GARCH or high-frequency estimates from yields data. This challenges the traditional DTSMs even more since these models can not be expected to capture the option implied volatilities if they can not even match the realized yield curve volatilities. Specifically, Collin-Dufresne, Goldstein, and Jones (2009, CDGJ) show that the LIBOR volatility implied by an affine multi-factor specification from the swap rate curve can be negatively correlated with the time series of volatility obtained from a standard GARCH approach. In response, they argue that an affine four-factor USV model delivers both realistic volatility estimates and a good cross-sectional fit. Andersen and Benzoni (2006), through the use of high-frequency data on bond yields, construct the model-free "realized yield volatility" measure by computing empirical quadratic yield variation for a cross-section of fixed maturities. They find that the yield curve fails to span yield volatility, as the systematic volatility factors are largely unrelated to the cross-section of yields. They claim that a broad class of affine diffusive, Gaussian-quadratic and affine jump-diffusive models is incapable of accommodating the observed yield volatility dynamics. An important implication is that the bond markets per se are incomplete and yield volatility risk cannot be hedged by taking positions solely in the Treasury bond market. They also advocate using the empirical realized yield volatility measures more broadly as a basis for specification testing and (parametric) model selection within the term structure literature. Thompson (2008), on the LIBOR swap data, argues when the affine models are estimated with the time-series filtered yield volatility they can pass on his newly proposed specification test, but not with the cross-sectional backed-out volatility. From these studies on the yields data alone, there may exist an alternative explanation for the failure of DTSMs in effectively pricing derivatives in that the bonds small convexity makes bonds not sensitive enough to identify the volatilities from measurement errors. Therefore efficient inference requires derivatives data as well.

The second question we study is how to incorporate USV into a term structure model so it can price wide spectrum of interest rate derivatives effectively. The existence of USV has profound implications for term structure modeling, in particular for the DTSMs. The presence of USV in the derivatives market implies that one fundamental assumption underlying all DTSMs does not hold and that these models need to be substantially extended to incorporate the unspanned factors before they can be applied to derivatives. However, as Collin-Dufresne and Goldstein (2002) show, it is rather difficult to introduce USV in traditional DTSMs: One must impose highly restrictive assumptions on model parameters to guarantee that certain factors that affect derivative prices do not affect bond prices. In contrast to the approach of adding USV restrictions to DTSMs, it is relatively easy to introduce USV in the Heath et al. (1992) (hereafter, HJM) class of models, which include the LIBOR models of Brace et al. (1997) and Miltersen et al. (1997), the random field models of Goldstein (2000), and the string models of Santa-Clara and Sornette (2001). Indeed, any HJM model in which the forward rate curve has stochastic volatility and the volatility and yield shocks are not perfectly correlated exhibits USV. Therefore, in addition to the commonly known advantages of HJM models (such as perfectly fitting the initial yield curve), they offer the additional advantage of easily accommodating USV. Of course, the trade-off here is that in an HJM model, the yield curve is an input rather than a prediction of the model.

Jarrow et al. (2007) develop multifactor HJM models with stochastic volatility and jumps in LIBOR forward rates to capture the smile in interest rate caps. The LIBOR rates follow the affine jump diffusions (hereafter, AJDs) of Duffie et al. (2000) and a closed-form solution for cap prices is provided. Given that a small number of factors can explain most of the variation of bond yields, we consider low-dimensional model specifications based on the first few (up to three) principal components of historical forward rates. Our model explicitly incorporates jumps in LIBOR rates, making it possible to differentiate between the importance of stochastic volatility versus jumps for pricing interest rate derivatives. Jarrow et al. (2007) provide one of the first empirical analyses of their model for capturing the volatility smile in the cap market. We show that although a three-factor stochastic volatility model can price at-the-money caps well, significant negative jumps in interest rates are needed to capture the smile.

Recently, several HJM models with USV have been developed and applied to price caps and swaptions. Collin-Dufresne and Goldstein (2003) develop a random field model with stochastic volatility and correlation in forward rates. Applying the transform analysis of Duffie et al. (2000), they obtain closed-form formulae for a wide variety of interest rate derivatives. However, they do not calibrate their models to market prices of caps and swaptions. Han (2007) extends the model of LSS (2001) by introducing stochastic volatility and correlation in forward rates. Han (2007) shows that stochastic volatility and correlation are important for reconciling the mispricing between caps and swaptions. Trolle and Schwartz (2009) develop a multifactor term structure model with unspanned stochastic volatility factors and correlation between innovations to forward rates and their volatilities.

The third question we study is what economic factors determine the shape of the volatility smile in interest rate caps. Li and Zhao (2009) provide one of the first nonparametric estimates of probability densities of LIBOR rates under forward martingale measures using caps with a wide range of strike prices and maturities.[1] The nonparametric estimates of LIBOR forward densities are conditional on the slope and volatility factors of LIBOR rates, while the level factor is automatically incorporated in existing methods.[2] They find that the forward densities depend significantly on the slope and volatility of LIBOR rates. In addition, they document important impacts of mortgage market activities on the LIBOR forward densities even after controlling for both the slope and volatility factors. For example, the forward densities at intermediate maturities (3, 4, and 5 years) are more negatively skewed when refinance activities, measured by the Mortgage Bankers Association

---

[1] The nonparametric forward densities estimated using caps, which are among the simplest and most liquid OTC interest rate derivatives, allow consistent pricing of more exotic and/or less liquid OTC interest rate derivatives based on the forward measure approach. The nonparametric forward densities can reveal potential misspecifications of most existing term structure models, which rely on strong parametric assumptions to obtain closed-form formula for interest rate derivative prices.

[2] Andersen and Benzoni (2006) show the "curvature" factor are not significantly correlated with the yield volatility and it is true in this paper as well, therefore the volatility effect here is not due to the "curvature" factor.

of America (MBAA) refinance index, are high. Demands for out-of–the-money (OTM) floors by investors in mortgage-backed securities (MBS) to hedge potential losses from prepayments could lead to more negatively skewed forward densities. These empirical results have important implications for the unspanned stochastic volatility puzzle by providing nonparametric and model-independent evidence of USV. The impacts of mortgage activities on the forward densities further suggest that the unspanned factors could be partially driven by activities in the mortgage markets. While Duarte (2008) shows mortgage-backed security (MBS) hedging activity affects interest rate volatility, Li and Zhao (2009) provide evidence on the impacts of mortgage market activities on the shape of the volatility smile.

The rest of the paper is organized as follows. In Sect. 7.2, we provide a comprehensive evidence on a volatility smile in interest rate cap markets. In Sect. 7.3, we present the main results of Li and Zhao (2006) on the pricing and hedging of interest rate caps under QTSMs. Section 7.4 contains the main results of Jarrow et al. (2007) on pricing the volatility smile in the cap markets using multifactor HJM model with stochastic volatility and jumps. In Sect. 7.5, we present the nonparametric evidence of Li and Zhao (2009) on the impacts of mortgage market activities on the shape of the volatility smile. Section 7.6 concludes, and the Appendix contains some mathematical details.

7.2 A Volatility Smile in the Interest Rate Cap Markets

In this section, using almost 4 years of cap price data we provide a comprehensive documentation of volatility smiles in the cap markets. The data come from SwapPX and include daily information on LIBOR forward rates (up to 10 years) and prices of caps with different strikes and maturities from August 1, 2000 to July 26, 2004. Jointly developed by GovPX and Garban-ICAP, SwapPX is the first widely distributed service delivering 24-hour real-time rates, data, and analytics for the world-wide interest rate swaps market. GovPX, established in the early 1990s by the major US fixed-income dealers in a response to regulators' demands for increased transparency in the fixed-income markets, aggregates quotes from most of the largest fixed-income dealers in the world. Garban-ICAP is the world's leading swap broker specializing in trades between dealers and trades between dealers and large customers. The data are collected every day the market is open between 3:30 and 4 p.m. To reduce noise and computational burdens, we use weekly data (every Tuesday) in our empirical analysis. If Tuesday is not available, we first use Wednesday followed by Monday. After excluding missing data, we have a total of 208 weeks in our sample. To our knowledge, our data set is the most comprehensive available for caps written on dollar LIBOR rates (see Gupta and Subrahmanyam 2005; Deuskar et al. 2003 for the only other studies that we are aware of in this area).

Interest rate caps are portfolios of call options on LIBOR rates. Specifically, a cap gives its holder a series of European call options, called caplets, on LIBOR forward rates. Each caplet has the same strike price as the others, but with different expiration dates. Suppose $L(t, T)$ is the 3-month LIBOR forward rate at $t \leq T$, for the interval from T to $T + \frac{1}{4}$. A caplet for the period $\left[T, T + \frac{1}{4}\right]$ struck at K pays $\frac{1}{4}(L(T, T) - K)$ at $T + \frac{1}{4}$. Note that although the cash flow of this caplet is received at time $T + \frac{1}{4}$, the LIBOR rate is determined at time T. Hence, there is no uncertainty about the caplet's cash flow after the LIBOR rate is set at time T. In summary, a cap is just a portfolio of caplets whose maturities are 3 months apart. For example, a 5-year cap on 3-month LIBOR struck at 6% represents a portfolio of 19 separately exercisable caplets with quarterly maturities ranging from 6 months to 5 years, where each caplet has a strike price of 6%.

The existing literature on interest rate derivatives mainly focuses on ATM contracts. One advantage of our data is that we observe prices of caps over a wide range of strikes and maturities. For example, every day for each maturity, there are 10 different strike prices: 4.0, 4.5, 5.0, 5.5, 6.0, 6.5, 7.0, 8.0, 9.0, and 10.0% between August 1, 2000 and October 17, 2001; 1.0, 1.5, 2.0, 2.5, 3.0, 3.5, 4.0, 4.5, 5.0 and 5.5% between October 18 and November 1, 2001; and 2.5, 3.0, 3.5, 4.0, 4.5, 5.0, 5.5, 6.0, 6.5, and 7.0% between November 2, 2001 and July 15, 2002; 2.0, 2.5, 3.0, 3.5, 4.0, 4.5, 5.0, 5.5, 6.0, and 6.5% between July 16, 2002 and April 14, 2003; 1.5, 2.0, 2.5, 3.0, 3.5, 4.0, 4.5, 5.0, 5.5, and 6.0% between April 15, 2003 and September 23, 2003; and 1.5, 2.0, 2.5, 3.0, 3.5, 4.0, 4.5, 5.0, 5.5, and 6.0% between April 15, 2003 and July 26, 2004. Moreover, caps have 15 different maturities throughout the whole sample period: 0.5, 1.0, 1.5, 2.0, 2.5, 3.0, 3.5, 4.0, 4.5, 5.0, 6.0, 7.0, 8.0, 9.0, and 10.0 years. This cross-sectional information on cap prices allows us to study the performance of existing term structure models in the pricing and hedging of caps for different maturity and moneyness.

Ideally, we would like to study caplet prices, which provide clear predictions of model performance across maturity. Unfortunately, we only observe cap prices. To simplify the empirical analysis, we consider the difference between the prices of caps with the same strike and adjacent maturities, which we refer to as difference caps. Thus, our analysis deals with the sum of the caplets between two neighboring maturities with the same strike. For example, 1.5-year difference caps with a specific strike represent the sum of the 1.25-year and 1.5-year caplets with the same strike.

Due to daily changes in LIBOR rates, difference caps realize different moneyness (defined as the ratio between the strike price and the average LIBOR forward rates underlying the caplets that form the difference cap) each day. Therefore, throughout our analysis, we focus on the prices of difference caps at given fixed moneyness. That is, each day we interpolate difference cap prices with respect to the strike price to obtain prices at fixed moneyness. Specifically, we use local cubic polynomials to preserve the shape of the original curves while smoothing over the grid points. We refrain from extrapolation and interpolation over grid points without nearby observations, and we eliminate all observations that violate various arbitrage restrictions. We also eliminate observations with zero prices, and observations that violate either monotonicity or convexity with respect to the strikes.

Figure 7.1a plots the average Black (1976)-implied volatilities of difference caps across moneyness and maturity, while Fig. 7.1b plots the average implied volatilities

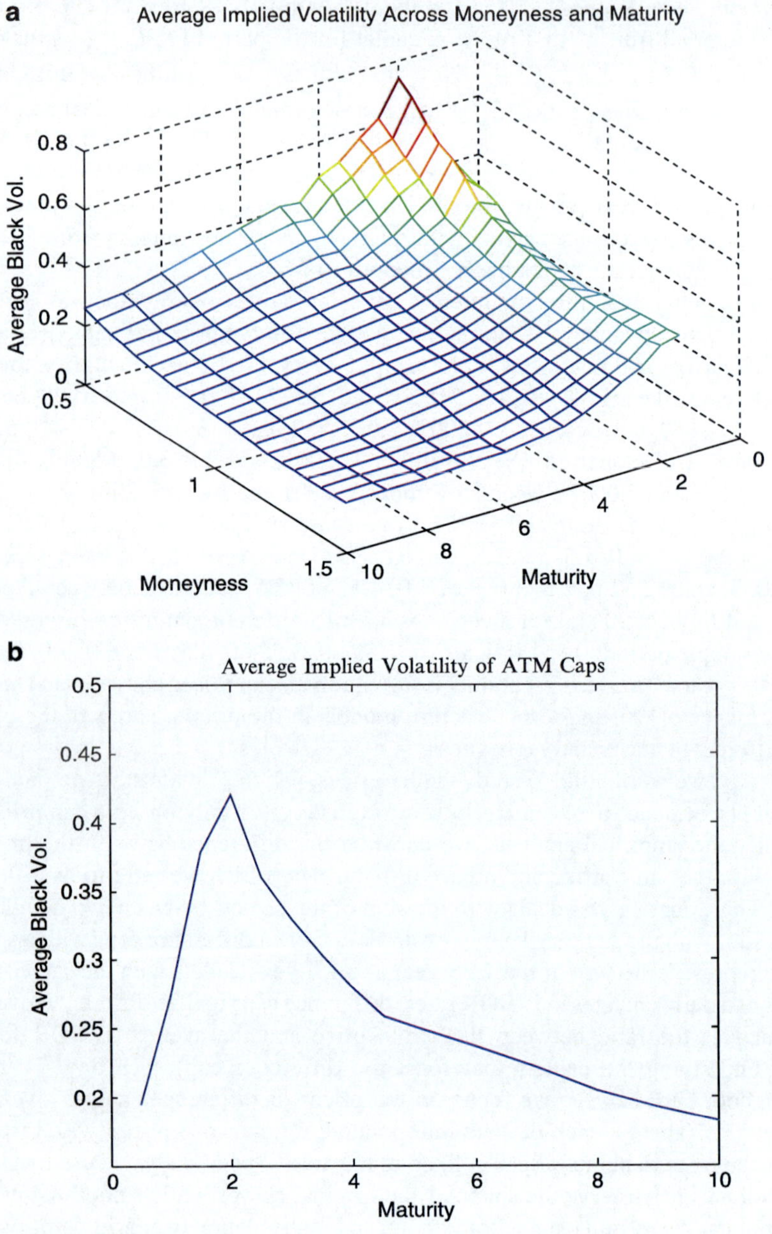

Fig. 7.1 Average black implied volatilities of difference caps between August 1, 2000 and July 26, 2004

of ATM difference caps over the whole sample period. Consistent with the existing literature, the implied volatilities of difference caps with a moneyness between 0.8 and 1.2 have a humped shape with a peak at around a maturity of 2 years. However, the implied volatilities of all other difference caps decline with maturity. There is also a pronounced volatility skew for difference caps at all maturities, with the skew being stronger for short-term difference caps. The pattern is similar to that of equity options: In-the-money (ITM) difference caps have higher implied volatilities than do out-of-the-money (OTM) difference caps. The implied volatilities of the very short-term difference caps are more like a symmetric smile than a skew.

Figure 7.2a–c, respectively, plots the time series of Black-implied volatilities for 2.5-, 5-, and 8-year difference caps across moneyness, while Fig. 7.2d plots the time series of ATM implied volatilities of the three contracts. It is clear that the implied volatilities are time varying and they have increased dramatically (especially for 2.5-year difference caps) over our sample period. As a result of changing interest rates and strike prices, there are more ITM caps in the later part of our sample.

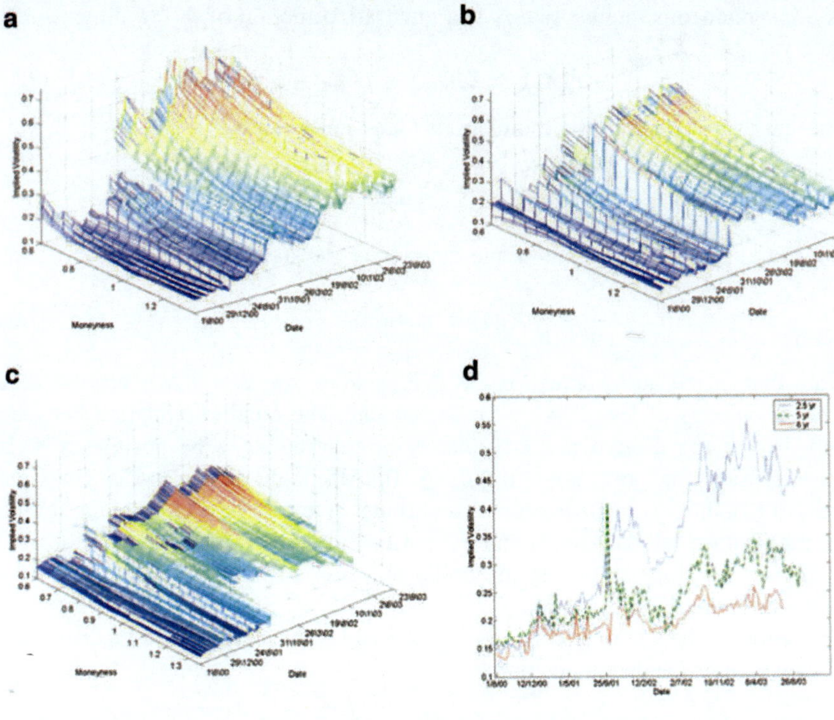

Fig. 7.2 (**a**) Black implied volatilities of 2.5-year difference caps. (**b**) Black implied volatilities of 5-year difference caps. (**c**) Black implied volatilities of 8-year difference caps. (**d**) Black implied volatilities of 2.5-, 5-, and 8-year ATM difference caps

7.3 Pricing and Hedging Interest Rate Caps Under QTSMs

In this section, we present the main results of Li and Zhao (2006), who study the pricing and hedging of interest rate caps under QTSMs. We first discuss the main ingredients of QTSMs and the method for model estimation. Then we discuss the main empirical findings of Li and Zhao (2006).

7.3.1 Model and Estimation

Suppose the economy is represented by a filtered probability space $(\Omega, \mathcal{F}, \{\mathcal{F}_t\}_{0\leq t\leq T}, P)$, where $\{\mathcal{F}_t\}_{0\leq t\leq T}$ is the augmented filtration generated by an N-dimensional standard Brownian motion, W, on this probability space. We assume $\{\mathcal{F}_t\}_{0\leq t\leq T}$ satisfies the usual hypothesis (see Protter 1990). The QTSMs rely on the following assumptions:

- The instantaneous interest rate r_t is a quadratic function of the N-dimensional state variables X_t,

$$r(X_t) = X_t'\Psi X_t + \beta' X_t + \alpha. \tag{7.1}$$

- The state variables follow a multivariate Gaussian process,

$$\mathrm{d}X_t = [\mu + \xi X_t]\,\mathrm{d}t + \Sigma \mathrm{d}W_t. \tag{7.2}$$

- The market price of risk is an affine function of the state variables,

$$\zeta(X_t) = \eta_0 + \eta_1 X_t. \tag{7.3}$$

Note that in the above equations Ψ, ξ, Σ, and η_1 are N-by-N matrices, β, μ and η_0 are vectors of length N and α is a scalar. The quadratic relation between r_t and X_t has the desired property that r_t is guaranteed to be positive if Ψ is positive semidefinite and $\alpha - \frac{1}{4}\beta'\Psi\beta \geq 0$. Although X_t follows a Gaussian process in (2), interest rate r_t exhibits conditional heteroskedasticity because of the quadratic relationship between r_t and X_t. As a result, the QTSMs are more flexible in modeling volatility clustering in bond yields and correlations among the state variables than the ATSMs.

To guarantee the stationarity of the state variables, we assume that ξ permits the following eigenvalue decomposition,

$$\xi = U\Lambda U^{-1},$$

where Λ is the diagonal matrix of the eigenvalues that take negative values, $\Lambda \equiv \mathit{diag}\,[\lambda_i]_N$, and U is the matrix of the eigenvectors of ξ, $U \equiv [u_1\, u_2\, \ldots\, u_N]$. The conditional distribution of the state variables X_t is multivariate Gaussian with conditional mean

$$E\left[X_{t+\Delta t}|X_t\right] = U\Lambda^{-1}\left[\Phi - I_N\right]U^{-1}\mu + U\Lambda^{-1}\left[\Phi - I_N\right]U^{-1}X_t \tag{7.4}$$

and conditional variance

$$\mathrm{var}\left[X_{t+\Delta t}|X_t\right] = U\Theta U', \tag{7.5}$$

where Φ is a diagonal matrix with elements $\exp(\lambda_i \Delta t)$ for $i = 1, \ldots, N$, Θ is a N-by-N matrix with elements $\left[\frac{v_{ij}}{\lambda_i + \lambda_j}\left(e^{\Delta t(\lambda_i+\lambda_j)} - 1\right)\right]$, where $\left[v_{ij}\right]_{N\times N} = U^{-1}\Sigma\Sigma' U'^{-1}$.

With the specification of market price of risk, we can relate the risk-neutral measure Q to the physical one P as follows:

$$E\left[\frac{\mathrm{d}Q}{\mathrm{d}P}|\mathcal{F}_t\right] = \exp\left[-\int_0^t \zeta(X_s)'\mathrm{d}W_s - \frac{1}{2}\int_0^t \zeta(X_s)'\zeta(X_s)\mathrm{d}s\right], \text{ for } t \le T.$$

Applying Girsanov's theorem, we obtain the risk-neutral dynamics of the state variables

$$\mathrm{d}X_t = [\delta + \gamma X_t]\,\mathrm{d}t + \Sigma \mathrm{d}W_t^Q$$

where $\delta = \mu - \Sigma\eta_0$, $\gamma = \xi - \Sigma\eta_1$, and W_t^Q is an N-dimensional standard Brownian motion under measure Q.

Under the above assumptions, a large class of fixed-income securities can be priced in (essentially) closed-form (see Leippold and Wu 2002). We discuss the pricing of zero-coupon bonds below and the pricing of caps in the Appendix. Let $V(t,\tau)$ be the time-t value of a zero-coupon bond that pays 1 dollar at time T $(\tau = T - t)$. In the absence of arbitrage, the discounted value process $\exp\left(-\int_0^t r\,(X_s)\,\mathrm{d}s\right)V(t,\tau)$ is a Q-martingale. Thus the value function must satisfy the fundamental PDE, which requires the bond's instantaneous return equals the risk-free rate,

$$\frac{1}{2}tr\left(\Sigma\Sigma'\frac{\partial^2 V(t,\tau)}{\partial X_t \partial X_t'}\right) + \frac{\partial V(t,\tau)}{\partial X_t'}(\delta + \gamma X_t) + \frac{\partial V(t,\tau)}{\partial t} = r_t V(t,\tau)$$

with the terminal condition $V(t,0) = 1$. The solution takes the form

$$V(t,\tau) = \exp\left[-X_t' A(\tau) X_t - b(\tau)' X_t - c(\tau)\right],$$

where $A(\tau), b(\tau)$ and $c(\tau)$ satisfy the following system of ordinary differential equations (ODEs),

$$\frac{\partial A(\tau)}{\partial \tau} = \Psi + A(\tau)\gamma + \gamma' A(\tau) - 2A(\tau)\Sigma\Sigma' A(\tau); \tag{7.6}$$

$$\frac{\partial b(\tau)}{\partial \tau} = \beta + 2A(\tau)\delta + \gamma' b(\tau) - 2A(\tau)\Sigma\Sigma' b(\tau); \tag{7.7}$$

$$\frac{\partial c(\tau)}{\partial \tau} = \alpha + b(\tau)'\delta - \frac{1}{2}b(\tau)'\Sigma\Sigma' b(\tau) + tr\left[\Sigma\Sigma' A(\tau)\right]; \tag{7.8}$$

with $A(0) = 0_{N\times N}; b(0) = 0_N; c(0) = 0.$

Consequently, the yield-to-maturity, $y(t,\tau)$, is a quadratic function of the state variables

$$y(t,\tau) = \frac{1}{\tau}\left[X_t' A(\tau) X_t + b(\tau)' X_t + c(\tau)\right]. \tag{7.9}$$

In contrast, in the ATSMs the yields are linear in the state variables and therefore the correlations among the yields are solely determined by the correlations of the state variables. Although the state variables in the QTSMs follow multivariate Gaussian process, the quadratic form of the yields helps to model the time varying volatility and correlation of bond yields.

To price and hedge caps in the QTSMs, we need to estimate both model parameters and latent state variables. Due to the quadratic relationship between bond yields and the state variables, the state variables are not identified by the observed yields even in the univariate case in the QTSMs. Previous studies, such as ADG (2002) have used the efficient method of moments (EMM) of Gallant and Tauchen (1996) to estimate the QTSMs. However, in our analysis, we need to estimate not only model parameters, but also the latent state variables. Hence, we choose the extended Kalman filter to estimate model parameters and extract the latent state variables. Duffee and Stanton (2004) show that the extended Kalman filter has excellent finite sample performance in estimating DTSMs. Previous studies that have used the extended Kalman filter in estimating the ATSMs include Duan and Simonato (1995), De Jong and Santa-Clara (1999), and Lund (1997), among many others.

To implement the extended Kalman filter, we first recast the QTSMs into a state-space representation. Suppose we have a time series of observations of yields of L zero-coupon bonds with maturities $\Gamma = (\tau_1, \tau_2, \ldots, \tau_L)$. Let Ξ be the set of parameters for QTSMs, $Y_k = f(X_k, \Gamma; \Xi)$ be the vector of the L observed yields at the discrete time points $k\Delta t$, for $k = 1, 2, \ldots, K$, where Δt is the sample interval (one day in our case). After the following change of variable,

$$Z_k = U^{-1}(\xi^{-1}\mu + X_k),$$

we have the state equation:

$$Z_k = \Phi Z_{k-1} + w_k, \qquad w_k \sim N(0, \Theta), \tag{7.10}$$

where Φ and Θ are first introduced in (4) and (5), and measurement equation:

$$Y_k = d_k + H_k Z_k + v_k, \quad v_k \sim N(0, R^v), \tag{7.11}$$

where the innovations in the state and measurement equations w_k and v_k follow serially independent Gaussian processes and are independent from each other. The time-varying coefficients of the measurement equation d_k and H_k are determined at the ex ante forecast of the state variables,

$$H_k = \frac{\partial f(Uz - \xi^{-1}\mu, \Gamma)}{\partial z} |_{z=Z_{k|k-1}}$$

$$d_k = f(UZ_{k|k-1} - \xi^{-1}\mu, \Gamma) - H_k Z_{k|k-1} + B_k,$$

where $Z_{k|k-1} = \Phi Z_{k-1}$.

In the QTSMs, the transition density of the state variables is multivariate Gaussian under the physical measure. Thus the transition equation in the Kalman filter is exact. The only source of approximation error is due to the linearization of the quadratic measurement equation. As our estimation uses daily data, the approximation error, which is proportional to one-day ahead forecast error, is likely to be minor.[3] The Kalman filter starts with the initial state variable $Z_0 = E(Z_0)$ and the variance-covariance matrix P_0^Z,

$$P_0^Z = E\left[(Z_0 - E(Z_0))\,(Z_0 - E(Z_0))'\right].$$

These unconditional mean and variance have closed-form expressions that can be derived using (4) and (5) by letting Δt goes to infinity. Given the set of filtering parameters, $\{\Xi, R^v\}$, we can write down the log-likelihood of observations based on the Kalman filter

$$\begin{aligned}\log \mathcal{L}(Y;\Xi) &= \sum_{k=1}^{K} \log f(Y_k; \mathcal{Y}_{k-1}, \{\Xi, R^v\}) \\ &= -\frac{LK}{2}\log(2\pi) - \frac{1}{2}\sum_{k=1}^{K}\log\left|P_{k|k-1}^Y\right| \\ &\quad -\frac{1}{2}\sum_{k=1}^{K}\left[\left(Y_k - \hat{Y}_{k|k-1}\right)'\left(P_{k|k-1}^Y\right)^{-1}\left(Y_k - \hat{Y}_{k|k-1}\right)\right],\end{aligned}$$

with $\mathcal{Y}_{k-1}$ is the information set at time $(k-1)\Delta t$, and $P_{k|k-1}^Y$ is the time $(k-1)\Delta t$ conditional variance of Y_k,

$$\begin{aligned}P_{k|k-1}^Y &= H_k P_{k|k-1}^Z H_k' + R^v; \\ P_{k|k-1}^Z &= \Phi P_{k-1}^Z \Phi' + \Theta.\end{aligned}$$

[3]The differences between parameter estimates with and without the correction term are very small and we report those estimates with the correction term B_k.

Parameters are obtained by maximizing the above likelihood function. To avoid local minimum, in our estimation procedure, we use many different starting values and search for the optimal point using simplex method. Then we use gradient-based optimization method to refine those estimates, until they cannot be further improved. This is the standard technique in the literature (see e.g., Duffee 2002).

7.3.2 *Empirical Results*

Now we discuss the main empirical results of Li and Zhao (2006) on the performance of QTSMs in pricing and hedging interest rate caps. Even though the study is based on QTSMs, the empirical findings are common to ATSMs as well.[4] The pricing analysis can reveal two sources of potential model misspecification. One is on the number of factors in the model as a missing factor usually causes large pricing errors. An analogy is using Black-Scholes model while the stock price is generated from a stochastic volatility model. The other is on the assumption of the innovation process of each factor. If the innovation of the factor has a fat-tailed distribution, the convenient assumption of Gaussian distribution is going to deliver large pricing error as well. So from a pricing study, we can not conclude one or the other or both cause large pricing errors. On the other hand, hedging analysis focuses on the changes of the prices, so even if the marginal distribution of the prices can be highly non-Gaussian, the conditional distribution for a small time step can still be reasonably approximated with Gaussian distribution. As the result, a deficiency in hedging, especially at high frequency, reveals more about the potential missing factors than the distribution assumption in a model.

Li and Zhao (2006) show that the QTSMs can capture yield curve dynamics extremely well. First, given the estimated model parameters and state variables, they compute the one day ahead projection of yields based on the estimated model. Figure 7.3 shows that QTSM1 model projected yields are almost indistinguishable from the corresponding observed yields. Secondly, they examine the performance of the QTSMs in hedging zero-coupon bonds assuming that the filtered state variables are traded and use them as hedging instruments. The delta-neutral hedge is conducted for zero-coupon bonds of six maturities on a daily basis. Hedging

---

[4]In the empirical analysis of Li and Zhao (2006), the QTSMs are chosen for several reasons. First, since the nominal spot interest rate is a quadratic function of the state variables, it is guaranteed to be positive in the QTSMs. On the other hand, in the ATSMs, the spot rate, an affine function of the state variables, is guaranteed to be positive only when all the state variables follow square-root processes. Second, the QTSMs do not have the limitations facing the ATSMs in simultaneously fitting interest rate volatility and correlations among the state variables. That is, in the ATSMs, increasing the number of factors that follow square-root processes improves the modeling of volatility clustering in bond yields, but reduces the flexibility in modeling correlations among the state variables. Third, the QTSMs have the potential to capture observed nonlinearity in term structure data (see e.g., Ahn and Gao 1999). Indeed, ADG (2002) and Brandt and Chapman (2002) show that the QTSMs can capture the conditional volatility of bond yields better than the ATSMs.

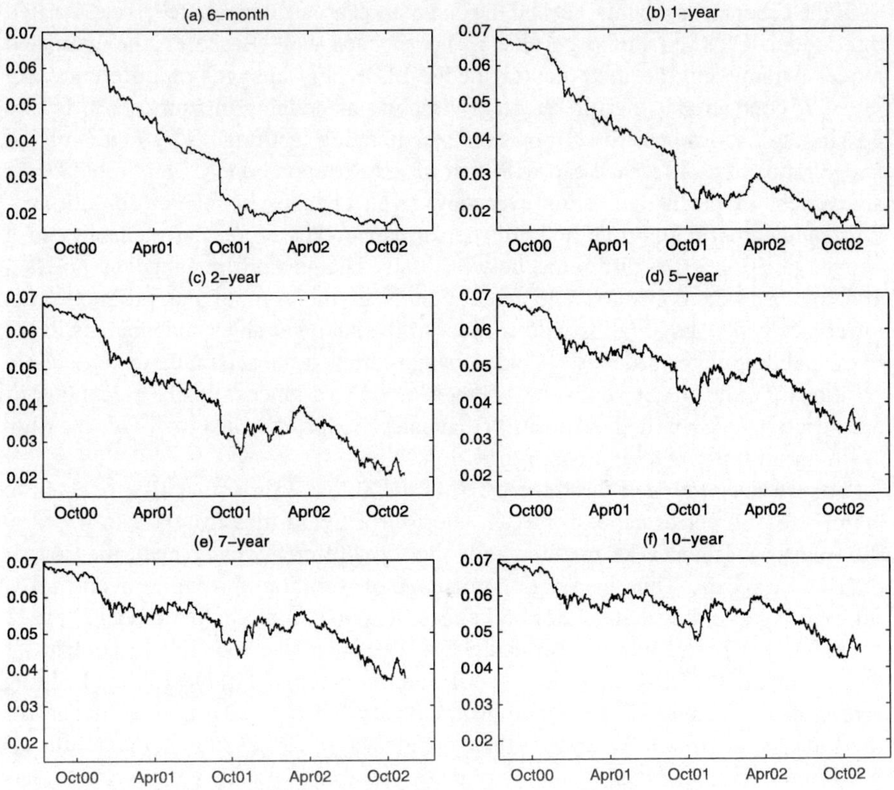

Fig. 7.3 The observed yields (*dot*) and the QTSM1 projected yields (*solid*)

Table 7.1 The performance of QTSMs in modeling bond yields. This table reports the performance of the three-factor QTSMs in capturing bond yields. Variance ratios of model-based hedging of zero-coupon bonds in QTSMs using filtered state variables as hedging instruments. Variance ratio measures the percentage of the variations of an unhedged position that can be reduced through hedging

	Maturity (yr)					
	0.5	1	2	5	7	10
QTSM3	0.717	0.948	0.982	0.98	0.993	0.93
QTSM2	0.99	0.956	0.963	0.975	0.997	0.934
QTSM1	0.994	0.962	0.969	0.976	0.997	0.932

performance is measured by variance ratio, which is defined as the percentage of the variations of an unhedged position that can be reduced by hedging. The results on the hedging performance in Table 7.1 show that in most cases the variance ratios are higher than 95%. This should not be surprising given the excellent fit of bond yields by the QTSMs.

If the Libor and swap market and the cap market are well integrated, the estimated three-factor QTSMs should be able to hedge caps well. Based on the estimated model parameters, the delta-neutral hedge of weekly changes of difference cap prices is conducted using filtered state variables as hedging instruments. It is also possible to use Libor zero-coupon bonds as hedging instruments by matching the hedge ratios of a difference cap with that of zero-coupon bonds. Daily rebalance – adjustment of the hedge ratios everyday given changes in market conditions – is implemented to improve hedging performance. Therefore, daily changes of a hedged position is the difference between daily changes of the unhedged position and the hedging portfolio. The latter equals to the sum of the products of a difference cap's hedge ratios with respect to the state variables and changes in the corresponding state variables. Weekly changes are just the accumulation over daily positions. The hedging effectiveness is measured by variance ratio, the percentage of the variations of an unhedged position that can be reduced by hedging. This measure is similar in spirit to R^2 in linear regression.

The variance ratios of the three QTSMs in Table 7.2 show that all models have better hedging performance for ITM, short-term (maturities from 1.5 to 4 years) difference caps than OTM, medium and long-term difference caps (maturities longer than 4 years) caps. There is a high percentage of variations in long-term and OTM difference cap prices that cannot be hedged. The maximal flexible model QTSM1 again has better hedging performance than the other two models. To control for the fact that the QTSMs may be misspecified, in Panel B of Table 7.2, the hedging errors of each moneyness/maturity group are further regressed on the changes of the three yield factors. While the three yield factors can explain some additional hedging errors, their incremental explanatory power is not very significant. Thus even excluding hedging errors that can be captured by the three yield factors, there is still a large fraction of difference cap prices that cannot be explained by the QTSMs.

Table 7.3 reports the performance of the QTSMs in hedging cap straddles. The difference floor prices are computed from difference cap prices using the put-call parity and construct weekly straddle returns. As straddles are highly sensitive to volatility risk, both delta and gamma neutral hedge is needed. Collin-Dufresne and Goldstein (2002) show that 80% of straddle regression residuals can be explained by one additional factor. Principle component analysis of ATM straddle hedging errors in Panel B of Table 7.3 shows that the first factor can explain about 60% of the total variations of hedging errors. The second and third factor each explains about 10% of hedging errors and two additional factors combined can explain about another 10% of hedging errors. The correlation matrix of the ATM straddle hedging errors across maturities in Panel C shows that the hedging errors of short-term (2, 2.5, 3, 3.5, and 4 years), medium-term (4.5 and 5 years) and long-term (8, 9, and 10 years) straddles are highly correlated within each group, suggesting that there could be multiple unspanned factors.

To further understand whether the unspanned factors are related to stochastic volatility, we study the relationship between ATM cap implied volatilities and straddle hedging errors. Principle component analysis in Panel A of Table 7.4 shows that the first component explains 85% of the variations of cap implied volatilities.

Table 7.2 The performance of QTSMs in hedging interest rate caps. This table reports the performance of the three QTSMs in hedging difference caps. Hedging effectiveness is measured by variance ratio, the percentage of the variations of an unhedged position that can be reduced through hedging. The bold entries represent moneyness/maturity groups that have less than 10% of missing values and the rest are the ones with 10–50% of missing values

	Maturity												
Money-ness (K/F)	1.5	2	2.5	3	3.5	4	4.5	5	6	7	8	9	10
Panel A. Variance ratio of QTSM1													
0.60	–	–	–	–	0.917	–	–	–	0.862	0.679	0.665	0.494	0.257
0.70	–	–	0.903	0.913	0.916	**0.862**	**0.666**	0.487	0.822	0.619	0.565	0.355	0.218
0.80	–	0.831	**0.890**	**0.900**	**0.876**	**0.864**	**0.670**	**0.504**	**0.785**	**0.594**	**0.537**	0.305	0.185
0.90	0.890	**0.810**	**0.853**	**0.872**	**0.851**	**0.832**	**0.631**	**0.514**	**0.748**	**0.577**	**0.491**	**0.314**	**0.171**
1.00	**0.875**	**0.677**	**0.803**	**0.824**	**0.826**	**0.815**	**0.575**	**0.456**	**0.695**	**0.533**	**0.476**	**0.287**	**0.164**
1.10	**0.851**	**0.575**	**0.737**	**0.779**	**0.763**	**0.773**	**0.523**	0.411	0.623	0.490	0.415	0.204	–
1.20	0.756	**0.489**	**0.645**	**0.692**	**0.654**	0.673	0.521	0.470	0.533	0.415	–	–	–
1.30	0.724	**0.393**	**0.534**	0.591	0.444	0.602	0.540	–	0.334	–	–	–	–
1.40	–	0.260	0.373	0.464	0.319	–	–	–	–	–	–	–	–
Panel B. Variance ratio of QTSM1 combined with the changes of the three yield factors													
0.60	–	–	–	–	0.921	–	–	–	0.912	0.788	0.815	0.658	0.579
0.70	–	–	0.927	0.922	0.919	**0.872**	**0.675**	0.507	0.847	0.679	0.666	0.462	0.353
0.80	–	0.886	**0.915**	**0.912**	**0.882**	**0.873**	**0.675**	**0.510**	**0.811**	**0.653**	**0.639**	0.417	0.324
0.90	0.942	**0.853**	**0.876**	**0.882**	**0.855**	**0.837**	**0.633**	**0.524**	**0.776**	**0.632**	**0.573**	**0.412**	**0.310**
1.00	**0.923**	**0.746**	**0.841**	**0.836**	**0.839**	**0.820**	**0.578**	**0.462**	**0.709**	**0.566**	**0.541**	**0.361**	**0.269**
1.10	**0.895**	**0.659**	**0.799**	**0.804**	**0.794**	**0.786**	**0.530**	0.421	0.630	0.508	0.480	0.278	–
1.20	0.848	**0.611**	**0.741**	**0.746**	**0.729**	0.701	0.530	0.486	0.572	0.427	–	–	–
1.30	0.796	**0.560**	**0.680**	0.687	0.626	0.679	0.559	–	0.378	–	–	–	–
1.40	–	0.455	0.582	0.638	0.573	–	–	–	–	–	–	–	–

Table 7.3 Hedging interest rate cap straddles

Panel A. The performance of QTSM1 in hedging difference cap straddles measured by variance ratio

	Maturity												
(K/F)	1.5	2	2.5	3	3.5	4	4.5	5	6	7	8	9	10
0.60	–	–	–	–	0.711	–	–	–	0.709	0.329	0.596	0.362	0.250
0.70	–	–	0.776	0.723	0.674	**0.557**	**0.250**	0.152	0.473	0.206	0.187	0.096	0.074
0.80	–	0.560	**0.652**	**0.615**	**0.437**	**0.488**	**0.179**	**0.093**	**0.293**	**0.126**	**0.113**	0.053	0.070
0.90	0.558	**0.278**	**0.405**	**0.339**	**0.248**	0.265	**0.066**	**0.049**	**0.138**	**0.052**	**0.032**	**0.016**	**0.060**
1.00	**0.364**	**0.081**	**0.210**	**0.142**	**0.149**	0.142	**0.024**	**0.006**	**0.045**	**0.006**	**0.047**	**0.009**	**0.002**
1.10	**0.622**	**0.212**	**0.368**	**0.226**	**0.314**	0.283	**0.146**	0.065	0.133	0.054	0.091	0.018	–
1.20	0.788	**0.527**	**0.633**	**0.515**	**0.593**	0.481	0.368	0.201	0.343	0.256	–	–	–
1.30	0.851	**0.727**	**0.808**	0.728	0.781	0.729	0.525	–	0.454	–	–	–	–
1.40	–	0.817	0.894	0.863	0.880	–	–	–	–	–	–	–	–

Panel B. Percentage of variance of ATM straddles hedging errors explained by the principle components

Principle component					
1	2	3	4	5	6
59.3%	12.4%	9.4%	6.7%	4.0%	2.8%

Panel C. Correlation matrix of ATM straddles hedging errors across maturity

	Maturity												
Maturity	1.5	2	2.5	3	3.5	4	4.5	5	6	7	8	9	10
1.5	1.00	–	–	–	–	–	–	–	–	–	–	–	–
2	0.38	1.00	–	–	–	–	–	–	–	–	–	–	–
2.5	0.28	0.66	1.00	–	–	–	–	–	–	–	–	–	–
3	0.03	0.33	0.73	1.00	–	–	–	–	–	–	–	–	–
3.5	0.27	0.52	0.63	0.59	1.00	–	–	–	–	–	–	–	–
4	0.13	0.44	0.37	0.37	0.77	1.00	–	–	–	–	–	–	–
4.5	0.20	0.21	−0.04	−0.08	−0.05	−0.06	1.00	–	–	–	–	–	–
5	0.10	0.11	−0.12	−0.13	−0.16	−0.15	0.96	1.00	–	–	–	–	–
6	0.21	0.16	0.19	0.13	0.25	0.05	0.27	0.23	1.00	–	–	–	–
7	0.30	0.34	0.33	0.35	0.46	0.38	0.28	0.22	0.08	1.00	–	–	–
8	0.10	0.12	0.30	0.30	0.25	0.11	0.36	0.34	0.29	0.29	1.00	–	–
9	0.14	0.11	0.25	0.29	0.26	0.12	0.39	0.37	0.32	0.38	0.83	1.00	–
10	0.08	−0.01	0.17	0.14	0.12	0.01	0.32	0.35	0.26	0.28	0.77	0.86	1.00

Table 7.4 Straddle hedging errors and cap implied volatilities. This table reports the relation between straddle hedging errors and ATM Cap implied volatilities

Panel A. Percentage of variance of ATM Cap implied volatilities explained by the principle components

Principle component					
1	2	3	4	5	6
85.73%	7.91%	1.85%	1.54%	0.72%	0.67%

Panel B. R^2s of the regressions of ATM straddles hedging errors on changes of the three yield factors (row one); changes of the three yield factors and the first four principle components of the ATM Cap implied volatilities (row two); and changes of the three yield factors and maturity-wise ATM Cap implied volatility (row three)

Maturity												
1.5	2	2.5	3	3.5	4	4.5	5	6	7	8	9	10
0.10	0.06	0.02	0.01	0.01	0.04	0.00	0.00	0.01	0.01	0.00	0.01	0.04
0.29	0.49	0.54	0.43	0.63	0.47	0.95	0.96	0.21	0.70	0.68	0.89	0.96
0.68	0.70	0.81	0.87	0.85	0.90	0.95	0.98	0.95	0.98	0.97	0.98	0.99

In Panel B, we regress straddle hedging errors on changes of the three yield factors and obtain R^2s that are close to zero. However, if we include the weekly changes of the first few principle components of cap implied volatilities, the R^2s increase significantly: for some maturities, the R^2s are above 90%. Although the time series of implied volatilities are very persistent, their differences are not and we do not suffer from the well-known problem of spurious regression. In the extreme case in which we regress straddle hedging errors of each maturity on changes of the yield factors and cap implied volatilities with the same maturity, the R^2s in most cases are above 90%. These results show that straddle returns are mainly affected by volatility risk but not term structure factors.

As ATM straddles are mainly exposed to volatility risk, their hedging errors can serve as a proxy of the USV. Panels A and B of Table 7.5 report the R^2s of regressions of hedging errors of difference caps and cap straddles across moneyness and maturity on changes of the three yield factors and the first five principle components of straddle hedging errors. In contrast to the regressions in Panel B of Table 2, which only include the three yield factors, the additional factors from straddle hedging errors significantly improve the R^2s of the regressions: for most moneyness/maturity groups, the R^2s are above 90%. Interestingly for long-term caps, the R^2s of ATM and OTM caps are actually higher than that of ITM caps. Therefore, a combination of the yield factors and the USV factors can explain cap prices across moneyness and maturity very well.

While the above analysis is mainly based on the QTSMs, the evidence on USV is so compelling that the results should be robust to potential model misspecification. The fact that the QTSMs provide excellent fit of bond yields but can explain only a small percentage of the variations of ATM straddle returns is a strong indication that the models miss some risk factors that are important for the cap market. While we estimate the QTSMs using only bond prices, we could also include cap prices in model estimation. We do not choose the second approach for several reasons. First, the current approach is consistent with the main objective of our study to use risk factors extracted from the swap market to explain cap prices. Second, it is not clear that modifications of model parameters without changing the fundamental structure of the model could remedy the poor cross-sectional hedging performance of the QTSMs. In fact, if the QTSMs indeed miss some important factors, then no matter how they are estimated (using bonds or derivatives data), they are unlikely to have good hedging performance. Finally, Jagannathan et al. (2003) do not find significant differences between parameters of ATSMs estimated using LIBOR/swap rates and cap/swaption prices. The existence of USV strongly suggests that existing DTSMs need to relax their fundamental assumption that derivatives are redundant securities by explicitly incorporating USV factors. Therefore, the DTSMs in their current form may not be directly applicable to derivatives because they all rely on the fundamental assumption that bonds and derivatives are driven by the same set of risk factors.

Table 7.5 ATM Straddle Hedging Error As a Proxy of Systematic USV. This table reports the contribution of USV proxied by the first few principle components of ATM straddle hedging errors in explaining the hedging errors of caps and cap straddles across moneyness and maturity. It reports the R^2s of regressions of hedging errors of caps across moneyness and maturity on changes of the three yield factors and the first five principle components of straddle hedging errors. The bold entries represent moneyness/maturity groups that have less than 10% of missing values and the rest are the ones with 10–50% of missing values

Money-ness	Maturity												
(K/F)	1.5	2	2.5	3	3.5	4	4.5	5	6	7	8	9	10
Panel A: R^2s of regressions of cap hedging errors													
0.60	–	–	–	–	0.945	–	–	–	0.948	0.880	0.884	0.786	0.880
0.70	–	–	0.934	0.944	0.943	**0.911**	**0.934**	0.936	0.940	0.885	0.839	0.791	0.776
0.80	–	0.917	**0.934**	**0.938**	**0.935**	**0.909**	**0.950**	**0.946**	**0.951**	**0.898**	**0.862**	0.821	0.840
0.90	0.949	**0.900**	**0.908**	**0.922**	**0.924**	**0.896**	**0.961**	**0.969**	**0.969**	**0.920**	**0.871**	**0.856**	**0.871**
1.00	**0.932**	**0.859**	**0.905**	**0.909**	**0.939**	**0.902**	**0.988**	**0.989**	**0.984**	**0.973**	**0.910**	**0.894**	**0.907**
1.10	**0.913**	**0.793**	**0.890**	**0.894**	**0.928**	**0.880**	**0.979**	0.976	0.974	0.967	0.913	0.921	–
1.20	0.881	**0.749**	**0.844**	**0.848**	**0.894**	0.846	0.966	0.963	0.954	0.957	–	–	–
1.30	0.861	**0.702**	**0.802**	0.808	0.836	0.802	0.920	–	0.908	–	–	–	–
1.40	–	0.640	0.725	0.743	0.761	0.536	–	–	–	–	–	–	–
Panel B: R^2s of regressions of cap straddle hedging errors													
0.60	–	–	–	–	0.851	–	–	–	0.839	0.589	0.724	0.712	0.744
0.70	–	–	0.874	0.897	0.825	**0.775**	**0.854**	0.886	0.800	0.621	0.688	0.749	0.730
0.80	–	0.812	**0.839**	**0.828**	**0.800**	**0.734**	**0.872**	**0.886**	**0.751**	**0.607**	**0.740**	0.805	0.805
0.90	0.883	**0.718**	**0.801**	**0.791**	**0.763**	**0.690**	**0.919**	**0.930**	**0.721**	**0.622**	**0.777**	**0.859**	**0.840**
1.00	**0.901**	**0.745**	**0.861**	**0.822**	**0.851**	**0.745**	**0.971**	**0.976**	**0.738**	**0.703**	**0.849**	**0.922**	**0.894**
1.10	**0.896**	**0.757**	**0.883**	**0.769**	**0.880**	**0.735**	**0.953**	0.960	0.758	0.788	0.892	0.931	–
1.20	0.920	**0.831**	**0.883**	**0.785**	**0.890**	0.808	0.947	0.935	0.809	0.843	–	–	–
1.30	0.945	**0.890**	**0.913**	0.859	0.921	0.872	0.920	–	0.830	–	–	–	–
1.40	–	0.924	0.941	0.912	0.946	–	–	–	–	–	–	–	–

7.4 LIBOR Market Models with Stochastic Volatility and Jumps: Theory and Evidence

The existence of USV factors suggests that it might be more convenient to consider derivative pricing in the forward rate models of HJM (1992) or the random field models of Goldstein (2000) and Santa-Clara and Sornette (2001) because it is generally very difficult to introduce USV in DTSMs. For example, Collin-Dufresne and Goldstein (2002) show that highly restrictive assumptions on model parameters need to be imposed to guarantee that some state variables that are important for derivative pricing do not affect bond prices. In contrast, they show that it is much easier to introduce USV in the HJM and random field class of models: Any HJM or random field model in which the forward rate has a stochastic volatility exhibits USV. While it has always been argued that HJM and random field models are more appropriate for pricing derivatives than DTSMs, the reasoning given here is quite different. That is, in addition to the commonly known advantages of these models (such as they can perfectly fit the initial yield curve while DTSMs generally cannot), another advantage of HJM and random field models is that they can easily accommodate USV (see Collin-Dufresne and Goldstein 2002 for illustration).

In this section, we discuss the multifactor HJM models with stochastic volatility and jumps in LIBOR forward rates developed in Jarrow et al. (2007) and their performance in capturing the volatility smile in interest rate cap markets. Instead of modeling the unobservable instantaneous forward rates as in standard HJM models, we focus on the LIBOR forward rates which are observable and widely used in the market.

7.4.1 *Model and Estimation*

Throughout our analysis, we restrict the cap maturity T to a finite set of dates $0 = T_0 < T_1 < \ldots < T_K < T_{K+1}$, and assume that the intervals $T_{k+1} - T_k$ are equally spaced by δ, a quarter of a year. Let $L_k(t) = L(t, T_k)$ be the LIBOR forward rate for the actual period $[T_k, T_{k+1}]$, and similarly let $D_k(t) = D(t, T_k)$ be the price of a zero-coupon bond maturing on T_k. Thus, we have

$$L(t, T_k) = \frac{1}{\delta}\left(\frac{D(t, T_k)}{D\,(t, T_{k+1})} - 1\right), \quad \text{for } k = 1, 2, \ldots, K. \tag{7.12}$$

For LIBOR-based instruments, such as caps, floors and swaptions, it is convenient to consider pricing under the forward measure. Thus, we will focus on the dynamics of the LIBOR forward rates $L_k(t)$ under the forward measure $\mathbb{Q}^{k+1}$, which is essential for pricing caplets maturing at T_{k+1}. Under this measure, the discounted price of any security using $D_{k+1}(t)$ as the numeraire is a martingale. Therefore, the time t price of a caplet maturing at T_{k+1} with a strike price of X is

$$Caplet\,(t, T_{k+1}, X) = \delta D_{k+1}(t) E_t^{\mathbb{Q}^{k+1}} \left[(L_k(T_k) - X)^+ \right],$$

where $E_t^{\mathbb{Q}^{k+1}}$ is taken with respect to $\mathbb{Q}^{k+1}$ given the information set at t. The key to valuation is modeling the evolution of $L_k(t)$ under $\mathbb{Q}^{k+1}$ realistically and yet parsimoniously to yield closed-form pricing formula. To achieve this goal, we rely on the flexible AJDs of Duffie et al. (2000) to model the evolution of LIBOR rates.

We assume that under the physical measure $\mathbb{P}$, the dynamics of LIBOR rates are given by the following system of SDEs, for $t \in [0, T_k)$ and $k = 1, \ldots, K$,

$$\frac{\mathrm{d}L_k(t)}{L_k(t)} = \alpha_k(t)\mathrm{d}t + \sigma_k(t)\mathrm{d}Z_k(t) + \mathrm{d}J_k(t), \tag{7.13}$$

where $\alpha_k(t)$ is an unspecified drift term, $Z_k(t)$ is the k-th element of a Kdimensional correlated Brownian motion with a covariance matrix $\Psi(t)$, and $J_k(t)$ is the k-th element of a Kdimensional independent pure jump process assumed independent of $Z_k(t)$ for all k. To introduce stochastic volatility and correlation, we could allow the volatility of each LIBOR rate $\sigma_k(t)$ and each individual element of $\Psi(t)$ to follow a stochastic process. But, such a model is unnecessarily complicated and difficult to implement. Instead, we consider a low dimensional model based on the first few principal components of historical LIBOR forward rates. We assume that the entire LIBOR forward curve is driven by a small number of factors $N < K$ ($N \leq 3$ in our empirical analysis). By focusing on the first N principal components of historical LIBOR rates, we can reduce the dimension of the model from K to N.

Following LSS (2001) and Han (2007), we assume that the instantaneous covariance matrix of changes in LIBOR rates share the same eigenvectors as the historical covariance matrix. Suppose that the historical covariance matrix can be approximated as $H = U\Lambda_0 U'$, where Λ_0 is a diagonal matrix whose diagonal elements are the first N largest eigenvalues in descending order, and the N columns of U are the corresponding eigenvectors.[5] Our assumption means that the instantaneous covariance matrix of changes in LIBOR rates with fixed time-to-maturity, Ω_t, share the same eigenvectors as H. That is

$$\Omega_t = U\Lambda_t U', \tag{7.14}$$

where Λ_t is a diagonal matrix whose i-th diagonal element, denoted by $V_i(t)$, can be interpreted as the instantaneous variance of the i-th common factor driving the yield curve evolution at t. We assume that $V(t)$ follows the square-root process that

[5] We acknowledge that with jumps in LIBOR rates, both the historical and instantaneous covariance matrix of LIBOR rates contain a component that is due to jumps. Our approach implicitly assumes that the first three principal components from the historical covariance matrix captures the variations in LIBOR rates due to continuous shocks and that the impact of jumps is only contained in the residuals.

has been widely used in the literature for modeling stochastic volatility (see, e.g., Heston 1993):

$$dV_i(t) = \kappa_i\,(\bar{v}_i - V_i(t))\,dt + \xi_i\,\sqrt{V_i(t)}d\widetilde{W}_i(t) \tag{7.15}$$

where $\widetilde{W}_i(t)$ is the i-th element of an N-dimensional independent Brownian motion assumed independent of $Z_k(t)$ and $J_k(t)$ for all k.[6]

While (14) and (15) specify the instantaneous covariance matrix of LIBOR rates with fixed time-to-maturity, in applications we need the instantaneous covariance matrix of LIBOR rates with fixed maturities Σ_t. At $t = 0$, Σ_t coincides with Ω_t; for $t > 0$, we obtain Σ_t from Ω_t through interpolation. Specifically, we assume that $U_{s,j}$ is piecewise constant,[7] i.e., for time to maturity $s \in (T_k, T_{k+1})$,

$$U_s^2 = \frac{1}{2}\left(U_k^2 + U_{k+1}^2\right). \tag{7.16}$$

We further assume that $U_{s,j}$ is constant for all caplets belonging to the same *difference cap*. For the family of the LIBOR rates with maturities $T = T_1, T_2, \ldots T_K$, we denote U_{T-t} the time-t matrix that consists of rows of U_{T_k-t}, and therefore we have the time-t covariance matrix of the LIBOR rates with fixed maturities,

$$\Sigma_t = U_{T-t}\Lambda_t U'_{T-t}. \tag{7.17}$$

To stay within the family of AJDs, we assume that the random jump times arrive with a constant intensity λ_J, and conditional on the arrival of a jump, the jump size follows a normal distribution $N(\mu_J, \sigma_J^2)$. Intuitively, the conditional probability at time t of another jump within the next small time interval Δt is $\lambda_J \Delta t$ and, conditional on a jump event, the mean relative jump size is $\mu = \exp(\mu_J + \frac{1}{2}\sigma_J^2) - 1$.[8] We also assume that the shocks driving LIBOR rates, volatility, and jumps (both jump time and size) are mutually independent from each other.

Given the above assumptions, we have the following dynamics of LIBOR rates under the physical measure $\mathbb{P}$,

[6]Many empirical studies on interest rate dynamics have shown that correlation between stochastic volatility and interest rates is close to zero. That is, there is not a strong "leverage" effect for interest rates as for stock prices. The independence assumption between stochastic volatility and LIBOR rates in our model captures this stylized fact.

[7]Our interpolation scheme is slightly different from that of Han (2007) for the convenience of deriving closed-form solution for cap prices.

[8]For simplicity, we assume that different forward rates follow the same jump process with constant jump intensity. It is not difficult to allow different jump processes for individual LIBOR rates and the jump intensity to depend on the state of the economy within the AJD framework.

$$\frac{dL_k(t)}{L_k(t)} = \alpha_k(t)dt + \sum_{j=1}^{N} U_{T_{k-t},j}\sqrt{V_j(t)}dW_j(t) + dJ_k(t), k = 1, 2, \ldots, K. \quad (7.18)$$

To price caps, we need the dynamics of LIBOR rates under the appropriate forward measure. The existence of stochastic volatility and jumps results in an incomplete market and hence the non-uniqueness of forward martingale measures. Our approach for eliminating this nonuniqueness is to specify the market prices of both the volatility and jump risks to change from the physical measure $\mathbb{P}$ to the forward measure $\mathbb{Q}^{k+1}$.[9] Following the existing literature, we model the volatility risk premium as $\eta_j^{k+1}\sqrt{V_j(t)}$, for $j = 1, \ldots, N$. For the jump risk premium, we assume that under the forward measure $\mathbb{Q}^{k+1}$, the jump process has the same distribution as that under P, except that the jump size follows a normal distribution with mean μ_J^{k+1} and variance σ_J^2. Thus, the mean relative jump size under $\mathbb{Q}^{k+1}$ is $\mu^{k+1} = \exp\left(\mu_J^{k+1} + \frac{1}{2}\sigma_J^2\right) - 1$. Our specification of the market prices of jump risks allows the mean relative jump size under $\mathbb{Q}^{k+1}$ to be different from that under $\mathbb{P}$, accommodating a premium for jump size uncertainty. This approach, which is also adopted by Pan (2002), artificially absorbs the risk premium associated with the timing of the jump by the jump size risk premium. In our empirical analysis, we make the simplifying assumption that the volatility and jump risk premiums are linear functions of time-to-maturity, i.e., $\eta_j^{k+1} = c_{jv}(T_k - 1)$ and $\mu_J^{k+1} = \mu_J + c_J(T_k - 1)$.[10] Due to the no arbitrage restriction, the risk premiums of shocks to LIBOR rates for different forward measures are intimately related to each other. If shocks to volatility and jumps are also correlated with shocks to LIBOR rates, then both volatility and jump risk premiums for different forward measures should also be closely related to each other. However, in our model shocks to LIBOR rates are independent of that to volatility and jumps, and as a result, the change of measure of LIBOR shocks does not affect that of volatility and jump shocks. Due to stochastic volatility and jumps, the underlying LIBOR market is no longer complete and there is no unique forward measure. This gives us the freedom to choose the functional forms of η_j^{k+1} and μ_J^{k+1}. See Andersen and Brotherton-Ratcliffe (2001) for similar discussions.

Given the above market prices of risks, we can write down the dynamics of $\log(L_k(t))$ under forward measure $\mathbb{Q}^{k+1}$,

[9]The market prices of interest rate risks are defined in such a way that the LIBOR rate is a martingale under the forward measure.

[10]In order to estimate the volatility and jump risk premiums, we need a joint analysis of the dynamics of LIBOR rates under both the physical and forward measure as in Pan (2002), and Eraker (2004). In our empirical analysis, we only focus on the dynamics under the forward measure. Therefore, we can only identify the differences in the risk premiums between forward measures with different maturities. Our specifications of both risk premiums implicitly use the 1-year LIBOR rate as a reference point.

$$\mathrm{d}\log(L_k(t)) = -\left(\lambda_J \mu^{k+1} + \frac{1}{2}\sum_{j=1}^{N} U_{T_k-t,j}^2 V_j(t)\right)\mathrm{d}t + \sum_{j=1}^{N} U_{T_k-t,j}\sqrt{V_j(t)}\mathrm{d}W_j^{\mathbb{Q}^{k+1}}(t) + \mathrm{d}J_k^{\mathbb{Q}^{k+1}}(t). \quad (7.19)$$

For pricing purpose, the above process can be further simplified to the following one which has the same distribution,

$$\mathrm{d}\log(L_k(t)) = -\left(\lambda_J \mu^{k+1} + \frac{1}{2}\sum_{j=1}^{N} U_{T_k-t,j}^2 V_j(t)\right)\mathrm{d}t + \sqrt{\sum_{j=1}^{N} U_{T_k-t,j}^2 V_j(t)}\mathrm{d}Z_k^{\mathbb{Q}^{k+1}}(t) + \mathrm{d}J_k^{\mathbb{Q}^{k+1}}(t), \quad (7.20)$$

where $Z_k^{\mathbb{Q}^{k+1}}(t)$ is a standard Brownian motion under $\mathbb{Q}^{k+1}$. Now the dynamics of $V_i(t)$ under $\mathbb{Q}^{k+1}$ becomes

$$\mathrm{d}V_i(t) = \kappa_i^{k+1}\left(\bar{v}_i^{k+1} - V_i(t)\right)\mathrm{d}t + \xi_i\sqrt{V_i(t)}\mathrm{d}\widetilde{W}_i^{\mathbb{Q}^{k+1}}(t) \quad (7.21)$$

where $\widetilde{W}^{\mathbb{Q}^{k+1}}$is independent of $Z^{\mathbb{Q}^{k+1}}$, $\kappa_j^{k+1} = \kappa_j - \xi_j \eta_j^{k+1}$, and $\bar{v}_j^{k+1} = \frac{\kappa_j \bar{v}_j}{\kappa_j - \xi_j \eta_j^{k+1}}$, $j = 1, \ldots, N$. The dynamics of $L_k(t)$ under the forward measure $\mathbb{Q}^{k+1}$ are completely captured by (20) and (21).

Given that LIBOR rates follow AJDs under both the physical and forward measure, we can directly apply the transform analysis of Duffie et al. (2000) to derive closed-form formula for cap prices. Denote the state variables at t as $Y_t = (\log(L_k(t)), V_t)'$ and the time-t expectation of $e^{u \cdot Y_{T_k}}$ under the forward measure $\mathbb{Q}^{k+1}$ as $\psi(u, Y_t, t, T_k) \triangleq E_t^{\mathbb{Q}^{k+1}}[e^{u \cdot Y_{T_k}}]$. Let $u = (u_0, 0_{1\times N})'$, then the time-$t$ expectation of LIBOR rate at T_k equals,

$$E_t^{\mathbb{Q}^{k+1}}\{\exp\left[u_0 \log\left(L_k(T_k)\right)\right]\} = \psi\left(u_0, Y_t, t, T_k\right) \quad (7.22)$$
$$= \exp\left[a(s) + u_0 \log(L_k(t)) + B(s)'V_t\right], \quad (7.23)$$

where $s = T_k - t$ and closed-form solutions of $a(s)$ and $B(s)$ (an N-by-1 vector) are obtained by solving a system of Ricatti equations in the Appendix.

Following Duffie et al. (2000), we define

$$G_{a,b}(y; Y_t, T_k, \mathbb{Q}^{k+1}) = E_t^{\mathbb{Q}^{k+1}}\left[e^{a \cdot \log(L_k(T_k))} 1_{\{b \cdot \log(L_k(T_k)) \leq y\}}\right], \quad (7.24)$$

and its Fourier transform,

$$\mathcal{G}_{a,b}(v; Y_t, T_k, \mathbb{Q}^{k+1}) = \int_R e^{ivy} \mathrm{d}G_{a,b}(y) \tag{7.25}$$

$$= E_t^{\mathbb{Q}^{k+1}} \left[e^{(a+ivb)\cdot\log(L_k(T_k))} \right]$$

$$= \psi\left(a + ivb, Y_t, t, T_k\right). \tag{7.26}$$

Levy's inversion formula gives

$$G_{a,b}(y; Y_t, T_k, \mathbb{Q}^{k+1}) = \frac{\psi\left(a + ivb, Y_t, t, T_k\right)}{2} - \frac{1}{\pi}\int_0^\infty \frac{\mathrm{Im}\left[\psi\left(a + ivb, Y_t, t, T_k\right) e^{-ivy}\right]}{v} \mathrm{d}v. \tag{7.27}$$

The time-0 price of a caplet that matures at T_{k+1} with a strike price of X equals

$$Caplet(0, T_{k+1}, X) = \delta D_{k+1}(0)\, E_0^{\mathbb{Q}^{k+1}} \left[(L_k(T_k) - X)^+ \right], \tag{7.28}$$

where the expectation is given by the inversion formula,

$$E_0^{\mathbb{Q}^{k+1}} \left[L_k(T_k) - X\right]^+ = G_{1,-1}(-\ln X; Y_0, T_k, \mathbb{Q}^{k+1}) \tag{7.29}$$

$$-XG_{0,-1}(-\ln X; Y_0, T_k, \mathbb{Q}^{k+1}). \tag{7.30}$$

The new models developed in this section nest some of the most important models in the literature, such as LSS (2001) (with constant volatility and no jumps) and Han (2007) (with stochastic volatility and no jumps). The closed-form formula for cap prices makes an empirical implementation of our model very convenient and provides some advantages over existing methods. For example, Han (2007) develops approximations of ATM cap and swaption prices using the techniques of Hull and White (1987). However, such an approach might not work well for away-from-the-money options. In contrast, our method would work well for all options, which is important for explaining the volatility smile.

In addition to introducing stochastic volatility and jumps, our multifactor HJM models also has advantages over the standard LIBOR market models of Brace et al. (1997), Miltersen et al. (1997), and their extensions often applied to caps in practice.[11] While our models provide a unified multifactor framework to characterize the evolution of the whole yield curve, the LIBOR market models typically make separate specifications of the dynamics of LIBOR rates with different maturities.

[11] Andersen and Brotherton-Ratcliffe (2001) and Glasserman and Kou (2003) develop LIBOR models with stochastic volatility and jumps, respectively.

As suggested by LSS (2001), the standard LIBOR models are "more appropriately viewed as a collection of different univariate models, where the relationship between the underlying factors is left unspecified." In contrast, the dynamics of LIBOR rates with different maturities under their related forward measures are internally consistent with each other given their dynamics under the physical measure and the market prices of risks. Once our models are estimated using one set of prices, they can be used to price and hedge other fixed-income securities.

We estimate our new market model using prices form a wide cross section of *difference caps* with different strikes and maturities. Every week we observe prices of *difference caps* with ten moneyness and thirteen maturities. However, due to changing interest rates, we do not have enough observations in all moneyness/maturity categories throughout the sample. Thus, we focus on the 53 moneyness/maturity categories that have less than ten percent of missing values over the sample estimation period. The moneyness and maturity of all *difference caps* belong to the following sets $\{0.7, 0.8, 0.9, 1.0, 1.1\}$ and $\{1.5, 2.0, 2.5, 3.0, 3.5, 4.0, 4.5, 5.0, 6.0, 7.0, 8.0, 9.0, 10.0\}$ (unit in years), respectively. The *difference caps* with time-to-maturity less than or equal to 5 years represent portfolios of two caplets, while those with time-to-maturity longer than 5 years represent portfolios of four caplets.

We estimate the model parameters by minimizing the sum of squared percentage pricing errors (SSE) of all relevant *difference caps*.[12] Consider the time series observations $t = 1, \ldots, \mathcal{T}$, of the prices of 53 *difference caps* with moneyness m_i and time-to-maturities τ_i, $i = 1, \ldots, M = 53$.Let θ represent the model parameters which remain constant over the sample period. Let $C(t, m_i, \tau_i)$ be the observed price of a *difference cap* with moneyness m_i and time-to-maturity τ_i and let $\hat{C}(t, \tau_i, m_i, V_t(\theta), \theta)$ denote the corresponding theoretical price under a given model, where $V_t(\theta)$is the model implied instantaneous volatility at t given model parameters θ. For each i and t, denote the percentage pricing error as

$$u_{i,t}(\theta) = \frac{C(t, m_i, \tau_i) - \hat{C}(t, m_i, \tau_i, V_t(\theta), \theta)}{C(t, m_i, \tau_i)}, \tag{7.31}$$

where $V_t(\theta)$ is defined as

$$V_t(\theta) = \arg\min_{\{V_t\}} \sum_{i=1}^{M} \left[\frac{C(t, m_i, \tau_i) - \hat{C}(t, m_i, \tau_i, V_t, \theta)}{C(t, m_i, \tau_i)} \right]^2. \tag{7.32}$$

[12]Due to the wide range of moneyness and maturities of the difference caps involved, there could be significant differences in the prices of difference caps. Using percentage pricing errors helps to mitigate this probelem.

7.4.2 Empirical Results

In this section, we provide empirical evidence on the performance of six different models in capturing the cap volatility smile. The first three models, denoted as SV1, SV2 and SV3, allow one, two, and three principal components to drive the forward rate curve, respectively, each with its own stochastic volatility. The next three models, denoted as SVJ1, SVJ2 and SVJ3, introduce jumps in LIBOR rates in each of the previous SV models. SVJ3 is the most comprehensive model and nests all the others as special cases. We first examine the separate performance of each of the SV and SVJ models, then we compare performance across the two classes of models. The estimation of all models is based on the principal components extracted from historical LIBOR forward rates between June 1997 and July 2000.[13]

The SV models contribute to cap pricing in four important ways. First, the three principal components capture variations in the levels of LIBOR rates caused by innovations in the "level", "slope", and "curvature" factors. Second, the stochastic volatility factors capture the fluctuations in the volatilities of LIBOR rates reflected in the Black implied volatilities of ATM caps.[14] Third, the stochastic volatility factors also introduce fatter tails in LIBOR rate distributions than implied by the log-normal model, which helps capture the volatility smile. Finally, given our model structure, innovations of stochastic volatility factors also affect the covariances between LIBOR rates with different maturities. The first three factors, however, are more important for our applications, because *difference caps* are much less sensitive to time varying correlations than swaptions.[15] Our discussion of the performance of the SV models focuses on the estimates of the model parameters and the latent volatility variables, and the time series and cross-sectional pricing errors of *difference caps*.

A comparison of the parameter estimates of the three SV models in Table 7.6 shows that the "level" factor has the most volatile stochastic volatility, followed, in decreasing order, by the "curvature" and "slope" factor. The long-run mean ($\bar{v}_1$) and volatility of volatility (ξ_1) of the first volatility factor are much bigger than that of the other two factors. This suggests that the fluctuations in the volatilities of LIBOR rates are mainly due to the time varying volatility of the "level" factor. The estimates of the volatility risk premium of the three models are significantly negative, suggesting that the stochastic volatility factors of longer maturity LIBOR rates under the forward measure are less volatile with lower long-run mean and faster speed of mean reversion. This is consistent with the fact that the Black implied volatilities of longer maturity *difference caps* are less volatile than that of short-term *difference caps*.

---

[13]The LIBOR forward curve is constructed from weekly LIBOR and swap rates from Datastream following the bootstrapping procedure of LSS (2001).

[14]Throughout our discussion, volatilities of LIBOR rates refer to market implied volatilities from cap prices and are different from volatilities estimated from historical data.

[15]See Han (2002) for more detailed discussions on the impact of time varying correlations for pricing swaptions.

Table 7.6 Parameter estimates of stochastic volatility models. This table reports parameter estimates and standard errors of the one-, two-, and three-factor stochastic volatility models. The estimates are obtained by minimizing sum of squared percentage pricing errors (SSE) of difference caps in 53 moneyness and maturity categories observed on a weekly frequency from August 1, 2000 to September 23, 2003. The objective functions reported in the table are re-scaled SSEs over the entire sample at the estimated model parameters and are equal to RMSE of difference caps. The volatility risk premium of the ith stochastic volatility factor for forward measure Q^{k+1} is defined as $\eta_i^{k+1} = c_{iv}(T_k - 1)$

	SV1		SV2		SV3	
Parameter	Estimate	Std. err	Estimate	Std. err	Estimate	Std. err
κ_1	0.0179	0.0144	0.0091	0.0111	0.0067	0.0148
κ_2			0.1387	0.0050	0.0052	0.0022
κ_3					0.0072	0.0104
$\bar{v}_1$	1.3727	1.1077	1.7100	2.0704	2.1448	4.7567
$\bar{v}_2$			0.0097	0.0006	0.0344	0.0142
$\bar{v}_3$					0.1305	0.1895
ζ_1	1.0803	0.0105	0.8992	0.0068	0.8489	0.0098
ζ_2			0.0285	0.0050	0.0117	0.0065
ζ_3					0.1365	0.0059
c_{1v}	−0.0022	0.0000	−0.0031	0.0000	−0.0015	0.0000
c_{2v}			−0.0057	0.0010	−0.0007	0.0001
c_{3v}					−0.0095	0.0003
Objective function	0.0834		0.0758		0.0692	

Our parameter estimates are consistent with the volatility variables inferred from the prices of *difference caps*. The volatility of the "level" factor is the highest among the three (although at lower absolute levels in the more sophisticated models). It starts at a low level and steadily increases and stabilizes at a high level in the later part of the sample period. The volatility of the "slope" factor is much lower and relatively stable during the whole sample period. The volatility of the "curvature" factor is generally between that of the first and second factors. The steady increase of the volatility of the "level" factor is consistent with the increase of Black implied volatilities of ATM *difference caps* throughout our sample period. In fact, the correlation between the Black implied volatilities of most *difference caps* and the implied volatility of the "level" factor are higher than 0.8. The correlation between Black implied volatilities and the other two volatility factors is much weaker. The importance of stochastic volatility is obvious: the fluctuations in Black implied volatilities show that a model with constant volatility simply would not be able to capture even the general level of cap prices.

The other aspects of model performance are the time series and cross-sectional pricing errors of *difference caps*. The likelihood ratio tests in Panel A of Table 7.7 overwhelmingly reject SV1 and SV2 in favor of SV2 and SV3, respectively. The Diebold-Mariano statistics in Panel A of Table 7 also show that SV2 and SV3 have significantly smaller SSEs than SV1 and SV2, respectively, suggesting that the more sophisticated SV models improve the pricing of all caps. The time series of RMSEs

Table 7.7 Comparison of the performance of stochastic volatility models. This table reports model comparison based on likelihood ratio and Diebold-Mariano statistics. The total number of observations (both cross sectional and time series), which equals 8,545 over the entire sample, times the difference between the logarithms of the SSEs between two models follows a χ^2distribution asymptotically. We treat implied volatility variables as parameters. Thus the degree of freedom of the χ^2distribution is 168 for the pairs of SV2/SV1 and SV3/SV2, because SV2 and SV3 have four more parameters and 164 additional implied volatility variables than SV1 and SV2, respectively. The 1% critical value of $\chi^2(168)$ is 214. The Diebold-Mariano statistics are calculated according to equation (14) with a lag order q of 40 and follow an asymptotic standard Normal distribution under the null hypothesis of equal pricing errors. A negative statistic means that the more sophisticated model has smaller pricing errors. Bold entries mean that the statistics are significant at the 5% level

Panel A. Likelihood ratio and Diebold-Mariano statistics for overall model performance based on SSEs

Models	D-M Stats	Likelihood ratio stats $\chi^2(168)$
SV2–SV1	−1.931	**1624**
SV3–SV2	**−6.351**	**1557**

Panel B. Diebold-Mariano statistics between SV2 and SV1 for individual difference caps based on squared percentage pricing errors

Moneyness	2yr	3yr	4yr	5yr	6yr	7yr	8yr	9yr	10yr
0.7	–	–	**2.895**	**5.414**	**4.107**	**5.701**	**2.665**	−1.159	−1.299
0.8	–	0.928	1.840	**6.676**	**3.036**	**2.274**	−0.135	−1.796	−1.590
0.9	−1.553	**−2.218**	−1.222	−1.497	0.354	−0.555	−1.320	−1.439	−1.581
1.0	**−5.068**	−1.427	−1.676	**−3.479**	**−2.120**	−1.734	−1.523	−0.133	**−2.016**
1.1	**−4.347**	0.086	**−3.134**	**−3.966**	–	–	–	–	–

Panel C. Diebold-Mariano statistics between SV3 and SV2 for individual difference caps based on squared percentage pricing errors

Moneyness	2yr	3yr	4yr	5yr	6yr	7yr	8yr	9yr	10yr
0.7	–	–	1.379	−0.840	**−3.284**	**−5.867**	**−4.280**	−0.057	**−2.236**
0.8	–	−1.212	1.682	−0.592	**−3.204**	**−6.948**	**−4.703**	1.437	−1.079
0.9	**−2.897**	**−3.211**	1.196	−1.570	−1.932	**−6.920**	−1.230	**−2.036**	−1.020
1.0	**−3.020**	−0.122	**−3.288**	**−3.103**	1.351	1.338	0.139	**−4.170**	−0.193
1.1	**−2.861**	0.315	**−3.523**	**−2.853**	–	–	–	–	–

of the three SV models over our sample period[16] suggest that except for two special periods where all models have extremely large pricing errors, the RMSEs of all models are rather uniform with the best model (SV3) having RMSEs slightly above 5%. The two special periods with high pricing errors cover the period between the second half of December of 2000 and the first half of January of 2001, and the first half of October 2001, and coincide with high prepayments in mortgage-backed securities (MBS). Indeed, the MBAA refinancing index and prepayment speed (see Fig. 7.3 of Duarte 2004) show that after a long period of low prepayments between the middle of 1999 and late 2000, prepayments dramatically increased at the end of 2000 and the beginning of 2001. There is also a dramatic increase of prepayments at the beginning of October 2001. As widely recognized in the fixed-income market,[17] excessive hedging demands for prepayment risk using interest rate derivatives may push derivative prices away from their equilibrium values, which could explain the failure of our models during these two special periods.[18]

In addition to overall model performance as measured by SSEs, we also examine the cross-sectional pricing errors of *difference caps* with different moneyness and maturities. We first look at the squared percentage pricing errors, which measure both the bias and variability of the pricing errors. Then we look at the average percentage pricing errors (the difference between market and model prices divided by the market price) to see whether SV models can on average capture the volatility smile in the cap market.

The Diebold-Mariano statistics of squared percentage pricing errors of individual *difference caps* between SV2 and SV1 in Panel B of Table 7 show that SV2 reduces the pricing errors of SV1 for some but not all *difference caps*. SV2 has the most significant reductions in pricing errors of SV1 for mid- and short-term around-the-money *difference caps*. On the other hand, SV2 has larger pricing errors for deep ITM *difference caps*. The Diebold-Mariano statistics between SV3 and SV2 in Panel C of Table 7 show that SV3 significantly reduces the pricing errors of many short- (2–3 years) and mid-term around-the-money, and long-term (6–10 years) ITM *difference caps*.

Table 7.8 reports the average percentage pricing errors of all *difference caps* under the three SV models. Panel A of Table 7.8 shows that, on average, SV1 underprices short-term and overprices mid- and long-term ATM *difference caps*, and underprices ITM and overprices OTM *difference caps*. This suggests that SV1 cannot generate enough skewness in the implied volatilities to be consistent with the

[16]RMSE of a model at t is calculated as $\sqrt{u_t'\left(\hat{\theta}\right) u_t\left(\hat{\theta}\right)/M}$. We plot RMSEs instead of SSEs because the former provides a more direct measure of average percentage pricing errors of *difference caps*.

[17]We would like to thank Pierre Grellet Aumont from Deutsche Bank for his helpful discussions on the influence of MBS markets on OTC interest rate derivatives.

[18]While the prepayments rates were also high in later part of 2002 and for most of 2003, they might not have come as surprises to participants in the MBS markets given the two previous special periods.

Table 7.8 Average percentage pricing errors of stochastic volatility models. This table reports average percentage pricing errors of difference caps with different moneyness and maturities of the three stochastic volatility models. Average percentage pricing errors are defined as the difference between market price and model price divided by the market price

Moneyness	2yr	3yr	4yr	5yr	6yr	7yr	8yr	9yr	10yr
Panel A. Average percentage pricing errors of SV1									
0.7	–	–	0.0258	0.0339	0.0361	0.0503	0.0344	0.0297	0.0402
0.8	–	0.0412	0.018	0.0332	0.0322	0.0468	0.0299	0.0244	0.0325
0.9	0.1092	0.0433	0.01	0.0208	0.0186	0.0348	0.0101	0.0062	0.0158
1.0	0.1217	0.0378	−0.0081	−0.0073	−0.0079	0.0088	−0.0114	−0.0192	−0.0062
1.1	0.0604	−0.0229	−0.0712	−0.0562	–	–	–	–	–
Panel B. Average percentage pricing errors of SV2									
0.7	–	–	0.0425	0.0524	0.0544	0.0663	0.0456	0.0304	0.0378
0.8	–	0.051	0.032	0.0486	0.0472	0.0586	0.0344	0.0138	0.0202
0.9	0.1059	0.0421	0.0145	0.0284	0.0265	0.0392	0.0054	−0.0184	−0.008
1.0	0.0985	0.0231	−0.0123	−0.005	−0.0042	0.008	−0.024	−0.0572	−0.0403
1.1	0.0584	−0.026	−0.0653	−0.0454	–	–	–	–	–
Panel C. Average percentage pricing errors of SV3									
0.7	–	–	0.0437	0.0494	0.0431	0.0466	0.031	0.03	0.028
0.8	–	0.0476	0.0378	0.0506	0.0367	0.0365	0.0226	0.0249	0.0139
0.9	0.0917	0.0379	0.0288	0.0377	0.0178	0.0145	−0.0026	0.0068	−0.0109
1.0	0.0782	0.0194	0.0105	0.011	−0.0129	−0.0221	−0.0299	−0.0192	−0.0432
1.1	0.0314	−0.0336	−0.0397	−0.0292	–	–	–	–	–

data. Panel B shows that SV2 has some improvements over SV1, mainly for some short-term (less than 3.5 years) ATM, and mid-term (3.5–5 years) slightly OTM *difference caps*. But SV2 has worse performance for most deep ITM ($m = 0.7$ and 0.8) *difference caps*: it actually worsens the underpricing of ITM caps. Panel C of Table 8 shows that relative to SV1 and SV2, SV3 has smaller average percentage pricing errors for most long-term (7–10 years) ITM, mid-term (3.5–5 years) OTM, and short-term (2–2.5 years) ATM *difference caps*, and bigger average percentage pricing errors for mid-term (3.5–6 years) ITM *difference caps*. There is still significant underpricing of ITM and overpricing of OTM *difference caps* under SV3.

Overall, the results show that stochastic volatility factors are essential for capturing the time varying volatilities of LIBOR rates. The Diebold-Mariano statistics in Table 7 shows that in general more sophisticated SV models have smaller pricing errors than simpler models, although the improvements are more important for close-to-the-money *difference caps*. The average percentage pricing errors in Table 8 show that, however, even the most sophisticated SV model cannot generate enough volatility skew to be consistent with the data. While previous studies, such as Han (2007), have shown that a three-factor stochastic volatility model similar to ours performs well in pricing ATM caps and swaptions, our analysis shows that the model fails to completely capture the volatility smile in the cap markets. Our findings highlight the importance of studying the relative pricing of caps with different moneyness to reveal the inadequacies of existing term structure models, the same inadequacies cannot be obtained from studying only ATM options.

One important reason for the failure of SV models is that the stochastic volatility factors are independent of LIBOR rates. As a result, the SV models can only generate a symmetric volatility smile, but not the asymmetric smile or skew observed in the data. The pattern of the smile in the cap market is rather similar to that of index options: ITM calls (and OTM puts) are overpriced, and OTM calls (and ITM puts) are underpriced relative to the Black model. Similarly, the smile in the cap market could be due to a market expectation of dramatically declining LIBOR rates. In this section, we examine the contribution of jumps in LIBOR rates in capturing the volatility smile. Our discussion of the performance of the SVJ models parallels that of the SV models.

Parameter estimates in Table 7.9 show that the three stochastic volatility factors of the SVJ models resemble that of the SV models closely. The "level" factor still has the most volatile stochastic volatility, followed by the "curvature" and the "slope" factor. With the inclusion of jumps, the stochastic volatility factors in the SVJ models, especially that of the "level" factor, tend to be less volatile than that of the SV models (lower long run mean and volatility of volatility). Negative estimates of the volatility risk premium show that the volatility of the longer maturity LIBOR rates under the forward measure have lower long-run mean and faster speed of mean-reversion.

Most importantly, we find overwhelming evidence of strong negative jumps in LIBOR rates under the forward measure. To the extend that cap prices reflect market expectations of future evolutions of LIBOR rates, the evidence suggests that the

Table 7.9 Parameter estimates of stochastic volatility and jumps models. This table reports parameter estimates and standard errors of the one-, two-, and three-factor stochastic volatility and jump models. The estimates are obtained by minimizing sum of squared percentage pricing errors (SSE) of difference caps in 53 moneyness and maturity categories observed on a weekly frequency from August 1, 2000 to September 23, 2003. The objective functions reported in the table are re-scaled SSEs over the entire sample at the estimated model parameters and are equal to RMSE of difference caps. The volatility risk premium of the ith stochastic volatility factor and the jump risk premium for forward measure Q^{k+1} is defined as $\eta_i^{k+1}=c_{iv}(T_k-1)$ and $\mu_J^{k+1} = \mu_J+c_J(T_k-1)$, respectively

	SVJ1		SVJ2		SVJ3	
Parameter	Estimate	Std. err	Estimate	Std. err	Estimate	Std. Err
κ_1	0.1377	0.0085	0.0062	0.0057	0.0069	0.0079
κ_2			0.0050	0.0001	0.0032	0.0000
κ_3					0.0049	0.0073
$\bar{v}_1$	0.1312	0.0084	0.7929	0.7369	0.9626	1.1126
$\bar{v}_2$			0.3410	0.0030	0.2051	0.0021
$\bar{v}_3$					0.2628	0.3973
ζ_1	0.8233	0.0057	0.7772	0.0036	0.6967	0.0049
ζ_2			0.0061	0.0104	0.0091	0.0042
ζ_3					0.1517	0.0035
c_{1v}	−0.0041	0.0000	−0.0049	0.0000	−0.0024	0.0000
c_{2v}			−0.0270	0.0464	−0.0007	0.0006
c_{3v}					−0.0103	0.0002
λ	0.0134	0.0001	0.0159	0.0001	0.0132	0.0001
μ_J	−3.8736	0.0038	−3.8517	0.0036	−3.8433	0.0063
c_J	0.2632	0.0012	0.3253	0.0010	0.2473	0.0017
σ_J	0.0001	3.2862	0.0003	0.8723	0.0032	0.1621
Objective function	0.0748		0.0670		0.0622	

market expects a dramatic declining in LIBOR rates over our sample period. Such an expectation might be justifiable given that the economy has been in recession during a major part of our sample period. This is similar to the volatility skew in the index equity option market, which reflects investors fear of the stock market crash such as that of 1987. Compared to the estimates from index options (see, e.g., Pan 2002), we see lower estimates of jump intensity (about 1.5% per annual), but much higher estimates of jump size. The positive estimates of a jump risk premium suggest that the jump magnitude of longer maturity forward rates tend to be smaller. Under SVJ3, the mean relative jump size, $\exp\left(\mu_J + c_J\,(T_k - 1) + \sigma_J^2/2\right) - 1$, for 1, 5, and 10 year LIBOR rates are −97%, −94%, and −80%, respectively. However, we do not find any incidents of negative moves in LIBOR rates under the physical measure with a size close to that under the forward measure. This big discrepancy between jump sizes under the physical and forward measures resembles that between the physical and risk-neutral measure for index options (see, e.g., Pan 2002). This could be a result of a huge jump risk premium.

The likelihood ratio tests in Panel A of Table 7.10 again overwhelmingly reject SVJ1 and SVJ2 in favor of SVJ2 and SVJ3, respectively. The Diebold-Mariano

Table 7.10 Comparison of the performance of stochastic volatility and jump models. This table reports model comparison based on likelihood ratio and Diebold-Mariano statistics. The total number of observations (both cross sectional and time series), which equals 8,545 over the entire sample, times the difference between the logarithms of the SSEs between two models follows a χ^2distribution asymptotically. We treat implied volatility variables as parameters. Thus the degree of freedom of the χ^2distribution is 168 for the pairs of SVJ2/SVJ1 and SVJ3/SVJ2, because SVJ2 and SVJ3 have four more parameters and 164 additional implied volatility variables than SVJ1 and SVJ2, respectively. The 1% critical value of $\chi^2(168)$ is 214. The Diebold-Mariano statistics are calculated according to equation (14) with a lag order q of 40 and follow an asymptotic standard Normal distribution under the null hypothesis of equal pricing errors. A negative statistic means that the more sophisticated model has smaller pricing errors. Bold entries mean that the statistics are significant at the 5% level

Panel A. Likelihood Ratio and Diebold-Mariano statistics for overall model performance based on SSEs

Models	D-M stats	Likelihood ratio stats $\chi^2(168)$
SVJ2–SVJ1	−2.240	**1886**
SVJ3–SVJ2	−7.149	**1256**

Panel B. Diebold-Mariano statistics between SVJ2 and SVJ1 for individual difference caps based on squared percentage pricing errors

Moneyness	2yr	3yr	4yr	5yr	6yr	7yr	8yr	9yr	10yr
0.7	–	–	−0.308	−0.467	−0.188	0.675	−0.240	−0.774	−0.180
0.8	–	−0.537	−1.031	−1.372	−0.684	−0.365	−0.749	−1.837	−1.169
0.9	−1.530	−0.934	−1.463	**−3.253**	−0.920	−1.588	**−2.395**	**−3.287**	−0.686
1.0	**−3.300**	−1.265	−1.647	**−2.020**	−0.573	−1.674	−1.396	**−2.540**	−0.799
1.1	**−5.341**	0.156	**−3.141**	**−2.107**	–	–	–	–	–

Panel C. Diebold-Mariano statistics between SVJ3 and SVJ2 for individual difference caps based on squared percentage pricing errors

Moneyness	2yr	3yr	4yr	5yr	6yr	7yr	8yr	9yr	10yr
0.7	–	–	0.690	−1.023	−1.133	**−2.550**	−1.469	−0.605	−1.920
0.8	–	−0.159	1.609	−1.898	−0.778	**−3.191**	**−3.992**	**−2.951**	**−3.778**
0.9	−1.235	−0.328	1.183	−1.361	−0.249	**−2.784**	−1.408	**−3.411**	**−2.994**
1.0	−1.245	−0.553	−0.463	−1.317	2.780	0.182	−0.551	−1.542	−1.207
1.1	−1.583	−0.334	−2.040	−1.259	–	–	–	–	–

statistics in Panel A of Table 7.10 show that SVJ2 and SVJ3 have significantly smaller SSEs than SVJ1 and SVJ2, respectively, suggesting that the more sophisticated SVJ models significantly improve the pricing of all *difference caps*. The Diebold-Mariano statistics of squared percentage pricing errors of individual *difference caps* in Panel B of Table 7.10 show that SVJ2 significantly improves the performance of SVJ1 for long-, mid-, and short-term around-the-money *difference caps*. The Diebold-Mariano statistics in Panel C of Table 7.10 show that SVJ3 significantly reduces the pricing errors of SVJ2 for long-term ITM, and some mid- and short-term around-the-money *difference caps*. Table 7.11 shows the average percentage pricing errors also improve over the SV models.

Table 7.12 compares the performance of the SVJ and SV models. During the first 20 weeks of our sample, the SVJ models have much higher RMSEs than the SV models. As a result, the likelihood ratio and Diebold-Mariano statistics between the three pairs of SVJ and SV models over the entire sample are somewhat smaller than that of the sample period without the first 20 weeks. Nonetheless, all the SV models are overwhelmingly rejected in favor of their corresponding SVJ models by both tests. The Diebold-Mariano statistics of individual *difference caps* in Panels B, C, and D show that the SVJ models significantly improve the performance of the SV models for most *difference caps* across moneyness and maturity. The most interesting results are in Panel D, which show that SVJ3 significantly reduces the pricing errors of most ITM *difference caps* of SV3, strongly suggesting that the negative jumps are essential for capturing the asymmetric smile in the cap market.

Our analysis shows that a low dimensional model with three principal components driving the forward rate curve, stochastic volatility of each component, and strong negative jumps captures the volatility smile in the cap markets reasonably well. The three yield factors capture the variations of the levels of LIBOR rates, while the stochastic volatility factors are essential to capture the time varying volatilities of LIBOR rates. Even though the SV models can price ATM caps reasonably well, they fail to capture the volatility smile in the cap market. Instead, significant negative jumps in LIBOR rates are needed to capture the smile. These results highlight the importance of studying the pricing of caps across moneyness: the importance of negative jumps is revealed only through the pricing of always-from-the-money caps. Excluding the first 20 weeks and the two special periods, SVJ3 has a reasonably good pricing performance with an average RMSEs of 4.5%. Given that the bid-ask spread is about 2–5% in our sample for ATM caps, and because ITM and OTM caps tend to have even higher percentage spreads,[19] this cam be interpreted as a good performance.

Despite its good performance, there are strong indications that SVJ3 is misspecified and the inadequacies of the model seem to be related to MBS markets. For example, while SVJ3 works reasonably well for most of the sample period, it has large pricing errors in several special periods coinciding with high prepayment activities in the MBS markets. Moreover, even though we assume that the stochastic

---

[19]See, for example, Deuskar et al. (2003).

Table 7.11 Average percentage pricing errors of stochastic volatility and jump models. This table reports average percentage pricing errors of difference caps with different moneyness and maturities of the three stochastic volatility and jump models. Average percentage pricing errors are defined as the difference between market price and model price divided by market price

Moneyness	2yr	3yr	4yr	5yr	6yr	7yr	8yr	9yr	10yr
Panel A. Average percentage pricing errors of SVJ1									
0.7	–	–	0.0073	0.01	0.0102	0.0209	−0.0001	−0.0061	0.0077
0.8	–	0.0167	−0.0014	0.0111	0.007	0.0207	−0.0009	−0.0076	0.0053
0.9	0.0682	0.0132	−0.0035	0.0104	0.0038	0.0204	−0.0062	−0.0114	0.0042
1.0	0.0839	0.016	−0.0004	0.0105	0.0062	0.0194	0.0013	−0.0083	0.0094
1.1	0.0625	−0.0144	−0.0255	0.0094	–	–	–	–	–
Panel B. Average percentage pricing errors of SVJ2									
0.7	–	–	0.0148	0.0188	0.0175	0.0279	0.0116	0.0106	0.0256
0.8	–	0.0271	0.0062	0.0172	0.0137	0.0255	0.0081	0.0061	0.0139
0.9	0.0698	0.0205	−0.0012	0.0068	0.0039	0.0198	−0.0041	−0.0047	−0.002
1.0	0.0668	0.0131	−0.0058	−0.0047	−0.0054	0.0127	−0.0058	−0.0112	−0.0128
1.1	0.0612	−0.0094	−0.0215	−0.0076	–	–	–	–	–
Panel C. Average percentage pricing errors of SVJ3									
0.7	–	–	0.0176	0.017	0.0085	0.0167	0.0008	−0.0049	−0.0021
0.8	–	0.0249	0.0115	0.0185	0.0016	0.0131	0.004	−0.0008	−0.0063
0.9	0.0713	0.0155	0.0073	0.0129	−0.0108	0.0072	0.0044	0.0048	−0.0092
1.0	0.0657	0.0054	0.0033	0.0047	−0.0232	−0.001	0.019	0.0206	−0.0058
1.1	0.0528	−0.0242	−0.0199	−0.0028	–	–	–	–	–

Table 7.12 Comparison of the Performance of SV and SVJ Models. This table reports model comparison based on likelihood ratio and Diebold-Mariano statistics. The total number of observations (both cross sectional and time series), which equals 8,545 over the entire sample and 7,485 without the first twenty weeks, times the difference between the logarithms of the SSEs between two models follows a χ^2distribution asymptotically. We treat implied volatility variables as parameters. Thus the degree of freedom of the χ^2distribution is 4 for the pairs of SVJ/SV1, SVJ2/SV2, and SVJ3/SV3, because SVJ models have four more parameters and equal number of additional implied volatility variables as the corresponding SV models. The 1% critical value of $\chi^2(4)$ is 13. The Diebold-Mariano statistics are calculated according to equation (14) with a lag order q of 40 and follow an asymptotic standard Normal distribution under the null hypothesis of equal pricing errors. A negative statistic means that the more sophisticated model has smaller pricing errors. Bold entries mean that the statistics are significant at the 5% level

Panel A. Likelihood ratio and Diebold-Mariano statistics for overall model performance based on SSEs

Models	D-M stats (whole sample)	D-M stats (without first 20 weeks)	Likelihood ratio stats $\chi^2(4)$ (whole sample)	Likelihood ratio stats $\chi^2(4)$ (without first 20 weeks)
SVJ1–SV1	**−2.972**	−3.006	**1854**	**2437**
SVJ2–SV2	**−3.580**	−4.017	**2115**	**2688**
SVJ3–SV3	**−3.078**	−3.165	**1814**	**2497**

Panel B. Diebold-Mariano statistics between SVJ1 and SV1 for individual difference caps based on squared percentage pricing errors (without first 20 weeks)

Moneyness	2yr	3yr	4yr	5yr	6yr	7yr	8yr	9yr	10yr
0.7	–	–	**−3.504**	−1.904	**−4.950**	**−3.506**	**−3.827**	**−2.068**	**−2.182**
0.8	–	**−12.68**	−1.520	−1.195	**−2.245**	**−1.986**	−1.920	−1.353	−1.406
0.9	**−7.162**	**−2.773**	−1.030	−0.923	−0.820	−0.934	−1.176	−1.109	−1.166
1.0	**−5.478**	−1.294	**−4.001**	−1.812	**−2.331**	−1.099	−1.699	**−2.151**	**−2.237**
1.1	−0.435	−0.261	**−2.350**	−0.892	–	–	–	–	–

(continued)

Table 7.12 (continued)

Panel C. Diebold-Mariano statistics between SVJ2 and SV2 for individual difference caps based on squared percentage pricing errors (without first 20 weeks)

Moneyness	2yr	3yr	4yr	5yr	6yr	7yr	8yr	9yr	10yr
0.7	–	–	**−7.714**	−3.914	**−14.01**	**−7.387**	**−7.865**	**−3.248**	−1.637
0.8	–	**−5.908**	**−2.917**	−2.277	**−7.591**	**−4.610**	**−4.182**	0.397	2.377
0.9	**−7.013**	**−2.446**	−1.246	−1.309	**−2.856**	−1.867	−0.183	0.239	3.098
1.0	**−6.025**	−1.159	**−4.478**	−3.754	−0.404	−0.416	−0.881	**−2.504**	0.023
1.1	−0.308	−1.284	**−3.148**	−2.267	–	–	–	–	–

Panel D. Diebold-Mariano statistics between SVJ3 and SV3 for individual difference caps based on squared percentage pricing errors (without first 20 weeks)

Moneyness	2yr	3yr	4yr	5yr	6yr	7yr	8yr	9yr	10yr
0.7	–	–	**−8.358**	−10.49	**−7.750**	**−5.817**	**−5.140**	**−3.433**	**−3.073**
0.8	–	**−7.373**	**−7.655**	−4.774	**−6.711**	**−3.030**	**−2.650**	**−2.614**	−1.239
0.9	**−1.980**	**−2.501**	**−3.622**	−2.384	−1.938	−1.114	−0.768	**−4.119**	−1.305
1.0	−1.124	−1.353	−0.880	0.052	0.543	**−2.110**	−0.359	−0.492	**−2.417**
1.1	1.395	**−2.218**	**−2.151**	−0.337	–	–	–	–	–

Table 7.13 Correlations between LIBOR rates and implied volatility variables. This table reports the correlations between LIBOR rates and implied volatility variables from SVJ3. Given the parameter estimates of SVJ3 in Table 4, the implied volatility variables are estimated at t by minimizing the SSEs of all difference caps at t

	L(t,1)	L(t,3)	L(t,5)	L(t,7)	L(t,9)	$V_1(t)$	$V_2(t)$	$V_3(t)$
$V_1(t)$	−0.8883	−0.8772	−0.8361	−0.7964	−0.7470	1	−0.4163	0.3842
$V_2(t)$	0.1759	0.235	0.2071	0.1545	0.08278	−0.4163	1	−0.0372
$V_3(t)$	−0.5951	−0.485	−0.4139	−0.3541	−0.3262	0.3842	−0.0372	1

volatility factors are independent of LIBOR rates, Table 7.13 shows strong negative correlations between the implied volatility variables of the first factor and the LIBOR rates. This result suggests that when interest rate is low, cap prices become too high for the model to capture and the implied volatilities have to become abnormally high to fit the observed cap prices. One possible explanation of the "leverage" effect is that higher demands for caps to hedge prepayments from MBS markets in low interest rate environments could artificially push up cap prices and implied volatilities. Therefore, extending our models to incorporate factors from MBS markets seems to be a promising direction of future research.

7.5 Nonparametric Estimation of LIBOR Forward Density

The studies presented so far have shown the importance of USV factors for pricing interest rate derivatives and have developed models that explicitly incorporate USV factors to capture the volatility smile. In this section, we try to identify economic factors that influence the shape of the volatility smile. In particular, we discuss the nonparametric analysis of LIBOR forward densities in Li and Zhao (2009), in which they identify the impacts of mortgage market activities on the volatility smile.

7.5.1 Nonparametric Method

For LIBOR-based instruments such as caps, floors, and swaptions, it is convenient to consider pricing using the forward measure approach. We will therefore focus on the dynamics of LIBOR forward rate $L_k(t)$ under the forward measure $\mathbb{Q}^{k+1}$, which is essential for pricing caplets maturing at T_{k+1}. Under this measure, the discounted price of any security using $D_{k+1}(t)$ as the numeraire is a martingale. Thus, the time-t price of a caplet maturing at T_{k+1} with a strike price of X is

$$C\left(L_k(t), X, t, T_k\right) = \delta D_{k+1}(t) \int_X^{\infty} (y - X)\, p^{\mathbb{Q}^{k+1}}\left(L_k(T_k) = y | L_k(t)\right) \mathrm{d}y, \tag{7.33}$$

where $p^{\mathbb{Q}^{k+1}}(L_k(T_k) = y|L_k(t))$ is the conditional density of $L_k(T_k)$ under forward measure $\mathbb{Q}^{k+1}$. Once we know the forward density, we can price any security whose payoff on T_{k+1} depends only on $L_k(t)$ by discounting its expected payoff under $\mathbb{Q}^{k+1}$ using $D_{k+1}(t)$.

Existing term structure models rely on parametric assumptions on the distribution of $L_k(t)$ to obtain closed-form pricing formulae for caplets. For example, the standard LIBOR market models of Brace et al. (1997) and Miltersen et al. (1997) assume that $L_k(t)$ follows a log-normal distribution and price caplet using the Black formula. The models of Jarrow et al. (2007) assume that $L_k(t)$ follows affine jump-diffusions of Duffie et al. (2000).

We estimate the distribution of $L_k(t)$ under $\mathbb{Q}^{k+1}$ using the prices of a cross section of caplets that mature at T_{k+1} and have different strike prices. Following Breeden and Litzenberger (1978), we know that the density of $L_k(t)$ under $\mathbb{Q}^{k+1}$ is proportional to the second derivative of $C(L_k(t), t, T_k, X)$ with respect to X,

$$p^{\mathbb{Q}^{k+1}}(L_k(T_k)|L_k(t)) = \frac{1}{\delta D_{k+1}(t)} \frac{\partial^2 C(L_k(t), t, T_k, X)}{\partial X^2}|_{X=L_k(T_k)}. \qquad (7.34)$$

In standard LIBOR market models, it is assumed that the conditional density of $L_k(T_k)$ depends only on the current LIBOR rate. This assumption, however, can be overly restrictive given the multifactor nature of term structure dynamics. For example, while the level factor can explain a large fraction (between 80 and 90%) of the variations of LIBOR rates, the slope factor still has significant explanatory power of interest rate variations. Moreover, there is overwhelming evidence that interest rate volatility is stochastic, and it has been suggested that interest rate volatility is unspanned in the sense that it can not be fully explained by the yield curve factors such as the level and slope factors.

One important innovation of our study is that we allow the volatility of $L_k(t)$ to be stochastic and the conditional density of $L_k(T_k)$ to depend on not only the level, but also the slope and volatility factors of LIBOR rates. Denote the conditioning variables as $Z(t) = \{s(t), v(t)\}$, where $s(t)$ (the slope factor) is the difference between the 10- and 2-year LIBOR forward rates and $v(t)$ (the volatility factor) is the first principal component of EGARCH-filtered spot volatilities of LIBOR rates across all maturities. Under this generalization, the conditional density of $L_k(T_k)$ under the forward measure $\mathbb{Q}^{k+1}$ is given by

$$p^{\mathbb{Q}^{k+1}}(L_k(T_k)|L_k(t), Z(t)) = \frac{1}{\delta D_{k+1}(t)} \frac{\partial^2 C(L_k(t), X, t, T_k, Z(t))}{\partial X^2}|_{X=L_k(T_k)}. \qquad (7.35)$$

Next we discuss how to estimate the SPDs by combining the forward and physical densities of LIBOR rates. Denote a SPD function as π. In general, π depends on multiple economic factors, and it is impossible to estimate it using interest rate caps alone. Given the available data, all we can estimate is the projection of π onto the future spot rate $L_k(T_k)$:

$$\pi_k \left(L_k(T_k); L_k(t), Z(t)\right) = E_t^{\mathbb{P}} \left[\pi | L_k(T_k); L_k(t), Z(t)\right], \tag{7.36}$$

where the expectation is taken under the physical measure. Then the price of the caplet can be calculated as

$$\begin{aligned} C\left(L_k(t), X, t, T_k, Z(t)\right) &= \delta E_t^{\mathbb{P}} \left[\pi \cdot \left(L_k(T_k) - X\right)^+\right] \\ &= \delta \int_X^{\infty} \pi_k(y)\,(y - X)\, p^{\mathbb{P}}\left(L_k(T_k)\right. \\ &= \left. y | L_k(t), Z(t)\right) \mathrm{d}y, \end{aligned} \tag{7.37}$$

where the second equality is due to iterated expectation and $p^{\mathbb{P}}(L_k(T_k) = y | L_k(t), Z(t))$ is the conditional density of $L_k(T_k)$ under the physical measure.

Comparing (33) and (37), we have

$$\pi_k \left(L_k(T_k); L_k(t), Z(t)\right) = D_{k+1}(t) \frac{p^{\mathbb{Q}^{k+1}}\left(L_k(T_k) | L_k(t), Z(t)\right)}{p^{\mathbb{P}}\left(L_k(T_k) | L_k(t), Z(t)\right)}. \tag{7.38}$$

Therefore, by combining the densities of $L_k(T_k)$ under $\mathbb{Q}^{k+1}$ and $\mathbb{P}$, we can estimate the projection of π onto $L_k(T_k)$. The SPDs contain rich information on how risks are priced in financial markets. While Ait-Sahalia and Lo (1998, 2000), Aït-Sahalia and Duarte (2003), Jackwerth (2000), Rosenberg and Engle (2002), and others estimate the SPDs using index options (i.e., the projection of π onto index returns), our analysis based on interest rate caps documents the dependence of the SPDs on term structure factors.

Similar to many existing studies, to reduce the dimensionality of the problem, we further assume that the caplet price is homogeneous of degree 1 in the current LIBOR rate:

$$C\left(L_k(t), X, t, T_k, Z(t)\right) = \delta D_{k+1}(t) L_k(t) C_M\left(M_k(t), t, T_k, Z(t)\right), \tag{7.39}$$

where $M_k(t) = X/L_k(t)$ represents the moneyness of the caplet. Hence, for the rest of the paper we estimate the forward density of $L_k(T_k)/L_k(t)$ as the second derivative of the price function C_M with respect to M :

$$p^{\mathbb{Q}^{k+1}}\left(\frac{L_k(T_k)}{L_k(t)} | Z(t)\right) = \frac{1}{\delta D_{k+1}(t)} \frac{\partial^2 C_M\left(M_k(t), t, T_k, Z(t)\right)}{\partial M^2} |_{M = L_k(T_k)/L_k(t)}. \tag{7.40}$$

7.5.2 Empirical Results

In this section, we present nonparametric estimates of the probability densities of LIBOR rates under physical and forward martingale measures. In particular, we document the dependence of the forward densities on the slope and volatility factors of LIBOR rates.

Figure 7.4 presents nonparametric estimates of the forward densities at different levels of the slope and volatility factors at 2, 3, 4, and 5 year maturities. The two levels of the slope factor correspond to a flat and a steep forward curve, while the two levels of the volatility factor represent low and high volatility of LIBOR rates. The 95% confidence intervals are obtained through simulation. The forward densities should have a zero mean since LIBOR rates under appropriate forward measures are martingales. The expected log percentage changes of the LIBOR rates are slightly negative due to an adjustment from the Jensen's inequality. We normalize the forward densities so that they integrate to one. However, we do not have enough data at the right tail of the distribution at 4 and 5 year maturities. We do not extrapolate the data to avoid potential biases.

Figure 7.4 documents three important features of the nonparametric LIBOR forward densities. First, the log-normal assumption underlying the popular LIBOR market models is grossly violated in the data, and the forward densities across all maturities are significantly negatively skewed. Second, all the forward densities depend significantly on the slope of the term structure. For example, moving from a flat to a steep term structure, the forward densities across all maturities become much more dispersed and more negatively skewed. Third, the forward densities also depend on the volatility factor. Under both flat and steep term structures, the forward densities generally become more compact when the volatility factor increases. This is consistent with a mean reverting volatility process: High volatility right now leads to low volatility in the future and more compact forward densities.

To better illustrate the dependence of the forward densities on the two conditioning variables, we also regress the quantiles of the forward densities on the two factors. We choose quantiles instead of moments of the forward densities in our regressions for two reasons. First, quantiles are much easier to estimate. While quantiles can be obtained from the CDF function, which is the first derivative of the price function, moments require integrations of the forward density, which is the second derivative of the price function. Second, a wide range of quantiles provide a better characterization of the forward densities than a few moments, especially for the tail behaviors of the densities.

Suppose we consider I and J levels of the transformed slope and volatility factors in our empirical analysis. For a given level of the two conditioning variables (s_i, v_j), we first obtain a nonparametric estimate of the forward density at a given maturity and its quantiles $Q_x(s_i, v_j)$, where x can range from 0 to 100%. Then we consider the following regression model

$$Q_x(s_i, v_j) = b_{0x} + b_{1x} \cdot s_i + b_{2x} \cdot v_j + b_{3x} \cdot s_i \cdot v_j + \varepsilon_x, \tag{7.41}$$

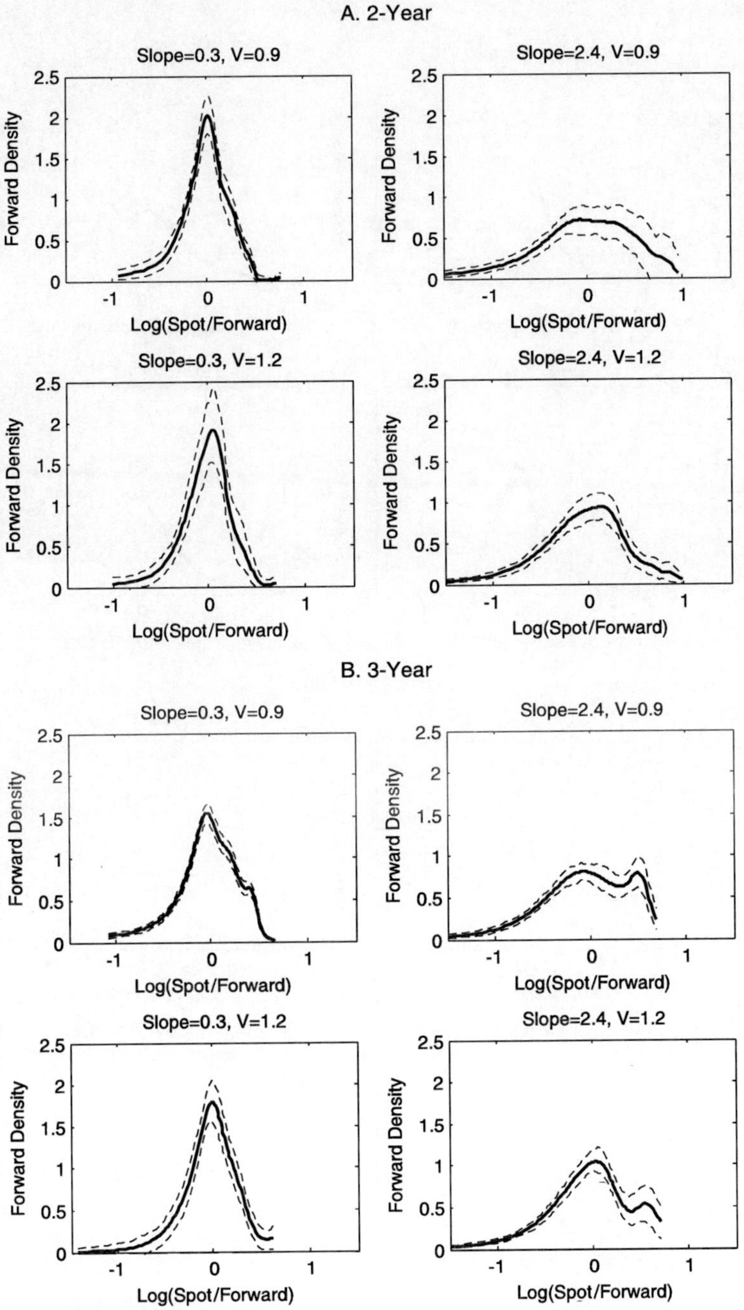

Fig. 7.4 Nonparametric estimates of the LIBOR forward densities at different levels of the slope and volatility factors. The slope factor is defined as the difference between the 10- and 2-year three-month LIBOR forward rates. The volatility factor is defined as the first principal component of EGARCH-filtered spot volatilities and has been normalized to a mean that equals one.

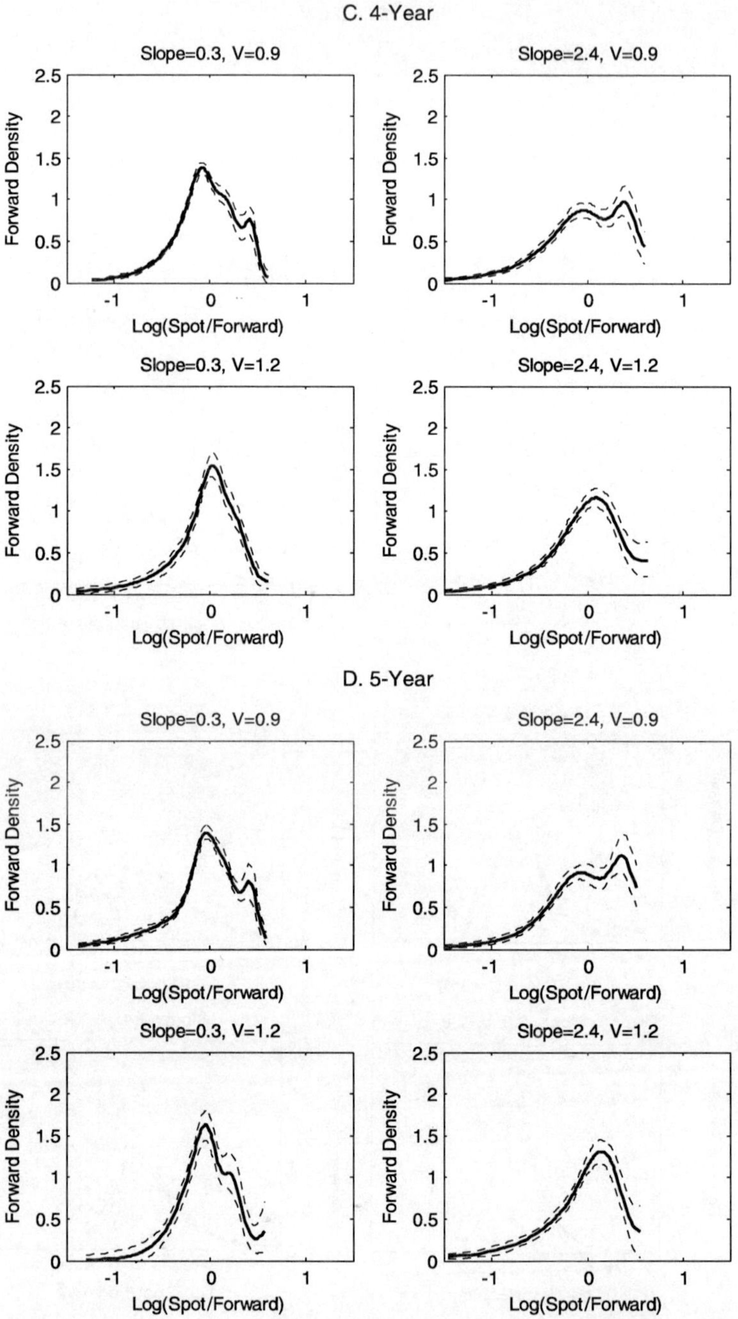

Fig. 7.4 (Continued) The two levels of the slope factor correspond to flat and steep term structures, while the two levels of the volatility factor corresponds to low and high levels of volatility

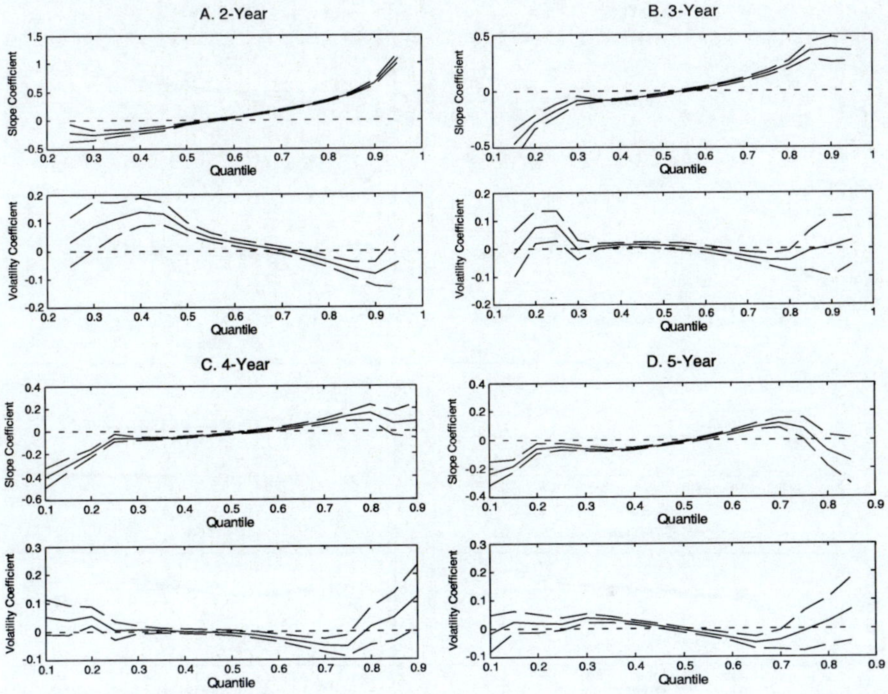

Fig. 7.5 Impacts of the slope and volatility factors on LIBOR forward densities. This figure reports regression coefficients of different quantiles of the forward densities at 2, 3, 4, and 5 year maturities on the slope and volatility factors of LIBOR rates in (27) without the interaction term

where $i = 1, 2, \ldots, I$, and $j = 1, 2, \ldots, J$. We include the interaction term to capture potential nonlinear dependence of the forward densities on the two conditioning variables.

Figure 7.5 reports regression coefficients of the slope and volatility factors for the most complete range of quantiles at each maturity, i.e., b_{1x} and b_{2x} as a function of x. While Fig. 7.4 includes only the slope and volatility factors as explanatory variables, Fig. 7.6 contains their interaction term as well. Though in results not reported here we also include lagged conditioning variables in our regressions, their coefficients are generally not statistically significant.

The regression results in Fig. 7.5 are generally consistent with the main findings in Fig. 7.4. The slope coefficients are generally negative (positive) for the left (right) half of the distribution and become more negative or positive at both tails. Consistent with Fig. 7.4, this result suggests that when the term structure steepens, the forward densities become more dispersed and the effect is more pronounced at both tails. One exception to this result is that the slope coefficients become negative and statistically insignificant at the right tail at 5-year maturity. The coefficients of the volatility factor are generally positive (negative) for the left (right) half of the distribution. Although the volatility coefficients start to turn positive at the right tail

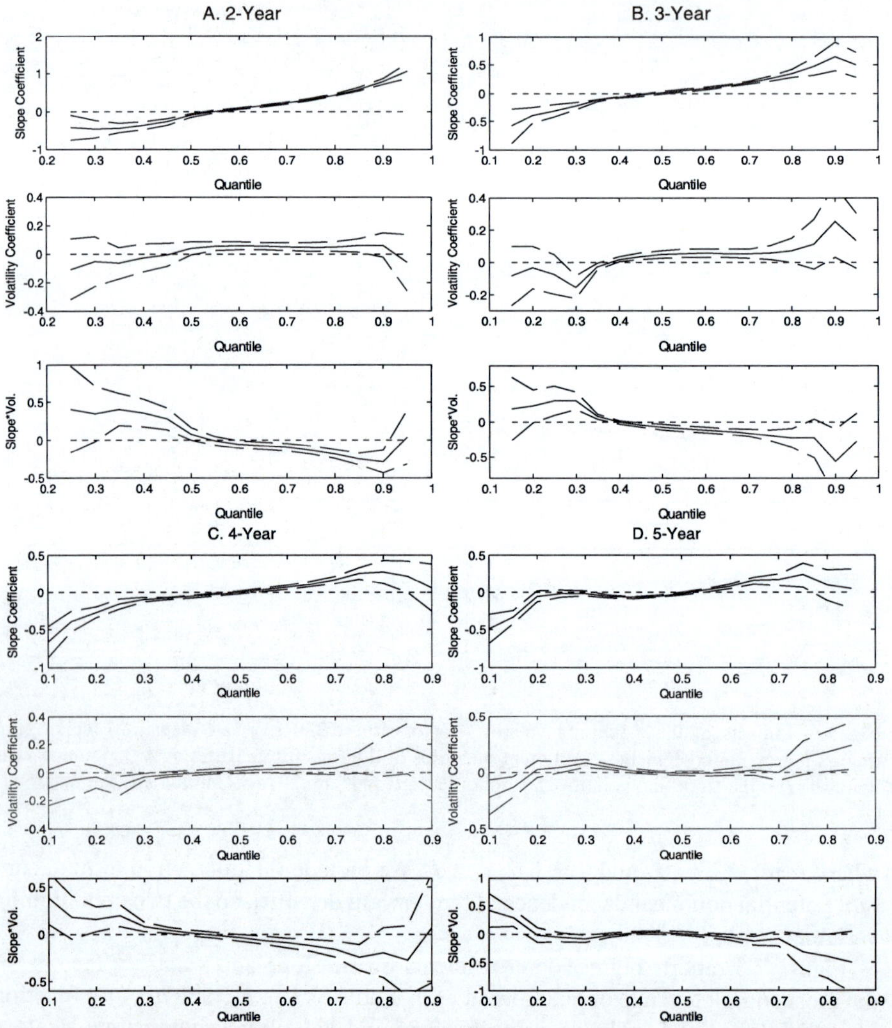

Fig. 7.6 Impacts of the slope and volatility factors (with their interaction term) on LIBOR forward densities. This figure reports regression coefficients of different quantiles of the forward densities at 2, 3, 4, and 5 year maturities on the slope and volatility factors of LIBOR rates and their interaction term in (27)

of the distribution, they are not statistically significant. These results suggest that higher volatility leads to more compact forward densities, a result that is generally consistent with that in Fig. 7.4.

In Fig. 7.6, although the slope coefficients exhibit similar patterns as that in Fig. 7.5, the interaction term changes the volatility coefficients quite significantly. The volatility coefficients become largely insignificant and exhibit quite different

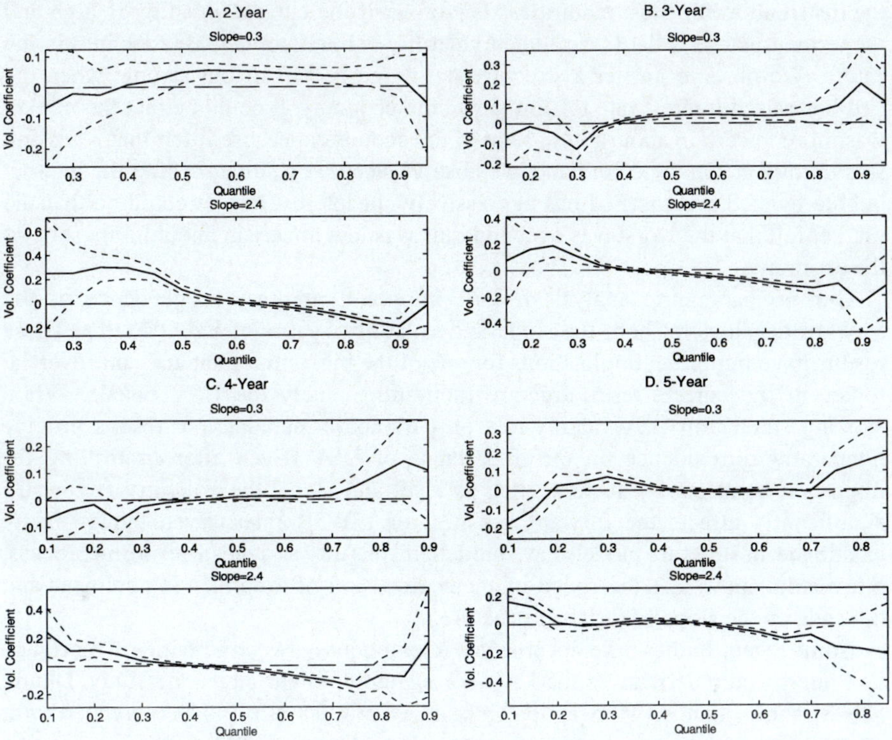

Fig. 7.7 Nonlinear dependence of LIBOR forward densities on the volatility factor of LIBOR rates. This figure presents regression coefficients of quantiles of LIBOR forward densities on the volatility factor at different levels of the slope factor. The two levels of the slope factor represent flat and steep term structures

patterns than those in Fig. 7.5. For example, the volatility coefficients at 2- and 3-year maturities are largely constant across different quantiles. At 4- and 5-year maturities, they even become negative (positive) for the left (right) half of the distribution. On the other hand, the coefficients of the interaction term exhibit similar patterns as that of the volatility coefficients in Fig. 7.5. These results suggest that the impacts of volatility on the forward densities depend on the slope of the term structure.

Figure 7.7 presents the volatility coefficients at different levels of the slope factor (i.e., $\hat{b}_{2x} + \hat{b}_{3x} \cdot s_i$, where $s_i = 0.3$ or 2.4). We see clearly that the impact of volatility on the forward densities depends significantly on the slope factor. With a flat term structure, the volatility coefficients generally increase with the quantiles, especially at 3-, 4-, and 5-year maturities. The volatility coefficients are generally negative (positive) for the left (right) tail of the distribution, although not all of them are statistically significant. However, with a steep term structure, the volatility coefficients are generally positive (negative) for the left (right) half of the distribution for most maturities. Therefore, if the current volatility is high and the term structure is flat (steep), then volatility is likely to increase (decline) in the future. We observe flat term structure during early part of our sample when the Fed has raised interest rate to slow down the economy. It could be that the market was more uncertain about future state of the economy because it felt that recession was imminent. On the other hand, we observe steep term structure after the internet bubble bursted and the Fed has aggressively cut interest rate. It could be that the market felt that the worst was over and thus was less uncertain about future state of the economy.

Our nonparametric analysis reveals important nonlinear dependence of the forward densities on both the slope and volatility factors of LIBOR rates. These results have important implications for one of the most important and controversial topics in the current term structure literature, namely the USV puzzle. While existing studies on USV mainly rely on parametric methods, our results provide nonparametric evidence on the importance of USV: Even after controlling for important bond market factors, such as level and slope, the volatility factor still significantly affects the forward densities of LIBOR rates. Even though many existing term structure models have modelled volatility as a mean-reverting process, our results show that the speed of mean reversion of volatility is nonlinear and depends on the slope of the term structure.

Some recent studies have documented interactions between activities in mortgage and interest rate derivatives markets. For example, in an interesting study, Duarte (2008) shows that ATM swaption implied volatilities are highly correlated with prepayment activities in the mortgage markets. Duarte (2008) extends the string model of Longstaff et al. (2001) by allowing the volatility of LIBOR rates to be a function of the prepayment speed in the mortgage markets. He shows that the new model has much smaller pricing errors for ATM swaptions than the original model with a constant volatility or a CEV model. Jarrow et al. (2007) also show that although their LIBOR model with stochastic volatility and jumps can price caps across moneyness reasonably well, the model pricing errors are unusually large during a few episodes with high prepayments in MBS. These findings suggest that if activities in the mortgage markets, notably the hedging activities of government sponsored enterprises, such as Fannie Mae and Freddie Mac, affect the supply/demand of interest rate derivatives, then this source of risk may not be fully spanned by the factors driving the evolution of the term structure.[20]

In this section, we provide nonparametric evidence on the impact of mortgage activities on LIBOR forward densities. Our analysis extends Duarte (2008) in several important dimensions. First, by considering caps across moneyness, we examine the impacts of mortgage activities on the entire forward densities. Second, by explicitly allowing LIBOR forward densities to depend on the slope and volatility factors of LIBOR rates, we examine whether prepayment still has incremental

[20]See Jaffee (2003) and Duarte (2008) for excellent discussions on the use of interest rate derivatives by Fannie Mae and Freddie Mac in hedging interest rate risks.

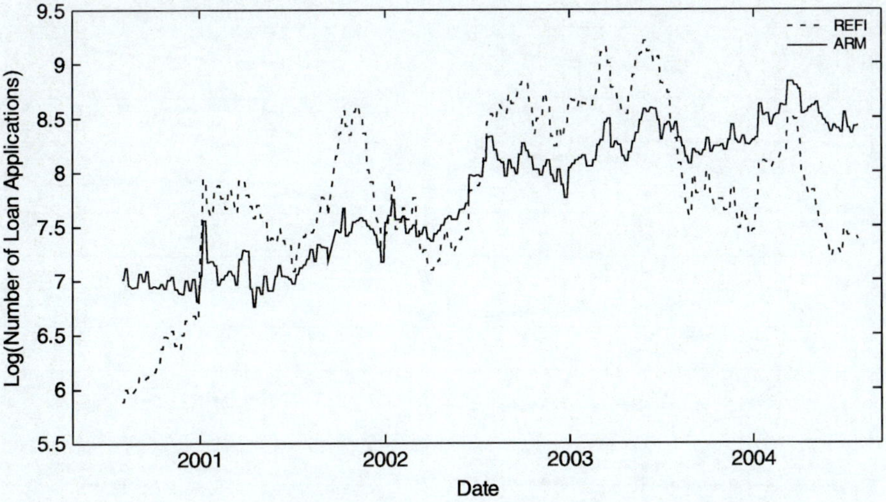

Fig. 7.8 Mortgage Bankers Association of America (MBAA) weekly refinancing and ARMs indexes. This figure reports the logs of the refinance and ARMs indexes obtained by weekly surveys at the Mortgage Bankers Association of America (MBAA)

contributions in explaining interest rate option prices in the presence of these two factors.[21] Finally, in addition to prepayment activities, we also examine the impacts of ARMs origination on the forward densities. Implicit in any ARM is an interest rate cap, which caps the mortgage rate at a certain level. Since lenders of ARMs implicitly sell a cap to the borrower, they might have incentives to hedge such exposures.[22]

Our measures of prepayment and ARMs activities are the weekly refinancing and ARMs indexes based on the weekly surveys conducted by MBAA, respectively. The two indexes, as plotted in Fig. 7.8, tend to be positively correlated with each other. There is an upward trend in ARMs activities during our sample period, which is consistent with what happened in the housing market in the past few years.

To examine the impacts of mortgage activities on LIBOR forward densities, we repeat the above regressions by including two additional explanatory variables that measure refinance and ARMs activities. Specifically, we refer to the top 20% of the observations of the refinance (ARMs) index as the high prepayment (ARMs) group. After obtaining a nonparametric forward density at a particular level of the two conditioning variables, we define two new variables "Refi" and "ARMs," which measure the percentages of observations used in estimating the forward density

[21]While the slope factor can have nontrivial impact on prepayment behavior, the volatility factor is crucial for pricing interest rate options.

[22]We thank the referee for the suggestion of examining the effects of ARMs origination on the forward densities.

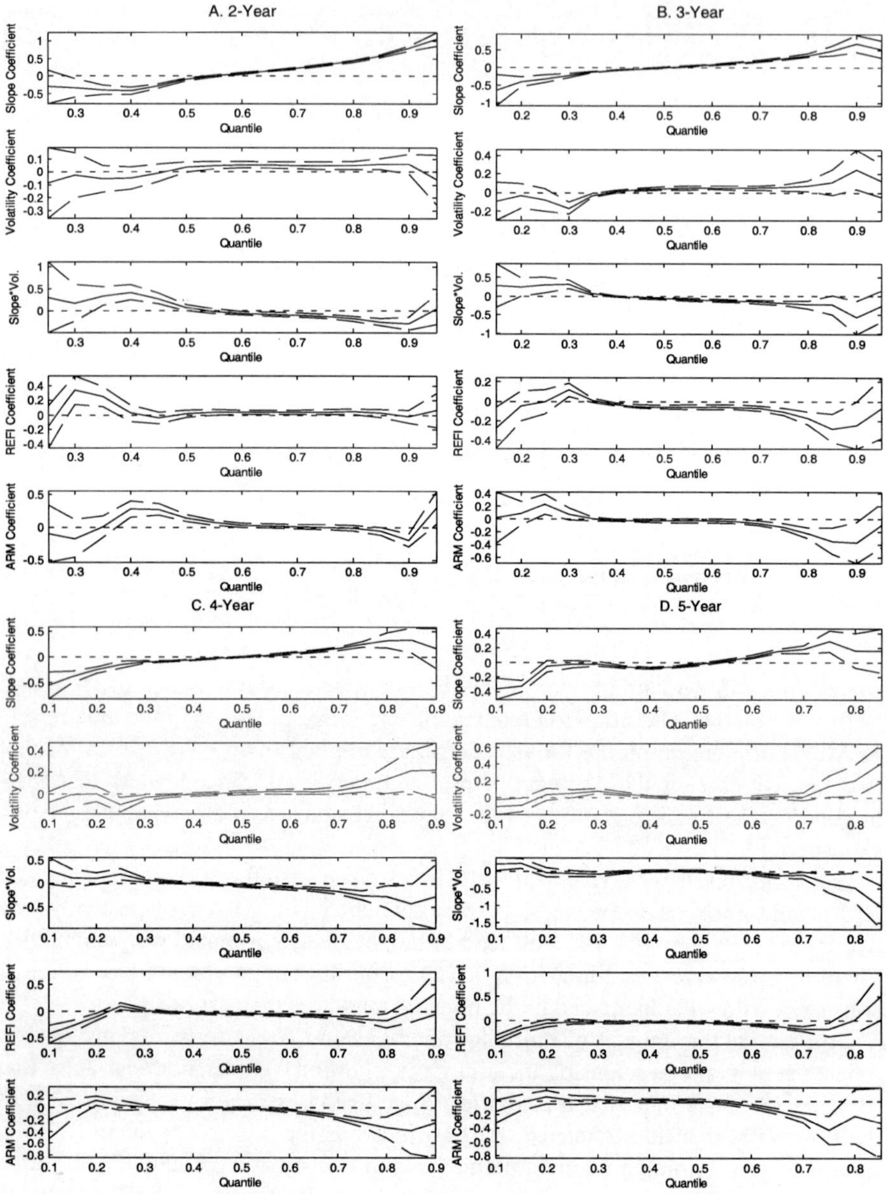

Fig. 7.9 Impacts of refinance and ARMs activities on LIBOR forward densities. In this figure, for each quantile of LIBOR forward densities at 2, 3, 4, and 5 year maturities, we report regression coefficients of the quantile on (1) the slope and volatility factors and their interaction term as in (27); and (2) refinance and ARMs activities

that belong to the high prepayment and ARMs groups, respectively. These two variables allow us to test whether the forward densities behave differently when prepayment/ARMs activities are high. To control for potential collinearity among the explanatory variables, we have orthogonalized any new explanatory variable with respect to existing ones.

Figure 7.9 contain the new regression results with "Refi" and "ARMs" for the four maturities. The coefficients of the slope, volatility, and the interaction term exhibit similar patterns as that in Fig. 7.6.[23]

The strongest impacts of ARMs on the forward densities occur at 2-year maturity, as shown in Panel A of Fig. 7.9. Therefore, high ARMs origination shifts the median and the right tail of the forward densities at 2-year maturity toward the right. This finding is consistent with the notion that hedging demands from ARMs lenders for the cap they have shorted might increase the price of OTM caps. One possible reason that the effects of ARMs are more pronounced at 2-year maturity than at 3-, 4-, and 5-year maturities is that most ARMs get reset within the first 2 years.

While high ARMs activities shift the forward density at 2-year maturity to the right, high refinance activities shift the forward densities at 3-, 4-, and 5-year maturities to the left. We see that the coefficients of Refi at the left tail are significantly negative. While the coefficients also are significantly negative for the middle of the distribution (40–70% quantiles), the magnitude of the coefficients are much smaller. These can be seen in Panels B, C, and D of Fig. 7.9. Therefore, high prepayment activities lead to much more negatively skewed forward densities. This result is consistent with the notion that investors in MBS might demand OTM floors to hedge their potential losses from prepayments. The coefficients of Refi are more significant at 4- and 5-year maturities because the duration of most of MBS are close to 5 years.

Our results confirm and extend the findings of Duarte (2008) by showing that mortgage activities affect the entire forward density and consequently the pricing of interest rate options across moneyness. While prepayment activities affect the left tail of the forward densities at intermediate maturities, ARMs activities affect the right tail of the forward densities at short maturity. Our findings hold even after controlling for the slope and volatility factors and suggest that part of the USV factors could be driven by activities in the mortgage markets.

7.6 Conclusion

In this paper, we have provided a review of some recent developments in the academic literature on interest rate derivatives. Our discussions have revolved around the new evidence of volatility smile in interest rate derivatives markets.

[23]In results not reported, we find that the nonlinear dependence of the forward densities on the volatility factor remain the same as well.

Studies on the pricing and hedging interest rate derivatives in the presence of volatility smile have provided rich insights on term structure modeling. For example, Li and Zhao (2006) have shown that one fundamental assumption of DTSMs that bonds and derivatives are driven by the same set of risk factors is violated in the data. Therefore, existing DTSMs, which have been popular and successful in pricing bonds and swaps, need substantial extension to price interest rate derivatives. Jarrow et al. (2007) also show that stochastic volatility and negative jumps are essential for pricing the smile in LIBOR market models. Finally, Li and Zhao (2009) provide nonparametric evidence on the impacts of mortgage refinance activities on the shape of the volatility smile. Given that volatility smile has guided the development of equity option pricing literature since Black and Scholes (1973) and Merton (1973), we hope that the volatility smile documented here will help the development of term structure models in the years to come.

Appendix

Derivatives Pricing Under QTSMs

Leippold and Wu (2002) show that a large class of fixed-income securities can be priced in closed-form in the QTSMs using the transform analysis of Duffie et al. Duffie et al. (2000). They show that the time-t value of a contract that has an exponential quadratic payoff structure at terminal time T, i.e.

$$\exp\left(-q(X_T)\right) = \exp\left(-X_T'\overline{A}X_T - \overline{b}'X_T - \overline{c}\right)$$

has the following form

$$\begin{aligned} \psi\left(q, X_t, t, T\right) &= E_Q\left(e^{-\int_t^T r(X_s)\mathrm{d}s} e^{-q(X_T)} | \mathcal{F}_t\right) \\ &= \exp\left[-X_t A(T-t) X_t - b(T-t)' X_t - c(T-t)\right]. \end{aligned} \tag{7.42}$$

where $A(.)$, $b(.)$ and $c(.)$ satisfy the ODEs (4)–(6) with the initial conditions $A(0) = \overline{A}, b(0) = \overline{b}$ and $c(0) = \overline{c}$.

The time-t price a call option with payoff $\left(e^{-q(X_T)} - y\right)^+$ at $T = t + \tau$ equals

$$\begin{aligned} C\left(q, y, X_t, \tau\right) &= E_Q\left(e^{-\int_t^T r(X_s)\mathrm{d}s}\left(e^{-q(X_T)} - y\right)^+ | \mathcal{F}_t\right) \\ &= E_Q\left(e^{-\int_t^T r(X_s)\mathrm{d}s}\left(e^{-q(X_T)} - y\right) \mathbf{1}_{\{-q(X_T) \geq \ln(y)\}} | \mathcal{F}_t\right) \\ &= G_{q,q}\left(-\ln(y), X_t, \tau\right) - y G_{0,q}\left(-\ln(y), X_t, \tau\right), \end{aligned}$$

where $G_{q_1,q_2}(y, X_t, \tau) = E_Q\left[e^{-\int_t^T r(X_s)ds} e^{-q_1(X_T)} \mathbf{1}_{\{q_2(X_T)\le y\}} | \mathcal{F}_t\right]$ and can be computed by the inversion formula,

$$G_{q_1,q_2}(y, X_t, \tau) = \frac{\psi(q_1, X_t, t, T)}{2} - \frac{1}{\pi}\int_0^\infty \frac{e^{ivy}\psi(q_1 + ivq_2) - e^{-ivy}\psi(q_1 - ivq_2)}{iv} dv. \quad (7.43)$$

Similarly, the price of a put option is

$$P(q, y, \tau, X_t) = yG_{0,-q}(\ln(y), X_t, \tau) - G_{q,-q}(\ln(y), X_t, \tau).$$

We are interested in pricing a cap which is portfolio of European call options on future interest rates with a fixed strike price. For simplicity, we assume the face value is one and the strike price is $\overline{r}$. At time 0, let $\tau, 2\tau, \ldots, n\tau$ be the fixed dates for future interest payments. At each fixed date $k\tau$, the $\overline{r}$-capped interest payment is given by $\tau(\mathcal{R}((k-1)\tau, k\tau) - \overline{r})^+$, where $\mathcal{R}((k-1)\tau, k\tau)$ is the τ-year floating interest rate at time $(k-1)\tau$, defined by

$$\frac{1}{1 + \tau\mathcal{R}((k-1)\tau, k\tau)} = \varrho((k-1)\tau, k\tau) = E^{\mathcal{Q}}\left(\exp\left(-\int_{(k-1)\tau}^{k\tau} r(X_s)\,ds\right) | \mathcal{F}_{(k-1)\tau}\right).$$

The market value at time 0 of the caplet paying at date $k\tau$ can be expressed as

$$\begin{aligned} Caplet(k) &= E^{\mathcal{Q}}\left[\exp\left(-\int_0^{k\tau} r(X_s)\,ds\right)\tau(\mathcal{R}((k-1)\tau, k\tau) - \overline{r})^+\right] \\ &= (1 + \tau\overline{r})E^{\mathcal{Q}}\left[\exp\left(-\int_0^{(k-1)\tau} r(X_s)\,ds\right)\right. \\ &\quad \times \left.\left(\frac{1}{(1+\tau\overline{r})} - \varrho((k-1)\tau, k\tau)\right)^+\right]. \end{aligned}$$

Hence, the pricing of the $k-th$ caplet is equivalent to the pricing of an $(k-1)\tau$-for-τ put struck at $K = \frac{1}{(1+\tau\overline{r})}$. Therefore,

$$\begin{aligned} Caplet(k) &= G_{0,-q_\tau}\left(\ln K, X_{(k-1)\tau}, (k-1)\tau\right) \\ &\quad - \frac{1}{K}G_{q_\tau,-q_\tau}\left(\ln K, X_{(k-1)\tau}, (k-1)\tau\right). \end{aligned} \quad (7.44)$$

Similarly for the $k-th$ floorlet

$$Floorlet(k) = -G_{0,q_\tau}\left(-\ln K, X_{(k-1)\tau}, (k-1)\tau\right) + \frac{1}{K} G_{q_\tau, q_\tau}\left(-\ln K, X_{(k-1)\tau}, (k-1)\tau\right). \quad (7.45)$$

Derivation of the Characteristic Function in Jarrow et al. (2007)

The solution to the characteristic function of $\log(L_k(T_k))$,

$$\psi\left(u_0, Y_t, t, T_k\right) = \exp\left[a(s) + u_0 \log(L_k(t)) + B(s)' V_t\right],$$

$a(s)$ and $B(s), 0 \le s \le T_k$ satisfy the following system of Ricatti equations:

$$\frac{dB_j(s)}{ds} = -\kappa_j^{k+1} B_j(s) + \frac{1}{2} B_j^2(s)\xi_j^2 + \frac{1}{2}\left[u_0^2 - u_0\right] U_{s,j}^2, \quad 1 \le j \le N,$$

$$\frac{da(s)}{ds} = \sum_{j=1}^{N} \kappa_j^{k+1} \theta_j^{k+1} B_j(s) + \lambda_J \left[\Gamma(u_0) - 1 - u_0 \left(\Gamma(1) - 1\right)\right],$$

where the function Γ is

$$\Gamma(c) = \exp(\mu_J^{k+1} c + \frac{1}{2}\sigma_J^2 c^2).$$

The initial conditions are $B(0) = 0_{N\times 1}$, $a(0) = 0$, and κ_j^{k+1} and θ_j^{k+1} are the parameters of $V_j(t)$ process under $\mathbb{Q}^{k+1}$.

For any $l < k$, Given that $B\left(T_l\right) = B_0$ and $a\left(T_l\right) = a_0$, we have the closed-form solutions for $B\left(T_{l+1}\right)$ and $a\left(T_{l+1}\right)$. Define constants $p = \left[u_0^2 - u_0\right] U_{s,j}^2$, $q = \sqrt{\left(\kappa_j^{k+1}\right)^2 + p\xi_j^2}$, $c = \frac{p}{q-\kappa_j^{k+1}}$ and $d = \frac{p}{q+\kappa_j^{k+1}}$. Then we have

$$B_j(T_{l+1}) = c - \frac{(c+d)(c-B_{j0})}{(d+B_{j0})\exp(-q\delta) + (c-B_{j0})}, 1 \le j \le N,$$

$$a(T_{l+1}) = a_0 - \sum_{j=1}^{N}\left[\kappa_j^{k+1}\theta_j^{k+1}\left(d\delta + \frac{2}{\xi_j^2}\ln\left(\frac{(d+B_{j0})\exp(-q\delta) + (c-B_{j0})}{c+d}\right)\right)\right] + \lambda_J \delta\left[\Gamma(u_0) - 1 - u_0\left(\Gamma(1) - 1\right)\right],$$

if $p \ne 0$ and $B_j(T_{l+1}) = B_{j0}, a(T_{l+1}) = a_0$ otherwise. $B(T_k)$ and $a(T_k)$ can be computed via iteration.

References

Ahn, D.-H., Dittmar, R. F., & Gallant, A. R. (2002). Quadratic term structure models: theory and evidence. *Review of Financial Studies, 15*, 243–288.

Aït-Sahalia, Y., & Lo, A. (1998). Nonparametric Estimation of State-Price Densities implicit in financial asset prices, *Journal of Finance, 53*, 499–547.

Aït-Sahalia, Y., & Lo, A. (2000). Nonparametric Risk Management and Implied Risk Aversion, *Journal of Econometrics, 94*, 9–51.

Aït-Sahalia, Y., & Duarte, J. (2003). Nonparametric option pricing under shape restrictions. *Journal of Econometrics, 116*, 9–47.

Andersen, T. G., & Benzoni, L. (2006). Do bonds span volatility risk in the U.S. reasury market? A specification test of affine term structure models. Working paper, Northwestern University.

Andersen, L., & Brotherton-Ratcliffe, R. (2001). Extended LIBOR market models with stochastic volatility. working paper, Gen Re Securities.

Bakshi, G., Cao, C., & Chen, Z. (1997). Empirical performance of alternative option pricing models. *Journal of Finance, 52*, 2003–2049.

Black, F. (1976). The pricing of commodity contracts. *Journal of Financial Economics, 3*, 167–179.

Black, F., & Scholes, M. (1973). The pricing of options and corporate liabilities. *Journal of Political Economy, 81*, 637–654.

Brace, A., Gatarek, D., & Musiela, M. (1997). The market model of interest rate dynamics. *Mathematical Finance, 7*, 127–155.

Breeden, D., & Litzenberger, R. (1978). Prices of State-Contingent Claims Implicit in Option Prices, *Journal of Business, 51*, 621–651.

Campbell, J., Lo, A., & MacKinlay, C. (1997). *The econometrics of financial markets*. New Jersey: Princeton University Press.

Chacko, George, and Sanjiv Das (2002). Pricing interest rate derivatives: A general approach, *Review of Financial Studies, 15*, 195–241.

Collin-Dufresne, P., & Goldstein, R. S. (2002). Do bonds span the fixed income markets? Theory and evidence for unspanned stochastic volatility. *Journal of Finance, 57*, 1685–1729.

Collin-Dufresne, P., & Goldstein, R. S. (2003). Stochastic Correlation and the relative pricing of caps and swaptions in a generalized affine framework. Working paper, Carnegie Mellon University.

Collin-Dufresne, P., Goldstein, R., & Jones, C. (2009). Can interest rate volatility be extracted from the cross section of bond yields? An investigation of unspanned stochastic volatility, *Journal of Financial Economics 94*, 47–66.

Dai, Q., & Singleton, K. (2003). Term structure dynamics in theory and reality. *Review of Financial Studies, 16*, 631–678.

De Jong, Frank, & Pedro Santa-Clara (1999). The dynamics of the forward interest rate curve: A formulation with state variables, *Journal of Financial and Quantitative Analysis, 34*, 131–157.

Deuskar, P., Gupta, A., & Subrahmanyam, M. (2003). Liquidity effects and volatility smiles in interest rate option markets. Working paper, New York University.

Duan, Jin-Chuan, and Jean-Guy Simonato (1999). Estimating and testing exponential-affine term structure models by Kalman filter, *Review of Quantitative Finance and Accounting, 13*, 111–135.

Duarte, J. (2004). Evaluating an alternative risk preference in affine term structure models. *Review of Financial Studies, 17*, 379–404.

Duarte, J. (2008). Mortgage-backed securities refinancing and the arbitrage in the swaption market. *Review of Financial Studies, 21*, 1689–1731.

Duffie, D. (2002). *Dynamic asset pricing theory*. New Jersey: Princeton University Press.

Duffie, D., & Kan, R. (1996). A yield-factor model of interest rates. *Mathematical Finance, 6*, 379–406.

Duffie, D., Pan, J., & Singleton, K. (2000). Transform analysis and asset pricing for affine jump-diffusions. *Econometrica, 68*, 1343–1376.

Duffee, G. R. (2002). Term premia and interest rate forecasts in affine models. *Journal of Finance, 57*, 405–443.

Duffee, G., & Stanton, R. (2004). Estimation of Dynamic Term Structure Models, Working Paper, University of California, Berkeley.

Fan, R., Gupta, A., & Ritchken, P. (2003). Hedging in the possible presence of unspanned stochastic volatility: evidence from swaption markets. *Journal of Finance, 58*, 2219–2248.

Gallant, A. Ronald, & George Tauchen (1996). Which moments to match? *Econometric Theory, 12*, 657–681.

Goldstein, R. S. (2000). The term structure of interest rates as a random field. *Review of Financial Studies, 13*, 365–384.

Gupta, A., & Subrahmanyam, M. (2001). An examination of the static and dynamic performance of interest rate option pricing models in the dollar cap-floor markets. Working paper, Case Western Reserve University.

Gupta, A., & Subrahmanyam, M. (2005). Pricing and Hedging Interest Rate Options: Evidence from Cap-Floor Markets, *Journal of Banking and Finance, 29*, 701–733.

Han, B. (2007). Stochastic volatilities and correlations of bond yields. *Journal of Finance, 62*, 1491–1524.

Heath, D., Jarrow, R., & Morton, A. (1992). Bond pricing and the term structure of interest rates: a new methodology. *Econometrica, 60*, 77–105.

Heidari, M., & Wu, L. (2003). Are interest rate derivatives spanned by the term structure of interest rates? *Journal of Fixed Income, 13*, 75–86.

Heston, S. (1993). A closed-form solution for options with stochastic volatility with applications to bond and currency options. *Review of Financial Studies, 6*, 327–343.

Hull, J., & White, A. (1987). The pricing of options on assets with stochastic volatilities. *Journal of Finance, 42*, 281–300.

Jackwerth, J. (2000). Recovering Risk Aversion from Option Prices and Realized Returns, *Review of Financial Studies, 13*, 433–451.

Jagannathan, R., Kaplin, A., & Sun, S. (2003). An evaluation of multi-factor CIR models using LIBOR, swap rates, and cap and swaption prices. *Journal of Econometrics, 116*, 113–146.

Jarrow, R., Li, H., & Zhao, F. (2007). Interest rate caps "smile" too! but can the LIBOR market models capture it? *Journal of Finance, 62*, 345–382.

Leippold, M., & Wu, L. (2002). Asset pricing under the quadratic class. *Journal of Financial and Quantitative Analysis, 37*, 271–295.

Li, H., & Zhao, F. (2006). Unspanned stochastic volatility: evidence from hedging interest rate derivatives. *Journal of Finance, 61*, 341–378.

Li, H., & Zhao, F. (2009). Nonparametric estimation of state-price densities implicit in interest rate cap prices. *Review of Financial Studies, 22*, 4335–4376.

Longstaff, F., Santa-Clara, P., & Schwartz, E. (2001). The relative valuation of caps and swaptions: theory and evidence. *Journal of Finance, 56*, 2067–2109.

Lund & Jesper (1997). Non-linear Kalman filtering techniques for term-structure models, Working paper, Aarhus School of Business.

Merton, R. (1973). The theory of rational option pricing. *Bell Journal of Economics and Management Science, 4*, 141–183.

Miltersen, M., Sandmann, K., & Sondermann, D. (1997). Closed-form solutions for term structure derivatives with lognormal interest rates. *Journal of Finance, 52*, 409–430.

Pan, J. (2002). The jump-risk premia implicit in options: evidence from an integrated time-series study. *Journal of Financial Economics, 63*, 3-50.

Piazzesi, M. (2009). Affine term structure models. *Handbook of financial econometrics*. North Holland.

Piazzesi, M. (2005). Bond yields and the federal reserve. *Journal of Political Economy, 113*, 311–344.

Protter, P. (1990). Stochastic Integration and Differential Equations (Springer, New York).

Rosenberg, J., & R. Engle. (2002). Empirical Pricing Kernels, *Journal of Financial Economics, 64*, 341–372.

Santa-Clara, P., & Sornette, D. (2001). The dynamics of the forward interest rate curve with stochastic string shocks. *Review of Financial Studies, 14*, 149–185.

Thompson, S. (2008). Identifying term structure volatility from the LIBOR-swap curve. *Review of Financial Studies, 21*, 819–854.

Trolle, A. B., & Schwartz, E. S. (2009). A general stochastic volatility model for the pricing of interest rate derivatives. *Review of Financial Studies*, 2007–2057.

Chapter 8
Volatility Investing with Variance Swaps

Wolfgang Karl Härdle and Elena Silyakova

Abstract Traditionally volatility is viewed as a measure of variability, or risk, of an underlying asset. However, recently investors began to look at volatility from a different angle. It happened due to emergence of a market for new derivative instruments - variance swaps. In this chapter, first we introduce the general idea of the volatility trading using variance swaps. Then we describe valuation and hedging methodology for vanilla variance swaps as well as for the third generation volatility derivatives: gamma swaps, corridor variance swaps, conditional variance swaps. Finally, we show the results of the performance investigation of one of the most popular volatility strategies - dispersion trading. The strategy was implemented using variance swaps on DAX and its constituents during the 5-year period from 2004 to 2008.

8.1 Introduction

Traditionally volatility is viewed as a measure of variability, or risk, of an underlying asset. However recently investors have begun to look at volatility from a different angle, variance swaps have been created.

The first variance swap contracts were traded in late 1998, but it was only after the development of the replication argument using a portfolio of vanilla options that

W.K. Härdle
Ladislaus von Bortkiewicz Chair of Statistics of Humboldt-Universität zu Berlin, Center for Applied Statistics and Economics and Department of Finance National Central University, Taipei, Taiwan, R.O.C.
e-mail: stat@wiwi.hu-berlin.de

E. Silyakova (✉)
Ladislaus von Bortkiewicz Chair of Statistics of Humboldt-Universität zu Berlin and Center for Applied Statistics and Economics, Spandauer Straße 1, 10178 Berlin, Germany
e-mail: silyakoe@cms.hu-berlin.de

J.-C. Duan et al. (eds.), *Handbook of Computational Finance*, Springer Handbooks of Computational Statistics, DOI 10.1007/978-3-642-17254-0_8,

variance swaps became really popular. In a relatively short period of time these over-the-counter (OTC) derivatives developed from simple contracts on future variance to more sophisticated products. Recently we have been able to observe the emergence of 3G volatility derivatives: gamma swaps, corridor variance swaps, conditional variance swaps and options on realised variance.

Constant development of volatility instruments and improvement in their liquidity allows for volatility trading almost as easily as traditional stocks and bonds. Initially traded OTC, now the number of securities having volatility as underlying are available on exchanges. Thus the variance swaps idea is reflected in volatility indices, also called "fear" indices. These indices are often used as a benchmark of equity market risk and contain option market expectations on future volatility. Among those are VIX – the Chicago Board Options Exchange (CBOE) index on the volatility of S&P500, VSTOXX on Dow Jones EURO STOXX 50 volatility, VDAX – on the volatility of DAX. These volatility indices represent the theoretical prices of one-month variance swaps on the corresponding index. They are calculated daily and on an intraday basis by the exchange from the listed option prices. Also, recently exchanges started offering derivative products, based on these volatility indices – options and futures.

8.2 Volatility Trading with Variance Swaps

Variance swap is a forward contract that at maturity pays the difference between realised variance σ_R^2 (floating leg) and predefined strike K_{var}^2 (fixed leg) multiplied by notional N_{var}.

$$(\sigma_R^2 - K_{\text{var}}^2) \cdot N_{\text{var}} \tag{8.1}$$

When the contract expires the realised variance σ_R^2 can be measured in different ways, since there is no formally defined market convention. Usually variance swap contracts define a formula of a final realised volatility σ_R. It is a square root of annualized variance of daily log-returns of an underlying over a swap's maturity calculated in percentage terms:

$$\sigma_R = \sqrt{\frac{252}{T} \sum_{t=1}^{T} \left(\log \frac{S_t}{S_{t-1}} \right)^2} \cdot 100 \tag{8.2}$$

There are two ways to express the variance swap notional: variance notional and vega notional. Variance notional N_{var} shows the dollar amount of profit (loss) from difference in one point between the realised variance σ_R^2 and the strike K_{var}^2. But since market participants usually think in terms of volatility, vega notional N_{vega} turns out to be a more intuitive measure. It shows the profit or loss from 1% change in volatility. The two measures are interdependent and can substitute each other:

$$N_{\text{vega}} = N_{\text{var}} \cdot 2K_{\text{var}}. \tag{8.3}$$

Example 1. Variance notional $N_{\text{var}} = 2500$. If K_{var} is 20% ($K^2_{\text{var}} = 400$) and the subsequent variance realised over the course of the year is $(15\%)^2$ (quoted as $\sigma^2_R = 225$), the investor will make a loss:

$$\text{Loss} = N_{\text{var}} \cdot (\sigma^2_R - K^2_{\text{var}})$$
$$437500 = 2500 \cdot (400 - 225).$$

Marking-to-market of a variance swap is straightforward. If an investor wishes to close a variance swap position at some point t before maturity, he needs to define a value of the swap between inception 0 and maturity T. Here the additivity property of variance is used. The variance at maturity $\sigma^2_{R,(0,T)}$ is just a time-weighted sum of variance realised before the valuation point $\sigma^2_{R,(0,t)}$ and variance still to be realised up to maturity $\sigma^2_{R,(t,T)}$. Since the later is unknown yet, we use its estimate $K^2_{\text{var},(t,T)}$. The value of the variance swap (per unit of variance notional) at time t is therefore:

$$T^{-1}\left\{t\sigma^2_{R,(0,t)} - (T-t)K^2_{\text{var},(t,T)}\right\} - K^2_{\text{var},(0,T)} \tag{8.4}$$

8.3 Replication and Hedging of Variance Swaps

The strike K^2_{var} of a variance swap is determined at inception. The realised variance σ^2_R, on the contrary, is calculated at expiry (8.2). Similar to any forward contract, the future payoff of a variance swap (8.1) has zero initial value, or $K^2_{\text{var}} = E[\sigma^2_R]$. Thus the variance swap pricing problem consists in finding the fair value of K^2_{var} which is the expected future realised variance.

To achieve this, one needs to construct a trading strategy that captures the realised variance over the swap's maturity. The cost of implementing this strategy will be the fair value of the future realised variance.

One of the ways of taking a position in future volatility is trading a delta-hedged option. The P&L from delta-hedging (also called hedging error) generated from buying and holding a vanilla option up to maturity and continuously delta-hedging it, captures the realised volatility over the holding period.

Some assumptions are needed:

- The existence of futures market with delivery dates $T' \geq T$
- The existence of European futures options market, for these options all strikes are available (market is complete)
- Continuous trading is possible
- Zero risk-free interest rate ($r = 0$)
- The price of the underlying futures contract F_t following a diffusion process with no jumps:

$$\frac{dF_t}{F_t} = \mu_t dt + \sigma_t dW_t \tag{8.5}$$

We assume that the investor does not know the volatility process σ_t, but believes that the future volatility equals σ_{imp}, the implied volatility prevailing at that time on the market. He purchases a claim (for example a call option) with σ_{imp}. The terminal value (or payoff) of the claim is a function of F_T. For a call option the payoff is denoted: $f(F_T) = (F_T - K)^+$. The investor can define the value of a claim $V(F_t, t)$ at any time t, given that σ_{imp} is predicted correctly. To delta-hedge the long position in V over $[0, T]$ the investor holds a dynamic short position equal to the option's delta: $\Delta = \partial V/\partial F_t$. If his volatility expectations are correct, then at time t for a delta-neutral portfolio the following relationship holds:

$$\Theta = -\frac{1}{2}\sigma_{imp}^2 F_t^2 \Gamma \tag{8.6}$$

subject to terminal condition:

$$V(F_T, T) = f(F_T) \tag{8.7}$$

$\Theta = \partial V/\partial t$ is called the option's theta or time decay and $\Gamma = \partial^2 V/\partial F_t^2$ is the option's gamma. Equation (8.6) shows how the option's value decays in time (Θ) depending on convexity (Γ).

Delta-hedging of V generates the terminal wealth:

$$P\&L_\Delta = -V(F_0, 0, \sigma_{imp}) - \int_0^T \Delta dF_t + V(F_T, T) \tag{8.8}$$

which consists of the purchase price of the option $V(F_0, 0, \sigma_{imp})$, P&L from delta-hedging at constant implied volatility σ_{imp} and final pay-off of the option $V(F_T, T)$.

Applying Itô's lemma to some function $f(F_t)$ of the underlying process specified in (8.5) gives:

$$f(F_T) = f(F_0) + \int_0^T \frac{\partial f(F_t)}{\partial F_t} dF_t + \frac{1}{2}\int_0^T F_t^2 \sigma_t^2 \frac{\partial^2 f(F_t)}{\partial F_t^2} dt + \int_0^T \frac{\partial f(F_t)}{\partial t} dt \tag{8.9}$$

For $f(F_t) = V(F_t, t, \sigma_t)$ we therefore obtain:

$$V(F_T, T) = V(F_0, 0, \sigma_{imp}) + \int_0^T \Delta dF_t + \frac{1}{2}\int_0^T F_t^2 \Gamma \sigma_t^2 dt + \int_0^T \Theta dt \tag{8.10}$$

Using relation (8.6) for (8.10) gives:

$$V(F_T, T) - V(F_0, 0, \sigma_{imp}) = \int_0^T \Delta dF_t + \frac{1}{2}\int_0^T F_t^2 \Gamma (\sigma_t^2 - \sigma_{imp}^2) dt \tag{8.11}$$

Finally substituting (8.11) into (8.8) gives $P\&L_\Delta$ of the delta-hedged option position:

$$P\&L_\Delta = \frac{1}{2}\int_0^T F_t^2 \Gamma(\sigma_t^2 - \sigma_{\text{imp}}^2)\mathrm{d}t \tag{8.12}$$

Thus buying the option and delta-hedging it generates P&L (or hedging error) equal to differences between instantaneous realised and implied variance, accrued over time $[0, T]$ and weighed by $F_t^2\Gamma/2$ (dollar gamma).

However, even though we obtained the volatility exposure, it is path-dependent. To avoid this one needs to construct a portfolio of options with path-independent P&L or in other words with dollar gamma insensitive to F_t changes. Figure 8.1 represents the dollar gammas of three option portfolios with an equal number of vanilla options (puts or calls) and similar strikes lying in a range from 20 to 200. Dollar gammas of individual options are shown with thin lines, the portfolio's dollar gamma is a bold line.

First, one can observe, that for every individual option dollar gamma reaches its maximum when the option is ATM and declines with price going deeper out of the money. One can make a similar observation by looking at the portfolio's dollar gamma when the constituents are weighted equally (first picture). However, when we use the alternative weighting scheme $(1/K)$, the portfolio's dollar gamma becomes flatter (second picture). Finally by weighting options with $1/K^2$ the

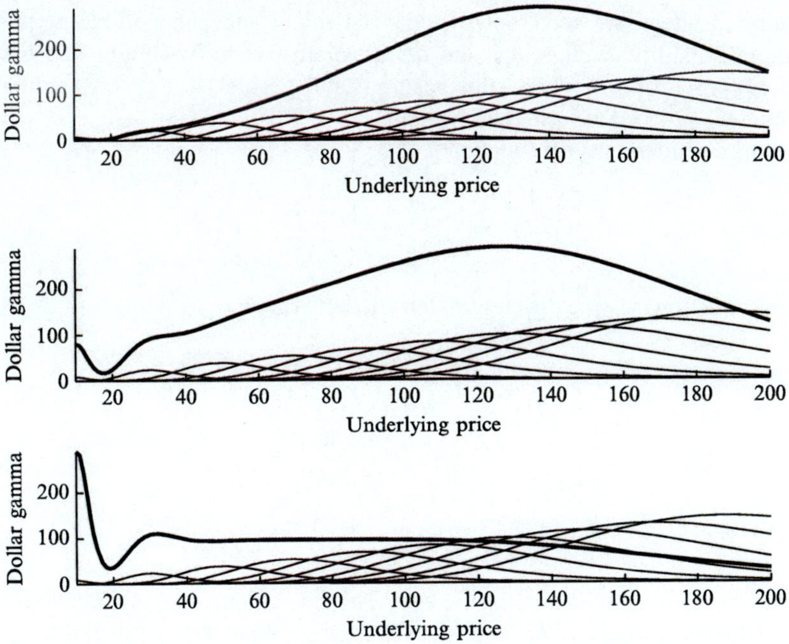

Fig. 8.1 Dollar gamma of option portfolio as a function of stock price. Weights are defined: equally, proportional to $1/K$ and proportional to $1/K^2$

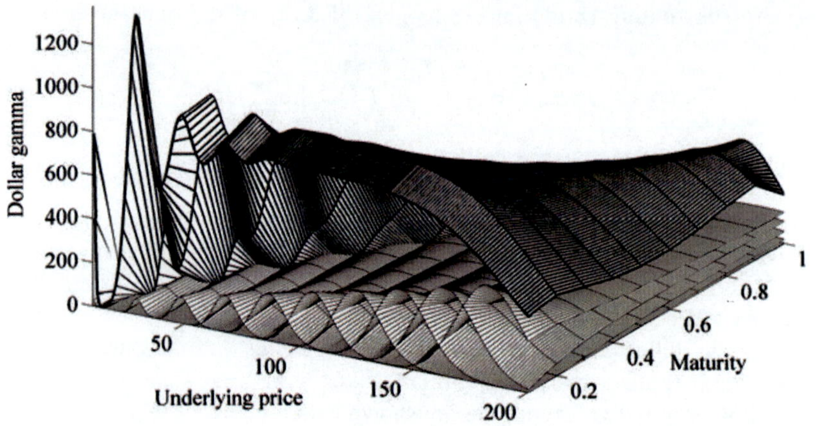

Fig. 8.2 Dollar gamma of option portfolio as a function of stock price and maturity. Weights are defined proportional to $1/K^2$

portfolio's dollar gamma becomes parallel to the vertical axis (at least in 20–140 region), which suggests that the dollar gamma is no longer dependent on the F_t movements.

We have already considered a position in a single option as a bet on volatility. The same can be done with the portfolio of options. However the obtained exposure is path-dependent. We need, however the static, path-independent trading position in future volatility. Figures 8.1 and 8.2 illustrate that by weighting the options' portfolio proportional to $1/K^2$ this position can be achieved. Keeping in mind this intuition we proceed to formal derivations.

Let us consider a payoff function $f(F_t)$:

$$f(F_t) = \frac{2}{T}\left(\log\frac{F_0}{F_t} + \frac{F_t}{F_0} - 1\right) \tag{8.13}$$

This function is twice differentiable with derivatives:

$$f'(F_t) = \frac{2}{T}\left(\frac{1}{F_0} - \frac{1}{F_t}\right) \tag{8.14}$$

$$f''(F_t) = \frac{2}{TF_t^2} \tag{8.15}$$

and

$$f(F_0) = 0 \tag{8.16}$$

One can give a motivation for the choice of the particular payoff function (8.13). The first term, $2\log F_0/TF_t$, is responsible for the second derivative of the payoff $f(F_t)$ w.r.t. F_t, or gamma (8.15). It will cancel out the weighting term in (8.12) and therefore will eliminate path-dependence. The second term $2/T(F_t/F_0 - 1)$ guarantees the payoff $f(F_t)$ and will be non-negative for any positive F_t.

Applying Itô's lemma to (8.13) (substituting (8.13) into (8.9)) gives the expression for the realised variance:

$$\frac{1}{T}\int_0^T \sigma_t^2 \mathrm{d}t = \frac{2}{T}\left(\log\frac{F_0}{F_T} + \frac{F_T}{F_0} - 1\right) - \frac{2}{T}\int_0^T\left(\frac{1}{F_0} - \frac{1}{F_t}\right)\mathrm{d}F_t \tag{8.17}$$

Equation (8.17) shows that the value of a realised variance for $t \in [0, T]$ is equal to

- A continuously rebalanced futures position that costs nothing to initiate and is easy to replicate:

$$\frac{2}{T}\int_0^T\left(\frac{1}{F_0} - \frac{1}{F_t}\right)\mathrm{d}F_t \tag{8.18}$$

- A *log contract*, static position of a contract that pays $f(F_T)$ at expiry and has to be replicated:

$$\frac{2}{T}\left(\log\frac{F_0}{F_T} + \frac{F_T}{F_0} - 1\right) \tag{8.19}$$

Carr and Madan (2002) argue that the market structure assumed above allows for the representation of any twice differentiable payoff function $f(F_T)$ in the following way:

$$f(F_T) = f(k) + f'(k)\left[\{(F_T - k)^+ - (k - F_T)^+\}\right] + \tag{8.20}$$

$$+\int_0^k f''(K)(K - F_T)^+\mathrm{d}K + \int_k^\infty f''(K)(F_T - K)^+\mathrm{d}K$$

Applying (8.20) to payoff (8.19) with $k = F_0$ gives:

$$\log\left(\frac{F_0}{F_T}\right) + \frac{F_T}{F_0} - 1 = \int_0^{F_0}\frac{1}{K^2}(K - F_T)^+\mathrm{d}K + \int_{F_0}^\infty \frac{1}{K^2}(F_T - K)^+\mathrm{d}K \tag{8.21}$$

Equation (8.21) represents the payoff of a log contract at maturity $f(F_T)$ as a sum of:

- The portfolio of OTM puts (strikes are lower than forward underlying price F_0), inversely weighted by squared strikes:

$$\int_0^{F_0}\frac{1}{K^2}(K - F_T)^+\mathrm{d}K \tag{8.22}$$

- The portfolio of OTM calls (strikes are higher than forward underlying price F_0), inversely weighted by squared strikes:

$$\int_{F_0}^{\infty} \frac{1}{K^2}(F_T - K)^+ \mathrm{d}K \tag{8.23}$$

Now coming back to equation (8.17) we see that in order to obtain a constant exposure to future realised variance over the period 0 to T the trader should, at inception, buy and hold the portfolio of puts (8.22) and calls (8.23). In addition he has to initiate and roll the futures position (8.18).

We are interested in the costs of implementing the strategy. Since the initiation of futures contract (8.18) costs nothing, the cost of achieving the strategy will be defined solely by the portfolio of options. In order to obtain an expectation of a variance, or strike K_{var}^2 of a variance swap at inception, we take a risk-neutral expectation of a future strategy payoff:

$$K_{\text{var}}^2 = \frac{2}{T} e^{rT} \int_0^{F_0} \frac{1}{K^2} P_0(K) \mathrm{d}K + \frac{2}{T} e^{rT} \int_{F_0}^{\infty} \frac{1}{K^2} C_0(K) \mathrm{d}K \tag{8.24}$$

8.4 Constructing a Replication Portfolio in Practice

Although we have obtained the theoretical expression for the future realised variance, it is still not clear how to make a replication in practice. Firstly, in reality the price process is discrete. Secondly, the range of traded strikes is limited. Because of this the value of the replicating portfolio usually underestimates the true value of a log contract.

One of the solutions is to make a discrete approximation of the payoff (8.19). This approach was introduced by Demeterfi et al. (Summer 1999).

Taking the logarithmic payoff function, whose initial value should be equal to the weighted portfolio of puts and calls (8.21), we make a piecewise linear approximation. This approach helps to define how many options of each strike investor should purchase for the replication portfolio.

Figure 8.3 shows the logarithmic payoff (dashed line) and the payoff of the replicating portfolio (solid line). Each linear segment on the graph represents the payoff of an option with strikes available for calculation. The slope of this linear segment will define the amount of options of this strike to be put in the portfolio.

For example, for the call option with strike K_0 the slope of the segment would be:

$$w(K_0) = \frac{f(K_{1,c}) - f(K_0)}{K_{1,c} - K_0} \tag{8.25}$$

where $K_{1,c}$ is the second closest call strike.

The slope of the next linear segment, between $K_{1,c}$ and $K_{2,c}$, defines the amount of options with strike $K_{1,c}$. It is given by

$$w(K_{1,c}) = \frac{f(K_{2,c}) - f(K_{1,c})}{K_{2,c} - K_{1,c}} - w(K_0) \tag{8.26}$$

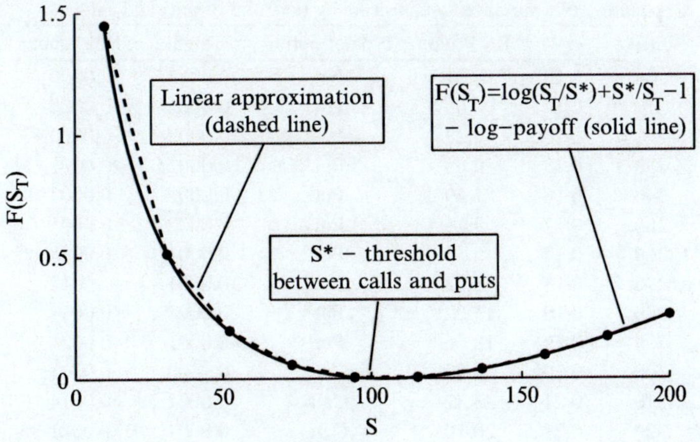

Fig. 8.3 Discrete approximation of a log payoff

Finally for the portfolio of n calls the number of calls with strike $K_{n,c}$:

$$w(K_{n,c}) = \frac{f(K_{n+1,c}) - f(K_{n,c})}{K_{n+1,c} - K_{n,c}} - \sum_{i=0}^{n-1} w(K_{i,c}) \tag{8.27}$$

The left part of the log payoff is replicated by the combination of puts. For the portfolio of m puts the weight of a put with strike $K_{m,p}$ is defined by

$$w(K_{m,p}) = \frac{f(K_{m+1,p}) - f(K_{m,p})}{K_{m,p} - K_{m+1,p}} - \sum_{j=0}^{m-1} w(K_{j,p}) \tag{8.28}$$

Thus constructing the portfolio of European options with the weights defined by (8.27) and (8.28) we replicate the log payoff and obtain value of the future realised variance.

Assuming that the portfolio of options with narrowly spaced strikes can produce a good piecewise linear approximation of a log payoff, there is still the problem of capturing the "tails" of the payoff. Figure 8.3 illustrates the effect of a limited strike range on replication results. Implied volatility is assumed to be constant for all strikes ($\sigma_{\text{imp}} = 25\%$). Strikes are evenly distributed one point apart. The strike range changes from 20 to 1,000. With increasing numbers of options the replicating results approach the "true value" which equals to σ_{imp} in this example. For higher maturities one needs a broader strike range than for lower maturities to obtain the value close to actual implied volatility.

Table 8.1 Replication of a variance swaps strike by portfolio of puts and calls

Strike	IV	BS Price	Type of option	Weight	Share value
200	0.13	0.01	Put	0.0003	0.0000
210	0.14	0.06	Put	0.0002	0.0000
220	0.15	0.23	Put	0.0002	0.0000
230	0.15	0.68	Put	0.0002	0.0001
240	0.16	1.59	Put	0.0002	0.0003
250	0.17	3.16	Put	0.0002	0.0005
260	0.17	5.55	Put	0.0001	0.0008
270	0.18	8.83	Put	0.0001	0.0012
280	0.19	13.02	Put	0.0001	0.0017
290	0.19	18.06	Put	0.0001	0.0021
300	0.20	23.90	Call	0.0000	0.0001
310	0.21	23.52	Call	0.0001	0.0014
320	0.21	20.10	Call	0.0001	0.0021
330	0.22	17.26	Call	0.0001	0.0017
340	0.23	14.91	Call	0.0001	0.0014
350	0.23	12.96	Call	0.0001	0.0011
360	0.24	11.34	Call	0.0001	0.0009
370	0.25	9.99	Call	0.0001	0.0008
380	0.25	8.87	Call	0.0001	0.0006
390	0.26	7.93	Call	0.0001	0.0005
400	0.27	7.14	Call	0.0001	0.0005
				K_{var}	0.1894

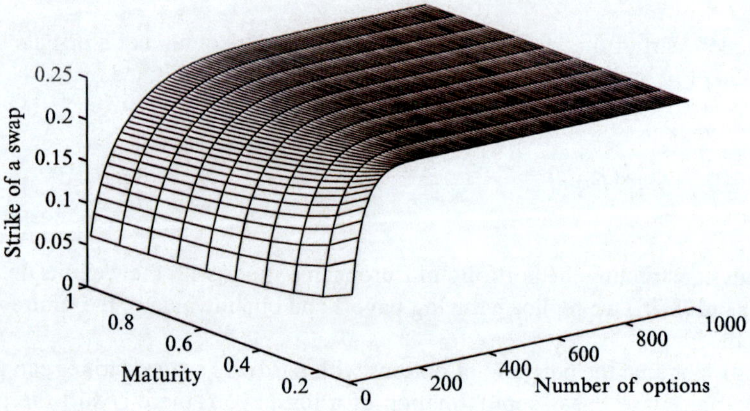

Fig. 8.4 Dependence of replicated realised variance level on the maturity of the swap and the number of options

Table 8.1 shows the example of the variance swap replication. The spot price of $S^* = 300$, riskless interest rate $r = 0$, maturity of the swap is one year $T = 1$, strike range is from 200 to 400 (Fig. 8.4). The implied volatility is 20% ATM and changes linearly with the strike (for simplicity no smile is assumed). The weight of each option is defined by (8.27) and (8.28).

8.5 3G Volatility Products

If we need to capture some particular properties of realised variance, standard variance swaps may not be sufficient. For instance by taking asymmetric bets on variance. Therefore, there are other types of swaps introduced on the market, which constitute the third-generation of volatility products. Among them are: gamma swaps, corridor variance swaps and conditional variance swaps.

By modifying the floating leg of a standard variance swap (8.2) with a weight process w_t we obtain a *generalized variance swap*.

$$\sigma_R^2 = \frac{252}{T}\sum_{t=1}^{T} w_t \left(\log \frac{F_t}{F_{t-1}}\right)^2 \tag{8.29}$$

Now, depending on the chosen w_t we obtain different types of variance swaps: Thus $w_t = 1$ defines a *standard variance swap*.

8.5.1 Corridor and Conditional Variance Swaps

The weight $w_t = w(F_t) = \mathbf{I}_{F_t \in C}$ defines a *corridor variance swap* with corridor C. $\mathbf{I}$ is the indicator function, which is equal to one if the price of the underlying asset F_t is in corridor C and zero otherwise.

If F_t moves sideways, but stays inside C, then the corridor swap's strike is large, because some part of volatility is accrued each day up to maturity. However if the underlying moves outside C, less volatility is accrued resulting the strike to be low. Thus corridor variance swaps on highly volatile assets with narrow corridors have strikes K_C^2 lower than usual variance swap strike K_{var}^2.

Corridor variance swaps admit model-free replication in which the trader holds statically the portfolio of puts and calls with strikes within the corridor C. In this case we consider the payoff function with the underlying F_t in corridor $C = [A, B]$

$$f(F_t) = \frac{2}{T}\left(\log \frac{F_0}{F_t} + \frac{F_t}{F_0} - 1\right)\mathbf{I}_{F_t \in [A,B]} \tag{8.30}$$

The strike of a corridor variance swap is thus replicated by

$$K_{[A,B]}^2 = \frac{2}{T}e^{rT}\int_A^{F_0} \frac{1}{K^2} P_0(K)\mathrm{d}K + \frac{2}{T}e^{rT}\int_{F_0}^{B} \frac{1}{K^2} C_0(K)\mathrm{d}K \tag{8.31}$$

$C = [0, B]$ gives a *downward variance swap*, $C = [A, \infty]$ – an *upward variance swap*.

Since in practice not all the strikes $K \in (0, \infty)$ are available on the market, corridor variance swaps can arise from the imperfect variance replication, when just strikes $K \in [A, B]$ are taken to the portfolio.

Similarly to the corridor, realised variance of conditional variance swap is accrued only if the price of the underlying asset in the corridor C. However the accrued variance is averaged over the number of days, at which F_t was in the corridor (T) rather than total number of days to expiry T. Thus ceteris paribus the strike of a conditional variance swap $K^2_{C,cond}$ is smaller or equal to the strike of a corridor variance swap K^2_C.

8.5.2 Gamma Swaps

As it is shown in Table 8.2, a standard variance swap has constant dollar gamma and vega. It means that the value of a standard swap is insensitive to F_t changes. However it might be necessary, for instance, to reduce the volatility exposure when the underlying price drops. Or in other words, it might be convenient to have a derivative with variance vega and dollar gamma, that adjust with the price of the underlying.

The weight $w_t = w(F_t) = F_t/F_0$ defines a price-weighted variance swap or *gamma swap*. At maturity the buyer receives the realised variance weighted to each t, proportional to the underlying price F_t. Thus the investor obtains path-dependent exposure to the variance of F_t. One of the common gamma swap applications is equity dispersion trading, where the volatility of a basket is traded against the volatility of basket constituents.

The realised variance paid at expiry of a gamma swap is defined by

$$\sigma_{\text{gamma}} = \sqrt{\frac{252}{T}\sum_{t=1}^{T}\frac{F_t}{F_0}\left(\log\frac{S_t}{S_{t-1}}\right)^2} \cdot 100 \tag{8.32}$$

Table 8.2 Variance swap greeks

Greeks		Call	Put	Standard variance swap	Gamma swap
Delta	$\frac{\partial V}{\partial F_t}$	$\Phi(d_1)$	$\Phi(d_1)-1$	$\frac{2}{T}(\frac{1}{F_0}-\frac{1}{F_t})$	$\frac{2}{TF_0}\log\frac{F_t}{F_0}$
Gamma	$\frac{\partial^2 V}{\partial F_t^2}$	$\frac{\phi(d_1)}{F_t\sigma\sqrt{\tau}}$	$\frac{\phi(d_1)}{F_t\sigma\sqrt{\tau}}$	$\frac{2}{F_t^2 T}$	$\frac{2}{TF_0F_t}$
Dollar gamma	$\frac{F_t^2\partial^2 V}{2\partial F_t^2}$	$\frac{F_t\phi(d_1)}{2\sigma\sqrt{\tau}}$	$\frac{F_t\phi(d_1)}{2\sigma\sqrt{\tau}}$	$\frac{1}{T}$	$\frac{F_t}{TF_0}$
Vega	$\frac{\partial V}{\partial \sigma_t}$	$\phi(d_1)F_t\sqrt{\tau}$	$\phi(d_1)F_t\sqrt{\tau}$	$\frac{2\sigma\tau}{T}$	$\frac{2\sigma\tau}{T}\frac{F_t}{F_0}$
Variance vega	$\frac{\partial V}{\partial \sigma_t^2}$	$\frac{F_t\phi(d_1)}{2\sigma\sqrt{\tau}}$	$\frac{F_t\phi(d_1)}{2\sigma\sqrt{\tau}}$	$\frac{\tau}{T}$	$\frac{\tau}{T}\frac{F_t}{F_0}$

One can replicate a gamma swap similarly to a standard variance swap, by using the following payoff function:

$$f(F_t) = \frac{2}{T}\left(\frac{F_t}{F_0}\log\frac{F_t}{F_0} - \frac{F_t}{F_0} + 1\right) \tag{8.33}$$

$$f'(F_t) = \frac{2}{TF_0}\log\frac{F_t}{F_0} \tag{8.34}$$

$$f''(F_t) = \frac{2}{TF_0F_t} \tag{8.35}$$

$$f(F_0) = 0 \tag{8.36}$$

Applying Itô's formula (8.9) to (8.33) gives

$$\frac{1}{T}\int_0^T \frac{F_t}{F_0}\sigma_t^2 \mathrm{d}t = \frac{2}{T}\left(\frac{F_T}{F_0}\log\frac{F_T}{F_0} - \frac{F_T}{F_0} + 1\right) - \frac{2}{TF_0}\int_0^T \log\frac{F_t}{F_0}\mathrm{d}F_t \tag{8.37}$$

Equation (8.37) shows that accrued realised variance weighted each t by the value of the underlying is decomposed into payoff (8.33), evaluated at T, and a continuously rebalanced futures position $\frac{2}{TF_0}\int_0^T \log\frac{F_t}{F_0}\mathrm{d}F_t$ with zero value at $t = 0$. Then applying the Carr and Madan argument (8.20) to the payoff (8.33) at T we obtain the $t = 0$ strike of a gamma swap:

$$K^2_{\text{gamma}} = \frac{2}{TF_0}e^{2rT}\int_0^{F_0}\frac{1}{K}P_0(K)\mathrm{d}K + \frac{2}{TF_0}e^{2rT}\int_{F_0}^{\infty}\frac{1}{K}C_0(K)\mathrm{d}K \tag{8.38}$$

Thus gamma swap can be replicated by the portfolio of puts and calls weighted by the inverse of strike $1/K$ and rolling the futures position.

8.6 Equity Correlation (Dispersion) Trading with Variance Swaps

8.6.1 Idea of Dispersion Trading

The risk of the portfolio (or basket of assets) can be measured by the variance (or alternatively standard deviation) of its return. Portfolio variance can be calculated using the following formula:

$$\sigma^2_{\text{Basket}} = \sum_{i=1}^{n} w_i^2\sigma_i^2 + 2\sum_{i=1}^{n}\sum_{j=i+1}^{n} w_i w_j \sigma_i \sigma_j \rho_{ij} \tag{8.39}$$

where σ_i – standard deviation of the return of an i-th constituent (also called volatility), w_i – weight of an i-th constituent in the basket, ρ_{ij} – correlation coefficient between the i-th and the j-th constituent.

Let's take an arbitrary market index. We know the index value historical development as well as price development of each of index constituent. Using this information we can calculate the historical index and constituents' volatility using, for instance, formula (8.2). The constituent weights (market values or current stock prices, depending on the index) are also known to us. The only parameter to be defined are correlation coefficients of every pair of constituents ρ_{ij}. For simplicity assume $\rho_{ij} = const$ for any pair of i, j and call this parameter $\overline{\rho}$ - *average index correlation, or dispersion*. Then having index volatility σ_{index} and volatility of each constituent σ_i, we can express the average index correlation:

$$\overline{\rho} = \frac{\sigma^2_{\text{index}} - \sum_{i=1}^{n} w_i^2 \sigma_i^2}{2\sum_{i=1}^{n}\sum_{j=i+1}^{n} w_i w_j \sigma_i \sigma_j} \tag{8.40}$$

Hence it appears the idea of *dispersion trading*, consisting of buying the volatility of index constituents according to their weight in the index and selling the volatility of the index. Corresponding positions in variances can be taken by buying (selling) variance swaps.

By going short index variance and long variance of index constituents we go short dispersion, or enter the direct dispersion strategy.

Why can this strategy be attractive for investors? This is due to the fact that index options appear to be more expensive than their theoretical Black-Scholes prices, in other words investors will pay too much for realised variance on the variance swap contract expiry. However, in the case of single equity options one observes no volatility distortion. This is reflected in the shape of implied volatility smile. There is growing empirical evidence that the index option skew tends to be steeper then the skew of the individual stock option. For instance, this fact has been studied in Bakshi et al. (2003) on example of the S&P500 and Branger and Schlag (2004) for the German stock index DAX.

This empirical observation is used in dispersion trading. The most widespread dispersion strategy, direct strategy, is a long position in constituents' variances and short in variance of the index. This strategy should have, on average, positive payoffs. Hoverer under some market conditions it is profitable to enter the trade in the opposite direction. This will be called – the inverse dispersion strategy.

The payoff of the direct dispersion strategy is a sum of variance swap payoffs of each of i-th constituent

$$(\sigma^2_{R,i} - K^2_{\text{var},i}) \cdot N_i \tag{8.41}$$

and of the short position in index swap

$$(K^2_{\text{var,index}} - \sigma^2_{R,\text{index}}) \cdot N_{\text{index}} \tag{8.42}$$

where

$$N_i = N_{\text{index}} \cdot w_i \tag{8.43}$$

The payoff of the overall strategy is:

$$N_{\text{index}} \cdot \left(\sum_{i=1}^{n} w_i \sigma_{R,i}^2 - \sigma_{R,\text{index}}^2 \right) - \text{ResidualStrike} \tag{8.44}$$

The residual strike

$$\text{ResidualStrike} = N_{\text{index}} \cdot \left(\sum_{i=1}^{n} w_i K_{\text{var},i}^2 - K_{\text{var,index}}^2 \right) \tag{8.45}$$

is defined by using methodology introduced before, by means of replication portfolios of vanilla OTM options on index and all index constituents.

However when implementing this kind of strategy in practice investors can face a number of problems. Firstly, for indices with a large number of constituent stocks (such as S&P500) it would be problematic to initiate a large number of variance swap contracts. This is due to the fact that the market for some variance swaps did not reach the required liquidity. Secondly, there is still the problem of hedging vega-exposure created by these swaps. It means a bank should not only virtually value (use for replication purposes), but also physically acquire and hold the positions in portfolio of replicating options. These options in turn require dynamic delta-hedging. Therefore, a large variance swap trade (as for example in case of S&P500) requires additional human capital from the bank and can be associated with large transaction costs. The remedy would be to make a stock selection and to form the offsetting variance portfolio only from a part of the index constituents.

It has already been mentioned that, sometimes the payoff of the strategy could be negative, in order words sometimes it is more profitable to buy index volatility and sell volatility of constituents. So the procedure which could help in decisions about trade direction may also improve overall profitability.

If we summarize, the success of the volatility dispersion strategy lies in correct determining:

- The direction of the strategy
- The constituents for the offsetting variance basket

The next sections will present the results of implementing the dispersion trading strategy on DAX and DAX constituents' variances. First we implement its classical variant meaning short position in index variance against long positions in variances of all 30 constituents. Then the changes to the basic strategy discussed above are implemented and the profitability of these improvements measured.

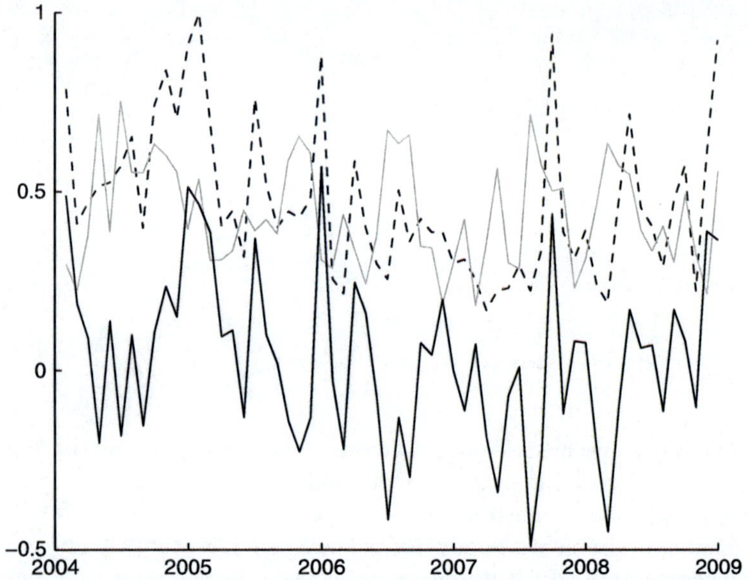

Fig. 8.5 Average implied correlation (*dotted*), average realized correlation (*gray*), payoff of the direct dispersion strategy (*solid black*)

8.6.2 Implementation of the Dispersion Strategy on DAX Index

In this section we investigate the performance of a dispersion trading strategy over the 5 years period from January 2004 to December 2008. The dispersion trade was initiated at the beginning of every moth over the examined period. Each time the 1-month variance swaps on DAX and constituents were traded.

First we implement the basic dispersion strategy, which shows on average positive payoffs over the examined period (Fig. 8.5). Descriptive statistics shows that the average payoff of the strategy is positive, but close to zero. Therefore in the next section several improvements are introduced.

It was discussed already that index options are usually overestimated (which is not the case for single equity options), the future volatility implied by index options will be higher than realized volatility meaning that the direct dispersion strategy is on average profitable. However the reverse scenario may also take place. Therefore it is necessary to define whether to enter a direct dispersion (short index variance, long constituents variance) or reverse dispersion (long index variance and short constituents' variances) strategy.

This can be done by making a forecast of the future volatility with GARCH (1,1) model and multiplying the result by 1.1, which was implemented in the paper of Deng (2008) for S&P500 dispersion strategy. If the variance predicted by GARCH is higher than the variance implied by the option market, one should enter the reverse

Table 8.3 Comparison of basic and improved dispersion strategy payoffs for the period from January 2004 to December 2008

Strategy	Mean	Median	SD	Skewness	Kurtosis	J-B	Probability
Basic	0.032	0.067	0.242	0.157	2.694	0.480	0.786
Improved	0.077	0.096	0.232	−0.188	3.012	0.354	0.838

dispersion trade (long index variance and short constituents variances). After using the GARCH volatility estimate the average payoff increased by 41.7% (Table 8.3).

The second improvement serves to decrease transaction cost and cope with market illiquidity. In order to decrease the number of stocks in the offsetting portfolio the Principal Components Analysis (PCA) can be implemented. Using PCA we select the most "effective" constituent stocks, which help to capture the most of index variance variation. This procedure allowed us to decrease the number of offsetting index constituents from 30 to 10. According to our results, the 1-st PC explains on average 50% of DAX variability. Thereafter each next PC adds only 2–3% to the explained index variability, so it is difficult to distinguish the first several that explain together 90%. If we take stocks, highly correlated only with the 1-st PC, we can significantly increase the offsetting portfolio's variance, because by excluding 20 stocks from the portfolio we make it less diversified, and therefore more risky.

However it was shown that one still can obtain reasonable results after using the PCA procedure. Thus in the paper of Deng (2008) it was successfully applied to S&P500.

References

Bakshi, G., Kapadia, N. & Madan, D. (2003). Stock return characteristics, skew laws, and the differential pricing of individual equity options. *Review of Financial Studies*, *16*(1):101–143.

Branger, N. & Schlag, C. (2004). Why is the index smile so steep? *Review of Finance*, *8*(1):109–127.

Carr, P., & Madan, D. (1998). Towards a theory of volatility trading. *Volatility*, 417–427.

Carr, P. & Madan, D. (2002). *Towards a Theory of Volatility Trading*, in: Volatility, Risk Publications, Robert Jarrow, ed.

Demeterfi, K., Derman, E., Kamal, M. & Zou, J. (Summer 1999). A guide to volatility and variance swaps. *The Journal of Derivatives*, *6*(4), 9–32.

Deng, Q. (2008). Volatility dispersion trading, *SSRN eLibrary*.

Franke, J., Härdle, W., & Hafner, C. M. (2008). *Statistics of financial markets: an introduction* (2nd ed.). Berlin Heidelberg: Springer.

Neuberger, A. (1990). Volatility trading, London business school Working paper.

Part III
Statistical Inference in Financial Models

Chapter 9
Evaluation of Asset Pricing Models Using Two-Pass Cross-Sectional Regressions

Raymond Kan and Cesare Robotti

Abstract This chapter provides a review of the two-pass cross-sectional regression methodology, which over the years has become the most popular approach for estimating and testing linear asset pricing models. We focus on some of the recent developments of this methodology and highlight the importance of accounting for model misspecification in estimating risk premia and in comparing the performance of competing asset pricing models.

9.1 Introduction

Since Black et al. (1972) and Fama and MacBeth (1973), the two-pass cross-sectional regression (CSR) methodology has become the most popular approach for estimating and testing linear asset pricing models. Although there are many variations of this two-pass methodology, the basic approach always involves two steps. In the first pass, the betas of the test assets are estimated from ordinary least squares (OLS) time series regressions of returns on some common factors. In the second pass, the returns on the test assets are regressed on the betas estimated from the first pass. The intercept and the slope coefficients from the second-pass CSR are then used as estimates of the zero-beta rate and factor risk premia. In addition, the R^2 from the second-pass CSR is a popular measure of goodness-of-fit and is often used to compare the performance of competing asset pricing models.

R. Kan (✉)
Joseph L. Rotman School of Management, University of Toronto, 105 St. George Street, Toronto, ON, Canada M5S 3E6
e-mail: kan@chass.utoronto.ca

C. Robotti
Research Department, Federal Reserve Bank of Atlanta, 1000 Peachtree St. N.E., Atlanta, GA 30309, USA
e-mail: cesare.robotti@atl.frb.org

J.-C. Duan et al. (eds.), *Handbook of Computational Finance*, Springer Handbooks of Computational Statistics, DOI 10.1007/978-3-642-17254-0_9,

Although the two-pass CSR approach is easy to implement, conducting robust statistical inference under this method is not trivial. In this article, we survey the existing asymptotic techniques and provide some new results. While we are not the first to review the CSR methodology (see Shanken 1996; Jagannathan et al. 2010), our summary of this approach is more current and emphasizes the role played by model misspecification in estimating risk premia and in comparing the performance of competing asset pricing models.

The remainder of the article is organized as follows. Section 9.2 presents the notation and introduces the two-pass CSR methodology. Section 9.3 discusses statistical inference under correctly specified models. Section 9.4 shows how to conduct statistical inference under potentially misspecified models. Section 9.5 reviews some popular measures of model misspecification and analyzes their statistical properties. Section 9.6 discusses some subtle issues associated with the two-pass CSR methodology that are often overlooked by researchers. The focus of Sects. 9.7 and 9.8 is on pairwise and multiple model comparison tests, respectively. Section 9.9 concludes and discusses several avenues for future research.

9.2 The Two-Pass Cross-Sectional Regression Methodology

Let f_t be a K-vector of factors at time t and R_t a vector of returns on N test assets at time t. We define $Y_t = [f_t', \ R_t']'$ and its unconditional mean and covariance matrix as

$$\mu = E[Y_t] \equiv \begin{bmatrix} \mu_1 \\ \mu_2 \end{bmatrix}, \tag{9.1}$$

$$V = \text{Var}[Y_t] \equiv \begin{bmatrix} V_{11} & V_{12} \\ V_{21} & V_{22} \end{bmatrix}, \tag{9.2}$$

where V is assumed to be positive definite. The multiple regression betas of the N assets with respect to the K factors are defined as $\beta = V_{21}V_{11}^{-1}$. These are measures of systematic risk or the sensitivity of returns to the factors. In addition, we denote the covariance matrix of the residuals of the N assets by $\Sigma = V_{22} - V_{21}V_{11}^{-1}V_{12}$. Throughout the article, we assume that the time series Y_t is jointly stationary and ergodic, with finite fourth moment.

The proposed K-factor beta pricing model specifies that asset expected returns are linear in the betas, i.e.,

$$\mu_2 = X\gamma, \tag{9.3}$$

where $X = [1_N, \ \beta]$ is assumed to be of full column rank, 1_N is an N-vector of ones, and $\gamma = [\gamma_0, \ \gamma_1']'$ is a vector consisting of the zero-beta rate (γ_0) and risk premia on the K factors (γ_1). In general, asset pricing models only require the linear relationship in (9.3) to hold conditionally. However, most empirical studies estimate an unconditional version of (9.3). This can be justified on the following grounds. First, the stochastic process of the conditional betas could be specified such that the K-factor beta pricing model holds unconditionally. See, for example, Caflisch and Chaudhary (2004) and Jagannathan and Wang (1996). Second, one could let γ be linear in a set of instruments. This will then lead to an expanded unconditional beta pricing model, which includes the instruments and the original factors multiplied by the instruments as additional factors.

Suppose that we have T observations on Y_t and denote the sample mean and covariance matrix of Y_t by

$$\hat{\mu} = \begin{bmatrix} \hat{\mu}_1 \\ \hat{\mu}_2 \end{bmatrix} = \frac{1}{T}\sum_{t=1}^{T} Y_t, \tag{9.4}$$

$$\hat{V} = \begin{bmatrix} \hat{V}_{11} & \hat{V}_{12} \\ \hat{V}_{21} & \hat{V}_{22} \end{bmatrix} = \frac{1}{T}\sum_{t=1}^{T}(Y_t - \hat{\mu})(Y_t - \hat{\mu})'. \tag{9.5}$$

The popular two-pass method first estimates the betas of the N assets by running the following multivariate regression:

$$R_t = \alpha + \beta f_t + \epsilon_t, \quad t = 1, \ldots, T. \tag{9.6}$$

The estimated betas from this first-pass time-series regression are given by the matrix $\hat{\beta} = \hat{V}_{21}\hat{V}_{11}^{-1}$.

In the second pass, we run a single CSR of $\hat{\mu}_2$ on $\hat{X} = [1_N,\ \hat{\beta}]$ to estimate γ. Note that some studies allow $\hat{\beta}$ to change throughout the sample period. For example, in the original Fama and MacBeth (1973) study, the betas used in the CSR for month t were estimated from data prior to that month. We do not study this case here mainly because the estimator of γ from this alternative procedure is generally not consistent. The second-pass CSR estimators will depend on the weighting matrix W. Popular choices of W in the literature are $W = I_N$ (OLS), $W = V_{22}^{-1}$ (generalized least squares, GLS), and $W = \Sigma_d^{-1}$ (weighted least squares, WLS), where $\Sigma_d = \text{Diag}(\Sigma)$ is a diagonal matrix containing the diagonal elements of Σ.

When W is known (say OLS CSR), we can estimate γ in (9.3) by

$$\hat{\gamma} = (\hat{X}'W\hat{X})^{-1}\hat{X}'W\hat{\mu}_2. \tag{9.7}$$

In the feasible GLS and WLS cases, W contains unknown parameters and one needs to substitute a consistent estimate of W, say $\hat{W}$, in (9.7). This is typically the corresponding matrix of sample moments, $\hat{W} = \hat{V}_{22}^{-1}$ for GLS and $\hat{W} = \hat{\Sigma}_d^{-1}$ for WLS. As pointed out by Lewellen et al. (2010), the estimates of γ are the same regardless of whether we use $W = V_{22}^{-1}$ or $W = \Sigma^{-1}$ as the weighting matrix for the GLS CSR. However, it should be noted that the cross-sectional R^2s are different for $W = V_{22}^{-1}$ and $W = \Sigma^{-1}$. For the purpose of model comparison, it makes sense to use a common W across models, so we prefer to use $W = V_{22}^{-1}$ for the case of GLS CSR.

9.3 Statistical Inference Under Correctly Specified Models

In this section, we present the asymptotic distribution of $\hat{\gamma}$ when the model is correctly specified, i.e., (9.3) holds exactly.

We first consider the special case in which the true betas are used in the second-pass CSR. The estimate of γ is given by

$$\hat{\gamma} = A\hat{\mu}_2, \tag{9.8}$$

where $A = (X'WX)^{-1}X'W$. Equation (9.8) shows that the randomness of $\hat{\gamma}$ is entirely driven by the randomness of $\hat{\mu}_2$. Under the joint stationarity and ergodicity assumptions, we have

$$\sqrt{T}(\hat{\mu}_2 - \mu_2) \overset{A}{\sim} N\left(0_N, \sum_{j=-\infty}^{\infty} E[(R_t - \mu_2)(R_{t+j} - \mu_2)']\right). \tag{9.9}$$

It follows that

$$\sqrt{T}(\hat{\gamma} - \gamma) \overset{A}{\sim} N(0_{K+1}, V(\hat{\gamma})), \tag{9.10}$$

where

$$V(\hat{\gamma}) = \sum_{j=-\infty}^{\infty} E[h_t h_{t+j}'], \tag{9.11}$$

with

$$h_t = \gamma_t - \gamma, \tag{9.12}$$

and $\gamma_t \equiv [\gamma_{0t}, \gamma_{1t}']' = AR_t$ is the period-by-period estimate of γ from regressing R_t on X.

If R_t is serially uncorrelated, then h_t is serially uncorrelated and

$$V(\hat{\gamma}) = AV_{22}A'. \tag{9.13}$$

For statistical inference, we need a consistent estimator of $V(\hat{\gamma})$. This can be accomplished by replacing h_t with

$$\hat{h}_t = \gamma_t - \hat{\gamma}. \tag{9.14}$$

When h_t is serially uncorrelated, a consistent estimator of the asymptotic variance of $\hat{\gamma}$ is given by

$$\hat{V}(\hat{\gamma}) = \frac{1}{T}\sum_{t=1}^{T} \hat{h}_t \hat{h}_t'. \tag{9.15}$$

Equation (9.15) yields the popular standard error of $\hat{\gamma}$ due to Fama and MacBeth (1973), which is obtained using the standard deviation of the time series $\{\gamma_t\}$. When h_t is autocorrelated, one can use the method proposed by Newey and West (1987) to obtain a consistent estimator of $V(\hat{\gamma})$.

In the general case, the betas are estimated with error in the first-pass time series regression and an errors-in-variables (EIV) problem is introduced in the second-pass CSR. Measurement errors in the betas cause two problems. The first is that the estimated zero-beta rate and risk premia are biased, though Shanken (1992) shows that they are consistent as the length of the time series increases to infinity. The second problem is that the usual Fama-MacBeth standard errors for the estimated zero-beta rate and risk premia are inconsistent. Shanken (1992) addresses this by developing an asymptotically valid EIV adjustment of the standard errors. Jagannathan and Wang (1998) extend Shanken's asymptotic analysis by relaxing the assumption that the returns are homoskedastic conditional on the factors.

It turns out that one can easily deal with the EIV problem by replacing h_t in (9.12) with

$$h_t = (\gamma_t - \gamma) - (\phi_t - \phi)w_t, \tag{9.16}$$

where $\phi_t = [\gamma_{0t},\ (\gamma_{1t} - f_t)']'$, $\phi = [\gamma_0,\ (\gamma_1 - \mu_1)']'$, and $w_t = \gamma_1' V_{11}^{-1}(f_t - \mu_1)$. The second term, $(\phi_t - \phi)w_t$, is the EIV adjustment term that accounts for the estimation error in $\hat{\beta}$. To estimate $V(\hat{\gamma})$, we replace h_t with its sample counterpart

$$\hat{h}_t = (\hat{\gamma}_t - \hat{\gamma}) - (\hat{\phi}_t - \hat{\phi})\hat{w}_t, \tag{9.17}$$

where $\hat{\gamma}_t = [\hat{\gamma}_{0t},\ \hat{\gamma}_{1t}']' = (\hat{X}'W\hat{X})^{-1}\hat{X}'WR_t$, $\hat{\phi}_t = [\hat{\gamma}_{0t},\ (\hat{\gamma}_{1t} - f_t)']'$, $\hat{\phi} = [\hat{\gamma}_0,\ (\hat{\gamma}_1 - \hat{\mu}_1)']'$, and $\hat{w}_t = \hat{\gamma}_1' \hat{V}_{11}^{-1}(f_t - \hat{\mu}_1)$.

When h_t is serially uncorrelated and $\mathrm{Var}[R_t | f_t] = \Sigma$ (conditional homoskedasticity case), we can simplify $V(\hat{\gamma})$ to

$$V(\hat{\gamma}) = AV_{22}A' + \gamma_1' V_{11}^{-1} \gamma_1 A \Sigma A', \tag{9.18}$$

which is the expression given in Shanken (1992). Using the fact that

$$V_{22} = \Sigma + \beta V_{11} \beta' = \Sigma + X \begin{bmatrix} 0 & 0_K' \\ 0_K & V_{11} \end{bmatrix} X', \tag{9.19}$$

we can also write (9.18) as

$$V(\hat{\gamma}) = (1 + \gamma_1' V_{11}^{-1} \gamma_1) A \Sigma A' + \begin{bmatrix} 0 & 0_K' \\ 0_K & V_{11} \end{bmatrix}. \tag{9.20}$$

In the above analysis, we have treated W as a known weighting matrix. Under a correctly specified model, the asymptotic distribution of $\hat{\gamma}$ does not depend on whether we use W or its consistent estimator $\hat{W}$ as the weighting matrix. Therefore, the asymptotic results in this section also hold for the GLS CSR and WLS CSR cases.

Under a correctly specified model, it is interesting to derive the optimal (in the sense that it minimizes $V(\hat{\gamma})$) weighting matrix W in the second-pass CSR. Ahn et al. (2009) provide an analysis of this problem. Using the fact that $\gamma_t - \gamma = A(R_t - \mu_2)$ and $\phi_t - \phi = A\epsilon_t$, where $\epsilon_t = (R_t - \mu_2) - \beta(f_t - \mu_1)$, we can write

$$h_t = A l_t, \tag{9.21}$$

where $l_t \equiv R_t - \mu_2 - \epsilon_t w_t$. It follows that

$$V(\hat{\gamma}) = A V_l A' = (X'WX)^{-1} X'W V_l W X (X'WX)^{-1}, \tag{9.22}$$

where

$$V_l = \sum_{j=-\infty}^{\infty} E[l_t l_{t+j}']. \tag{9.23}$$

From this expression, it is obvious that we can choose $W = V_l^{-1}$ to minimize $V(\hat{\gamma})$ and we have $\min_W V(\hat{\gamma}) = (X' V_l^{-1} X)^{-1}$. However, it is important to note that V_l^{-1} is not the only choice of W that minimizes $V(\hat{\gamma})$. Using a lemma in Kan and Zhou (2004), it is easy to show that any W that is of the form $(a V_l + XCX')^{-1}$, where a is a positive scalar and C is an arbitrary symmetric matrix, will also yield the lowest $V(\hat{\gamma})$.

There are a few cases in which the GLS CSR will give us the lower bound of $V(\hat{\gamma})$. The first case arises when h_t is serially uncorrelated and $\text{Var}[R_t | f_t] = \Sigma$ (conditional homoskedasticity case). In this scenario, we have

$$V_l = E[l_t l_t'] = (1 + \gamma_1' V_{11}^{-1} \gamma_1) \Sigma + \beta V_{11} \beta' = V_{22} + \gamma_1' V_{11}^{-1} \gamma_1 \Sigma. \tag{9.24}$$

It can be readily shown that

$$(X' V_l^{-1} X)^{-1} = (X' V_{22}^{-1} X)^{-1} + \gamma_1' V_{11}^{-1} \gamma_1 (X' \Sigma^{-1} X)^{-1}. \tag{9.25}$$

The second case arises when $Y_t = [f_t', \; R_t']'$ is i.i.d. multivariate elliptically distributed with multivariate excess kurtosis parameter κ. In this case, we have

$$V_l = E[l_t l_t'] = [1 + (1+\kappa) \gamma_1' V_{11}^{-1} \gamma_1] \Sigma + \beta V_{11} \beta' = V_{22} + (1+\kappa) \gamma_1' V_{11}^{-1} \gamma_1 \Sigma \tag{9.26}$$

and

$$(X'V_l^{-1}X)^{-1} = (X'V_{22}^{-1}X)^{-1} + (1+\kappa)\gamma_1'V_{11}^{-1}\gamma_1(X'\Sigma^{-1}X)^{-1}. \quad (9.27)$$

In general, the GLS CSR is not the optimal CSR to be used in the second pass. The best choice of W is V_l^{-1}. To use the optimal two-pass CSR, one needs to obtain a consistent estimator of V_l. This can be accomplished with a two-step procedure: (1) Obtain a consistent estimate of γ_1 using, for example, the OLS CSR. (2) Estimate V_l using $\hat{l}_t = (R_t - \hat{\mu}_2) - \hat{\epsilon}_t\hat{w}_t$.

9.4 Statistical Inference Under Potentially Misspecified Models

Standard inference using the two-pass CSR methodology typically assumes that expected returns are exactly linear in the betas, i.e., the beta pricing model is correctly specified. It is difficult to justify this assumption when estimating many different models because some (if not all) of the models are bound to be misspecified. Moreover, since asset pricing models are, at best, approximations of reality, it is inevitable that we will often, knowingly or unknowingly (because of limited power), estimate an expected return relation that departs from exact linearity in the betas. In this section, we discuss how to conduct statistical inference on $\hat{\gamma}$ when the model is potentially misspecified. The results that we present here are mostly drawn from Kan et al. (2010), which generalizes the earlier results of Hou and Kimmel (2006) and Shanken and Zhou (2007) that are obtained under a normality assumption.

When the model is misspecified, the pricing-error vector, $\mu_2 - X\gamma$, will be nonzero for all values of γ. For a given weighting matrix W, we define the (pseudo) zero-beta rate and risk premia as the choice of γ that minimizes the quadratic form of pricing errors:

$$\gamma_W \equiv \begin{bmatrix} \gamma_{W,0} \\ \gamma_{W,1} \end{bmatrix} = \text{argmin}_\gamma (\mu_2 - X\gamma)'W(\mu_2 - X\gamma) = (X'WX)^{-1}X'W\mu_2. \quad (9.28)$$

The corresponding pricing errors of the N assets are then given by

$$e_W = \mu_2 - X\gamma_W. \quad (9.29)$$

It should be emphasized that unless the model is correctly specified, γ_W and e_W depend on the choice of W. To simplify the notation, we suppress the subscript W from γ_W and e_W when the choice of W is clear from the context.

Unlike the case of correctly specified models, the asymptotic variance of $\hat{\gamma}$ under a misspecified model depends on whether we use W or $\hat{W}$ as the weighting matrix. As a result, we need to separate these two cases when presenting the asymptotic distribution of $\hat{\gamma}$. For the known weighting matrix case, the asymptotic variance of $\hat{\gamma}$ is obtained by replacing h_t in (9.16) with

$$h_t = (\gamma_t - \gamma) - (\phi_t - \phi)w_t + (X'WX)^{-1}z_t u_t, \tag{9.30}$$

where $z_t = [0,\ (f_t - \mu_1)'V_{11}^{-1}]'$ and $u_t = e'W(R_t - \mu_2)$.

For the GLS case that uses $W = V_{22}^{-1}$ as the weighting matrix, h_t has the following expression:

$$h_t = (\gamma_t - \gamma) - (\phi_t - \phi)w_t + (X'V_{22}^{-1}X)^{-1}z_t u_t - (\gamma_t - \gamma)u_t. \tag{9.31}$$

For the WLS case, h_t is given by

$$h_t = (\gamma_t - \gamma) - (\phi_t - \phi)w_t + (X'\Sigma_d^{-1}X)^{-1}z_t u_t - A\Psi_t\Sigma_d^{-1}e, \tag{9.32}$$

where $\Psi_t = \text{Diag}(\epsilon_t\epsilon_t')$ and $\epsilon_t = (R_t - \mu_2) - \beta(f_t - \mu_1)$. As before, we can obtain a consistent estimator of $V(\hat{\gamma})$ by replacing h_t with its sample counterpart.

Note that model misspecification adds extra terms to h_t and this could have a serious impact on the standard error of $\hat{\gamma}$. For example, when h_t is serially uncorrelated and the conditional homoskedasticity assumption holds, we can show that for the GLS CSR

$$\begin{aligned} V(\hat{\gamma}) = {} & (X'V_{22}^{-1}X)^{-1} + \gamma_1'V_{11}^{-1}\gamma_1(X'\Sigma^{-1}X)^{-1} + e'V_{22}^{-1}e \\ & \times \left[(X'\Sigma^{-1}X)^{-1} \begin{bmatrix} 0 & 0_K' \\ 0_K & V_{11}^{-1} \end{bmatrix} (X'\Sigma^{-1}X)^{-1} + (X'\Sigma^{-1}X)^{-1} \right]. \end{aligned} \tag{9.33}$$

We call the last term in (9.33) the misspecification adjustment term. When $e'V_{22}^{-1}e > 0$, the misspecification adjustment term is positive definite since it is the sum of two matrices, the first positive semidefinite and the second positive definite. It can also be shown that the misspecification adjustment term crucially depends on the variance of the residuals from projecting the factors on the returns. For factors that have very low correlations with the returns (e.g., macroeconomic factors), the impact of the misspecification adjustment term on the asymptotic variance of $\hat{\gamma}_1$ can be very large.

9.5 Specification Tests and Measures of Model Misspecification

One of the earliest problems in empirical asset pricing has been to determine whether a proposed model is correctly specified or not. This can be accomplished by using various specification tests, which are typically aggregate measures of sample pricing errors. However, some of these specification tests aggregate the pricing errors using weighting matrices that are model dependent, and these test statistics cannot be used to perform model comparison. Therefore, researchers are often interested in a normalized goodness-of-fit measure that uses the same weighting matrix across models. One such measure is the cross-sectional R^2. Following Kandel and Stambaugh (1995), this is defined as

$$\rho^2 = 1 - \frac{Q}{Q_0}, \tag{9.34}$$

where

$$\begin{aligned} Q_0 &= \min_{\gamma_0}(\mu_2 - 1_N\gamma_0)'W(\mu_2 - 1_N\gamma_0) \\ &= \mu_2'W\mu_2 - \mu_2'W1_N(1_N'W1_N)^{-1}1_N'W\mu_2, \end{aligned} \tag{9.35}$$

$$\begin{aligned} Q &= e'We \\ &= \mu_2'W\mu_2 - \mu_2'WX(X'WX)^{-1}X'W\mu_2. \end{aligned} \tag{9.36}$$

In order for ρ^2 to be well defined, we need to assume that μ_2 is not proportional to 1_N (the expected returns are not all equal) so that $Q_0 > 0$. Note that $0 \le \rho^2 \le 1$ and it is a decreasing function of the aggregate pricing-error measure $Q = e'We$. Thus, ρ^2 is a natural measure of goodness of fit. However, it should be emphasized that unless the model is correctly specified, ρ^2 depends on the choice of W. Therefore, it is possible that a model with a good fit under the OLS CSR provides a very poor fit under the GLS CSR.

The sample measure of ρ^2 is similarly defined as

$$\hat{\rho}^2 = 1 - \frac{\hat{Q}}{\hat{Q}_0}, \tag{9.37}$$

where $\hat{Q}_0$ and $\hat{Q}$ are consistent estimators of Q_0 and Q in (9.35) and (9.36), respectively. When W is known, we estimate Q_0 and Q using

$$\hat{Q}_0 = \hat{\mu}_2'W\hat{\mu}_2 - \hat{\mu}_2'W1_N(1_N'W1_N)^{-1}1_N'W\hat{\mu}_2, \tag{9.38}$$

$$\hat{Q} = \hat{\mu}_2'W\hat{\mu}_2 - \hat{\mu}_2'W\hat{X}(\hat{X}'W\hat{X})^{-1}\hat{X}'W\hat{\mu}_2. \tag{9.39}$$

When W is not known, we replace W with $\hat{W}$ in the formulas above.

To test the null hypothesis of correct model specification, i.e., $e = 0_N$ (or, equivalently, $Q = 0$ and $\rho^2 = 1$), we typically rely on the sample pricing errors $\hat{e}$. Therefore, it is important to obtain the asymptotic distribution of $\hat{e}$ under the null hypothesis. For a given weighting matrix W (or $\hat{W}$ with a limit of W), let P be an $N \times (N - K - 1)$ orthonormal matrix with its columns orthogonal to $W^{\frac{1}{2}}X$. Kan et al. (2010) derive the asymptotic distribution of $\hat{e}$ under the null hypothesis:

$$\sqrt{T}\hat{e} \overset{A}{\sim} N(0_N, V(\hat{e})), \tag{9.40}$$

where

$$V(\hat{e}) = \sum_{j=-\infty}^{\infty} E[q_t q'_{t+j}], \tag{9.41}$$

with

$$q_t = W^{-\frac{1}{2}} P P' W^{\frac{1}{2}} \epsilon_t y_t, \tag{9.42}$$

and $y_t = 1 - \gamma_1' V_{11}^{-1}(f_t - \mu_1)$.

Remark 1. Under the correctly specified model, the asymptotic distribution of $\hat{e}$ does not depend on whether we use W or $\hat{W}$ as the weighting matrix.

Remark 2. Under the correctly specified model, q_t in (9.42) can also be written as

$$q_t = W^{-\frac{1}{2}} P P' W^{\frac{1}{2}} R_t y_t. \tag{9.43}$$

Remark 3. $V(\hat{e})$ is a singular matrix and some linear combinations of $\sqrt{T}\hat{e}$ are not asymptotically normally distributed. As a result, one has to be careful when relying on individual sample pricing errors to test the validity of a model because some of them may not be asymptotically normally distributed. Gospodinov et al. (2010b) provide a detailed analysis of this problem. For our subsequent analysis, it is easier to work with

$$\tilde{e} = P' W^{\frac{1}{2}} \hat{e}. \tag{9.44}$$

The reason is that the asymptotic variance of $\tilde{e}$ is given by

$$V(\tilde{e}) = \sum_{j=-\infty}^{\infty} E[\tilde{q}_t \tilde{q}'_{t+j}], \tag{9.45}$$

where

$$\tilde{q}_t = P' W^{\frac{1}{2}} \epsilon_t y_t, \tag{9.46}$$

and $V(\tilde{e})$ is nonsingular.

Given (9.40), we can obtain the asymptotic distribution of any quadratic form of sample pricing errors. For example, let Ω be an $N \times N$ positive definite matrix, and let $\hat{\Omega}$ be a consistent estimator of Ω. When the model is correctly specified, we have

$$T\hat{e}'\hat{\Omega}\hat{e} \overset{A}{\sim} \sum_{i=1}^{N-K-1} \xi_i x_i, \tag{9.47}$$

where the x_i's are independent χ^2_1 random variables, and the ξ_i's are the $N-K-1$ eigenvalues of

$$(P'W^{-\frac{1}{2}}\Omega W^{-\frac{1}{2}}P)V(\tilde{e}). \tag{9.48}$$

Using an algorithm due to Imhof (1961) and later improved by Davies (1980) and Lu and King (2002), one can easily compute the cumulative distribution function of a linear combination of independent χ^2 random variables. As a result, one can use (9.47) as a specification test of the model.

There are several interesting choices of $\hat{\Omega}$. The first one is $\hat{\Omega} = \hat{W}$, and the test statistic is simply given by $T\hat{e}'\hat{W}\hat{e} = T\hat{Q}$. In this case, the ξ_i's are the eigenvalues of $V(\tilde{e})$. The second one is $\hat{\Omega} = \hat{V}(\hat{e})^+$, where $\hat{V}(\hat{e})$ is a consistent estimator of $V(\hat{e})$ and $\hat{V}(\hat{e})^+$ stands for its pseudo-inverse. This choice of $\hat{\Omega}$ yields the following Wald test statistic:

$$J_W = T\hat{e}'\hat{V}(\hat{e})^+\hat{e} = T\tilde{e}'\hat{V}(\tilde{e})^{-1}\tilde{e} \overset{A}{\sim} \chi^2_{N-K-1}, \tag{9.49}$$

where $\hat{V}(\tilde{e})$ is a consistent estimator of $V(\tilde{e})$. The advantage of using J_W is that its asymptotic distribution is simply χ^2_{N-K-1} and does not involve the computation of the distribution of a linear combination of independent χ^2 random variables. The disadvantage of using J_W is that the weighting matrix is model dependent, making it problematic to compare the J_W's of different models.

When q_t is serially uncorrelated and $\text{Var}[R_t\,|\,f_t] = \Sigma$ (conditional homoskedasticity case), we can show that

$$V(\tilde{e}) = (1 + \gamma_1'V_{11}^{-1}\gamma_1)P'W^{\frac{1}{2}}\Sigma W^{\frac{1}{2}}P. \tag{9.50}$$

For the special case of $W = V_{22}^{-1}$ or $W = \Sigma^{-1}$, we have

$$V(\tilde{e}) = (1 + \gamma_1'V_{11}^{-1}\gamma_1)I_{N-K-1}. \tag{9.51}$$

If we estimate $V(\tilde{e})$ using $\hat{V}(\tilde{e}) = (1 + \hat{\gamma}_1'\hat{V}_{11}^{-1}\hat{\gamma}_1)I_{N-K-1}$, the Wald test in (9.49) becomes

$$J_W = T\tilde{e}'\hat{V}(\tilde{e})^{-1}\tilde{e} = \frac{T\hat{e}'\hat{V}_{22}^{-1}\hat{e}}{1 + \hat{\gamma}_1'\hat{V}_{11}^{-1}\hat{\gamma}_1} = \frac{T\hat{e}'\hat{\Sigma}^{-1}\hat{e}}{1 + \hat{\gamma}_1'\hat{V}_{11}^{-1}\hat{\gamma}_1} \overset{A}{\sim} \chi^2_{N-K-1}, \tag{9.52}$$

and J_W coincides with the cross-sectional regression test (CSRT) proposed by Shanken (1985). Better finite sample properties of the Wald test can be obtained, as suggested by Shanken (1985), by using the following approximate F-test:

$$J_W \overset{\text{app.}}{\sim} \frac{T(N-K-1)}{T-N+1}F_{N-K-1,T-N+1}. \tag{9.53}$$

Using the general result in (9.47), one can show that when the model is correctly specified,

$$T(\hat{\rho}^2 - 1) \overset{A}{\sim} \sum_{i=1}^{N-K-1} -\frac{\xi_i}{Q_0} x_i, \tag{9.54}$$

and the sample cross-sectional R^2 can be used as a specification test.

When the model is misspecified, i.e., $\rho^2 < 1$, there are two possible asymptotic distributions for $\hat{\rho}^2$. When $\rho^2 = 0$, we have

$$T\hat{\rho}^2 \overset{A}{\sim} \sum_{i=1}^{K} \tilde{\xi}_i x_i, \tag{9.55}$$

where the x_i's are independent χ_1^2 random variables and the $\tilde{\xi}_i$'s are the eigenvalues of

$$[\beta' W \beta - \beta' W 1_N (1_N' W 1_N)^{-1} 1_N' W \beta] V(\hat{\gamma}_1), \tag{9.56}$$

where $V(\hat{\gamma}_1)$ is the asymptotic covariance matrix of $\hat{\gamma}_1$ under potentially misspecified models (i.e., based on the expressions of h_t in (9.30)–(9.32)). This asymptotic distribution permits a test of whether the model has any explanatory power for expected returns. It can be shown that $\rho^2 = 0$ if and only if $\gamma_1 = 0_K$. Therefore, one can also test $H_0 : \rho^2 = 0$ using a Wald test of $H_0 : \gamma_1 = 0_K$.

When $0 < \rho^2 < 1$, the asymptotic distribution of $\hat{\rho}^2$ is given by

$$\sqrt{T}(\hat{\rho}^2 - \rho^2) \overset{A}{\sim} N\left(0, \sum_{j=-\infty}^{\infty} E[n_t n_{t+j}]\right), \tag{9.57}$$

where

$$n_t = 2\left[-u_t y_t + (1-\rho^2) v_t\right] / Q_0 \qquad \text{for known } W, \tag{9.58}$$

$$n_t = \left[u_t^2 - 2u_t y_t + (1-\rho^2)(2v_t - v_t^2)\right] / Q_0 \qquad \text{for } \hat{W} = \hat{V}_{22}^{-1}, \tag{9.59}$$

$$n_t = \left[e' \Gamma_t e - 2u_t y_t + (1-\rho^2)(2v_t - e_0 \Gamma_t e_0)\right] / Q_0 \quad \text{for } \hat{W} = \hat{\Sigma}_d^{-1}, \tag{9.60}$$

with $v_t = e_0' W (R_t - \mu_2)$ and $\Gamma_t = \Sigma_d^{-1} \text{Diag}(\epsilon_t \epsilon_t') \Sigma_d^{-1}$.

In the $0 < \rho^2 < 1$ case, $\hat{\rho}^2$ is asymptotically normally distributed around its true value. It is readily verified that the expressions for n_t approach zero when $\rho^2 \to 0$ or $\rho^2 \to 1$. Consequently, the standard error of $\hat{\rho}^2$ tends to be lowest when ρ^2 is close to zero or one, and thus it is not monotonic in ρ^2. Note that the asymptotic normal distribution of $\hat{\rho}^2$ breaks down for the two extreme cases ($\rho^2 = 0$ or 1). Intuitively, the normal distribution fails because, by construction, $\hat{\rho}^2$ will always be above zero (even when $\rho^2 = 0$) and below one (even when $\rho^2 = 1$).

9.6 Some Subtle Issues

In this section, we discuss two issues related to the two-pass CSR methodology that are worth clarifying. The first point is about testing whether the risk premium associated with an individual factor is equal to zero. The second point is about the assumption of full column rank on the matrix $X = [1_N, \beta]$.

While the betas are typically used as the regressors in the second-pass CSR, there is a potential issue with the use of multiple regression betas when $K > 1$: in general, the beta of an asset with respect to a particular factor depends on what other factors are included in the first-pass time series OLS regression. As a consequence, the interpretation of the risk premia in the context of model selection can be problematic.

For example, suppose that a model has two factors f_1 and f_2. We are often interested in determining whether f_2 is needed in the model. Some researchers have tried to answer this question by performing a test of $H_0 : \gamma_2 = 0$, where γ_2 is the risk premium associated with factor 2. When the null hypothesis is rejected by the data, they typically conclude that factor 2 is important, and when the null hypothesis is not rejected, they conclude that factor 2 is unimportant. In the following, we provide two numerical examples that illustrate that the test of $H_0 : \gamma_2 = 0$ does not answer the question of whether factor 2 helps to explain the cross-sectional differences in expected returns on the test assets.

In the first example, we consider two factors with

$$V_{11} = \begin{bmatrix} 15 & -10 \\ -10 & 15 \end{bmatrix}. \tag{9.61}$$

Suppose there are four assets and their expected returns and covariances with the two factors are

$$\mu_2 = [2,\ 3,\ 4,\ 5]', \qquad V_{12} = \begin{bmatrix} 1 & 2 & 3 & 4 \\ 3 & 5 & 2 & 1 \end{bmatrix}. \tag{9.62}$$

It is clear that the covariances (or simple regression betas) of the four assets with respect to the first factor alone can fully explain μ_2 because μ_2 is exactly linear in the first row of V_{12}. As a result, the second factor is irrelevant from a cross-sectional expected return perspective. However, when we compute the (multiple regression) beta matrix with respect to the two factors, we obtain:

$$\beta = V_{21}V_{11}^{-1} = \begin{bmatrix} 0.36 & 0.64 & 0.52 & 0.56 \\ 0.44 & 0.76 & 0.48 & 0.44 \end{bmatrix}'. \tag{9.63}$$

Simple calculations give $\gamma = [1,\ 15,\ -10]'$ and γ_2 is nonzero even though f_2 is irrelevant. This suggests that when the capital asset pricing model is true, it does not imply that the betas with respect to the other two Fama and French (1993) factors should not be priced. See Grauer and Janmaat (2009) for a discussion of this point.

In the second example, we change μ_2 to $[10,\ 17,\ 14,\ 15]'$. In this case, the covariances (or simple regression betas) with respect to f_1 alone do not fully explain μ_2 (in fact, the OLS R^2 for the model with just f_1 is only 28%). However, it is easy to see that μ_2 is linear in the first column of the beta matrix, implying that the R^2 of the full model is 100%. Simple calculations give us $\gamma = [1,\ 25,\ 0]'$ and $\gamma_2 = 0$, even though f_2 is needed in the factor model, along with f_1, to explain μ_2.

To overcome this problem, we propose an alternative second-pass CSR that uses the covariances V_{21} as the regressors. Let $C = [1_N,\ V_{21}]$ and λ_W be the choice of coefficients that minimizes the quadratic form of pricing errors:

$$\lambda_W \equiv \begin{bmatrix} \lambda_{W,0} \\ \lambda_{W,1} \end{bmatrix} = \text{argmin}_\lambda (\mu_2 - C\lambda)' W (\mu_2 - C\lambda) = (C'WC)^{-1} C' W \mu_2. \quad (9.64)$$

Given (9.28) and (9.64), there is a one-to-one correspondence between γ_W and λ_W:

$$\lambda_{W,0} = \gamma_{W,0}, \qquad \lambda_{W,1} = V_{11}^{-1} \gamma_{W,1}. \quad (9.65)$$

To simplify the notation, we suppress the subscript W from λ_W when the choice of W is clear from the context. It is easy to see that the pricing errors from this alternative second-pass CSR are the same as the pricing errors from the CSR that uses the betas as regressors. It follows that the ρ^2 for these two CSRs are also identical. However, it is important to note that unless V_{11} is a diagonal matrix, $\lambda_{1,i} = 0$ does not imply $\gamma_{1,i} = 0$, and vice versa. If interest lies in determining whether a particular factor i contributes to the explanatory power of the model, the correct hypothesis to test is $H_0 : \lambda_{1,i} = 0$ and not $H_0 : \gamma_{1,i} = 0$. This issue is also discussed in Jagannathan and Wang (1998) and Cochrane (2005, Chap. 13.4). Another solution to this problem is to use simple regression betas as the regressors in the second-pass CSR, as in Chen et al. (1986) and Jagannathan and Wang (1996, 1998). Kan and Robotti (2011) provide asymptotic results for the CSR with simple regression betas under potentially misspecified models.

Let $\hat{C} = [1_N,\ \hat{V}_{21}]$. The estimate of λ from the second-pass CSR is given by

$$\hat{\lambda} = (\hat{C}'W\hat{C})^{-1} \hat{C}' W \hat{\mu}_2. \quad (9.66)$$

For the GLS and WLS cases, one needs to replace W with $\hat{W}$ in the expression for $\hat{\lambda}$.

Under a potentially misspecified model, the asymptotic distribution of $\hat{\lambda}$ is given by

$$\sqrt{T}(\hat{\lambda} - \lambda) \stackrel{A}{\sim} N(0_{K+1}, V(\hat{\lambda})), \quad (9.67)$$

where

$$V(\hat{\lambda}) = \sum_{j=-\infty}^{\infty} E[\tilde{h}_t \tilde{h}'_{t+j}]. \tag{9.68}$$

To simplify the expressions for $\tilde{h}_t$, we define $\tilde{G}_t = V_{21} - (R_t - \mu_2)(f_t - \mu_1)'$, $\tilde{z}_t = [0,\ (f_t - \mu_1)']'$, $\tilde{A} = (C'WC)^{-1}C'W$, $\lambda_t = \tilde{A}R_t$, and $u_t = e'W(R_t - \mu_2)$. The $\tilde{h}_t$ expressions for the different cases are given by

$$\tilde{h}_t = (\lambda_t - \lambda) + \tilde{A}\tilde{G}_t\lambda_1 + (C'WC)^{-1}\tilde{z}_t u_t \quad \text{for known } W, \tag{9.69}$$

$$\tilde{h}_t = (\lambda_t - \lambda) + \tilde{A}\tilde{G}_t\lambda_1 + (C'V_{22}^{-1}C)^{-1}\tilde{z}_t u_t - (\lambda_t - \lambda)u_t \quad \text{for } \hat{W} = \hat{V}_{22}^{-1}, \tag{9.70}$$

$$\tilde{h}_t = (\lambda_t - \lambda) + \tilde{A}\tilde{G}_t\lambda_1 + (C'\Sigma_d^{-1}C)^{-1}\tilde{z}_t u_t - \tilde{A}\Psi_t\Sigma_d^{-1}e \quad \text{for } \hat{W} = \hat{\Sigma}_d^{-1} \tag{9.71}$$

where $\Psi_t = \text{Diag}(\epsilon_t\epsilon_t')$. Besides allowing us to test whether a given factor is important in a model, the asymptotic distribution of $\hat{\lambda}$ is necessary for the implementation of model comparison, a topic that will be discussed in Sects. 9.7 and 9.8. To test $H_0 : \lambda_{1,i} = 0$, one needs to obtain a consistent estimator of $V(\hat{\lambda})$. This can be easily accomplished by replacing $\tilde{h}_t$ with its sample counterpart.

The second issue that is often overlooked by researchers is related to the full column rank assumption on $X = [1_N,\ \beta]$. In the two-pass CSR methodology, we need to assume that X has full rank to ensure that γ_W (or λ_W) is uniquely defined. This assumption is often taken for granted and most researchers do not examine its validity before performing the two-pass CSR. When X does not have full column rank, the asymptotic results will break down, leading to misleading statistical inference on $\hat{\gamma}$ and $\hat{\rho}^2$, especially when the model is misspecified. For example, Kan and Zhang (1999) show that there is a high probability that the estimated risk premium on a useless factor is significantly different from zero. This happens because, when the factor is useless, $\beta = 0_N$ and $X = [1_N,\ \beta]$ does not have full column rank. As a result, γ_W is not uniquely defined and the usual asymptotic standard error of $\hat{\gamma}$ is no longer valid. Note that the useless factors scenario is not completely unreasonable since many macroeconomic factors exhibit very low correlations with asset returns. Even when the full column rank condition is satisfied in population, Kleibergen (2009) shows that there can still be serious finite sample problems with the asymptotic results if the factors have low correlations with the returns and the beta estimates are noisy.

When the factors have very low correlations with the returns, it is sensible to test whether X has full column rank before running the two-pass CSR. Note that testing $H_0 : \text{rank}(X) = K$ is the same as testing $H_0 : \text{rank}(\Pi) = K - 1$, where $\Pi = P'\beta$ and P is an $N \times (N-1)$ orthonormal matrix with its columns orthogonal to 1_N. When $K = 1$, it is easy to perform this test because the null hypothesis is simply $H_0 : P'\beta = 0_{N-1}$.

A simple Wald test of $H_0 : P'\beta = 0_{N-1}$ can be performed using the following test statistic:

$$J_1 = T\hat{\beta}'P(P'\hat{V}(\hat{\beta})P)^{-1}P'\hat{\beta} \overset{A}{\sim} \chi^2_{N-1}, \tag{9.72}$$

where $\hat{V}(\hat{\beta})$ is a consistent estimator of $V(\hat{\beta})$, the asymptotic covariance of $\hat{\beta}$. Under the stationarity and ergodicity assumptions on Y_t,

$$V(\hat{\beta}) = \sum_{j=-\infty}^{\infty} E[x_t x_{t+j}'], \tag{9.73}$$

where

$$x_t = V_{11}^{-1}(f_t - \mu_1)\epsilon_t. \tag{9.74}$$

If we further assume that x_t is serially uncorrelated and $\text{Var}[R_t | f_t] = \Sigma$ (conditional homoskedasticity case),

$$V(\hat{\beta}) = V_{11}^{-1}\Sigma, \tag{9.75}$$

and we can use the following Wald test:

$$J_2 = T\hat{V}_{11}\hat{\beta}' P(P'\hat{\Sigma}P)^{-1}P'\hat{\beta} \stackrel{A}{\sim} \chi^2_{N-1}. \tag{9.76}$$

When ϵ_t is i.i.d. multivariate normal, we have the following exact test:

$$F_2 = \frac{(T-N)J_2}{(N-1)T} \sim F_{N-1,T-N}. \tag{9.77}$$

When ϵ_t is not normally distributed, this F-test is only an approximate test but it generally works better than the asymptotic one.

When $K > 1$, the test of H_0 : rank$(\Pi) = K - 1$ is more complicated. Several tests of the rank of a matrix have been proposed in the literature. In the following, we describe the test of Cragg and Donald (1997). Let $\hat{\Pi} = P'\hat{\beta}$, we have

$$\sqrt{T}\text{vec}(\hat{\Pi} - \Pi) \stackrel{A}{\sim} N(0_{(N-1)K}, S_{\hat{\Pi}}), \tag{9.78}$$

where

$$S_{\hat{\Pi}} = \sum_{j=-\infty}^{\infty} E[\tilde{x}_t \tilde{x}_{t+j}'], \tag{9.79}$$

and

$$\tilde{x}_t = V_{11}^{-1}(f_t - \mu_1) \otimes P'\epsilon_t. \tag{9.80}$$

Denoting by $\hat{S}_{\hat{\Pi}}$ a consistent estimator of $S_{\hat{\Pi}}$, Cragg and Donald (1997) suggest that we can test H_0 : rank$(\Pi) = K - 1$ using

$$J_3 = T \min_{\Pi \in \Gamma(K-1)} \text{vec}(\hat{\Pi} - \Pi)' \hat{S}_{\hat{\Pi}}^{-1} \text{vec}(\hat{\Pi} - \Pi) \stackrel{A}{\sim} \chi^2_{N-K}, \tag{9.81}$$

where $\Gamma(K-1)$ is the space of an $(N-1)\times K$ matrix with rank $K-1$. This test is not computationally attractive in general since we need to optimize over $N(K-1)$ parameters. Gospodinov et al. (2010a) propose an alternative expression for J_3 that greatly reduces the computational burden. Their test is given by

$$J_3 = T \min_c [-1,\ c']\hat{\Pi}'(C\hat{S}_{\hat{\Pi}}C')^{-1}\hat{\Pi}[-1,\ c']' \overset{A}{\sim} \chi^2_{N-K}, \tag{9.82}$$

where c is a $(K-1)$-vector and $C = [-1,\ c'] \otimes I_{N-1}$. With this new expression, one can easily test whether X has full column rank even when K is large.

If we further assume that $\tilde{x}_t$ is serially uncorrelated and $\text{Var}[R_t|f_t] = \Sigma$ (conditional homoskedasticity case),

$$S_{\hat{\Pi}} = V_{11}^{-1} \otimes P'\Sigma P, \tag{9.83}$$

and we have the following simple test of H_0 : rank$(\Pi) = K-1$:

$$J_4 = T\xi_K \overset{A}{\sim} \chi^2_{N-K}, \tag{9.84}$$

where ξ_K is the smallest eigenvalue of $\hat{V}_{11}\hat{\beta}'P(P'\hat{\Sigma}P)^{-1}P'\hat{\beta}$.

9.7 Pairwise Model Comparison Tests

One way to think about pairwise model comparison is to ask whether two competing beta pricing models have the same population cross-sectional R^2. Kan et al. (2010) show that the asymptotic distribution of the difference between the sample cross-sectional R^2s of two models depends on whether the models are nested or non-nested and whether the models are correctly specified or not. In this section, we focus on the R^2 of the CSR with known weighting matrix W and on the R^2 of the GLS CSR that uses $\hat{W} = \hat{V}_{22}^{-1}$ as the weighting matrix. Since the weighting matrix of the WLS CSR is model dependent, it is not meaningful to compare the WLS cross-sectional R^2s of two or more models. Therefore, we do not consider the WLS cross-sectional R^2 in the remainder of the article. Our analysis in this section is based on the earlier work of Vuong (1989), Rivers and Vuong (2002), and Golden (2003).

Consider two competing beta pricing models. Let f_{1t}, f_{2t}, and f_{3t} be three sets of distinct factors at time t, where f_{it} is of dimension $K_i \times 1$, $i = 1, 2, 3$. Assume that model 1 uses f_{1t} and f_{2t}, while Model 2 uses f_{1t} and f_{3t} as factors. Therefore, model 1 requires that the expected returns on the test assets are linear in the betas or covariances with respect to f_{1t} and f_{2t}, i.e.,

$$\mu_2 = 1_N\lambda_{1,0} + \text{Cov}[R_t, f_{1t}']\lambda_{1,1} + \text{Cov}[R_t, f_{2t}']\lambda_{1,2} = C_1\lambda_1, \tag{9.85}$$

where $C_1 = [1_N,\ \mathrm{Cov}[R_t, f_{1t}'],\ \mathrm{Cov}[R_t, f_{2t}']]$ and $\lambda_1 = [\lambda_{1,0},\ \lambda_{1,1}',\ \lambda_{1,2}']'$. Model 2 requires that expected returns are linear in the betas or covariances with respect to f_{1t} and f_{3t}, i.e.,

$$\mu_2 = 1_N\lambda_{2,0} + \mathrm{Cov}[R_t, f_{1t}']\lambda_{2,1} + \mathrm{Cov}[R_t, f_{3t}']\lambda_{2,3} = C_2\lambda_2, \tag{9.86}$$

where $C_2 = [1_N,\ \mathrm{Cov}[R_t, f_{1t}'],\ \mathrm{Cov}[R_t, f_{3t}']]$ and $\lambda_2 = [\lambda_{2,0},\ \lambda_{2,1}',\ \lambda_{2,3}']'$.

In general, both models can be misspecified. The λ_i that maximizes the ρ^2 of model i is given by

$$\lambda_i = (C_i'WC_i)^{-1}C_i'W\mu_2, \tag{9.87}$$

where C_i is assumed to have full column rank, $i = 1, 2$. For each model, the pricing-error vector e_i, the aggregate pricing-error measure Q_i, and the corresponding goodness-of-fit measure ρ_i^2 are all defined as in Sects. 9.4 and 9.5.

When $K_2 = 0$, model 2 nests model 1 as a special case. Similarly, when $K_3 = 0$, model 1 nests model 2. When both $K_2 > 0$ and $K_3 > 0$, the two models are non-nested.

We study the nested models case next and deal with non-nested models later in the section. Without loss of generality, we assume $K_3 = 0$, so that model 1 nests model 2. Since $\rho_1^2 = \rho_2^2$ if and only if $\lambda_{1,2} = 0_{K_2}$ (this result is applicable even when the models are misspecified), testing whether the models have the same ρ^2 is equivalent to testing $H_0 : \lambda_{1,2} = 0_{K_2}$. Under the null hypothesis,

$$T\hat{\lambda}_{1,2}'\hat{V}(\hat{\lambda}_{1,2})^{-1}\hat{\lambda}_{1,2} \overset{A}{\sim} \chi^2_{K_2}, \tag{9.88}$$

where $\hat{V}(\hat{\lambda}_{1,2})$ is a consistent estimator of the asymptotic covariance of $\sqrt{T}(\hat{\lambda}_{1,2} - \lambda_{1,2})$ given in Sect. 9.6. This statistic can be used to test $H_0 : \rho_1^2 = \rho_2^2$. It is important to note that, in general, we cannot conduct this test using the usual standard error of $\hat{\lambda}$, which assumes that model 1 is correctly specified. Instead, we need to rely on the misspecification-robust standard error of $\hat{\lambda}$ given in Sect. 9.6.

Alternatively, one can derive the asymptotic distribution of $\hat{\rho}_1^2 - \hat{\rho}_2^2$ and use this statistic to test $H_0 : \rho_1^2 = \rho_2^2$. Partition $\tilde{H}_1 = (C_1'WC_1)^{-1}$ as

$$\tilde{H}_1 = \begin{bmatrix} \tilde{H}_{1,11} & \tilde{H}_{1,12} \\ \tilde{H}_{1,21} & \tilde{H}_{1,22} \end{bmatrix}, \tag{9.89}$$

where $\tilde{H}_{1,22}$ is $K_2 \times K_2$. Under the null hypothesis $H_0 : \rho_1^2 = \rho_2^2$,

$$T(\hat{\rho}_1^2 - \hat{\rho}_2^2) \overset{A}{\sim} \sum_{i=1}^{K_2} \frac{\xi_i}{Q_0} x_i, \tag{9.90}$$

where the x_i's are independent χ^2_1 random variables and the ξ_i's are the eigenvalues of $\tilde{H}_{1,22}^{-1}V(\hat{\lambda}_{1,2})$. Once again, it is worth emphasizing that the misspecification-robust version of $V(\hat{\lambda}_{1,2})$ should be used to test $H_0 : \rho_1^2 = \rho_2^2$. Model misspecification tends to create additional sampling variation in $\hat{\rho}_1^2 - \hat{\rho}_2^2$. Without taking this into account, one might mistakenly reject the null hypothesis when it is true. In actual testing, we replace ξ_i with its sample counterpart $\hat{\xi}_i$, where the $\hat{\xi}_i$'s are the eigenvalues of $\hat{\tilde{H}}_{1,22}^{-1}\hat{V}(\hat{\lambda}_{1,2})$, and $\hat{\tilde{H}}_{1,22}$ and $\hat{V}(\hat{\lambda}_{1,2})$ are consistent estimators of $\tilde{H}_{1,22}$ and $V(\hat{\lambda}_{1,2})$, respectively.

The test of $H_0 : \rho_1^2 = \rho_2^2$ is more complicated for non-nested models. The reason is that under H_0, there are three possible asymptotic distributions for $\hat{\rho}_1^2 - \hat{\rho}_2^2$, depending on why the two models have the same cross-sectional R^2. To see this, first let us define the normalized stochastic discount factors at time t for models 1 and 2 as

$$y_{1t} = 1 - (f_{1t} - E[f_{1t}])'\lambda_{1,1} - (f_{2t} - E[f_{2t}])'\lambda_{1,2}, \tag{9.91}$$

$$y_{2t} = 1 - (f_{1t} - E[f_{1t}])'\lambda_{2,1} - (f_{3t} - E[f_{3t}])'\lambda_{2,3}. \tag{9.92}$$

Kan et al. (2010) show that $y_{1t} = y_{2t}$ implies that the two models have the same pricing errors and hence $\rho_1^2 = \rho_2^2$. If $y_{1t} \neq y_{2t}$, there are additional cases in which $\rho_1^2 = \rho_2^2$. A second possibility is that both models are correctly specified (i.e., $\rho_1^2 = \rho_2^2 = 1$). This occurs, for example, if model 1 is correctly specified and the factors f_{3t} in model 2 are given by $f_{3t} = f_{2t} + \epsilon_t$, where ϵ_t is pure "noise" – a vector of measurement errors with mean zero, independent of returns. In this case, we have $C_1 = C_2$ and both models produce zero pricing errors. A third possibility is that the two models produce different pricing errors but the same overall goodness of fit. Intuitively, one model might do a good job of pricing some assets that the other prices poorly and vice versa, such that the aggregation of pricing errors is the same in each case ($\rho_1^2 = \rho_2^2 < 1$). As it turns out, each of these three scenarios results in a different asymptotic distribution for $\hat{\rho}_1^2 - \hat{\rho}_2^2$.

For non-nested models, Kan et al. (2010) show that $y_{1t} = y_{2t}$ if and only if $\lambda_{1,2} = 0_{K_2}$ and $\lambda_{2,3} = 0_{K_3}$. This result, which is applicable even when the models are misspecified, implies that we can test $H_0 : y_{1t} = y_{2t}$ by testing the joint hypothesis $H_0 : \lambda_{1,2} = 0_{K_2},\ \lambda_{2,3} = 0_{K_3}$. Let $\psi = [\lambda_{1,2}',\ \lambda_{2,3}']'$ and $\hat{\psi} = [\hat{\lambda}_{1,2}',\ \hat{\lambda}_{2,3}']'$. Under $H_0 : y_{1t} = y_{2t}$, the asymptotic distribution of $\hat{\psi}$ is given by

$$\sqrt{T}(\hat{\psi} - \psi) \overset{A}{\sim} N(0_{K_2+K_3}, V(\hat{\psi})), \tag{9.93}$$

where

$$V(\hat{\psi}) = \sum_{j=-\infty}^{\infty} E[\tilde{q}_t \tilde{q}_{t+j}'], \tag{9.94}$$

and $\tilde{q}_t$ is a $K_2 + K_3$ vector obtained by stacking up the last K_2 and K_3 elements of $\tilde{h}_t$ for models 1 and 2, respectively, where $\tilde{h}_t$ is given in Sect. 9.6.

Let $\hat{V}(\hat{\psi})$ be a consistent estimator of $V(\hat{\psi})$. Then, under the null hypothesis $H_0 : \psi = 0_{K_2+K_3}$,

$$T\hat{\psi}'\hat{V}(\hat{\psi})^{-1}\hat{\psi} \stackrel{A}{\sim} \chi^2_{K_2+K_3}, \tag{9.95}$$

and this statistic can be used to test $H_0 : y_{1t} = y_{2t}$. As in the nested models case, it is important to conduct this test using the misspecification-robust standard error of $\hat{\psi}$.

Alternatively, one can derive the asymptotic distribution of $\hat{\rho}_1^2 - \hat{\rho}_2^2$ given $H_0 : y_{1t} = y_{2t}$. Let $\tilde{H}_1 = (C_1'WC_1)^{-1}$ and $\tilde{H}_2 = (C_2'WC_2)^{-1}$, and partition them as

$$\tilde{H}_1 = \begin{bmatrix} \tilde{H}_{1,11} & \tilde{H}_{1,12} \\ \tilde{H}_{1,21} & \tilde{H}_{1,22} \end{bmatrix}, \qquad \tilde{H}_2 = \begin{bmatrix} \tilde{H}_{2,11} & \tilde{H}_{2,13} \\ \tilde{H}_{2,31} & \tilde{H}_{2,33} \end{bmatrix}, \tag{9.96}$$

where $\tilde{H}_{1,11}$ and $\tilde{H}_{2,11}$ are $(K_1+1)\times(K_1+1)$. Under the null hypothesis $H_0 : y_{1t} = y_{2t}$,

$$T(\hat{\rho}_1^2 - \hat{\rho}_2^2) \stackrel{A}{\sim} \sum_{i=1}^{K_2+K_3} \frac{\xi_i}{Q_0} x_i, \tag{9.97}$$

where the x_i's are independent χ^2_1 random variables and the ξ_i's are the eigenvalues of

$$\begin{bmatrix} \tilde{H}_{1,22}^{-1} & 0_{K_2\times K_3} \\ 0_{K_3\times K_2} & -\tilde{H}_{2,33}^{-1} \end{bmatrix} V(\hat{\psi}). \tag{9.98}$$

Note that we can think of the earlier nested models scenario as a special case of testing $H_0 : y_{1t} = y_{2t}$ with $K_3 = 0$. The only difference is that the ξ_i's in (9.90) are all positive whereas some of the ξ_i's in (9.97) are negative. As a result, we need to perform a two-sided test based on $\hat{\rho}_1^2 - \hat{\rho}_2^2$ in the non-nested models case.

If we fail to reject $H_0 : y_1 = y_2$, we are finished since equality of ρ_1^2 and ρ_2^2 is implied by this hypothesis. Otherwise, we need to consider the case $y_{1t} \neq y_{2t}$. As noted earlier, when $y_{1t} \neq y_{2t}$, the asymptotic distribution of $\hat{\rho}_1^2 - \hat{\rho}_2^2$ given $H_0 : \rho_1^2 = \rho_2^2$ depends on whether the models are correctly specified or not. A simple chi-squared statistic can be used for testing whether models 1 and 2 are both correctly specified. As this joint specification test focuses on the pricing errors, it can be viewed as a generalization of the CSRT of Shanken (1985), which tests the validity of the expected return relation for a single pricing model.

Let $n_1 = N - K_1 - K_2 - 1$ and $n_2 = N - K_1 - K_3 - 1$. Also let P_1 be an $N\times n_1$ orthonormal matrix with columns orthogonal to $W^{\frac{1}{2}}C_1$ and P_2 be an $N\times n_2$ orthonormal matrix with columns orthogonal to $W^{\frac{1}{2}}C_2$. Define

$$g_t(\theta) = \begin{bmatrix} g_{1t}(\lambda_1) \\ g_{2t}(\lambda_2) \end{bmatrix} = \begin{bmatrix} \epsilon_{1t}y_{1t} \\ \epsilon_{2t}y_{2t} \end{bmatrix}, \tag{9.99}$$

where ϵ_{1t} and ϵ_{2t} are the residuals of models 1 and 2, respectively, $\theta = (\lambda_1', \lambda_2')'$, and

$$S \equiv \begin{bmatrix} S_{11} & S_{12} \\ S_{21} & S_{22} \end{bmatrix} = \sum_{j=-\infty}^{\infty} E[g_t(\theta)g_{t+j}(\theta)']. \tag{9.100}$$

If $y_{1t} \neq y_{2t}$ and the null hypothesis $H_0 : \rho_1^2 = \rho_2^2 = 1$ holds, then

$$T \begin{bmatrix} \hat{P}_1'\hat{W}^{\frac{1}{2}}\hat{e}_1 \\ \hat{P}_2'\hat{W}^{\frac{1}{2}}\hat{e}_2 \end{bmatrix}' \begin{bmatrix} \hat{P}_1'\hat{W}^{\frac{1}{2}}\hat{S}_{11}\hat{W}^{\frac{1}{2}}\hat{P}_1 & \hat{P}_1'\hat{W}^{\frac{1}{2}}\hat{S}_{12}\hat{W}^{\frac{1}{2}}\hat{P}_2 \\ \hat{P}_2'\hat{W}^{\frac{1}{2}}\hat{S}_{21}\hat{W}^{\frac{1}{2}}\hat{P}_1 & \hat{P}_2'\hat{W}^{\frac{1}{2}}\hat{S}_{22}\hat{W}^{\frac{1}{2}}\hat{P}_2 \end{bmatrix}^{-1} \begin{bmatrix} \hat{P}_1'\hat{W}^{\frac{1}{2}}\hat{e}_1 \\ \hat{P}_2'\hat{W}^{\frac{1}{2}}\hat{e}_2 \end{bmatrix} \overset{A}{\sim} \chi^2_{n_1+n_2}, \tag{9.101}$$

where $\hat{e}_1$ and $\hat{e}_2$ are the sample pricing errors of models 1 and 2, and $\hat{P}_1$, $\hat{P}_2$, and $\hat{S}$ are consistent estimators of P_1, P_2, and S, respectively.

An alternative specification test makes use of the cross-sectional R^2s. If $y_{1t} \neq y_{2t}$ and the null hypothesis $H_0 : \rho_1^2 = \rho_2^2 = 1$ holds, then

$$T(\hat{\rho}_1^2 - \hat{\rho}_2^2) \overset{A}{\sim} \sum_{i=1}^{n_1+n_2} \frac{\xi_i}{Q_0} x_i, \tag{9.102}$$

where the x_i's are independent χ_1^2 random variables and the ξ_i's are the eigenvalues of

$$\begin{bmatrix} -P_1'W^{\frac{1}{2}}S_{11}W^{\frac{1}{2}}P_1 & -P_1'W^{\frac{1}{2}}S_{12}W^{\frac{1}{2}}P_2 \\ P_2'W^{\frac{1}{2}}S_{21}W^{\frac{1}{2}}P_1 & P_2'W^{\frac{1}{2}}S_{22}W^{\frac{1}{2}}P_2 \end{bmatrix}. \tag{9.103}$$

Note that the ξ_i's are not all positive because $\hat{\rho}_1^2 - \hat{\rho}_2^2$ can be negative. Thus, again, we need to perform a two-sided test of $H_0 : \rho_1^2 = \rho_2^2$.

If the hypothesis that both models are correctly specified is not rejected, we are finished, as the data are consistent with $H_0 : \rho_1^2 = \rho_2^2 = 1$. Otherwise, we need to determine whether $\rho_1^2 = \rho_2^2$ for some value less than one. As in the earlier analysis for $\hat{\rho}^2$, the asymptotic distribution of $\hat{\rho}_1^2 - \hat{\rho}_2^2$ changes when the models are misspecified. Suppose $y_{1t} \neq y_{2t}$ and $0 < \rho_1^2 = \rho_2^2 < 1$. Then,

$$\sqrt{T}(\hat{\rho}_1^2 - \hat{\rho}_2^2) \overset{A}{\sim} N\left(0, \sum_{j=-\infty}^{\infty} E[d_t d_{t+j}]\right). \tag{9.104}$$

When the weighting matrix W is known,

$$d_t = 2Q_0^{-1}(u_{2t}y_{2t} - u_{1t}y_{1t}), \tag{9.105}$$

where $u_{1t} = e_1'W(R_t - \mu_2)$ and $u_{2t} = e_2'W(R_t - \mu_2)$. With the GLS weighting matrix $\hat{W} = \hat{V}_{22}^{-1}$,

$$d_t = Q_0^{-1}(u_{1t}^2 - 2u_{1t}y_{1t} - u_{2t}^2 + 2u_{2t}y_{2t}). \tag{9.106}$$

Note that if $y_{1t} = y_{2t}$, then $\rho_1^2 = \rho_2^2$, $u_{1t} = u_{2t}$, and hence $d_t = 0$. Or, if $y_{1t} \neq y_{2t}$, but both models are correctly specified (i.e., $u_{1t} = u_{2t} = 0$ and $\rho_1^2 = \rho_2^2 = 1$), then again $d_t = 0$. Thus, the normal test cannot be used in these cases.

Given the three distinct cases encountered in testing $H_0 : \rho_1^2 = \rho_2^2$ for non-nested models, the approach we have described above entails a sequential test, as suggested by Vuong (1989). In our context, this involves first testing $H_0 : y_{1t} = y_{2t}$ using (9.95) or (9.97). If we reject $H_0 : y_{1t} = y_{2t}$, then we use (9.101) or (9.102) to test $H_0 : \rho_1^2 = \rho_2^2 = 1$. Finally, if this hypothesis is also rejected, we use the normal test in (9.104) to test $H_0 : 0 < \rho_1^2 = \rho_2^2 < 1$. Let α_1, α_2, and α_3 be the significance levels employed in these three tests. Then the sequential test has an asymptotic significance level that is bounded above by $\max[\alpha_1, \alpha_2, \alpha_3]$.

Another approach is to simply perform the normal test in (9.104). This amounts to assuming that $y_{1t} \neq y_{2t}$ and that both models are misspecified. The first assumption rules out the unlikely scenario that the additional factors are completely irrelevant for explaining cross-sectional variation in expected returns. The second assumption is sensible because asset pricing models are approximations of reality and we do not expect them to be perfectly specified.

9.8 Multiple Model Comparison Tests

Thus far, we have considered comparison of two competing models. However, given a set of models of interest, one may want to test whether one model, the "benchmark," has the highest ρ^2 of all models in the set. This gives rise to a common problem in applied work – if we focus on the statistic that provides the strongest evidence of rejection, without taking into account the process of searching across alternative specifications, there will be a tendency to reject the benchmark more often than the nominal size of the tests suggests. In other words, the true p-value will be larger than the one associated with the most extreme statistic.

Therefore, in this section we discuss how to perform model comparison when multiple models are involved. Suppose we have p models. Let ρ_i^2 denotes the cross-sectional R^2 of model i. We are interested in testing if model 1 performs as well as models 2 to p. Let $\delta = (\delta_2, \ldots, \delta_p)$, where $\delta_i = \rho_1^2 - \rho_i^2$. We are interested in testing $H_0 : \delta \geq 0_r$, where $r = p - 1$.

We consider two different tests of this null hypothesis. The first one is the multivariate inequality test developed by Kan et al. (2010). Numerous studies in statistics focus on tests of inequality constraints on parameters. The relevant work dates back to Bartholomew (1961), Kudo (1963), Perlman (1969), Gourieroux et al. (1982) and Wolak (1987, 1989). Following Wolak (1989), we state the null and alternative hypotheses as

$$H_0 : \delta \geq 0_r, \qquad H_1 : \delta \in \Re^r. \tag{9.107}$$

We also consider another test based on the reality check of White (2000) that has been used by Chen and Ludvigson (2009). Let $\delta_{\min} = \min_{2\leq i\leq p} \delta_i$. Define the null and alternative hypotheses as

$$H_0 : \delta_{\min} \geq 0, \qquad H_1 : \delta_{\min} < 0. \tag{9.108}$$

The null hypotheses presented above suggest that no other model outperforms model 1, whereas the alternative hypotheses suggest that at least one of the other models outperforms model 1.

Let $\hat{\delta} = (\hat{\delta}_2, \ldots, \hat{\delta}_p)$, where $\hat{\delta}_i = \hat{\rho}_1^2 - \hat{\rho}_i^2$. For both tests, we assume

$$\sqrt{T}(\hat{\delta} - \delta) \overset{A}{\sim} N(0_r, \Sigma_{\hat{\delta}}). \tag{9.109}$$

Starting with the multivariate inequality test, its test statistic is constructed by first solving the following quadratic programming problem

$$\min_{\delta} (\hat{\delta} - \delta)' \hat{\Sigma}_{\hat{\delta}}^{-1} (\hat{\delta} - \delta) \quad \text{s.t.} \quad \delta \geq 0_r, \tag{9.110}$$

where $\hat{\Sigma}_{\hat{\delta}}$ is a consistent estimator of $\Sigma_{\hat{\delta}}$. Let $\tilde{\delta}$ be the optimal solution of the problem in (9.110). The likelihood ratio test of the null hypothesis is given by

$$LR = T(\hat{\delta} - \tilde{\delta})' \hat{\Sigma}_{\hat{\delta}}^{-1} (\hat{\delta} - \tilde{\delta}). \tag{9.111}$$

For computational purposes, it is convenient to consider the dual problem

$$\min_{\lambda} \lambda' \hat{\delta} + \frac{1}{2} \lambda' \hat{\Sigma}_{\hat{\delta}} \lambda \quad \text{s.t.} \quad \lambda \geq 0_r. \tag{9.112}$$

Let $\tilde{\lambda}$ be the optimal solution of the problem in (9.112). The Kuhn-Tucker test of the null hypothesis is given by

$$KT = T \tilde{\lambda}' \hat{\Sigma}_{\hat{\delta}} \tilde{\lambda}. \tag{9.113}$$

It can be readily shown that $LR = KT$.

To conduct statistical inference, it is necessary to derive the asymptotic distribution of LR. Wolak (1989) shows that under $H_0 : \delta = 0_r$ (i.e., the least favorable value of δ under the null hypothesis), LR has a weighted chi-squared distribution:

$$LR \overset{A}{\sim} \sum_{i=0}^{r} w_i(\Sigma_{\hat{\delta}}^{-1}) X_i = \sum_{i=0}^{r} w_{r-i}(\Sigma_{\hat{\delta}}) X_i, \tag{9.114}$$

where the X_i's are independent χ^2 random variables with i degrees of freedom, $\chi_0^2 \equiv 0$, and the weights w_i sum up to one. To compute the p-value of LR, $\Sigma_{\hat{\delta}}^{-1}$ needs to be replaced with $\hat{\Sigma}_{\hat{\delta}}^{-1}$ in the weight functions.

The biggest hurdle in determining the p-value of this multivariate inequality test is the computation of the weights. For a given $r \times r$ covariance matrix $\Sigma = (\sigma_{ij})$, the expressions for the weights $w_i(\Sigma)$, $i = 0, \ldots, r$, are given in Kudo (1963). The weights depend on Σ through the correlation coefficients $\rho_{ij} = \sigma_{ij}/(\sigma_i \sigma_j)$. When $r = 1$, $w_0 = w_1 = 1/2$. When $r = 2$,

$$w_0 = \frac{1}{2} - w_2, \tag{9.115}$$

$$w_1 = \frac{1}{2}, \tag{9.116}$$

$$w_2 = \frac{1}{4} + \frac{\arcsin(\rho_{12})}{2\pi}. \tag{9.117}$$

When $r = 3$,

$$w_0 = \frac{1}{2} - w_2, \tag{9.118}$$

$$w_1 = \frac{1}{2} - w_3, \tag{9.119}$$

$$w_2 = \frac{3}{8} + \frac{\arcsin(\rho_{12\cdot 3}) + \arcsin(\rho_{13\cdot 2}) + \arcsin(\rho_{23\cdot 1})}{4\pi}, \tag{9.120}$$

$$w_3 = \frac{1}{8} + \frac{\arcsin(\rho_{12}) + \arcsin(\rho_{13}) + \arcsin(\rho_{23})}{4\pi}, \tag{9.121}$$

where

$$\rho_{ij\cdot k} = \frac{\rho_{ij} - \rho_{ik}\rho_{jk}}{[(1-\rho_{ik}^2)(1-\rho_{jk}^2)]^{\frac{1}{2}}}. \tag{9.122}$$

For $r > 3$, the computation of the weights is more complicated. Following Kudo (1963), let $P = \{1, \ldots, r\}$. There are 2^r subsets of P, which are indexed by M. Let $n(M)$ be the number of elements in M and M' be the complement of M relative to P. Define Σ_M as the submatrix that consists of the rows and columns in the set M, $\Sigma_{M'}$ as the submatrix that consists of the rows and columns in the set M', $\Sigma_{M,M'}$ the submatrix with rows corresponding to the elements in M and columns corresponding to the elements in M' ($\Sigma_{M',M}$ is similarly defined), and $\Sigma_{M\cdot M'} = \Sigma_M - \Sigma_{M,M'}\Sigma_{M'}^{-1}\Sigma_{M',M}$. Kudo (1963) shows that

$$w_i(\Sigma) = \sum_{M:\, n(M)=i} P(\Sigma_{M'}^{-1}) P(\Sigma_{M\cdot M'}), \tag{9.123}$$

where $P(A)$ is the probability for a multivariate normal distribution with zero mean and covariance matrix A to have all positive elements. In the above equation, we use the convention that $P[\Sigma_{\emptyset\cdot P}] = 1$ and $P[\Sigma_{\emptyset}^{-1}] = 1$. Using (9.123), we have $w_0(\Sigma) = P(\Sigma^{-1})$ and $w_r(\Sigma) = P(\Sigma)$.

Researchers have typically used a Monte Carlo approach to compute the positive orthant probability $P(A)$. However, the Monte Carlo approach is not efficient because it requires a large number of simulations to achieve the accuracy of a few digits, even when r is relatively small.

To overcome this problem, Kan et al. (2010) rely on a formula for the positive orthant probability due to Childs (1967) and Sun (1988a). Let $R = (r_{ij})$ be the correlation matrix corresponding to A. Childs (1967) and Sun (1988a) show that

$$P_{2k}(A) = \frac{1}{2^{2k}} + \frac{1}{2^{2k-1}\pi} \sum_{1\leq i<j\leq 2k} \arcsin(r_{ij}) + \sum_{j=2}^{k} \frac{1}{2^{2k-j}\pi^j} \sum_{1\leq i_1<\cdots<i_{2j}\leq 2k} I_{2j}\left(R_{(i_1,\ldots,i_{2j})}\right), \tag{9.124}$$

$$P_{2k+1}(A) = \frac{1}{2^{2k+1}} + \frac{1}{2^{2k}\pi} \sum_{1\leq i<j\leq 2k+1} \arcsin(r_{ij}) + \sum_{j=2}^{k} \frac{1}{2^{2k+1-j}\pi^j} \sum_{1\leq i_1<\cdots<i_{2j}\leq 2k+1} I_{2j}\left(R_{(i_1,\ldots,i_{2j})}\right), \tag{9.125}$$

where $R_{(i_1,\ldots,i_{2j})}$ denotes the submatrix consisting of the $(i_1,\ldots,i_{2j})$th rows and columns of R, and

$$I_{2j}(\Lambda) = \frac{(-1)^j}{(2\pi)^j} \int_{-\infty}^{\infty} \cdots \int_{-\infty}^{\infty} \left(\prod_{i=1}^{2j} \frac{1}{\omega_i}\right) \exp\left(-\frac{\omega'\Lambda\omega}{2}\right) \mathrm{d}\omega_1 \cdots \mathrm{d}\omega_{2j}, \tag{9.126}$$

where Λ is a $2j \times 2j$ covariance matrix and $\omega = (\omega_1,\ldots,\omega_{2j})'$. Sun (1988a) provides a recursive relation for $I_{2j}(\Lambda)$ that allows us to obtain I_{2j} starting from I_2. Sun's formula enables us to compute the $2j$th order multivariate integral I_{2j} using a $(j-1)$th order multivariate integral, which can be obtained numerically using the Gauss-Legendre quadrature method. Sun (1988b) provides a Fortran subroutine to compute $P(A)$ for $r \leq 9$. Kan et al. (2010) improve on Sun's program and are able to accurately compute $P(A)$ and hence $w_i(\Sigma)$ for $r \leq 11$.

Turning to the $\delta_{\min}$ test based on White (2000), one can use the sample counterpart of $\delta_{\min}$:

$$\hat{\delta}_{\min} = \min_{2\leq i\leq p} \hat{\delta}_i \tag{9.127}$$

to test (9.108). To determine the p-value of $\hat{\delta}_{\min}$, we need to identify the least favorable value of δ under the null hypothesis. It can be easily shown that the least favorable value of δ under the null hypothesis occurs at $\delta = 0_r$. It follows that asymptotically,

$$
\begin{aligned}
P[\sqrt{T}\delta_{\min} < c] \to P[\min_{1\le i\le r} z_i < c] \\
&= 1 - P[\min_{1\le i\le r} z_i > c] \\
&= 1 - P[z_1 > c, \ldots, z_r > c] \\
&= 1 - P[z_1 < -c, \ldots, z_r < -c], \qquad (9.128)
\end{aligned}
$$

where $z = (z_1, \ldots, z_r)' \sim N(0_r, \Psi)$, and the last equality follows from symmetry since $E[z] = 0_r$. Therefore, to compute the asymptotic p-value one needs to evaluate the cumulative distribution function of a multivariate normal distribution.

Note that both tests crucially depend on the asymptotic normality assumption in (9.109). Sufficient conditions for this assumption to hold are (1) $0 < \rho_i^2 < 1$, and (2) the implied stochastic discount factors of the different models are distinct. Even though the multivariate normality assumption may not always hold at the boundary point of the null hypothesis (i.e., $\delta = 0_r$), it is still possible to compute the p-value as long as we assume that the true δ is not at the boundary point of the null hypothesis. There are, however, cases in which this assumption does not hold. For example, if model 2 nests model 1, then we cannot have $\delta_2 > 0$. As a result, the null hypothesis $H_0 : \delta_2 \ge 0$ becomes $H_0 : \delta_2 = 0$. Under this null hypothesis, $\sqrt{T}\hat{\delta}_2$ no longer has a multivariate normal distribution and both the multivariate inequality test and the $\delta_{\min}$ test will break down.

Therefore, when nested models are involved, the two tests need to be modified. If model 1 nests some of the competing models, then those models that are nested by model 1 will not be included in the model comparison tests. The reason is that these models are clearly dominated by model 1 and we no longer need to perform tests in presence of these models. If some of the competing models are nested by other competing models, then the smaller models will not be included in the model comparison tests. This is reasonable since if model 1 outperforms a larger model, it will also outperform the smaller models that are nested by the larger model. With these two types of models being eliminated from the model comparison tests, the remaining models will not nest each other and the multivariate asymptotic normality assumption on $\sqrt{T}(\hat{\delta} - \delta)$ can be justified.

Finally, if model 1 is nested by some competing models, one should separate the set of competing models into two subsets. The first subset will include competing models that nest model 1. To test whether model 1 performs as well as the models in this subset, one can construct a model M that contains all the distinct factors in this subset. It can be easily verified that model 1 performs as well as the models in this subset if and only if $\rho_1^2 = \rho_M^2$. In this case, a test of $H_0 : \rho_1^2 = \rho_M^2$ can be simply performed using the model comparison tests for nested models described earlier. The second subset includes competing models that do not nest model 1. For this second subset, we can use the non-nested multiple model comparison tests as before. If we perform each test at a significance level of $\alpha/2$ and accept the null hypothesis if we fail to reject in both tests, then by the Bonferroni inequality, the size of the joint test is less than or equal to α.

9.9 Conclusion

In this review article, we provide an up-to-date summary of the asymptotic results related to the two-pass CSR methodology, with special emphasis on the role played by model misspecification in estimating risk premia and in comparing the performance of competing models. We also point out some pitfalls with certain popular usages of this methodology that could lead to erroneous conclusions.

There are some issues related to the two-pass CSR methodology that require further investigation. At the top of the list are the finite sample properties of the risk premium and cross-sectional R^2 estimates. At the current stage, we have little understanding of the finite sample biases of these estimates and, as a result, no good way to correct them. This is a serious concern especially when the number of assets is large relative to the length of the time series. An important related issue is how to implement the two-pass CSR methodology when the number of assets is large. In this respect, the standard practice is to simply run an OLS CSR since the GLS CSR becomes infeasible. However, in this scenario, relying on asymptotic results may not be entirely appropriate. Alternatively, one could form portfolios and use the potentially more efficient GLS CSR. How many portfolios should we consider and how should we form them are certainly open questions that we hope future research will address.

While most of the econometric results in this review article are relatively easy to program, some of them require specialized subroutines that may be time consuming to develop. To facilitate this task, a set of Matlab programs is available from the authors upon request.

References

Ahn, S., Gadarowski, C., & Perez, M. (2009). *Two-pass cross-sectional regression of factor pricing models: Minimum distance approach*. Working paper, Wilfrid Laurier University.

Bartholomew, D. J. (1961). A test of homogeneity of means under restricted alternatives. *Journal of the Royal Statistical Society, 23*, 239–281.

Black, F., Jensen, M. C., & Scholes, M. (1972). The capital asset pricing model: some empirical findings. In M. C. Jensen (Ed.), *Studies in the theory of capital markets*. New York: Praeger.

Chan, K. C., & Chen, N. F. (1988). An unconditional asset-pricing test and the role of firm size as an instrumental variable for risk. *Journal of Finance, 43*, 309–325.

Chen, X., & Ludvigson, S. C. (2009). Land of addicts? An empirical investigation of habit-based asset pricing models. *Journal of Applied Econometrics, 24*, 1057–1093.

Chen, N. F., Roll, R., & Ross, S. (1986). Economic forces and the stock market. *Journal of Business, 59*, 383–404.

Childs, D. R. (1967). Reduction of the multivariate normal integral to characteristic form. *Biometrika, 54*, 293–300.

Cochrane, J. H. (2005). *Asset pricing*. Princeton: Princeton University Press.

Cragg, J. G., & Donald, S. G. (1997). Inferring the rank of a matrix. *Journal of Econometrics, 76*, 223–250.

Davies, R. B. (1980). Algorithm as 155: The distribution of a linear combination of χ^2 random variables. *Applied Statistics*, *29*, 323–333.

Fama, E. F., & French, K. R. (1993). Common risk factors in the returns on stocks and bonds. *Journal of Financial Economics*, *33*, 3–56.

Fama, E. F., & MacBeth J. D. (1973). Risk, return, and equilibrium: Empirical tests. *Journal of Political Economy*, *71*, 607–636.

Golden, R. M. (2003). Discrepancy risk model selection test theory for comparing possibly misspecified or nonnested models. *Psychometrika*, *68*, 229–249.

Gospodinov, N., Kan, R., & Robotti, C. (2010a). *A simplified rank restriction test*. Working paper, Federal Reserve Bank of Atlanta.

Gospodinov, N., Kan, R., & Robotti, C. (2010b). *Further results on the limiting distribution of GMM sample moment conditions*. Working paper, Federal Reserve Bank of Atlanta.

Gourieroux, C., Holly, A., & Monfort, A. (1982). Likelihood ratio test, Wald Test, and Kuhn-Tucker test in linear models with inequality constraints on the regression parameters. *Econometrica*, *50*, 63–80.

Grauer, R. R., & Janmaat, J. J. (2009). On the power of cross-sectional and multivariate tests of the CAPM. *Journal of Banking and Finance*, *33*, 775–787.

Hou, K., & Kimmel, R. (2006). *On the estimation of risk premia in linear factor models*. Working paper, Ohio State University.

Imhof, J. P. (1961). Computing the distribution of quadratic forms in normal variables. *Biometrika*, *48*, 419–426.

Jagannathan, R., & Wang, Z. (1996). The conditional CAPM and the cross-section of expected returns. *Journal of Finance*, *51*, 3–53.

Jagannathan, R., & Wang, Z. (1998). An asymptotic theory for estimating beta-pricing models using cross-sectional regression. *Journal of Finance*, *53*, 1285–1309.

Jagannathan, R., Skoulakis, G., & Wang, R. (2010). The analysis of the cross-section of security returns. In Y. Aït-Sahalia & L. Hansen (Eds.), *Handbook of financial econometrics* (Vol. 2, pp. 73–134). Amsterdam: Elsevier Science BV.

Kan, R., & Robotti, C. (2011). On the estimation of asset pricing models using univariate betas. *Economics Letters*, 110, 117–121.

Kan, R., & Zhang, C. (1999) Two-pass tests of asset pricing models with useless factors. *Journal of Finance*, *54*, 203–235.

Kan, R., & Zhou, G. (2004). *Hansen-Jagannathan distance: Geometry and exact distribution*. Working paper, University of Toronto.

Kan, R., Robotti, C., & Shanken, J. (2010). *Pricing model performance and the two-pass cross-sectional regression methodology*. Working paper, University of Toronto.

Kandel, S., & Stambaugh, R. F. (1995). Portfolio inefficiency and the cross-section of expected returns. *Journal of Finance*, *50*, 157–184.

Kleibergen, F. (2009). Tests of risk premia in linear factor models. *Journal of Econometrics*, *149*, 149–173.

Kudo, A. (1963). A Multivariate analogue of the one-sided test. *Biometrika*, *50*, 403–418.

Lewellen, J. W., Nagel, S., & Shanken, J. (2010). A skeptical appraisal of asset-pricing tests. *Journal of Financial Economics*, *96*, 175–194.

Lu, Z. H., & King, M. L. (2002). Improving the numerical technique for computing the accumulated distribution of a quadratic form in normal variables. *Econometric Reviews*, *21*, 149–165.

Newey, W. K., & West, K. D. (1987). A simple positive definite heteroskedasticity and autocorrelation consistent covariance matrix. *Econometrica*, *55*, 703–708.

Perlman, M. D. (1969). One-sided testing problems in multivariate analyses. *Annals of Mathematical Statistics*, *40*, 549–567.

Rivers, D., & Vuong, Q. H. (2002). Model selection tests for nonlinear dynamic models. *Econometrics Journal*, *5*, 1–39.

Shanken, J. (1985). Multivariate tests of the zero-beta CAPM. *Journal of Financial Economics*, *14*, 327–348.

Shanken, J. (1992). On the estimation of beta-pricing models. *Review of Financial Studies*, *5*, 1–33.
Shanken, J. (1996). Statistical methods in tests of portfolio efficiency: A synthesis. In G. S. Maddala & C. R. Rao (Eds.), *Handbook of statistics* (Vol. 14, pp. 693–711). Amsterdam: Elsevier Science BV.
Shanken, J., & Zhou, G. (2007). Estimating and testing beta pricing models: alternative methods and their performance in simulations. *Journal of Financial Economics*, *84*, 40–86.
Sun, H. J. (1988a). A general reduction method for n-variate normal orthant probability. *Communications in Statistics – Theory and Methods*, *11*, 3913–3921.
Sun, H. J. (1988b). A fortran subroutine for computing normal orthant probabilities of dimensions up to nine. *Communications in Statistics – Simulation and Computation*, *17*, 1097–1111.
Vuong, Q. H. (1989). Likelihood ratio tests for model selection and non-nested hypotheses. *Econometrica*, *57*, 307–333.
White, H. L. (2000). A reality check for data snooping. *Econometrica*, *68*, 1097–1127.
Wolak, F. A. (1987). An exact test for multiple inequality and equality constraints in the linear regression model. *Journal of the American Statistical Association*, *82*, 782–793.
Wolak, F. A. (1989). Testing inequality constraints in linear econometric models. *Journal of Econometrics*, *31*, 205–235.

Chapter 10
Parametric Estimation of Risk Neutral Density Functions

Maria Grith and Volker Krätschmer

Abstract This chapter deals with the estimation of risk neutral distributions for pricing index options resulting from the hypothesis of the risk neutral valuation principle. After justifying this hypothesis, we shall focus on parametric estimation methods for the risk neutral density functions determining the risk neutral distributions. We we shall differentiate between the direct and the indirect way. Following the direct way, parameter vectors are estimated which characterize the distributions from selected statistical families to model the risk neutral distributions. The idea of the indirect approach is to calibrate characteristic parameter vectors for stochastic models of the asset price processes, and then to extract the risk neutral density function via Fourier methods. For every of the reviewed methods the calculation of option prices under hypothetically true risk neutral distributions is a building block. We shall give explicit formula for call and put prices w.r.t. reviewed parametric statistical families used for direct estimation. Additionally, we shall introduce the Fast Fourier Transform method of call option pricing developed in Carr and Madan [J. Comput. Finance 2(4):61–73, 1999]. It is intended to compare the reviewed estimation methods empirically.

M. Grith (✉)
Research Associate at Humboldt-Universität zu Berlin, Ladislaus von Bortkiewicz Chair of Statistics, Spandauer Straße 1, 10178 Berlin, Germany
e-mail: gritmari@wiwi.hu-berlin.de

V. Krätschmer
Weierstrass Institute of Applied Analysis and Stochastics, Mohrenstrasse 39, 10117 Berlin, Germany
e-mail: kraetsch@wias-berlin.de.

J.-C. Duan et al. (eds.), *Handbook of Computational Finance*, Springer Handbooks of Computational Statistics, DOI 10.1007/978-3-642-17254-0_10,

10.1 Introduction

It is usual practice of empirical studies on index option pricing in financial markets to start with the hypothesis of risk neutral valuation principle. That means it is assumed that prices of path independent derivatives with expiration at maturity may be represented as expected pay offs. The underlying distribution is referred as the risk neutral distribution. In the seminal paper (Black and Scholes (1973)), a stochastic model for financial markets has been established where this risk neutral distribution may be singled out by arbitrage arguments up to the volatility parameter. This Black Scholes model is nowadays very well understood, and widely used in financial industries due to the derived formula of risk neutral index call and put option prices.

Several empirical studies had come to the conclusion that the stochastic assumptions underlying the Black Scholes model does not fit very well the observed dynamics of asset prices. Therefore several alternative stochastic models have been proposed in the literature where typically risk neutral distributions may not be obtained by arbitrage arguments alone. However, within quite general stochastic frameworks one may identify theoretically risk neutral distributions compatible with observable liquid derivatives like call and put options. These risk neutral distributions are often called implied risk neutral distributions.

Compared to the risk neutral distribution according to the Black Scholes model implied risk neutral distributions generally do not have further specifications in advance. This complicates estimations in two directions. From the point of view of accuracy specification aspects like the choice of statistical families for the risk neutral distributions or the assumptions on stochastic models for the asset price processes have to be taken into account when selecting the estimation method and controlling the accuracy. Additionally the numerical problems associated with the implementation of the estimation method typically became more involved.

As a general assumption within the literature on estimation of risk neutral distributions they are considered as continuous distributions. The object is then to estimate related probability density functions called the risk neutral density functions, with a slight abuse of mathematical correctness. Two principal ways to estimate risk neutral density functions may be pointed out, parametric and nonparametric methods. This chapter deals with the parametric ones. One class of them is built upon parametric statistical families assumed to describe the risk neutral distribution accurately. The problem reduces to the estimation of the distribution parameters. The other group of methods estimate the probability density functions indirectly. A parametric stochastic model is assumed for the asset price processes, and the risk neutral density functions are extracted then after the calibration of the model to observed option prices. The chapter is organized as follows.

We shall start with the risk neutral valuation principle. There are controversial standpoints concerning the reasonability of this principle. Since the field of mathematical finance is mainly built upon the framework of arbitrage theory, many mathematicians accept risk neutral pricing for replicable options only. Instead non-linear pricing rules like superhedging are favoured which reduce to risk neutral pricing for replicable options. In Sect. 10.2 we shall present an argumentation which might reconcile the different views.

The general idea behind the estimation methods under considerations is to fit option prices calculated under hypothetically true risk neutral density distributions to respective observed ones. Therefore these calculations play an important role for the implementation of the estimation methods. In Sect. 10.4 we shall assume particular statistical families to model the risk neutral distributions. The considered families, namely log-normal distributions, mixtures of log-normal distributions and general gamma distributions, allow for explicit formula of call and put prices. Section 10.3 deals with calculations of call prices based on parametric stochastic models for the asset price processes. There the classical Black Scholes formula will be reviewed, and the Fast Fourier Transform method developed in Carr and Madan (1999) will be introduced. This method might be used as a tool for the model calibration as presented in Sect. 10.5. There it will also be shown how to extract the risk neutral density functions via Fourier methods. The whole line of reasoning will be explified by Merton's jump diffusion and Heston's volatility model. In the last section it is intended to compare the different reviewed estimation methods empirically.

10.2 The Risk Neutral Valuation Principle

Let $[0, T]$ be the time interval of investment in the financial market, where $t = 0$ denotes the present time and $t = T \in]0, \infty[$ the time of maturity.

Furthermore it is assumed that a riskless bond with constant interest rate $r > -1$ and a risky asset are traded in the financial market as basic underlyings. The evolution of the risky asset is expressed in terms of a state dependent nonnegative price process $(S_t)_{t \in [0,T]}$ with constant S_0. Notice that time discrete modelling may be subsumed under this framework.

For the pricing of nonnegative derivatives $\psi(S_T)$ it is often assumed that the *risk valuation principle* is valid. That means that there is a stochastic model for $(S_t)_{t \in [0,T]}$ by means of a probability measure Q such that the price of any $\psi(S_T)$ is characterized by

$$\mathbb{E}_{\mathrm{Q}}\left[e^{-rT}\psi(S_T)\right].$$

There exist many arguments supporting this principle. From the viewpoint of the arbitrage theory it is closely related to the condition that Q is a so called martingale measure, i.e.

$$\mathbb{E}_{\mathrm{Q}}\left[\, e^{-tr} S_t | S_\tau, \tau \leq \tilde{t}\,\right] = e^{-\tilde{t}r} S_{\tilde{t}} \text{ for } 0 \leq \tilde{t} < t \leq T. \tag{10.1}$$

In this case the financial market is arbitrage free in the sense that the value process $(V_t(H))_{t \in [0,T]}$ of a self-financing investment strategy $H = (H_t)_{t \in [0,T]}$ which is bounded from below by $-(\delta S_t)_{t\in[0,T]}$ for some $\delta > 0$, with $V_0(H) \leq 0$ the value at maturity $V_T(H)$ is vanishing almost surely if it is nonnegative. For a comprehensive account on the theory of arbitrage the reader is kindly referred to the monograph (Delbaen and Schachermayer (2006)).

The expectation $\mathbb{E}_{\mathrm{Q}}\left[e^{-rT}\psi(S_T)\right]$ is then a so called arbitrage free price of $\psi(S_T)$, meaning that Q remains a martingale measure for the new financial market with an additional underlying having price process

$$\{\mathbb{E}\left[\, e^{-r(T-t)}\psi(S_T) \mid S_\tau, \tau \leq t \,\right]\}_{t\in[0,T]}\,.$$

Unfortunately, arbitrage free prices vary over the martingal measures unless a derivative $\psi(S_T)$ is replicable by the terminal wealth $V_T(H)$ of a value process $(V_t(H))_{t\in[0,T]}$ of a self-financing investment strategy $H = (H_t)_{t\in[0,T]}$ satisfying boundness conditions as above. If every such derivative is replicable the financial market is called complete. An outstanding example is the famous Black-Scholes model (see below). However, at least in the special case of time discrete modelling complete financial markets are very exceptional, e.g. reducing directly to a binomial model within out setting (cf. Föllmer and Schied (2004), Theorem 5.38). Hence arbitrage arguments alone are not sufficient for a justification of the risk neutral valuation. Several suggestions have combined them with additional criteria. In Hugonnier et al. (2005) arbitrage free markets are embedded into a utility-based model for the terminal wealths of value processes of self-financing investment strategies that leads to risk neutral valuation of the derivatives $\psi(S_T)$. Another suggestion is built upon the observation that in organized markets call and put options are traded so often that they might be viewed as liquid derivatives. So the idea is to look for martingale measures Q consistent with observable prices $C(K)$ of call options with expiration T and strike price K in the sense

$$C(K) = \mathbb{E}_{\mathrm{Q}}\left[e^{-rT}\max\{0, S_T - K\}\right].$$

If consistency is required for all strike prices K, then for any pair $\mathrm{Q}_1, \mathrm{Q}_2$ of such martingale measures the marginal distributions of S_T w.r.t. $\mathrm{Q}_1, \mathrm{Q}_2$ coincide (see proof of Lemma 7.23 in Föllmer and Schied (2004)), implying $\mathbb{E}_{\mathrm{Q}_1}\left[e^{-rT}\psi(S_T)\right] = \mathbb{E}_{\mathrm{Q}_2}\left[e^{-rT}\psi(S_T)\right]$ for a derivative $\psi(S_T)$. Moreover, there exist axiomatizations for pricing rules in financial markets that guarantee the existence of martingale measures which are consistent with the observable call prices $C(K)$ for all strikes K (cf. e.g. Föllmer and Schied (2004), Proposition 7.26, Biagini and Cont (2006)).

If the risk neutral valuation principle is valid w.r.t. to some stochastic model in terms of a probability measure Q, we shall call it *risk neutral probability measure*. As discussed above marginal distributions of S_T are independent of the chosen risk neutral probability measure so that we may speak of *the risk neutral distribution* of S_T, henceforth denoted by Q_{S_T}. Of course the marginal distributions of $\ln(S_T)$ are independent of the choice of risk neutral probability measures too, suggesting the convention of *the log-price risk neutral distribution*. We shall further restrict considerations to continuous risk neutral distributions admitting a probability density function q, which we shall call *risk neutral density function*. So from now on the assumption of the risk valuation principle should mean that the price of a derivative $\psi(S_T)$ is expressed by

$$\int \psi(x)\, q(x)\, dx.$$

Since the risk neutral density function is unknown, the task is to estimate it upon observed prices for options $\psi(S_T)$ at time $t = 0$. Typically, prices for call and put options are used. We shall review some widely used parametric methods. There always computations of hypothetical prices for options w.r.t. candidates of the risk neutral density functions are involved. For some models like the Black Scholes model such hypothetical prices for call and put options are given in implementable analytical expressions, for others like several stochastic volatility models numerically efficient ways of calculations have been developed. These results and methods will be the subject of the next section.

10.3 Calculations of Risk Neutral Option Prices

Let us assume that the stock price process $(S_t)_{t\in[0,T]}$ is characterized by a parameter vector $\vartheta \in \Theta \subseteq \mathbb{R}^r$ under the risk neutral probability measures. In the special case of the Black Scholes model the famous Black Scholes formulas provide explicit formulas for parameter dependent call and put prices. They will be reviewed in the following subsection. Afterwards we shall introduce the Fast Fourier Transform method to calculate call option prices as proposed in Carr and Madan (1999). It relies on the additional assumption that the characteristic function of the log-price risk neutral distribution is known analytically.

10.3.1 The Black Scholes Formula

In the Black Scholes model the price process $(S_t)_{t\in[0,T]}$ is modelled under the risk neutral probability measure by

$$S_t = S_0 \exp\left\{\left(r - \frac{\sigma^2}{2}\right)t + \sigma W_t\right\},$$

where $\sigma > 0$, and $(W_t)_{t\in[0,\infty[}$ denotes a standard Brownian motion. In particular $\vartheta \overset{\text{def}}{=} \sigma \in \Theta \overset{\text{def}}{=}]0,\infty[$, the so called volatility, and the risk neutral distribution is a log-normal distributions with parameters $\mu \overset{\text{def}}{=} \left(r - \frac{\sigma^2}{2}\right)T + \ln(S_0)$ and $\sigma^2 T$.

As usual, let Φ denote the distribution function of the standard normal distribution, and let $M_K \stackrel{\text{def}}{=} \frac{S_T}{K}$ be the moneyness w.r.t. strike price $K > 0$. With these notations we may report the celebrated Black Scholes formula (cf. Black and Scholes (1973)) for the prices $C^{BS}(K,\sigma)$, $P^{BS}(K,\sigma)$ of respectively the call and put with expiration at T and strike price $K > 0$ dependent on the volatility σ:

$$C^{BS}(K,\sigma) = S_T\Phi(d_1) - Ke^{-rT}\Phi(d_2) \tag{10.2}$$

$$P^{BS}(K,\sigma) = C_{BS}(K,\sigma) - S_0 + e^{-rT}K \tag{10.3}$$

$$d_1 \stackrel{\text{def}}{=} \frac{-\ln(M) + T(r + \frac{\sigma^2}{2})}{\sigma\sqrt{T}}, \quad d_2 \stackrel{\text{def}}{=} d_1 - \sigma\sqrt{T} \tag{10.4}$$

10.3.2 Fast Fourier Transform Method to Calculate Call Option Prices

We shall follow the line of reasoning in Carr and Madan (1999), assuming that the characteristic function $\Phi_{T|\vartheta}$ of the log-price risk neutral distribution is known analytically. Prominent examples are some widely used stochastic volatility models with or without jumps (see below). The aim is to calculate the hypothetical price $C^{\vartheta}(K)$ for the call option with expiration at T and strike K if ϑ is the true parameter vector driving the risk neutral model for the stock price process.

Recall that for an integrable function $f : \mathbb{R} \to \mathbb{R}$ we may define the so called Fourier transform $\hat{f}$ of f via

$$\hat{f}(y) \stackrel{\text{def}}{=} \int fe^{iyv}\, dv.$$

Due to Plancherel's theorem (cf. Rudin (1974), Theorem 9.13) we may recover f from its fourier transform by

$$f(x) = \int \frac{e^{-ixy}\hat{f}(y)}{2\pi}\, dy$$

if f is in addition square integrable. Under the assumption

$$\mathbb{E}\left[S_T^{1+\alpha}\right] < \infty \text{ for some } \alpha > 0 \tag{10.5}$$

this relationship between functions and their fourier transforms may be applied to

$$C_\alpha^{\vartheta} : \mathbb{R} \to \mathbb{R}, x \mapsto e^{\alpha x} C^{\vartheta}(e^x) \ (\vartheta \in \Theta)$$

(see Carr and Madan (1999)). We obtain the following formulas for the fourier transforms $\widehat{C_\alpha^\vartheta}$ of C_α^ϑ ($\vartheta \in \Theta$)

$$\widehat{C_\alpha^\vartheta} = \frac{e^{-rT}\Phi_{T|\vartheta}(y-(1+\alpha)i)}{\alpha^2+\alpha-y^2+i(2\alpha+1)y}, \tag{10.6}$$

A derivation may be found in Carr and Madan (1999) or Cizek et al. (2005), p. 189. This yields

$$\begin{aligned} C^\vartheta(K) = K^{-\alpha}C_\alpha^\vartheta(\ln(K)) &= \int \frac{K^{-\alpha}e^{-iy\ln(K)}\widehat{C_\alpha^\vartheta}(y)}{2\pi}\,dy \\ &= \int_0^\infty \frac{K^{-\alpha}e^{-iy\ln(K)}\widehat{C_\alpha^\vartheta}(y)}{\pi}\,dy. \end{aligned} \tag{10.7}$$

The last equation holds because $C^\vartheta(K)$ is real, which implies that fourier transform $\widehat{C_\alpha^\vartheta}$ is odd in its imaginary part and even in its real part. Using the Trapezoid rule for the integral on the right hand side of (10.7), we may approximate the prices $C^\vartheta(K)$ by

$$C^\vartheta(K) \approx \frac{1}{K^\alpha\pi}\sum_{j=0}^{N-1} e^{-i\eta j}\widehat{C_\alpha^\vartheta}(\eta j)\eta, \tag{10.8}$$

where $\eta > 0$ is the distance between the points of the integration grid. Bounds for sampling and truncation errors of this approximation have been developed in Lee (2004).

Approximation (10.8) suggests to apply the Fast Fourier algorithm which is an efficient algorithm to compute sums of the form

$$w_u = \sum_{j=0}^{N-1} e^{-i\frac{2\pi}{N}ju} z_j \text{ for } u = 0,\ldots,N-1$$

(cf. Walker (1996)). In general, the strikes near the spot price S_0 are of interest because call options with such prices are traded most frequently. We thus consider an equidistant spacing of the log-strikes around the log spot price $\ln(S_0)$:

$$x_u = -\frac{1}{N}N\zeta + \zeta u + \ln(S_0) \text{ for } u = 0,\ldots,N-1, \tag{10.9}$$

where $\zeta > 0$ denotes the distance between the log-strikes. Inserting (10.9) into formula (10.8) yields

$$C^\vartheta\{\exp(x_u)\} \approx \frac{\exp(-\alpha x_u)}{\pi}\sum_{j=0}^{N-1} e^{-i\zeta\eta j u}\, e^{i\eta j\{\frac{1}{2}N\zeta-\ln(S_0)\}}\widehat{C_\alpha^\vartheta}(\eta j)\eta \tag{10.10}$$

for $u = 0, \ldots, N-1$. Now we may apply the Fast Fourier algorithm to

$$z_j \stackrel{\text{def}}{=} e^{i\eta j\{\frac{1}{2}N\zeta - \ln(S_0)\}} \widehat{C_\alpha^\vartheta}(\eta j)\eta \text{ for } j = 0, \ldots, N-1$$

provided $\zeta\eta = \frac{2\pi}{N}$ holds. This restriction means on one hand that if we choose η small in order to obtain a fine grid for the integration, we have a relatively large spacing between the log-strikes with few log-strikes lying around the desired region near $\ln(S_0)$. On the other hand a small ζ to catch many log-strikes near $\ln(S_0)$ a more rough grid for the integration is forced by the restriction. So we face a trade-off between accuracy and the number of interesting strikes. Accuracy may be improved for large η by using better numerical integration rules. Carr and Madan considered the Simpson rule leading to the approximation

$$\frac{\exp(-\alpha x_u)}{\pi} \sum_{j=0}^{N-1} e^{-i\zeta\eta j u} e^{i\eta j\{\frac{1}{2}N\zeta - \ln(S_0)\}} \widehat{C_\alpha^\vartheta}(\eta j)\frac{\eta}{3}\{3 + (-1)^j - \delta_0(j)\} \tag{10.11}$$

for $u = 0, \ldots, N-1$, instead of (10.10), where $\delta_0(0) \stackrel{\text{def}}{=} 1$ and $\delta_0(j) \stackrel{\text{def}}{=} 0$ for $j \neq 0$. The Fast Fourier algorithm may be applied to calculate

$$z_j \stackrel{\text{def}}{=} e^{i\eta j\{\frac{1}{2}N\zeta - \ln(S_0)\}} \widehat{C_\alpha^\vartheta}(\eta j)\frac{\eta}{3}\{3 + (-1)^j - \delta_0(j)\} \text{ for } j = 0, \ldots, N-1,$$

again taking into account the condition $\zeta\eta = \frac{2\pi}{N}$.

10.4 Direct Parametric Estimation of the Risk Neutral Density Function

The parametric approach to estimate the risk neutral density function directly starts with the assumption that the risk neutral distribution of S_T belongs to a parametric family W_Θ ($\Theta \subseteq \mathbb{R}^r$) of one-dimensional continuous distributions. For any parameter vector $\vartheta \in \Theta$ and every strike price K we may calculate the hypothetical prices for the call $C(K, \vartheta)$, the put $P(K, \vartheta)$ both with expiration T, and the forward F_η by

$$C(K|\vartheta) = e^{-rT} \int_K^\infty (x - K)\, q(x|\eta)\, dx \tag{10.12}$$

$$P(K|\vartheta) = e^{-rT} \int_0^K (K - x)\, q(x|\eta)\, dx \tag{10.13}$$

$$F_{\vartheta} = e^{-rT} \int_0^{\infty} x\, q(x|\vartheta)\, dx \tag{10.14}$$

Therein, $q(\cdot|\vartheta)$ denotes any probability density function of the distribution from W_{Θ} associated with ϑ.

The estimation of the risk neutral density function reduces to the estimation of the distribution parameter vector ϑ. The most common approach is based on S_0, observed prices $Y_1, \ldots, Y_n$ for calls with strikes $K_1, \ldots, K_m$, and $\tilde{Y}_1, \ldots, \tilde{Y}_m$ with strikes $\tilde{K}_1, \ldots, \tilde{K}_n$. Both, calls and puts with expiration T. The parameter vector ϑ is estimated by minimizing the sum of the squared differences between the observed call, put and forward price and the hypothetical ones. More precisely, the estimation involves the solution of the following minimization problem

$$\min \sum_{i=1}^{m} \{Y_i - C(K_i|\vartheta)\}^2 + \sum_{j=1}^{n} \{\tilde{Y}_i - P(\tilde{K}_i|\vartheta)\}^2 \tag{10.15}$$

$$+ \left(e^{-rT} S_0 - F_{\vartheta}\right)^2 \text{ s.t. } \vartheta \in \Theta.$$

The crucial step to implement this parametric approach is to find a proper statistical family W_{Θ} as a model for the risk neutral distribution. Usually, either a very general class a distribution is selected or mixtures of log-normal distributions are utilized. As general classes we shall discuss the benchmark case of log-normal distributions and the generalized Gamma distributions. Let us start with assumption of log-normal distributions.

10.4.1 Estimation Using Log-Normal Distributions

Closely related to the Black Scholes model the log-normal distributions are sometimes used for the risk neutral distribution, indicated as a benchmark case (cf. e.g. Jondeau and Rockinger (2000)). Recall that a probability density function $f_{LN(\mu,\sigma)}$ of a log-normal distribution with parameters $\mu \in \mathbb{R}$ and $\sigma > 0$ is given by

$$f_{LN(\mu,\sigma)}(x) \stackrel{\text{def}}{=} \begin{cases} \dfrac{1}{\sqrt{2\pi}\sigma x}\, e^{-\frac{\{\ln(x)-\mu\}^2}{2\sigma^2}} & : \quad x > 0 \\ 0 & : \text{ otherwise} \end{cases}.$$

For fixed σ^2 and different μ_1, μ_2 the respective probability density functions $f_{LN(\mu_1,\sigma)}$ and $f_{LN(\mu_2,\sigma)}$ are linked by

$$f_{LN(\mu_2,\sigma)} = e^{(\mu_1-\mu_2)}\, f_{LN(\mu_1,\sigma)}\left\{x\, e^{(\mu_1-\mu_2)}\right\}. \tag{10.16}$$

Then applying the change of variables theorem for integration we obtain the following relationships between the call and put prices

$$C(K|\mu_2,\sigma) = e^{(\mu_2-\mu_1)}\, C\left\{K\, e^{(\mu_1-\mu_2)}|\mu_1,\sigma\right\} \tag{10.17}$$

$$P(K|\mu_2,\sigma) = e^{(\mu_2-\mu_1)}\, P\left\{K\, e^{(\mu_1-\mu_2)}|\mu_1,\sigma\right\}. \tag{10.18}$$

The equations suggest to express prices $C(K|\mu,\sigma)$ and $P(K|\mu,\sigma)$ in terms of Black Scholes formulas, noticing that $C^{BS}(K,\sigma)=C\left\{K|\left(r-\frac{\sigma^2}{2}\right)T+\ln(S_0),\sigma\right\}$ and $P^{BS}(K,\sigma) = P\left\{K|\left(r-\frac{\sigma^2}{2}\right)T+\ln(S_0),\sigma\right\}$ holds for any strike K. For $\mu\in\mathbb{R}$ and $\sigma>0$ we obtain

$$C^{BS}(K|\mu,\sigma) \stackrel{\text{def}}{=} C(K|\mu,\sigma) \tag{10.19}$$

$$= e^{\left\{\mu-\left(r-\frac{\sigma^2}{2}\right)T+\ln(S_0)\right\}}\, C^{BS}\left\{K\, e^{\left(r-\frac{\sigma^2}{2}\right)T+\ln(S_0)-\mu},\sigma\right\}$$

$$P^{BS}(K|\mu,\sigma) \stackrel{\text{def}}{=} C(K|\mu,\sigma) \tag{10.20}$$

$$= e^{\left\{\mu-\left(r-\frac{\sigma^2}{2}\right)T+\ln(S_0)\right\}}\, P^{BS}\left\{K\, e^{\left(r-\frac{\sigma^2}{2}\right)T+\ln(S_0)-\mu},\sigma\right\}.$$

With a slight abuse of convention we shall call $C^{BS}(K|\mu,\sigma)$ and $P^{BS}(K|\mu,\sigma)$ Black Scholes call and put prices too.

Next we want to introduce the approach to substitute log-normal distributions for the risk neutral distributions by mixtures of them.

10.4.2 Estimation Using Log-Normal Mixtures

The use of log-normal mixtures to model the risk neutral distribution of S_T was initiated by Ritchey (1990) and became further popular even in financial industries by the studies Bahra (1997), Melick and Thomas (1997) and Söderlind and Swensson (1997). The idea is to model the risk neutral density function as a weighted sum of probability density functions of possibly different log-normal distribution. Namely, we set

$$q(x|\mu_1,\ldots,\mu_k,\sigma_1,\ldots,\sigma_k,\lambda_1,\ldots,\lambda_k) \stackrel{\text{def}}{=} \sum_{i=1}^{k} \lambda_i\, f_{\text{LN}(\mu_i,\sigma_i)}(x),$$

where $f_{\text{LN}(\mu_i,\sigma_i)}$ denotes a probability density function of the log-normal distribution with parameters $\mu_i \in \mathbb{R}$ as well as $\sigma_i > 0$, and nonnegative weights $\lambda_1, \ldots, \lambda_k$ summing up to 1.

This approach might be motivated w.r.t. two aspects. Firstly such density functions are flexible enough to model a great variety of potential shapes for the risk neutral density function. Secondly, we may compute easily the hypothetical call and put prices in terms of respective Black-Scholes formulas by

$$C(K|\mu_1, \ldots, \mu_k, \sigma_1, \ldots, \sigma_k, \lambda_1, \ldots, \lambda_k) = \sum_{i=1}^{k} \lambda_i C^{BS}(K|\mu_i, \sigma_i) \quad (10.21)$$

$$P(K|\mu_1, \ldots, \mu_k, \sigma_1, \ldots, \sigma_k, \lambda_1, \ldots, \lambda_k) = \sum_{i=1}^{k} \lambda_i P^{BS}(K|\mu_i, \sigma_i). \quad (10.22)$$

Additionally, drawing on well known formulas for the expectations of log-normal distributions, we obtain

$$F_{\mu_1,\ldots,\mu_k,\sigma_1,\ldots,\sigma_k,\lambda_1,\ldots,\lambda_k} = \sum_{i=1}^{k} \lambda_i e^{(\mu_i + \frac{\sigma_i^2}{2} - rT)}$$

Recalling that the parameter estimation is based on observations of m call and n put prices we have to take into account the problem of overfitting. More precisely, the number $3k - 1$ of parameters should not exceed $m + n$, the number of observations. Furthermore in order to reduce the numerical complexity of the minimization problem underlying the estimation it is often suggested to restrict estimation to the choice of $k \in \{2, 3\}$.

Empirical evidence (cf. e.g. Corrado and Su (1997); Savickas (2002, 2005)) shows that the implied skewness of the underlying used in options is often negative, in contrary to the skewness of log-normal distributions. In order to take into account negative skewness Savickas proposed to use Weibull distributions (cf. Savickas (2002, 2005)). In Fabozzi et al. (2009) this suggestion has been extended to the family of generalized gamma distributions that will be considered in the next subsection.

10.4.3 Estimation Using Generalized Gamma Distributions

According to $\vartheta \stackrel{\text{def}}{=} (\alpha, \beta, k) \in \Theta \stackrel{\text{def}}{=}]0, \infty[^3$ a respective probability density function $f_{\text{G}}(\cdot|\alpha, \beta, \delta)$ is given by

$$f_{\text{G}}(\cdot|\alpha, \beta, k) = \begin{cases} \frac{1}{\Gamma(k)} \left(\frac{\beta}{\alpha}\right) \left(\frac{x}{\alpha}\right)^{\beta k - 1} \exp\left\{-\left(\frac{x}{\alpha}\right)^{\beta}\right\} & : \quad x > 0 \\ 0 & : \text{otherwise} \end{cases},$$

where Γ denotes the Gamma function. The corresponding cumulative distribution function $G(\cdot|\alpha, \beta, k)$ is given by

$$G(x|\alpha, \beta, k) \stackrel{\text{def}}{=} I\left\{k, \left(\frac{x}{\alpha}\right)^{\beta}\right\},$$

where I denotes the incomplete gamma function defined as

$$I\,(k, y) \stackrel{\text{def}}{=} \frac{1}{\Gamma(k)} \int_0^y x^{k-1}\, e^{-x}\, dx.$$

It is known that $k = 1$ leads to a Weibull distribution, when $\beta = 1$ we get a gamma distribution, when $\beta = k = 1$ we obtain an exponential distribution and when $k \to \infty$ we arrive a log-normal distribution. Explicit formulas for the respective hypothetical prices $C(K|\alpha, \beta, k)$, $P(K|\alpha, \beta, k)$ and $F_{\alpha,\beta,k}$, the moment generating function, have been derived in Fabozzi et al. (2009) (pp. 58, 70). They read as follows.

$$F_{\alpha,\beta,k} = \alpha\, \frac{\Gamma(k + \frac{1}{\beta})}{\Gamma(k)} \tag{10.23}$$

$$\begin{aligned} C(K|\alpha, \beta, k) = {} & e^{-rT} F_{\alpha,\beta,k} \; - \; e^{-rT} K \\ & - \left[F_{\alpha,\beta,k}\; I\left\{k - \frac{1}{\beta}, \left(\frac{K}{\alpha}\right)^{\beta}\right\} \; + \; K\, I\left\{k, \left(\frac{K}{\alpha}\right)^{\beta}\right\} \right] \end{aligned} \tag{10.24}$$

$$P(K|\alpha, \beta, k) = e^{-rT} \left[K\, I\left\{k, \left(\frac{K}{\alpha}\right)^{\beta}\right\} - F_{\alpha,\beta,k}\; I\left\{k + \frac{1}{\beta}, \left(\frac{K}{\alpha}\right)^{\beta}\right\} \right]. \tag{10.25}$$

A different class of methods to estimate the risk neutral density start with a parametric model of the whole stock price process which determines in an analytic way the risk neutral distribution. Then the risk neutral density will be estimated indirectly via calibration of the stock price process.

10.5 Estimation via Calibration of the Stock Price Process

The starting point for the indirect estimation of the risk neutral density function via model calibration is the assumption that the risk neutral probability measure of the stock price process $(S_t)_{t\in[0,T]}$ is characterized by a parameter vector $\vartheta \in \Theta \subseteq \mathbb{R}^r$. Furthermore it is supposed that the characteristic functions $\Phi_{T|\vartheta}$ of $\ln(S_T)$ under ϑ is known analytically. Prominent examples are some widely used models (see below).

Based on observed prices $Y_1, \ldots, Y_m$ for call options with expiration T and strike prices $K_1, \ldots, K_m$ the stock price process is calibrated to obtain an estimated parameter vector $\hat{\vartheta}$. A popular way is to solve the following inverse problem (cf. e.g. Bates (1996); Andersen and Andreasen (2000))

$$\min \sum_{i=1}^{m} \{Y_i - C^{\vartheta}(K_i)\}^2 \tag{10.26}$$

$$\text{s.t. } \vartheta \in \Theta, \tag{10.27}$$

where $C^{\vartheta}(K_i)$ denotes the hypothetical call price with expiration T and strike price K_i if ϑ is the true characteristic parameter vector. These prices might be calculated via the Fast Fourier Transform method as introduced in Sect. 10.3.2. This approach has the attractive numerical feature that for implementation we may draw on the Fast Fourier algorithm.

Once we have solved the inverse problem some parameter vector say $\hat{\vartheta}$, we might extract the risk neutral density function in the following way. Firstly we obtain by Fourier inversion theorem (cf. Dudley (2002), 9.5.4) for probability density function $q_{\log|\hat{\vartheta}}$ of $\ln(S_T)$

$$q_{\log|\hat{\vartheta}}(x) = \int \frac{\Phi_{T|\hat{\vartheta}}(y)e^{-ity}}{2\pi}\, dy.$$

Then application of the transformation theorem for probability density functions yields the estimation

$$q_{\hat{\vartheta}}(x) = \begin{cases} \frac{q_{\log|\hat{\vartheta}}(x)}{x} & : \quad x > 0 \\ 0 & : \text{ otherwise} \end{cases}.$$

Let us now have a closer look at some special models where we shall identify the respective calibration parameter and characteristic functions. We shall consider refinements of the classical Black Scholes model. Namely Merton's jump diffusion model which incoporates possible large or sudden movement in prices, and Heston's volatility model which take into account state dependent changes in volatilities.

10.5.1 Merton's Jump Diffusion Model

The jumps of the log prices are usually modelled by a compound Poisson process $\sum_{i=1}^{N_t} Y_i$, consisting of a Poisson process $(N_t)_{t\in[0,\infty[}$ with intensity parameter $\lambda > 0$ independent of a sequence $(Y_i)_{i\in\mathbb{N}}$ of i.i.d. random variables. The N_t model the random number of jumps, whereas the respective jump sizes are expressed by the Y_i having a common distribution of typical jump size. Within the Merton's jump diffusion model a normal distribution $N(\mu, \delta^2)$ is assumed as the distribution of typical jump size. Then this compound Poisson process is added to classical Black Scholes model. As introduced in Merton (1976), the risk neutral price process within Merton's jump diffusion model may be described by

$$S_t = S_0 \exp\left(\mu^M t + \sigma W_t + \sum_{i=1}^{N_t} Y_i,\right),$$

where $\mu^M = r - \frac{\sigma^2}{2} - \lambda\left\{\exp\left(\mu + \frac{\delta^2}{2}\right) - 1\right\}$, $\sigma > 0$, and $(W_t)_{t\in[0,\infty[}$ denoting a standard Brownian motion which is independent of the compound Poisson process.

Drawing on well-known formulas for characteristic functions of normally distributed random variables (cf. Dudley (2002), Proposition 9.4.2), and that for compound Poisson processes (Cont and Tankov (2004), Proposition 3.4), we obtain the characteristic function $\Phi_{\ln(S_T)}$ of $\ln(S_T)$ by an easy calculation, yielding

$$\Phi_{\ln(S_T)}(z) = \exp\{iz\ln(S_0)\} \times \exp\left[T\left\{1 - \frac{\sigma^2 z^2}{2} + i\mu^M z + \lambda\left(e^{(-\frac{\delta^2 z^2}{2} + i\mu z)}\right)\right\}\right] \tag{10.28}$$

As parameter vector we may identify $\vartheta \stackrel{\text{def}}{=} (\sigma^2, \lambda, \mu, \delta^2) \in]0,\infty[^2 \times \mathbb{R} \times]0,\infty[\stackrel{\text{def}}{=} \Theta$.

10.5.2 Heston's Volatility Model

A popular approach to substitute the deterministic volatility in the Black Scholes model by a stochastic process $(v_t)_{t\in[0,\infty[}$. was proposed in Heston (1993). In this model the risk neutral dynamics of the log price $\ln(S_t)$ is expressed by the stochastic differential equations

$$d\ \ln(S_t) = \left(r - \frac{1}{2}v_t\right) dt + \sqrt{v_t}\, dW_t^S \tag{10.29}$$

$$d\ v_t = \kappa(\eta - v_t)\, dt + \theta\sqrt{v_t}\, dW^V, \tag{10.30}$$

where $(W_t^S)_{t\in[0,\infty[}, (W_t^V)_{t\in[0,\infty[}$ are correlated standard Brownian motion with rate ρ:

$$\mathrm{Cov}(d\ W_t^S, d\ W_t^V) = \rho\, dt.$$

An analytical expression of the characteristic function $\Phi_{\ln(S_T)}$ of $\ln(S_T)$ has been derived in Heston (1993) in the following way

$$\Phi_{\ln(S_T)}(z) = \frac{\exp\left[\frac{\kappa\eta T(\kappa - i\rho\theta z)}{\theta^2} + iz\{Tr + \ln(S_0)\}\right]}{\left\{\cosh(\frac{\gamma T}{2}) + \frac{\kappa - i\rho\theta z}{\gamma}\sinh(\frac{\gamma T}{2})\right\}^{\frac{2\kappa\eta}{\theta^2}}} \tag{10.31}$$

$$\times \exp\left\{-\frac{(z^2 + iz)v_0}{\gamma\coth(\frac{\gamma T}{2}) + \kappa - i\rho\theta z}\right\},$$

where $\gamma = \sqrt{\theta^2(z^2 + iz) + (\kappa - i\rho\theta z)^2}$. As parameter vector we obtain

$$\vartheta \stackrel{\text{def}}{=} (\theta, \rho, \kappa, \eta) \in \,]0, \infty[\times[-1, 1] \times [0, \infty[\times]0, \infty[\stackrel{\text{def}}{=} \Theta.$$

10.6 Empirical Study

In this section we will demonstrate the methods exposed in the theoretical part and address some aspects of concern for practitioners. Estimating the risk neutral density by direct methods involves the choice of parametric distribution family to which it belongs to. This introduces some arbitrariness in modelling because the distribution family must be selected a priori from a set of candidates. Indirect modelling relies on assumptions about the data generating process and the shape of the risk neutral density is intrinsically related to the parameter values of the underlying process.

Practitioners are interested in modelling the RND from observed data and therefore have to solve an inverse problem. Model parameters are often obtained by solving nonlinear least squares equations for which analytical solutions may be very difficult or impossible to derive. Therefore, one has to rely on numerical optimization algorithms in order to retrieve the unknown parameters. In addition, the approaches may suffer the drawbacks associated with the ill-posedness of some inverse problems in pricing models: there may exist no solution at all or an infinity of solutions. The last case means that there are many sets of parameters reproducing call prices with equal precision, which in turn may translate in pricing errors with many local minima or flat regions with low model sensitivity to variations in parameters. The solutions are often very sensitive to the numerical starting values; numerical instability may also occur if the dependence of solutions to the observed data is discontinuous. Uniqueness and stability may be achieved by introducing a regularization method: e.g. adding penalty to the linear least squares term. For further discussions on regularization methods see Cont and Tankov (2004).

In order to assess the shape of RND implied by different parametric approaches we use paired European call options written on the underlying DAX stock index which mature in 1 month (21 days) and strike prices observed on 20040121. The data is provided by Eurex – Deutsche Börse and collected from the Research Data Center (RDC) of the Collaborative Research Center 649. Strike prices have been transformed to account for intraday stock price movements; these have been computed from the futures prices following a methodology by Fengler (2005). The

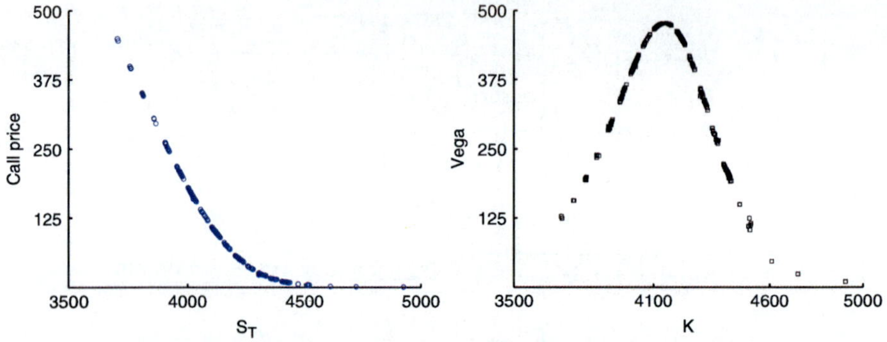

Fig. 10.1 *Left*: European call option vs. strike prices on 21 Jan 2004. *Right*: ν of the observed call prices

EURIBOR interpolated interest rate for this maturity is $r = 2.05\%$ per annum and the stock index value taken from the daily series DAX 30 Index is $S = 4{,}138$. The dividend rate is zero. Observations that do not respect general arbitrage conditions (see Jackwerth (2000)) have been excluded from the sample. We are left with 2,562 paired observations, which we display in Fig. 10.1. The counterpart representation of the observations in the implied volatility space (based on Black-Scholes valuation formula) will be further used to assess the quality of the estimates. Note that in practice, it is often more convenient to match implied volatilities which are of the same order of magnitude relative to call prices which display a much larger out-of-the-money variation.

Figure 10.2 depicts the estimation results for the RND by direct methods as well as the fit in the implied volatility space. In the upper left panel, the Black-Scholes log-normal RND depends on only one unknown for given risk free interest rate, the constant – across strikes – volatility parameter σ. It is contrasted with the implied volatility of the observed call prices in the right panel.

Next, we fit a mixture of log-normal densities. The parameter k is usually assumed to be unknown and one has to apply appropriate criteria to find the optimal parameter value. Here, we illustrate the method for fixed $k = 2$ in the central part of Fig. 10.2. Since μ_1 and μ_2 are known up to the volatility parameters σ_1 and σ_2 respectively of the components, the mixing distribution will have three unknown parameters. We have investigated the shape of the resulting density and found that it is robust with respect to the mixtures, in the sense that for known basic densities, the proportion parameter λ regulates the relative impact of each component. Conversely, one can fix λ and try to estimate σ_1 and σ_2. The mixture generates a rather symmetric smile, with a minimum different from that of the volatility skew of the observed prices. The higher kurtosis improves the fit at a price of higher skewness compared with the simple log-normal case. This shows that using mixtures of log-normals improves the fit especially by higher kurtosis. Every (central) moment of a linear combination of densities is given by the same combination of the corresponding

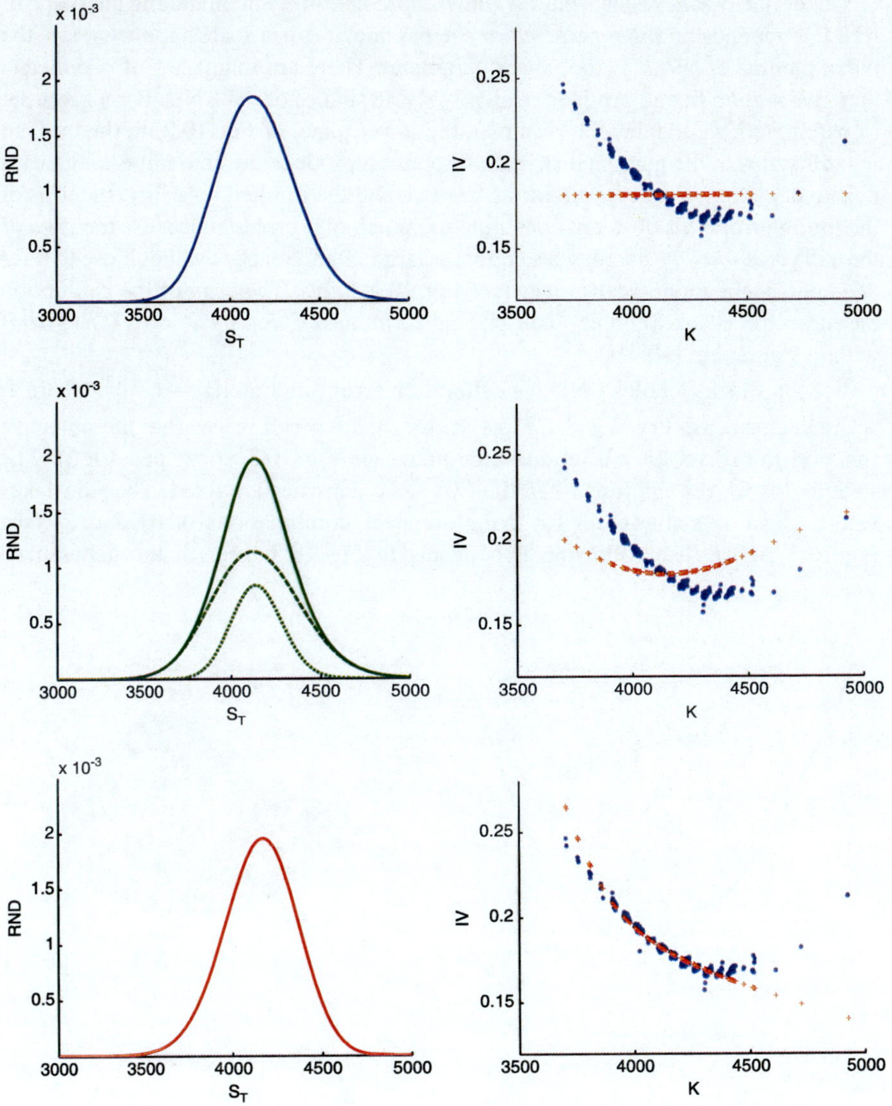

Fig. 10.2 *Left*: RND estimated by: log-normal distribution with $\sigma = 0.18$ (*top*), mixture of log-normal distributions for weighted components $\sigma_1 = 0.24$ – *dashed*, $\sigma_2 = 0.15$ – *dotted* with $\lambda = 0.31$ (*center*) and generalized gamma distribution with $\alpha_1 = 3033.03$, $\beta_2 = 6.86$ and $k = 9.05$ (*bottom*). *Right*: IV_{BS} for observed call prices (*asterisk*) and fitted call prices (*plus sign*)

moments. The third moment of a log-normal density is always positive, therefore a mixture of log-normal can never generate negative skewness. In order to generate a negative skew either other mixture types or other parametric models for the RND has to be considered.

Generalized gamma distribution allows more flexibility in modelling the shape of RND. It depends on three parameters: the parameter α is a scale parameter, k is the index parameter and β is the power parameter. There are many sets of parameters that give a good fit and produce relatively stable shapes of the RND. For a given set of parameters we display the results in the lower panel of Fig. 10.2. In the implied volatility space, the gamma distribution cannot reproduce the smile; it establishes a negative relationship between the strike price and the implied volatility. In terms of the fitting errors this does not constitute too much of a problem because the *vega* of the call price $\nu = \frac{\partial C}{\partial \sigma}$ decreases steeply for large K and reaches values close to 0 for deep out-of-the money call prices (see Fig. 10.1 right). The *vega* of the call option based on the Black-Scholes's call pricing formula is given by $\nu = S\sqrt{T}(\phi(d_1))$ with d_1 defined in (10.4).

For the Black-Scholes RND the calibration error function $||Y - C^{\hat{\vartheta}}||^2$, where Y is the vector of observed and $C^{\hat{\vartheta}}$ the vector of fitted call prices (i.e. the objective function in (10.15)) has a unique minimum (see Fig. 10.3 upper panel left). The same holds for the mixture when the two basic densities are fixed. The RSS takes values close to a minimum for a multitude of combinations of σ_1 and σ_2 (see Fig. 10.3 right). The following two panels in Fig. 10.3 refer to the generalized

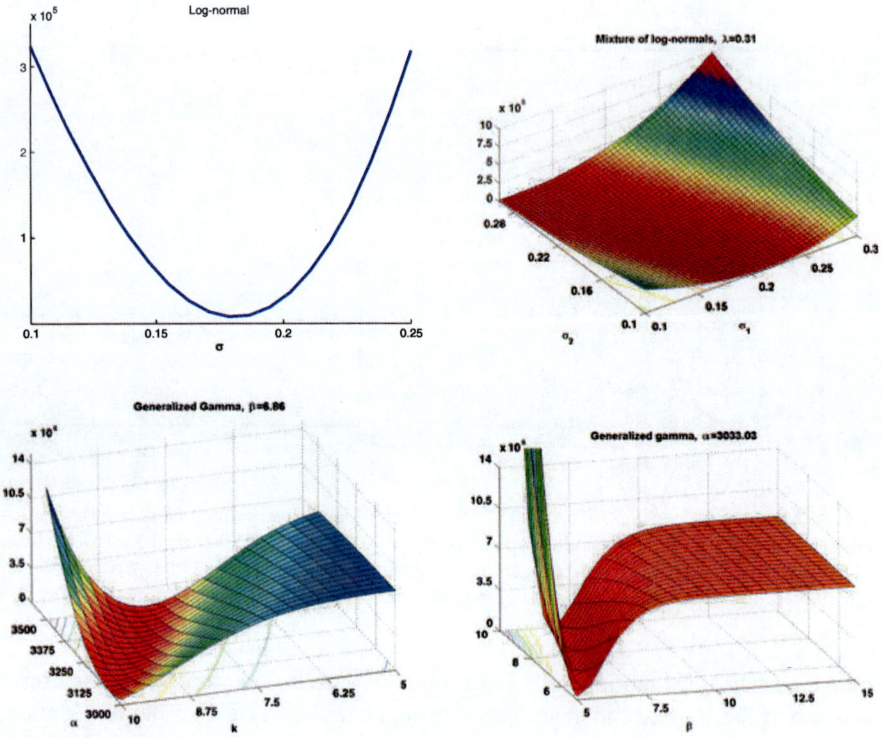

Fig. 10.3 Estimation error function by direct methods

Table 10.1 Comparison of the RND estimates by direct method in terms of moments and fit: log-normal for $\sigma = 0.18$ (*blue*), mixture of lognormals for $\sigma_1 = 0.24$, $\sigma_2 = 0.15$ and $\lambda = 0.31$, generalized gamma for $\alpha_1 = 3033.03$, $\beta_2 = 6.86$ and $k = 9.05$

Model	Mean	St. Dev.	Skewness	Kurtosis	RSS
Log-normal	4,145.11	216.99	0.15	3.04	7,693.08
Mixture	4,139.39	221.28	0.20	3.83	7,465.22
Generalized gamma	4,152.76	205.44	−0.18	3.06	351.07

gamma distribution. The objective function is a surface which forms a valley or rift of minimum values. This illustrate the ill-posed problem.

The pricing errors computed as a difference between observed and fitted call prices, display some regularities: RND-s estimated by the first two methods lead to underpriced calls for ITM options and overpriced calls for OTM options; the discrepancies diminish for deep ITM and OTM options. Generalized gamma distribution is flexible enough to give a good fit for a large range of strikes in the central part of the distribution. Since the observations in the tails are more sparse, the pricing errors will be higher for deep ITM call options. In this particular case, the estimated density will have fetter left tails resulting in overpriced options for small strike prices. However, the absolute pricing errors are smaller than for the other candidates. The resulting moments of the estimated densities are summarized in Table 10.1.

In the remaining of this section, we describe the results by the indirect approach for finding the RND. The calibration of the second type of models is further supported by advanced numerical methods available, such as Fast Fourier Transform (FFT). In order to apply the FFT-based algorithm we use the characteristic function of the risk neutral density as described in Sect. 5 for the Merton and Heston models and set the parameters $\alpha = 1.25$, $N = 4{,}096$, and $\eta = 0.25$. For OTM option prices the calibration error increases; therefore, we use the Fourier Transform of OTM option prices as described in Carr and Madan (1999). With the above parameters choice and pricing rules, we solve the problem of model calibration. This implies solving the minimization problem given in (10.26) and (10.27). We describe the results for both models in terms of the resulting RND and fit in the IV space in Fig. 10.4.

Merton model for pricing European options tries to capture the deviations from normality of log-returns by adding a compound Poisson jump process to the Black-Scholes model. Jump components add mass to the tails of the returns distribution. Increasing δ adds mass to both tails. The sign of μ determines the sign of the skewness: negative μ implies relatively more mass in the left (negative skew) and the other way around. Larger values of the intensity parameters λ (which means that the jumps are expected to occur more frequently) makes the density flatter tailed, i.e. increases kurtosis.

In the Merton model an implied volatility skew is attainable by the presence of jumps. By introducing a correlation parameter ρ between log-returns and volatility movements in the Heston model has a similar effect on the volatility smile. Varying the parameter ρ around 0 gives us asymmetric tails of RND. Intuitively, if $\rho > 0$, then volatility will increase as the asset price/return increases. This will spread

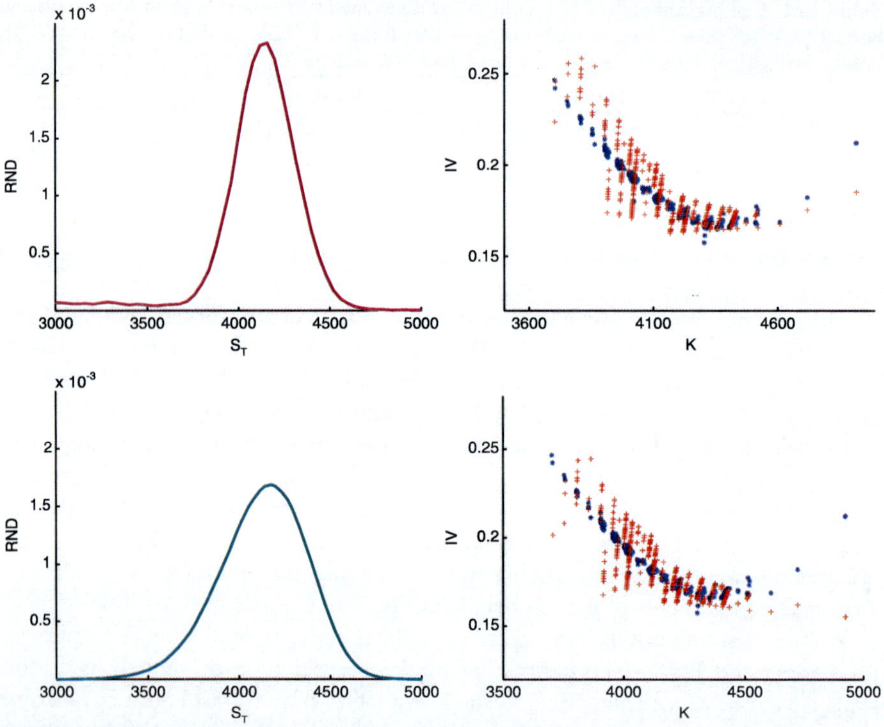

Fig. 10.4 *Left*: RND estimated for the Merton with $\sigma = 0.13$, $\lambda = 0.10$, $\mu = -0.23$, $\delta = 0.17$ and Heston with $\theta = 0.19$, $\rho = -0.61$, $\kappa = 1.18$, $\eta = 0.21$. *Right*: IV_{BS} for observed call prices (*asterisk*) and fitted call prices (*plus sign*)

the right tail and squeeze the left tail of the distribution creating a fat right-tailed distribution. Parameter κ measures the speed of mean reversion and can be interpreted as the degree of "volatility clustering" in the sense that large price variations are likely to be followed by large price variations and the other way around. η is the long run level of volatility and θ is the volatility of volatility. θ affects the kurtosis of the distribution: when it is 0 the log-returns will be normally distributed. Increasing θ will then increase the kurtosis only, creating heavy tails on both sides. Conversely, if $\theta < 0$, then volatility will increase when the asset price/return decreases, thus spreading the left tail and squeezing the right tail of the distribution and creating a fat left-tailed distribution.

Empirical results for the RND by both method indicate negative skewness: $\mu > 0$ in Merton model and $\rho < 0$ in Heston model. Negative correlation ρ is in line with the empirical studies of the financial returns which show that volatility is negatively correlated with the returns. Reproducing some of the essential features of asset dynamics can result in significant shape differences. We can see in Fig. 10.4 that RND implied by Merton has a much fatter left tail and a higher kurtosis than the

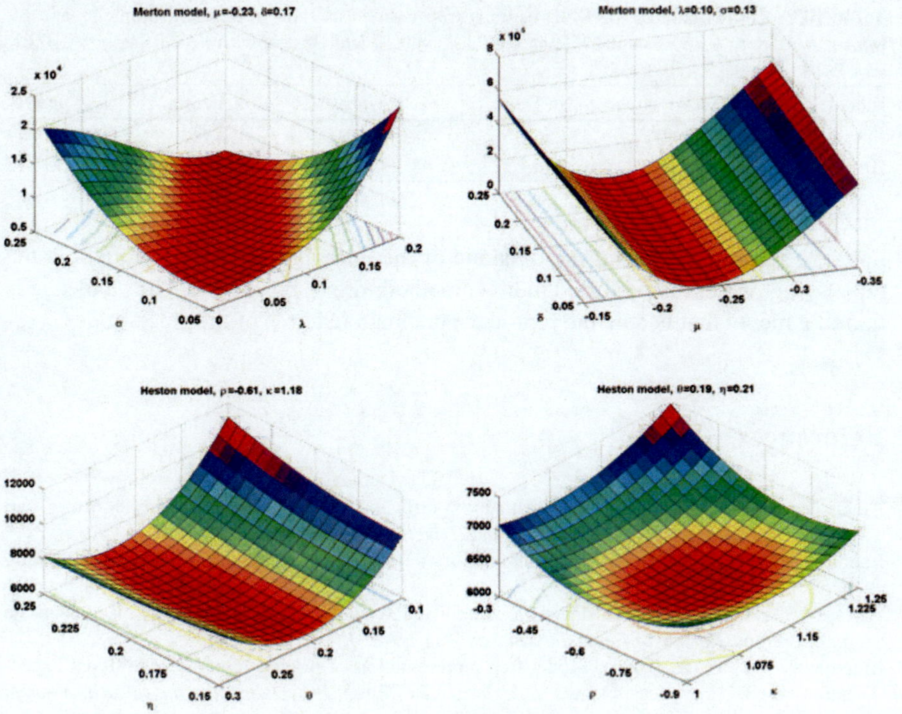

Fig. 10.5 Calibration error function by indirect methods

RND obtained from the Heston model. This shows how different models for the stock prices give various shapes of the risk neutral density. In terms of the implied volatility, Merton model seems more suitable to reproduce the skew in Fig. 10.4. Pricing errors have a very similar structure for the two models: they are almost symmetrical against the 0 line and decrease for high strike prices.

The graphs in Fig. 10.5 show the calibration error function in both models for pairs of parameters in each model. Three of the panels indicate that the calibration is ill-posed because there is a large, nearly flat region or a valley of minima for the objective function. It implies that there are many parameter sets for which the model prices match the observed prices. However, by using this approach the shape of RND for different set of parameters that give a comparable good fit may differ a lot. We do not report such graphs here, but one can easily vary two of the parameters along a valley in Fig. 10.5 to verify this. The right panel bottom indicate that the objective function has a clearly defined minimum so that the pairs (ρ, κ) in the Heston model are uniquely defined when keeping the other model parameters fixed.

In modelling the risk neutral densities based on option data the practitioners face a trade off between modelling aspects of the underlying's dynamics and reliability of calculations concerning the shape of the RND. While some distribution families allow for great flexibility in the shape of RND (e.g. generalized gamma) they are

Table 10.2 Comparison of the RND estimates by indirect method in terms of moments and fit: Merton with $\sigma = 0.13$, $\lambda = 0.10$, $\mu = -0.23$, $\delta = 0.17$ and Heston with $\theta = 0.19$, $\rho = -0.61$, $\kappa = 1.18$, $\eta = 0.21$

Model	Mean	St. Dev.	Skewness	Kurtosis	RSS
Merton	4,008.40	256.61	−0.09	4.88	6,468.49
Heston	4,130.12	240.20	−0.35	3.19	6,362.18

not very informative about the dynamic of the underlying asset. If modelling the underlying process is preferred indirect methods are to be chosen. The challenge is to find a model that is able the reproduce the main features of the stock prices.

References

Andersen, L., & Andreasen, J. (2000). Jump-diffusion models: Volatility smile fitting and numerical methods for pricing. *Review of Derivatives Research*, *4*(3), 231–262.

Bahra, B. (1997). Implied risk-neutral probability density functions from option prices. Working paper, Bank of England.

Bates, D. S. (1996). Jumps and stochastic volatility: Exchange rate processes implicit in deutsche mark options. *Review of Financial Studies*, *9*(1), 69–107.

Biagini, S., & Cont, R. (2006). Model-free representation of pricing rules as conditional expectations. In J. Akahori, S. Ogawa, & S. Watanabe (Eds.), *Proceedings of the 6th Ritsumeikan International Symposium – Stochastic Processes and Applications to Mathematical Finance* (pp. 53–66). Singapore: World Scientific.

Black, F., & Scholes, M. (1973). The pricing of options and corporate liabilities. *Journal of Political Economy*, *81*(3), 637–659.

Carr, P., & Madan, D. (1999). Option valuation using the fast fourier transform. *Journal of Computational Finance*, *2*(4), 61–73.

Cizek, P., Härdle, W., & Weron, R. (2005). *Statistical tools in finance and insurance*. Berlin: Springer

Cont, R., & Tankov, P. (2004). *Financial modelling with jump processes*. London: Chapman & Hall.

Corrado, C., & Su, T. (1997). Implied volatility skew and stock index skewness and kurtosis implied by s& p 500 index option prices. *Journal of Derivatives*, *4*, 8–19.

Delbaen, F., & Schachermayer, M. (2006). *Mathematics of arbitrage*. Berlin: Springer.

Dudley, R. M. (2002). *Real analysis and probability*. Cambridge: Cambridge University Press.

Fabozzi, F. J., Tunaru, R., & Albotas, G. (2009). Estimating risk-neutral density with parametric models in interest rate markets. *Quantitative Finance*, *9*(1), 55–70.

Fengler, M. (2005). Semiparametric modeling of implied volatility, Springer Finance

Föllmer, H., & Schied, A. (2004). *Stochastic finance* (2nd ed.). Berlin: de Gruyter

Heston, S. (1993). A closed-form solution for options with stochastic volatility with applications to bond and currency options. *Review of Financial Studies*, *6*(2), 327–343.

Hugonnier, J., Kramkov, D., & Schachermayer, W. (2005). On utility-based pricing of contingent claims in complete markets. *Mathematical Finance*, *15*(2), 203–212.

Jackwerth, J. (2000). Recovering risk aversion from option prices and realized returns. *Review of Financial Studies*, *1*(2), 433–451.

Jondeau, E., & Rockinger, M. (2000). Reading the smile: The message conveyed by methods which infer risk neutral densities. *Journal of International Money and Finance*, *19*, 885–915.

Lee, R. (2004). Option pricing by transform methods: extensions, unifications and error control. *Journal of Computational Finance*, *7*(3), 51–86.

Melick, W., & Thomas, C. (1997). Recovering an asset's implied pdf from option prices: An application to crude oil crisis. *Journal of Financial and Quantitative Analysis*, *32*, 91–115.

Merton, R. (1976). Option pricing when underlying stock returns are discontinuous. *Journal of Financial Economics*, *3*, 125–183.

Ritchey, R. J. (1990). Call option valuation for discrete normal mixtures. *Journal of Financial Research*, *13*(4), 285–295.

Rudin, W. (1974). *Real and complex analysis* (2nd ed.). New York: McGraw-Hill.

Savickas, R. (2002). A simple option-pricing formula. *The Financial Review*, *37*, 207–226.

Savickas, R. (2005). Evidence on delta hedging and implied volatilities for black-scholes, gamma and weibull option-pricing models. *Journal of Financial Research*, *28*, 299–317.

Söderlind, P., & Swensson, L. (1997). New techniques to extract market expectation from financial instruments. *Journal of Monetary Economics*, *40*, 383–429.

Walker, J. S. (1996). *Fast fourier transforms*. Boca Raton: CRC Press.

Chapter 11
Nonparametric Estimation of Risk-Neutral Densities

Maria Grith, Wolfgang Karl Härdle, and Melanie Schienle

Abstract This chapter deals with nonparametric estimation of the risk neutral density. We present three different approaches which do not require parametric functional assumptions on the underlying asset price dynamics nor on the distributional form of the risk neutral density. The first estimator is a kernel smoother of the second derivative of call prices, while the second procedure applies kernel type smoothing in the implied volatility domain. In the conceptually different third approach we assume the existence of a stochastic discount factor (pricing kernel) which establishes the risk neutral density conditional on the physical measure of the underlying asset. Via direct series type estimation of the pricing kernel we can derive an estimate of the risk neutral density by solving a constrained optimization problem. The methods are compared using European call option prices. The focus of the presentation is on practical aspects such as appropriate choice of smoothing parameters in order to facilitate the application of the techniques.

M. Grith (✉)
Ladislaus von Bortkiewicz Chair of Statistics, Humboldt-Universität zu Berlin,
Spandauer Straße 1, 10178 Berlin, Germany
e-mail: gritmari@wiwi.hu-berlin.de

W.K. Härdle
Ladislaus von Bortkiewicz Chair of Statistics and CASE – Center for Applied Statistics and Economics, Humboldt-Universität zu Berlin, Spandauer Straße 1, 10178 Berlin, Germany

Graduate Institute of Statistics, CDA – Centre for Complex Data Analysis, National Central University, No. 300, Jhongda Rd., Jhongli City, Taoyuan County 32001, Taiwan (R.O.C.)
e-mail: haerdle@wiwi.hu-berlin.de

M. Schienle
Chair of Econometrics and CASE – Center for Applied Statistics and Economics,
Humboldt-Universität zu Berlin, Spandauer Straße 1, 10178 Berlin, Germany
e-mail: melanie.schienle@wiwi.hu-berlin.de

J.-C. Duan et al. (eds.), *Handbook of Computational Finance*, Springer Handbooks of Computational Statistics, DOI 10.1007/978-3-642-17254-0_11,

11.1 Introduction

Most of our economic understanding of investment under uncertainty is based on pure Arrow–Debreu securities (Arrow 1964; Debreu 1959), which pay one unit of currency at the end of a period if a state of nature is realized and zero otherwise. Their theoretical state-contingent prices are the starting point for pricing any security in an economic equilibrium under uncertainty. In a continuum of states, the prices of the Arrow–Debreu securities can be characterized by the state-price density, which yields one dollar if the final state is in the interval $[x, x + \mathrm{d}x]$ when starting from any point x. The existence and form of a state-price density can be justified by preference-based equilibrium models (Lucas 1978) or reasoning from arbitrage-based models (Merton 1973). We focus on the latter, where the state-price density is denoted as risk neutral density (RND). It exists if the underlying market is dynamically complete, i.e. any position can be replicated by a cash-flow neutral (self-financing) trading strategy over subsequent trades. We assume this for the rest of the chapter. Then the RND also uniquely characterizes the equivalent martingale measure under which all asset prices discounted at the risk-free rate are martingales.

In standard option pricing models such as Merton (1976), Heston (1993) or Bates (1996), estimation of the risk neutral density crucially depends on underlying model assumptions such as the underlying asset price dynamics and the statistical family of distributions that the risk neutral density is assumed to belong to. Consumption based asset pricing models prespecify preferences of the representative agent and condition therefore the shape of the pricing kernel (Lucas 1978; Rubinstein 1976). Recent empirical findings, however, question the validity of these popular specifications which drive the overall result (Campbell et al. 1997). Nonparametric estimation offers an alternative by avoiding possibly biased parametric restrictions and therefore reducing the respective misspecification risk. Since nonparametric estimation techniques require larger sample sizes for the same accuracy as a parametric estimation procedure, increasing availability of large data sets as intraday traded option prices have raised their feasibility. On the other hand, due to their flexibility, many existing nonparametric risk neutral density estimation techniques are afflicted by irregularities such as data sparsity in the tails, negative probabilities and failure of integration to unity. We will address these problems by appropriate choices of smoothing parameters, by employing semiparametric techniques or imposing relevant constraints.

We present a thorough picture of nonparametric estimation strategies for the RND q: We study direct standard kernel based approaches (local polynomial regression) which are flexible and yield point estimates as opposed to series expansion, sieve methods or splines. Though shape constraints such as convexity or monotonicity of the call price are hard to incorporate directly in the estimation step. Therefore, in particular in small samples, they are not satisfied leading to problems with economic theory. Thus we also propose an indirect way of estimation by employing series methods for directly controlling constraints in the estimation.

In the following, we will briefly outline the main ideas for direct or indirect estimation of q.

In a dynamically complete market, the price of a European call is obtained by discounting the expected payoff, where the expectation is taken with respect to the risk neutral measure

$$C(X,\tau,r_{t,\tau},\delta_{t,\tau},S_t) = e^{-r_{t,\tau}\tau}\int_0^{\infty}(S_T - X)^{+} q(S_T|\tau,r_{t,\tau},\delta_{t,\tau},S_t)\,dS_T. \quad (11.1)$$

Here S_t is the underlying asset price at time t, X the strike price, τ the time to maturity, $T = t + \tau$ the expiration date, $r_{t,\tau}$ the deterministic risk free interest rate at t until maturity T, $\delta_{t,\tau}$ the corresponding dividend yield of the asset, and $q(S_T|\tau,r_{t,\tau},\delta_{t,\tau},S_t)$ is the conditional risk neutral density. We assume that these state variables contain all essential information needed for estimation of C and q while quantities such as stochastic market volatility, trading volumes, bid-ask spreads are negligible. We write $q(S_T)$ instead of $q(S_T|\cdot)$ to keep notation simple. The risk neutral density can be derived from (11.1) as

$$q(S_T) = e^{r_{t,\tau}\tau}\left\{\frac{\partial^2 C}{\partial X^2}\right\}_{X=S_T}, \quad (11.2)$$

see Breeden and Litzenberger (1978). It has been exploited to derive two standard nonparametric kernel estimation strategies for q: Either obtain an estimate of the RND from estimating a continuous twice-differentiable call function in all its arguments from traded options by smoothing in the call price, or alternatively, by smoothing in the implied volatility space.

In addition to these standard approaches, here we also introduce a third indirect way via series estimation of the empirical pricing kernel. Assuming that all the variables other than X are fixed, the price of the European call option with strike price X expiring in τ years under the historical measure p is given by

$$\begin{aligned} C(X) &= e^{-r_{t,\tau}\tau}\int_0^{\infty}(S_T - X)^{+}\frac{q(S_T)}{p(S_T)}p(S_T)dS_T \\ &= e^{-r_{t,\tau}\tau}\int_0^{\infty}(S_T - X)^{+} m(S_T)p(S_T)dS_T, \end{aligned} \quad (11.3)$$

where p is the subjective density of the stock price at the expiration of the option, at time T and m is the so called pricing kernel characterizing the change of measure from q to p.

The rest of this chapter is organized as follows: Section 11.2 describes kernel based regression methods for direct estimation of the RND from the call price function, Sect. 3 introduces the pricing kernel concept and explains the indirect method of estimating RND, Sect. 4 concludes. Throughout the chapter, empirical studies using EUREX DAX Index based European option data illustrate the methods and compare their performance.

11.2 Estimation of RND Based on the Second Derivative

The standard approach for a nonparametric estimator of the risk neutral density is by estimating the second derivative of the call price with respect to the strike X. Then an estimate for q is obtained by discounting according to (11.2). Therefore in the following we focus on estimation of the second derivative of a regression function.

Call the d-dimensional vector of covariates $\mathbf{Z}$ which comprises all estimation relevant variables of $(X, \tau, r_{t,\tau}, \delta_{t,\tau}, S_t)$ from (11.1) and denote call prices as response Y. From paired observations Y_i and $\mathbf{Z}_i = (Z_{ik})_{k=1}^d$, for $i = 1, ..., n$ we want to estimate the following general, possibly nonlinear relationship

$$Y_i = C(\mathbf{Z}_i) + \varepsilon_i, \quad i = 1, ..., n, \tag{11.4}$$

where $C(\cdot) : \mathbb{R}^d \to \mathbb{R}$ is a smooth function in all d dimensions and ε is i.i.d. with $\mathsf{E}[\varepsilon|\mathbf{Z}] = 0$.

Kernel based methods are local techniques for estimating the function C at any point $\mathbf{z}$ in its domain; they use a weighted average of the Y_i's to yield fitted values via

$$\widehat{C}(\mathbf{z}) = \sum_{i=1}^{n} w_i(\mathbf{z}) Y_i, \tag{11.5}$$

where the weights $w_i(\mathbf{z})$ assigned to each point of fit $\mathbf{z}$ decline with the distance $|\mathbf{Z}_i - \mathbf{z}|$ and satisfy $\frac{1}{n}\sum_{i=1}^{n} w_i(\mathbf{z}) = 1$. Kernel regression methods use kernel functions to construct weights. A univariate kernel is a smooth, symmetric real-valued squared integrable function $K(u) : \mathbb{R} \to \mathbb{R}$ which integrates to one. We can think of a standard kernel function as a probability density with potentially compact support. Examples of such K are presented in Table 11.1, that is an updated version of H"ardle (1990) Table 4.5.2.

Table 11.1 Kernel functions $K(u)$

Uniform	$\frac{1}{2}\,\mathrm{I}(\lvert u\rvert \le 1)$
Triangle	$(1 - \lvert u\rvert)\,\mathrm{I}(\lvert u\rvert \le 1)$
Epanechnikov	$\frac{3}{4}(1 - u^2)\,\mathrm{I}(\lvert u\rvert \le 1)$
Quartic (Biweight)	$\frac{15}{16}(1 - u^2)^2\,\mathrm{I}(\lvert u\rvert \le 1)$
Triweight	$\frac{35}{32}(1 - u^2)^3\,\mathrm{I}(\lvert u\rvert \le 1)$
Gaussian	$\frac{1}{\sqrt{2\pi}} \exp(-\frac{1}{2}u^2)$
Cosine	$\frac{\pi}{4} \cos(\frac{\pi}{2}u)\,\mathrm{I}(\lvert u\rvert \le 1)$

Furthermore, there exist more general types of kernel functions, so called higher order kernels, which can be used for bias refinements in the estimation, see Sect. 2.1. The order of a kernel $\upsilon > 0$ is defined as the first nonzero moment of the kernel, that is

$$\int u^l K(u)du = 0, \quad l = 1, ..., \upsilon - 1 \tag{11.6}$$

$$\int u^\upsilon K(u)du = \kappa_\upsilon \neq 0$$

and $\kappa_\upsilon < \infty$. Solving the system of (11.6) for kernel functions integrating to unity for a fixed υ, yields a υth order kernel. The larger υ, however, the more "wiggly" the resulting kernel becomes – covering more and more negative areas. Here we mostly consider standard second order kernels, which are nonnegative functions.

Set $K_h(u) = \frac{1}{h}K\left(\frac{u}{h}\right)$ for all $u \in \mathbb{R}$ where h is the bandwidth, the smoothing parameter. In a d-dimensional space, for each pair $\mathbf{z}$ and $\mathbf{Z}_i$ the multivariate kernel function $\mathcal{K}(\mathbf{z} - \mathbf{Z}_i) : \mathbb{R}^d \to \mathbb{R}$ must analogously fulfil

$$\mathcal{K}_H(\mathbf{z} - \mathbf{Z}_i) = \frac{1}{|H|}\mathcal{K}\{H^{-1}(\mathbf{z} - \mathbf{Z}_i)\},$$

where $H = \operatorname{diag}(\tilde{h})$ is the diagonal matrix of bandwidths $\tilde{h} = [h_1, ..., h_d]$. The matrix H can in general also contain off-diagonal elements – but in practice such generality is not needed. Define the multidimensional kernel $\mathcal{K}_H(\mathbf{z}-\mathbf{Z}_i)$ as a product of univariate kernels

$$\mathcal{K}_H(\mathbf{z} - \mathbf{Z}_i) = \prod_{k=1}^{d} K\left(\frac{z_k - Z_{ik}}{h_k}\right).$$

For expositional simplicity we let $h_1 = \ldots = h_d = h$. Details on how to choose the optimal bandwidths are addressed in the next section.

The simplest case of choosing w_i in (11.5) is to use Nadaraya-Watson weights

$$w_i(\mathbf{z}) = \frac{\mathcal{K}_h(\mathbf{z} - \mathbf{Z}_i)}{\sum_{i=1}^{n} \mathcal{K}_h(\mathbf{z} - \mathbf{Z}_i)}.$$

These are a special constant case of general local polynomial weights derived below. Besides, other choices of weights such as in the k-nearest neighbour or the Gasser-Müller estimator are possible.

Estimators of the second derivative of a function are constructed by twice differentiating the estimator of the function. Such estimators, however, have inferior statistical properties and are therefore not included here (see e.g. Härdle et al. 2004; Fan and Gijbels 1996 (for details)). We focus on local polynomial estimation which directly yields estimates of derivatives. The idea of local polynomial regression is based on Taylor expansion approximating an unknown function C at a point $\mathbf{z}$. In order to keep notation simple, we first illustrate the method for the univariate case. The multivariate case is systematically the same and will be sketched afterwards.

Locally, any sufficiently smooth function C can be approximated by a polynomial of degree p

$$C(Z_i) = \sum_{j=0}^{p} \frac{C^{(j)}(z)}{j!}(Z_i - z)^j + \mathcal{O}\{(Z_i - z)^{p+1}\} \tag{11.7}$$
$$= \sum_{j=0}^{p} \beta_j (Z_i - z)^j + \mathcal{O}\{(Z_i - z)^{p+1}\}$$

with $i = 1, ..., n$. Therefore by minimizing a locally weighted least squares regression

$$\min_{\beta} \sum_{i=1}^{n} \left\{ Y_i - \sum_{j=0}^{p} \beta_j(z)(z - Z_i)^j \right\}^2 \mathcal{K}_h(z - Z_i) \tag{11.8}$$

the solution $\hat{\beta}_0(z)$ provides an estimator of C at point z, while $j!\hat{\beta}_j(z)$, with $j = 1, ..., p$ are the estimated derivatives at that point. Closed forms for $\hat{\beta}(z) = (\hat{\beta}_0(z), \ldots, \hat{\beta}_p(z))$ can be obtained by solving (11.8) via equating the corresponding system of first order conditions to zero. As we are interested in an estimator for the second derivative of a function, we should choose $p \geq 2$. As will be outlined in the subsection below, for good statistical properties without requiring too much smoothness $p = 3$ will be a suitable choice.

In d-dimensional case, expansion (11.7) will include mixed terms which must be appropriately ordered. Then the interpretation of the coefficients is similar: $\widehat{\beta}_0(\mathbf{z})$ is the estimator of C at point $\mathbf{z}$, while $j!\widehat{\beta}_j(\mathbf{z}) = j!\left[\beta_{j1}(\mathbf{z}), \cdots, \beta_{j\mu}(\mathbf{z})\right]$ with $\mu = 1, ..., N_j$ is N_j-dimensional vector of jth order derivatives of C evaluated at point $\mathbf{z}$. It is obvious that $N_0 = 1$ (β_0 is the local constant) and $N_1 = d$ (β_1 is the vector of partial derivatives) but for $j \geq 2$ the expansion contains cross order derivatives and the general formula for N_j is

$$N_j = \binom{d + j - 1}{j - 1} = \frac{(d + j - 1)!}{d!(j - 1)!}.$$

For example, when $j = 2$ we have $N_2 = d(d + 1)/2$ distinct derivatives and, $\nabla^{(2)}\hat{C}(\mathbf{z}) = 2\hat{\beta}_2(\mathbf{z})$ is the estimate of

$$\nabla^{(2)}C(\mathbf{z}) = \begin{pmatrix} \frac{\partial^2 C(\mathbf{z})}{\partial z_1^2} \\ \frac{\partial^2 C(\mathbf{z})}{\partial z_1 \partial z_2} \\ \vdots \\ \frac{\partial^2 C(\mathbf{z})}{\partial z_d^2} \end{pmatrix}.$$

For the estimation of the RND we are interested in the second derivative of the call price with respect to the strike price X. In our notation with $\mathbf{Z} = (X, \tau, r_{t,\tau}, \delta_{t,\tau}, S_t)$, this is $2\beta_{21}$. Thus

$$\hat{q}(S_T) = 2e^{r_{t,\tau}\tau}\hat{\beta}_{21}(S_T, \mathbf{z}_{-1}) = e^{r_{t,\tau}\tau}\left\{\frac{\widehat{\partial^2 C}(\mathbf{z})}{\partial z_1^2}\right\}_{X=S_T}$$

with $\mathbf{z}_{-1} = (\tau, r_{t,\tau}, \delta_{t,\tau}, S_t)$.

11.2.1 Statistical Properties

Assume for simplicity that C is univariate and has continuous derivatives of total order $(p+1)$. The probability density function f of Z is continuous, it is $f \geq 0$, and f is $(p+1)$ times continuously differentiable. The kernel K is a bounded second order kernel with compact support and the $\mathsf{E}[\varepsilon^2|Z = z]$ exists and is continuous in z. Let $\widehat{C}^{(j)}$ denote the estimator of $C^{(j)}$ based on a pth order local polynomial fit $(j \leq p)$. The results below are standard and can be found for instance in Li and Racine (2007).

Theorem 1. *When $p - j$ is odd, the bias is*

$$\mathsf{E}\left[\widehat{C}^{(j)}(z)\right] - C^{(j)}(z) = h^{p-j+1}c_{1,j,p}\left\{\frac{\omega^{(p+1)}(z)}{(p+1)!}\right\} + o(h^{p-j+1}). \tag{11.9}$$

When $p - j$ is even, the bias is

$$\mathsf{E}\left[\widehat{C}^{(j)}(z)\right] - C^{(j)}(z) = h^{p-j+2}c_{2,j,p}\left\{\frac{\omega^{(p+2)}(z)}{(p+2)!}\right\}\int u^{p+2}K(u)du \tag{11.10}$$
$$+ h^{p-j+2}c_{3,j,p}\left\{\frac{\omega^{(p+1)}(z)f^{(1)}(z)}{f(z)(p+1)!}\right\},$$

where $\omega(z) = \left\{\widehat{C}(z) - C(z)\right\}f(z)$. In either case

$$\mathsf{Var}\left(\widehat{C}^{(j)}(z)\right) = \left\{\frac{c_{4,j,p}\sigma^2(z)}{nh^{2j+1}}\right\} + o\left\{(nh^{2j+1})^{-1}\right\}, \tag{11.11}$$

where $\sigma^2(z) = \mathsf{E}[\varepsilon^2|Z = z]$ is the residual variance. The exact form of the constants $c_{a,j,p}$ for $a = 1, 2, 3, 4$ can be found in Ruppert and Wand (1994).

Theorem 1 provides asymptotic bias and variance expressions of local polynomial estimators of degree p for a general jth derivative. For illustration consider the special case $p = 0$ and $j = 0$ – local constant estimation of a function. The bias is

$$\frac{h^2}{2}\left\{C^{(2)}(z) + 2\frac{C^{(1)}(z)f^{(1)}(z)}{f(z)}\right\}\mu_2(K)\,, \tag{11.12}$$

with $\mu_2(K) = \int u^2 K(u)\,du$. For $p = 1$ and $j = 0$ – local linear estimation of a function – the bias becomes

$$\frac{h^2}{2}\{C^{(2)}(z)\}\,\mu_2(K). \tag{11.13}$$

Observe in general from (11.9) and (11.10) that the bias for $p-j$ even contains an additional design dependent term with factor $\frac{f^{(1)}(z)}{f(z)}$ as opposed to the odd case. Sign and size of this quantity, however, depend on the shape of the underlying estimated function and the shape of f_Z. In particular at the boundary of the support of Z, small values of f inflate the entire term. Therefore odd values of $p-j$ are preferable avoiding such boundary bias problems and pertaining the same variance. In our case, we are interested in the second derivative. We therefore choose the polynomial order $p = 3$ and not $p = 2$ according to Theorem 1.

With higher order kernels (11.6) of order υ and corresponding higher smoothness assumptions the bias in Theorem 1 can be further reduced to be of order h^{υ} for fixed p and j with $\upsilon > p - j + 2$ without changing the rate in the variance. In practice the order υ, however, cannot be chosen too large as with increasing υ the estimates have robustness problems in finite samples due to negative weights associated with the kernels (Müller 1988).

Observe from Theorem 1 that kernel estimation of a derivative is harder than of the function itself. While the variance in the function estimation decreases with $\mathcal{O}(1/(nh))$ the corresponding rate in the second derivative is only $\mathcal{O}_P(1/(nh^5))$ which is much slower. Therefore the finite sample performance of second derivatives lacks the precision of the fit achieved for the function.

Rates of convergence can be judged according to the mean squared error (MSE). Assuming that $p - j$ is odd, it is

$$\begin{aligned} \mathrm{MSE}(z,h,j) &= \mathsf{E}\left[\hat{C}^{(j)}(z) - C^{(j)}(z)\right]^2 \\ &= \mathcal{O}\{\underbrace{h^{2(p-j+1)}}_{bias^2} + \underbrace{(nh^{2j+1})^{-1}}_{var}\}. \end{aligned} \tag{11.14}$$

For constructing confidence intervals of the nonparametric estimates use the following normality result

Theorem 2. *Under some additional moment assumptions it is for $p - j$ odd*

$$\sqrt{nh^{2j+1}}\{\hat{C}^{(j)}(z) - C^{(j)}(z)\} \to \mathsf{N}(0, V_j) \tag{11.15}$$

with V_j as in Theorem 1.

For a precise statement of the standard moment conditions see Li and Racine (2007). Analogous results to Theorem 1 and 2 hold for d-dimensional functions. The only remarkable systematic difference is that the dimension of the regressors enters in the rate of the variance which is then $\mathcal{O}_P\{(nh^{2j+d})^{-1}\}$. Likewise the rate of convergence to the asymptotic distribution also deteriorates with d and is nh^{2j+d}. This phenomenon is known in the literature as the curse of dimensionality capturing the fact that finite sample performance of nonparametric estimators decreases with an increasing number of regressors. Therefore in practice, appropriate semiparametric dimension reduction techniques are used. They keep high modeling flexibility but yield better finite sample properties in regression settings with more than three regressors. See Sect. 11.2.3 for details.

11.2.2 Selection of the Smoothing Parameter

In practice, most important for good nonparametric estimation results is an appropriate choice of bandwidth. Other parameters like the selection of a suitable kernel K only have little influence on the final result in practice. Asymptotically the choice of K has no effect, and in finite samples its impact is negligible (see Marron and Nolan 1988). For the choice of order p of the employed local polynomial estimator it is sufficient to follow the logic outlined above.

An optimal bandwidth should minimize both bias and variance of the estimator. Though according to Theorem 1 there is a tradeoff between these quantities as smaller bandwidths would reduce the bias but inflate the variance. Therefore selecting h by minimizing $\mathrm{MSE}(\mathbf{z}, h, j)$ (multivariate analogue to (11.14)) balances bias and variance (see Fig. 11.1 for an illustration in averages). However, such a choice depends on the location $\mathbf{z}$. For a global choice, use an integrated criterion like the weighted integrated mean square error (WIMSE)

$$\mathrm{WIMSE}(h, j) = \int \mathrm{MSE}(\mathbf{z}, h, j)\psi(\mathbf{z})d\mathbf{z} = \int \mathsf{E}[\hat{C}^{(j)}(\mathbf{z}) - C^{(j)}(\mathbf{z})]^2\psi(\mathbf{z})d\mathbf{z}, \tag{11.16}$$

where $\psi(\mathbf{z})$ is a nonnegative weight function which ensures that WIMSE is well defined. Instead of an integrated criterion also an averaged criterion like the mean average squared error (MASE) can be used which replaces integration with summation in (11.16). When using a second order kernel straightforward calculations yield

$$h^{\star} = \begin{cases} cn^{-1/(2p+d+2)} & \text{for } p - j \text{ odd} \\ c'n^{-1/(2p+d+4)} & \text{for } p - j \text{ even} \end{cases} \tag{11.17}$$

for the optimal bandwidth $h^{\star}$ in a multivariate setting with constants $c, c' > 0$ depending on kernel constants and higher derivatives of C and the density f of regressors at $\mathbf{z}$ as we can see from (11.9)–(11.11). In our case of interest for

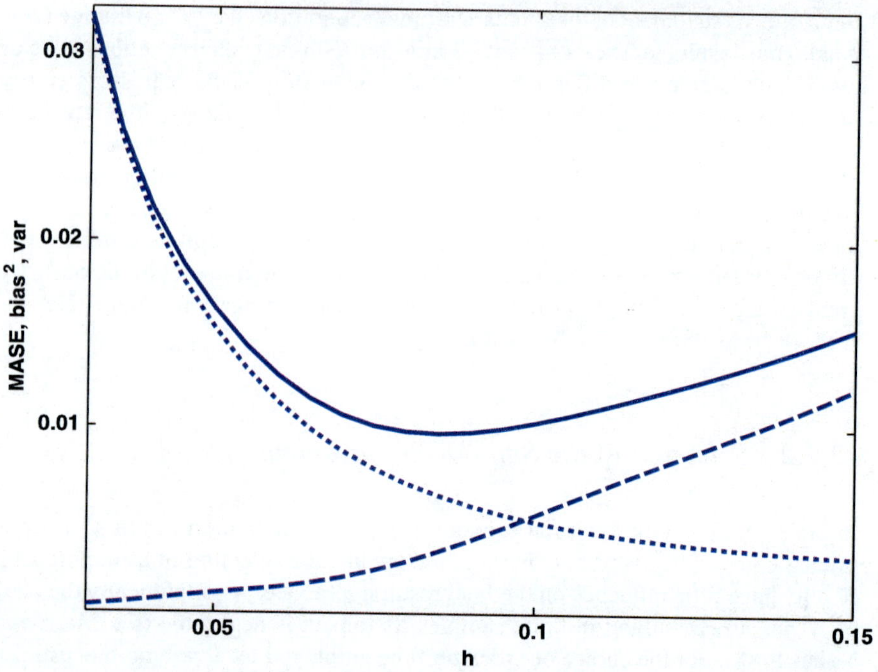

Fig. 11.1 MASE (*solid line*), squared bias (*dashed line*) and variance part (*dotted line*) for simulated data, weights $w(x) = \mathrm{I}(x \in [0.05, 0.95])$

$p = 3, j = 2$ and $d = 1$, it is $h^\star = n^{-1/9}$. As for larger j also p must be enlarged, the optimal bandwidth for estimating the jth derivative decreases in j. Note, however, that $h^\star$ is not feasible in practice because the constants c, c' in (11.17) contain unknown quantities such as higher derivatives of C and the density f of regressors.

A way to operationalize these are plug-in methods. They replace unknown quantities by pilot estimates and then calculate $h^\star$ via (11.17). The rule-of-thumb additionally uses normality assumptions in the distribution of the regressors and for the kernel to calculate exact constants. For $p = j = 0$ it is $h_k \approx s_k n^{-1/(4+d)}$ for $h = (h_1, \ldots, h_d)$ with s_k the standard deviation of observations of covariate Z_k. It is an easy and fast way to obtain a rough estimate and can be used for pilot estimates in plug-in procedures. Nevertheless, a bandwidth choice based on these procedures yields only an asymptotically optimal selection as the employed criteria are asymptotic ones.

In small samples, however, there are better choices which can be made by data driven cross-validation (CV) methods. In general, these procedures yield valid finite sample bandwidth choices, but do not have closed form solutions. Therefore computation intensive numerical methods must be used in order to obtain such an automatic bandwidth h_{CV}. This can amount to a feasibility issue in particular for time series. We present a least squares cross-validation for local cubic estimation as our interest is in estimating the second derivative of C. Here, we select h_{CV} as minimizer of the sum of squared errors between obtained local cubic fit and observed response used as cross-validation criterion.

$$\mathrm{CV}(\tilde{h}) = \sum_{i=1}^{n}\sum_{j\neq i}^{n}\Bigg\{ Y_i - \widehat{C}_{\tilde{h},-i}(\mathbf{Z}_i) - \widehat{C}^{(1)}_{\tilde{h},-i}(\mathbf{Z}_i)(\mathbf{Z}_j - \mathbf{Z}_i) \tag{11.18}$$

$$- \frac{1}{2}\widehat{C}^{(2)}_{\tilde{h},-i}(\mathbf{Z}_i)(\mathbf{Z}_j - \mathbf{Z}_i)^2 \Bigg\}^2 M(\mathbf{Z}_i),$$

where $0 \leq M(\mathbf{Z}_i) \leq 1$ is a weight function that ensures existence of the limit for n large. and $(\widehat{C}_{\tilde{h},-i}, \widehat{C}^{(1)}_{\tilde{h},-i}, \widehat{C}^{(2)}_{\tilde{h},-i})$ denote the local cubic regression estimate obtained without using the ith observation $(\mathbf{Z}_i, C_i)$. This way we ensure that the observations used for calculating $\widehat{C}_{\tilde{h},-i}(\cdot)$ are independent of $\mathbf{Z}_i$. It can be shown that asymptotically h_{CV} converges to the corresponding theoretical bandwidth obtained from (11.17). This design driven choice of bandwidth is completely free of functional form assumptions and therefore most appealing in finite samples at the expense of potentially long computation time.

11.2.3 Dimension Reduction Techniques

While flexible, high-dimensional kernel regression requires large data samples for precise results in terms of tight pointwise confidence intervals. Aït-Sahalia and Lo (1998), for example, use 1 year option data to empirically derive the call function based on five-dimensional kernel regression. Asymptotically, rates of convergence of nonparametric estimators decrease the more regressors are included in the model. This is referred to as the "curse of dimensionality" (see Sect. 2.1. for theoretical details). Hence, there is a need to keep the dimension or equivalently the number of regressors low.

There exists a vast literature on methods which reduce the complexity of high dimensional regression problems resulting in better feasibility. In particular, the reduction of dimensionality is achieved by putting some structure on the model by e.g. imposing a parametric model or an additive or partially linear structure. The resulting models are so-called semiparametric models, among which the additive models are the most flexible kind requiring the least structural assumptions. In additive models, the regression function additively separates the influence of each univariate regressor. Thus estimation is restricted to a surface of the full-dimensional space of regressors $\mathbf{Z}$, which allows to construct estimators with univariate nonparametric rates of convergence and thus substantially improved finite sample properties. We refer to Mammen et al. (1999) and Linton and Nielsen (1995) for detailed methods in this case. Here, however, we will focus on suitable parametric assumptions tailored to financial modeling.

One way is to use no-arbitrage arguments and collapse S_t, $r_{t,\tau}$ and $\delta_{t,\tau}$ into the forward price $F_t = S_t e^{(r_{t,\tau}-\delta_{t,\tau})\tau}$ in order to express the call pricing function as

$$C(S_t, X, \tau, r_{t,\tau}, \delta_{t,\tau}) = C(F_{t,\tau}, X, \tau, r_{t,\tau})$$

Alternatively use the non-arbitrage relation to estimate dividends and express the function in terms of the discounted stock price, that is either by $S_t^0 = S_t e^{-\delta_{t,\tau}} = S_t - D_{t,\tau}$ where $D_{t,\tau}$ is the present value of the dividends to be paid before the expiration. Thus it is

$$C(S_t, X, \tau, r_{t,\tau}, \delta_{t,\tau}) = C(S_t^0, X, \tau, r_{t,\tau}) \ .$$

A further reduction of the number of regressors is achieved by assuming that the call option function is homogeneous of degree one in S_t and X so that

$$C(S_t, X, \tau, r_{t,\tau}, \delta_{t,\tau}) = XC(S_t/X, \tau, r_{t,\tau}, \delta_{t,\tau}).$$

Combining the assumptions of the last two equations, the call price function can be further reduced to a function of three variables: moneyness $M_t = \frac{S_t^0}{K}$, maturity τ and risk free interest rate $r_{t,\tau}$. Notice that by smoothing with respect to moneyness, rather than with respect to the dividend adjusted index level we implicitly assume the theoretical option function is homogeneous of degree one with respect to the index and strike price. The basic Black and Scholes (1973) formula is an example of such a function, and as shown by Merton (1973) and discussed in Ingersoll (1987), a call price is homogeneous of degree one in the asset price and strike price if the asset's return distribution is independent of the level of the underlying index. We use these dimension reduction techniques in the empirical study in both settings, direct estimation of the RND from the call prices and but also in indirect estimation via implied volatility.

11.2.4 Application

We use tick data on the DAX index based European options prices maturing in 1 month (21 trading days), provided by EUREX for 20040121. The transformed data according to a methodology by Fengler (2005) contain date stamp, implied volatility, type of the option, maturity, strike price, option price, interest rate, intraday future price, average dividend rate.

The index stock price varies within 1 day and one needs to identify the price at which a certain transaction has taken place. Intraday DAX index prices are available on EUREX. Several authors report that the change of the index price is stale and for every pair option/strike we use instead the prices of futures contracts closest to the time of the registered trade, see e.g. Jackwerth (2000).

Original strike prices are given on an equidistant grid and in order to account for movements in the intraday price we use the following transformation $\frac{X_i}{F_i} S_t e^{r_{t,\tau} - \delta_{t,\tau}}$, where X_i and F_i are paired observations and S_t is the median intraday stock price, $r_{t,\tau}$ is the 1 month interest rate (linearly interpolated EURIBOR rates, for the desired maturity) and $\delta_{t,\tau}$ the average dividend (see Fengler 2005). Conditional on these values we estimate q and interpret it as an average curve for the estimation date.

We use only at-the-money and out-of-the-money call options and in-the-money puts translated in call prices by using the put call parity

$$C_t - P_t = S_t e^{-\delta_{t,\tau}\tau} - X e^{-r_{t,\tau}\tau}$$

This guarantees that unreliable observations (high volatility) will be removed from our estimation samples. Since, as mentioned before, the intraday stock price varies, we use its median to compute the risk neutral density. For this price, we verify if our observations satisfy the arbitrage condition and delete for our sample those who do not satisfy it

$$S_t \geq C_i \geq \max(S_t - X_i e^{-r_{t,\tau}\tau}, 0).$$

Finally, if we have different call price observations for the same strike price we take their median at that point. In this way we ensure that we have a one to one relationship between every call and strike price.

11.2.4.1 Smoothing in Call Option Space

As described in Sect. 2.1 local polynomial estimation allows to compute the second derivative of the call price directly, in a single step. We use local polynomial smoothing of degree three and a quartic kernel to reduce finite sample bias. In the first step, we rescale the call price by dividing it by S_t and we smooth in this direction. We use cross-validation to choose the optimal bandwidth; however, this bandwidth appears to undersmooth in extreme areas around the mode and in the tails yielding a wiggly estimator in these regions (see in Fig. 11.2). Therefore we decide to gradually increase the bandwidth to "stabilize" the estimate in the extremes. However, Fig. 11.2 also illustrates that this should be done with care, as too much of oversmoothing can easily cause a huge bias at the mode.

11.2.4.2 Smoothing in Implied Volatility Space

In practice, the smoothing is mainly done in the implied volatility span because call prices respond asymmetrically to changes in the strike prices. In the present context, implied volatility is the volatility that yields a theoretical value for the option equal to the observed market price of that option, when using the Black-Scholes pricing model. We then estimate a smooth function $\hat{\sigma}$ and recover the call price by a bijective

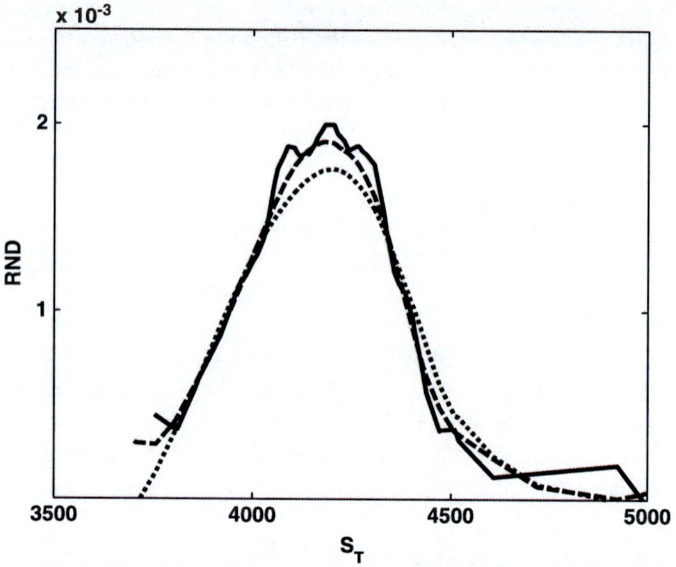

Fig. 11.2 $\hat{q}(S_T)$ by local polynomial smoother for the optimally chosen bandwidth $h = 114.34$ by cross-validation (*solid line*) and oversmoothing bandwidths $h = 227.59$ (*dashed line*) and $h = 434.49$ (*dotted line*)

function evaluated at some fixed values of the regressors and variable σ

$$\hat{C}(S_t, X, \tau, r_{t,\tau}, \delta_{t,\tau}) = C_{BS}(\cdot\,; \hat{\sigma}(S_t, X, \tau, r_{t,\tau}, \delta_{t,\tau}))$$
$$= e^{-\delta_{t,\tau}\tau} S_t \Phi(y + \sigma\sqrt{\tau}) - e^{-r_{t,\tau}\tau} X \Phi(y),$$

where Φ is the distribution function of the standard normal distribution and

$$y = \frac{\log(\frac{S_t}{K}) + (r_{t,\tau} - \frac{1}{2}\sigma^2)\tau}{\sigma\sqrt{\tau}}.$$

In this chapter we use a method based on Rookley (1997) who shows how to improve the efficiency of the estimator by estimating σ and its first two derivatives by local polynomial regression and plugging them into a modified version of the Black–Scholes formula. Below we describe the method for fixed maturity of 1 month.

For each pair $\{C_i, X_i\}_{i=1}^n$ we define the rescaled call option $c_i = C_i / S_t$ in terms of moneyness $M_i = S_t / X_i$ so that starting from the Black–Scholes formula for the call price we can write

$$C_i = C\{M_i; \sigma(M_i)\} = \Phi(d_1) - \frac{e^{-r\tau}\Phi(d_2)}{M_i}$$

$$d_1 = \frac{\log(M_i) + \left\{r_{t,\tau} + \frac{1}{2}\sigma(M_i)^2\right\}\tau}{\sigma(M_i)\sqrt{\tau}}$$

$$d_2 = d_1 - \sigma(M_i)\sqrt{\tau}.$$

For simplification we drop the indices. The risk neutral density can be expressed in terms of rescaled call price

$$q(\cdot) = e^{r\tau}\frac{\partial^2 C}{\partial X^2} = e^{r\tau} S \frac{\partial^2 c}{\partial X^2}$$

with

$$\frac{\partial^2 C}{\partial X^2} = \frac{\mathrm{d}^2 c}{\mathrm{d}M^2}\left(\frac{M}{X}\right)^2 + 2\frac{\mathrm{d}c}{\mathrm{d}M}\frac{M}{X^2}$$

and

$$\begin{aligned}\frac{\mathrm{d}^2 C}{\mathrm{d}M^2} &= \varphi(d_1)\left\{\frac{\mathrm{d}^2 d_1}{\mathrm{d}M^2} - d_1\left(\frac{\mathrm{d}d_1}{\mathrm{d}M}\right)^2\right\}\\ &\quad - \frac{e^{-r\tau}\varphi(d_2)}{M}\left\{\frac{\mathrm{d}^2 d_2}{\mathrm{d}M^2} - \frac{2}{M}\frac{\mathrm{d}d_2}{\mathrm{d}M} - d_2\left(\frac{\mathrm{d}d_2}{\mathrm{d}M}\right)^2\right\}\\ &\quad - \frac{2e^{-r\tau}\Phi(d_2)}{M^3},\end{aligned}$$

where φ is the probability density function of the standard normal distribution. The results depend further on the following quantities, where $\sigma(M)$, $\sigma'(M)$, $\sigma''(M)$ are smooth functions in moneyness direction

$$\begin{aligned}\frac{\mathrm{d}^2 d_1}{\mathrm{d}M^2} &= -\frac{1}{M\sigma(M)\sqrt{\tau}}\left\{\frac{1}{M} + \frac{\sigma'(M)}{\sigma(M)}\right\}\\ &\quad + \sigma''(M)\left\{\frac{\sqrt{\tau}}{2} - \frac{\log(M) + r\tau}{\sigma(M)^2\sqrt{\tau}}\right\}\\ &\quad + \sigma'(M)\left\{2\sigma'(M)\frac{\log(M) + r\tau}{\sigma(M)^3\sqrt{\tau}} - \frac{1}{M\sigma(M)^2\sqrt{\tau}}\right\}\\ \frac{\mathrm{d}^2 d_2}{\mathrm{d}M^2} &= -\frac{1}{M\sigma(M)\sqrt{\tau}}\left\{\frac{1}{M} + \frac{\sigma'(M)}{\sigma(M)}\right\}\\ &\quad - \sigma''(M)\left\{\frac{\sqrt{\tau}}{2} + \frac{\log(M) + r\tau}{\sigma(M)^2\sqrt{\tau}}\right\}\\ &\quad + \sigma'(M)\left\{2\sigma'(M)\frac{\log(M) + r\tau}{\sigma(M)^3\sqrt{\tau}} - \frac{1}{M\sigma(M)^2\sqrt{\tau}}\right\}.\end{aligned}$$

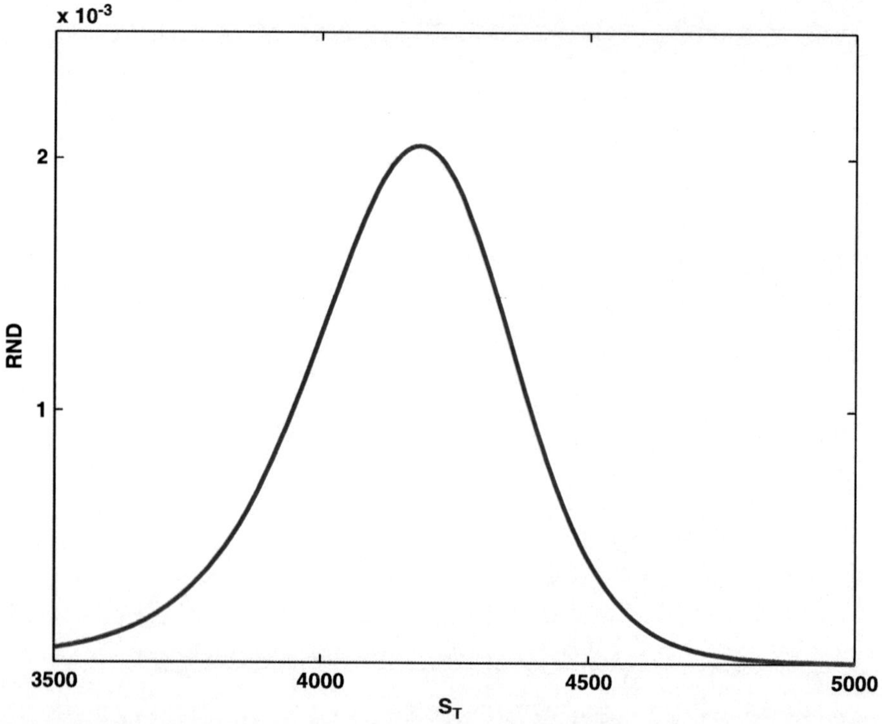

Fig. 11.3 $\hat{q}(S_T)$ by Rookley method with oversmoothing bandwidth $h = 372.42$

In order to estimate $\sigma(M)$ and its associated first and second derivatives with respect to moneyness we use univariate local polynomial kernel regression of degree three and a quartic kernel. The optimal bandwidth has been computed using cross-validation criteria (11.18) for the implied volatility. Oversmoothing bandwidths, see Fig. 11.3, improve the fit in the tails because they allow for more observations to be included in the estimators while having little effects on the values of $\hat{q}$ situated in the middle of the distribution, where the estimates by different bandwidths overlap almost perfectly. It follows that smoothing in implied volatility yields a more stable estimator in terms of shape in finite sample, for varying bandwidths. This can be well seen in Figs. 11.2 and 11.3. It is because the implied volatility responds with a fairly constant magnitude to the changes in the strike price over the estimation domain.

11.2.4.3 Problems and Refinements

In applications the support of strike prices is mostly compact and thus bounded. As shown in Sect. 2.1. the quality of estimates in regions close to the boundary might be low due to small values of the regressors' density when using even order polynomials. By using a polynomial of order three, estimation is design adaptive for the second derivative avoiding this problem.

Furthermore, associated with the boundary, option data is characterized by scarce observations close to the bounds. In general, nonparametric techniques do not perform well in regions with sparse data and other methods are required. Parametrization of the tails using Pareto type distributions might be advantageous leaving however the question of how to join the two regions in order to assure that the resulting distribution integrates to one. Alternatively, Rookley (1997) proposes to further parametrize these distributions by matching them with an Edgeworth expansion type density

$$q(S_T) = \frac{1}{S_T \sigma} \Phi(Z)\{1 + \beta(Z^3 - 3Z) + \gamma(Z^4 - 6Z^2 + 3)\}$$

for $Z = \frac{\log(S_T) - \tilde{\mu}}{\tilde{\sigma}}$, where $\tilde{\mu}$ and $\tilde{\sigma}$ are the conditional mean and standard deviation of $\log(S_T)$ implied by the risk neutral measure, and β and γ are coefficients related to the higher moments of $\log(S_T)$.

In order for the risk neutral density to be well defined, an estimate of the call price function C must satisfy certain high-level conditions (see e.g. Aït-Sahalia and Duarte 2003): It should be (1) positive, (2) decreasing in X, (3) convex, and (4) its second derivative should exist, be nonnegative and integrable. Given that the first derivative of C with respect to X is the (negative) discounted cumulative density function of q conditions (2) and condition (3) can be summarized by the following inequality

$$-e^{r_{t,\tau}} \leq \frac{\partial C(S_t, X, \tau, r_{t,\tau}, \delta_{t,\tau})}{\partial X} \leq 0.$$

Convexity requires

$$\frac{\partial^2 C(S_t, X, \tau, r_{t,\tau}, \delta_{t,\tau})}{\partial^2 X} \geq 0.$$

Nonparametric kernel estimates may violate these constraints, unless we deal with large samples of observations. Imposing constraints like monotonicity or convexity directly in the estimation leads to nontrivial optimization problems in topological cones. If it is crucial for the outcome to fulfill the shape restrictions in small samples, it is recommended to use series type estimation methods which easily allow to incorporate them directly in the estimation. In general, these constrains must be applied directly to the call price, because theoretical properties of the implied volatility are not well known. For further references see Ait-Sahalia (2003). This will be illustrated in the next section.

11.3 Estimation of the RND via Empirical Pricing Kernel

In the previous section, we studied nonparametric kernel methods for estimating q as the discounted second derivative of the call price function and discussed the problems associated with kernel type estimators in this setting. Now, we propose a new approach, based on series expansion of the pricing kernel.

In financial mathematics the relationship between the physical measure p and RND q of a financial asset can be represented via the pricing kernel m. Also called stochastic discount factor, the pricing kernel is the quotient of the Arrow security prices and the objective probability measure and summarizes information related to asset pricing. Thus it is

$$q(S_T) = m(S_T)p(S_T). \tag{11.19}$$

From a behavioral economics perspective m describes risk preferences of a representative agent in an exchange economy. In many applications, the empirical pricing kernel is the object of interest. In most of the studies Aït-Sahalia and Lo (2000), Brown and Jackwerth (2004), Grith et al. (2010) it has been estimated as a ratio of two estimated densities: $\hat{q}$ computed as the second derivative of a smooth call function (as described in Sect. 2) and $\hat{p}$ based on historical returns. This approach leads to difficulties in deriving the statistical properties of the estimator. In particular, the sample sizes for estimating p and q may differ substantially: p uses daily observations, whereas q is based on intraday high-frequency observations. On the other hand, methods for estimating p are in general much simpler and more stable compared to those for q for which typically nonparametric kernel estimation of a second derivative is required. Direct estimation of the pricing kernel can be seen as an improvement in this sense. Engle and Rosenberg (2002), for instance, specify the pricing kernel using a polynomial expansion.

For estimating q, however, a series approach is additionally appealing, as high-level shape constraints are straightforward to incorporate in finite samples. Recall that for kernel type estimators this is not the case, see the end of Sect. 2.4.

We introduce the series expansion for the pricing kernel in (11.3). With an estimate of the physical measure from historical data and the pricing kernel m from option prices, these indirectly imply an estimate of q via (11.19). In statistical theory and also in practice, this indirect way of estimating q has a faster rate of convergence than using series methods directly for q in (11.1) which require the choice of an additional regularization parameter to guarantee invertibility of an ill-posed statistical problem. In particular, in (11.19) large values of S_T are downweighted by integrating over the physical measure, while they enter undamped in (11.1) leading to unreliable results.

11.3.1 Direct Estimation of Empirical Pricing Kernel via Series Methods

As for q, there are several factors which drive the form of the pricing kernel. Here, however, we focus on the projection of the pricing kernel on the set of available payoff functions m^*, which allows us to represent m in terms of S_T only. In practice this is a reasonable assumption. Thus we require that m and m^* are close in the following sense

$$||m - m^*||^2 = \int |m(x) - m^*(x)|^2 \mathrm{d}x < \epsilon \tag{11.20}$$

with ϵ small. Further we assume that m^* has a Fourier series expansion

$$m^*(S_T) = \sum_{l=1}^{\infty} \alpha_l g_l(S_T), \tag{11.21}$$

where $\{\alpha_l\}_{l=1}^{\infty}$ are Fourier coefficients and $\{g_l\}_{l=1}^{\infty}$ is a fixed collection of basis functions. The functions g_l are chosen as orthonormal with respect to a particular norm. Such a representation is possible if the function is absolutely integrable.

Based on (11.21), we can construct an estimator for m^* and thus m. If a finite number L of basis functions is sufficient for a good approximation of m then

$$\widehat{m}(S_T) = \sum_{l=1}^{L} \hat{\alpha}_l g_l(S_T). \tag{11.22}$$

Estimates $\hat{\alpha}_l$ for the coefficients α_l could be obtained by least squares for fixed basis functions g_l if a direct response was observable. Clearly the choice of L controls the quality of the estimate. The larger L, the better the fit but the higher the computing cost and less robust the result. See Sect. 3.3 for a sophisticated way of selecting the smoothing parameter.

In financial applications the following polynomial basis functions are frequently used: e.g. Laguerre, Legendre, Chebyshev polynomials, see Fig. 11.4 and Sect. 3.5. While asymptotically equivalent, in finite samples their form will influence the size of L. In general, one would prefer to have g_l such that L small is sufficient. For a formal criterion on how to select between different basis options see Li and Racine (2007). They assess different candidate basis functions by comparing a CV-type criterion for fixed L.

Though the form of m is only indirectly determined by relating observable call prices Y_i to strike prices X_i for given T, τ via (11.3). A response to observed payoffs via the pricing kernel is not directly observable. In sample an estimate of m should fulfill

$$\begin{aligned} Y_i &= e^{-r_{t,\tau}\tau} \int_0^{\infty} (S_T - X_i)^+ \sum_{l=1}^{L} \hat{\alpha}_l g_l(S_T) p_t(S_T) dS_T + \varepsilon_i \\ &= \sum_{l=1}^{L} \hat{\alpha}_l \left\{ e^{-r_{t,\tau}\tau} \int_0^{\infty} (S_T - X_i)^+ g_l(S_T) p_t(S_T) dS_T \right\} + \varepsilon_i \end{aligned} \tag{11.23}$$

with error ε such that $\mathsf{E}[\varepsilon|X] = 0$. Set

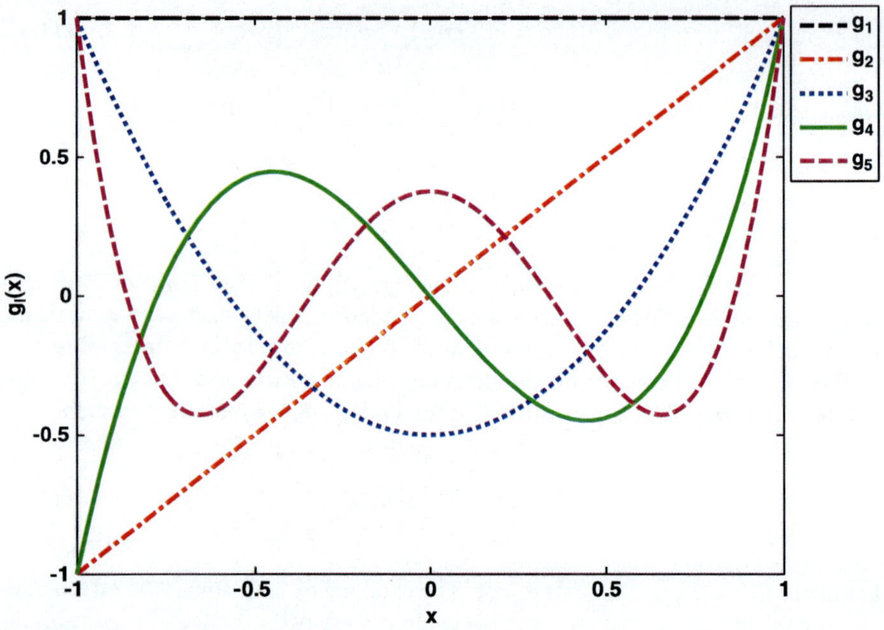

Fig. 11.4 First five terms of the Legendre polynomials

$$\psi_{il} = \psi_l(X_i) = e^{-r_{t,\tau}\tau} \int_0^\infty (S_T - X_i)^+ g_l(S_T) p_t(S_T) dS_T. \tag{11.24}$$

Then for known p and fixed basis functions and fixed L, the vector $\hat{\alpha} = (\hat{\alpha}_1, ..., \hat{\alpha}_L)^\top$ is obtained as

$$\arg\min_{\alpha} \sum_{i=1}^{n} \left\{ Y_i - \sum_{l=1}^{L} \alpha_l \psi_l(X_i) \right\}^2 \tag{11.25}$$

In practice, however, p is not known and can only be estimated. Therefore instead of ψ_l in (11.24) we have only estimates $\hat{\psi}_l$ of the basis functions. We consider two possible ways for constructing them. First, regard ψ as an expectation which can be estimated by sample averaging over J different payoffs at time T for fixed τ and given X

$$\hat{\psi}_{il} = e^{-r_{t,\tau}\tau} J^{-1} \sum_{s=1}^{J} (S_T^k - X_i)^+ g_l(S_T^k) \quad \text{with} \tag{11.26}$$

How $(S_T^k)_{k=1}^J$ are obtained is explained in detail in the following section. Alternatively, replace p by an estimator, e.g. a kernel density estimator. Then it is

$$\hat{\psi}_{il} = e^{-r_{t,\tau}\tau} \int_0^\infty (S_T - X_i)^+ g_l(S_T) \hat{p}(S_T) dS_T. \tag{11.27}$$

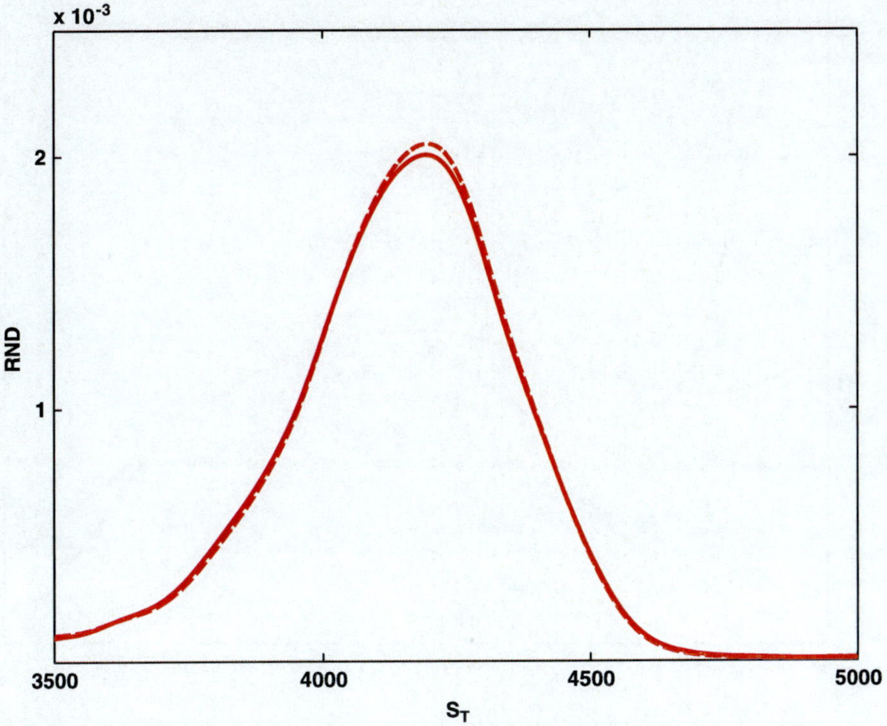

Fig. 11.5 $\hat{q}(S_T)$ in Legendre basis with $L = 5$ based on approximation (11.26) (*simple line*) and (11.27) of Ψ (*dashed line*)

Here some care is needed in numerical integration to keep discretization errors negligible. Furthermore, for an appropriate choice of bandwidth in $\hat{p}$, both approaches are asymptotically equivalent. In finite samples, however, estimates for q might differ (see Fig. 11.5, for $J = 4{,}500$, and S_T^k simulated based on historical log-returns).

In total we obtain a feasible estimator of α based on a feasible version of (11.25) as

$$\check{\alpha} = (\hat{\Psi}^\top \hat{\Psi})^- \hat{\Psi}^\top Y. \tag{11.28}$$

The elements of $\hat{\Psi}_{(n \times L)}$ are given either by (11.26) or (11.27) and $Y = (Y_1, \cdots, Y_n)^\top$.

Then an estimate of the pricing kernel at observation s of S_T is given by

$$\hat{m}(s) = g^L(s)^\top \check{\alpha}, \tag{11.29}$$

where $g^L(s) = (g_1(s), \ldots, g_L(s))^\top$. We see in Fig. 11.6 that the estimator of m is less stable for different approximations of ψ_{il}. Finally, the risk neutral density is estimated as

$$\hat{q}(s) = \hat{m}(s)\hat{p}(s). \tag{11.30}$$

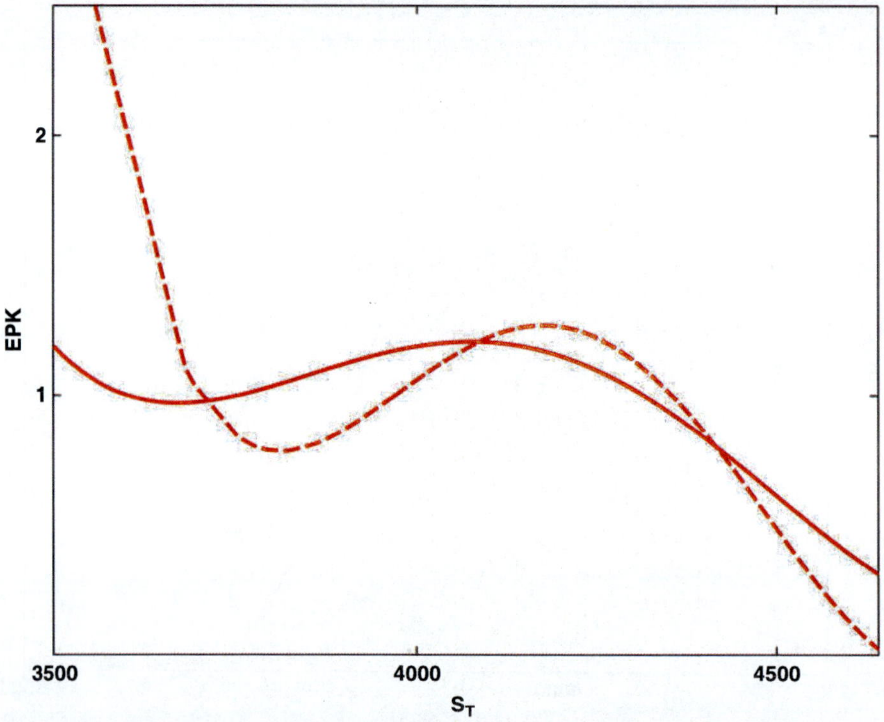

Fig. 11.6 $\hat{m}^*(S_T)$ by Legendre basis expansion with $L = 5$ based on approximation (11.26) (*solid line*) and (11.27) of Ψ (*dashed line*)

11.3.2 Estimation of the PDF of S_T

In the empirical study we use two different ways of obtaining $\hat{p}$ from the DAX Index prices at time T. And we look at the sensitivity of $\hat{q}$ w.r.t. $\hat{p}$. First, we extrapolate possible realizations of S_T in the future from historical log-returns. Based on a window of historical DAX Index values of length J we get

$$S_T^k = S_t e^{r_T^k}, \quad \text{for} \quad r_T^k = \log(S_{t-k}/S_{t-(k+1)}). \tag{11.31}$$

Alternatively, we use a GARCH(1,1) specification for the log-returns to account for slowly decaying autocorrelation in the data. The model is specified as follows

$$\log(S_t/S_{t-1}) = \mu + u_t\,, \qquad u_t \sim f(0, \sigma_t^r). \tag{11.32}$$

In (11.32), the returns consist of a simple constant, plus an uncorrelated, non-Gaussian disturbance. The conditional variance $(\sigma_t^r)^2$ follows an ARMA(1,1) type specification

$$(\sigma_t^r)^2 = a_1 + a_2 r_{t-1}^2 + a_3 (\sigma_{t-1}^r)^2 . \tag{11.33}$$

We can estimate the parameters of the model (μ, a_1, a_2, a_3) and retrieve a time series of stochastic volatilities $\{\sigma_{t-k}^r\}_{k=1}^J$. The simulated index prices at time T are obtained as in (11.31) above for

$$r_T^k = r_{t-k} \frac{\sigma_T^r}{\sigma_{t-k}^r},$$

where we use for the forecasted volatility σ_T^r today's volatility σ_t based on GARCH.

Then the probability density p of S_T is estimated at each point S_T using a kernel density estimator

$$\widehat{p}_h(S_T) = \frac{1}{Jh} \sum_{k=1}^{J} K\left(\frac{S_T^k - S_T}{h}\right), \tag{11.34}$$

where K is a kernel function and the bandwidth is selected similarly to the criteria introduced in Sect. 11.2.2. Resulting estimates of the two approaches are illustrated in Fig. 11.7 for $J = 5{,}000$. We observe that they differ significantly in spread and

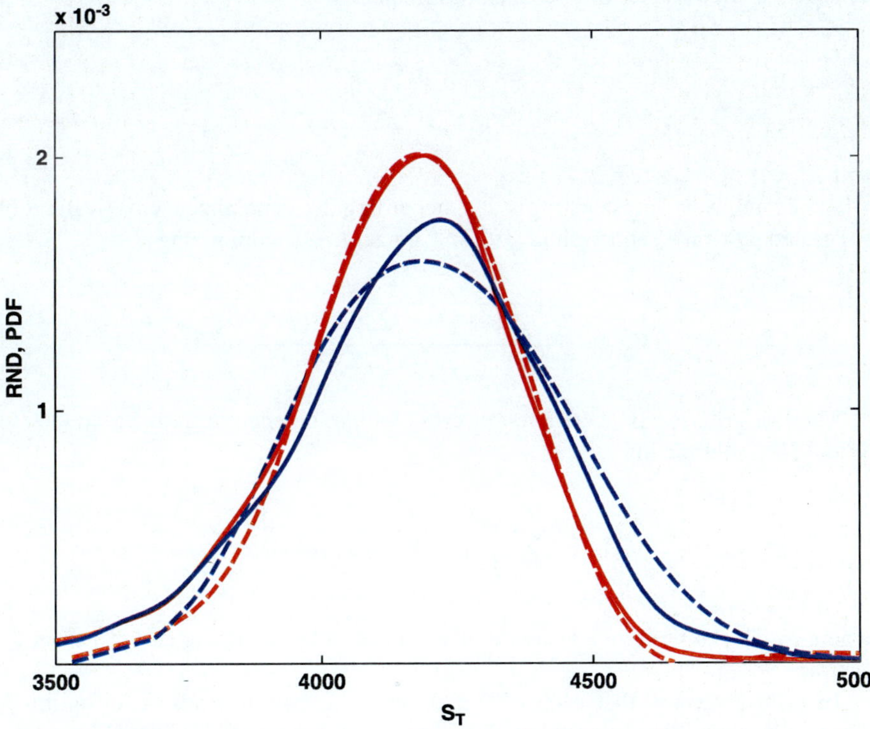

Fig. 11.7 $\hat{q}$ in Legendre basis with $L = 5$ and $\hat{p}$ based on log-returns (*blue*) and weighted log-returns (*red*). *Solid* and *dashed lines* correspond to the specifications in the Fig. 11.5

the mode. However, the differences depend on the window length J of returns used to estimate the parameters of the GARCH model, as well as on the choice of the bandwidth used to estimate p, which carries over to q via (11.30) directly or indirectly.

11.3.3 Choice of the Tuning Parameters

The quality of the obtained series estimators (11.29) and (11.30) depends on a suitable choice of the number $L(n) \to \infty$ for $n \to \infty$ for given basis functions. Note that the role of L (or L/n) is similar to that played by the smoothing parameter h for the kernel methods. There are three well-known procedures for a data-driven optimal selection of L, see Wahba (1985). The first one is Mallows's C_L as proposed in Mallows (1973): Select L_M such that it minimizes

$$C_L = n^{-1} \sum_{i=1}^{n} \left\{ Y_i - \sum_{l=1}^{L} \breve{\alpha}_l \hat{\psi}_l(X_i) \right\}^2 + 2\sigma^2 (L/n),$$

where σ^2 is the variance of ε. One can estimate σ^2 by

$$\hat{\sigma}^2 = n^{-1} \sum_{i=1}^{n} \hat{\varepsilon}_i^2$$

with $\hat{\varepsilon}_i = Y_i - \sum_l \breve{\alpha}_l \hat{\psi}_l(X_i)$.

A second way for selecting L is according to generalized cross-validation suggested by Craven and Wahba (1979). Choose L_{GCV} minimizing

$$CV_L^G = \frac{n^{-1} \sum_{i=1}^{n} \left\{ Y_i - \sum_{l=1}^{L} \breve{\alpha}_l \hat{\psi}_l(X_i) \right\}^2}{\{1 - (L/n)\}^2}.$$

The last criterion is leave-one-out cross-validation according to Stone (1974): Select L_{CV} minimizing

$$CV_L = \sum_{i=1}^{n} \left\{ Y_i - \sum_{l=1}^{L} \breve{\alpha}_l^{-i} \hat{\psi}_l(X_i) \right\}^2,$$

where $\breve{\alpha}_l^{-i}$ is the leave one estimate of α_l obtained by removing (X_i, Y_i) from the sample.

Li (1987) showed that each of the above three criteria leads to an optimally selected L in the sense that they all minimize the asymptotic weighted integrated squared error (see (11.16)). In this sense the obtained L are asymptotically equivalent.

11.3.4 Statistical Properties

Series type estimators are designed to provide good approximations in an L_2 sense, see (11.20). Therefore asymptotic properties as consistency and rates of convergence should be derived from the asymptotic mean squared error. The rate of convergence for the indirect estimator of q via the pricing kernel depends on the two smoothing parameters h and L.

$$\int_0^\infty \{\hat{q}(S_T) - q(S_T)\}^2 dS_T = \int_0^\infty \{\hat{m}(S_T)\hat{p}(S_T) - m(S_T)p(S_T)\}^2 dS_T$$

$$= \int_0^\infty [\hat{m}(S_T)\{\hat{p}(S_T) - p(S_T)\}]^2 dS_T + \int_0^\infty [p(S_T)\{\hat{m}(S_T) - m(S_T)\}]^2 dS_T$$

$$+ \int_0^\infty 2\hat{m}(S_T)\{\hat{p}(S_T) - p(S_T)\}p(S_T)\{\hat{m}(S_T) - m(S_T)\} dS_T$$

It easily follows from the law of iterated expectations that the third term equals zero. Consequently, the convergence of $\hat{q}(S_T)$ depends only on the first two terms. Since $\sup \hat{m}(s) = \mathcal{O}_P(1)$ under Assumption 1 given below, the order of convergence for the first term is dominated by $\{\hat{p}(S_T) - p(S_T)\}^2$.

Assumption 1. Suppose that p is twice continuously differentiable, K is a second order kernel and the bandwidth is chosen optimally as $h = cn^{-1/5}$, for a known constant c.

Then the asymptotic mean squared error for the kernel density estimator is

$$||\widehat{p}_h(x) - p(x)||_2^2 = \mathcal{O}_P(n^{-4/5}) \tag{11.35}$$

This follows along the same logic as the results for local polynomials in Sect. 2.1. For further details see e.g. Härdle et al. (2004).

The order of convergence for the second term only depends on $\{\hat{m}(S_T) - m(S_T)\}^2$ since $\sup p(s) \leq 1$. The next assumption establishes consistency of $\hat{m}(S_T)$.

Assumption 2. $\{X_i, Y_i\}$ are i.i.d. observations of (X, Y), $\mathsf{Var}(Y|X)$ is bounded on S, the compact connected interval of support of X. Furthermore p is bounded away from zero and m is ν-times continuously differentiable on S. Choose L such that $L^3/n \to 0$ as $n \to \infty$.

Under Assumption 2 it is

$$\int_0^\infty \{\hat{m}(S_T) - m(S_T)\}^2 dS_T = \mathcal{O}_p(L/n + L^{-2\nu}). \tag{11.36}$$

This result is from Newey (1997) for fixed basis functions ψ_l. With estimated basis $\hat{\psi}_l$ the result still goes through as the convergence of $\hat{\psi}_l$ to the true ψ_l is at parametric rate. The i.i.d. assumption is for simplicity of the exposition only. It can be easily relaxed to mixing type of observations.

The theorem below puts (11.35) and (11.36) together for an asymptotic result for q.

Theorem 3. *Assume that Assumptions 1 and 2 hold. Then the integrated square error* (ISE) *converges as*

$$\int_0^\infty \{\hat{q}(S_T) - q(S_T)\}^2 dS_T = \mathcal{O}_p(n^{-4/5} + L/n + L^{-2\nu}) . \qquad (11.37)$$

11.3.5 Implementation

We illustrate the method using the data described in Sect. 11.2.4. We consider the univariate regression of C on the strike price X for fixed maturity and fixed interest rate. We estimate q using three different systems of orthogonal basis: Laguerre, Legendre and Chebyshev. We found that the fit of the call price is almost identical for fixed L, while $\hat{\alpha}$ varies obviously with the series. There is little sensitivity with respect to the choice of the basis functions that holds also for the empirical pricing kernel and the implied risk neutral density. Based on the selection criteria for L from Sect. 3.3, we have chosen $L = 5$. We exemplify the method with Legendre polynomials. Estimation results are displayed in Figs. 11.5–11.8.

11.4 Conclusions

We have studied three nonparametric approaches for estimating the risk neutral density. They are based on fundamentally different techniques: two of them use local features and the third one is based on global curve fitting. For these approaches we have described the estimation methodology and their performance in finite sample, in terms of robustness and stability. Statistical properties of all procedures have been derived and illustrated focusing on practically most relevant aspects.

Figure 11.8 shows estimates of q using the three methods we discussed in this article for suitable choices of tuning parameters. While for the given sample size, all three nonparametric methods yield similar results, there still are some peculiarities. Our empirical results suggest that kernel methods for the estimation of q in implied volatility space work much better than those which smooth in the call price space in our sample. Local polynomial methods applied to call prices yield estimates which are highly sensitive to the choice of the bandwidth: A globally optimal bandwidth might in fact severely undersmooth around the mode or in the tails,

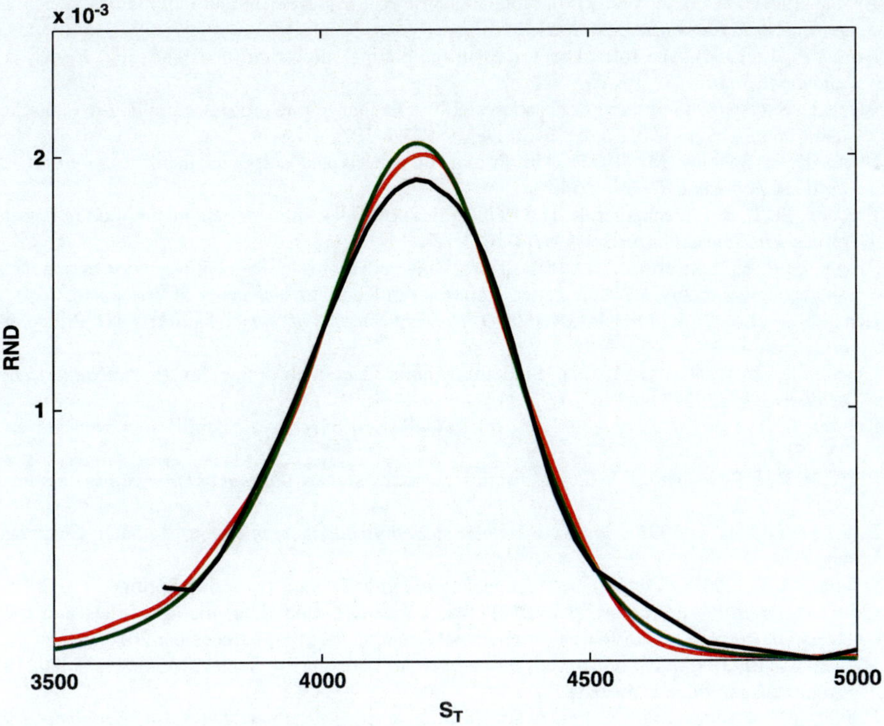

Fig. 11.8 $\hat{q}(S_T)$ by local polynomial regression with $h = 227.59$ in call space (*black*), by Rookley method $h = 372.42$ in IV space (*green*), indirect estimation of the pricing kernel as Legendre basis expansion with $L = 5$ (*green*)

resulting in wiggly estimates in this areas. In comparison to this, when we smooth in the implied volatility space, the Rookley method yields a much smoother estimate directly without additional oversmoothing and performs better in regions of sparse data, see Figs. 11.2 and 11.3. Estimation of the risk neutral density based on the pricing kernel is not affected by the choice of the basis functions in small samples – differences occur only in the tails due to scarcity of observations at the boundaries in our empirical findings. Generally, series type methods allow for direct incorporation of shape constraints. Thus resulting estimates are consistent with economic theory even in finite samples.

References

Aït-Sahalia, Y., & Duarte, J. (2003). Nonparametric option pricing under shape restrictions. *Journal of Econometrics*, *116*(1), 9–47.

Aït-Sahalia, Y., & Lo, A. W. (1998). Nonparametric estimation of state-price densities implicit in financial asset prices. *Journal of Finance*, *53*, 499–547.

Aït-Sahalia, Y., & Lo, A. W. (2000). Nonparametric risk management and implied risk aversion. *Journal of Econometrics, 94*, 9–51.
Arrow, K. J. (1964). The role of securities in the optimal allocation of risk-bearing. *Review of Economic Studies, 31*, 91–96.
Bates, D. S. (1996). Jumps and stochastic volatility: Exchange rate processes implicit in deutsche mark options. *Review of Financial Studies, 9*(1), 69–107.
Black, F., & Scholes, M. (1973). The Pricing of options and corporate liabilities. *Journal of Political Economy, 81*, 637–654.
Breeden, D. T., & Litzenberger, R. H. (1978). Prices of state-contingent claims implicit in option prices. *The Journal of Business, 51*(4), 621–651.
Brown, D. P., & Jackwerth, J. C. (2004). The pricing kernel puzzle: Reconciling index option data and economic theory, Working Paper, University of Konstanz/University of Wisconsin.
Campbell, J., Lo, A., & McKinlay, A. (1997). *The econometrics of financial markets*. NJ: Princeton University Press.
Craven, P., & Wahba, G. (1979). Smoothing noisy data with spline functions, *Numerische Mathematik, 31*, 377–403
Debreu, G. (1959). *Theory of value: An axiomatic analysis of economic equilibrium*. New Haven: Yale University Press.
Engle, R. F., & Rosenberg, J. V. (2002). Empirical pricing kernels. *Journal of Financial Economics, 64*, 341–372.
Fan, J., & Gijbels, I. (1996). *Local polynomial modelling and its applications*. London: Chapman and Hall.
Fengler, M. R. (2005). *Semiparametric modeling of implied volatility*. Berlin: Springer
Grith, M., Härdle, W., & Park, J. (2010). Shape invariant modelling pricing kernels and risk aversion. Resubmitted to Journal of Financial Econometrics on 17 December 2010
Härdle, W. (1990). *Applied nonparametric regression*. Econometric Society Monographs No. 19. London: Cambridge University Press
Härdle, W., & Hlavka, Z. (2009). Dynamics of state price densities. *Journal of Econometrics, 150*(1), 1–15.
Härdle, W., Müller, M., Sperlich, S., & Werwatz, A. (2004). *Nonparametric and semiparametric models*. Heidelberg: Springer.
Härdle, W., Okhrin, Y., & Wang, W. (2009). Uniform confidence for pricing kernels. SFB649DP2010-003. *Econometric Theory* (Submitted).
Heston, S. (1993). A closed-form solution for options with stochastic volatility with applications to bond and currency options. *Review of Financial Studies, 6*(2), 327–343.
Ingersoll, J.E. (1987). Theory of financial decision making. Rowman & Littlefield
Jackwerth, J. C. (2000). Recovering risk aversion from option prices and realized returns. *Review of Financial Studies, 13*(2), 433–451.
Li, K. C. (1987). Asymptotic optimality for c_p, c_l, cross-validation and generalized cross-validation: Discrete index set. *Annals of Statistics, 15*, 958–975.
Li, Q., & Racine, J. S. (2007). *Nonparametric econometrics: Theory and practice*. NJ: Princeton University Press.
Linton, O., & Nielsen, J. P. (1995). A kernel method of estimating structured nonparametric regression based on marginal integration. *Biometrika, 82*(1), 93–100.
Lucas, R. E. (1978). Asset prices in an exchange economy. *Econometrica, 46*, 1429–1445.
Mallows, C. L. (1973). Some comments on c_p. *Technometrics, 15*, 661–675.
Mammen, E., Linton, O., & Nielsen, J. (1999). The existence and asymptotic properties of a backfitting projection algorithm under weak conditions. *Annals of Statistics, 27*(5), 1443–1490.
Marron, J. S., & Nolan, D. (1988). Canonical kernels for density estimation. *Statistics and Probability Letters, 7*(3), 195–199.
Merton, R. C. (1973). Theory of rational option pricing. *Bell Journal of Economics, 4*(1), 141–183.
R. Merton (1976) Option pricing when underlying stock returns are discontinuous. *Journal of Financial Economics, 3*, 125–183.

Müller, H. G. (1988). *Nonparametric regression analysis of longitudinal data*. Lecture Notes in Statistics (Vol. 46). New York: Springer.

Newey, W. K. (1997). Convergence rates and asymptotic normality for series estimators. *Journal of Econometrics*, *79*(1), 147–168.

Rookley, C. (1997). Fully exploiting the information content of intra day option quotes: Applications in option pricing and risk management, Working paper, University of Arizona.

Rubinstein. (1976). The valuation of uncertain income streams and the pricing of options. *Bell Journal of Economics*, *7*(2), 407–425.

Ruppert, D., & Wand, M. P. (1994). Multivariate locally weighted least squares regression. *Annals of Statistics*, *22*(3), 1346–1370.

Stone, M. (1974). Cross-validatory choice and assessment of statistical predictions. *J. Roy. Statist. Soc. Ser. B*, *36*, 111–147.

Wahba, G. (1985). A comparison of GCV and GML for choosing the smoothing parameter in the generalized spline smoothing problem. *Annals of Statistics*, *13*(4), 1378–1402.

Chapter 12
Value at Risk Estimation

Ying Chen and Jun Lu

Abstract This chapter reviews the recent developments of Value at Risk (VaR) estimation. In this survey, the most available univariate and multivariate methods are presented. The robustness and accuracy of these estimation methods are investigated based on the simulated and real data. In the backtesting procedure, the conditional coverage test (Christoffersen, Int. Econ. Rev. 39:841–862, 1998), the dynamic quantile test (Engle and Manganelli, J. Bus. Econ. Stat. 22(4):367–381, 2004) and Ljung-Box test (Berkowitz and O'Brien, J. Finance 57(3):1093–1111, 2002) are used to justify the performance of the methods.

12.1 Introduction

Value-at-Risk (VaR) is a standard risk measure, which indicates the possible loss of a financial portfolio at a certain risk level over a certain time horizon. The introduction of VaR dated back to the late 1980s, when stock market crashed and immediately measuring market risk became overwhelmingly necessary. In 1994 Morgan launched RiskMetrics with a free access to VaR estimation, making the analysis of VaR simple and standard. The Basel Accord in 1996 allowed financial institutions to use internal models to calculate VaR, which further prompted the development of VaR estimation. After that, many VaR estimation methods have been proposed and widely used in risk management. Hence, it was believed that VaR estimation has been well developed and can provide reliable risk measure. Early studies even show that many large financial institutions adopt conservative internal

Y. Chen and J. Lu (✉)
Department of Statistics and Applied Probability, National University of Singapore, Singapore 117546
e-mail: stacheny@nus.edu.sg; g0700712@nus.edu.sg

J.-C. Duan et al. (eds.), *Handbook of Computational Finance*, Springer Handbooks of Computational Statistics, DOI 10.1007/978-3-642-17254-0_12,

VaR models and report over-estimated risks in the US and Canada (see Berkowitz et al. 2006; Pérignon et al. 2008).

The 2007–2009 financial crisis however called the entire risk management system into question, when so many large and reputable financial firms either bankrupted or faced survival problem, e.g. Lehman fell, Merrill Lynch was sold, AIG was saved by the US government. The VaR, as an industrial standard risk measure, and its estimation have again attracted global attention. It is necessary to at least investigate the robustness and accuracy of the most available VaR estimation methods under different market conditions, e.g. with and without financial crisis.

From the statistical point of view, VaR is in fact a certain quantile of a portfolio's returns. Given a probability level $\alpha \in (0, 1)$, VaR for period $t + h$ is defined as a threshold value such that the probability that the mark-to-market loss on the portfolio over the given time horizon exceeds this value is α:

$$\text{VaR}_{t+h}^{\alpha} = -\inf\{c \in \mathbb{R} : P(r_{t+h} \leq c|\mathcal{F}_t) \geq \alpha\}, \tag{12.1}$$

where $\mathcal{F}_t$ represents the past information at time t. Even before the subprime mortgage crisis, VaR has been criticized over its mathematical properties and over its potential destabilizing impact on financial activities by e.g. Artzner et al. (1999) and Bibby and Sørensen (2001). It has been well-known that VaR is not a coherent risk measure, i.e. it is not necessarily subadditive. For example, the VaR value for a portfolio may exceed the summation of the individual VaR values of its component assets, which contradicts the principle of diversification. Moreover, VaR provides less information about the potential size of the loss that exceeds it. Despite these criticisms, VaR remains the industrial benchmark for measuring market risk. In addition, the coherent risk measures such as conditional VaR, also depends on VaR. Therefore, it is still meaningful to discuss VaR and its estimation in risk management.

To date, there are many methods used for VaR estimation, see Jorion (2001) and the references therein. In addition, Kuester et al. (2006) gives an extensive review on the VaR estimation methods with a focus on the univariate financial time series. In accordance with the definition (12.1), the initial focus of VaR estimation is to estimate the distributional quantile of the portfolio's returns. The simplest way is the historical simulation (HS) method, where the sample quantile of returns conditional on past information is used as VaR. Another widely-used method is based on the extreme value theory (EVT). With a focus on the distributional behavior in tails, EVT is expected to provide accurate VaR estimates (see e.g. Embrechts et al. 1999b; McNeil 2000). Alternatively, the quantile regression and its implementation in VaR estimation is also used. For example, the class of conditional autoregressive value at risk (CAViaR) models estimates VaR using the quantile regression minimization (Engle and Manganelli 2004).

These methods are based on the assumption that the returns are independently and identically distributed (IID). Financial returns are unfortunately not IID. As

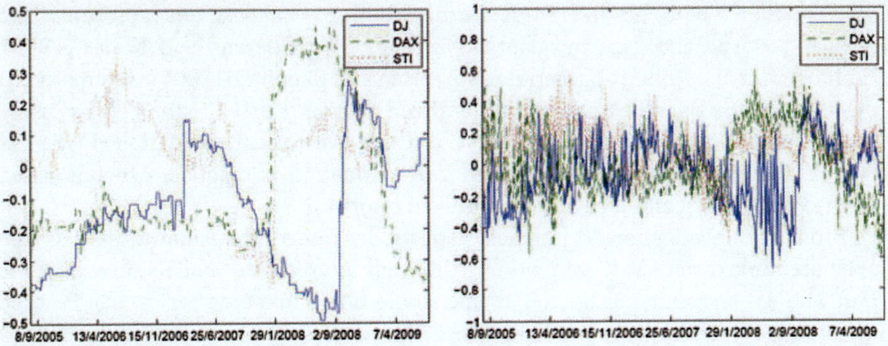

Fig. 12.1 The EVT shape parameter fluctuation for the returns (*left*) and the standardized returns (*right*). EVT is presented in Sect. 12.2

noted by Mandelbrot, the variance of financial returns changes and clusters over time. This heteroscedasticity (non IID) of financial returns can be illustrated in Fig. 12.1. Based on three real financial data, the Dow Jones Industry Average 30, the German DAX index and the Singapore Straight Time Index from 18/08/2003 to 31/07/2009, the distributional parameters of the returns (left panel) are re-estimated using a 500-day rolling window for each time point. If the returns are IID, the parameters should be constant over time and straight lines are expected. However, the fitted parameters for the return series fluctuates, especially around January 2008 (financial crisis). On the other hand, the fitted parameters against the standardized returns (returns are filtered by volatility estimates) on the right panel display relative stable (IID) pattern. To account for this heteroscedasticity, financial returns are modeled without loss of generality as:

$$r_t = \sigma_t \varepsilon_t, \tag{12.2}$$

where σ_t denotes the conditional variance of returns r_t. The standardized returns, i.e. the residuals ε_t, are assumed to be IID with $E(\varepsilon_t) = 0$ and $E(\varepsilon_t^2) = 1$. In other words, the probability distribution of the residuals is assumed to be invariant over time.

Clearly, volatility estimation is an important issue in VaR estimation. Among volatility models, the GARCH model (see Engle 1982; Bollerslev 1986) and its extensions are widely-used. Their success stems from the ability to capture the stylized facts of financial time series, such as time-varying volatility and volatility clustering, see among others Poon and Granger (2003) and the references therein. Therefore, we estimate volatility with the GARCH(1,1) model using a rolling window in this survey. The density of the standardized returns can be estimated with or without distributional assumption. In the parametric literature, financial returns are typically assumed to be Gaussian distributed for simplicity, e.g. in RiskMetrics. However, this assumption contradicts the empirical observation – daily financial time series are heavy-tailed distributed. Therefore, heavy-tailed distribution families such as the hyperbolic and Student-*t* distributions and the Lévy process have been introduced and used in quantitative finance by Eberlein and Keller (1995), Embrechts et al. (1999a) and Barndorff-Nielsen and Shephard (2001). Alternatively, the density (or quantile) of the standardized returns can be estimated by using nonparametric methods. For example, the filtered historical simulation (FHS) has been considered as the most successful VaR estimation in practice, which is based on the empirical quantile of the standardized returns.

From both academic and practical aspects, it is also interesting to discuss and evaluate multivariate VaR estimation. Although the reduced models, in which the portfolio's returns are considered as one single univariate time series, can be used to calculate VaR for large portfolio, the reduced models (univariate method) may yield low accuracy by ignoring the complicated correlation among the individual returns. Consequently, reduced models provide less detailed information on the source of risk. Therefore, multivariate methods are necessary to be investigated. Moreover, the multivariate VaR estimation will be at least a useful complement to reduced models and could help in evaluating the accuracy of univariate methods. Nevertheless, there are few contributions in multivariate VaR estimation, due to the numerical complexity in the covariance estimation and the joint density identification.

In this chapter, we will present three workable multivariate VaR estimation methods. The DCC-VaR method estimates the covariance matrix by using the dynamic conditional covariance (DCC) model (see Engle 2002; Tse and Tsui 2002) and assumes the standardized returns are Gaussian distributed. The Copula-VaR method calculate VaR based on the fitted copula function, in which the joint distribution of portfolios is estimated by linking all the marginal distributions with a fixed form (see e.g. Nelsen 1999; Embrechts and Dias 2003; Giacomini et al. 2009). Moreover, we present a multivariate VaR estimation based on the independent component analysis (ICA) method that converts the high dimensional analysis to univariate study with a simple linear transformation (Hyvärinen et al. 2001). It is worth noting that the first two approaches are numerically cumbersome when the number of assets involved is large, while the ICA based method significantly speeds up the VaR calculation even for a large portfolio (Chen et al. 2009).

The chapter is organized as follows. We will first give a definition of VaR. The most available VaR estimation methods, including the volatility/covariance estimation and the calculation of the residual quantile position, will be presented in Sect. 12.2. In this survey, three tests are used in the backtesting procedure to evaluate the robustness and accuracy of the VaR methods. They will be discussed in Sect. 12.3. In particular, the conditional coverage test (Christoffersen 1998), the dynamic quantile test (Engle and Manganelli 2004) and Ljung-box test (Berkowitz and O'Brien 2002) are considered. The implementation of various VaR methods and the backtesting results are illustrated based on simulated and real data with and without financial crisis. Finally we will give a short conclusion.

12.2 VaR and Methodology

Value at Risk (VaR) is so far the most widely used risk measure by financial institutions. With a target probability $\alpha \in (0,1)$ and time horizon $[t, t+h]$, VaR is defined as a threshold value such that the probability that the mark-to-market loss on the portfolio over the given time horizon exceeds this value is the given probability level:

$$\text{VaR}^{\alpha}_{t+h} = -\inf\{c \in \mathbb{R} : P(r_{t+h} \leq c|\mathcal{F}_t) \geq \alpha\},$$

where $\mathcal{F}_t$ represents the past information at time t. The target level α is often to be set between 0.1 and 5% for different purposes such as regulatory requirement and internal supervisory.

Despite its simple definition, the calculation of VaR is a very challenging statistical problem. From a statistical point of view, VaR is a certain quantile of the distribution of the portfolio's future returns. Hence VaR is tightly linked to estimating the (joint) distribution of returns. A direct estimation of the returns' quantile, with or without distributional assumption, is however insufficient. The reason is that the variable of interest, the financial returns, is not IID. To take into account this heteroscedasticity, financial returns are modeled as:

$$r_t = \mu_t + \sigma_t \varepsilon_t, \tag{12.3}$$

where μ_t and σ_t denote the conditional mean and variance of returns r_t and the residuals (also known as standardized returns) ε_t are assumed to be IID with $E(\varepsilon_t) = 0$ and $E(\varepsilon_t^2) = 1$. In literature, the conditional mean of financial returns plays a relatively trivial role since $\mu_t = 0$ holds under the efficient market hypothesis. Therefore, we illustrate the VaR estimation with focus on (1) estimating the conditional variance and (2) identifying the distributional behavior of the residuals.

The forecast of VaR at the future time point $t + h$, based on the fitted heteroscedastic model, can be formulated as:

$$\widehat{\text{VaR}}^{\alpha}_{t+h} = \widehat{\sigma}_{t+h} Q_{\alpha}, \tag{12.4}$$

where Q^{α} denotes the αth quantile of the residuals ε_t.

In the following, we firstly discuss the volatility estimation and then show how to calculate the quantile of the standardized returns. Some multivariate VaR estimation methods will be presented after that.

12.2.1 Volatility Estimation

Volatility plays an important role in VaR estimation and other financial activities. It is a latent variable and not directly observed in markets. Therefore, many volatility models and volatility proxies have been proposed in the quantitative finance literature. Among others, the generalized autoregressive conditional heteroscedasticity (GARCH) model and its extension can capture with success the volatility clustering (see e.g. Bollerslev 1995; Nelson 1990, 1991) and hence the ARCH-type models dominate the modeling and estimation of variance. We refer to Engle (1995), Franke et al. (2008) and Tsay (2005) for a comprehensive review. In general, GARCH(1,1) has a good performance in estimation as well as prediction of volatility. Based on the comparison of 330 ARCH-type models, Lunde and Hansen (2005) finds no evidence that a GARCH(1,1) is outperformed for exchange rate series, although the GARCH(1,1) is inferior when there is leverage effect in the data. In addition, Andersen and Bollerslev (1998) demonstrates that based on the realized variance (RV) – a consistent estimator of volatility calculated from ultra-high frequency data (see Andersen et al. 2001; Barndorff-Nielsen and Shephard 2002; Zhang et al. 2005; McAleer and Medeiros 2008) GARCH models produce accurate forecasts. Motivated by these works, the GARCH(1,1) set up is used to estimate the latent volatility variable in our study.

GARCH(1,1) model is defined as (see Engle, 1982; Bollerslev, 1986):

$$\sigma_t^2 = \omega + \beta_1 r_{t-1}^2 + \beta_2 \sigma_{t-1}^2, \tag{12.5}$$

where the unconditional variance $\sigma^2 = \omega/(1 - \beta_1 - \beta_2)$ exists if $1 - \beta_1 - \beta_2 \neq 0$. (Very often, it is observed that the sum of the estimated parameters $\tilde{\beta}_1$ and $\tilde{\beta}_2$ is close to 1, partially due to the nonstationarity or persistence of volatility process, Nelson 1990.)

It is worth noting that the dynamic of variance may change over time, especially when market shifts. To achieve accuracy in estimation, one can use the most recent observations to adapt the estimation, which is referred to as rolling window average method. Basel accord has suggested to use a window size of 2 years (roughly 500 days) in VaR estimation, we here follow the suggestion. Figure 12.2 displays the fitted GARCH parameters for three stock indices by using a rolling window for

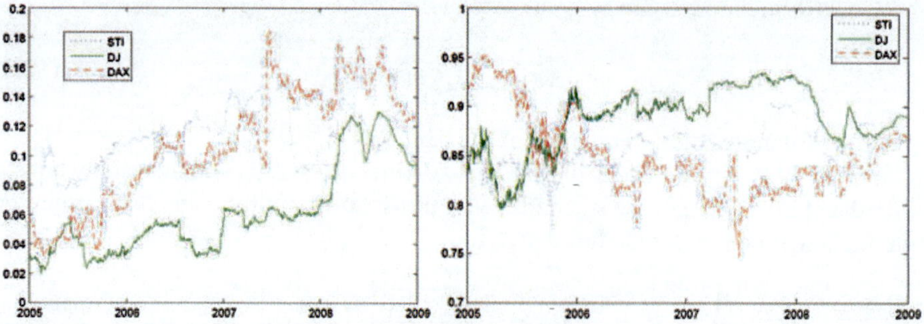

Fig. 12.2 The GARCH(1,1) fitted parameter α (*left*) and β (*right*) by using a 500-day rolling window for each time point

each time point. It shows that the parameters β_1 and β_2 change over time, especially during the sub-prime financial crisis 2007–2009. The results support the use of rolling window. The selection of 500 days, however, raises the question whether the experience-based value is really better than others. Clearly, a large window size often corresponds to low variation of estimate but rises the risk of modeling bias. On the contrary, a small window size delivers estimates that are sensitive to model changes but have high variation. As a consequence, it is suggested to choose a large window size when the markets are stationary and to reduce the window size to a small value when the markets change. For selecting a local optimal window size, we refer to the recently developed adaptive approach (see e.g. Chen and Spokoiny, 2009; Čížek et al., 2009).

12.2.2 *Quantile of (Standardized) Returns*

The quantile of residuals can be estimated in either nonparametric (without distributional assumption, e.g. historical simulation) or parametric way. In parametric way, for reasons of stochastic and numerical simplicity, it is often assumed that the (standardized) returns are normally distributed e.g. in the Morgan's RiskMetrics framework. Although returns will converge to normality under temporal aggregation, it is observed that most concerned short-term returns, e.g. daily returns, obviously deviate from the assumption of normality (Andersen et al. 2005). The heavy-tailed distribution families such as the normal inverse Gaussian (NIG) and Student-t distributions, on the other hand, have been used in VaR models, see e.g. Eberlein and Keller (1995) and Embrechts et al. (1999a). In particular, the density of NIG random variable has a form of:

$$f_{\mathrm{NIG}}(x;\varpi,\beta,\delta,\mu) = \frac{\varpi\delta}{\pi}\frac{K_1\left\{\varpi\sqrt{\delta^2+(x-\mu)^2}\right\}}{\sqrt{\delta^2+(x-\mu)^2}}\exp\{\delta\sqrt{\varpi^2-\beta^2}+\beta(x-\mu)\},$$

where the distributional parameters fulfill $\mu \in \mathbb{R}$, $\delta > 0$ and $|\beta| \leq \varpi$. The modified Bessel function of the third kind $K_\lambda(\cdot)$ with an index $\lambda = 1$ has a form of:

$$K_\lambda(x) = \frac{1}{2}\int_0^\infty x^{\lambda-1}\exp\{-\frac{x}{2}(x+x^{-1})\}\,dy$$

It is worth noting that these parametric approaches tend to fit the density curves that accommodate the mass of central observations. On the contrary, extreme value theory (EVT) provides a natural approach to VaR estimation, which projects tail out from a data over a certain threshold, see Pickands (1975a) and Embrechts et al. (1997). The details will be presented in Sect. 12.2.2.1.

The simplest nonparametric VaR estimation is the historical simulation (HS) method. The VaR at $t+1$ for instance is given by the empirical α-quantile of the past K observations up to date t:

$$\widehat{\text{VaR}}_{t+1} = Q_\alpha(r_t, \cdots, r_{t-K+1})$$

Empirically, filtered HS (FHS) shows improved accuracy and is more adopted than HS method. In FHS, VaRs are calculated based on the empirical quantiles of residuals and the estimated volatility:

$$\widehat{\text{VaR}}^{\alpha}_{t+1} = \widehat{\sigma}_{t+1} Q_\alpha(\widehat{\varepsilon}_t, \cdots, \widehat{\varepsilon}_{t-K+1})$$

These quantile estimation methods only based on the portfolio's (standardized) returns and assume that returns contain sufficient information for forecasting. Recently, Engle and Manganelli (2004) opens a new door to VaR estimation, in which the dynamic of VaR depends not only on returns but also on other covariates, for example, the returns. The details are given in the following.

12.2.2.1 Extreme Value Theory

There are two kinds of models for extreme values; the block maxima models that are for the largest observations collected from large samples of identically distributed observations, and the peaks-over-threshold (POT) models that are for all large observations which exceed a high threshold. The POT models are often considered to be more useful for practical usage, especially in VaR estimation. In our study we will focus on POT model based on the generalized Pareto distribution. The other models based on the Hill estimator and its relatives are referred to Beirlant et al. (1996).

Let $z_1, z_2, \cdots$ be IID random variables representing financial losses ($z_t = -r_t$) and having distribution function F. Let u be the high threshold and defines the excess distribution above the threshold u as:

$$F_u(z) = P\{Z - u \leq z | Z > u\} = \frac{F(z+u) - F(u)}{1 - F(u)}, \quad z \geq 0 \tag{12.6}$$

For a wide class of distributions, the distribution of the excess over a sufficiently large threshold u converges to generalized Pareto distribution (GPD), see Pickands (1975b):

$$G_{\xi,\beta}(z) = \begin{cases} 1 - (1 + \frac{\xi z}{\beta})^{-1/\xi} & \text{if} \quad \xi \neq 0 \\ 1 - \exp(-z/\beta) & \text{if} \quad \xi = 0 \end{cases}, \tag{12.7}$$

where ξ is the shape parameter and β is the scale parameter. In addition, we have:

$$\begin{cases} z \geq 0 & \text{if} \quad \xi \geq 0 \\ 0 \leq z < -\beta/\xi - u & \text{if} \quad \xi < 0. \end{cases}$$

The GDP distribution nests certain other distributions. For example, if $\xi > 0$ then GPD is in fact the ordinary Pareto distribution. If $\xi = 0$ then GPD corresponds to the exponential distribution and $\xi < 0$ is known as a Pareto type II distribution.

Given IID $z > u$, (12.6) can be reformulated to

$$\bar{F}(z+u) = \bar{F}(u)\bar{F}_u(z).$$

$\bar{F}(u)$ is estimated by its empirical counterpart k/n, i.e. proportion of the k observations that are larger than the threshold u to the total n observations. On the basis of EVT, $\bar{F}_u(z)$ for $z > u$ is estimated using a GPD approximation to obtain the tail estimator, i.e. $F_u(z) \approx G_{\xi,\beta}(z)$ and

$$\widehat{\bar{F}(z)} = \frac{k}{n}\left(1+\hat{\xi}\frac{z-u}{\hat{\beta}}\right)^{-1/\hat{\xi}}.$$

By inverting the above formula we get the quantile estimator, i.e. VaR:

$$\widehat{\mathrm{VaR}}_\alpha = z_{k+1} + \frac{\hat{\beta}}{\hat{\xi}}\left\{1-\left(\frac{1-\alpha}{k/n}\right)^{-\hat{\xi}}\right\},$$

recalling $u = z_{k+1}$. It is important to note that the tail estimator is only valid for $z > u$. Needless, the threshold u plays an important role in the estimation. A high u reduces bias in estimating the excess function since the approximation (12.7) only works well on the tails. Simultaneously, choosing a high u leads to very few exceedances and hence a high variance of the estimator. On the other hand, a low u induces a large bias but a small variance for estimator. For practical use we must trade off bias and variance. Data-driven tools, e.g. mean excess plot, can help to choose a suitable threshold value u, see Embrechts et al. (1997).

A direct application of EVT is possibly inappropriate for most financial assets returns, since EVT assumes IID of random variables. It is suggested to use filtered EVT (FEVT), in which the time varying volatility is estimated and EVT is applied to the residuals, see e.g. McNeil and Frey (2000), Franke et al. (2008).

12.2.2.2 Quantile Regression: CAViaR Model

Given a standard linear regression model:

$$r_t = x_t^\top \beta + \varepsilon_t,$$

the median regression is concerned with the estimation of the conditional median given $X = x$, which corresponds to the minimization of the mean absolute error (MAE). Let $u = r - x^\top\beta$, we denote MAE as $f(u) = |u|$ and rewrite the optimization problem for the median – the 50% quantile as:

$$\min 0.5 f(u) = 0.5 u I_{[0,\infty)}(u) - (1-0.5) u I_{(-\infty,0))}(u),$$

where $I_A(u) = 1$ if $u \in A$ and 0 otherwise is the usual indicator function of the set A. This definition has been generalized by replacing 0.5 by α to obtain an α-quantile regression (Koenker and Bassett 1978):

$$\beta(\alpha) = \operatorname{argmin} \left\{ \sum_{r_t \geq x_t^\top \beta} \alpha |r_t - x_t^\top \beta| + \sum_{r_t < x_t^\top \beta} (1-\alpha)|r_t - x_t^\top \beta| \right\} \tag{12.8}$$

One desirable feature of the quantile regression is that there is no distributional assumptions for the portfolio's returns.

Motivated by the good performance of quantile regression, Engle and Manganelli (2004) proposed the CAViaR models, which estimates VaR by using both the (standardized) returns and the past values of VaR. Given the observation that financial returns tends to be autocorrelated and have clustering phenomenon, the quantile of portfolio's returns – VaR with a natural link to the distribution – is expected to exhibit a similar behavior. One version of the CAViaR models with a focus on absolute value of past returns (Symmetric absolute value CAViaR) is defined as:

$$\text{VaR}_t = \beta_0 + \beta_1 \text{VaR}_{t-1} + \beta_2 |r_{t-1}|$$

The CAViaR model measures the relationship between VaRs and the past value of returns r_{t-1}. Moreover, the autoregressive term (past values of VaR) ensures that the VaR changes smoothly over time.

12.2.3 Multivariate VaR Models

Along with the remarkable development of univariate VaR estimation, there are few contributions to multidimensional VaR estimation. Although the univariate VaR estimation methods, which are based on the simple modeling structure and assumption, can be extended to multivariate time series, the performance is poor due to the unrealistic assumptions. On the other hand, the VaR estimation methods, which are based on the realistic but complex modeling structure and assumption are infeasible or inappropriate for solving high-dimensional problem. In this section, we will introduce three methods that balance the numerical calculation and the estimation accuracy.

Let $X_t \in \mathbb{R}^d$ denote the vector of individual returns in a portfolio. The portfolio's return is mapped by the trading strategy $b_t \in \mathbb{R}^d$ as:

$$r_t = b_t^\top X_t = b_t^\top \Sigma_t^{1/2} \varepsilon_t,$$

where the portfolio's returns rely on the trading allocation $b_t = (b_{1,t}, \cdots, b_{d,t})^\top$, the covariance matrix Σ_t for returns of individual components in the portfolio and the residuals ε_t. The VaR estimation provides the future VaR values based on the fitted model:

$$\widehat{\text{VaR}}^{\alpha}_{t+h} = Q_\alpha(r_{t+h}) = Q_\alpha(b^\top_{t+h}\widehat{\Sigma}^{1/2}_{t+h}\varepsilon_{t+h}) \tag{12.9}$$

12.2.3.1 DCC-VaR

We introduce one multivariate GARCH volatility models, the dynamic conditional correlation (DCC) model (see Engle, 2002; Tse and Tsui, 2002) and incorporate the covariance estimator into the VaR estimation.

$$\Sigma_t = D_t R_t D_t,$$

where Σ_t denotes the covariance matrix at time point t, $D_t = diag(h^{1/2}_{11,t} \dots h^{1/2}_{dd,t})$ and $h_{ii,t}$ denotes the variance of the ith component. Its dynamic can be modeled in a univariate GARCH framework. In addition, the dynamic conditional correlation is formulated as (Tse and Tsui 2002):

$$R_t = (1 - \theta_1 - \theta_2)R + \theta_1\Psi_{t-1} + \theta_2 R_{t-1},$$

where θ_1 and θ_2 are non-negative parameters satisfying $\theta_1 + \theta_2 < 1$, R is a symmetric $d \times d$ positive definite matrix with $\rho_{ii} = 1$. Ψ_{t-1} is the $d \times d$ correlation matrix of ϵ_τ for $\tau = t - M, t - M + 1, \dots, t - 1$, in which the i, jth elements is given by:

$$\Psi_{i,j,t-1} = \frac{\sum_{m-1}^{M} u_{i,t-m}u_{j,t-m}}{\sqrt{(\sum_{m-1}^{M} u^2_{i,t-m})(\sum_{m-1}^{M} u^2_{j,t-m})}}$$

with $u_{it} = \varepsilon_{it}/\sqrt{h_{ii,t}}$. The covariance estimation is however numerically cumbersome when the dimension of portfolio is high. The distribution of the filtered series ($\varepsilon_t = \Sigma_t^{-1/2}x_t$) is assumed to be Gaussian distributed for simplicity. The VaR estimator is obtained as the empirical quantile of the portfolio.

12.2.3.2 Copula-VaR

Copula is a function that links a multidimensional joint distribution F to its one-dimensional margins F_j:

$$F(x_1, \cdots, x_d) = C\{F_1(x_1), \cdots, F_d(x_d); \theta\},$$

where C is a copula function and the parameter θ measures the dependence of the variables. According to Sklar's theorem, given any joint distribution function and respective marginal distribution functions, there exists a copula C to bind the margins and give the joint distribution (Nelsen, 1999). Many families of copulas have been proposed in the literature, which differ in the detail of the dependence they represent. For example, one elliptical copula – the Gaussian copula function – is defined:

$$C_\rho(u_1, \cdots, u_d) = \Phi_\rho\left(\Phi^{-1}(u_1), \cdots, \Phi^{-1}(u_d)\right),$$

where $u_1, \cdots, u_d \in [0, 1]$ and Φ denotes the standard normal cumulative distribution function. The parameter ρ measures the linear dependence among the variables. In the later calculation, we will use the Gaussian copula. Note that the elliptical copula group is rich in parameters – parameter for each pair of variables and hence it is easy for simulation.

For a sample $\{x_t\}_{t=1}^T$, the procedure of VaR estimation with copula can be defined as (Giacomini et al. 2009):

1. Identify of marginal distributions $F_{x_j}(x_j; \delta_j)$ and select copula $C(u_1, \cdots, u_d; \theta)$
2. Fit the copula C
3. Generate Monte Carlo data $x_{T+h} \sim C\{F_1(x_1), \cdots, F_d(x_d); \hat{\theta}\}$
4. Generate a sample of portfolio returns
5. Estimate $\widehat{\text{VaR}}_{t+h}^{\alpha}$, the empirical quantile at level α from the generated returns

It is worth noting that the Copula-VaR approach is also numerically cumbersome when the dimension of portfolio is high, since Monte Carlo simulation is necessary in the VaR estimation. In the following, we introduce an advanced statistical method, with which a fast multivariate VaR model is derived that is even applicable to very high dimensional problem.

12.2.3.3 ICA-VaR

Independent component analysis (ICA) is to retrieve, out of high dimensional time series, stochastically independent components (ICs) through a linear transformation:

$$y(t) = Wx(t),$$

where the transformation matrix $W = (w_1, \cdots, w_d)^\top$ is nonsingular. Chen et al. (2009) proposes an ICA-VaR model (named as GHICA in the reference), where the high-dimensional risk factors are modeled as:

$$x(t) = W^{-1}y(t) = W^{-1}\Sigma_y^{1/2}(t)\varepsilon_y(t).$$

Clearly, the covariance of the ICs $\Sigma_y(t)$ is a diagonal matrix because of independence. The diagonal elements are variances of the ICs at time point t. The stochastic innovations $\varepsilon_y(t) = \{\varepsilon_{y_1}(t), \cdots, \varepsilon_{y_d}(t)\}^\top$ are cross independent and can be individually identified in any univariate distributional framework. Based on Jacobian transformation (see e.g. Härdle and Simar, 2003), the joint distribution can be straightforwardly derived from the identified marginals. In other words, ICA-VaR converts the multivariate problem to univariate problems, where the realistic and complex univariate approaches and assumptions can be easily used. Hence, the ICA-VaR estimation provides a solution to balance the numerical tractability and the realistic distributional assumption on the risk factors.

Various methods have been proposed to compute the transformation matrix W, see Hyvärinen et al. (2001) and references therein. In this study, we use the fastICA method. The idea is to find ICs by maximizing nongaussianity, in particular, to maximize the negentropy:

$$J(y_j) = \int f_{y_j}(u) \log f_{y_j}(u) du - \int \varphi_{0,1}(u) \log \varphi_{0,1}(u) du,$$

where $\varphi_{0,1}$ is the density function of a standard Gaussian random variable (Hyvärinen, 1998). This optimization problem is solved by using the symmetric FastICA algorithm.

The formal procedure of the ICA-VaR method is defined as:

1. Apply ICA to the given risk factors to get ICs.
2. Estimate the variance of each IC and identify the distribution of every IC's innovation in the normal inverse Gaussian (NIG) or other distributional framework.
3. Estimate the density of the portfolio return using the FFT technique.
4. Calculate risk measures.

12.3 Backtesting

Empirical study shows that VaR models sometimes provide quite different VaR values for the same portfolio data (Beder, 1995). Therefore, it is important to justify the VaR models in the backtesting procedure. Backtesting involves a systematical comparison of the historical VaR forecasts with the associated portfolio returns. By far, several statistical tests have been proposed in the framework of backtesting procedure. In our study, three tests are considered: Christoffersen (1998) 's test based on the use of a Markov chain, dynamic quantile test of Engle and Manganelli (2004) derived from a linear auto-regressive model, Berkowitz and O'Brien (2002)'s test – a portmanteau test of weak white noise.

12.3.1 Conditional Coverage Test

Let us define exceedance as a violation when VaR is exceeded by the actual losses. The simplest way to verify the accuracy of VaR methods is to record the failure rate – the proportion of occurrence of exceedances in a given sample. In the baktesting procedure, each day is marked to 0 if VaR is not exceeded, i.e. no exceedance, and to 1 otherwise:

$$I_t = I(r_t < -\text{VaR}_t)$$

The test of unconditional coverage is initially developed by Kupiec (1995), where the failure rate with n_1 exceedances over T points is equal to:

$$E[I_t|\mathcal{F}_{t-1}] = \frac{n_1}{T}$$

According to the definition of VaR (12.1), the failure rate should be close to the expected risk level α. The test of unconditional coverage is formulated as:

$$H_0 : E[I_t] = \alpha \qquad H_1 : E[I_t] \neq \alpha \tag{12.10}$$

The sequence of I_t is naturally Bernoulli distributed with parameter α (Christoffersen, 1998). Under the null hypothesis, the likelihood ratio test statistic is constructed:

$$\text{LRuc} = -2\log\left[\frac{(1-\alpha)^{T-n_1}\alpha^{n_1}}{(1-n_1/T)^{T-n_1}(n_1/T)^{n_1}}\right] \xrightarrow{\mathscr{L}} \chi_1^2$$

It is worth noting that the unconditional coverage test ignores the temporal dependence of exceedances, e.g. the clustering of exceedances. The fact that extreme losses follow extreme losses, will lead to bankruptcy and hence invalidate VaR methods. To overcome this limitation, Christoffersen (1998) extends the unconditional coverage test (12.10) with a focus on the serial correlation of exceedances. In particular, the process of exceedances $\{I_t\}$ is modeled by a binary first-order Markov chain with a transition probability matrix:

$$\Pi = \begin{pmatrix} \pi_{00} & \pi_{01} \\ \pi_{10} & \pi_{11} \end{pmatrix}$$

where $\pi_{ij} = P(I_t = j|I_{t-1} = i)$ denotes the probability of observing a state j in 1 day given a state i in the previous day. For example, π_{01} records the conditional failure rate with no exceedance (state $= 0$) followed by an exceedance (state $= 1$). The null hypothesis of the temporal dependence is formulated as:

$$H_0 : \Pi = \Pi_\alpha = \begin{pmatrix} 1-\alpha & \alpha \\ 1-\alpha & \alpha \end{pmatrix} \qquad H_1 : \Pi \neq \Pi_\alpha$$

Under the null hypothesis, the occurrence of exceedances should not contain information for future state. In other words, the hypothesis is $\pi_{ij} = \pi_{jj}$. Moreover, the probability of observing exceedances is equivalent to the target probability α. Overall, it leads to the likelihood ratio test statistic:

$$\mathrm{LR_{td}} = -2\log\left[\frac{(1-\alpha)^{T-n_1}\alpha^{n_1}}{(1-\pi_{01})^{n_{00}}\pi_{01})^{n_{01}}(1-\pi_{11})^{n_{10}}\pi_{11})^{n_{11}}}\right] \xrightarrow{\mathscr{L}} \chi_1^2,$$

where n_{ij} represents the number of transitions from state i to state j, i.e. $n_{ij} = \sum_{t=2}^{T}(I_t = j\,|\,I_{t-1} = i)$, and the number of days with exceedances is $n_1 = n_{0j} + n_{1j}$.

The combination of the two tests yields the conditional coverage test statistic (Christoffersen, 1998):

$$\mathrm{LR_{cc}} = \mathrm{LR_{uc}} + \mathrm{LR_{td}} \tag{12.11}$$

Although the conditional coverage test has been widely used in verifying VaR methods, it is criticized for two limitations. First the temporal dependence test only take into account the dependence of order one (two consecutive days). Secondly, the Markov chain only measures the influence of past exceedances and not that of any other exogenous variable. Next we introduce the tests proposed by Engle and Manganelli (2004) and Berkowitz and O'Brien (2002) that overcome the limitations.

12.3.2 Dynamic Quantile Test

Engle and Manganelli (2004) proposes a conditional coverage test by using a linear regression model based on the process of hit function:

$$H_t = I_t - \alpha = \begin{cases} 1-\alpha, & \text{if } r_t < -\mathrm{VaR}_t \\ -\alpha, & \text{else} \end{cases},$$

where $\{H_t\}$ is a centered process on the target probability α. The dynamic of the hit function is modeled as:

$$H_t = \beta_0 + \sum_{j=1}^{p} \beta_j H_{t-j} + \sum_{k=1}^{K} \gamma_k g_k(z_t) + \varepsilon_t, \tag{12.12}$$

where ε_t is an IID process with mean of zero and $g(\cdot)$ is a function of past exceedances and of variable z_t.

Under the hypothesis that the VaR estimation can deliver accurate VaR estimates and also the occurrence of p consecutive exceedances is uncorrelated, the regressors should have no explanatory power. Hence, the dynamic quantile (DQ) test is defined as:

$$H_0 : \Psi = (\beta_0, \beta_1, \cdots, \beta_p, \gamma_1, \cdots, \gamma_K)^\top = 0.$$

It is easy to show that the DQ test statistic, in association with the Wald statistic, is:

$$\mathrm{DQ} = \frac{\hat{\Psi}^\top \mathbf{X}^\top \mathbf{X} \hat{\Psi}}{\alpha(1-\alpha)} \xrightarrow{\mathcal{L}} \chi^2_{1+p+K}, \tag{12.13}$$

where $\mathbf{X}$ denotes the covariates matrix in (12.12). In our study, we select $p = 4$, $K = 1$ and $g(z_t) = \widehat{\mathrm{VaR}}_t$ to account for the influence of past exceedances up to 4 days and that of the VaR estimates.

12.3.3 *Ljung-Box Test*

Berkowitz and O'Brien (2002) suggests to use the Ljung-Box test in checking the temporal dependence of exceedances. The Ljung-Box is used to assess the temporal dependence of time series. Here it is again motivated by the hypothesis of absence of autocorrelation in the centered process of exceedances $\{H_t = I_t - \alpha\}$. The null hypothesis is:

$$H_0 : \rho_1(H_t) = \cdots = \rho_p(H_t) = 0,$$

where ρ_j is the jth autocorrelations of the exceedances process: Under the null hypothesis, the test statistic is:

$$\mathrm{LB} = (T)(T+2) \sum_{j=1}^{p} \frac{\hat{\rho}_j^2}{T-j} \xrightarrow{\mathcal{L}} \chi^2_p, \tag{12.14}$$

where p is the order of the autocorrelation for the hit function process. In practice, $p = \log(T)$ is recommended. As same as the dynamic quantile test, the Ljung-Box test also overcomes the limitation of the traditional conditional coverage test (12.11) by considering temporal dependence with order higher than one.

12.4 Simulation Study

In this section, we will demonstrate the performance of the discussed VaR estimation methods based on simulated data. The initial focus is the accuracy of these methods. In addition, it is interesting to investigate the effect of financial crisis on the accuracy of the VaR estimation. In particular, we will simulate the processes with and without financial crisis respectively. The backtesting tests will help to justify the validity of the VaR methods, see Sect. 12.3.

The return series is generated based on the heteroscedastic model:

$$r_t = \sigma_t \varepsilon_t,$$

where the volatility index VIX is chosen as the "actual" value of market volatility σ_t. VIX is the implied volatility based on S&P 500 index options and has been adopted across financial communities. To investigate the accuracy of the discussed VaR models to various market situations, we use the daily VIX observations from 18/08/2003 to 31/06/2009 (1,500 days) that covers financial crisis time period (whole sample period) and the observations from 18/08/2003 to 30/04/2007 (pre-financial crisis period) respectively in data generation. The innovations or shocks ε_t are assumed to be either standard normal distributed, Student-t distributed or normal inverse gaussian (NIG) distributed. The assumption of the last two distributions (Student-t and NIG) helps us to mimic the fat-tails of the financial return series. The degree of freedoms of Student-t distribution is set to be 5, which is enough to show the fat-tail feature of the return series. The NIG parameters are empirically estimated based on the VIX-filtered S&P series with $\alpha = 1.34$, $\beta = -0.015$, $\delta = 1.337$ and $\mu = 0.01$. For each type of distribution, we generate 200 processes with 1,500 observations (whole sample) or 930 observations (pre-financial crisis) respectively.

For each scenario, the first 500 observations are used as training set. Note that a rolling window with a window size of 500 observations is used in the dynamic estimation. The 1 day ahead forecasts of volatility based on the GARCH(1,1) set up are reasonable, see Fig. 12.3.

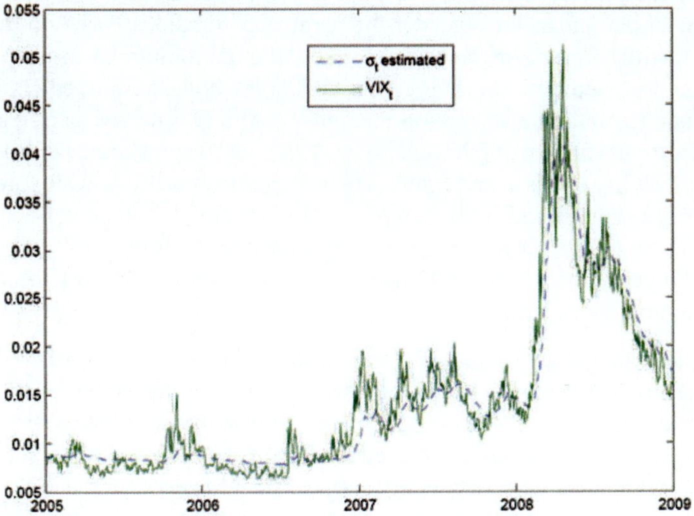

Fig. 12.3 VIX and the average value of 1 day ahead volatility forecasts

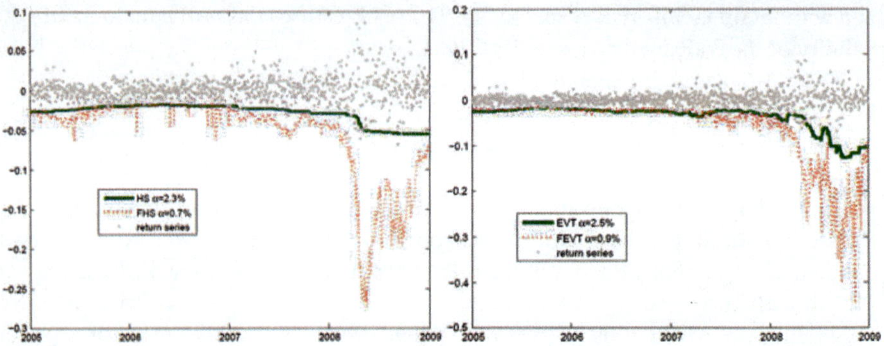

Fig. 12.4 The 1-day ahead VaR plots using HS vs FHS (*left*) and EVT vs filtered EVT (*right*)

At the beginning, we show the performance of the historical simulation (HS) and the filtered HS based on the whole time period, see Fig. 12.4. The VaR forecasts at risk level $\alpha = 1\%$ and $h = 1$ day are reported. It shows that the overall performance of the filtered HS is better than HS, where HS obviously underestimates the risk with many exceedances. In fact, the empirical failure rate – the probability of exceedances' occurrence – is larger than 2% and all the three tests reject the validation of the HS method in backtesting. The similar results are observed for EVT without filtration (right panel). As mentioned before, the poor performance of the direct implementation of HS and EVT on the return series is due to the heteroscedasticity of financial data. In the following, we will justify the VaR methods accounting for heteroscedasticity.

After filtering the generated return series by using the estimated volatility in Fig. 12.3, various VaR methods are used to calculate the quantile position of the residuals. In particular, we consider the parametric methods based on the normal, Student-t with degrees of freedom 5 and NIG distributional assumptions, the nonparametric method – the filtered HS (FHS) method, the filtered EVT (FEVT) method and the quantile regression method – CAViaR. The one-period and multi-period VaR with forecasting horizon $h = 1$, 5 and 10 steps are calculated for two target probability levels 1 and 0.5%. The average values of the VaR forecasts and backtesting results based on the conditional coverage test (Christoffersen, 1998), the dynamic quantile (DQ) test (Engle and Manganelli, 2004) and Ljung-Box (LB) test (Berkowitz and O'Brien, 2002) are reported in Tables 12.1 and 12.2. The results can be summarized as follows:

1. The performance of some VaR estimation methods is very sensitive to market conditions. For example, under the normal market condition in the pre-financial crisis period up to 30/04/2007, the VaR estimation based on the *normal distributional assumption* yields quite accurate VaR forecasts over both short interval ($h = 1$) and relatively long interval ($h = 10$). However, if the financial crisis period is included, the normality based estimation has a low accuracy. On the other hand, the quantile regression method – *CAViaR* – outperforms many

Table 12.1 Backtesting results – the conditional coverage test, Ljung-Box test and the dynamic quantile test – applied to the simulated data with three distributional types before and over the global financial crisis 2007–2009. The forecasting horizon of VaRs at level $\alpha = 1.0\%$ is respectively $h = 1, 5$ and 10. The VaR estimation methods are ordered according to the average value of accuracy measured by $|\hat{\alpha} - \alpha|$, where $\hat{\alpha}$ is the empirical failure rate. The significant testing statistics (5%) are labeled by $*$

	$h = 1$, Pre-FC: 11/08/2005 - 30/04/2007					$h = 1$, Whole sample 11/08/2005 - 31/07/2009				
ε_t	Method	$\hat{\alpha}$%	C's test	LB	DQ	Method	$\hat{\alpha}$%	C's test	LB	DQ
N	N(0,1)	1.1	1.299	7.895	6.495	FEVT	0.9	2.337	10.622	*13.520
	FEVT	1.1	1.946	9.926	8.467	NIG	0.7	2.117	7.082	9.480
	FHS	0.8	1.078	5.424	6.035	FHS	0.7	2.012	7.396	*11.160
	NIG	0.4	1.786	3.097	3.802	CAViaR	0.6	3.534	7.434	9.767
	CAViaR	0.3	1.520	4.778	6.256	N(0,1)	1.6	4.952	*13.334	6.424
	t(5)	0.1	1.115	1.109	1.433	t(5)	0.2	7.595	6.228	*14.185
t(5)	NIG	0.9	1.187	4.144	3.463	CAViaR	1.0	1.978	6.734	8.031
	FHS	0.8	1.027	4.794	5.419	FHS	0.9	1.251	5.923	7.760
	CAViaR	0.6	1.576	4.008	5.703	NIG	1.4	2.984	9.324	5.600
	N(0,1)	1.5	2.428	8.607	6.221	t(5)	0.6	3.075	4.431	6.748
	FEVT	1.5	3.508	10.405	5.653	FEVT	1.6	7.208	*16.336	8.944
	t(5)	0.4	1.697	2.227	2.538	N(0,1)	0.2	*12.946	*22.153	6.789
NIG	FHS	0.9	0.976	3.841	3.817	CAViaR	1.0	1.855	6.096	6.567
	FEVT	1.2	2.718	9.343	5.489	NIG	1.2	1.740	8.506	7.526
	NIG	0.8	0.997	4.611	4.929	FHS	0.8	1.235	5.789	8.204
	CAViaR	0.6	1.414	3.860	3.819	FEVT	1.2	3.651	*12.320	10.220
	N(0,1)	1.6	2.919	9.717	5.193	t(5)	0.5	3.510	4.582	6.469
	t(5)	0.3	1.683	1.983	1.572	N(0,1)	2.1	*14.014	*23.495	6.255
	$h = 5$, Pre-FC: 11/08/2005 - 30/04/2007					$h = 5$, Whole sample 11/08/2005 - 31/07/2009				
ε_t	Method	$\hat{\alpha}$%	C's test	LB	DQ	Method	$\hat{\alpha}$%	C's test	LB	DQ
N	N(0,1)	1.0	1.593	10.710	9.890	NIG	1.0	2.290	*11.777	*12.884
	FEVT	1.0	2.586	*12.063	11.04	FEVT	1.0	3.669	*17.638	*21.629
	NIG	1.1	1.785	*12.077	*11.713	FHS	0.7	2.602	*12.629	*19.063
	FHS	0.6	1.437	7.042	7.036	CAViaR	0.4	5.988	*11.114	*26.424
	CAViaR	0.2	1.436	2.719	3.149	t(5)	0.3	6.722	10.240	*21.946
	t(5)	0.1	0.937	0.964	0.839	N(0,1)	1.9	*14.820	*32.983	*13.564
t(5)	NIG	0.8	1.490	6.409	6.567	CAViaR	0.9	2.023	9.188	*12.488
	CAViaR	0.8	1.445	5.148	5.350	FHS	0.9	2.195	*11.132	*14.755
	N(0,1)	1.3	1.857	8.851	7.213	NIG	1.4	5.142	*15.064	9.358
	FEVT	1.3	3.367	10.73	6.909	t(5)	0.7	2.832	*7.935	10.807
	FHS	0.7	1.368	5.888	8.103	FEVT	1.6	10.39	*24.290	*13.860
	t(5)	0.3	1.797	3.000	3.359	N(0,1)	2.2	*21.199	*37.144	*12.821
NIG	FEVT	1.2	3.228	11.03	7.420	FHS	0.9	2.279	*12.633	*15.598
	FHS	0.8	1.432	6.223	7.134	CAViaR	0.7	2.991	10.121	*14.258
	NIG	0.7	1.334	6.372	7.283	t(5)	0.7	3.338	10.596	*16.221
	N(0,1)	1.4	2.989	*12.031	7.641	FEVT	1.4	7.207	*21.003	*15.708
	CAViaR	0.5	1.728	4.530	5.125	NIG	1.6	8.646	*23.816	*12.575
	t(5)	0.3	1.692	3.641	4.459	N(0,1)	2.4	*26.726	*43.899	*12.122
	$h = 10$, Pre-FC: 11/08/2005 - 30/04/2007					$h = 10$, Whole sample 11/08/2005 - 31/07/2009				
ε_t	Method	$\hat{\alpha}$%	C's test	LB	DQ	Method	$\hat{\alpha}$%	C's test	LB	DQ
N	FEVT	0.8	2.076	10.00	*11.555	FEVT	1.1	4.543	*20.409	*23.763
	N(0,1)	0.8	1.469	8.302	8.667	FHS	0.8	3.022	*15.685	*25.028
	FHS	0.5	1.565	5.201	5.863	t(5)	0.5	5.207	*14.398	*29.401
	NIG	0.4	1.964	4.710	4.735	CAViaR	0.5	5.450	*12.910	*29.478
	CAViaR	0.2	1.439	2.721	3.149	NIG	1.8	*14.482	*42.473	*25.996
	t(5)	0.1	0.817	1.029	1.286	N(0,1)	2.4	*31.890	*60.810	*25.380
t(5)	N(0,1)	1.2	1.934	8.190	6.819	CAViaR	1.0	2.457	*11.750	*14.560
	FEVT	1.2	3.361	10.17	6.458	FHS	0.9	2.645	*13.280	*18.650
	CAViaR	0.8	1.428	5.168	5.419	t(5)	0.8	2.941	10.744	*14.670
	FHS	0.6	1.361	5.337	7.188	NIG	1.4	6.881	*24.930	19.628
	t(5)	0.3	1.732	2.841	3.836	FEVT	1.7	*13.160	*30.720	*19.310
	NIG	0.3	1.893	2.287	2.410	N(0,1)	2.4	*31.430	*53.390	*19.700
NIG	FEVT	1.0	2.473	9.158	7.746	FHS	1.0	2.619	*15.510	*18.630
	FHS	0.7	1.442	6.299	7.223	t(5)	0.9	2.619	*13.890	*18.450
	N(0,1)	1.3	2.604	10.46	7.350	CAViaR	0.8	2.600	*11.500	*19.219
	NIG	1.5	3.186	*12.758	7.762	FEVT	1.5	9.373	*27.487	*19.122
	CAViaR	0.5	1.738	4.594	5.175	NIG	1.6	8.769	*21.379	*12.303
	t(5)	0.3	1.702	3.356	4.156	N(0,1)	2.8	*47.559	*72.862	*20.917

others in most cases, if the financial crisis period is included in the estimation. Especially, CAViaR yields accurate VaR values for the heavy-tailed (realistic) t(5) and NIG distributed data.

2. Some methods are robust to both normal and "bad" market conditions. For example, the *FHS* method and the *NIG based method* in general deliver reasonable VaR estimation. The observation is consistent to the popularity of the FHS method in industry and also explains why the NIG based method attract much attention of researchers, although the NIG based method does not perform very well for the long term ($h = 10$) prediction.

Table 12.2 Backtesting results – the conditional coverage test, LB test and the dynamic quantile test – applied to the simulated data with three distributional types before and over the global financial crisis 2007–2009. The forecasting horizon of VaRs at level $\alpha = 0.5\%$ is respectively $h = 1$, 5 and 10. The VaR estimation methods are ordered according to the average value of accuracy measured by $|\hat{\alpha} - \alpha|$, where $\hat{\alpha}$ is the empirical failure rate. The significant testing statistics (5%) are labeled by $*$

	$h = 1$, Pre-FC: 11/08/2005 - 30/04/2007					$h = 1$, Whole sample 11/08/2005 - 31/07/2009				
ε_t	Method	$\hat{\alpha}$%	C's test	LB	DQ	Method	$\hat{\alpha}$%	C's test	LB	DQ
N	NIG	0.4	0.932	6.423	5.929	N(0,1)	0.6	1.065	9.906	5.789
	FHS	0.3	1.742	8.027	*12.690	FHS	0.3	0.490	4.963	4.137
	CAViaR	0.2	2.071	6.834	8.575	CAViaR	0.1	0.242	2.456	2.094
	t(5)	0.1	1.878	1.649	1.310	FEVT	0.9	6.358	*24.820	*13.730
	N(0,1)	1.0	5.085	*17.920	6.796	NIG	0.0	0.097	0.166	0.001
	FEVT	1.1	4.616	*22.742	8.514	t(5)	0.0	0.062	0.098	0.000
t(5)	NIG	0.4	0.503	2.325	1.236	CAViaR	0.5	0.964	4.799	6.269
	FHS	0.3	0.392	3.098	3.424	FHS	0.4	1.133	6.106	10.160
	CAViaR	0.3	0.376	2.494	1.816	t(5)	0.3	1.467	2.444	2.681
	t(5)	0.2	0.323	0.566	0.552	NIG	0.9	4.662	*14.180	5.733
	N(0,1)	1.1	3.493	*14.650	6.175	N(0,1)	1.5	*16.810	*35.530	6.301
	FEVT	1.5	9.140	*26.931	5.773	FEVT	1.6	*24.095	*48.784	*9.677
NIG	FHS	0.4	4.207	*16.240	5.184	CAViaR	0.5	1.443	3.890	4.933
	CAViaR	0.3	0.321	2.346	2.204	FHS	0.4	0.875	6.236	8.500
	NIG	0.1	0.268	0.496	0.009	NIG	0.8	2.428	10.500	6.575
	t(5)	0.1	0.222	1.096	0.546	t(5)	0.2	1.832	3.293	1.978
	N(0,1)	1.1	3.541	*13.908	4.520	FEVT	1.2	*13.110	*33.620	10.630
	FEVT	1.2	6.764	*22.823	5.497	N(0,1)	1.5	*16.774	*36.761	6.520
	$h = 5$, Pre-FC: 11/08/2005 - 30/04/2007					$h = 5$, Whole sample 11/08/2005 - 31/07/2009				
ε_t	Method	$\hat{\alpha}$%	C's test	LB	DQ	Method	$\hat{\alpha}$%	C's test	LB	DQ
N	N(0,1)	0.5	1.034	*11.940	6.964	NIG	0.5	1.955	*16.240	*14.040
	FHS	0.3	0.527	4.652	2.835	FHS	0.3	1.564	*14.830	*19.000
	NIG	0.3	0.527	4.652	2.835	CAViaR	0.2	2.346	5.926	10.274
	CAViaR	0.1	0.142	0.784	0.542	t(5)	0.1	2.369	6.136	9.355
	FEVT	1.0	5.034	*25.690	10.94	FEVT	1.0	9.518	*38.510	*21.930
	t(5)	0.0	0.054	0.055	0.000	N(0,1)	1.3	*13.490	*45.220	*13.830
t(5)	CAViaR	0.4	0.621	3.878	1.928	CAViaR	0.5	1.442	10.14	10.15
	NIG	0.6	1.214	10.63	7.082	FHS	0.4	1.386	*12.150	*13.690
	FHS	0.3	0.386	2.920	3.047	t(5)	0.4	1.449	7.578	9.018
	t(5)	0.2	0.311	1.111	1.089	NIG	0.8	3.739	*16.760	9.321
	N(0,1)	0.9	2.828	*14.970	6.817	FEVT	1.6	*30.270	*65.330	*14.980
	FEVT	1.3	8.392	*26.260	6.981	N(0,1)	1.6	*26.740	*58.880	*11.280
NIG	NIG	0.5	1.022	9.684	5.747	FHS	0.4	1.609	*13.400	*16.190
	FHS	0.3	0.471	5.378	4.946	CAViaR	0.4	1.921	*11.760	*14.820
	CAViaR	0.2	0.472	3.262	2.200	t(5)	0.3	1.810	9.756	9.211
	t(5)	0.1	0.224	1.181	0.5447	NIG	0.9	5.561	*27.100	*14.290
	N(0,1)	1.0	3.408	*15.790	6.257	FEVT	1.4	*21.410	*54.270	*16.480
	FEVT	1.2	7.038	*25.750	7.395	N(0,1)	1.7	*30.320	*69.580	*12.800
	$h = 10$, Pre-FC: 11/08/2005 - 30/04/2007					$h = 10$, Whole sample 11/08/2005 - 31/07/2009				
ε_t	Method	$\hat{\alpha}$%	C's test	LB	DQ	Method	$\hat{\alpha}$%	C's test	LB	DQ
N	N(0,1)	0.4	0.861	7.989	4.476	FHS	0.4	1.847	*17.560	*25.920
	FHS	0.2	0.475	4.549	3.569	t(5)	0.5	2.073	*17.630	*19.170
	FEVT	0.8	3.269	*20.940	*11.610	NIG	0.7	3.958	*28.410	*23.580
	NIG	0.1	0.172	1.065	1.016	t(5)	0.2	2.719	*12.690	*18.630
	CAViaR	0.1	0.142	0.783	0.5419	FEVT	1.1	*12.130	*47.150	*24.220
	t(5)	0.0	0.0425	0.546	0.540	N(0,1)	1.7	*31.340	*86.250	*25.700
t(5)	CAViaR	0.4	0.663	4.008	2.043	CAViaR	0.5	1.843	*13.140	*14.420
	NIG	0.4	0.631	5.972	5.375	FHS	0.4	1.618	*14.620	*18.340
	FHS	0.3	0.365	3.041	2.870	t(5)	0.4	1.689	9.480	9.835
	N(0,1)	0.8	2.800	*12.910	5.518	NIG	1.1	10.57	*38.050	*17.050
	t(5)	0.2	0.282	1.416	1.325	FEVT	1.7	*35.860	*80.100	*20.920
	FEVT	1.2	7.724	*24.300	6.428	N(0,1)	1.8	*38.500	*85.510	*19.190
NIG	FHS	0.3	0.456	5.341	4.706	FHS	0.5	1.611	*16.360	*18.520
	NIG	0.3	0.564	6.073	4.712	t(5)	0.5	2.073	*17.630	*19.160
	CAViaR	0.2	0.4704	3.254	2.202	CAViaR	0.4	1.792	*11.960	*15.350
	N(0,1)	0.9	2.810	*15.120	6.804	NIG	1.2	*14.950	*48.950	*21.220
	t(5)	0.1	0.220	0.368	0.005	FEVT	1.5	*26.650	*69.920	*20.540
	FEVT	1.0	5.305	*20.830	7.522	N(0,1)	2.1	*52.720	*108.700	*20.610

3. The exceedance clustering becomes serious if the forecasting horizon is getting long or if the financial crisis happens. Among these methods, *FEVT* is very easy to generate the clustering phenomenon, supported by the rejection of LB or DQ test. It is possibly a limitation of FEVT, although FEVT can pass the conditional coverage test and furthermore provide accurate VaR forecasts at $\alpha = 1\%$ level in most cases.
4. The Student-t based estimation method is not attractive in terms of accuracy. However we do observe that the performance of the t(5) method is getting improved as the financial crisis happens and the long forecasting interval is considered in the VaR estimation.

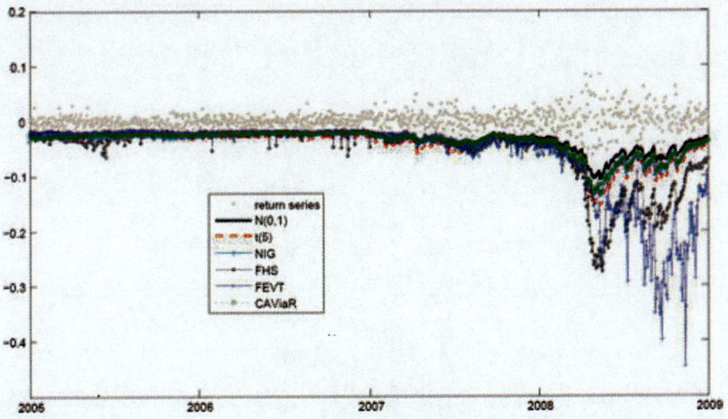

Fig. 12.5 The VaR plot based on different VaR estimation methods. The financial crisis is in 2007–2009

Figure 12.5 shows the time plot of 1-day ahead VaR forecasts at 1% level based on one generated process. It demonstrates that VaR estimation may yield quite different VaR values by using various methods, especially during the financial crisis period. It is observed that the VaR estimation based on the NIG assumption delivers small and accurate VaRs that follow the fluctuations of return series closely, especially when market is volatile. It motivates us to use NIG distributional assumption in the ICA-VaR estimation.

12.5 Empirical Analysis

For the real data analysis, we consider three stock indices, the Dow Jones (DJ) Industry Average 30, the German DAX index and the Singapore Straight Time Index (STI) from 18/08/2003 to 31/07/2009 (each with 1,500 observations). The log returns of these indices are displayed in Fig. 12.6. For these 3 indices, the VaRs can be estimated either by using the reduced models (univariate methods) or by using the multivariate methods based on the returns of the component stocks. As an illustration, we apply the multivariate methods to DJ30 and investigate the accuracy of the multivariate methods to reduced models. In particular, the multivariate VaR estimation methods – by using DCC, ICA and Copula are respectively considered. For a fair comparison, the actual weights used for composing the indices are assigned to the multivariate stocks.

The $h = 1, 5$ and 10-day ahead VaR forecasts are calculated given two risk levels $\alpha = 1\%$ and $\alpha = 0.5\%$. The volatility of the univariate variables is estimated based on the GARCH(1,1) model. The rolling window approach with 500-day window size is again used to adapt the estimation for each time point. Figure 12.7 shows that VaR plot by using different estimation methods. The plots against DAX and STI are similar and omitted.

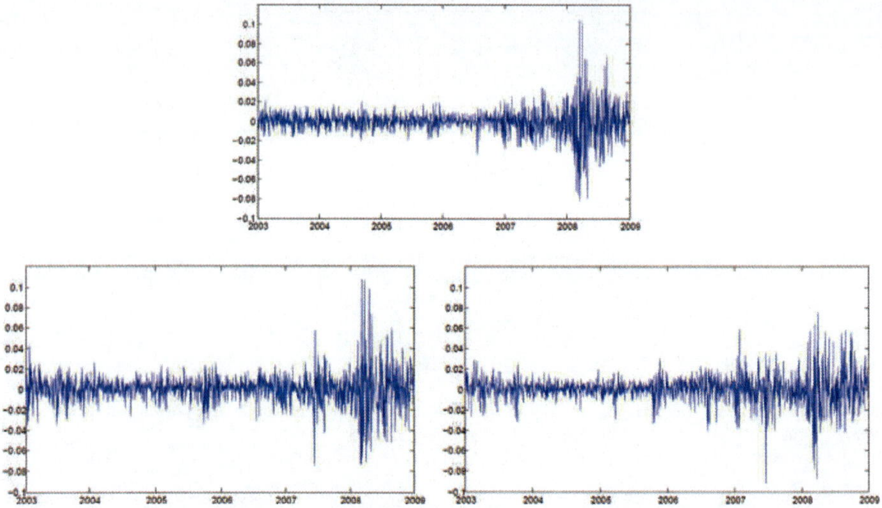

Fig. 12.6 Dow Jones Industry average 30 log return series from 18/08/2003 to 31/07/2009 (*upper*). DAX index log return series from 12/09/2003 to 31/07/2009 (*bottom left*). Singapore STI index log return series from 7/08/2003 to 31/07/2009 (*bottom right*)

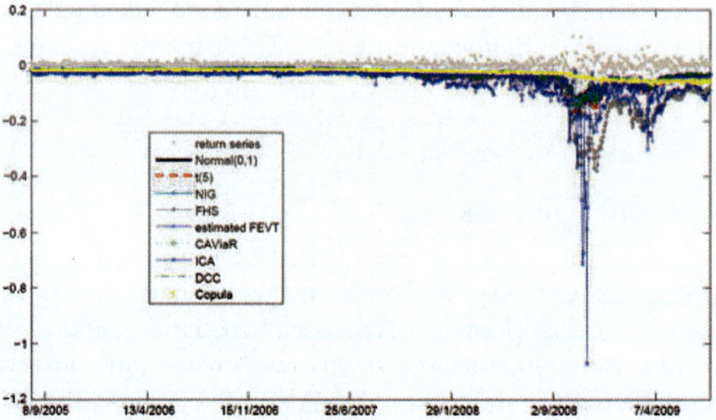

Fig. 12.7 The VaR plot against the Dow Jones Index

The backtesting results are reported in Tables 12.3 and 12.4, where the methods are ordered according to the estimation accuracy $|\hat{\alpha} - \alpha|$ with $\hat{\alpha}$ denoting the empirical failure rate. The observations are summarized as follows:

1. In general, *CAViaR* and the *NIG-based* estimation are robust and deliver accurate VaR estimation, if the short forecasting interval is considered, i.e. $h = 1$. However, if the forecasting horizon is relatively long, e.g. $h = 5$, these estimation are sensitive to market condition, especially the NIG-based method. To be more specifically, NIG works well if the market is normal, i.e. without financial crisis,

Table 12.3 Backtesting results – the conditional coverage test, Ljung-Box test and the dynamic quantile test – applied to the three indices: DJ30, DAX and STI. The forecasting horizon of VaRs at level $\alpha = 1.0\%$ is respectively $h = 1, 5$ and 10. The univariate (and multivariate for DJ30) VaR estimation methods are ordered according to the average value of accuracy measured by $|\hat{\alpha} - \alpha|$, where $\hat{\alpha}$ is the empirical failure rate. The significant testing statistics (5%) are labeled by ∗

	$h = 1$, Pre-FC: crisis: 11/08/2005 - 30/04/2007					$h = 1$, Whole sample 11/08/2005 - 31/07/2009				
ε_t	Method	$\hat{\alpha}$%	C's test	LB	DQ	Method	$\hat{\alpha}$%	C's test	LB	DQ
DJ30	NIG	0.9	0.022	0.275	0.232	CAViaR	1.0	0.000	1.157	0.620
	FEVT	0.9	0.022	1.637	0.232	NIG	1.3	0.831	8.237	5.086
	FHS	0.7	0.444	1.738	0.088	ICA	1.4	4.629	*13.440	*11.130
	CAViaR	0.7	0.444	1.894	0.088	FHS	0.6	1.886	2.867	0.070
	Copula	1.4	0.605	*15.350	10.950	FEVT	0.6	1.886	3.657	0.070
	DCC	0.5	1.551	1.622	0.004	t(5)	0.3	6.826	5.214	1.026
	ICA	0.5	1.551	2.648	0.005	DCC	2.1	9.284	*22.030	3.137
	N(0,1)	1.6	1.439	*15.073	7.881	N(0,1)	2.4	*14.220	*49.970	14.54
	t(5)	0.2	3.708	3.470	1.281	Copula	4.0	*104.780	*138.800	*54.030
DAX	FHS	0.9	0.022	*27.170	*25.870	NIG	1.2	0.380	8.076	5.985
	NIG	1.2	0.109	*20.270	*16.120	FEVT	0.7	1.016	2.221	0.203
	CAViaR	0.7	0.444	*34.530	*47.290	CAViaR	0.7	1.016	*14.390	*19.280
	FEVT	0.7	0.044	5.864	0.088	FHS	0.7	1.016	*15.470	*19.280
	t(5)	0.5	1.551	1.578	0.005	t(5)	0.4	4.706	3.690	0.144
	N(0,1)	1.6	1.439	*14.724	7.873	N(0,1)	1.9	6.473	*16.690	5.136
STI	CAViaR	0.7	0.444	0.601	0.088	NIG	1.1	0.098	8.670	*13.640
	FHS	0.7	0.444	1.116	*47.490	FEVT	0.8	0.434	6.102	*28.440
	t(5)	0.7	0.444	0.593	0.088	CAViaR	0.5	3.094	3.255	0.000
	NIG	1.4	0.605	1.674	10.970	t(5)	0.5	3.094	3.188	0.000
	FEVT	1.6	1.440	4.261	15.300	FHS	0.4	4.706	4.284	*64.890
	N(0,1)	2.1	3.947	*60.180	*31.820	N(0,1)	2.2	*22.140	*33.030	*14.860
	$h = 5$, Pre-FC: crisis: 11/08/2005 - 30/04/2007					$h = 5$, Whole sample 11/08/2005 - 31/07/2009				
ε_t	Method	$\hat{\alpha}$%	C's test	LB	DQ	Method	$\hat{\alpha}$%	C's test	LB	DQ
DJ30	NIG	0.9	0.022	1.574	0.232	FHS	1.0	0.000	*39.240	*37.400
	FEVT	0.9	0.022	1.355	0.232	ICA	1.1	0.106	7.968	*13.580
	FHS	1.2	0.109	*19.680	*16.150	FEVT	0.8	0.418	4.090	0.352
	Copula	1.4	0.605	*15.350	10.950	t(5)	0.7	0.992	*15.130	19.210
	N(0,1)	1.6	1.439	*13.820	7.888	CAViaR	0.7	0.992	*58.750	*118.000
	CAViaR	0.2	3.708	2.533	1.281	DCC	0.5	3.044	2.514	0.000
	ICA	0.2	3.708	2.530	1.281	NIG	1.8	5.290	*25.900	*12.230
	DCC	0.2	3.708	2.663	1.281	N(0,1)	3.3	*67.130	*107.500	*23.250
	t(5)	0.2	3.708	3.603	1.281	Copula	4.0	*104.780	*138.800	*54.030
DAX	NIG	1.2	0.109	*18.400	*16.120	CAViaR	1.0	0.000	*39.240	*37.400
	FHS	0.7	0.444	*33.760	*47.290	FEVT	1.1	0.106	*37.430	*43.300
	FEVT	0.7	0.444	2.780	0.088	FHS	0.8	0.418	*33.210	42.210
	CAViaR	0.7	0.444	0.804	0.088	t(5)	0.8	0.418	*110.400	*151.300
	t(5)	0.5	1.551	1.851	0.005	NIG	1.7	4.418	*54.420	*33.690
	N(0,1)	1.6	1.439	*54.500	*40.880	N(0,1)	2.5	*16.170	*100.200	*38.290
STI	FEVT	1.2	0.109	*40.070	*63.540	FEVT	0.7	0.992	*77.390	*136.500
	NIG	1.4	0.605	*19.360	*21.850	t(5)	0.5	3.054	*22.770	*39.960
	t(5)	0.5	1.551	1.232	0.005	FHS	0.5	3.054	*23.020	*80.010
	FHS	0.5	1.551	1.775	*109.500	NIG	1.6	3.125	*47.100	*30.720
	N(0,1)	1.6	1.439	*81.060	*48.840	CAViaR	1.6	7.585	*44.940	*35.220
	CAViaR	1.6	1.439	*16.200	*15.520	N(0,1)	2.0	7.901	*117.500	*57.050
	$h = 10$, Pre-FC: crisis: 11/08/2005 - 30/04/2007					$h = 10$, Whole sample 11/08/2005 - 31/07/2009				
ε_t	Method	$\hat{\alpha}$%	C's test	LB	DQ	Method	$\hat{\alpha}$%	C's test	LB	DQ
DJ30	Copula	1.4	0.605	*15.350	10.950	CAViaR	0.8	0.398	*51.012	*133.030
	FHS	1.4	0.605	*15.717	10.951	ICA	1.3	0.886	1.987	5.069
	NIG	1.6	1.439	*13.921	7.888	DCC	0.5	2.994	2.477	0.000
	FEVT	1.6	1.439	*30.799	*15.241	FHS	1.6	3.187	*57.505	*36.363
	ICA	0.2	3.708	2.537	1.281	FEVT	1.8	*11.723	*69.152	*40.248
	DCC	0.2	3.708	2.663	1.281	t(5)	1.8	5.373	*53.229	*27.870
	CAViaR	0.2	3.708	2.543	1.281	NIG	2.7	*20.243	*79.064	*22.761
	t(5)	0.2	3.708	3.950	1.281	Copula	4.0	*104.780	*138.800	*54.030
	N(0,1)	1.9	2.566	*13.330	5.949	N(0,1)	5.3	*181.790	*259.780	*43.490
DAX	CAViaR	0.7	0.444	1.982	0.088	FEVT	0.9	3.559	*75.946	*170.100
	N(0,1)	1.6	1.439	*54.240	*66.672	CAViaR	1.2	3.132	*116.290	*189.420
	FHS	0.2	3.708	3.313	1.381	NIG	0.7	0.962	*145.620	236.720
	t(5)	0.0	0.000	0.000	0.000	FHS	0.5	3.004	*34.860	*79.519
	NIG	0.0	0.000	0.000	0.000	t(5)	0.4	4.598	*54.605	*128.140
	FEVT	0.0	0.000	0.000	0.000	N(0,1)	2.8	*44.752	*95.923	*40.001
STI	t(5)	0.9	0.022	1.983	*25.975	FHS	1.1	0.117	*16.524	*49.541
	FHS	1.6	0.605	5.050	*21.524	FEVT	1.2	0.417	*67.513	*84.053
	FEVT	1.6	1.439	*72.387	*55.606	t(5)	1.4	1.511	*77.803	*83.141
	CAViaR	1.6	1.439	*14.391	*15.240	CAViaR	1.7	9.583	*53.977	*64.059
	NIG	2.6	*15.987	*47.235	*24.985	NIG	3.6	*84.635	*181.690	*82.434
	N(0,1)	2.8	*19.732	*46.836	*20.272	N(0,1)	4.2	*118.750	*222.910	*85.107

whereas the method provides low accurate VaR values if the financial crisis happens.

2. The widely used methods, *FHS* and *FEVT*, on the other hand, display robust and good performance in terms predictability over e.g. 1 week $h = 5$, if the risk level is $\alpha = 1.0\%$. However it is not necessary applied to extreme risks e.g. $\alpha = 0.5\%$.
3. The other two univariate estimation methods, based on normal and Student-t distributional assumption respectively, are often out-performed.

Table 12.4 Backtesting results – the conditional coverage test, Ljung-Box test and the dynamic quantile test – applied to the three indices: DJ30, DAX and STI. The forecasting horizon of VaRs at level $\alpha = 0.5\%$ is respectively $h = 1, 5$ and 10. The univariate (and multivariate for DJ30) VaR estimation methods are ordered according to the average value of accuracy measured by $|\hat{\alpha} - \alpha|$, where $\hat{\alpha}$ is the empirical failure rate. The significant testing statistics (5%) are labeled by $*$

	$h = 1$, Pre-FC: crisis: 11/08/2005 - 30/04/2007					$h = 1$, Whole sample 11/08/2005 - 31/07/2009				
ε_t	Method	$\hat{\alpha}$%	C's test	LB	DQ	Method	$\hat{\alpha}$%	C's test	LB	DQ
DJ30	NIG	0.5	0.011	2.570	0.058	ICA	0.9	8.584	*23.660	*22.230
	CAViaR	0.5	0.011	2.920	0.058	FEVT	0.6	0.189	4.325	0.209
	DCC	0.5	0.011	0.970	0.058	NIG	0.6	0.189	0.913	0.209
	Copula	1.1	1.843	0.058	0.110	FHS	0.4	0.216	1.298	0.087
	FEVT	0.7	0.301	4.994	0.110	CAViaR	0.3	0.939	1.682	0.017
	FHS	0.7	0.301	1.394	0.110	t(5)	0.2	2.344	2.200	0.036
	ICA	0.2	0.772	0.656	0.001	N(0,1)	1.4	10.910	*30.820	4.431
	t(5)	0.2	0.772	2.492	0.011	DCC	1.4	10.910	*33.630	4.431
	N(0,1)	1.2	2.759	4.472	0.164	Copula	2.4	*77.660	*152.200	45.680
DAX	NIG	0.5	0.011	0.679	0.058	NIG	0.6	0.189	0.419	0.209
	CAViaR	0.5	0.011	0.646	0.058	FHS	0.4	0.216	*50.180	*61.470
	FHS	0.7	0.301	*68.060	*47.180	CAViaR	0.3	0.939	1.381	0.017
	FEVT	0.7	0.301	*11.270	0.110	FEVT	0.7	0.715	3.475	0.254
	t(5)	0.2	0.772	1.457	0.001	t(5)	0.2	2.059	0.036	0.144
	N(0,1)	1.4	4.650	*42.622	*11.253	N(0,1)	1.3	8.908	*27.942	5.175
STI	CAViaR	0.5	0.011	0.215	0.058	NIG	0.5	0.000	1.374	0.153
	FHS	0.5	0.011	1.430	0.058	CAViaR	0.4	0.216	1.361	0.087
	NIG	0.7	0.301	0.778	0.110	FHS	0.3	0.939	1.556	0.017
	t(5)	0.2	0.772	0.771	0.001	FEVT	0.8	1.529	*13.210	*28.693
	N(0,1)	1.4	4.650	8.980	*11.300	t(5)	0.2	2.344	1.975	0.036
	FEVT	1.6	6.881	*16.131	*16.201	N(0,1)	1.3	8.908	*38.644	*14.727
	$h = 5$, Pre-FC: crisis: 11/08/2005 - 30/04/2007					$h = 5$, Whole sample 11/08/2005 - 31/07/2009				
ε_t	Method	$\hat{\alpha}$%	C's test	LB	DQ	Method	$\hat{\alpha}$%	C's test	LB	DQ
DJ30	NIG	0.5	0.011	2.516	0.058	FHS	0.5	0.000	1.802	0.154
	FHS	0.7	0.301	1.990	0.110	DCC	0.4	0.206	0.305	0.088
	t(5)	0.2	0.772	2.758	0.001	ICA	0.7	0.731	1.159	0.254
	CAViaR	0.2	0.772	0.942	0.001	FEVT	0.8	1.553	9.269	0.290
	ICA	0.2	0.772	0.692	0.001	CAViaR	0.2	2.320	1.803	0.035
	DCC	0.2	0.772	1.385	0.001	t(5)	0.2	2.320	2.188	0.034
	FEVT	0.9	1.275	4.316	0.143	NIG	0.9	2.629	4.492	0.320
	Copula	1.1	1.843	0.058	0.110	N(0,1)	2.3	*34.670	*127.200	*16.600
	N(0,1)	1.2	2.759	4.472	0.164	Copula	2.4	*77.660	*152.200	*45.680
DAX	FHS	0.5	0.011	*100.510	*107.800	t(5)	0.5	0.000	*64.300	*76.970
	FEVT	0.7	0.301	5.142	0.110	FHS	0.4	0.208	*104.500	*122.300
	CAViaR	0.7	0.301	6.672	0.110	CAViaR	0.7	0.731	*54.690	37.920
	t(5)	0.2	0.772	0.962	0.001	NIG	1.0	3.929	*176.100	*95.780
	NIG	0.2	0.772	0.882	0.001	FEVT	1.1	5.431	*81.640	*45.050
	N(0,1)	1.2	2.759	*40.180	*16.450	N(0,1)	1.9	*23.040	*151.300	*35.250
STI	NIG	0.5	0.011	0.428	0.058	t(5)	0.4	0.208	*52.650	*61.150
	t(5)	0.2	0.772	1.283	0.001	NIG	0.7	0.731	*55.260	*37.890
	FEVT	1.2	2.759	*83.360	*65.460	FEVT	0.7	0.731	*153.100	*135.800
	CAViaR	1.2	2.759	*52.910	*16.590	FHS	0.1	4.765	3.192	1.280
	N(0,1)	1.4	4.650	*44.500	*22.590	CAViaR	1.2	7.117	*81.590	*31.660
	FHS	0.0	0.000	0.000	0.000	N(0,1)	1.7	*17.850	*159.800	*46.120
	$h = 10$, Pre-FC: crisis: 11/08/2005 - 30/04/2007					$h = 10$, Whole sample 11/08/2005 - 31/07/2009				
ε_t	Method	$\hat{\alpha}$%	C's test	LB	DQ	Method	$\hat{\alpha}$%	C's test	LB	DQ
DJ30	NIG	0.5	0.011	3.835	0.058	CAViaR	0.5	0.000	*38.440	*38.401
	FHS	0.7	0.301	0.597	0.110	DCC	0.4	0.196	0.298	0.089
	t(5)	0.2	0.772	3.448	0.001	ICA	0.7	0.751	1.172	*19.016
	CAViaR	0.2	0.772	1.125	0.001	FHS	0.8	1.584	*26.554	*14.338
	ICA	0.2	0.772	0.736	0.001	t(5)	1.0	3.980	*36.471	*17.522
	DCC	0.2	0.772	1.394	0.001	FEVT	1.8	*42.022	*158.92	*42.912
	Copula	1.1	1.843	0.058	0.110	NIG	2.3	*34.857	*202.590	*50.187
	N(0,1)	1.4	4.650	*36.251	*11.278	Copula	2.4	*77.660	*152.200	*45.680
	FEVT	1.6	6.881	*68.935	*16.127	N(0,1)	4.1	*102.520	*423.330	*43.199
DAX	NIG	0.5	0.011	0.791	0.058	CAViaR	0.6	0.208	*33.371	*26.203
	CAViaR	0.7	0.301	9.823	0.110	FHS	0.2	2.290	2.298	0.033
	N(0,1)	0.2	0.772	2.591	0.001	t(5)	0.1	4.725	3.178	1.253
	FHS	0.2	0.772	2.351	0.001	FEVT	0.9	8.724	*154.340	*172.910
	FEVT	0.0	0.000	0.000	0.000	NIG	1.0	3.980	*196.200	*112.49
	t(5)	0.0	0.000	0.000	0.000	N(0,1)	1.0	10.936	*211.230	*158.510
STI	t(5)	0.7	0.301	2.752	*47.398	t(5)	0.7	0.751	*46.907	*75.249
	CAViaR	1.4	4.650	*39.285	*22.371	FHS	0.2	2.290	1.761	0.033
	FEVT	1.6	6.881	*151.70	*58.121	FEVT	1.2	7.188	*143.710	*88.124
	NIG	1.9	9.404	*36.249	*18.206	CAViaR	1.2	*16.676	*91.455	*42.897
	N(0,1)	2.3	*31.926	*68.697	*18.476	NIG	2.2	*64.068	*151.670	*63.245
	FHS	0.0	0.000	0.000	0.000	N(0,1)	3.2	*134.860	*258.440	*70.950

4. As an illustration, we applied the multivariate VaR estimation methods to DJ30. The results show that the *ICA-VaR* method performs better than some univariate methods as well as the other two multivariate methods in most cases. This observation is consistent to the study of Chen et al. (2009), where the ICA-VaR method shows superior performance.

12.6 Conclusion

In this chapter we reviewed and implemented several popular or recently developed VaR estimation methods. The robustness and accuracy of the univariate and multivariate methods are investigated and demonstrated based on simulated data and real data. Backtesting is used as a standard tool to evaluate the performance. We tested both the unconditional and conditional coverage properties of all the models we covered using the Christofferson's test, the Ljung-Box test and the dynamic quantile test.

In the simulation study, we generated three types of processes with normal, Student-t and NIG distributional assumption, and applied the discussed univariate VaRestimation methods. Backtesting results show that FHS and the NIG-based method are robust to market conditions and deliver reasonable VaR estimation. The CAViaR model outperforms many others in most cases, however it is sensitive to market condition. In addition, FEVT is very easy to generate the clustering phenomenon, although it provides accurate VaR forecasts at $\alpha = 1\%$ level in most cases. Last but not least, the filtered HS and EVT methods overall outperform the non-filtered counter parties.

For empirical analysis, three composite indices DJ30, DAX and STI are used to illustrate the performance of the VaR estimation. In general, CAViaR and the NIG-based estimation are robust and deliver accurate VaR estimation, if the short forecasting interval is considered, i.e. $h = 1$. FHS and FEVT, on the other hand, display robust and good performance in terms predictability over e.g. 1 week $h = 5$, if the risk level is $\alpha = 1.0\%$. One multivariate estimation based on ICA performs better than many univariate methods in most cases.

References

Andersen, T., Bollerslev, T., Diebold, F., & Ebens, H. (2001). The distribution of realized stock return volatility. *Journal of Financial Economics*, *61*, 43–76.

Andersen, T., Bollerslev, T., Christoffersen, P., & Diebold, F. (2005). Volatility and correlation forecasting. In G. Elliott, C. Granger, and A. Timmermann (Eds.), *Handbook of economic forecasting*. Amsterdam: North-Holland.

Andersen, T. G., & Bollerslev, T. (1998). Answering the skeptics: Yes, standard volatility models do provide accurate forecasts. *International Economic Review*, *39*(4), 885–905.

Artzner, P., Delbaen, F., Eber, J., & Heath, D. (1999). Coherent measures of risk. *Mathematical finance*, *9*(3), 203–228.

Barndorff-Nielsen, O., & Shephard, N. (2001). Modelling by lévy processes for financial econometrics. In O. Barndorff-Nielsen, T. Mikosch, and S. Resnik (Eds.), *Lévy processes: Theory and applications*. Boston: Birkhauser.

Barndorff-Nielsen, O. E., & Shephard, N. (2002). Econometric analysis of realized volatility and its use in estimating stochastic volatility models. *Journal of the Royal Statistical Society B*, *64*, 253–280.

Beder, T. (1995). VaR: Seductive but dangerous. *Financial Analysts Journal*, *51*(5), 12–24.

Beirlant, J., Teugels, J., & Vynckier, P. (1996). *Practical analysis of extreme values*. Leuven: Leuven University Press.

Berkowitz, J., & O'Brien, J. (2002). How accurate are value-at-risk models at commercial banks? *Journal of Finance*, *57*(3), 1093–1111.

Berkowitz, J., Christoffersen, P., & Pelletier, D. (2006). Evaluating value-at-risk models with desk-level data. Working Paper Series 010, Department of Economics, North Carolina State University.

Bibby, B. M., & Sørensen, M. (2001). Hyperbolic processes in finance, Technical Report 88, University of Aarhus, Aarhus School of Business.

Bollerslev, T. (1986). Generalized autoregressive conditional heteroskedasticity. *Journal of Econometrics*, *31*, 307–327.

Bollerslev, T. (1995). Generalied autoregressive conditional heteroskedasticity. In R. Engle (Ed.), *ARCH, selected readings* (pp. 42–60). New York: Oxford University Press.

Chen, Y., & Spokoiny, V. (2009). Modeling and estimation for nonstationary time series with applications to robust risk management (submitted).

Chen, Y., Härdle, W., & Spokoiny, V. (2010). GHICA risk analysis with GH distributions and independent components. *Journal of Empirical Finance*, *17*(2), 255–269.

Christoffersen, P. F. (1998). Evaluating interval forecast. *International Economic Review*, *39*, 841–862.

Čížek, P., Härdle, W., & Spokoiny, V. (2009). Adaptive pointwise estimation in time-inhomogeneous conditional heteroscedasticity models, *Econom. J.*, *12*, 248–271.

Eberlein, E., & Keller, U. (1995). Hyperbolic distributions in finance. *Bernoulli*, *1*, 281–299.

Embrechts, P., & Dias, A. (2003). Dynamic copula models for multivariate high-frequency data in finance. Working paper.

Embrechts, P., Klüppelberg, C., & Mikosch, T. (1997). *Modelling extremal events for insurance and finance*. Berlin: Springer.

Embrechts, P., McNeil, A., & Straumann, D. (1999a). Correlation: Pitfalls and alternatives. *Risk*, *12*, 69–71.

Embrechts, P., Resnick, S., & Samorodnitsky, G. (1999b). Extreme value theory as a risk management tool. *North American Actuarial Journal*, *3*, 30–41.

Engle, R. (1982). Autoregressive conditional heteroscedasticity with estimates of the variance of uk inflation. *Econometrica*, *50*, 987–1008.

Engle, R., & Manganelli, S. (2004). CAViaR: Conditional autoregressive value at risk by regression quantiles. *Journal of Business and Economic Statistics*, *22*(4), 367–381.

Engle, R. F. (1995). Autoregressive conditional heteroscedasticity with estimates of the variance of united kingdom inflation. In *ARCH*. New York: Oxford University Press.

Engle, R. F. (2002). Dynamic conditional correlation: a simple class of multivariate garch models. *Journal of Business and Economic Statistics*, *20*, 339–350.

Franke, J., Härdle, W., & Hafner, C. (2008). *Statistics of Financial Markets*. Berlin: Springer.

Giacomini, E., Härdle, W., & Spokoiny, V. (2009). Inhomogeneous dependence modeling with time-varying copulae. *Journal of Business and Economic Statistics*, *27*(2), 224–234.

Härdle, W., & Simar, L. (2003). *Applied multivariate statistical analysis*. Berlin: Springer.

Hyvärinen, A. (1998). *New approximations of differential entropy for independent component analysis and projection pursuit* (pp. 273–279). Cambridge: MIT.

Hyvärinen, A., Karhunen, J., & Oja, E. (2001). *Independent component analysis*. New York: Wiley.

Jorion, P. (2001). *Value at risk*. NY: McGraw-Hill.

Koenker, R., & Bassett, G. (1978). Regression quantiles. *Econometrica*, *46*, 33–50.

Kuester, K., Mittnik, S., & Paolella, M. S. (2006). Value-at-risk prediction: A comparison of alternative strategies. *Journal of Financial Econometrics*, *4*(1), 53–89.

Kupiec, P. (1995). Techniques for verifying the accuracy of risk measurement models. *Journal of Derivatives*, *3*, 73–84.

Lunde, A., & Hansen, P. R. (2005). A forecast comparison of volatility models: Does anything beat a garch(1,1)?. *Journal of Applied Econometrics*, *20*(7), 873–889.

McAleer, M., & Medeiros, M. (2008). Realized volatility: A review. *Econometric Reviews*, *27*(1), 10–45.

McNeil, A. (2000). Extreme value theory for risk managers. Internal Modeling and CAD II published by RISK Books, 93–113.
McNeil, A., & Frey, R. (2000). Estimation of tail-related risk measures for heteroscedastic financial time series: An extreme value approach. *Journal of Empirical Finance*, *7*, 271–300.
Nelsen, R. (1999). *An introduction to copulas*. Berlin: Springer.
Nelson, D. (1990). Stationarity and persistence in the garch (1, 1) model. *Econometric Theory*, *6*(3), 318–334.
Nelson, D. (1991). Conditional heteroskedasticity in asset returns: A new approach. *Econometrica*, *59*(2), 347–370.
Pérignon, C., Deng, Z. Y., & Wang, Z. J. (2008). Do banks overstate their value-at-risk?. *Journal of Banking and Finance*, *32*(5), 783–794.
Pickands, J. (1975a). Statistical inference using extreme order statistics. *The Annals of Statistics*, *3*, 119–131.
Pickands, J. I. (1975b). Statistical inference using extreme order statistics. *The Annals of Statistics*, *3*, 119–131.
Poon, S., & Granger, C. (2003). Forecasting volatility in financial markets: A review. *Journal of Economic Literature*, *XLI*, 478–539.
Tsay, R. (2005). *Analysis of financial time series*. New Jersey: Wiley.
Tse, Y. K., & Tsui, A. K. C. (2002). A multivariate garch model with time-varying correlations. *Journal of Business and Economic Statistics*, *20*, 351–362.
Zhang, L., Mykland, P. A., & Aït-Sahalia, Y. (2005). A tale of two time scales: Determining integrated volatility with noisy high-frequency data. *Journal of the American Statistical Association*, *100*, 1394–1411.

Chapter 13
Volatility Estimation Based on High-Frequency Data

Christian Pigorsch, Uta Pigorsch, and Ivaylo Popov

Abstract With the availability of high-frequency data ex post daily (or lower frequency) nonparametric volatility measures have been developed, that are more precise than conventionally used volatility estimators, such as squared or absolute daily returns. The consistency of these estimators hinges on increasingly finer sampled high-frequency returns. In practice, however, the prices recorded at the very high frequency are contaminated by market microstructure noise. We provide a theoretical review and comparison of high-frequency based volatility estimators and the impact of different types of noise. In doing so we pay special focus on volatility estimators that explore different facets of high-frequency data, such as the price range, return quantiles or durations between specific levels of price changes.The various volatility estimators are applied to transaction and quotes data of the S&P500 E-mini and of one stock of Microsoft using different sampling frequencies and schemes. We further discuss potential sources of the market microstructure noise and test for its type and magnitude. Moreover, due to the volume of high-frequency financial data we focus also on computational aspects, such as data storage and retrieval.

C. Pigorsch
Department of Economics, University of Bonn, Adenauerallee 24-42, 53113 Bonn, Germany
e-mail: christian.pigorsch@uni-bonn.de

U. Pigorsch (✉)
Department of Economics, University of Mannheim, L7, 3-5, 68131 Mannheim, Germany
e-mail: uta.pigorsch@vwl.uni-mannheim.de

I.Popov
Business School of the University of Mannheim, L5, 5, 68131 Mannheim, Germany
e-mail: ipopov@mail.uni-mannheim.de

J.-C. Duan et al. (eds.), *Handbook of Computational Finance*, Springer Handbooks of Computational Statistics, DOI 10.1007/978-3-642-17254-0_13,

13.1 Introduction

This chapter presents a review and empirical illustration of nonparametric volatility estimators that exploit the information contained in high-frequency financial data. Such ex-post volatility measures can be directly used for the modelling and forecasting of the (future) volatility dynamics, which in turn may be essential for an adequate risk management or hedging decisions. Moreover, volatility constitutes the main ingredient in asset pricing and the knowledge of this quantity therefore plays a major role in most financial applications.

One of the most recent milestones in financial econometrics is therefore probably the introduction of the concept of realized volatility, which allows to consistently estimate the price variation accumulated over some time interval, such as 1 day, by summing over squared (intraday) high-frequency returns. The consistency of this estimator hinges on increasingly finer sampled high-frequency returns. In practice, however, the sampling frequency is limited by the actual quotation or transaction frequency and prices are contaminated by market microstructure effects, so-called noise. We discuss different types and potential sources of the noise and its impact on realized volatility. We further review two of the probably most popular approaches to estimate volatility based on squares or products of high-frequency returns, i.e. the two time scales estimators and kernel-based approaches. However, our main focus in this chapter is on volatility estimators that explore different facets of high-frequency data, such as the price range, return quantiles or durations between specific levels of price changes. Our review thus differs from the one provided in McAleer and Medeiros (2008). A theoretical summary and comparison of the estimators is given. Moreover, as the high-frequency financial data exceeds the amount of data usually encountered by financial econometricians we provide a discussion on data storage and retrieval, i.e. computational aspects that may be of interest to anybody dealing with such high-frequency data. In our empirical application we estimate and illustrate realized volatility over various frequencies for different sampling schemes and price series of one future (S&P500 E-mini) and one stock (Microsoft). We test for the magnitude and type of market microstructure noise and implement the discussed volatility estimators.

13.2 Realized Volatility

Assume that the logarithmic price of a financial asset is given by the following diffusion process

$$p_t = \int_0^t \mu(s)\mathrm{d}s + \int_0^t \sigma(s)\mathrm{d}W(s), \tag{13.1}$$

where the mean process μ is continuous and of finite variation, $\sigma(t) > 0$ denotes the càdlàg instantaneous volatility and W is a standard Brownian motion. The object of interest is the *integrated variance* (IV), i.e. the amount of variation at time point t

accumulated over a past time interval Δ:

$$IV_t = \int_{t-\Delta}^{t} \sigma^2(s)\mathrm{d}s.$$

In the sequel, our focus is on the estimation of IV over one period, e.g. 1 day. For the ease of exposition we, thus, normalize $\Delta = 1$ and drop the time subscript. Suppose there exist m intraday returns, the ith intraday return is then defined as:

$$r_i^{(m)} = p_{i/m} - p_{(i-1)/m},\ i = 1, 2, \ldots, m.$$

The sum of the squared intraday returns:

$$RV^{(m)} = \sum_{i=1}^{m} r_i^{(m)2} \tag{13.2}$$

provides a natural estimator of IV. In fact, based on the theory of quadratic variation, Andersen et al. (2003) show that $RV^{(m)} \xrightarrow{p} IV$ as $m \to \infty$. Following the recent literature we will refer to this ex-post measure of IV as the *realized volatility*, see e.g. Andersen and Bollerslev (1998).

Barndorff-Nielsen and Shephard (2002a) show the consistency of this estimator and that its asymptotic distribution is normal:

$$\frac{\sqrt{m}\left(RV^{(m)} - IV\right)}{\sqrt{2IQ}} \xrightarrow{d} \mathcal{N}(0, 1),$$

with $IQ = \int_0^1 \sigma^4(s)\mathrm{d}s$ denoting the *integrated quarticity*. An application of this asymptotic result, e.g. the construction of confidence intervals, however, is complicated by the unobservability of IQ. A solution is offered in Barndorff-Nielsen and Shephard (2004), who propose the concept of realized power variation, that allows to estimate IQ via the *realized quarticity*:

$$RQ^{(m)} = \frac{m}{3}\sum_{i=1}^{m} r_i^{(m)^4},$$

such that

$$\frac{RV^{(m)} - IV}{\sqrt{\frac{2}{3}\sum_{i=1}^{m} r_i^{(m)^4}}} \xrightarrow{d} \mathcal{N}(0, 1)$$

can be used for large m. In practice, however, the sampling frequency is limited by the actual transaction or quotation frequency. Moreover, the very high-frequency prices are contaminated by market microstructure effects (noise), such as bid-ask bounce effects, price discreteness etc., leading to biases in realized volatility, see e.g. Andersen et al. (2001) and Barndorff-Nielsen and Shephard (2002b). The next section discusses typical assumptions on the structure of the market microstructure noise and its implications for realized volatility. Section 13.4 presents modifications of the realized volatility estimator, while Sect. 13.5 focuses on estimators that exploit other data characteristics for measuring *IV*. Section 13.6 provides a comparison of the various estimators and discusses the situation where the price process (13.1) additionally exhibits finite active jumps.

13.3 Market Microstructure Noise: Assumptions and Implications

Assume that the observed (log) price is contaminated by market microstructure noise u (or measurement error), i.e.:

$$p_{i/m} = p^*_{i/m} + u_{i/m}, \quad i = 1, 2, \ldots m,$$

where $p^*_{i/m}$ is the latent true, or so-called efficient, price that follows the semimartingale given in (13.1). In this case, the observed intraday return is given by:

$$r_i^{(m)} = r_i^{*(m)} + \epsilon_i^{(m)}, \quad i = 1, 2, \ldots m,$$

i.e. by the efficient intraday return $r_i^{*(m)} = p^*_{i/m} - p^*_{(i-1)/m}$ and the intraday noise increment $\epsilon_i^{(m)} = u_{i/m} - u_{(i-1)/m}$. As a consequence, the observed RV can be decomposed as:

$$RV^{(m)} = RV^{*(m)} + 2\sum_{i=1}^{m} r_i^{*(m)} \epsilon_i{}^{(m)} + \sum_{j=1}^{m} \epsilon_j^{(m)2},$$

where the last term on the right-hand side can be interpreted as the (unobservable) realized variance of the noise process, while the second term is induced by potential dependence between the efficient price and the noise. Based on this decomposition and the assumption of covariance stationary noise with mean zero, Hansen and Lunde (2006) show that *RV* is a biased estimator of *IV*. Interestingly, this bias is positive if the noise increments and the returns are uncorrelated, but may become negative in the case of negative correlation. One possible explanation for such negative correlation is given in Hansen and Lunde (2006), who show that in price series compiled from mid-quotes (see Sect. 13.7 for a definition), this can be caused by non-synchronous revisions of the bid and the ask prices, leading to a temporary widening of the spread. Another source of negative correlation may be the staleness of the mid-quote prices.

Obviously, the precise implications of the presence of noise for the properties of the *RV* estimator depend on the assumed structure of the noise process. In the following we focus on the most popular noise assumption.

Assumption 1: Independent noise.

(a) The noise process u is independent and identically distributed with mean zero and finite variance ω^2 and finite fourth moment.
(b) The noise is independent of the efficient price.

The independent noise assumption implies that the intraday returns have an MA(1) component. Such a return specification is well established in the market microstructure literature and is usually justified by the existence of the bid-ask bounce effect, see e.g. Roll (1984). However, as shown in Hansen and Lunde (2006) and Zhang et al. (2005) the iid noise introduces a bias into the RV estimator:

$$E\left[RV^{(m)}\right] = IV + 2m\omega^2 \tag{13.3}$$

that diverges to infinity as $m \to \infty$. Moreover, the asymptotic distribution of *RV* is given by:

$$\frac{\left(RV^{(m)} - IV - 2m\omega^2\right)}{2\sqrt{mE(u^4)}} \xrightarrow{d} \mathcal{N}(0,1).$$

Sampling at lower frequencies, i.e. sparse sampling, reduces the bias but leads to an increase in the variance (see e.g. Barndorff-Nielsen and Shephard 2002b), which is usually referred to as the bias-variance trade-off.

The independent noise assumption seems restrictive. In fact, Hansen and Lunde (2006) provide some evidence of serial dependence in the noise process and correlation with the efficient price, i.e. *time-dependent* and *endogenous* noise, respectively. Alternative estimators of *IV* have been developed and are shown to be robust to some dependence in the noise process, but they are in no way developed around a universally accepted dependence specification like Assumption 1. The next section discusses the probably most popular alternatives to the RV estimator that are asymptotically unbiased and consistent under iid and under dependent noise types.

13.4 Subsampling and Realized Kernels

In the following we briefly present two more elaborate, but under specific noise assumptions consistent procedures for estimating *IV*.

13.4.1 Averaging and Subsampling

The subsampling approach originally suggested by Zhang et al. (2005) builds on the idea of averaging over various *RV*s constructed by sampling sparsely over high-frequency subsamples. To this end the intraday observations are allocated to K subsamples. Using a regular allocation, 5 min returns can for example be sampled at the time points 9:30, 9:35, 9:40, …; and at the time points 9:31, 9:36, 9:41, … and so forth. Averaging over the subsample *RV*s yields the so-called *average RV* estimator: $(1/K)\sum_{k=1}^{K} RV^{(k,m_k)}$ with m_k denoting the sampling frequency used in the RV computation for subsample k. Usually, m_k is equal across all subsamples. The average *RV* estimator is still biased, but the bias now depends on the average size of the subsamples rather than on the total number of observations. RV constructed from all observations, $RV^{(all)}$ can be used for bias correction yielding the estimator:

$$TTSRV^{(m,m_1,\ldots,m_K,K)} = \frac{1}{K}\sum_{k=1}^{K} RV^{(k,m_k)} - \frac{\bar{m}}{m} RV^{(all)}, \tag{13.4}$$

where $\bar{m} = (1/K)\sum_{k=1}^{K} m_k$. As the estimator (13.4) consists of a component based on sparsely sampled data and one based on the full grid of price observations, the estimator is also called *two time scales estimator*.

Under the independent noise assumption, the estimator is consistent. Furthermore, under equidistant observations and under regular allocation to the grids, the asymptotic distribution is given by:

$$\frac{m^{1/6}(TTSRV^{(m,m_1,\ldots,m_K,K)} - IV)}{\sqrt{\frac{8}{c^2}(\omega^2)^2 + c\frac{4}{3}IQ}} \xrightarrow{d} \mathcal{N}(0,1).$$

for $K = cm^{2/3}$. The optimal value of K, i.e. minimizing the expected asymptotic variance, can be obtained by estimating $c_{opt} = \left(12\omega^2/IQ\right)^{1/3}$ based on data prior to the day under consideration (see Zhang et al. 2005).

A generalization of *TTSRV* was introduced by Aït-Sahalia et al. (2010) and Zhang (2006), which is consistent and asymptotically unbiased also under time-dependent noise. To account for serial correlation in the noise, the *RV*s are based on overlapping J-period intraday returns. Using these so-called *average-lag-J RV*s the estimator becomes:

$$TTSRV_{adj}^{(m,K,J)} = s\left(\frac{1}{K}\sum_{i=0}^{m-K}(p_{(i+K)/m} - p_{i/m})^2 - \frac{\bar{m}^{(K)}}{\bar{m}^{(J)}}\frac{1}{J}\sum_{l=0}^{m-J}(p_{(l+J)/m} - p_{l/m})^2\right) \tag{13.5}$$

with $\bar{m}^{(K)} = (m - K + 1)/K$, $\bar{m}^{(J)} = (m - J + 1)/J$, $1 \leq J < K < m$ and the small sample adjustment factor $s = \left(1 - \bar{m}^{(K)}/\bar{m}^{(J)}\right)^{-1}$. Note that K and J now basically denote the slow and fast time scales, respectively. The asymptotic distribution is given by:

$$\frac{m^{1/6}\left(TTSRV_{adj}^{(m,K,J)} - IV\right)}{\sqrt{\frac{1}{c^2}\xi^2 + c\frac{4}{3}IQ}} \overset{d}{\approx} \mathcal{N}(0,1)$$

with $\xi^2 = 16(\omega^2)^2 + 32\sum_{l=1}^{\infty}(E(u_0, u_l))^2$ and $\overset{d}{\approx}$ denotes that when multiplied by a suitable factor, then the convergence is in distribution.

Obviously, $TTSRV_{adj}$ converges to IV at rate $m^{1/6}$, which is below the rate of $m^{1/4}$, established as optimal in the fully parametric case in Aït-Sahalia et al. (2005). As a consequence, Aït-Sahalia et al. (2010) introduced the multiple time scale estimator, *MTSRV*, which is based on the weighted average of average-lag-J *RV*s computed over different multiple scales. It is computationally more complex, but for suitably selected weights it attains the optimal convergence rate $m^{1/4}$.

13.4.2 Kernel-Based Estimators

Given the similarity to the problem of estimating the long-run variance of a stationary time series in the presence of autocorrelation, it is not surprising that kernel-based methods have been developed for the estimation of *IV*. Such an approach was first adopted in Zhou (1996) and generalized in Hansen and Lunde (2006), who propose to estimate *IV* by:

$$KRV_{Z\&HL}^{(m,H)} = RV^{(m)} + 2\sum_{h=1}^{H}\frac{m}{m-h}\gamma_h$$

with $\gamma_h = \sum_{i=1}^{m} r_i^{(m)} r_{i+h}^{(m)}$. As the bias correction factor $m/(m-h)$ increases the variance of the estimator, Hansen and Lunde (2006) replaced it by the Bartlett kernel. Nevertheless, all three estimators are inconsistent.

Recently, Barndorff-Nielsen et al. (2008) proposed a class of consistent kernel based estimators, *realized kernels*. The *flat-top realized kernel*:

$$KRV_{FT}^{(m,H)} = RV^{(m)} + \sum_{h=1}^{H} k\left(\frac{h-1}{H}\right)(\gamma_h + \gamma_{-h}),$$

where $k(x)$ for $x \in [0,1]$ is a deterministic weight function. If $k(0) = 1$, $k(1) = 0$ and $H = cm^{2/3}$ the estimator is asymptotically mixed normal and converges at rate $m^{1/6}$. The constant c is a function of the kernel and the integrated quarticity, and is chosen such that the asymptotic variance of the estimator is minimized. Note that for the flat-top Bartlett kernel, where $k(x) = 1 - x$, and the cubic kernel, $k = 1 - 3x^2 + 2x^3$, $KRV_{FT}^{(m,H)}$ has the same asymptotic distribution as the TTSRV and the MTSRV estimators, respectively.

Furthermore, if $H = cm^{1/2}$, $k'(0) = 0$ and $k'(1) = 0$ (called *smooth* kernel functions), the convergence rate becomes $m^{1/4}$ and the asymptotic distribution is given by:

$$\frac{m^{1/4}\left(KRV_{FT}^{(m,H)} - IV\right)}{\sqrt{4ck_\circ IQ + \frac{8}{c}k'_\circ \omega^2 IV + \frac{4}{c^3}k''_\circ \omega^4}} \xrightarrow{d} \mathcal{N}(0,1)$$

with $k_\circ = \int_0^1 k(x)^2 \mathrm{d}x$, $k'_\circ = \int_0^1 k'(x)^2 \mathrm{d}x$ and $k''_\circ = \int_0^1 k''(x)^2 \mathrm{d}x$.

For practical applications, Barndorff-Nielsen et al. (2009) consider the *non-flat-top realized kernels*, which are robust to serial dependent noise and to dependence between noise and efficient price. The estimator is defined as:

$$KRV_{NFT}^{(m,H)} = RV^{(m)} + \sum_{h=1}^{H} k\left(\frac{h}{H}\right)(\gamma_h + \gamma_{-h}). \tag{13.6}$$

However, the above mentioned advantages of this estimator come at the cost of a lower convergence rate, i.e. $m^{1/5}$, and a small asymptotic bias:

$$m^{1/5}\left(KRV_{NFT}^{(m,H)} - IV\right) \xrightarrow{ds} \mathcal{MN}\left(c^{-2}\left|k''(0)\right|\omega^2,\ 4ck_\circ IQ\right),$$

where ds denotes stable convergence and $\mathcal{MN}$ a mixed normal distribution. Barndorff-Nielsen et al. (2009) recommend the use of the Parzen kernel as it is smooth and always produces non-negative estimates. The kernel is given by:

$$k(x) = \begin{cases} 1 - 6x^2 + 6x^3 & \text{for } 0 \leq x < 1/2 \\ 2(1-x)^3 & \text{for } 1/2 \leq x \leq 1 \\ 0 & \text{for } x > 1 \end{cases}. \tag{13.7}$$

For non-flat-top realized kernels, the bandwidth H can be optimally selected as:

$$H^* = c^* \xi^{4/5} m^{3/5}, \quad c^* = \left(\frac{k''(0)^2}{k_\circ}\right)^{1/5} \quad \text{and} \quad \xi^2 = \frac{\omega^2}{\sqrt{IQ}}.$$

For the Parzen kernel $c^* = 3.5134$. Obviously, the optimal value of H is larger if the variance of the microstructure noise is large in comparison to the integrated quarticity. The estimation of this signal-to-noise ratio ξ^2 is discussed in Barndorff-Nielsen et al. (2008, 2009), see also Sect. 13.8.

Realized kernels are subject to the so-called *end effects*, caused by the missing sample size adjustment of the autocovariance terms. This can be accounted for by using local averages of returns in the beginning and the end of the sample. However, Barndorff-Nielsen et al. (2009) argue that for actively traded assets these effects can be ignored in practice.

Further refinements of the realized kernels in the spirit of the subsampling approach adopted in the *TTSRV* and *MTSRV* estimators are considered in Barndorff-Nielsen et al. (2010) by using averaged covariance terms in the realized kernel estimators.

13.5 Alternative Volatility Estimators

All of the realized variance measures discussed so far are based on squared intraday returns. In the following we present estimators of the quadratic variation that exploit other aspects of high-frequency financial data.

13.5.1 Range-Based Estimation

In volatility estimation, the usage of the range, i.e. the difference between high and low (log) prices, is appealing, as it is based on extremes from the entire price path and, thus, provides more information than returns sampled at fixed time intervals. The range-based estimator has therefore attracted researcher's interest, see e.g. Feller (1951), Garman and Klass (1980), Parkinson (1980), and it has been found that using the squared range based on the daily high and low is about five times more efficient than the daily squared return. Nevertheless, it is less efficient than *RV* based on a sampling frequency higher than two hours.

Recently, Christensen and Podolskij (2007) proposed a *realized range-based estimator*, that replaces the squared intraday returns by normalized squared ranges. Assume that the (log) price process follows a continuous semimartingale and that $m_K K + 1$ equidistant prices are observed discretely over a day. Decomposing the daily time interval into K *non-overlapping* intervals of size m_K, the estimator is given by:

$$RRV^{(m_K,K)} = \frac{1}{\lambda_{2,m_K}} \sum_{i=1}^{K} s_i^{(m_K)2}, \tag{13.8}$$

where the range of the price process over the ith interval is defined as:

$$s_i^{(m_K)} = \max_{0 \le h,l \le m_K} \left(p_{\frac{i-1+h/m_K}{K}} - p_{\frac{i-1+l/m_K}{K}} \right), \quad i = 1, \dots K,$$

and $\lambda_{r,m_K} = E\left[\max_{0\leq h,l\leq m_K}\left(W_{h/m_K} - W_{l/m_K}\right)^r\right]$. I.e. λ_{2,m_K} is the second moment of the range of a standard Brownian motion over the unit interval with m_K observed increments. This factor corrects for the downward bias arising from discretely observed data. In particular, the observed high and low prices may under- and overestimate the true ones, respectively, such that the true range is underestimated.

The estimator is asymptotically distributed according to:

$$\frac{\sqrt{K}\left(RRV^{(m_K,K)} - IV\right)}{\sqrt{\Lambda_c IQ}} \xrightarrow{d} \mathcal{N}(0,1)$$

as $K \to \infty$, where it is sufficient that m_K converges to a natural number c, i.e. $m_K \to c \in \mathbf{N} \cup \infty$, $\Lambda_c = \lim_{m_K \to c} \Lambda_{m_K}$ and $\Lambda_{m_K} = \left(\lambda_{4,m_K} - \lambda_{2,m_K}^2\right)/\lambda_{2,m_K}^2$. The efficiency of the *RRV* estimator obviously depends on the variance factor Λ. Christensen and Podolskij (2007) illustrate that for $m_K = 10$, which is a reasonable choice for moderately liquid assets, the factor is about 0.7. Its asymptotic value, i.e. for continuously observed prices, the factor is 0.4, such that *RRV* is five times more efficient than *RV*. For $m_K = 1$ the efficiency of *RV* is obtained. Notably, *IQ* can also be estimated based on the range, i.e. via the so-called *realized range-based quarticity* $RRQ^{(m_K,K)} = (1/\lambda_{4,m_K}) \sum_{i=1}^{K} s_i^{(m_K)^4}$.

Market microstructure noise corrections of range-based volatility estimators have been proposed by Martens and van Dijk (2007), who focus particularly on the effect of the bid-ask bounce, and by Christensen et al. (2009a). The latter address bias correction under iid noise. However, bias correction is not as straightforward as in the case of using squared returns as the extreme value theory of *RRV* depends on the distribution of the noise. Moreover, *RRV* is more sensitive towards price outliers than squared returns. Nevertheless, the empirical results reported in Christensen et al. (2009a) indicate that bias reduction can be achieved by imposing simple parametric assumptions on the distribution of the noise process and sampling at a 1–2 min frequency.

13.5.2 Quantile-Based Estimation

An approach that is very similar to the range-based estimators is to consider quantiles of the return rather than of the (log) price. We refer to these estimators as the *quantile-based estimators*. This idea dates back at least to David (1970) and Mosteller (1946) and was generalized in Christensen et al. (2009b) by combining multiple quantiles for each of the m_K intraday subintervals yielding the so-called *quantile-based realized variance* (*QRV*) estimator.

The setup is similar to the one used in *RRV*, i.e. the sample is again split into K non-overlapping blocks with m_K returns, where we denote the set of returns contained in block j by $r_{[(j-1)m_K+1:jm_K]}$. For a vector of p return quantiles $\bar{\lambda} = (\lambda_1, \ldots \lambda_p)'$ the QRV estimator is given by:

$$QRV^{(m_K,K,\bar{\lambda})} = \frac{1}{K}\sum_{i=1}^{p}\alpha_i \sum_{j=0}^{K}\frac{q_j^{(m_K,\lambda_i)}}{\nu_1^{(m_K,\lambda_i)}} \quad \text{for } \lambda_i \in (1/2, 1) \tag{13.9}$$

with the realized squared symmetric λ_i-quantile

$$\begin{aligned} q_j^{(m_K,\lambda_i)} &= g^2_{\lambda_i m_K}\left(\sqrt{m_K K}\, r_{[(j-1)m_K+1:jm_K]}\right) \\ &\quad + g^2_{m_K-\lambda_i m_K+1}\left(\sqrt{m_K K}\, r_{[(j-1)m_K+1:jm_K]}\right), \end{aligned} \tag{13.10}$$

where the function $g_l(x)$ extracts the lth order statistic from a vector x, $\alpha = (\alpha_1, \ldots \alpha_p)'$ is a non-negative vector of quantile weights, summing to unity, and

$$\nu_r^{(m_K,\lambda)} = E\left[\left(\left|U_{(\lambda m_K)}\right|^2 + \left|U_{(m_K-\lambda m_K+1)}\right|^2\right)^r\right]$$

with $U_{(\lambda m_K)}$ denoting the (λm_K)th order statistic of an independent standard normal sample $\{U_i\}_{i=1}^{m_K}$. For m_K fixed and as the number of blocks is increasing, i.e. $m = m_K K \to \infty$, $q_j^{(m_K,\lambda_i)}/\nu_1^{(m_K,\lambda_i)}$ is an estimator of the (scaled) return variance over the jth block. Summing across all blocks naturally yields a consistent estimator of the integrated variance. Christensen et al. (2009b) derive the asymptotic distribution of QRV:

$$\frac{\sqrt{m}\left(QRV^{(m_K,K,\bar{\lambda})} - IV\right)}{\sqrt{\theta^{(m_K,\bar{\lambda},\alpha)} IQ}} \xrightarrow{d} \mathcal{N}(0,1),$$

where $\theta^{(m_K,\bar{\lambda},\alpha)} = \alpha' \Theta^{(m_K,\bar{\lambda})}\alpha$ and the i,jth element of the $p \times p$ matrix $\Theta^{(m_K,\bar{\lambda})}$ is given by

$$\Theta_{i,j}^{(m_K,\bar{\lambda})} = m_K \frac{\nu_1^{(m_K,\lambda_i\lambda_j)} - \nu_1^{(m_K,\lambda_i)}\nu_1^{(m_K,\lambda_j)}}{\nu_1^{(m_K,\lambda_i)}\nu_1^{(m_K,\lambda_j)}}$$

with

$$\begin{aligned} \nu_1^{(m_K,\lambda_i\lambda_j)} &= E\left[\left(\left|U_{(\lambda_i m_K)}\right|^2 + \left|U_{(m_K-\lambda_i m_K+1)}\right|^2\right)\right. \\ &\quad \left.\times \left(\left|U_{(\lambda_j m_K)}\right|^2 + \left|U_{(m_K-\lambda_j m_K+1)}\right|^2\right)\right]. \end{aligned}$$

The fourth power of the realized quantiles can be used to construct a quantile-based estimator of IQ.

Christensen et al. (2009b) further propose a subsampled version of the QRV estimator that yields improvements in the efficiency of the above estimator by using overlapping subintervals.

The implementation of the estimator involves the choice of several hyperparameters, i.e. the selection of the quantiles λ, the block length m_K, and the assignment of the optimal weights α. For a fixed set of quantiles and a fixed block size, the weights α can be chosen to minimize the asymptotic variance of *QRV* estimators, i.e.minimizing θ yields the optimal weights:

$$\alpha^* = \frac{\Theta^{(m,\bar{\lambda})^{-1}}\iota}{\iota' \Theta^{(m,\bar{\lambda})^{-1}}\iota},$$

where ι is a $(p \times 1)$ vector of ones. Comparing the efficiency of the estimator, Christensen et al. (2009b) conclude that the gains from optimizing α for finite samples, instead of using the asymptotic optimal values, are only minor.

For the quantile selection, Christensen et al. (2009b) find that the 90–95% quantiles are most informative. The quantiles around the median are uninformative and those around the extremes are too erratic and less robust to potential jumps in the price process or to outliers. Nevertheless, quantiles outside the most informative region may be used to exploit the covariances structure of the order statistics for $p > 1$. Smaller block sizes deliver slightly more efficient estimators, as they achieve better locality of volatility. Also, the subsampled version is shown to be slightly more efficient than the blocked version for multiple quantiles. Finally, the efficiency constant θ can be reduced to around 2.5 for one quantile and is close to 2 for multiple quantiles, achieving the efficiency constant of *RV*.

Christensen et al. (2009b) propose a modification of the *QRV* estimator that makes it robust to iid noise. Based on a pre-averaging technique similar to Podolskij and Vetter (2009), the robust estimator is obtained by applying the *QRV* methodology to a weighted average of the observed returns. In particular, define the averaged data by:

$$\bar{y}_j = \sum_{i=1}^{L-1} h\left(\frac{i}{L}\right) r_{j+i}^{(m)}$$

with $L = c\sqrt{m} + o(m^{1/4})$ for some constant c and weight function h on $[0, 1]$. Further conditions of h are given in Christensen et al. (2009b), who use in their simulation and application the weight function $h(x) = \min(x, 1 - x)$. The *QRV* estimator is then given by:

$$QRV_{\bar{y}}^{(L,m_K,K,\bar{\lambda})} = \frac{1}{c\psi_2(m - m_K(L-1)+1)} \sum_{i=1}^{p} \alpha_i \sum_{j=0}^{m_K(K-L+1)} \frac{q_{\bar{y};j}^{(m_K,\lambda_i)}}{\nu_1^{(m_K,\lambda_i)}}$$

with

$$\begin{aligned} q_{\bar{y};j}^{(m_K,\lambda_i)} &= g^2_{\lambda_i m_K}\left(m^{1/4}\bar{y}_{[j:j+m_K(L-1)]}\right) \\ &\quad + g^2_{m_K-\lambda_i m_K+1}\left(m^{1/4}\bar{y}_{[j:j+m_K(L-1)]}\right). \end{aligned}$$

The problem of $QRV_{\bar{y}}^{(L,m_K,K,\bar{\lambda})}$ is that it is biased. Incorporating a bias-correction finally yields the iid noise-robust estimator:

$$QRV_{iid}^{(L,m_K,K,\bar{\lambda})} = QRV_{\bar{y}}^{(L,m_K,K,\bar{\lambda})} - \frac{\psi_1}{c^2\psi_2}\omega^2, \tag{13.11}$$

where ψ_1 and ψ_2 can be computed by

$$\psi_2 = L\sum_{j=1}^{L}\left(h\left(\frac{j}{L}\right) - h\left(\frac{j-1}{L}\right)\right)^2$$

and

$$\psi_1 = \frac{1}{L}\sum_{j=1}^{L-1} h^2\left(\frac{j}{L}\right).$$

Under some further mild assumptions, Christensen et al. (2009b) show that this estimator converges at rate $m^{-1/4}$ to the IV. However, in contrast to the other volatility estimators its asymptotic variance has no explicit expression in terms of IQ. Nevertheless, it can be estimated based on the estimates of the q_i, ψ_2 and ν_1 terms. For $h(x) = \min(x, 1-x)$ and the constant volatility setting the estimator achieves a lower bound of $8.5\sigma^3\omega$. This is close to the theoretical bound of the variance of the realized kernel approach discussed in Sect. 13.4.2, which is $8\sigma^3\omega$. The behavior of the noise robust estimator will of course depend on the choice of L, which trades-off between the noise reduction and the efficiency loss due to pre-averaging. A simulation study suggests that a conservative choice, e.g. a larger value of L, such as $L = 20$, may be preferable. In applications the estimated signal-to-noise ratio can be used to determine L based on the mean-square error (MSE) criterion.

13.5.3 Duration-Based Estimation

While the return- and range-based volatility estimators make use of a functional of the price path between fixed points in time, the duration-based approach focuses on the time it takes the price process to travel between fixed price levels. Such an approach was first investigated by Cho and Frees (1988) for the constant volatility case. Recently, Andersen et al. (2009) provide a more comprehensive treatment of this concept in the case of constant volatility and for stochastic volatility evolving without drift. They consider three different *passage times*, i.e. three different ways to measure the time a Brownian motion needs to travel a given distance r:

$$\tau^{(r)} = \begin{cases} \inf\{t : t > 0 \; \& \; |W_t| > r\} & \text{(first exit time)} \\ \inf\{t : (\max_{0<s\le t} W_s - \min_{0<s^*\le t} W_{s^*}) > r\} & \text{(first range time)} \\ \inf\{t : t > 0 \; \& \; W_t = r\} & \text{(first hitting time)} \end{cases}$$

In the constant volatility case, the moments of these passage times are available in closed-form:

$$E\left[\tau^{(r)}\right] = \begin{cases} \frac{r^2}{\sigma^2} & \text{(first exit time)} \\ \frac{1}{2}\frac{r^2}{\sigma^2} & \text{(first range time)} \\ \infty & \text{(first hitting time)} \end{cases} \tag{13.12}$$

Interestingly, comparing these moments to the expected value of a squared Brownian increment over the interval τ, which is $\sigma^2\tau$, illustrates the duality between *RV* and the range-based volatility approaches and the duration-based one.

The moment conditions (13.12) suggest to estimate σ^2 via the method of moments using either an observed sample of first exit times or of first range times with fixed r. However, as the expected passage times are inversely proportional to the instantaneous variance, these estimators will suffer from severe small sample biases induced by Jensen's inequality. For this reason, Andersen et al. (2009) propose small sample unbiased estimators based on the reciprocal passage times:

$$E\left[\frac{r^2}{\tau^{(r)}}\right] = \mu_1\sigma^2 = \begin{cases} 2C\sigma^2 & \text{(first exit time)} \\ (4\log 2)\sigma^2 & \text{(first range time)} \\ \sigma^2 & \text{(first hitting time)} \end{cases},$$

where $C \approx 0.916$ is the Catalan constant. Interestingly, moment based estimators for the reciprocal hitting time are now also feasible.

The concept also allows to define a *local* volatility estimator for a single passage time at (intraday) time point i:

$$\left(\hat{\sigma}_i^{(r)}\right)^2 = \frac{1}{\mu_1}\frac{r^2}{\tau_i^{(r)}},$$

such that *IV* can also be estimated in the case of stochastic volatility by applying the Riemann sum. The resulting *duration-based realized variance* estimator is given by:

$$DRV^{(m,r)} = \sum_{i=1}^{m}\left(\hat{\sigma}_i^{(r)}\right)^2 \delta_i \tag{13.13}$$

with δ_i denoting the times between the intraday observations.

Based on the time-reversibility of the Brownian motion, the local volatility estimates can be constructed using either the *previous passage time* or the *next passage time*, i.e. the time is determined by the path of the Brownian motion either prior to or after the time point i, respectively. In practice, market closures will thus induce the problem of censoring. In particular, the expected next passage time is affected by the time left until the market closes, *(right censoring)*, while the expected previous passage time is limited by the time the market opened, *(left censoring)*. Andersen et al. (2009) show that a one-directional approach may be

Table 13.1 Efficiency constant of *DRV* for different types of passage time

	Bi-directional	Uni-directional
First exit time	0.3841	0.7681
First range time	0.2037	0.4073
First hitting time	1.0000	2.0000

preferred, although combining both schemes, so-called *bi-directional* local volatility estimation, has the potential of reducing the variance of the duration-based estimator by a factor of two, see also Table 13.1. More precisely, to account for censoring effects, they suggest to construct the *DRV* estimator based on the next passage time scheme during the first half of the day and to use the previous passage time scheme over the second half of the day. The suggestion is motivated by their simulation results for exit and range passage times showing that left and right censoring can be ignored, if the difference in time to the market opening and closing is 2–3 times longer than the expected passage times.

The duration-based approach can also be used to construct estimators of the integrated quarticity:

$$DRQ^{(m,r)} = \sum_{i=1}^{m} \left(\hat{\sigma}^{(r)}\right)^4 \delta_i,$$

which allows the construction of confidence intervals using the asymptotic result for *DRV*:

$$\frac{\sqrt{m}\left(DRV^{(m,r)} - IV\right)}{\sqrt{\nu IQ}} \xrightarrow{d} \mathcal{N}(0,1),$$

where ν is a constant that is specific to the type of passage time used in the estimation and that is independent of the choice of r. Table 13.1 presents the respective values of this efficiency constant.

The asymptotic efficiency is much higher compared to the return-based estimators, especially if the dataset allows the usage of bi-directional passage times through non-interrupted trading, suggesting the use of trade data from FOREX or GLOBEX. However, similarly to the other estimators, the *DRV* not only suffers from the problem that the price process is observed only at m discrete time points, but also that the number of observed price changes is even less, see Sect. 13.7. Andersen et al. (2009) therefore suggest to sample sparsely in order to avoid this potentially more pronounced discreetness effect. Moreover, similarly to the range-based estimator the *DRV* based on first range time and on first exit time may be biased, as the observed times may not coincide with the true ones.

A formal noise-robust *DRV* estimator has not been developed so far, however, Andersen et al. (2009) investigate the impact of market microstructure noise on the *DRV* estimator within a simulation study with independent and serial dependent noise assumptions. The results indicate that the estimator is nevertheless sufficiently robust to independent noise with moderate levels of noise-to-signal ratio even in the case of first range and first exit times. Also, as may be naturally expected, higher threshold values r make the estimator more robust to noise.seems to be very robust to the higher persistent levels typically encountered for quote data as argued in Andersen et al. (2009).

13.6 Theoretical Comparison of Volatility Estimators and Price Jumps

So far we have discussed and presented the most popular and the most recent approaches to estimate IV based on various characteristics of high-frequency financial data. In the following we provide a brief summary of the main large sample properties of these estimators in order to facilitate their comparison. An empirical evaluation and illustration of the estimators is given in Sect. 13.8.

Table 13.2 summarizes the estimators, which are grouped according to the underlying assumption on the market microstructure noise under which they achieve consistency. We further report the asymptotic variances (based on rounded and optimally chosen parameter values) and the convergence rates of the various estimators. Note that due to the unavailability of a closed-form expression of the asymptotic variance of QRV_{iid} we report here only its lower bound in the setting of constant volatility. The reported asymptotic variance of KRV_{NFT} is based on the Parzen kernel. Moreover, the complexity and the performance of the estimators often depend on the choice of hyperparameters as is indicated in the table. The exact impact of those parameters and their determination have been discussed in the previous sections. Note that with the exception of the non-flat-top realized kernel all the estimators are unbiased in large samples and we, thus do not comment on this property in the table.

The table also reports the robustness of the various estimators to the presence of jumps. So far we have assumed that the log price follows a pure diffusion process. However, recent empirical evidence suggests that jumps may have a non-trivial contribution to the overall daily price variation, see e.g. Andersen et al. (2007), Eraker et al. (2003) and Huang and Tauchen (2005). Suppose the log price follows in fact a continuous-time jump diffusion process:

$$p_t = \int_0^t \mu(s)\mathrm{d}s + \int_0^t \sigma(s)\mathrm{d}W(s) + \sum_{j=1}^{N(t)} \kappa(s_j),$$

where the $N(t)$ process counts the number of jumps occurring with possibly time-varying intensity $\lambda(t)$ and jump size $\kappa(s_j)$. Given the presence of jumps in the price process, the question arises, whether the proposed approaches still deliver estimators of the integrated variance, i.e. the object of interest in many financial applications?

From the theory of quadratic variation it follows that the basic RV estimator converges uniformly in probability to the quadratic variation as the sampling frequency of the underlying returns approaches infinity:

$$RV \xrightarrow{p} IV + \sum_{j=N(t-1)+1}^{N(t)} \kappa^2(s_j).$$

In other words, the realized variance provides an ex-post measure of the true *total* price variation, i.e. including the discontinuous jump part.

In order to distinguish the continuous variation from the jump component, Barndorff-Nielsen and Shephard (2004) first proposed the so-called *Bipower variation* measure, defined by:

$$BPV^{(m)} = \frac{\pi}{2} \sum_{j=2}^{m} |r_j||r_{j-1}|,$$

which becomes immune to jumps and consistently estimates the integrated variance as $m \to \infty$. A central limit theory for Bipower variation has just recently been derived in Vetter (2010). He also provides a brief review of alternative jump-robust estimators of *IV* including multipower variations, see Barndorff-Nielsen et al. (2006), and a threshold-based realized variance estimator, see Mancini (2009).

Table 13.2 shows, that only a few of the previously discussed approaches deliver consistent estimators of *IV* in the presence of (finite active) jumps. In the quantile-based estimation, the jump robustness is due to the exclusion of the extreme quantiles in the construction of the estimator. Similarly, the *DRV* estimator can be made robust by the choice of the price threshold r, i.e. limiting the impact of

Table 13.2 Asymptotic properties of the IV estimators

Estimator	Equation	Asymptotic variance	Convergence rate	Jump robust	Parameters
No microstructure noise					
RV	(13.2)	$2IQ$	$m^{1/2}$	No	m
RRV	(13.8)	$0.4IQ$	$K^{1/2}$	No	m_K, K
QRV	(13.9)	$2.3IQ$	$m^{1/2}$	Yes	$m_K, K, \bar{\lambda}$
DRV first exit	(13.13)	$0.77IQ$	$m^{1/2}$	Yes[a]	m, r
iid noise					
TTSRV	(13.4)	$1.33\frac{K}{m^{2/3}}IQ + 8\frac{m^{4/3}}{K^2}\omega^2$	$m^{1/6}$	No	$m, m_1, \ldots, m_K, K$
QRV_{iid}	(13.11)	$8.5\sigma^3\omega$	$m^{1/4}$	Yes	$m_K, K, L, \bar{\lambda}$
Time-dependent noise					
$TTSRV_{adj}$	(13.5)	$1.33\frac{K}{m^{2/3}}IQ + \frac{m^{4/3}}{K^2}\xi^2$	$m^{1/6}$	No	m, K, J
Time-dependent and endogenous noise					
KRV_{NFT}	(13.6)+(13.7)	$3.78IQ$	$m^{1/5}$	No	m, k, H

[a]Explanation is given in the text

jumps that exceed this threshold. The asterisk in Table 13.2 indicates that the jump robustness of this estimator has been shown only within a Monte Carlo simulation for a modified version of the estimator that utilizes the threshold corresponding to the observed log price at the tick time prior to the crossing of the target threshold. A jump robust range-based estimator is proposed by Klößner (2009).

13.7 High-Frequency Financial Data: Characteristics and Computational Aspects

In the following we briefly review some of the main characteristics of high-frequency financial data and of the existing sampling schemes. Moreover, as the volume of the high-frequency dataset exceeds the one usually encountered in financial statistics or econometrics, we discuss also the computational aspects concerning data storage and retrieving, which will be useful not only for the reader interested in implementing volatility estimators, but also to those planning to work with high-frequency financial data in general.

13.7.1 Price Series and Sampling Schemes

Electronic trading systems have lead to the availability of detailed price and trade information at the ultrahigh frequency. In particular, information on the arrival and volume of the sell and buy orders is stored along with the ask and bid quotes. A trade takes place if buy and sell orders could be matched and the corresponding price of this transaction, i.e. the transaction price, is recorded. As the underlying type of trading mechanism differs across exchanges, we refer the interested reader to Hasbrouck (2007) and Gouriéroux and Jasiak (2001) for a more detailed discussion on order books and existing types of markets.

An important feature of an exchange market is that prices at which one can send buy (bid) and sell (ask) quotations and at which transactions take place must be multiples of a predetermined number, called *tick size*. As a consequence, markets with a tick size relatively large in comparison to the price level of the asset, *large tick markets*, often exhibit a *spread*, i.e. the difference between the price of the highest available bid and the lowest available ask quote, that equals most of the time exactly one tick. The S&P500 future is an example for such a market, see Hasbrouck (2007). Obviously, such price discreteness or round-off errors represent one source of market microstructure noise that will affect the performance of the *IV* estimators, especially of *DRV*.

Given the availability of transaction, bid and ask prices, the question arises on which of these price series should be used in the construction of the estimators. Financial theory of course suggests to use the price at which the asset trades. However, assuming a random flow of alternating buying and selling market orders, the trading mechanism and the discrete prices will cause transaction prices to randomly fluctuate between the best bid and ask price. This effect is called *bid-ask bounce* and was first described in Roll (1984). It induces a strong negative autocorrelation in the returns and, thus, violates the assumption of a semimartingale for the price process.

This has lead to the consideration of the *mid-quotes*, i.e. the average of the best bid and ask price. However, the mid-quotes are also not immune to microstructure effects. In fact, they suffer from the so-called *price staleness*. They change rather rarely, and are subject to non-synchronous adjustments of the bid and ask quotes. Alternatively, the bid and ask prices can be used, which in large tick markets contain a similar amount of information as the mid-quotes, but do not suffer from non-synchronous adjustment effects.

Apart from deciding upon the price series used in the empirical implementation of the *IV* estimators, one also has to choose the scheme at which prices are sampled. The literature basically distinguishes between four types of sampling schemes, see Oomen (2006): calendar time sampling, transaction time sampling, business time sampling and tick time sampling. The most obvious one is *calendar time sampling*, CTS, which samples at equal intervals in physical time. As high-frequency observations are irregularly spaced in physical time, an artificial construction of CTS from the full record of prices is necessary. A natural approach is given by the *previous tick method*, see Wasserfallen and Zimmermann (1985), which uses the last record observed prior to the sampling point. The *linear interpolation* method instead interpolates between the previous and the next observed price. At ultra high-frequencies this implies, however, that $RV \to 0$ as $m \to \infty$, see Hansen and Lunde (2006).

Alternatively, one can sample whenever a transactions takes place, i.e. the so-called *transaction time sampling*, TTS, or whenever prices change, so-called *tick time sampling*, TkTS. The latter can further be distinguish according to the type of price series, yielding tick-time sampling for transactions TkTS(T), mid-quotes TkTS(MQ), and bid and ask prices, TkTS(B) and TkTS(A), respectively. A generalization of TTS is *event time sampling*, ETS, where sampling takes place at all market events including transactions and quotations. Thus, TTS, TkTS and ETS are only based on observed prices and time points. This is not the case in *business time sampling*, BTS, which samples data such that *IV* of the intraday intervals are all equal, i.e. $IV_i = \frac{IV}{m}$.

BTS is infeasible as it depends on *IV*. However, in practice it can be approximated by prior estimates of *IV* or by standard non-parametric smoothing methods using the transaction times, see Oomen (2006). The *φ-sampling* scheme introduced by Dacorogna et al. (1993) is similar to BTS, but also removes seasonalities in the volatility across days, while the BTS just operates within the day. Empirical results of Andersen and Bollerslev (1997) and Curci and Corsi (2006) suggest that BTS can be well approximated by TTS. In the setting of Oomen (2006) random transaction times are generated by a quantity related to IV such that TTS can be directly interpreted as a feasible variant of BTS. Hansen and Lunde (2006) show that BTS,

Table 13.3 Overview of the number of observations (ticks) in different sampling schemes (01/01/2008–03/31/2008)

	S&P500 E-mini (8:30–15:15)				MSFT (9:30–16:00)			
	Total ticks	Ticks/day	Δ s	Ticks/s	Total ticks	Ticks/day	Δ s	Ticks/s
CTS (1 s)	1,496,576	23,755	1.00	1.00	1,427,277	22,655	1.05	0.95
TTS	9,466,209	150,257	0.16	6.33	8,452,679	134,170	0.18	5.65
ETS	44,646,176	708,669	0.03	29.83	–	–	–	–
QTS	–	–	–	–	22,342,994	354,651	0.07	14.93
TkTS (T)	2,772,594	44,009	0.54	1.85	1,191,310	18,910	1.26	0.80
TkTS (MQ)	1,935,415	30,721	0.77	1.29	1,893,741	30,059	0.79	1.27
TkTS (B)	968,666	15,376	1.54	0.65	831,659	13,201	1.80	0.56

by construction, minimizes *IQ*. Thus, using BTS and TTS rather than CTS may reduce the variance of RV.

Moreover, the results in Griffin and Oomen (2010), who introduce a model for transaction time patterns for analyzing the effects of TkTS and TTS, suggest that TkTS is equivalent to TTS for high levels of noise and is superior for low levels. However, once a first-order bias correction is applied, TTS is preferable.

Table 13.3 illustrates the impact of the various sampling schemes on the number of ticks available for the construction of the *IV* estimators. The numbers are based on the two datasets used in our empirical illustration, see Sect. 13.8. Generally, the number of ticks as well as the time scales are quite different across the sampling schemes. For example, for the S&P500 E-mini the one minute CTS corresponds to sampling about every 380 transactions in TTS and 1,750 events in ETS. For MSFT we obtain 340 in TTS and 900 in QTS (see Sect. 13.8). For both assets the markets have become more active, i.e. there are more quotes and trades in 2008 than in 2006. As the tick number for the Bid and Ask are similar we just report here TkTS(B).

13.7.2 Computational Aspects

A unique feature of high-frequency datasets is the vast mounds of data. In comparison to datasets commonly used in financial econometrics, e.g. daily financial data, high-frequency data requires a different approach to data storage and retrieval. The full Trade and Quote, TAQ, dataset contains for example around 10 million records per day in November 2004 and around 150 million records per day in November 2008. Obviously, the mere storage, fast retrieval and processing of this amount of data requires advanced information technology, which is discussed in this paragraph. In addition to the established row-oriented database systems we also discuss column-oriented systems and perform a comparison in terms of storage requirements and query execution speed. All computations are performed on the Microsoft Windows.Net framework but can also be replicated in econometric/statistics packages, e.g. Matlab or R, given a suitable interface to the database.

Structured data is usually stored in database management systems, i.e. software packages that offer convenient, flexible and fast read and write access to it. There is a high number of mature general purpose databases, including Microsoft SQL Server, Oracle Database, IBM DB2 and others. They are all *row-oriented*, i.e. they store entire records one after another, which may be highly disadvantageous for analytical data. Only recently have *column-oriented* databases attracted more attention, see Abadi et al. (2008). column-oriented databases store all the attributes from different records belonging to the same column contiguously and densely packed, which allows for more efficient read access, when few columns but many rows are required. column-oriented storage can be traced back to the 1970s, when transposed files and vertical partitioning clustering techniques were first studied. The interest in these techniques accelerated during the 2000s, partially because of the exponentially growing data volumes, which have become increasingly hard to handle by general purpose row-oriented databases. Another factor, which necessitates a rethinking of the design of database systems, is the increasing discrepancy between processor and physical memory speeds (Boncz 2002). While over the last decades transistor density in chips, affecting processor speed and storage density, has closely followed Moore's law – postulating a doubling of transistor chip density every 18 months – external and internal memory latency have been lagging, creating a growing bottleneck. Modern column-oriented databases are designed considering this bottleneck. Each column is stored separately, typically using large disk read units to amortize head seeks. Columnar values are stored densely, in sequence, which especially on sparse data types, can deliver astonishing levels of compression (see e.g. Stonebraker et al. 2005). Consider the storage of a bid price column, for instance. There is approx. one price change per 50 quote size changes and these price changes are mostly within a narrow band. Obviously the information entropy of the column is quite low. Furthermore, to partially avoid the processing cost involved in the decompression, column-oriented databases usually have query executors which work directly on the compressed data Abadi et al. (2006). The benefits of compression are not entirely limited to column-oriented stores but are a lot bigger, considering that the information entropy of the values within a column is almost always lower than the information entropy of the values within a record.

The biggest disadvantages of column-oriented storage manifest themselves during tuple (record) reconstruction and write operations. Write operations are problematic as inserted records have to be broken into columns and stored separately and as densely packed data makes moving records almost impossible. Some of the techniques used to mitigate the write issues are in-memory buffering and partition-merging. The problem with tuple reconstruction is again that the data for a single row is scattered in different locations on the disk. Most database interface standards (e.g. ODBC) access the results of a query on a row basis, not per columns. Thus, at some point of the query plan of a column-oriented database, the data from multiple columns must be combined in records. Abadi et al. (2008) consider several techniques, which can be used to minimize this reconstruction overhead.

A list of the currently available commercial column-oriented databases includes Sybase IQ, Vertica, VectorWise, InfoBright, Exasol, ParAccel, SAP BI Accelerator,

Kickfire and others. Not all of them are general purpose databases, e.g. InfoBright is actually an MySQL storage engine and Kickfire is offered as an hardware appliance. The most mature academic system is MonetDB/X100, developed at Centrum Wiskund & Informaticas.

The MySQL/InfoBright solution can be referred to as a hybrid system, as MySQL can simultaneously handle a number of different engines, including both column- and row-oriented stores, which can be selected on a per table basis. The usage of the highly mature MySQL database platform and the fact that InfoBright is freely available in an open-source edition (InfoBright ICE), make it a good candidate for academic comparison.

In the following we compare the retrieval speed and the compression levels of the row-oriented Microsoft SQL Server, which is a mature database system, introduced in 1989 and very well integrated into the whole palette of development tools from Microsoft, and the column-oriented database MySQL/InfoBright. The dataset used for the test comprises all transactions and quote updates in the first quarter of 2008, a total of 97.9 million records.

The sizes of the sample in the form of: raw data, flat file, uncompressed Microsoft SQL database, compressed Microsoft SQL database and compressed InfoBright database are given in Fig. 13.1. Indeed, the compression rate of InfoBright is astonishing – 1–20 compared to the raw data size. The implications are huge – a raw dataset of e.g. 10 Terabyte (TB) can be stored on an off-the-shelf hard drive with 500 GB capacity. On the contrary, Microsoft SQL manages to achieve only compression ratios of 1–2, compared to the raw data size, and 1–3, compared to the uncompressed Microsoft SQL database. The benefits of these rates may become questionable in the light of the processor overhead caused by decompression.

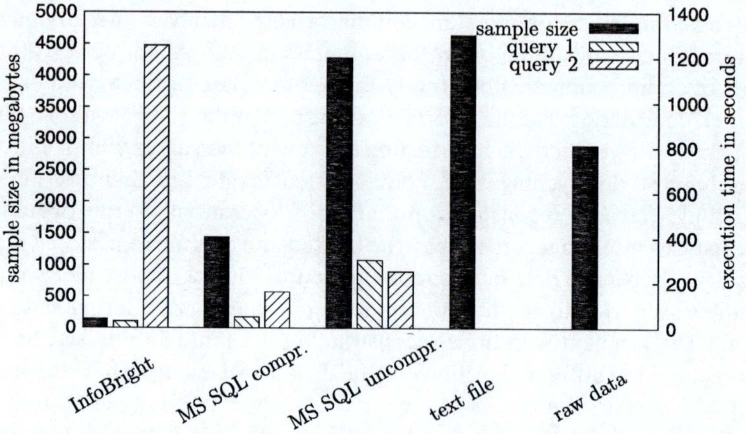

Fig. 13.1 Sample size in megabytes (*left scale*) in the form of InfoBright database, Microsoft SQL compressed database, Microsoft SQL uncompressed database, comma-separated text file and theoretical size as an in-memory structure (raw data); and query execution speed in seconds (*right scale*) for the InfoBright, Microsoft SQL compressed and Microsoft SQL uncompressed databases

The performance of the two database systems will be compared with the help of two queries, the first of which will test the speed of retrieval of aggregated and filtered information, performing an *in-database* full table scan:

```
SELECT SecurityID, DateID, MIN(Timestamp),
MAX(Timestamp), SUM(Count), SUM(Size), MIN(Price),
MAX(Price) FROM tblTickES WHERE FieldID = 3 GROUP
BY SecurityID, DateID
```

The second query will test the sequential retrieval of all the information in the table from an *external* environment:

```
SELECT SecurityID, DateID, Timestamp, FieldID, Price,
Size, Count FROM tblTickES
```

Both query types are important in analytical work (e.g. in econometrics) but the performance of the second is especially relevant, as it is used on a more regular basis and requires the transfer of huge amounts of data between applications, a process which can quickly become a bottleneck for the whole system.

It is important to note that these tests were not performed under ideal conditions and are in no way representative for the general and even optimal performance of the two database systems. The tests are designed to assess the performance which can be expected from the systems in a normal working environment by an analytical worker, who is not able or willing to spend considerable amounts of time on learning and optimizing the systems. The results of the tests on the second run are reported in Fig. 13.1. The results of the first run are ignored because they can unfairly penalize systems which make use of cache and memory to optimize their performance. The results of runs after the second one, on the other side, can be too hardware specific, since some systems could manage to cache large amount of data from hard drive media in the memory.

The speed of retrieving all data from InfoBright is low. The number in Fig. 13.1 reports the result for the general ODBC driver for MySQL. Changes of diverse settings in the ODBC driver did not improve the situation. The in-database query speed of InfoBright is satisfactory. Overall, the compressed Microsoft SQL variant offers a promising improvement over the uncompressed one – a factor of 1–6 for the in-database query and slightly less than 1–2 for the external query.

To conclude, column-oriented database systems provide a comfortable way to achieve a high comparison of the data and fast in-database queries. On the other side a sequential retrieval of all records is significantly slower than for row-oriented database systems. Thus, the preferred choice of the database system may depend whether good compression or fast sequential retrieval of all records is important.

13.8 Empirical Illustration

Our empirical application aims at illustrating the impact of market microstructure noise on the estimation of IV in finite samples and on an empirical comparison of the various estimators. To this end, we consider two high-frequency datasets over the first quarters of 2008: data on the highly liquid futures S&P500 E-mini and on an individual stock, i.e. of Microsoft, MSFT. The reason for this choice is, that the data sources and the type of asset are quite different allowing for a more detailed analysis of the market microstructure noise and the performance of the estimators.

The S&P500 E-mini, is traded on the CME Globex electronic trading platform and the dataset consists of all transaction and quotation updates in correct order and with time-stamps given in milliseconds. The quality of the data is very high and no filtering or cleaning is required, except for a trivial removal of any non-positive prices or volume. Note that in our application we roll-over between the most liquid contracts.

Such highly accurate information is not available for MSFT, which is obtained from the (monthly) TAQ dataset, disseminated by the NYSE for all listed stocks. The dataset includes quotation updates and transactions provided in separate files and the time-stamps are available only up to the precision of a second. This requires a more involved data filtering and cleaning. As the TAQ is probably the most popular high-frequency dataset, we give here a few more details on the data manipulations conducted for our analysis. In particular, we focus on the TAQ data coming from the NASDAQ, such that we filter all records with exchange identifiers being different from T, D or Q, as specified in the TAQ 3 User's Guide (2004–2008). For transactions we have additionally removed records with a CORR attribute different from 0 or 1. The resulting data contains numerous outliers, such as prices equal to 0.01$ or 2,000$ right next to regular prices varying around the usual trading price range of MSFT, i.e. around 30$. Such outliers have been removed by first dismissing records with non-positive price or size and by discarding records with a price that deviates from the last one by more than 10%. More advanced methods involve filtering based on rolling windows and a deviation threshold adapted to the current volatility of the price, see Brownlees and Gallo (2006).

One of the major problems of the TAQ dataset, however, is the separate distribution of transaction and quote data and the lack of millisecond precision in the time-stamps, such that the exact order of the generation of transaction prices and quotes over the trading day cannot be deduced. An approach to match transactions at least to the corresponding bid and ask quotes has been proposed in Lee and Ready (1991). For volatility estimation such synchronicity is only required for sampling schemes involving price type pairs. For MSFT we have, thus, limited TTS to transaction prices and introduce a modification of ETS, which only regards quote updates as events, called *quote-time sampling* (QTS). This sampling scheme can of course be applied to mid-quotes, bid and ask prices avoiding any mixing of transaction and quote series.

13.8.1 Type and Magnitude of the Noise Process

A common tool to visualize the impact of market microstructure noise on the high-frequency based volatility estimators are the so-called volatility signature plots made popular in Andersen et al. (2000). Depicted are usually the average estimates of daily volatility as a function of the sampling frequency, where the average is taken across multiple days, i.e. in our case all trading days in the first quarter of 2008. Figure 13.2 shows the volatility signature plots for RV based on different sampling schemes and different prices.

Overall, it seems that CTS is most strongly affected by market microstructure noise, compared to the alternative sampling schemes. This is important, as CTS is probably the most commonly applied sampling method. Moreover, under the assumption of a pure jump diffusion price process, Oomen (2006) shows theoretically that TTS is superior to CTS, if the in a MSE sense optimal sampling frequency is used.

Interestingly, the biases observed in the *RV* estimates for both of our datasets are all positive, irrespective of the sampling scheme and the employed price series. Moreover, we find across all sampling schemes that transaction prices produce the most severe bias in the case of the S&P500 E-mini (note that all quotes series yield identical *RV*s in both CTS and TTS and we thus only display the estimates based on the mid-quotes), but are preferable for MSFT. Using the same stock but a different sample period, Hansen and Lunde (2006) instead observe a superiority of quotation data in terms of bias reduction and a negative bias if quote data is used. The latter may be induced by the non-synchronous quote revisions or price staleness. Recall that another source of a negative bias may be given by the dependence between the efficient price and the noise. Obviously, the potential presence of these different types of market microstructure effects make it difficult to draw general statements on the expected sign and size of the biases in the *RV* estimator and the preferable sampling method/price series. Negative and positive biases may be present at the same time, leading to overall small biases or to non-monotone patterns like the one observed for the S&P500 E-mini under ETS. Volatility signature plots based on estimators that are robust to particular types of microstructure noise, allow to shed more light on the noise effects. Using a kernel-based approach, Hansen and Lunde (2006) for example find, that the iid robust *RV* based on transaction prices also exhibits a negative bias and that this may be due to endogenous noise.

Instead of using pure visual inspections to judge the presence and potential type of market microstructure noise, Awartani et al. (2009) propose statistical tests on the no noise assumption and on noise with constant variance. The no market microstructure noise test builds on the idea, that *RV* sampled at two different frequencies, e.g. very frequently and sparsely, should both converge in probability to *IV*. The test therefore evaluates, whether the difference of both estimators is zero. Asymptotically the test statistic is normally distributed. The implementation of the

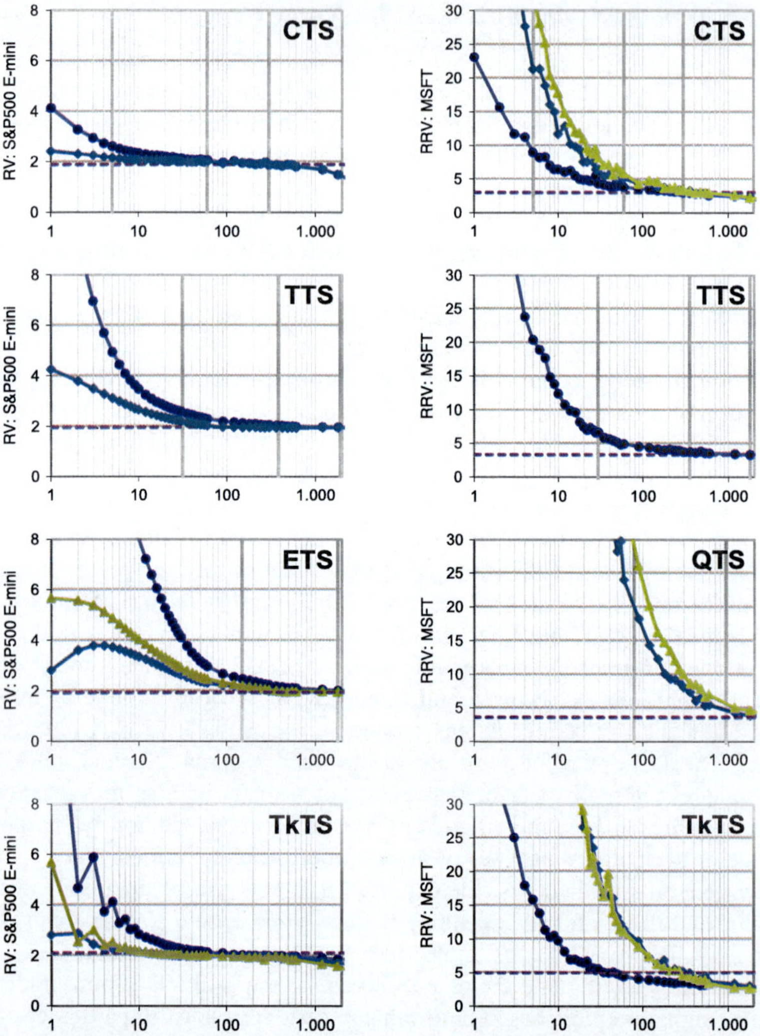

Fig. 13.2 Volatility signature plots for RV of S&P500 E-mini (*left*) and MSFT (*right*), first quarter 2008, based on different sampling schemes and different price series: transaction prices (*circles*), mid-quotes (*rhombuses*) and bid/ask prices (*triangles*). The *bold vertical lines* represent the frequency equal, on average, to 5 s, 1 and 5 min in the respective sampling scheme, the *horizontal line* refers to the average RV estimate based on a 30 min frequency

test of course depends on the choice of both sampling frequencies. As an alternative, Awartani et al. (2009) suggest to exploit the autocovariance structure of the intraday returns. Focusing on the first lag, the scaled autocovariance estimator over e.g. n days can be expressed by

$$\bar{m}^{3/2}\widehat{cov}(r_i^{(m)}, r_{i-1}^{(m)}) = \sqrt{\bar{m}}\left(\sum_{i=2}^{\bar{m}} r_i^{(m)} r_{i-1}^{(m)} + \sum_{i=2}^{\bar{m}} \epsilon_{i/m}\epsilon_{(i-1)/m} + \sum_{i=2}^{\bar{m}} r_i^{(m)}\epsilon_{(i-1)/m} + \sum_{i=2}^{\bar{m}} \epsilon_{i/m} r_{i-1}^{(m)}\right),$$

where $\bar{m} = nm$. Under the null of no noise the last three terms converge to zero almost surely. The first term therefore drives the asymptotic distribution of the test statistic which is given by:

$$\frac{\sqrt{\bar{m}}\sum_{i=2}^{\bar{m}} r_i^{(m)} r_{i-1}^{(m)}}{\sqrt{IQ}} \xrightarrow{d} \mathcal{N}(0, 1).$$

After some rearrangement and for large $\bar{m}$, the feasible test statistic can also be computed in terms of the sample autocorrelation coefficient of the intraday returns $\hat{\rho}_1(r_i^{(m)}) = \sum_{i=2}^{\bar{m}} r_i^{(m)} r_{i-1}^{(m)} / \sum_{i=2}^{\bar{m}} (r_i^{(m)})^2$:

$$z_{AC,1} = \frac{\hat{\rho}_1(r_i^{(m)})}{\sqrt{\frac{1}{3}\sum_{i=2}^{\bar{m}} (r_i^{(m)})^4 / \sum_{i=2}^{\bar{m}} (r_i^{(m)})^2}} \xrightarrow{d} \mathcal{N}(0, 1).$$

Figure 13.3 presents the test statistic and corresponding confidence intervals as a function of the sampling frequency over the first quarter of 2008. The results indicate that the noise "kicks in" at frequencies exceeding approximately 1 and 3 min for the S&P500 E-mini and MSFT data, respectively. Moreover, if quote data is used in the case of MSFT, then the noise robust sampling frequency should be lower than approx. every 5 min.

Most of the noise robust estimators have been derived under the assumption of iid noise, implying also that the noise has a constant noise variance, irrespective of the sampling frequency. Awartani et al. (2009) therefore propose a test for the null of constant noise variance. To this end it is instructive to first consider feasible estimators of the noise variance. Based on the bias of *RV* in the presence of iid noise, see (13.3), the noise variance can be estimated by $\tilde{\omega}^2 = RV/2m$ using sparse sampling. However, Hansen and Lunde (2006) show that this estimator will overestimate the true noise variance whenever $IV/2m$ is negligible. They therefore suggest the following estimator:

$$\hat{\omega}^2 = \frac{RV^{(m_K)} - RV^{(m_J)}}{2(m_K - m_J)}, \tag{13.14}$$

where m_J denotes a lower sampling frequency, such that $RV^{(m_J)}$ is an unbiased estimator of *IV*. However, both variance estimators may be inadequate if the iid noise assumption is not appropriate, which may be the case at very high frequencies.

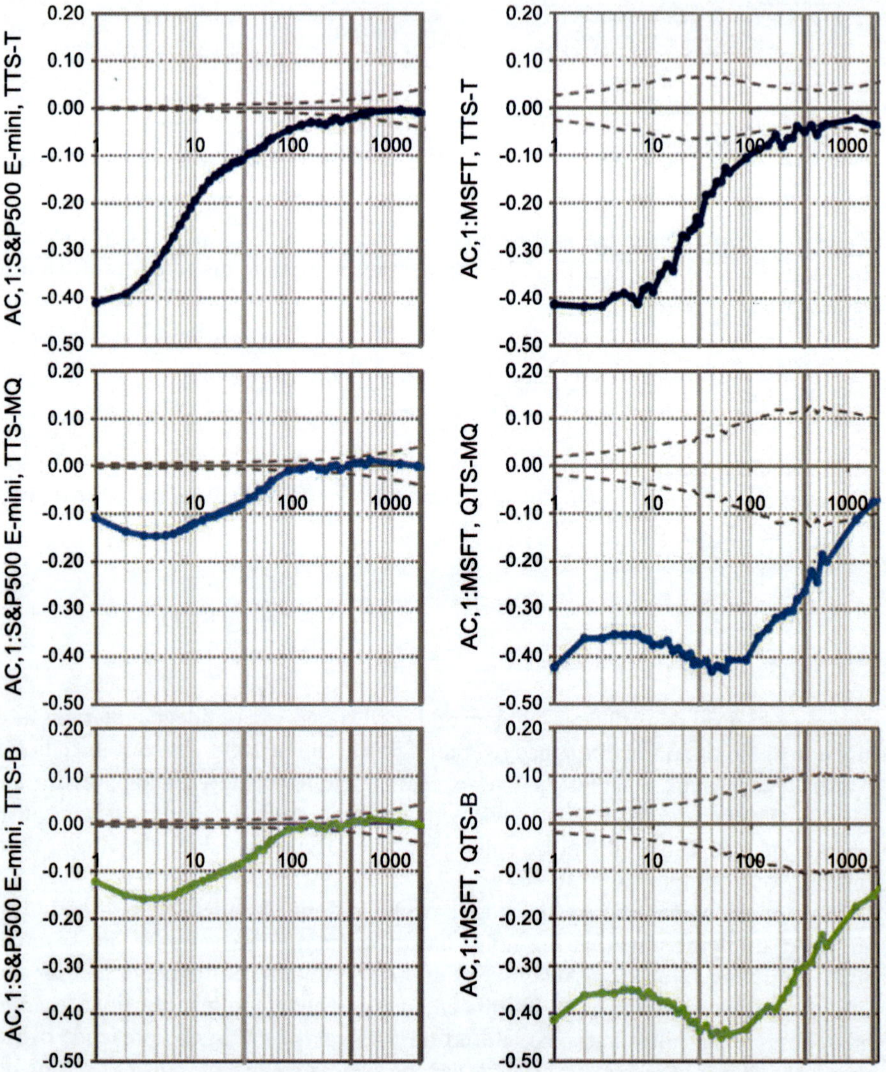

Fig. 13.3 Tests on no noise. Depicted are the $z_{AC,1}$ statistics and corresponding confidence intervals (*dashed*) based on different sampling frequencies for the S&P500 E-mini (*left*) and MSFT (*right*) using TTS/QTS and transaction (*top*), mid-quote (*middle row*) and bid/ask quote (*bottom*) price series. The *bold vertical lines* give the frequency equal, on average, to 5 s, 1 and 5 min

The constant noise variance test of Awartani et al. (2009) considers the difference between two noise variances estimated at different sampling frequencies:

$$z_{IND} = \sqrt{\bar{m}_J}\,\frac{\frac{RV^{(\bar{m}_K)}-RV^{(\bar{m}_L)}}{2\bar{m}_K} - \frac{RV^{(\bar{m}_J)}-RV^{(\bar{m}_L)}}{2\bar{m}_J}}{\sqrt{3\left(\frac{IQ^{(\bar{m}_J)}}{2\bar{m}_J{}^2} - \left(\frac{RV^{(\bar{m}_J)}}{2\bar{m}_J}\right)^2\right)}} \sim \mathcal{N}(0,1)\,,$$

Table 13.4 Test on constant noise variance. Reported are the test statistics z_{IND}. Asterisks denote rejections at the 5% significance level

S&P500 E-mini, TTS, B					MSFT, TTS, T				
	m_J					m_J			
m_K	40	90	120	240	m_K	90	180	360	420
60	20.71*	–	–	–	180	8.86*	–	–	–
120	15.43*	2.61*	–	–	360	9.09*	5.78*	–	–
300	12.38*	1.2	0.85	−0.61	540	6.65*	4.11*	2.73*	1.6

where the third frequency $\bar{m}_L$ should be unaffected by the noise. Moreover, $m_J < m_K$.

Table 13.4 presents the test results for some pairs of m_K and m_J, where the choice of m_L is conditional on the no noise test results. Obviously, the constant variance assumption is rejected only at the very high frequencies. The results are very much in line with those reported in Awartani et al. (2009) and we conclude that noise seems to be statistically significant at frequencies higher than 1–5 min (depending on the dataset and the price series used) and that the iid noise assumption is violated only at the ultrahigh frequencies, e.g. at approximately 0.5 min TTS, B sampling for S&P500 E-mini and at 1.5 min TTS, transaction prices for MSFT.

The noise test results may serve as a guidance for the selection of the sampling frequencies in the noise variance estimation, see (13.14). In particular, m_J should be set to a frequency where no noise is present, while m_K should correspond to a very high frequency, at which, however, the iid assumption is not violated. The procedure should produce reliable noise variance estimates. Applying this method to the S&P500 E-mini, for example, yields an estimated signal-to-noise ratio $2m_J\omega^2/IV$ of about 8% using TTS, B. In contrast, $\tilde{\omega}^2$ yields a signal-to-noise ratio of about 45%.

13.8.2 Volatility Estimates

In the following we provide an empirical illustration of the various volatility estimators. To this end we first need to determine the values of the hyperparameters, which can be partly guided by the findings of the previous section. In the computation of $TTSRV_{adj}$, for example, we can set the return horizon of the slow time scale (K) to the highest frequencies without significant noise and the horizon of the fast time scale returns (J) to the highest frequencies at which the iid assumption is not violated. For the KRV estimator we implement the non-flat-top Parzen kernel. The optimal bandwidth H is selected to minimize the variance of the estimator, as described in Sect. 13.4. Estimates for ω^2 and IV are derived from the full sample period to obtain constant values $H(m)$ for all days.

In the implementation of the RRV estimator we vary K and sample at every observation in each interval. The DRV estimator is implemented in its first exit time variant. Specifically, in the first half of the day the next exit time is used, while the second half of the day is based on the previous exit time. We depict the results for 8 ticks on the original scale, which is converted to the logarithmic scale at the beginning of each day using the current price level. Note that we have also experimented with alternative numbers of ticks ranging from 1 to 15 and we found that the resulting *IV* estimates are quite stable for values of $r = 3$–14.

Following Christensen et al. (2009b), we compute the subsampled version of the *QRV* estimators for three different block lengths, $m_K = 20, 40, 100$, for a fixed set of quantiles, $\bar{\lambda} = (0.80, 0.85, 0.90, 0.95)'$, and asymptotically optimal weights. These parameters are also adopted for the QRV_{iid} estimator. While the optimal value of L can be determined by a data-driven simulation of the MSE loss function, we set here L to a conservative value of 20 and $c = 0.02$, which is motivated by the finite sample performance study of Christensen et al. (2009b). Nevertheless, note that ideally L and c should be chosen at each sampling frequency m. Thus, our volatility signature plots of *QRV* should be interpreted with care.

Figures 13.4 and 13.5 depict the resulting volatility estimates over the period from 01/01/2006 to 05/31/2008 with respect to various sampling frequencies m. Clearly, the estimators that have been derived under the no noise assumption seem to be subject to severe positive biases at high frequencies, with the exception of the *BPV* for the S&P500 E-mini, which seems to be negatively biased. Interestingly, the two estimators that are robust to time-dependent noise specifications, i.e. *TTSRV* and *KRV*, appear to be unbiased. In contrast, the QRV_{iid} is negatively biased at ultrahigh frequencies, pointing towards the presence of time-dependent noise at those frequencies. Overall, the results are thus in line with our findings from the noise specification tests. Moreover, although the *DRV* estimator has been formally derived under the no noise assumption, we find empirical support for the simulation results of Andersen et al. (2009), indicating that *DRV* is robust to iid and serial dependent noise. (Note that we do not report *DRV* for MSFT due to the coarse time stamping.)

Another aspect that should be kept in mind when interpreting the volatility estimates is that some estimators do not only measure *IV*, but additionally the variation due to jumps. From a closer, i.e. zoomed-in, look at the volatility signature plots (not presented here), however, we cannot observe systematically lower volatility estimates based on the jump robust estimators. Testing for the relative contribution of jumps to total price variation based on the ratio *BPV* to *RV*, see e.g. Huang and Tauchen (2005), we do not find significant evidence for jumps at lower frequencies, e.g. lower than 5 min for the S&P500 data, respectively. Computing the tests at various sampling frequencies, similarly to the volatility signature plots, we could, however, observe that the relative jump contribution seems to be increasing strongly at the very high frequencies (e.g. for the S&P500 E-mini we observe that 20% of the total price variation is due to jumps at a sampling frequency of 1 s and reaches up to 80% for ultrahigh frequencies). Still it is interesting to understand the behavior of the statistic at higher frequencies, which, at a first glance, points to significant presence of discontinuities. *BPV* and the jump statistic are not derived under microstructure noise. We know that bid prices are rather stale, with long series

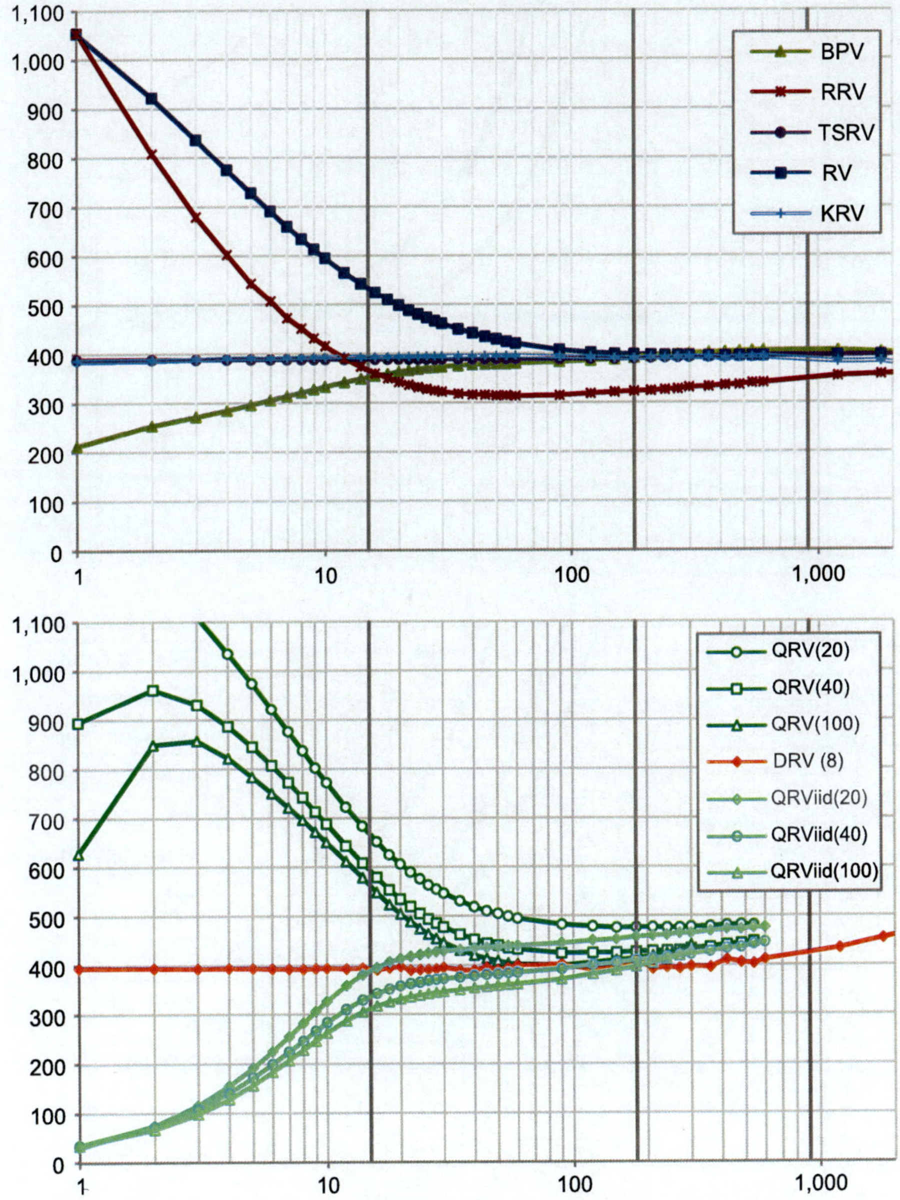

Fig. 13.4 Volatility signatures of the various high-frequency based estimators for the S&P500 E-mini based on TTS with bid prices over the period from 01/01/2006 to 05/31/2008. The *bold vertical lines* represent the frequency equal, on average, to 5 s, 1 and 5 min

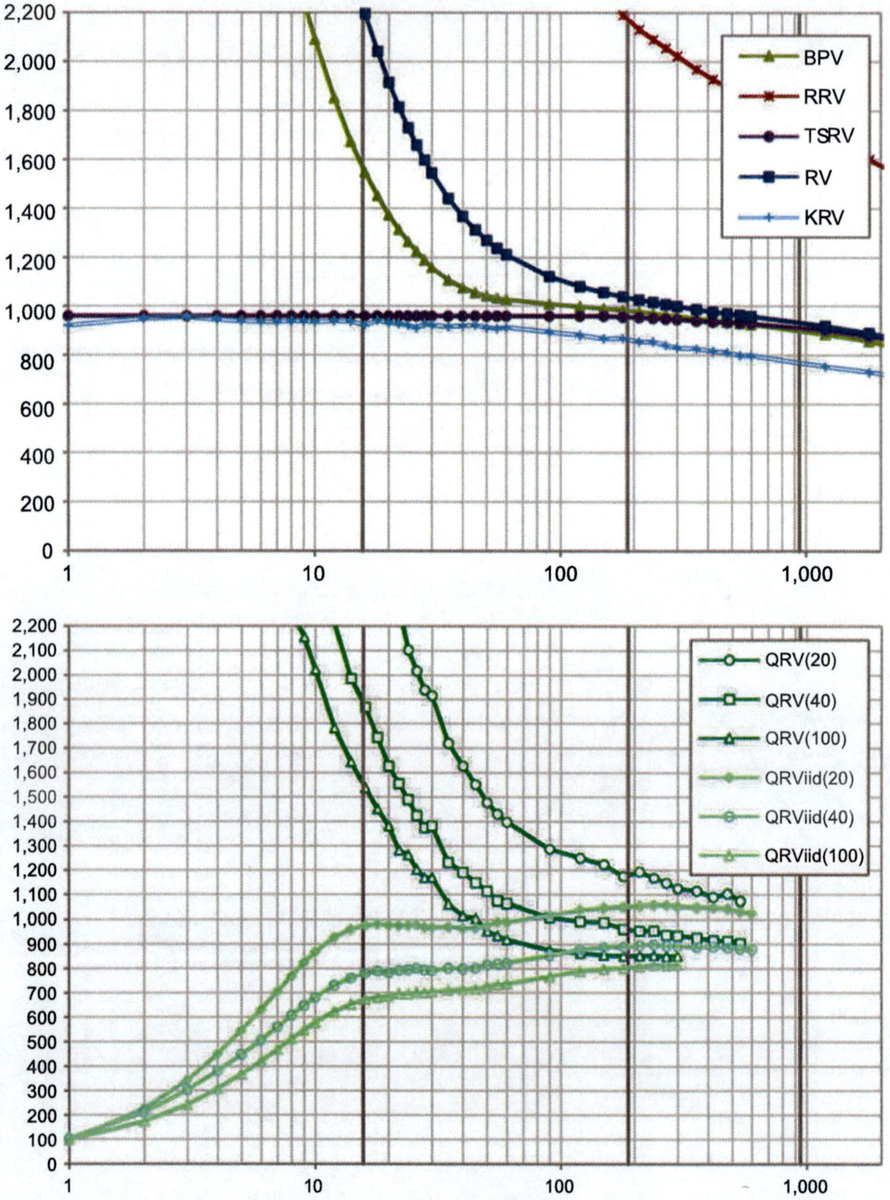

Fig. 13.5 Volatility signatures of the various high-frequency based estimators for MSFT based on QTS with transaction prices over the period from 01/01/2006 to 05/31/2008. The *bold vertical lines* represent the frequency equal, on average, to 5 s, 1 and 5 min

of zero-return observations and infrequent returns of at least one tick. If we imagine a scenario, in which we sample in TTS and there are at least two transactions between every two consecutive price moves, actually not so far from reality, then there will be no two consecutive returns, both $\neq 0$, and therefore *BPV* goes to zero. Thus, from the perspective of *BPV*, high-frequency data approaches a pure jump process as the frequency of the observations increases.

13.9 Conclusion

In this chapter we have reviewed the rapidly growing literature on volatility estimation based on high-frequency financial data. We have paid particular attention to estimators that exploit different facets of such data. We have provided a theoretical and empirical comparison of the discussed estimators. Moreover, statistical tests indicated that for the series considered in this chapter market microstructure noise can be neglected at sampling frequencies lower than 5 min, and that the common assumption of iid noise is only violated at the very high frequencies. The specific type of noise at these ultra-high frequencies is still an open question. Interestingly, estimators that are robust to serial dependent and/or endogenous noise (*TSRV*, *KRV*) seem to provide plausible estimates at all frequencies. Nevertheless, understanding the properties of estimators under different noise types could be considered in more detail within a simulation study, allowing also for a more thorough comparison of the various estimators in terms of their finite sample performance.

References

Abadi, D. J., Madden, S. R., & Ferreira, M. (2006). Integrating compression and execution in column-oriented database systems. In *Proceedings of the 2006 ACM SIGMOD International Conference on Management of Data* (pp. 671–682).

Abadi, D. J., Madden, S. R., & Hachem, N. (2008). Column-stores vs. row-stores: How different are they really? In *Proceedings of the 2008 ACM SIGMOD International Conference on Management of Data* (pp. 967–980).

Aït-Sahalia, Y., Mykland, P. A., & Zhang, L. (2005). How often to sample a continuous-time process in the presence of market microstructure noise. *Review of Financial Studies, 18*, 351–416.

Aït-Sahalia, Y., Mykland, P. A., & Zhang, L. (2010). Ultra high frequency volatility estimation with dependent microstructure noise. *Journal of Econometrics, 160*(1), 2011, 160–175.

Andersen, T. G., & Bollerslev, T. (1997). Intraday periodicity and volatility persistence in financial markets. *Journal of Empirical Finance, 4*, 115–158.

Andersen, T. G., & Bollerslev, T. (1998). Answering the skeptics: Yes, standard volatility models do provide accurate forecasts. *International Economic Review, 39*, 885–905.

Andersen, T. G., Bollerslev, T., Diebold, F. X., & Labys, P. (2000). Great realisations. *Risk, 13*, 105–108.

Andersen, T. G., Bollerslev, T., Diebold, F. X., & Ebens, H. (2001). The distribution of realized stock return volatility. *Journal of Financial Economics, 61*, 43–76.

Andersen, T. G., Bollerslev, T., Diebold, F. X., & Labys, P. (2003). Modeling and forecasting realized volatility. *Econometrica, 71*, 579–625.

Andersen, T. G., Bollerslev, T., & Diebold, F. X. (2007). Roughing it up: Including jump components in the measurement, modeling, and forecasting of return volatility. *The Review of Economics and Statistics, 89*, 701–720.

Andersen, T. G., Dobrev, D., & Schaumburg, E. (2009). Duration-based volatility estimation. Working Paper.

Awartani, B., Corradi, V., & Distaso, W. (2009). Assessing market microstructure effects via realized volatility measures with an application to the dow jones industrial average stocks. *Journal of Business & Economic Statistics, 27*, 251–265.

Barndorff-Nielsen, O. E., & Shephard, N. (2002a). Econometric analysis of realised volatility and its use in estimating stochastic volatility models. *Journal of the Royal Statistical Society, Series B, 64*, 253–280.

Barndorff-Nielsen, O. E., & Shephard, N. (2002b). Estimating quadratic variation using realized variance. *Journal of Applied Econometrics, 17*(5), 457–477.

Barndorff-Nielsen, O. E., & Shephard, N. (2004). Power and bipower variation with stochastic volatility and jumps. *Journal of Financial Econometrics, 2*, 1–37.

Barndorff-Nielsen, O. E., Shephard, N., & Winkel, M. (2006). Limit theorems for multipower variation in the presence of jumps. *Stochastic Processes and their Applications, 116*, 796–806.

Barndorff-Nielsen, O. E., Hansen, P. R., Lunde, A., & Shephard, N. (2008). Designing realised kernels to measure the ex-post variation of equity prices in the presence of noise. *Econometrica, 76*, 1481–1536.

Barndorff-Nielsen, O. E., Hansen, P. R., Lunde, A., & Shephard, N. (2009). Realized kernels in practice: Trades and quotes. *Econometrics Journal, 12*, C1–C32.

Barndorff-Nielsen, O. E., Hansen, P. R., Lunde, A., & Shephard, N. (2010). Subsampled realised kernels. *Journal of Econometrics, 160*(1), 2011, 204–219.

Boncz, P. A. (2002). *Monet: A Next-Generation DBMS Kernel for Query-Intensive Applications*. PhD thesis, Universiteit van Amsterdam: Netherlands.

Brownlees, C. T., & Gallo, G. M. (2006). Financial econometric analysis at ultra-high frequency: Data handling concerns. *Computational Statistics & Data Analysis, 51*, 2232–2245.

Cho, C., & Frees, F. (1988). Estimating the volatility of discrete stock prices. *Journal of Finance, 43*, 451–466.

Christensen, K., & Podolskij, M. (2007). Realized range-based estimation of integrated variance. *Journal of Econometrics, 141*, 323–349.

Christensen, K., Podolskij, M.. & Vetter, M. (2009a). Bias-correcting the realized range-based variance in the presence of market microstructure noise. *Finance and Stochastics, 13*, 239–268.

Christensen, K., Oomen, R., & Podolskij, M. (2009b). Realised quantile-based estimation of the integrated variance. Working Paper.

Curci, G., & Corsi, F. (2006). A discrete sine transform approach for realized volatility measurement. Working Paper.

Dacorogna, M. M., Müller, U. A., Nagler, R. J., Olsen, R. B., & Puctet, O. V. (1993). A geographical model for the daily and weekly seasonal volatility in the foreign exchange market. *Journal of International Money and Finance, 12*, 413–438.

David, H. A. (1970). *Order statistics*. New York: Wiley.

Eraker, B., Johannes, M., & Polson, N. (2003). The impact of jumps in volatility and returns. *Journal of Finance, 58*, 1269–1300.

Feller, W. (1951). The asymptotic distribution of the range of sums of independent random variables. *The Annals of Mathematical Statistics, 22*, 427–432.

Garman, M. B., & Klass, M. J. (1980). On the estimation of security price volatilities from historical data. *Journal of Business, 53*, 67–78.

Gouriéroux, C., & Jasiak, J. (2001). *Financial Econometrics: Problems, Models, and Methods*. Princeton, NJ: Princeton University Press.

Griffin, J. E., & Oomen, R. C. A. (2010). Covariance measurement in the presence of non-synchronous trading and market microstructure noise. *Journal of Econometrics*, *160*(1), 2011, 58–68.

Hansen, P. R., & Lunde, A. (2006). Realized variance and market microstructure noise. *Journal of Business & Economic Statistics*, *24*(2), 127–161.

Hasbrouck, J. (2007). *Empirical market microstructure: The institutions, economics, and econometrics of securities trading*. UK: Oxford University Press

Huang, X., & Tauchen, G. (2005). The relative contribution of jumps to total price variance. *Journal of Financial Econometrics*, *3*(4), 456–499.

Klößner, S. (2009). Estimating volatility using intradaily highs and lows. Working Paper.

Lee, C. M. C., & Ready, M. J. (1991). Inferring trade direction from intraday data. *Journal of Finance*, *46*, 733–746.

Mancini, C. (2009). Non-parametric threshold estimation for models with stochastic diffusion coefficient and jumps. *Scandinavian Journal of Statistics*, *36*, 270–296.

Martens, M., & van Dijk, D. (2007). Measuring volatility with the realized range. *Journal of Econometrics*, *138*, 181–207.

McAleer, M., & Medeiros, M. (2008). Realized volatilty: A review. *Econometric Reviews*, *26*, 10–45.

Mosteller, F. (1946). On some useful "inefficient" statistics. *The Annals of Mathematical Statistics*, *17*, 377–408.

Oomen, R. C. (2006). Properties of realized variance under alternative sampling schemes. *Journal of Business & Economic Statistics*, *24*(2), 219–237.

Parkinson, M. (1980). The extreme value method for estimating the variance of the rate of return. *Journal of Business*, *53*, 61–65.

Podolskij, M., & Vetter, M. (2009). Estimation of volatility functionals in the simultaneous presence of microstructure noise and jumps. *Bernoulli*, *15*, 634–658.

Roll, R. (1984). A simple implicit measure of the effective bid-ask spread in an efficient market. *Journal of Finance*, *39*, 1127–1139.

Stonebraker, M., Abadi, D., Batkin, A., Chen, X., Cherniack, M., Ferreiraa, M., Lau, E., Lin, A., Madden, S., ONeil, E., ONeil, P., Rasin, A., Tran, N., & Zdonik, S. (2005). C-Store: A column-oriented DBMS. In *Proceedings of the 31st International Conference on Very Large Data Bases* (pp. 553–564). NY: ACM.

Vetter, M. (2010). Limit theorems for bipower variation of semimartingales. *Stochastic Processes and their Applications*, *120*, 22–38.

Wasserfallen, W., & Zimmermann, H. (1985). The behavior of intra-daily exchange rates. *Journal of Banking and Finance*, *9*, 55–72.

Zhang, L. (2006). Efficient estimation of stochastic volatility using noisy observations: A multi-scale approach. *Bernoulli*, *12*, 1019–1043.

Zhang, L., Mykland, P. A., & Aït-Sahalia, Y. (2005). A tale of two time scales: Determining integrated volatility with noisy high-frequency data. *Journal of the American Statistical Association*, *100*, 1394–1411.

Zhou, B. (1996). High-frequency data and volatility in foreign-exchange rates. *Journal of Business & Economic Statistics*, *14*, 45–52.

Chapter 14
Identifying Jumps in Asset Prices

Johan Bjursell and James E. Gentle

14.1 Jump/Diffusion Models of Asset Prices

For over a hundred years, diffusion differential equations have been used to model the changes in asset prices. Despite obvious fundamental problems with these equations, such as the requirement of continuity, they often provide adequate local fits to the observed asset price process. There are, however, several aspects of the empirical process that are not fit by simple diffusion equations.

Direct observation indicates that the standard deviation of an asset's price changes is not constant. A diffusion model can be modified so that this term is not constant. There are, of course, several ways that this can be done, for example, by coupling the basic diffusion model with another stochastic differential equation that models changes in the standard deviation.

Direct observation indicates that the distribution of an asset's price changes, even over local periods in which parameters can be assumed constant, do not follow a Gaussian distribution. The observations clearly cannot be realizations of a random process with a kurtosis of 0. Again, it might be possible to modify a simple diffusion equation so that the stochastic component has heavy tails. Various continuous probability distributions with varying tail weights, or with other non-Gaussian moment properties such as skewness can be used instead of a simple Brownian process.

Finally, direct observation indicates that asset prices often suddenly change by a very large amount. While a diffusion model could perhaps account for this kind of change if the random component has extremely heavy tails, this approach has the problems of dealing with infinite moments. Many statistical procedures such as those based on least-squares cannot be used. Many statistics of interest, such as

J. Bjursell (✉) · J.E. Gentle
George Mason University, Braddock, VA, USA
e-mail: cbjursel@gmu.edu; jgentle@gmu.edu

J.-C. Duan et al. (eds.), *Handbook of Computational Finance*, Springer Handbooks of Computational Statistics, DOI 10.1007/978-3-642-17254-0_14,

confidence intervals, cannot be computed. In addition, the fundamental disconnect between a continuous-time model and real world markets remains. An alternative approach to the problem of using diffusion differential equations to model a process that may suddenly change by a very large amount is to superimpose a discrete jump process on the continuous diffusion process.

Let $X_t = \log S_t$ denote the logarithmic price where S_t is the observed price at time t. Assume that the logarithmic price process, X_t, follows a continuous-time diffusion process coupled with a discrete process defined as,

$$\mathrm{d}X_t = \mu_t \mathrm{d}t + \sigma_t \mathrm{d}W_t + \kappa_t \mathrm{d}q_t, \tag{14.1}$$

where μ_t is the instantaneous drift process and σ_t is the diffusion process; W_t is the standard Wiener process; q_t is a counting process with intensity λ_t, that is, $P(\mathrm{d}q_t = 1) = \lambda_t \mathrm{d}t$; and κ_t is the size of the price jump at time t if a jump occurred. If X_{t-} denotes the price immediately prior to the jump at time t, then $\kappa_t = X_t - X_{t-}$.

This model has been used in various applications of financial modeling, such as options pricing; see Cont and Tankov (2004) for general background and methodology. Use of this model raises two interesting questions:

when has a jump occurred and how large was the jump?

That is, in an observed sequence of prices, $\{X_t\}$, identify t_j when the counting process fired, and determine $\kappa_{t_j} = X_{t_j} - X_{t_j -}$.

In Sect. 14.2, we study some tests that have been proposed for identification of jumps in a jump/diffusion process, and Sect. 14.3, we consider applications of this model in the context of the U.S. energy futures market. This is an interesting setting for a model with jumps because of the effects of the regular release of data on petroleum commodities by the U.S. Energy Information Administration.

14.2 Identification of Jumps

An increase in the availability of high-frequency data has produced a growing literature on nonparametric methods to identify jumps such as Barndorff-Nielsen and Shephard (2004, 2006), Fan and Wang (2007), Jiang and Oomen (2008), Lee and Mykland (2008) and Sen (2008) This section introduces the work by Barndorff-Nielsen and Shephard, which many others have built on and used as a benchmark.

14.2.1 Theoretical Framework

Define the *intraday return*, r_{t_j}, as the difference between two logarithmic prices,

$$r_{t_j} = X_{t_j} - X_{t_{j-1}}, \tag{14.2}$$

where t_j denotes the jth intraday observation on the tth day. Let Δ denote the discrete intraday sample period of length, $t_j - t_{j-1}$. The *realized variance* is defined as the sum of squared intraday returns,

$$\mathrm{RV}_t = \sum_{j=1}^{m_t} r_{t_j}^2, \tag{14.3}$$

where m_t is the number of Δ-returns during the tth time horizon (such as a trading day) and is assumed to be an integer. Jacod and Shiryaev (1987) show that the realized (quadratic) variation converges to the integrated variation assuming that the underlying process follows (14.1) without jumps ($\lambda = 0$). Furthermore, in the presence of jumps ($\lambda > 0$), the realized volatility converges in probability to the total variation as $\Delta \to 0$,

$$\mathrm{RV}_t \xrightarrow{p} \int_{t-1}^{t} \sigma_s^2 \mathrm{d}s + \sum_{t<s<t+1} \kappa^2(s). \tag{14.4}$$

Hence, the realized variation captures the effects of both the continuous and the discrete processes where the first term in (14.4) is the return variation from the diffusion process and the second term is due to the jump component.

A second estimator of the integrated variance is the *realized bipower variation*, which is defined as,

$$\mathrm{BV}_t = \mu_1^{-1} \frac{m_t}{m_t - 1} \sum_{j=2}^{m_t} |r_{t_j}||r_{t_{j-1}}|, \tag{14.5}$$

where μ_1 is a constant given by,

$$\mu_k = \frac{2^{k/2}}{\sqrt{\pi}} \Gamma\left(\frac{k+1}{2}\right), \tag{14.6}$$

where Γ is the Gamma function. Barndorff-Nielsen and Shephard (2004) show that as $\Delta \to 0$,

$$\mathrm{BV}_t \xrightarrow{p} \int_{t-1}^{t} \sigma_s^2 \mathrm{d}s. \tag{14.7}$$

The result follows from that only a finite number of terms in the sum in (14.5) are affected by jumps while the remaining returns go to zero in probability. Since the probability of jumps goes to zero as $\Delta \to 0$, those terms do not impact the limiting probability. Hence, the asymptotic convergence of the bipower variation captures only the effects of the continuous process even in the presence of jumps. By combining the results from (14.4) and (14.7), the contribution of the jump process in the total quadratic variation can be estimated by the difference between these two variations where,

$$\mathrm{RV}_t - \mathrm{BV}_t \xrightarrow{p} \sum_{t<s<t+1} \kappa^2(s), \tag{14.8}$$

as $\Delta \rightarrow 0$. Hence, (14.8) estimates the integrated variation due to the jump component and, as such, provides the basis for a nonparametric statistic for identifying jumps.

Barndorff-Nielsen and Shephard (2004, 2006) and Barndorff-Nielsen et al. (2006) show that in the absence of jumps in the price process,

$$\Delta^{-1/2} \frac{\mathrm{RV}_t - \mathrm{BV}_t}{\left(\left(\nu_{bb} - \nu_{qq}\right) \int_{t-1}^{t} \sigma^4(s) \mathrm{d}s\right)^{1/2}} \xrightarrow{p} \mathrm{N}(0, 1), \tag{14.9}$$

as $\Delta \rightarrow 0$ where RV_t and BV_t are defined in (14.3) and (14.5) and $\nu_{bb} = \pi^2/2 + \pi - 3$ and $\nu_{qq} = 2$. The integral in the denominator, called the *integrated quarticity*, is unobservable. From the work by Barndorff-Nielsen and Shephard (2004) on multipower variations, Andersen et al. (2007) propose to estimate the integrated quarticity using the *realized tripower quarticity*, TP_t, which is defined as,

$$\mathrm{TP}_t = m_t \mu_{4/3}^{-3} \frac{m_t}{m_t - 2} \sum_{j=3}^{m_t} \prod_{i=0}^{2} |r_{t_{j-i}}|^{4/3}, \tag{14.10}$$

where $\mu_{4/3}$ is defined in (14.6). Asymptotically, as $\Delta \rightarrow 0$,

$$\mathrm{TP}_t \xrightarrow{p} \int_{t-1}^{t} \sigma_s^4 \mathrm{d}s. \tag{14.11}$$

Hence, a test statistic based on (14.9) is given by,

$$\Delta^{-1/2} \frac{\mathrm{RV}_t - \mathrm{BV}_t}{\left((\nu_{bb} - \nu_{qq})\mathrm{TP}_t\right)^{1/2}}. \tag{14.12}$$

Barndorff-Nielsen and Shephard (2004, 2006) propose a number of variations of the statistic in (14.12), all of which asymptotically have a standard normal distribution. See Huang and Tauchen (2005) for a list of these statistics and finite sample studies of their properties. Empirical studies favor the statistic,

$$Z_{\mathrm{TPRM},t} = \frac{\mathrm{RJ}_t}{\sqrt{\left(\nu_{bb} - \nu_{qq}\right) \frac{1}{m_t} \max\left\{1, \frac{\mathrm{TP}_t}{\mathrm{BV}_t^2}\right\}}}, \tag{14.13}$$

where

$$\mathrm{RJ}_t = \frac{\mathrm{RV}_t - \mathrm{BV}_t}{\mathrm{RV}_t}. \tag{14.14}$$

The ratio, RJ_t, is an estimator of the relative contribution of the jump component to the total variance.

14.2.2 Market MicroStructure Noise

The test statistic relies on estimates of integrated variations, which are obtained with model-free methods on high-frequency intraday data. The asymptotic results hinge on efficient (noise-free) price processes. Observed prices, however, are noisy due to market microstructure. Thus, the variation in intraday returns can be attributed to two components: the efficient price returns and the microstructure frictions. The variance generated by market frictions is the result of price formation under specific trade mechanisms and rules, such as discrete price grids and bid-ask bounce effects. Such noise introduces bias in the variance estimates, which becomes particularly severe at high sampling rates. The variance due to noise rather than the integrated variance will dominate the estimate as the sampling interval goes to zero.

One approach that is used in the applied literature to alleviate the bias is simply to sample the price process at lower frequencies than what the data permits. The sampling intervals are typically arbitrarily chosen and commonly in the range of 5–30 min. Bandi and Russell (2006) and Zhang et al. (2005) propose methods that finds an optimal sampling rate for estimating the realized volatility. Andersen et al. (2007) take another approach to reduce the bias. These methods are introduced in this section.

14.2.2.1 Optimal Sampling Rate

Define a noisy logarithmic price process, Y_{t_i}, which is observed in the market by,

$$Y_{t_j} = X_{t_j} + \epsilon_{t_j}, \tag{14.15}$$

where ϵ_{t_j} denotes the microstructure noise process. The observed returns, $\widetilde{r}_{t_j}$, are then given by,

$$\widetilde{r}_{t_j} = Y_{t_j} - Y_{t_{j-1}} = r_{t_j} + \eta_{t_j}, \tag{14.16}$$

where as before r_{t_j} denotes the efficient returns,

$$r_{t_j} = X_{t_j} - X_{t_{j-1}}. \tag{14.17}$$

The microstructure noise in the observed return process is given by,

$$\eta_{t_j} = \epsilon_{t_j} - \epsilon_{t_{j-1}}. \tag{14.18}$$

The random shocks, ϵ_{t_j}, are assumed to be iid with mean zero and variance σ_ϵ^2. Furthermore, the true price return process, r_{t_j}, and the noise process, ϵ_{t_j}, are assumed to be independent.

The efficient returns are then of order $O(\sqrt{\Delta})$, which follows from the definition of the true price returns in (14.17) and the properties of the standard Brownian motion. Meanwhile, the microstructure noise, η_{t_j}, is of order O(1). The independence from the time duration in the microstructure noise component is motivated by that adjustments of observed prices (such as the bid-ask spread) are fixed in size regardless of how short the time interval is. Hence, the accumulated noise dominates the realized variance at high sampling rates, whereas at lower sample rates the variance of the efficient price process is proportionally larger compared to the component due to noise.

An optimal sampling rate is obtained by minimizing the conditional mean-square error (MSE), which Bandi and Russell (2006) show can be written as,

$$E\left(\sum_{j=1}^{m_t} \widetilde{r}_{t_i}^2 - \int_{t-1}^{t} \sigma_s^2 \mathrm{d}s\right)^2 = 2\frac{1}{m_t}(Q_t + o(1)) + m_t\beta + {m_t}^2\alpha + \gamma, \tag{14.19}$$

where Q_t denotes the quarticity, $\int_{t-1}^{t} \sigma^4 \mathrm{d}s$. The three other parameters are defined as,

$$\begin{aligned}
\alpha &= \left(E\left(\eta_t^2\right)\right)^2, \\
\beta &= 2E\left(\eta_t^4\right) - 3\left(E\left(\eta_t^2\right)\right)^2, \\
\gamma &= 4E\left(\eta_t^2\right)\int_{t-1}^{t} \sigma_s^2 \mathrm{d}s - E\left(\eta_t^4\right) - 2\left(E\left(\eta_t^2\right)\right)^2.
\end{aligned}$$

The optimal number of samples, m_0, is obtained by minimizing the MSE in (14.19). Bandi and Russell (2006) show that m_0 can be approximated by,

$$m_0 \sim \left(\frac{Q_t}{(E(\eta^2))^2}\right)^{1/3}, \tag{14.20}$$

when the optimal sampling frequency is high. Notice that the approximation does not depend on the fourth moment and has a closed-form solution. Intuitively, the approximation seems reasonable since for large estimates of the second moment of the microstructure noise component, η_{t_j} (that is, the more contaminated the series is), the lower the sampling frequency should be.

14.2.2.2 Staggered Returns

Andersen et al. (2007), Barndorff-Nielsen and Shephard (2006) and Huang and Tauchen (2005) evaluate a different approach to reduce the impact of microstructure noise. Specifically, the method addresses the bias generated by spurious correlations in the returns due to noise, such as the bid-ask spread, which generates negative correlations. Any correlation structure in the returns may bias the bipower, tripower and quadpower estimators since these are functions of adjacent returns. The method, referred to as *staggered returns*, attempts to break up or at least reduce the correlation structure by skipping one or more returns when computing the estimators. The bipower variation using staggered returns becomes,

$$\mathrm{BV}_{t+i} = \frac{\pi}{2}\frac{m_t}{m_t - 1 - i}\sum_{j=2+i}^{m_t} |r_{t_j}||r_{t_{j-1-i}}|. \tag{14.21}$$

The offset, i, is chosen based on the order of the autocorrelation in the return process. Similarly, the staggered version of the tripower quarticity is defined by,

$$\mathrm{TP}_t = m_t \mu_{4/3}^{-3} \frac{m_t}{m_t - 2(1+i)} \sum_{j=1+2(1+i)}^{m_t} \prod_{k=0}^{2} |r_{t_{j-k(1+i)}}|^{4/3}. \tag{14.22}$$

14.2.3 Empirical Results

The following section presents finite sample results of the statistics and examines the implications of noise.

14.2.3.1 Design of Simulation Study

The setup follows Huang and Tauchen (2005), who consider a one-factor stochastic volatility jump-diffusion model written as,

$$\begin{aligned} \mathrm{d}X_t &= \mu \mathrm{d}t + \mathrm{e}^{\beta_0 + \beta_1 v_t} \mathrm{d}w_{p,t} + \kappa_t \mathrm{d}q_t, \\ \mathrm{d}v_t &= \alpha_v v_t \mathrm{d}t + \mathrm{d}w_{v,t}, \end{aligned} \tag{14.23}$$

where v_t is a stochastic volatility factor; α_v is the mean reversion parameter; and $\mathrm{d}w_p$ and $\mathrm{d}w_v$ are standard Brownian motions with correlation, ρ. q_t is a discontinuous jump process where jumps occur at a rate denoted by λ. κ_t is the size of the jumps. In the following, we refer to the model defined in (14.23) as SV1F for $\lambda_t = 0$, that is, when no jumps are simulated, and SV1FJ otherwise.

Table 14.1 presents values of the parameters in the data-generating processes that we consider. The values are obtained from Huang and Tauchen (2005), who select the values based on empirical studies reported in literature.

We simulate observed prices per second. The number of simulated prices per interval t is equivalent to six hours and a half of trading, that is, t corresponds to a typical trading day. We compute intraday price returns for time intervals ranging from 1 s to 30 min. We assume that the number of jumps in the SV1FJ model has

Table 14.1 The experimental design for SV1F and SV1FJ (14.23) where the jump rate, λ, is set to zero for SV1F

Parameter	Value
μ	0.030
β_0	0.000
β_1	0.125
α_v	−0.100
ρ	−0.620
λ	{0.000, 0.014, 0.118, 1.000, 2.000}
σ_{jmp}	{0.500, 1.000, ..., 2.500}

a Poisson distribution; hence, the interarrival times have an exponential distribution with parameter λ. The size of the jumps, κ, has a normal distribution with zero mean and variance, σ^2_{jmp}. This jump model produces the asymmetric leptokurtic features of the return distribution which is typical for market data.

Figure 14.1 graphs realizations of 10000 simulated days from the SV1FJ model. The parameters λ and σ_{jmp} are 0.014 and 1.50, respectively. The top panel plots daily closing prices; the second panel plots daily price returns; the third panel plots the volatility factor, v_t; and the bottom panel plots the jump component, $\kappa_t dq_t$.

14.2.3.2 Optimal Sampling Rates

Bandi and Russell (2006) derive optimal sampling frequencies for estimating the integrated variance, $\int_{t-1}^{t} \sigma_s^2 ds$, using the realized variation estimator, RV_t. The jump statistics, however, also require the bipower, BV_t, and tripower, TP_t estimators, all of which are based on intraday sampling. We evaluate how well the optimal sampling rates apply to these power variations.

For this study, we assume that the return process, X_t, follows the geometric Brownian motion with constant drift and volatility so that the bias and mean-square error can be computed without any error. Thus, let the data-generating process be,

$$dX_t = \mu dt + \sigma dW_t, \tag{14.24}$$

where W_t is a standard Wiener process and the drift, μ, and volatility, σ, parameters are constant. Let Y_{t_i} denote the observed noisy price process given by,

$$Y_{t_i} = X_{t_i} + \epsilon_{t_i}, \tag{14.25}$$

where ϵ_{t_i} is normally distributed. The estimates are mean values of 1000 realized trading days per data point, where each trading day is equivalent to six and a half hours. The drift rate, μ, is zero and the volatility, σ, is one. The sampling intervals range from 1 to 60 min in increments of 1 min.

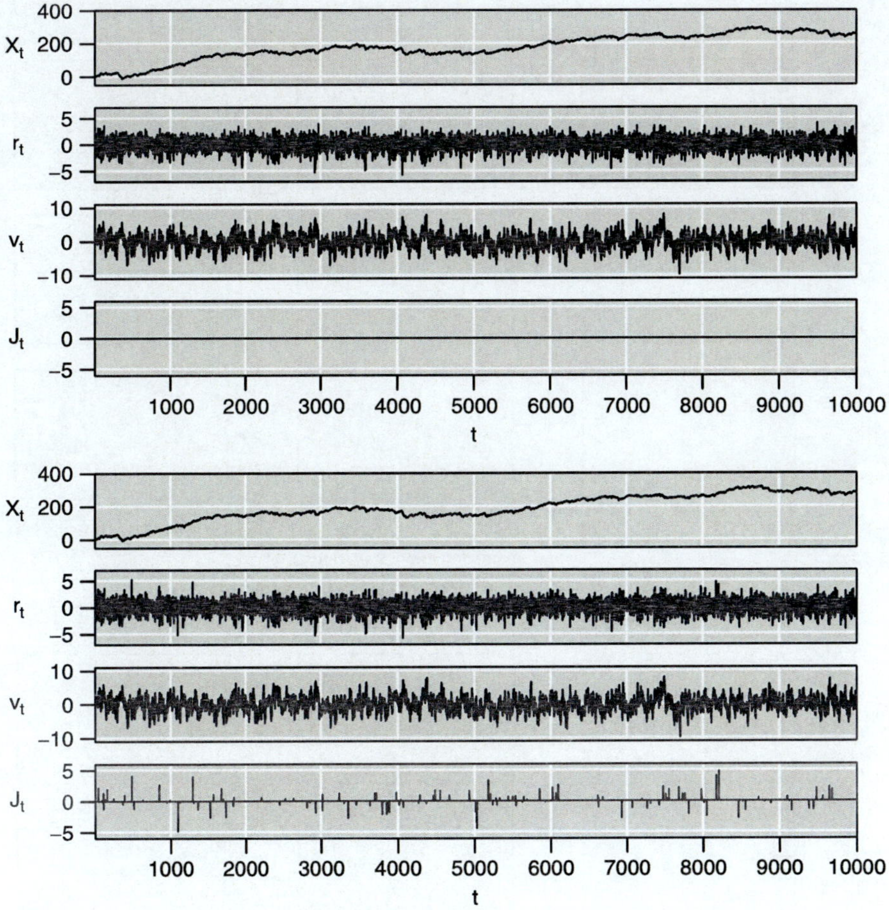

Fig. 14.1 The figure plots results based on realizations of 10000 simulated days of the SV1FJ model, (14.23). The experimental design is described in Table 14.1 with $\lambda = 0.014$ and $\sigma_{\text{jmp}} = 1.50$. The top panel is the daily closing price; the second panel is the daily price returns given by the logarithmic difference between the last and first price; the third panel plots the volatility factor, v_t; and the bottom panel plots the jump process

Panel A in Fig. 14.2 plots the bias (first column), variance (second column) and mean square error (third column) for the realized variance (RV_t), bipower variation (BV_t) and tripower variation (TP_t) for a price process without noise. Under these conditions, the asymptotic theory states that the price process should be sampled as frequently as possible. Consistent with this, the MSE obtains its minimum at the highest frequency, that is, 1 min and increases linearly with the sampling interval. This is expected since the variance is negatively related to the sampling frequency.

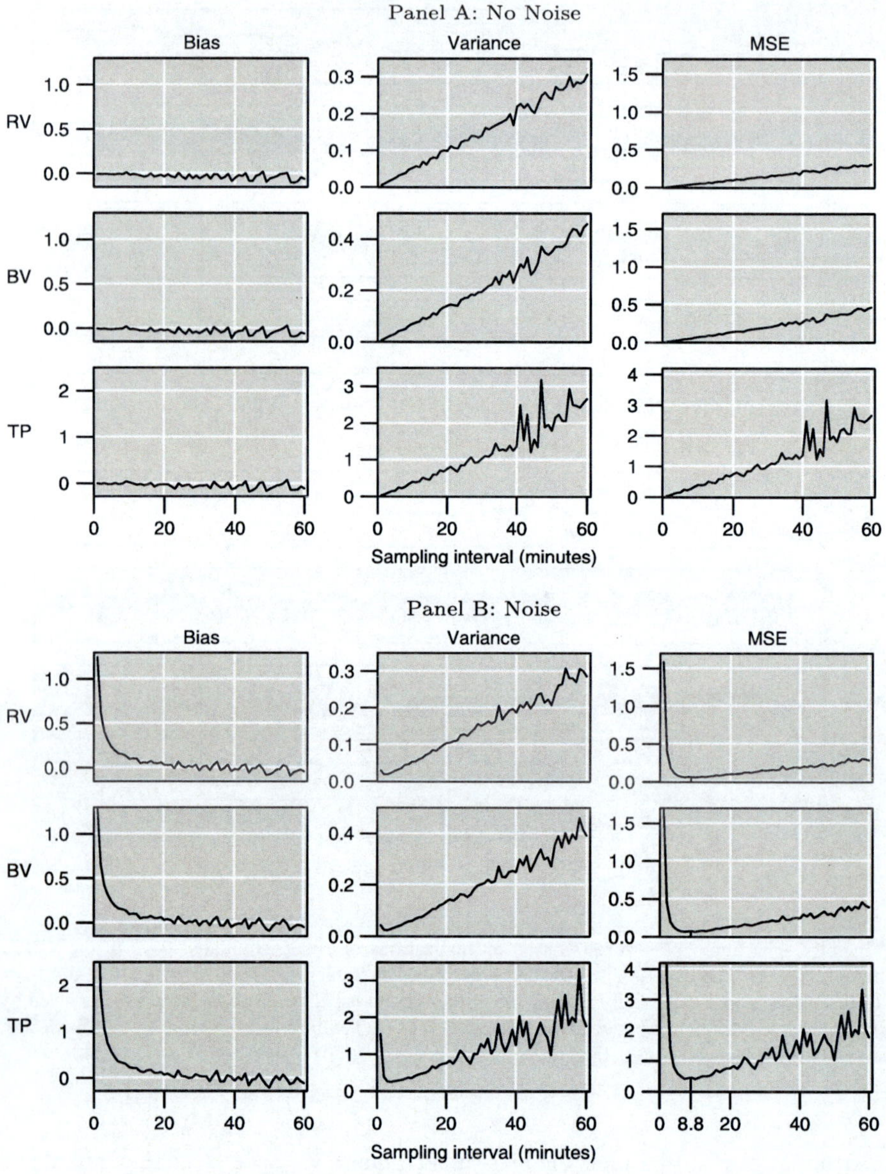

Fig. 14.2 The figure plots bias, variance and mean-square error for three estimators: realized variance (RV_t), bipower (BV_t), and tripower (TP_t) variations. Prices are generated from the geometric Brownian motion model, (14.24), with $\mu = 0$ and $\sigma = 1$. Panel A plots results for efficient price series; Panel B presents results where an iid $N(0, 0.040^2)$ noise component is added to the price process

Panel B graphs the equivalent results with the standard deviation of the noise component equal to 0.040. The pattern is consistent across all three estimators. The bias is large at high sampling frequencies but drops as the sampling interval increases by a few minutes and thereafter flattens out after about 10 min. Similarly, as in the previous case with no noise, the variance is low at the highest sampling frequencies but increases nearly linearly as the sampling frequency drops. As a result, the mean-square error peaks at the shortest sampling interval but drops rapidly and reaches its minimum around 7–10 min.

We estimate the optimal sampling rates by Bandi and Russell (2006) (BR) and compare with the minimum point of the MSE. For the first set of results without noise, the optimal sampling intervals are about 20 s for BR. Once we add noise to the observed prices, the MSE in Panel B in Fig. 14.2 suggests that the optimal sampling interval is in the range of 7–10 min for all three estimators. The mean (standard deviation) of the sampling interval based on 1000 simulations using BR is 8.8 (1.8) min. The vertical line in the MSE plots represents the BR estimate. The sampling rate given by BR coincides with the minimum of the MSE for all three estimators. In sum, these results suggest that the optimal sampling rates derived for the realized variance also are appropriate for the bipower and tripower estimators.

14.2.3.3 Asymptotics

This section documents the convergence to the asymptotics of the jump statistics. We examine the distribution of the Z_{TPRM} statistic as $\Delta \to 0$ by generating 100 trading days from the SV1F model and compute the jump statistics per day. Thereafter, we calculate the p-value from the Kolmogorov–Smirnov test of normality. The results are equivalent for the Anderson-Darling and Shapiro-Francia tests. We repeat these steps 100 times and examine the distribution of the p-values, which is uniform under the asymptotic theory. We produce these results for sampling intervals, Δ, ranging from 1 s to about 30 min; specifically, $\Delta = 1, 2, 4, \ldots, 2048$ s.

Panel A in Fig. 14.3 plots histograms for prices without noise. The labels specifies the sampling interval, Δ, in seconds. The distributions appear to be uniform with the exception for the longest sampling intervals (1024 and 2048). Panel B graphs histograms for a price process with a relatively severe noise component, $\text{NIID}(0, 0.160)$, which show that there are proportionally too many small p-values at the lower sampling frequencies. In spite of the severe noise process, the distribution converges to a normal distribution for the high frequency estimates. We examine the mean and standard deviations of the jump statistic which asymptotically are zero and one, respectively. Panel A in Table 14.2 presents the mean and standard deviation of 10000 realizations of the Z_{TPRM} statistic. The row labels denote the sampling interval in seconds, and the column labels denote the standard deviation of the noise process. The means remain close to zero for efficient prices (first column); however, even for small levels of noise, the estimates become negatively biased at high sampling frequencies. For the lowest levels of noise ($\sigma_{\text{mn}} = 0.027$), a sampling interval around 3 min or longer seems appropriate. The

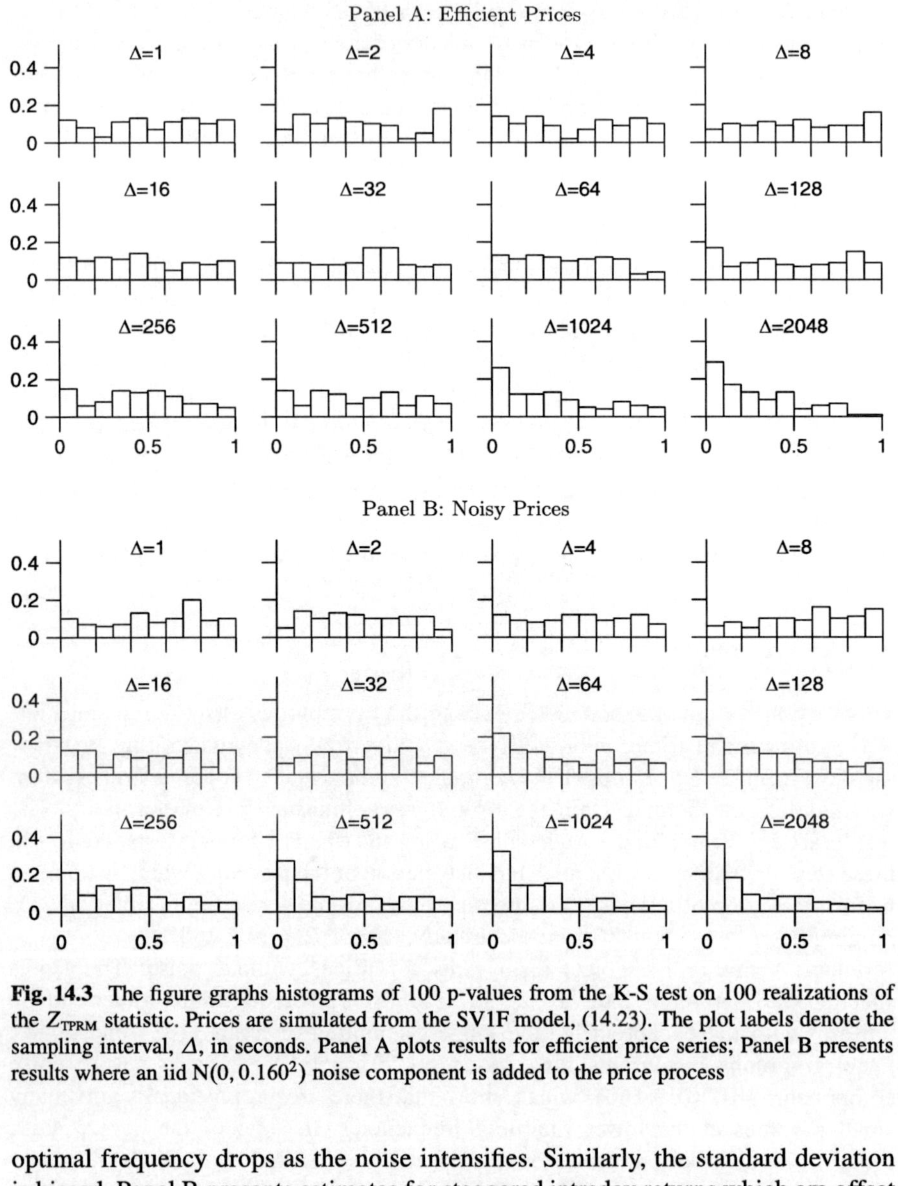

Fig. 14.3 The figure graphs histograms of 100 p-values from the K-S test on 100 realizations of the Z_{TPRM} statistic. Prices are simulated from the SV1F model, (14.23). The plot labels denote the sampling interval, Δ, in seconds. Panel A plots results for efficient price series; Panel B presents results where an iid $\text{N}(0, 0.160^2)$ noise component is added to the price process

optimal frequency drops as the noise intensifies. Similarly, the standard deviation is biased. Panel B presents estimates for staggered intraday returns which are offset by one lag. The means are close to zero at any noise level and sampling frequency. Similarly, the standard deviations are close to one. Hence, offsetting the intraday returns appears to adequately address the impact of noise.

The results without offsetting the intraday returns presented in Panel A in Table 14.2 suggest that for each noise variance, there exists a range of sampling intervals that produces estimates of the moments that are consistent with the asymptotic properties. The objective of the optimal sampling rate method introduced

Table 14.2 The table presents means and standard deviations (in parentheses) of the Z_{TPRM} statistic for 10000 price paths. The prices are generated from the SV1F model, (14.23). Panel A and Panel B report estimates without and with staggered returns by one lag

	0.000	0.020	0.040	0.080	0.160	0.320
Panel A: No Staggering						
1	0.01	−21.04	−23.00	−23.53	−23.66	−23.69
	(1.00)	(1.61)	(0.84)	(0.73)	(0.72)	(0.72)
2	0.01	−13.37	−15.77	−16.50	−16.68	−16.73
	(1.01)	(1.77)	(0.91)	(0.73)	(0.72)	(0.71)
4	−0.01	−7.85	−10.54	−11.49	−11.76	−11.83
	(0.99)	(1.77)	(1.01)	(0.76)	(0.72)	(0.72)
8	0.01	−4.10	−6.71	−7.89	−8.25	−8.34
	(0.98)	(1.52)	(1.09)	(0.79)	(0.73)	(0.72)
16	−0.00	−1.81	−3.94	−5.27	−5.73	−5.87
	(1.00)	(1.24)	(1.12)	(0.83)	(0.73)	(0.72)
32	0.00	−0.65	−2.06	−3.36	−3.92	−4.11
	(1.00)	(1.04)	(1.06)	(0.87)	(0.75)	(0.72)
64	0.01	−0.19	−0.91	−1.97	−2.63	−2.87
	(0.99)	(1.00)	(1.00)	(0.90)	(0.78)	(0.74)
128	0.02	−0.03	−0.33	−1.01	−1.65	−1.95
	(0.98)	(0.98)	(0.98)	(0.93)	(0.83)	(0.76)
256	−0.01	0.00	−0.11	−0.45	−0.93	−1.27
	(0.98)	(0.97)	(0.98)	(0.95)	(0.86)	(0.79)
512	−0.02	−0.01	−0.02	−0.15	−0.47	−0.79
	(0.98)	(0.98)	(0.98)	(0.96)	(0.91)	(0.83)
1024	0.00	0.01	−0.01	−0.06	−0.19	−0.43
	(1.00)	(0.99)	(1.00)	(0.98)	(0.94)	(0.89)
2048	0.01	0.00	0.01	−0.01	−0.07	−0.21
	(0.99)	(1.00)	(0.99)	(0.99)	(0.98)	(0.95)
Panel B: Staggering						
1	−0.01	−0.02	−0.01	−0.01	−0.02	−0.02
	(1.00)	(1.00)	(0.99)	(0.99)	(0.99)	(0.99)
2	0.00	0.01	0.00	0.00	0.01	0.01
	(1.00)	(0.98)	(0.99)	(0.99)	(0.98)	(0.98)
4	−0.01	−0.00	0.00	0.00	0.00	0.00
	(1.00)	(1.00)	(0.99)	(0.98)	(0.99)	(0.99)
8	0.02	0.00	−0.01	−0.01	0.00	0.00
	(1.00)	(1.00)	(0.99)	(0.99)	(0.99)	(0.99)
16	−0.00	−0.00	0.01	0.01	0.00	0.00
	(1.01)	(0.99)	(1.00)	(0.99)	(0.98)	(0.98)
32	−0.00	0.02	−0.01	−0.02	0.02	0.01
	(1.00)	(0.98)	(0.99)	(1.00)	(0.98)	(0.98)
64	0.02	0.01	0.01	−0.00	−0.01	−0.01
	(0.99)	(1.00)	(1.00)	(0.99)	(0.99)	(0.99)
128	0.02	0.02	0.01	0.01	0.01	0.01
	(0.99)	(0.99)	(1.01)	(0.99)	(1.00)	(1.00)
256	−0.02	−0.00	−0.01	−0.01	0.01	0.01
	(0.99)	(0.99)	(0.99)	(0.99)	(0.99)	(0.98)
512	−0.00	−0.01	0.01	0.01	−0.00	0.01
	(1.01)	(1.01)	(1.01)	(1.01)	(1.00)	(1.00)
1024	0.01	0.01	0.00	0.00	0.01	0.01
	(1.04)	(1.04)	(1.04)	(1.03)	(1.04)	(1.03)
2048	0.01	0.01	0.00	0.00	0.01	0.02
	(1.07)	(1.07)	(1.07)	(1.07)	(1.08)	(1.08)

in the previous section is to find these ranges. We calculate the rates given by Bandi and Russell (2006) and find that these correspond well with the sampling intervals in Panel A that produces values close to the asymptotics. The estimated optimal sampling intervals for the first five noise processes are 31, 242, 569, 1227, and 2125 s, which all seem appropriate. The interval for the most severe noise component is 2827, and thus goes beyond the range covered in the table.

14.2.4 Size

The following section evaluates the size of the statistic for different sampling rates. The left panel in Fig. 14.4 plots the size of the Z_{TPRM} statistics against the sampling intervals: 1, 3, 5, 10, 15 and 30 min. The nominal size is 0.05. The estimates are for 10000 simulated trading days from the SV1F model without any noise process.

The rejection rates remain near the nominal size for all sizes and sampling intervals. The right figure plots the size for prices with noise. We add an iid $\mathrm{N}(0, \sigma_{mn})$ process to the simulated prices with $\sigma_{mn} = 0.080$. The statistic becomes conservative at higher sampling frequencies but approaches the nominal size as the sample rate increases and reaches 0.05 for sampling intervals at about 10 min or longer. These findings agree with the applied literature on high-frequency data where the sampling interval typically is chosen in the 5–30-min range.

Table 14.3 reports the rejection frequencies under the null hypothesis; the significance level is set to $\alpha = 0.99$. The columns report rejection rates for different values of the standard deviation of the noise process, σ_{mn}. The sampling rates are kept constant at 1, 3, 5 and 30 min. We include results for the Z_{TPRM} statistic. The three panels tabulate the rejection rates for the statistic where the bipower (14.21) and tripower (14.22) are computed using staggered returns with offset zero (panel $i = 0$), one (panel $i = 1$).

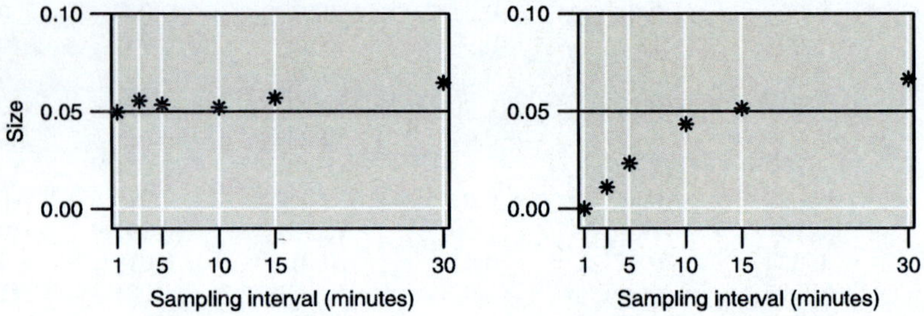

Fig. 14.4 The figure plots the size of the Z_{TPRM} statistic based on 10000 simulated days of the SV1F model, (14.23). The estimates are based on efficient prices in the left panel and noisy prices to the right where an iid $\mathrm{N}(0, 0.080^2)$ process is added to the simulated prices. The sizes are plotted against sampling intervals which range from 1 to 30 min. The horizontal lines denote the nominal size 0.05

Table 14.3 The size of the Z_{TPRM} statistic is tabulated based on 10000 days simulated from the SV1F model, (14.23). An iid $N(0, \sigma_{mn}^2)$ noise process is added to the simulated prices; σ_{mn} is set to 0.000, 0.027, 0.040, 0.052, 0.065, 0.080. The panel labels $i = 0, 1$ denote the staggered offset. The return horizons are 1, 3, 5 and 30 min. The test size is 0.01

	σ_{mn}					
Interval	0.000	0.027	0.040	0.052	0.065	0.080
$(i = 0)$						
1 minutes	0.010	0.003	0.001	0.000	0.000	0.000
3 minutes	0.015	0.011	0.008	0.006	0.004	0.002
5 minutes	0.013	0.013	0.012	0.010	0.007	0.004
30 minutes	0.016	0.015	0.016	0.017	0.017	0.018
$(i = 1)$						
1 minutes	0.014	0.014	0.013	0.013	0.013	0.011
3 minutes	0.017	0.015	0.015	0.015	0.016	0.015
5 minutes	0.016	0.016	0.017	0.017	0.018	0.017
30 minutes	0.033	0.033	0.034	0.033	0.034	0.034

The first panel clearly shows that the noise has a considerable impact on the test sizes, particularly at high sampling frequencies. For 1-min sampling intervals, the statistic becomes biased against identifying jumps, which is consistent with the convergence results above. In fact, the rejection rates are less than 0.000 for the three largest values of the noise variations although the nominal test size is 0.01. As the sampling interval increases to 3 min, the test size approaches the nominal size yet remains conservative for the larger values of σ_{mn}. Notice, however, that the statistic is becoming increasingly anti-conservative for no or minor noise at this sampling rate. The same patterns hold for 5-min sampling. Thus, we confirm that the optimal constant sampling rate is highly dependent on the noise variance. A high sampling rate yields test sizes that are closer to the true size without noise while the appropriate sampling frequency drops as the noise variance increases.

Applying staggered returns reduces the impact of noise considerably. The estimated sizes are nearly constant across all values of the noise variations, and thus alleviate the user from having to gauge the level of noise in order to select an appropriate sampling rate.

The rejection rates for Z_{TPRM} at the highest sampling frequency when offsetting the returns by one lag is analogous to the 30-min sampling interval without staggering. That is, the former uses thirty times more data. We investigate below whether this translates into a more powerful test.

Table 14.4 presents results based on the method by Bandi and Russell (2006). We compute sampling rates per day using their exact and approximate equations, see (14.19) and (14.20), which we refer to as BR1 and BR0, respectively. Notice that the optimal sampling rates are computed per day; that is, the sampling rate is adjusted per day. The benefit is that if the price process is noisier during certain periods, the sampling rate is appropriately adjusted.

Table 14.4 The size is tabulated for the Z_{TPRM} statistics based on 10000 days simulated from the SV1F model, (14.23). An iid $N(0, \sigma_{mn}^2)$ noise process is added to the simulated prices; σ_{mn} is set to 0.000, 0.027, 0.040, 0.052, 0.065, 0.080. BR1 and BR0 denote sampling rates that are obtained by solving (14.19) and by (14.20), respectively. The test size is 0.01

	σ_{mn}					
Interval	0.000	0.027	0.040	0.052	0.065	0.080
BR0	0.011	0.013	0.013	0.014	0.013	0.014
BR1	0.011	0.013	0.013	0.012	0.013	0.012

In contrast to the results for constant sampling without staggered returns, the sizes stay effectively constant near the nominal size across all standard deviations for the Z_{TPRM} statistic. That is, the noise does not bias the Z_{TPRM} statistic against rejecting the null hypothesis, which is remarkable considering the large bias resulting from sampling at constant sampling rates, see Table 14.3. The application of staggered returns combined with BR makes the test statistics anti-conservative and thus invalidates the test. Further analysis shows that the mean values of the statistics becomes positively biased, which results in too many rejection, when both methods are applied.

Bandi and Russell (2006) evaluate two methods for estimating the optimal sampling rate, one exact and one approximate. The tables shows that the results based on the exact (BR1) and approximate (BR0) rates are equivalent.

We document the sampling intervals estimated by BR1 to further explore what causes the difference between applying constant and optimal sampling rates with no staggering. For prices without noise, the optimal sampling interval predicted by BR1 is around 30 s. The interval gradually increases with the noise and reaches about 30 min for the largest noise variance. Interestingly, even though BR1 on average gives a 30-min sampling rate for the largest noise variance, holding the sampling interval constant at that rate across the whole sample period yields worse results (compare with the first panel in Table 14.3 for results with constant 30-min sampling.) This suggests that estimating the sampling rate per trading day rather than across the full sample is beneficial since some intervals are more (or less) noisy and thus require longer (shorter) sampling intervals.

14.2.5 Power

In this section, we add the jump component to the data-generating process and evaluate the power of the test statistics for different values of the jump intensity, λ, and the standard deviation of the jump size, σ_{jmp}.

We generate prices from the jump-diffusion model, SV1FJ, as specified in (14.23) (page 377). The experimental design is described in Table 14.1 (page 378) with $\lambda = 0.014$ and $\sigma_{\text{jmp}} = 1.50$. We initially consider a price process without noise.

Table 14.5 Confusion matrices are tabulated for the Z_{TPRM} statistic based on 10000 days simulated from the SV1FJ model, (14.23). The jump rates, λ, are 0.014, 0.118, 1.000, and 2.000. Results are presented for four return horizons: 1, 3, 5 and 30 min. The labels, NJ and J, denote days without and with a jump, respectively. The rows correspond to the actual event of a jump or no jump while the columns denote the statistical inference. The test size is 0.01

		$\lambda = 0.014$		$\lambda = 0.118$		$\lambda = 1.000$		$\lambda = 2.000$	
		(NJ)	(J)	(NJ)	(J)	(NJ)	(J)	(NJ)	(J)
1 minutes	(NJ)	0.990	0.010	0.989	0.011	0.990	0.010	0.984	0.016
	(J)	0.239	0.761	0.208	0.792	0.211	0.789	0.211	0.789
3 minutes	(NJ)	0.985	0.015	0.985	0.015	0.986	0.014	0.990	0.010
	(J)	0.319	0.681	0.323	0.677	0.307	0.693	0.305	0.695
5 minutes	(NJ)	0.987	0.013	0.987	0.013	0.988	0.012	0.989	0.011
	(J) 0.377	0.623	0.404	0.596	0.375	0.625	0.373	0.627	
30 minutes	(NJ)	0.984	0.016	0.984	0.016	0.980	0.020	0.985	0.015
	(J)	0.754	0.246	0.761	0.239	0.734	0.266	0.734	0.266

Table 14.5 presents confusion matrices for the Z_{TPRM} statistic for different values of the jump intensity, λ. The labels, NJ and J, denote days without and with a jump, respectively. The rows represent the true events while the columns denote the statistical inference. Hence, the rows for the 2×2 matrices add up to one where the 1×1 element is the fraction of correct non-rejections of the null (no-jump) hypothesis and the 1×2 element is the false rejection rate. Meanwhile, the 2×1 element is the false non-rejection of the null hypothesis and the 2×2 element is the correct rejection. The jump intensity, λ, is set to 0.014, 0.118, 1.000 and 2.000, respectively, while the standard deviation of the jump size, σ_{jmp}, is held constant at 1.50. The significance level, α, is 0.99.

Since the underlying prices are efficient, theory states that the price series should be sampled as frequently as possible. Consistently, the type I error is smallest and near the nominal test size for the highest frequency, that is, for the 1-min sampling interval. Furthermore, the test correctly rejects the null hypothesis at relatively high rates. As the sampling interval increases, the statistic is computed on fewer data points. Consequently, the test properties deteriorate as the variance increases. The type I error holds up reasonable well for the Z_{TPRM} statistic as the sampling rate decreases; the type II error, however, increases significantly. Hence, for efficient prices there is a considerable loss in power at low sampling rates. In fact, there is an evident drop in the power already at the 5-min sampling rate compared to the highest frequency.

Moreover, the observed patterns are nearly constant across the different jump intensities, which is anticipated since the nonparametric statistic is applied to each day individually. If the jump arrival rate were large enough to generate multiple jumps per day, the power should increase as the statistics would accrue the effects of several jumps. We expect the variance of the jump size, σ_{jmp}, however, to be positively related to the power of the test since larger jumps are easier to identify. This is confirmed in Table 14.6 where we explore the relationship between the

Table 14.6 The power is tabulated for Z_{TPRM} per jump intensity and standard deviation of the jump size based on 10000 days simulated from the SV1FJ model, (14.23). The jump rates, λ, are 0.5, 1.0, 1.5 and 2.0. The standard deviation of the jumps, σ_{jmp}, ranges from 0.5 to 2.5 by increments of 0.5. The return horizons are 1, 3, 5 and 30 min. The test size is 0.01

		σ_{jmp}				
	λ	0.5	1.0	1.5	2.0	2.5
1 minutes	0.5	0.432	0.678	0.780	0.837	0.868
	1.0	0.439	0.691	0.789	0.842	0.871
	1.5	0.443	0.690	0.789	0.839	0.872
	2.0	0.438	0.692	0.789	0.841	0.871
3 minutes	0.5	0.278	0.552	0.679	0.756	0.806
	1.0	0.288	0.559	0.693	0.765	0.812
	1.5	0.289	0.569	0.698	0.768	0.812
	2.0	0.290	0.562	0.695	0.768	0.811
5 minutes	0.5	0.197	0.465	0.615	0.700	0.749
	1.0	0.211	0.477	0.625	0.712	0.765
	1.5	0.206	0.484	0.634	0.715	0.768
	2.0	0.205	0.485	0.627	0.714	0.769
30 minutes	0.5	0.038	0.138	0.252	0.349	0.426
	1.0	0.037	0.139	0.266	0.368	0.448
	1.5	0.041	0.147	0.273	0.377	0.462
	2.0	0.041	0.149	0.266	0.368	0.455

power and the magnitude of the jump size, by simulating price processes for different values of the jump size. It is remarkable how low the power drops for the 30-min sampling intervals. Even going from one to 5-min sampling leads to a considerable reduction in power, which is significant since 5-min sampling intervals are commonplace in the applied empirical literature.

To examine the impact of noise on the power of the test, we tabulate confusion matrices for different sampling intervals and noise variances based on 10000 simulated days from the SV1FJ model, (14.23). The jump intensity, λ, is 0.014 and the standard deviation of the jump size is 1.50. Table 14.7 presents matrices for constant 1, 3, 5 and 30-min sampling intervals. For σ_{mn} equal to 0.052 and 0.080, the type I errors are less than 0.0005 at the highest sampling frequency. For the 30-min sampling interval, the type I errors are near 0.01 for all values of σ_{mn}. The power decreases with the sampling frequency. Staggering the returns, however, increases the power. The type I errors remain nearly constant only narrowly exceeding 0.01. Without noise, the test rejects the false null about 75% of the time while the percentage drops to 50% for the largest noise variance for 1-min sampling.

Table 14.8 presents the confusion matrices for the method by Bandi and Russell (2006). The rates for BR without applying staggered returns are equivalent to the values for constant sampling at the highest frequency with staggering the returns at one lag. Moreover, the results for BR0 and BR1 are equivalent.

Table 14.7 Confusion matrices are tabulated for Z_{TPRM} based on 10000 days simulated from the SV1FJ model, (14.23), with $\lambda = 0.014$, and $\sigma_{\text{jmp}} = 1.50$. An iid $N(0, \sigma_{mn})$ noise process is added to the simulated prices; σ_{mn} is set to 0.000, 0.027, 0.052, 0.080. Results are presented for four return horizons: 1, 3, 5 and 30 min. The panel label i denotes the staggered offset. The labels, NJ and J, denote days without and with a jump, respectively. The rows correspond to the actual event of a jump or no jump while the columns denote the statistical inference. The test size is 0.01

		$\{\sigma_{mn}\}$							
		0.000		0.027		0.052		0.080	
		(NJ)	(J)	(NJ)	(J)	(NJ)	(J)	(NJ)	(J)
$(i = 0)$									
1 min	(NJ)	0.990	0.010	0.997	0.003	1.000	0.000	1.000	0.000
	(J)	0.239	0.761	0.297	0.703	0.493	0.507	0.623	0.377
3 min	(NJ)	0.985	0.015	0.988	0.012	0.994	0.006	0.998	0.002
	(J)	0.319	0.681	0.399	0.601	0.471	0.529	0.572	0.428
5 min	(NJ)	0.987	0.013	0.987	0.013	0.990	0.010	0.996	0.004
	(J)	0.377	0.623	0.377	0.623	0.464	0.536	0.536	0.464
30 min	(NJ)	0.984	0.016	0.984	0.016	0.983	0.017	0.982	0.018
	(J)	0.754	0.246	0.775	0.225	0.761	0.239	0.768	0.232
$(i = 1)$									
1 min	(NJ)	0.986	0.014	0.986	0.014	0.987	0.013	0.989	0.011
	(J)	0.246	0.754	0.297	0.703	0.413	0.587	0.500	0.500
3 min	(NJ)	0.983	0.017	0.985	0.015	0.985	0.015	0.985	0.015
	(J)	0.348	0.652	0.355	0.645	0.435	0.565	0.536	0.464
5 min	(NJ)	0.984	0.016	0.984	0.016	0.983	0.017	0.983	0.017
	(J)	0.362	0.638	0.391	0.609	0.449	0.551	0.493	0.507
30 min	(NJ)	0.968	0.032	0.967	0.033	0.967	0.033	0.966	0.034
	(J)	0.674	0.326	0.703	0.297	0.725	0.275	0.754	0.246

Table 14.8 Confusion matrices are tabulated for the Z_{TPRM} statistic based on 10000 days simulated from the SV1FJ model, (14.23). The experimental design is described in Table 14.1 with $\lambda = 0.014$ and $\sigma_{jmp} = 1.50$. BR1 and BR0 denote sampling rates that are obtained by solving (14.19) and by (14.20), respectively. The labels, NJ and J, denote days without and with a jump, respectively. The rows correspond to the actual event of a jump or no jump while the columns denote the statistical inference. The test size is 0.01

		$\{\sigma_{mn}\}$							
		0.000		0.027		0.052		0.080	
		(NJ)	(J)	(NJ)	(J)	(NJ)	(J)	(NJ)	(J)
BR0	(NJ)	0.989	0.011	0.987	0.013	0.986	0.014	0.985	0.015
	(J)	0.203	0.797	0.370	0.630	0.493	0.507	0.572	0.428
BR1	(NJ)	0.989	0.011	0.987	0.013	0.988	0.012	0.988	0.012
	(J)	0.210	0.790	0.399	0.601	0.493	0.507	0.609	0.391

14.3 Volatility in U.S. Energy Futures Market

14.3.1 Introduction

Observers of energy futures markets have long noted that energy futures prices are very volatile and often exhibit jumps (price spikes) during news event periods. Thus, the assumption of a continuous diffusion process for asset price behavior is often violated in practice. Since volatility behavior is the central topic for option pricing, risk management and asset allocation strategies, market participants, regulators and academics have a strong interest in the identification of jumps over time and measuring the relative importance of the jump component versus the smooth sample path component as contributors to total volatility. Motivated by the increase in the availability of high-frequency data (tick by tick data), Barndorff-Nielsen and Shephard (2004, 2006) and Jiang and Oomen (2008) have developed nonparametric procedures for detecting the presence of jumps in high-frequency intraday financial time series.

This section shows how a method based on the statistic by Barndorff-Nielsen and Shephard from Sect. 14.2 can be applied to study the jump process. We examine the realized volatility behavior of natural gas, heating oil and crude oil futures contracts traded on the New York Mercantile Exchange (NYMEX) using high-frequency intraday data from January 1990 to January 2008.

14.3.2 Background of Statistical Methodology

In Sect. 14.2, we described the method by Barndorff-Nielsen and Shephard. In this applied study, we follow Jiang et al. (2008) and combine the Z_{TPRM} statistic with another jump statistic by Jiang and Oomen (2008).

14.3.2.1 Swap Variance

Jiang and Oomen (2008) base a statistic to test for jumps in asset prices on the variance swap replication strategy (Neuberger 1994). This strategy allows traders to hedge their exposure to volatility risk more effectively than by using traditional put or call options. The hedge portfolio is based on that the accumulated difference between the simple return and the logarithmic return is one half of the integrated variance under the assumption that the asset price process is continuous. The relation between the two return measures breaks down, however, if the data-generating process has discontinuities in the price process, which Jiang and Oomen (2008) use to develop a test for jumps.

The price process in (14.1) with $S_t = \exp(X_t)$ can be written as,

$$\frac{dS_t}{S_t} = \left(\mu_t + \frac{1}{2}\sigma^2\right) dt + \sigma_t dW_t + (e^{\kappa_t} - 1)\, dq_t, \tag{14.26}$$

which can be shown to be,

$$2\int_0^1 \left(\frac{dS_t}{S_t} - dX_t\right) = \sigma^2_{(0,1)} + 2\int_0^1 (e^{\kappa_T} - \kappa_t - 1)\, dq_t. \tag{14.27}$$

In the discrete case, the left-hand side of (14.27) is the *swap variance*, which can be estimated by,

$$\text{SwV}_t = 2\sum_{i=1}^{m_t} (R_{t_i} - r_{t_i}), \tag{14.28}$$

where $R_{t_i} = (S_{t_i} - S_{t_{i-1}})/S_{t_{i-1}}$ is the ith intraday *simple return*, r_{t_i} is the *logarithmic return*, and m_t is the number of intraday returns. Asymptotically, as $m_t \to \infty$,

$$\text{SwV}_t - \text{RV}_t \xrightarrow{p} \begin{cases} 0, & \text{if no jump;} \\ 2\int_{t-1}^{t} \left(e^{\kappa_t} - \frac{1}{2}\kappa_t^2 - \kappa_t - 1\right) dq, & \text{if jump,} \end{cases} \tag{14.29}$$

where RV_t is the realized variation (14.3). The result in (14.29) follows from (14.27) and that $\text{RV}_t \to \int_{t-1}^{t} \sigma_s ds + \sum_{t<s<t+1} \kappa^2(s)$ (Jacod and Shiryaev 1987). Jiang and Oomen (2008) uses these results to derive a statistic which is defined as,

$$Z_{\text{swv},t} = \frac{\widehat{\sigma}_t^2 m_t}{\sqrt{\widehat{\Omega}_t}} \left(1 - \frac{\text{RV}_t}{\text{SwV}_t}\right). \tag{14.30}$$

$\widehat{\Omega}_t$ is an estimator of,

$$\Omega_t = \frac{\mu_6}{9} \int_{t-1}^{t} (\sigma_u^2)^3\, du, \tag{14.31}$$

where σ_t is the volatility term in the data-generating process defined in (14.1) (page 372) and μ_6 is a constant given by (14.6). The estimator, $\widehat{\Omega}_t$, is defined by,

$$\widehat{\Omega}_t^{(p)} = \frac{\mu_6}{9} \frac{m_t^3 \mu_{6/p}^{-p}}{m_t - p + 1} \sum_{j=0}^{N-p} \prod_{k=1}^{p} |r_{t+k}|^{6/p}. \tag{14.32}$$

Jiang and Oomen (2008) conclude in simulations studies that four and six are appropriate choices for p.

Simulation studies (see for example Huang and Tauchen (2005) and Jiang et al. (2008)) on these two statistics have shown that both methods may become anti-conservative. Jiang et al. (2008) propose to address this by only rejecting the null hypothesis when both tests reject. They provide empirical evidence suggesting that the combined version is conservative and powerful. We apply this combined method to the energy market data to increase the validity of the results.

The daily variance due to the jump component is estimated by the difference between RV_t and BV_t, (14.8), where RV_t estimates the total variation including the contribution due to the jump component, whereas BV_t is robust to jumps and only captures the variation due to the continuous component. Hence, the difference is zero in absence of jumps and greater than zero otherwise. However, due to measurement errors, the difference can be negative. Barndorff-Nielsen and Shephard (2004) suggest imposing a lower bound at zero by letting the variance due to the jump component be given by,

$$J_t = \max[\mathrm{RV}_t - \mathrm{BV}_t, 0]. \tag{14.33}$$

Furthermore, since small values of J_t may be due to noise rather than discontinuities in the price process. We identify the variance contributed by *significant* jumps as,

$$J_{t,\alpha} = (\mathrm{RV}_t - \mathrm{BV}_t)\,\mathrm{I}_{(p<1-\alpha)}, \tag{14.34}$$

where p is the p-value which is set to the maximum value of the p-values based on the Z_{TPRM} and $Z_{\mathrm{SWV},t}$ statistics; α is the significance level; and I is the indicator function, which is equal to one if the test rejects the null hypothesis and zero otherwise. The variation that is contributed by the continuous sample path component can then be estimated by,

$$C_{t,\alpha} = \mathrm{I}_{(p<1-\alpha)}\mathrm{RV}_t + \mathrm{I}_{(p\geq 1-\alpha)}\mathrm{BV}_t. \tag{14.35}$$

By this definition, the sum of $J_{t,\alpha}$ and $C_{t,\alpha}$ adds up to the total variation, RV_t.

14.3.3 Contract Specifications and Data

We examine price series for three contracts from the U.S. energy futures markets. The contracts are on natural gas, crude oil, and heating oil, all of which are traded on the New York Mercantile Exchange (NYMEX).

The natural gas futures contract is commonly cited as the benchmark for the spot market, which accounts for nearly 25% of the energy consumption in the U.S. The futures contract began trading on April 3, 1990 and is based on delivery at the Henry Hub in Louisiana. The futures contract on crude oil began trading in 1983 and, according to NYMEX, is the world's most liquid futures contract on a physical commodity. The contract calls for delivery of both domestic as well as international crude oils of different grades in Cushing, Oklahoma. The heating oil futures contract began trading on November 14, 1978 and calls for deliver of heating oil in New York Harbor. Heating oil currently accounts for about a fourth of the yield of a barrel of crude oil, second only to gasoline.

The intraday data series for crude oil and heating oil range from January 1, 1990 to December 31, 2007 and the series for the natural gas contract span from January 1, 1993 to January 1, 2008.

14.3.4 Empirical Results

14.3.4.1 Realized Variations and Jump Dynamics

The time series behavior of daily closing prices (top panel) and log-returns (bottom panel) for natural gas are presented in Fig. 14.5. It clearly exhibits that the closing prices of the three energy markets have generally increased since around 1999.

The Augmented Dickey-Fuller (ADF) test is used to test for the presence of a unit root in realized variance, realized volatility (realized variance in standard deviation form), and log transformation of realized variance and the same forms of the jump component. The first row of Table 14.9 reports the ADF test statistics which indicate that the null hypothesis of unit root is rejected at the 1% level of significance for all series.

The top panel in Fig. 14.2 shows daily volatilities (realized variance in standard deviation form) for the natural gas series. Each of the three series exhibits strong autocorrelation. This is confirmed by the Ljung-Box statistic (LB_{10}), which is equal to 10,926 for crude oil, 9,263 for heating oil and 6,184 for natural gas (see the bottom row of Panel A-C in Table 14.9). A cross-market comparison shows that the natural gas market is the most volatile market; the annualized realized volatilities are 39.4% for natural gas, 26.5% for heating oil and 26.0% for crude oil. The equivalent values for the S&P 500 and the thirty-year U.S. Treasury bond futures over the sample period 1990–2002 are 14.7% and 8.0%, respectively (Andersen et al. 2007). Based on the skewness and excess kurtosis, the logarithmic form appears to be the

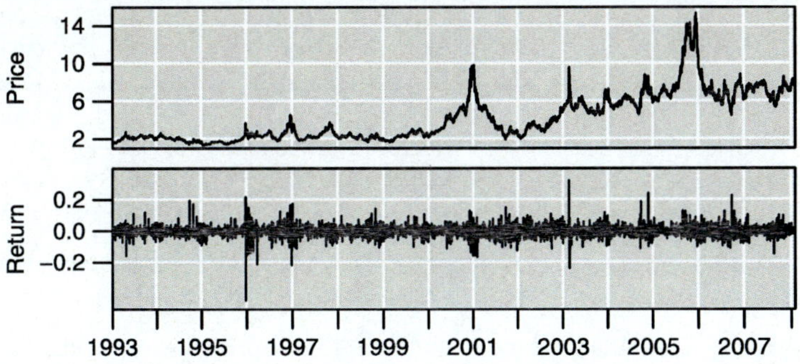

Fig. 14.5 The figure graphs closing prices (*top panel*) and daily returns (*bottom panel*) for futures contracts on natural gas. The returns are computed as the logarithmic price difference between the last and first transactions per day

Table 14.9 Daily summary statistics for futures contracts on crude oil (Panel A), heating oil (Panel B) and natural gas (Panel C) for realized variance, RV_t (equation (14.3)) and jump component, J_t (equation (14.33)). Statistics are also computed in standard deviation form, $\mathrm{RV}_t^{1/2}$ ($J_t^{1/2}$), and logarithmic form, $\log(\mathrm{RV}_t)$ ($\log(J_t + 1)$). ADF denotes the augmented Dickey-Fuller statistic. The lag orders are determined by Schwartz criterion. Only intercepts are included in the level series. The critical value for the ADF test for the 1% (5%) significance level is −3.4393 (−2.8654). Min and Max are minimum and maximum daily values. JB is the Jarque-Bera test statistic for normality. LB_{10} denotes the Ljung-Box tenth-order serial correlation test statistic. Kurtosis denotes excess kurtosis. The realized variations are computed based on 5-min intraday returns and staggered returns with one lag offset

	RV_t	$\mathrm{RV}_t^{1/2}$	$\log(\mathrm{RV}_t)$	J_t	$\mathrm{J}_t^{1/2}$	$\log(\mathrm{J}_t + 1)$
Panel A: Crude Oil						
ADF[1]	−16.04	−6.75	−5.67	−33.99	−19.34	−33.96
Mean	0.0003	0.0164	−8.3774	0.0000	0.0033	0.0000
Std Dev	0.0007	0.0072	0.7718	0.0003	0.0045	0.0003
Skewness	44.30	4.87	0.04	59.82	7.34	59.71
Kurtosis	2534.14	88.62	1.06	3835.89	175.34	3825.63
Min	0.0000	0.0030	−11.6462	0.0000	0.0000	0.0000
Max	0.0381	0.1953	−3.2664	0.0188	0.1370	0.0186
JB	$1.21E+09$	$1.49E+06$	$2.13E+02$	$2.77E+09$	$5.82E+06$	$2.76E+09$
LB_{10}	968	10926	16947	91	283	93
Panel B: Heating Oil						
ADF[1]	−15.35	−6.80	−4.95	−27.02	−23.85	−27.00
Mean	0.0003	0.0167	−8.3128	0.0000	0.0038	0.0000
Std Dev	0.0004	0.0064	0.6897	0.0002	0.0044	0.0002
Skewness	27.81	3.24	0.17	50.19	3.83	50.07
Kurtosis	1286.59	39.97	0.83	2998.58	56.75	2988.21
Min	0.0000	0.0034	−11.3906	0.0000	0.0000	0.0000
Max	0.0207	0.1439	−3.8779	0.0103	0.1017	0.0103
JB	$3.08E+08$	$3.04E+05$	$1.50E+02$	$1.67E+09$	$6.08E+05$	$1.66E+09$
LB_{10}	1873	9263	13033	137	193	138
Panel C: Natural Gas						
ADF[1]	−10.91	−8.63	−7.53	−21.34	−13.62	−21.33
Mean	0.0007	0.0248	−7.5419	0.0001	0.0061	0.0001
Std Dev	0.0008	0.0105	0.7556	0.0003	0.0075	0.0003
Skewness	6.73	2.16	0.22	11.83	2.74	11.80
Kurtosis	81.85	10.01	0.66	207.69	14.62	206.60
Min	0.0000	0.0038	−11.1209	0.0000	0.0000	0.0000
Max	0.0165	0.1286	−4.1015	0.0073	0.0852	0.0072
JB	$1.06E+06$	$1.82E+04$	$9.60E+01$	$6.70E+06$	$3.74E+04$	$6.63E+06$
LB_{10}	2912	6184	8503	194	231	194

most normally distributed, which is consistent with previous empirical findings in the equity and foreign exchange markets (Andersen et al. 2007) although the Jarque-Bera test statistic rejects normality for all forms and markets at the 1% significance level.

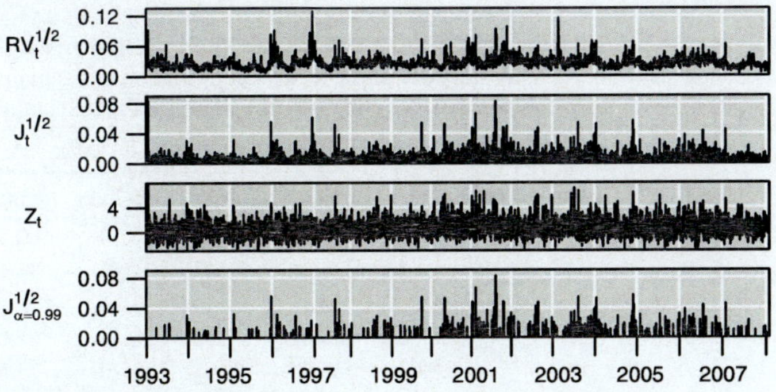

Fig. 14.6 The figure graphs time-series of the realized volatility and jump component for futures contracts natural gas. The top panel for respective contract graphs the daily realized volatility, $\mathrm{RV}_t^{1/2}$ (14.3); the second panel plots the jump component $J_t^{1/2}$ (14.33); the third panel shows the jump statistic, Z_{TPRM} (14.13); and the bottom panel plots the significant jump component, $J_{t,\alpha=0.99}^{1/2}$ (14.34). The realized variations are computed based on 5-min intraday returns and staggered returns with one lag offset

The second panel in Fig. 14.6 plots the separate measurement of the jump components in standard deviation form. The jump component is defined as the difference between the realized and bipower variations with a lower bound at zero (14.33). The mean of the daily volatility due to the jump component is equivalent for crude and heating oil at 0.0033 and 0.0038, respectively, while it is larger for natural gas at 0.0061; the corresponding annualized volatilities are 5.2%, 6.0% and 9.7%, respectively. The jump component is highly positively skewed with a large kurtosis in all three markets. The Ljung-Box test statistics reported in the bottom row are significant although considerably smaller than for the total volatility. The Ljung-Box statistics for the standard deviation form of the jump components are between 190 and 290 for the three markets while the corresponding statistics are greater than 6,000 for the realized volatility of each of the three series. Hence, the smooth component appears to contribute more to the persistency in the total volatility.

Since the jump component in Table 14.9 is computed by the difference defined in (14.33), the properties and in particular the prevalence of autocorrelation may partially be due to that the estimator captures some of the smooth process on days without jumps. Hence, to alleviate such potential bias, we examine the properties for significant jumps as defined by (14.34). The significant jumps are determined by the combined statistic where the bipower and tripower estimators are obtained using staggered returns with one lag offset to reduce the impact of market microstructure noise. The significant jump components based on the test level α set to 0.99 are plotted in the last panel in Fig. 14.6 which clearly exhibits that large volatility often can be associated with a large jump component.

Table 14.10 reports yearly statistics of the significant jump components for α equal to 0.99. There are significant jumps in all three price series. The number of days with a jump ranges from 5 to 34 for natural gas, 5–28 for heating oil and 4–20

Table 14.10 Yearly estimates for natural gas. No. Days denotes the number of trading days, No. Jumps denotes the number of days with jumps, and Prop denotes the proportion of days with jumps. Min, Mean, Median and Max are daily statistics of the relative contribution of the jump component to the total realized variance (14.14) computed for days with a significant jump component

				RJ on Jump Days (%)			
	No. Days	No. Jumps	Prop	Min	Mean	Median	Max
1993	250	5	0.020	31.72	46.17	46.58	60.52
1994	248	11	0.044	25.18	34.49	34.53	54.62
1995	250	8	0.032	26.42	39.34	33.76	75.23
1996	248	15	0.060	26.62	37.22	36.43	61.08
1997	213	8	0.038	28.84	38.65	33.20	73.60
1998	240	11	0.046	26.47	42.90	37.51	78.50
1999	232	12	0.052	25.32	33.53	32.12	55.07
2000	235	17	0.072	28.23	48.46	48.03	87.47
2001	236	34	0.144	23.56	45.76	44.06	85.92
2002	245	17	0.069	28.12	46.05	43.43	72.97
2003	249	25	0.100	25.89	38.51	34.75	77.15
2004	249	26	0.104	26.45	42.05	37.26	69.19
2005	251	19	0.076	26.50	42.05	40.37	68.96
2006	250	23	0.092	25.47	41.88	42.09	62.96
2007	258	14	0.054	23.39	33.81	32.18	52.13

days for crude oil. The proportion of days with jumps in natural gas is higher during the second half of the sample period; the other markets do not reveal the same trend. The table also includes daily summary statistics per year for the relative contribution for days with a significant jump. The relative contribution of the jump component to the total variance ranges from 23% to 87% for natural gas futures, 23%–64% for crude oil futures and 23%–74% for heating oil futures for days with jumps. Hence, jumps have a significant impact in all three markets.

To further examine the jump dynamics, we consider different levels of α ranging from 0.5 to 0.9999. The empirical results are reported in Table 14.11. The first row tabulates the number of days with a significant jump. As a comparison, the total number of trading days for the complete sample period for natural gas is 3,676, for crude oil is 4,510, and for heating oil is 4,449. As expected, the proportion of days with significant jumps declines from 0.49 to 0.02 for natural gas, 0.49 to 0.01 for heating oil, and from 0.44 to 0.01 for crude oil, as the level of α increases from 0.5 to 0.9999. Andersen et al. (2007) report that the equivalent values for S&P 500 futures and thirty-year U.S. Treasury bond futures are 0.737–0.051 and 0.860–0.076, respectively; thus, jumps are more frequent in the latter markets. Andersen et al. (2007) identifies jumps by the Barndorff-Nielsen and Shephard framework which partially explain the differences. The rates using this statistic for the energy markets are 0.64 to 0.02 for natural gas and heating oil and from 0.44 to 0.01 for crude oil. Based on the proportions of days with a jump for the energy futures markets, the test statistic consistently rejects the null hypothesis too frequently for the larger test sizes had the underlying data generating process been a continuous

Table 14.11 Summary statistics for significant jumps, $J_{t,\alpha}^{1/2}$ (14.34), for futures contracts on crude oil, heating oil and natural gas. No. Jumps denotes the number of jumps in the complete sample. Proportion denotes the ratio of days with a jump. The sample consists of 4, 510 trading days for crude oil, 4, 449 for heating oil, and 3, 676 for natural gas. Mean and Std Dev are the mean and standard deviation of the daily jump component, $J_{t,\alpha}^{1/2}$. $LB_{10}, J_{t,\alpha}^{1/2}$ denotes the Ljung-Box tenth-order autocorrelation test statistic. The realized variations are computed based on 5-min intraday returns and staggered returns with one lag offset

α	0.500	0.950	0.990	0.999	0.9999
Panel A: Crude Oil					
No. Jumps	1993	440	197	80	37
Proportion	0.44	0.10	0.04	0.02	0.01
Mean ($J_{t,\alpha}^{1/2}$)	0.0061	0.0100	0.0121	0.0152	0.0144
Std Dev	0.0051	0.0082	0.0109	0.0159	0.0079
$LB_{10}, J_{t,\alpha}^{1/2}$	75	71	59	58	0
Panel B: Heating Oil					
No. Jumps	2161	502	272	115	66
Proportion	0.49	0.11	0.06	0.03	0.01
Mean ($J_{t,\alpha}^{1/2}$)	0.0063	0.0103	0.0121	0.0144	0.0157
Std Dev	0.0046	0.0064	0.0077	0.0096	0.0116
$LB_{10}, J_{t,\alpha}^{1/2}$	124	101	105	41	0
Panel C: Natural Gas					
No. Jumps	1816	470	246	121	75
Proportion	0.49	0.13	0.07	0.03	0.02
Mean ($J_{t,\alpha}^{1/2}$)	0.0101	0.0171	0.0207	0.0263	0.0297
Std Dev	0.0079	0.0103	0.0120	0.0137	0.0135
$LB_{10}, J_{t,\alpha}^{1/2}$	179	241	216	222	38

diffusion process. For natural gas, 13% of the days are identified as having a jump for $\alpha = 0.95$ and 7% for $\alpha = 0.99$. Similar percentages hold for the other markets. The sample mean and standard deviations are daily values of the volatility due to significant jumps where the estimates are computed only over days with significant jumps. Hence, the average jump size increases as the significance level increases. The annualized estimates range from 16.0% to 47.1% for natural gas, 9.68%–22.6% for crude oil and 10.0%–24.9% for heating oil. The Ljung-Box test statistics for significant jumps ($LB_{10}, J_{\alpha}^{1/2}$) are lower than the equivalent values for jumps defined by (14.33) reported in Table 14.9. Consistently, the Ljung-Box statistics decrease as the size of α increases. Yet, even as the number of jumps declines, the Ljung-Box statistics indicate that some persistency remains in the jump component. The p-values are less than 0.01 for $\alpha = 0.999$ for all markets and less than 0.01 for $\alpha = 0.9999$ for natural gas. The time series plot of the significant jump component is graphed in the fourth panel of Fig. 14.6.

Finally, Table 14.12 presents summary statistics for jump returns conditioned on the sign of the returns. Since the test statistic does not provide the direction of the price change, we define the largest (in magnitude) intraday price return as the

Table 14.12 Summary statistics for jump returns for days with significant jumps ($\alpha = 0.99$) for crude oil, heating oil and natural gas. N denotes the number of jumps. The largest (in magnitude) 5-min intraday return per day with a significant jump is identified as the jump return. The statistics are computed for positive and negative returns, respectively

Contract	N	Mean	Median	StdDev	Max	Min
Positive Jumps						
Crude Oil	89	0.012	0.009	0.015	0.137	0.003
Heating Oil	101	0.012	0.010	0.010	0.102	0.005
Natural Gas	89	0.021	0.016	0.014	0.083	0.007
Negative Jumps						
Crude Oil	107	0.012	0.011	0.005	0.031	0.003
Heating Oil	165	0.012	0.011	0.005	0.033	0.004
Natural Gas	153	0.020	0.018	0.011	0.067	0.006

jump for each day for which the test rejects the null hypothesis and thus obtain the size and sign of the jump return. We observe that there are more negative than positive jumps for all three energy futures markets. The mean and median values are equivalent, however.

In summary, using high-frequency data, we have applied a nonparametric statistical procedure to decompose total volatility into a smooth sample path component and a jump component for three markets. We find that jump components are less persistent than smooth components and large volatility is often associated with a large jump component. Across the three markets, natural gas futures is the most volatile, followed by heating oil and then by crude oil futures.

References

Andersen, T. G., Bollerslev, T., & Diebold, F. X. (2007). Roughing it up: Including jump components in the measurement, modeling and forecasting of return volatility. *Review of Economics and Statistics, 89*(4), 701–720.

Bandi, F. M., & Russell, J. R. (2006). Separating microstructure noise from volatility. *Journal of Financial Economics, 79*, 655–692.

Barndorff-Nielsen, O. E., Graversen, S. E., Jacod, J., Podolskij, M., & Shephard, N. (2006). *From stochastic analysis to mathematical finance, festschrift for Albert Shiryaev*. New York: Springer.

Barndorff-Nielsen, O. E., & Shephard, N. (2004). Power and bipower variation with stochastic volatility and jumps. *Journal of Financial Econometrics, 2*, 1–48.

Barndorff-Nielsen, O. E., & Shephard, N. (2006). Econometrics of testing for jumps in financial economics using bipower variation. *Journal of Financial Econometrics, 4*(1), 1–30.

Cont, R., & Tankov, P. (2004). *Financial modelling with jump processes*. Chapman & Hall/CRC.

Fan, J., & Wang, Y. (2007). Multi-scale jump and volatility analysis for high-frequency financial data. *Journal of the American Statistical Association, 102*(480), 1349–1362.

Huang, X., & Tauchen, G. (2005). The relative contribution of jumps to total price variation. *Journal of Financial Econometrics, 3*, 456–499.

Jacod, J., & Shiryaev, A. N. (1987). *Limit theorems for stochastic processes*. New York: Springer.
Jiang, G. J., Lo, I., & Verdelhan, A. (2008). Information shocks and bond price jumps: Evidence from the U.S. treasury market. Working Paper.
Jiang, G. J., & Oomen, R. C. (2008). Testing for jumps when asset prices are observed with noise - A "Swap Variance" approach. *Journal of Econometrics, 144*(2), 352–370.
Lee, S. S., & Mykland, P. A. (2008). Jumps in financial markets: A new nonparametric test and jump dynamics. *Review of Financial Studies, 21*(6), 2535–2563.
Neuberger, A. (1994). The log contract - A new instrument to hedge volatility. *Journal of Portfolio Management, 20*(2), 74–80.
Sen, R. (2008). Jumps and microstructure noise in stock price volatility. In G. N. Gregoriou (Ed.), *Stock market volatility*. Chapman Hall-CRC/Taylor and Francis.
Zhang, L., Mykland, P. A., & Aït-Sahalia, Y. (2005). A tale of two time scales: Determining integrated volatility with noisy high-frequency data. *Journal of the American Statistical Association, 100*, 1394–1411.

Chapter 15
Simulation-Based Estimation Methods for Financial Time Series Models

Jun Yu

Abstract This chapter overviews some recent advances on simulation-based methods of estimating financial time series models that are widely used in financial economics. The simulation-based methods have proven to be particularly useful when the likelihood function and moments do not have tractable forms and hence the maximum likelihood (ML) method and the generalized method of moments (GMM) are difficult to use. They are also useful for improving the finite sample performance of the traditional methods. Both frequentist and Bayesian simulation-based methods are reviewed. Frequentist's simulation-based methods cover various forms of simulated maximum likelihood (SML) methods, simulated generalized method of moments (SGMM), efficient method of moments (EMM), and indirect inference (II) methods. Bayesian simulation-based methods cover various MCMC algorithms. Each simulation-based method is discussed in the context of a specific financial time series model as a motivating example. Empirical applications, based on real exchange rates, interest rates and equity data, illustrate how to implement the simulation-based methods. In particular, we apply SML to a discrete time stochastic volatility model, EMM to estimate a continuous time stochastic volatility model, MCMC to a credit risk model, the II method to a term structure model.

15.1 Introduction

Relative to other fields in economics, financial economics has a relatively short history. Over the last half century, however, there has been an explosion of theoretical work in financial economics. At the same time, more and more complex

J. Yu (✉)
School of Economics, Lee Kong Chian School of Economics and Sim Kee Boon Institute for Financial Economics, Singapore Management University, 90 Stamford Road, Singapore 178903
e-mail: yujun@smu.edu.sg

J.-C. Duan et al. (eds.), *Handbook of Computational Finance*, Springer Handbooks of Computational Statistics, DOI 10.1007/978-3-642-17254-0_15,

financial products and services have been created. The size of financial markets has exponentially increased and the quality of database is hugely advanced. The major developments in theoretical finance and the availability of high quality data provide an extremely rich framework for empirical work in financial economics.

How to price financial assets has been a driving force for much of the research on financial asset pricing. With the growth in complexity in financial products and services, the challenges faced by the financial economists naturally grow accordingly, one of which is the computing cost. Another driving force for research in financial economics is to bring finance theory to data. Empirical analysis in financial economics often involves calculating the likelihood function or solving a set of moment conditions.

Traditional econometric methods for analyzing models in financial economics include maximum likelihood (ML), quasi-ML, generalized method of moments (GMM), and classical Bayesian methods. When the model is fully specified and the likelihood function has a tractable form, ML and Bayesian methods provide the full likelihood-based inference. Under mild regularity conditions, it is well recognized that the ML estimator (MLE) is consistent, asymptotically normally distributed and asymptotically efficient. Due to the invariance principle, a function of MLE is a MLE and hence inherits all the nice asymptotic properties (e.g., Zehna 1966). These features greatly facilitate applications of ML in financial economics. When the model is not fully specified but certain moments exist, GMM can be applied. Relative to ML, GMM trades off efficiency with robustness.

Financial data are typically available in the time series format. Consequently, time series methods are of critical importance to empirical research in financial economics. Historically, financial economists restricted themselves to a small class of time series models so that the setups were simple enough to permit an analytical solution for asset prices. Moreover, empirical analysis was often done based a small set of financial assets, so that the computational cost is kept low. The leading example is perhaps the geometric Brownian motion, which was used by Black and Scholes to price European options (Black and Scholes 1973) and by Merton to price corporate bonds (Merton 1974). In recent years, however, many alternative models and financial products have been proposed so that asset prices do not have analytical solutions any more. As a result, various numerical solutions have been proposed, one class of which is based on simulations. Although the use of simulation-based methods for asset pricing is sufficient important and merits a detailed review, it is beyond the scope of the present chapter. We refer readers to McLeish (2005) for a textbook treatment on asset pricing via simulation methods.

Even if the pricing formula of a financial asset has a tractable form, estimation of the underlying time series model is not always feasible by standard econometric methods. For many important financial time series models, the likelihood function or the moment conditions cannot be evaluated analytically and may be numerically formidable so that standard econometric methods, such as ML, GMM and Bayesian, are not feasible. For example, Heston (1993) derived a closed-form expression for the European option price under the square root specification for volatility. It is known that the ML estimation of Heston's stochastic volatility (SV) model from stock prices is notoriously difficult. For more complicated models where asset prices do not have a closed-form expression, it is almost always the case that standard estimation methods are difficult to use.

Parameter estimation is important for asset pricing. For example, in order to estimate the theoretical price of a contingent claim implied by the underlying time series model, one has to estimate the parameters in the time series model and then plug the estimates into the pricing formula. In addition, parameter estimates in financial time series models are necessary inputs to many other financial decision makings, such as asset allocation, value-at-risk, forecasting, estimation of the magnitude of microstructure noise, estimation of transaction cost, specification analysis, and credit risk analysis. For example, often alternative and sometimes competing time series specifications co-exist. Consequently, it may be important to check the validity of a particular specification and to compare the relative performance of alternative specifications. Obviously, estimation of these alternative specifications is an important preliminary step to the specification analysis. For another example, in order to estimate the credit spread of a risky corporate bond over the corresponding Treasury rate and the default probability of a firm, the parameters in the underlying structural model have to be estimated first.

In some cases where ML or GMM or Bayesian methods are feasible but financial time series are highly persistent, classical estimators of certain parameters may have poor finite sample statistical properties, due to the presence of a large finite sample bias. The bias in parameter estimation leads to a bias in other financial decision making. Moreover, the large finite sample bias often leads to a poor approximation to the finite sample distribution by the asymptotic distribution. As a result, statistical inference based on the asymptotic distribution may be misleading. Because many financial variables, such as interest rates and volatility, are highly persistence, this finite sample problem may be empirically important.

To overcome the difficulties in calculating likelihood and moments and to improve the finite sample property of standard estimators, many simulation-based estimation methods have been proposed in recent years. Some of them are methodologically general; some other are specially tailored to deal with a particular model structure. In this chapter, we review some simulation-based estimation methods that have been used to deal with financial time series models.

Stern (1997) is an excellent review of the simulation-based estimation methods in the cross-sectional context while Gouriéroux and Monfort (1995) reviewed the simulation-based estimation methods in the classical framework. Johannes and Polson (2009) reviewed the Bayesian MCMC methods used in financial econometrics. Our present review is different from these reviews in several important aspects. First, our review covers both the classical and Bayesian methods whereas Johannes and Polson (2009) only reviewed the Bayesian methods. Second, relative to Stern (1997) and Gouriéroux and Monfort (1995), more recently developed classical methods are discussed in the present chapter. Moreover, only our review discuss the usefulness of simulation-based methods to improve finite sample performances.

We organize the rest of this chapter by collecting the methods into four categories: simulation-based ML (SML), simulation-based GMM (SGMM), Bayesian

Markov chain Monte Carlo (MCMC) methods, and simulation-based resampling methods. Each method is discussed in the context of specific examples and an empirical illustration is performed using real data correspondingly. Section 15.2 overviews the classical estimation methods and explains why they may be difficult to use in practice. Section 15.3 discusses discrete time stochastic volatility models and illustrates the implementation of a SML method. Section 15.4 discusses continuous time models and illustrates the implementation of EMM. Section 15.5 discusses structure credit risk models and illustrates the implementation of a Bayesian MCMC method. Section 15.6 discusses continuous time models with a linear and persistent drift function and illustrates the implementation of the indirect inference (II) method in the context of Vasicek model for the short term interest rate. Finally, Sect. 15.7 concludes.

15.2 Problems with Traditional Estimation Methods

In many cases the likelihood function of a financial time series model can be expressed as:

$$L(\theta) = p(\mathbf{X};\theta) = \int p(\mathbf{X},\mathbf{V};\theta)d\mathbf{V}, \tag{15.1}$$

where $\mathbf{X} = (X_1, \cdots, X_n) := (X_h, \cdots, X_{nh})$ is the data observed by econometricians, h the sampling interval, $p(\mathbf{X})$ the joint density of $\mathbf{X}$, $\mathbf{V}$ a vector of latent variables, θ a set of K parameters that econometricians wish to estimate. As $X(t)$ often represents the annualized data, when daily (weekly or monthly) data are used, h is set at 1/252 (1/52 or 1/12). Assume $T = nh$ is the time span of the data and the true values for θ is θ_0.

MLE maximizes the log-likelihood function over θ in a certain parameter space:

$$\hat{\theta}_n^{ML} := \text{argmax}_{\theta\in\Theta}\ell(\theta)),$$

where $\ell(\theta) = \ln L(\theta) = \ln p(\mathbf{X};\theta)$. The first order condition of the maximization problem is:

$$\frac{\partial \ell}{\partial \theta} = 0.$$

Under mild regularity conditions, the ML estimator (MLE) has desirable asymptotic properties of consistency, normality and efficiency. Moreover, the invariance property of MLE ensures that a smoothed transformation of MLE is a MLE of the same transformation of the corresponding parameters (Zehna 1966). This property has proven very useful in financial applications.

Unfortunately, when the integration in (15.1) is not analytically available and the dimension of $\mathbf{V}$ is high, numerical evaluation of (15.1) is difficult. If $p(\mathbf{X};\theta)$ is difficult to calculate, ML is not easy to implement.

Instead of maximizing the likelihood function, Bayesian methods update the prior density to the posterior density using the likelihood function, based on the Bayes theorem:

$$p(\theta|\mathbf{X}) \propto p(\mathbf{X};\theta)p(\theta),$$

where $p(\theta)$ is the prior density and $p(\theta|\mathbf{X})$ the posterior distribution. As in ML, if $p(\mathbf{X};\theta)$ is difficult to calculate, the posterior density $p(\theta|\mathbf{X})$ is generally difficult to evaluate.

Unlike ML or Bayesian methods that rely on the distributional assumption of the model, GMM only requires a set of moment conditions to be known. Let g be a set of q moment conditions, i.e.

$$E[g(\mathbf{X};\theta_0)] = 0$$

GMM minimizes a distance measure, i.e.

$$\hat{\theta}_n^{GMM} := \text{argmin}_{\theta\in\Theta} \left(\frac{1}{n}\sum_{t=1}^{n} g(X_t;\theta)\right)' W_n \left(\frac{1}{n}\sum_{t=1}^{n} g(X_t;\theta)\right)',$$

where W_n is a certain positive definite weighting matrix of $q \times q$-dimension ($q \geq K$), which may depend on the sample but not θ. Obviously, the implementation of GMM requires the moments to be known analytically or easy to calculate numerically. Since a fixed set of moments contain less information than a density, in general GMM uses less information than ML and hence is statistically less efficient. In the case where the moment conditions are selected based on the score functions (in which case $q = K$), GMM and ML are equivalent. However, sometimes moment conditions are obtained without a distributional assumption and hence GMM may be more robust than the likelihood-based methods. Under mild regularity conditions, Hansen (1982) obtained the asymptotic distributions of GMM estimators. Unfortunately, many financial time series models do not have an analytical expression for moments and moments are difficult to evaluate numerically, making GMM not trivial to implement.

Even if ML is applicable, MLE is not necessarily the best estimator in finite sample. Phillips and Yu (2005a,b, 2009a,b) have provided numerous examples to demonstrate the poor finite sample properties of MLE. In general there are three reasons for this. First, many financial variables (such as interest rates and volatility) are very persistent. When a linear time series model is fitted to these variables, ML and GMM typically lead to substantial finite sample bias for the mean reversion parameter even in very large samples. For example, when 2,500 daily observations are used to estimate the square root model of the short term interest rate, ML estimates the mean reversion parameter with nearly 300% bias. Second, often financial applications involve non-linear transformation of estimators of the system parameters. Even if the system parameters are estimated without any bias, insertion of even unbiased estimators into the nonlinear functions will not assure unbiased estimation of the quantity of interest. A well known example is the MLE of a deep out-of-money option which is highly nonlinear in volatility. In general, the more pronounced the nonlinearity, the worse the finite sample performance is. Third, even if a long-span sample is available for some financial variables and hence asymptotic properties of econometric estimators is more relevant, full data sets are not always employed in estimation because of possible structural changes in long-span data. When short-span samples are used in estimation, finite sample distributions can be far from the asymptotic theory.

A natural way to improve the finite sample performance of classical estimators is to obtain the bias in an analytical form and then remove the bias from the biased estimator, with the hope that the variance of the bias-corrected estimator does not increase or only increases slightly so that the mean square error becomes smaller. Unfortunately, the explicit analytical bias function is often not available, except in very simple cases.

When the likelihood function and moments are difficult to calculate or traditional estimators perform poorly in finite sample, one can resort to simulation methods. There has been an explosion of theoretical and empirical work using simulation methods in financial time series analysis over the last 15 years. In the following sections we will consider some important examples in financial economics and financial econometrics. Simulated-based methods are discussed in the context of these examples and an empirical illustration is provided in each case.

15.3 Simulated ML and Discrete Time SV Models

To illustrate the problem in ML, we first introduce the basic lognormal (LN) SV model of Taylor (1982) defined by

$$\begin{cases} X_t &= \sigma e^{h_t/2}\epsilon_t,\ t = 1, \ldots, n, \\ h_{t+1} &= \phi h_t + \gamma \eta_t,\ t = 1, \ldots, n-1, \end{cases} \tag{15.2}$$

where X_t is the return of an asset, $|\phi| < 1$, $\epsilon_t \overset{iid}{\sim} N(0,1)$, $\eta_t \overset{iid}{\sim} N(0,1)$, $corr(\epsilon_t, \eta_t) = 0$, and $h_1 \sim N(0, \gamma^2/(1-\phi^2))$. The parameters of interest are $\theta = (\sigma, \phi, \gamma)'$. This model is proven to be a powerful alternative to ARCH-type models (Geweke 1994; Danielsson 1994). Its continuous time counterpart has been used to pricing options contracts (Hull and White 1987).

Let $\mathbf{X} = (X_1, \ldots, X_n)'$ and $\mathbf{V} = (h_1, \ldots, h_n)'$. Only $\mathbf{X}$ is observed by the econometrician. The likelihood function of the model is given by

$$p(\mathbf{X}; \theta) = \int p(\mathbf{X}, \mathbf{V}; \theta) d\mathbf{V} = \int p(\mathbf{X}|\mathbf{V}; \theta) p(\mathbf{V}; \theta) d\mathbf{V}. \tag{15.3}$$

To perform the ML estimation to the SV model, one must approximate the high-dimensional integral (15.3) numerically. Since a typical financial time series has at least several hundreds observations, using traditional numerical integration methods, such as quadratures, to approximate the high-dimensional integral (15.3) is numerically formidable. This is the motivation of the use of Monte Carlo integration methods in much of the SV literature.

The basic LN-SV model has been found to be too restrictive empirically for many financial time series and generalized in various dimensions to accommodate stylized facts. Examples include the leverage effect (Harvey and Shephard 1996; Yu 2005), SV-t (Harvey et al. 1994), super-position (Pitt and Shephard 1999b), jumps (Duffie et al. 2000), time varying leverage effect (Yu 2009b). An widely used specification, alternative to the LN-SV model, is the Heston model (Heston 1993).

In this section, we will review several approaches to do simulated ML estimation of the basic LN-SV model. The general methodology is first discussed, followed by a discussion of how to use the method to estimate the LN-SV model and then by an empirical application.

15.3.1 Importance Sampler Based on the Laplace Approximation

Taking the advantage that the integrand is a probability distribution, a widely used SML method evaluates the likelihood function numerically via simulations. One method matches the integrand with a multivariate normal distribution, draws a sequence of independent variables from the multivariate normal distribution, and approximates the integral by the sample mean of a function of the independent draws. Namely, a Monte Carlo method is used to approximate the integral numerically and a carefully selected multivariate normal density is served as an importance function in the Monte Carlo method. The technique in the first stage is known as the Laplace approximation while the technique in the second stage is known as the importance sampler. In this chapter the method is denoted LA-IS.

To fix the idea, in Stage 1, we approximate $p(\mathbf{X}, \mathbf{V}; \theta)$ by a multivariate normal distribution for $\mathbf{V}$, $N(\cdot; \mathbf{V}^*, -\Omega^{-1})$, where

$$\mathbf{V}^* = \arg\max_{\mathbf{V}} \ln p(\mathbf{X}, \mathbf{V}; \theta) \tag{15.4}$$

and

$$\Omega = \frac{\partial^2 \ln p(\mathbf{X}, \mathbf{V}^*; \theta)}{\partial \mathbf{V} \partial \mathbf{V}'}. \tag{15.5}$$

For the LN-SV model $\mathbf{V}^*$ does not have the analytical expression and hence numerical methods are needed. For example, Shephard and Pitt (1997), Durham (2006), Skaug and Yu (2007) proposed to use Newton's method, which involves recursive calculations of $\mathbf{V} = \mathbf{V}_- - \Omega^{-1}\mathbf{V}_-$, based on a certain initial vector of log-volatilities, $\mathbf{V}_0$.

Based on the Laplace approximation, the likelihood function can be written as

$$p(\mathbf{X}; \theta) = \int p(\mathbf{X}, \mathbf{V}; \theta) d\mathbf{V} = \int \frac{p(\mathbf{X}, \mathbf{V}; \theta)}{N(\mathbf{V}; \mathbf{V}^*, -\Omega^{-1})} N(\mathbf{V}; \mathbf{V}^*, -\Omega^{-1}) d\mathbf{V}. \tag{15.6}$$

The idea of importance sampling is to draw samples $\mathbf{V}^{(1)}, \ldots, \mathbf{V}^{(S)}$ from $N(\cdot; \mathbf{V}^*, -\Omega^{-1})$ so that $p(\mathbf{X}; \theta)$ is approximated by

$$\frac{1}{S} \sum_{s=1}^{S} \frac{p(\mathbf{X}, \mathbf{V}^{(s)}; \theta)}{N(\mathbf{V}^{(s)}; \mathbf{V}^*, -\Omega^{-1})}. \tag{15.7}$$

After the likelihood function is obtained, a numerical optimization procedure, such as the quasi Newton method, can be applied to obtain the ML estimator.

The convergence of (15.7) to the likelihood function $p(\mathbf{X}; \theta)$ with $S \to \infty$ is ensured by Komogorov's strong law of large numbers. The square root rate of convergence is achieved if and only if the following condition holds

$$Var\left(\frac{p(\mathbf{X}, \mathbf{V}^{(s)}; \theta)}{N(\mathbf{V}^{(s)}; \mathbf{V}^*, -\Omega^{-1})}\right) < \infty.$$

See Koopman et al. (2009) for further discussions on the conditions and a test to check the convergence.

The idea of the LA-IS method is quite general. The approximation error is determined by the distance between the integrant and the multivariate normal distribution and the size of S. The Laplace approximation does not have any error if $p(\mathbf{X}, \mathbf{V}; \theta)$ is the Gaussianity in $\mathbf{V}$. In this case, $S = 1$ is big enough to obtain the exact value of the integral. The further $p(\mathbf{X}, \mathbf{V}; \theta)$ away from Gaussian in $\mathbf{V}$, the less precise the Laplace approximation is. In this case, a large value is needed for S.

For the LN-SV model, the integrand in (15.3) can be written as

$$p(\mathbf{X}, \mathbf{V}; \theta) = N\left(h_1, 0, \frac{\gamma^2}{1-\phi^2}\right) \prod_{t=2}^{n} N\left(h_t, \phi h_{n-1}, \gamma^2\right) \prod_{t=1}^{n} N\left(X_t, 0, \sigma^2 e^{h_t}\right), \tag{15.8}$$

and hence

$$\ln p(\mathbf{X}, \mathbf{V}; \theta) = \ln N\left(h_1, 0, \frac{\gamma^2}{1-\phi^2}\right) + \sum_{t=2}^{n} \ln N\left(h_t, \phi h_{n-1}, \gamma^2\right) + \sum_{t=1}^{n} \ln N\left(X_t, 0, \sigma^2 e^{h_t}\right). \tag{15.9}$$

It is easy to show that

$$\frac{\partial N(x; \mu, \sigma^2)/\partial x}{N(x; \mu, \sigma^2)} = -\frac{x-\mu}{\sigma^2}, \quad \frac{\partial N(x; \mu, \sigma^2)/\partial \mu}{N(x; \mu, \sigma^2)} = -\frac{\mu - x}{\sigma^2},$$

$$\frac{\partial N(x; \mu, \sigma^2)/\partial \sigma^2}{N(x; \mu, \sigma^2)} = -\frac{1}{\sigma^2}\left(1 - \frac{(x-\mu)^2}{\sigma^2}\right),$$

Using these results, we obtain the gradient of the log-integrand:

$$\begin{pmatrix} \frac{\partial \ln p(\mathbf{X},\mathbf{V};\theta)}{\partial h_1} \\ \frac{\partial \ln p(\mathbf{X},\mathbf{V};\theta)}{\partial h_2} \\ \vdots \\ \frac{\partial \ln p(\mathbf{X},\mathbf{V};\theta)}{\partial h_{n-1}} \\ \frac{\partial \ln p(\mathbf{X},\mathbf{V};\theta)}{\partial h_n} \end{pmatrix} = \begin{pmatrix} \frac{\phi h_2 - h_1}{\gamma^2} - \frac{1}{2} + \frac{1}{2}\epsilon_1^2 \\ \frac{\phi h_3 - \phi^2 h_2 + \phi h_1}{\gamma^2} - \frac{1}{2} + \frac{1}{2}\epsilon_2^2 \\ \vdots \\ \frac{\phi h_n - \phi^2 h_{n-1} + \phi h_{T-2}}{\gamma^2} - \frac{1}{2} + \frac{1}{2}\epsilon_{n-1}^2 \\ \frac{h_n - \phi h_{n-1}}{\gamma^2} - \frac{1}{2} + \frac{1}{2}\epsilon_n^2 \end{pmatrix}, \tag{15.10}$$

and the Hessian matrix of the log-integrand:

$$\Omega = \begin{pmatrix} -\frac{1}{\gamma^2} - \frac{1}{2}\epsilon_1^2 & \frac{\phi}{\gamma^2} & \cdots & 0 & 0 \\ \frac{\phi}{\gamma^2} & -\frac{1+\phi^2}{\gamma^2} - \frac{1}{2}\epsilon_2^2 & \cdots & 0 & 0 \\ \vdots & \vdots & \ddots & \vdots & \vdots \\ 0 & 0 & \cdots & -\frac{1+\phi^2}{\gamma^2} - \frac{1}{2}\epsilon_{n-1}^2 & \frac{\phi}{\gamma^2} \\ 0 & 0 & \cdots & \frac{\phi}{\gamma^2} & -\frac{1}{\gamma^2} - \frac{1}{2}\epsilon_n^2 \end{pmatrix}. \tag{15.11}$$

Durham (2006, 2007), Koopman et al. (2009), Skaug and Yu (2007) and Yu (2009b) applied the SML method to estimate generalized SV models and documented the reliable performance in various contexts.

15.3.2 Monte Carlo Likelihood Method

Durbin and Koopman (1997) proposed a closely related SML method which is termed Monte Carlo likelihood (MCL) method. MCL was originally designed to evaluate the likelihood function of a linear state-space model with non-Gaussian errors. The basic idea is to decompose the likelihood function into the likelihood of a linear state-space model with Gaussian errors and that of the remainder. It is known that the likelihood function of a linear state-space model with Gaussian errors can be calculated by the Kalman filter. The likelihood of the remainder is calculated by simulations using LA-IS.

To obtain the linear state-space form for the LN-SV model, one can apply the log-squared transformation to X_t:

$$\begin{cases} Y_t &= \ln X_t^2 = \ln \sigma^2 + h_t + \varepsilon_t, \; t = 1, \ldots, n, \\ h_{t+1} &= \phi h_t + \gamma \eta_t, \; t = 1, \ldots, n-1, \end{cases} \tag{15.12}$$

where $\varepsilon_t \stackrel{iid}{\sim} \ln \chi^2_{(1)}$ (i.e. no-Gaussian), $\eta_t \stackrel{iid}{\sim} N(0,1)$, $corr(\varepsilon_t, \eta_t) = 0$, and $h_1 \sim N(0, \gamma^2/(1-\phi^2))$. For any linear state-space model with non-Gaussian measurement errors, Durbin and Koopman (1997) showed that the log-likelihood function can be expressed as

$$\ln p(\mathbf{X};\theta) = \ln L_G(\mathbf{X};\theta) + \ln E_G\left[\frac{p_\varepsilon(\varepsilon;\theta)}{p_G(\varepsilon;\theta)}\right], \tag{15.13}$$

where $\ln L_G(\mathbf{X};\theta)$ is the the log-likelihood function of a carefully chosen approximating Gaussian model, $p_\varepsilon(\varepsilon;\theta)$ the true density of $\varepsilon(:= (\varepsilon_1,\ldots,\varepsilon_n)')$, $p_G(\varepsilon;\theta)$ the Gaussian density of the measurement errors of the approximating model, E_G the expectation with respect to the importance density in connection to the approximating model.

Relative to (15.3), (15.13) has the advantage that simulations are only needed to estimate the departure of the likelihood from the Gaussian likelihood, rather than the full likelihood. For the LN-SV model, $\ln L_G(\mathbf{X};\theta)$ often takes a much larger value than $\ln E_G\left[\frac{p_\varepsilon(\varepsilon;\theta)}{p_G(\varepsilon;\theta)}\right]$. As a result, MCL is computationally efficient than other simulated-based ML methods because it only needs a small number of simulations to achieve the desirable accuracy when approximating the likelihood. However, the implementation of the method requires a linear non-Gaussian state-space representation. Jungbacker and Koopman (2007) extended the method to deal with nonlinear non-Gaussian state-space models. Sandmann and Koopman (1998) applied the method to estimate the LN-SV model and the SV-t model. Broto and Ruiz (2004) compared the performance of alternative methods for estimating the LN-SV model and found supporting evidence for of the good performance of MCL.

15.3.3 Efficient Importance Sampler

Richard and Zhang (2007) developed an alternative simulated ML method. It is based on a particular factorization of the importance density and termed as Efficient Importance Sampling (EIS). Relative to the two SML methods reviewed in Sects 3.1 and 3.2, EIS minimizes locally the Monte Carlo sampling variance of the approximation to the integrand by factorizing the importance density. To fix the idea, assume $g(\mathbf{V}|\mathbf{X})$ is the importance density which can be constructed as

$$g(\mathbf{V}|\mathbf{X}) = \prod_{t=1}^{n} g(h_t|h_{t-1},\mathbf{X}) = \prod_{t=1}^{n}\left\{C_t e^{c_t h_t + d_t h_t^2} p(h_t|h_{t-1})\right\}, \tag{15.14}$$

where c_t, C_t and d_t depend on $\mathbf{X}$ and h_{t-1} with $\{C_t\}$ be a normalization sequence so that g is a normal distribution. The sequences $\{c_t\}$ and $\{d_t\}$ should be chosen to match $p(\mathbf{X},\mathbf{V};\theta)$ and $g(\mathbf{V}|\mathbf{X})$ which, as we shown in Sect. 15.3.1, requires a high-dimensional non-linear regression. The caveat of EIS is to match each component in $g(\mathbf{V}|\mathbf{X})$ (i.e. $C_t e^{c_t h_t + d_t h_t^2} p(h_t|h_{t-1})$), to the corresponding element in the integrand

$p(\mathbf{X};\mathbf{V})$ (ie $p(X_t|h_t)p(h_t|h_{t-1})$) in a backward manner, with $t = n, n-1, \cdots, 1$. It is easy to show that C_t depends only on h_{t-1} but not on h_t. As a result, the recursive matching problem is equivalent to running the following linear regression backward:

$$\ln p(X_t|h_t^{(s)}) - \ln C_{t+1} = a + c_t h_t^{(s)} + d_t (h_t^{(s)})^2, \; s = 1, \cdots, S, \tag{15.15}$$

where $h_t^{(1)}, \ldots, h_t^{(S)}$ are drawn from the importance density and $h_t^{(s)}$ and $(h_t^{(s)})^2$ are treated as the explanatory variables in the regression model with $C_{n+1} = 1$.

The method to approximate the likelihood involves the following procedures:

1. Draw initial $\mathbf{V}^{(s)}$ from (15.2) with $s = 1, \cdots, S$.
2. Estimate c_t and d_t from (15.15) and do it backward with $C_{n+1} = 1$.
3. Draw $\mathbf{V}^{(s)}$ from importance density $g(\mathbf{V}|\mathbf{X})$ based on c_t and d_t.
4. Repeat Steps 2-3 until convergence. Denote the resulting sampler by $\mathbf{V}^{(s)}$.
5. Approximate the likelihood by

$$\frac{1}{S}\sum_{s=1}^{S}\left\{\prod_{t=1}^{n}\frac{p(X_t|h_t^{(s)})}{C_t \exp\left(c_t h_t^{(s)} + d_t (h_t^{(s)})^2\right)}\right\}.$$

The EIS algorithm relies on the user to provide a problem-dependent auxiliary class of importance samplers. An advantage of this method is that it does not rely on the assumption that the latent process is Gaussian. Liesenfeld and Richard (2003, 2006) applied this method to estimate a number of discrete SV models while Kleppe et al. (2009) applied this method to estimate a continuous time SV model. Lee and Koopman (2004) compared the EIS method with the LA-IS method and found two methods are comparable in the context of the LN-SV model and the SV-t model. Bauwens and Galli (2008) and Bauwens and Hautsch (2006) applied EIS to estimate a stochastic duration model and a stochastic conditional intensity model, respectively.

15.3.4 An Empirical Example

For the purposes of illustration, we fit the LN-SV model to a widely used dataset (namely svpd1.txt). The dataset consists of 945 observations on daily pound/dollar exchange rate from 01/10/1981 to 28/06/1985. The same data were used in Harvey et al. (1994), Shephard and Pitt (1997), Meyer and Yu (2000), and Skaug and Yu (2007).

Matlab code (namely LAISLNSV.m) is used to implement the LA-IS method. Table 15.1 reports the estimates and the likelihood when $S = 32$. In Skaug and Yu (2007) the same method was used to estimate the same model but S was set at 64. The estimates and the log-likelihood value based on $S = 32$ are very similar to those based on $S = 64$, suggesting that a small number of random samples can approximate the likelihood function very well.

Table 15.1 SMLE of the LN-SV model

	σ	γ	ϕ	Log-likelihood
$S = 32$	0.6323	0.1685	0.9748	917.845
$S = 64$	0.6305	0.1687	0.9734	917.458

15.4 Simulated GMM and Continuous Time Models

Many models that are used to describe financial time series are written in terms of a continuous time diffusion $X(t)$ that satisfies the stochastic differential equation

$$dX(t) = \mu(X(t);\theta)dt + \sigma(X(t);\theta)dB(t), \tag{15.16}$$

where $B(t)$ is a standard Brownian motion, $\sigma(X(t);\theta)$ a diffusion function, $\mu(X(t);\theta)$ a drift function, and θ a vector of unknown parameters. The target here is to estimate θ from a discrete sampled observations, $X_h, \ldots, X_{nh}$ with h being the sampling interval. This class of parametric models has been widely used to characterize the temporal dynamics of financial variables, including stock prices, interest rates, and exchange rates.

Many estimation methods are based on the construction of the likelihood function derived from the transition probability density of the discretely sampled data. This approach is explained as follows. Suppose $p(X_{ih}|X_{(i-1)h},\theta)$ is the transition probability density. The Markov property of model (15.16) implies the following log-likelihood function for the discrete sample

$$\ell(\theta) = \sum_{i=1}^{n} \ln(p(X_{ih}|X_{(i-1)h},\theta)). \tag{15.17}$$

To perform exact ML estimation, one needs a closed form expression for $\ell(\theta)$ and hence $\ln(p(X_{ih}|X_{(i-1)h},\theta))$. In general, the transition density p satisfies the forward equation:

$$\frac{\partial p}{\partial t} = \frac{1}{2}\frac{\partial^2 p}{\partial y^2}$$

and the backward equation:

$$\frac{\partial p}{\partial s} = -\frac{1}{2}\frac{\partial^2 p}{\partial x^2},$$

where $p(y,t|x,s)$ is the transition density. Solving the partial differential equation numerically at $y = X_{ih}, x = X_{(i-1)h}$ yields the transition density. This approach was proposed by Lo (1988).

Unfortunately, only in rare cases, does the transition density $p(X_{ih}|X_{(i-1)h},\theta)$ have a closed form solution. Phillips and Yu (2009) provide a list of examples in which $\ln(p(X_{ih}|X_{(i-1)h},\theta))$ have a closed form analytical expression. These examples include the geometric Brownian Motion, Ornstein-Uhlenbeck (OU) process, square-root process, and inverse square-root process. In general solving the forward/backward equations is computationally demanding.

A classical and widely used estimation method is via the Euler scheme, which approximates a general diffusion process such as equation (15.16) by the following discrete time model

$$X_{ih} = X_{(i-1)h} + \mu(X_{(i-1)h}, \theta)h + \sigma(X_{(i-1)h}, \theta)\sqrt{h}\epsilon_i, \tag{15.18}$$

where $\epsilon_i \sim$ i.i.d. $N(0,1)$. The transition density for the Euler discrete time model (15.18) has the following closed form expression:

$$X_{ih}|X_{(i-1)h} \sim N\left(X_{(i-1)h} + \mu(X_{(i-1)h}, \theta)h, \sigma^2(X_{(i-1)h}, \theta)h\right). \tag{15.19}$$

Obviously, the Euler scheme introduces a discretization bias. The magnitude of the bias introduced by Euler scheme is determined by h, which cannot be controlled econometricians. In general, the bias becomes negligible when h is close to zero. One way to use the full likelihood analysis is to make the sampling interval arbitrarily small by partitioning the original sampling interval so that the new subintervals are sufficiently fine for the discretization bias to be negligible. By making the subintervals smaller, one inevitably introduces latent variables between the two original consecutive observations $X_{(i-1)h}$ and X_{ih}. While our main focus is SGMM in this section, SML is possible and is discussed first.

15.4.1 SML Methods

To implement ML estimation, one can integrate out these latent observations. When the partition becomes finer, the discretization bias is approaching 0 but the required integration becomes high dimensional. In general, the integral does not have a closed-form expression and hence simulation-based methods can be used, leading to simulated ML estimators. To fix the idea, suppose that $M-1$ auxiliary points are introduced between $(i-1)h$ and ih, i.e.

$$((i-1)h \equiv)\tau_0, \tau_1, \cdots, \tau_{M-1}, \tau_M(\equiv ih).$$

Thus

$$\begin{aligned} p(X_{ih}|X_{(i-1)h};\theta) &= \int \cdots \int p(X_{\tau_M}, X_{\tau_{M-1}}, \cdots, X_{\tau_1}|X_{\tau_0};\theta) dX_{\tau_1} \cdots dX_{\tau_{M-1}} \\ &= \int \cdots \int \prod_{m=1}^{M} p(X_{\tau_m}|X_{\tau_{m-1}};\theta) dX_{\tau_1} \cdots dX_{\tau_{M-1}}. \end{aligned} \tag{15.20}$$

The second equality follows from the Markov property. The idea behind the simulated ML method is to approximate the densities $p(X_{\tau_m}|X_{\tau_{m-1}};\theta)$ (step 1), evaluate the multidimensional integral using importance sampling techniques (step 2) and then maximize the likelihood function numerically. To the best of my knowledge, Pedersen (1995) was the first study that suggested the idea in this context.

Pedersen's method relies on the Euler scheme, namely, approximates the latent transition densities $p(X_{\tau_m}|X_{\tau_{m-1}};\theta)$ based on the Euler scheme and approximates the integral by drawing samples of $(X_{\tau_{M-1}},\cdots,X_{\tau_1})$ via simulations from the Euler scheme. That is, the importance sampling function is the mapping from $(\epsilon_1,\epsilon_2,\cdots,\epsilon_{M-1})\mapsto(X_{\tau_1},X_{\tau_2},\cdots,X_{\tau_{M-1}})$ given by the Euler scheme:

$$X_{\tau_{m+1}} = X_{\tau_m} + \mu(X_{\tau_m};\theta)h/M + \sigma(X_{\tau_m},\theta)\sqrt{h/M}\epsilon_{m+1},\ m=0,\cdots,M-2,$$

where $(\epsilon_1,\cdots,\epsilon_{M-1})$ is a multivariate standard normal.

Durham and Gallant (2002) noted two sources of approximation error in Pedersen's method, the discretization bias in the Euler scheme and the errors due to the Monte Carlo integration. A number of studies have provided methods to reduce these two sources of error. For example, to reduce the discretization bias in step 1, Elerian (1998) used the Milstein scheme instead of the Euler scheme while Durham and Gallant advocated using a variance stablization transformation, i.e. applying the Lamperti transform to the continuous time model. Certainly, other methods that can reduce the discretization bias may be used. Regarding step 2, Elerian et al. (2001) argued that the importance sampling function of Pedersen ignores the end-point information, X_{τ_M}, and Durham and Gallant (2002) showed that Pedersen's importance function draws most samples from regions where the integrand has little mass. Consequently, Pedersen's method is simulation-inefficient.

To improve the efficiency of the importance sampler, Durham and Gallant (2002) considered the following importance sampling function

$$X_{\tau_{m+1}} = X_{\tau_m} + \frac{X_{ih}-X_{\tau_m}}{ih-\tau_m}h/M + \sigma(X_{\tau_m},\theta)\sqrt{h/M}\epsilon_{m+1},\ m=0,\cdots,M-2,$$

where $(\epsilon_1,\cdots,\epsilon_{M-1})$ is a multivariate standard normal. Loosing speaking, this is a Brownian bridge because it starts from $X_{(i-1)h}$ at $(i-1)h$ and is conditioned to terminate with X_{ih} at ih. Another importance sampling function proposed by Durham and Gallant (2002) is to draw $X_{\tau_{m+1}}$ from the density $N(X_{\tau_m}+\tilde{\mu}_m h/M, \tilde{\sigma}_m^2 h/M)$ where $\tilde{\mu}_m=(X_{\tau_M}-X_{\tau_m})/(ih-\tau_m)$, $\tilde{\sigma}_m^2=\sigma^2(X_{\tau_m})(M-m-1)/(M-m)$. Elerian et al. (2001) suggested the following tied-down process:

$$p(X_{\tau_1},\cdots,X_{\tau_{M-1}}|X_{\tau_0},X_{\tau_M}),$$

as the importance function and proposed using the Laplace approximation to the tied-down process. Durham and Gallant (2002) compared the performance of these three importance functions relative to Pedersen (1995) and found that all these methods deliver substantial improvements.

15.4.2 Simulated GMM (SGMM)

Not only is the likelihood function for (15.16) difficult to construct, but also the moment conditions; see, for example, Duffie and Singleton (1993) and He (1990). While model (15.16) is difficult to estimate, data can be easily simulated from it. For example, one can simulate data from the Euler scheme at an arbitrarily small sampling interval. With the interval approaches to zero, the simulated data can be regarded as the exact simulation although the transition density at the coarser sampling interval is not known analytically. With simulated data, moments can be easily constructed, facilitating simulation-based GMM estimation. Simulated GMM (SGMM) methods have been proposed by McFadden (1989), Pakes and Pollard (1989) for iid environments, and Lee and Ingram (1991), Duffie and Singleton (1993) for time series environments.

Let $\{\widetilde{\mathbf{X}}_t^{(s)}(\theta)\}_{t=1}^{\mathcal{N}(n)}$ be the data simulated from (15.16) when parameter is θ using random seed s. Therefore, $\{\widetilde{X}_t^{(s)}(\theta_0)\}$ is drawn from the same distribution as the original data $\{\mathbf{X}_t\}$ and hence share the same moment characteristic. The parameter θ is chosen so as to "match moments", that is, to minimize the distance between sample moments of the data and those of the simulated data. Assuming H represents K-moments, SGMM estimator is defined as:

$$\hat{\theta}_n^{SGMM} := \text{argmin}_{\theta \in \Theta} \left(\frac{1}{n} \sum_{t=1}^{n} g(X_t) - \frac{1}{\mathcal{N}(n)} \sum_{t=1}^{\mathcal{N}(n)} g(\tilde{X}_t^{(s)}; \theta) \right)' W_n$$

$$\left(\frac{1}{n} \sum_{t=1}^{n} g(X_t) - \frac{1}{\mathcal{N}(n)} \sum_{t=1}^{\mathcal{N}(n)} g(\tilde{X}_t^{(s)}; \theta) \right)',$$

where W_n is a certain positive definite weighting matrix of $q \times q$-dimension ($q \geq K$), which may depend on the sample but not θ, $\mathcal{N}(n)$ is the number of number of observations in a simulated path. Under the ergodicity condition,

$$\frac{1}{\mathcal{N}(n)} \sum_{t=1}^{\mathcal{N}(n)} g(\tilde{X}_t^{(s)}; \theta_0) \xrightarrow{p} E(g(X_t; \theta_0))$$

and

$$\frac{1}{n} \sum_{t=1}^{n} g(X_t) \xrightarrow{p} E(g(X_t; \theta_0)),$$

justifying the SGMM procedure.

The SGMM procedure can be made optimal with a careful choice of the weighting function, given a set of moments. However, the SGMM estimator is in general asymptotically less efficient than SML for the reason that moments are less informative than the likelihood. Gallant and Tauchen (1996a,b) extended the SGMM technique so that the GMM estimator is asymptotically as efficient as SML. This approach is termed efficient method of moments (EMM), which we review below.

15.4.3 Efficient Method of Moments

EMM is first introduced by Gallant and Tauchen (1996a,b) and has now found many applications in financial time series; see Gallant and Tauchen (2001a,c) for the detailed account of the method and a review of the literature. While it is closely related to the general SGMM, there is one important difference between them. Namely, GMM relies on an ad hoc chosen set of moment conditions, EMM is based on a judiciously chosen set of moment conditions. The moment conditions that EMM is based on are the expectation of the score of an auxiliary model which is often referred to as the score generator.

For the purpose of illustration, let a SV model be the structural model. The SV model is the continuous time version of the Box-Cox SV model of Yu et al. (2006), which contains many classical continuous SV models as special cases, and is of the form:

$$dS(t) = \alpha_{10}S(t)dt + S(t)[1 + \delta(\beta_{10} + \beta_{12}h(t))]^{1/(2\delta)}dB_1(t),$$

$$dh(t) = -\alpha_{22}h(t)dt + dB_2(t).$$

Let the conditional density of the structural model (the Box-Cox SV model in this case) is defined by

$$p_t(X_t|Y_t, \theta),$$

where $X_t = \ln S(t)$, the true value of θ is θ_0, $\theta_0 \in \Theta \subset \Re^{\ell_\theta}$ with ℓ_θ being the length of θ_0 and Y_t is a vector of lagged X_t. Denote the conditional density of an auxiliary model by

$$f_t(X_t|Y_t, \beta), \beta \in R \subset \Re^{\ell_\beta}.$$

Further define the expected score of the auxiliary model under the structural model as

$$m(\theta, \beta) = \int \cdots \int \frac{\partial}{\partial \beta} \ln f(x|y, \beta) p(x|y, \theta) p(y|\theta) dx dy.$$

Obviously, in the context of the SV model, the integration cannot be solved analytically since neither $p(x|y, \theta)$ nor $p(y|\theta)$ has a closed form expression. However, it is easy to simulate from an SV model so that one can approximate the integral by Monte Carlo simulations. That is

$$m(\theta,\beta) \approx m_N(\theta,\beta) \equiv \frac{1}{N}\sum_{\tau=1}^{N}\frac{\partial}{\partial\beta}\ln f(\hat{X}_\tau(\theta)|\hat{Y}_\tau(\theta),\beta),$$

where $\{\hat{X}_\tau, \hat{Y}_\tau\}$ are simulated from the structural model. The EMM estimator is a minimum chi-squared estimator which minimizes the following quadratic form,

$$\hat{\theta}_n = \arg\min_{\theta\in\Theta} m'_N(\theta,\hat{\beta}_n)(I_n)^{-1}m_N(\theta,\hat{\beta}_n),$$

where $\hat{\beta}_n$ is a quasi maximum likelihood estimator of the auxiliary model and I_n is an estimate of

$$I_0 = \lim_{n\to\infty} Var\left(\frac{1}{\sqrt{n}}\sum_{t=1}^{n}\left\{\frac{\partial}{\partial\beta}\ln f_t(x_t|y_t,\beta^*)\right\}\right)$$

with β^* being the pseudo true value of β. Under regularity conditions, Gallant and Tauchen (1996a,b) show that the EMM estimator is consistent and has the following asymptotic normal distribution,

$$\sqrt{n}(\hat{\theta}_n - \theta_0) \xrightarrow{d} N\left(0, \frac{\partial}{\partial\theta}m(\theta_0,\beta^*)(I_0)^{-1}\frac{\partial}{\partial\theta'}m(\theta_0,\beta^*)\right).$$

For specification testing, we have

$$J_n = nm'_N(\hat{\theta}_n,\hat{\beta}_n)(I_n)^{-1}m_N(\hat{\theta}_n,\hat{\beta}_n) \xrightarrow{d} \chi^2_{\ell_\beta-\ell_\theta}$$

under the null hypothesis that the structural model is correct. When a model fails the above specification test one may wish to examine the quasi-t-ratios and/or t-ratios to look for some suggestion as to what is wrong with the structural model. The quasi-t-ratios are defined as

$$\hat{T}_n = S_n^{-1}\sqrt{n}m_N(\hat{\theta}_n,\hat{\beta}_n),$$

where $S_n = [diag(I_n)]^{1/2}$. It is well known that the elements of $\hat{T}_n$ are downward biased in absolute value. To correct the bias one can use the t-ratios defined by

$$\tilde{T}_n = Q_n^{-1}\sqrt{n}m_N(\hat{\theta}_n,\hat{\beta}_n),$$

where

$$Q_n = \left(diag\{I_n - \frac{\partial}{\partial\theta'}m_N(\hat{\theta}_n,\hat{\beta}_n)[m'_N(\hat{\theta}_n,\hat{\beta}_n)(I_n)^{-1}m_N(\hat{\theta}_n,\hat{\beta}_n)]^{-1}\frac{\partial}{\partial\theta}m_N(\hat{\theta}_n,\hat{\beta}_n)\}\right)^{1/2}.$$

Large quasi-t-ratios and t-ratios reveal the features of the data that the structural model cannot approximate.

Furthermore, Gallant and Tauchen (1996a,b) show that if the auxiliary model nests the data generating process, under regularity conditions the EMM estimator has the same asymptotic variance as the maximum likelihood estimator and hence is fully efficient. If the auxiliary model can closely approximate the data generating process, the EMM estimator is nearly fully efficient (Gallant and Long 1997; Tauchen 1997).

To choose an auxiliary model, the seminonparametric (SNP) density proposed by Gallant and Tauchen (1989) can be used since its success has been documented in many applications. As to SNP modeling, six out of eight tuning parameters are to be selected, namely, L_u, L_g, L_r, L_p, K_z, and K_y. The other two parameters, I_z and I_x, are irrelevant for univariate time series and hence set to be 0. L_u determines the location transformation whereas L_g and L_r determine the scale transformation. Altogether they determine the nature of the leading term of the Hermite expansion. The other two parameters K_z and K_y determine the nature of the innovation. To search for a good auxiliary model, one can use the Schwarz BIC criterion to move along an upward expansion path until an adequate model is found, as outlined in Bansal et al. (1995). To preserve space we refer readers to Gallant and Tauchen (2001b) for further discussion about the role of the tuning parameters and how to design an expansion path to choose them.

While EMM has found a wide range of applications in financial time series, Duffee and Stanton (2008) reported finite sample evidence against EMM when financial time series is persistent. In particular, in the context of simple term structure models, they showed that although EMM has the same asymptotic efficiency as ML, the variance of EMM estimator in finite sample is too large and cannot be accepted in practice.

15.4.4 An Empirical Example

For the purposes of illustration, we fit the continuous time Box-Cox SV model to daily prices of Microsoft. The stock price data consist of 3,778 observations on the daily price of a share of Microsoft, adjusted for stock split, for the period from March 13, 1986 to February 23, 2001. The same data have been used in Gallant and Tauchen (2001a) to fit a continuous time LN-SV model. For this reason, we use the same sets of tuning parameters in the SNP model as in Gallant and Tauchen (2001a), namely,

$$(L_u, L_g, L_r, L_p, K_z, I_z, K_y, I_y) = (1, 1, 1, 1, 6, 0, 0, 0).$$

Fortran code and the date can be obtained from an anonymous ftp site at ftp.econ.duke.edu. A EMM User Guide by Gallant and Tauchen (2001a) is available from the same site. To estimate the Box-Cox SV model, we only needed to change the specification of the diffusion function in the subroutine difuse in the fortran file emmuothr.f, i.e. "tmp1 = DEXP(DMIN1 (tmp1,bnd))" is changed to

Table 15.2 EMM estimate of the continuous time box-cox SV model

α_{10}	α_{22}	β_{10}	β_{12}	δ	χ^2_6
0.4364	0.5649	−0.1094	0.2710	0.1367	13.895

"tmp1 = (1+ delta* DMIN1 (tmp1,bnd))**(0.5/delta)". Table 15.2 reports the EMM estimates. Obviously, the volatility of Microsoft is very persistent since the estimated mean reversion parameter is close to zero and the estimate value of δ is not far away from 0, indicating that the estimated Box-Cox SV is not very different from the LN-SV model model.

15.5 Bayesian MCMC and Credit Risk Models

Credit derivatives market had experienced a fantastic growth before the global financial meltdown in 2007. The size of the market had grew so much and the credit risk management had been done so poorly in practice that the impact of the financial crisis is so big. Not surprisingly, how to estimate credit risk has received an increasing attention from academic researchers, industry participants, policy makers and regulators.

A widely used approach to credit risk modelling in practice is the so-called structural method. All structural credit risk models specify a dynamic structure for the underlying firm's asset and default boundary. Let V be the firm's asset process, r the risk-free interest rate, F the face value of a zero-coupon debt that the firm issues with the time to maturity T. Merton (1974) is the simplest structural model where V_t is assumed to follow a geometric Brownian motion:

$$d\ln V_t = (\mu - \sigma^2/2)dt + \sigma dB_t, \; V_0 = c, \tag{15.21}$$

The exact discrete time model, sampled with the step size h, is

$$\ln V_{t+1} = (\mu - \sigma^2/2)h + \ln V_t + \sigma\sqrt{h}\epsilon_t, \; V_0 = c, \tag{15.22}$$

which contains a unit root.

There are two types of outstanding claims faced by a firm that is listed in a stock exchange, an equity and a zero-coupon debt whose face value is F maturing at T. The default occurs at the maturity date of debt in the event that the issuer's assets are less than the face value of the debt (i.e. $V_T < F$). Under the assumption of (15.21) the firm's equity can be priced with the Black-Scholes formula as if it is a call option on the total asset value V of the firm with the strike price of F and the maturity date T. Namely, the equity claim, denoted by S_t, is

$$S_t \equiv S(V_t;\sigma) = V_t\Phi(d_{1t}) - Fe^{-r(T-t)}\Phi(d_{2t}), \tag{15.23}$$

where $\Phi(\cdot)$ is the cumulative distribution function of the standard normal variate,

$$d_{1t} = \frac{\ln(V_t/F) + (r + \sigma^2/2)(T - t)}{\sigma\sqrt{T - t}},$$

and

$$d_{2t} = \frac{\ln(V_t/F) + (r - \sigma^2/2)(T - t)}{\sigma\sqrt{T - t}}.$$

Merton's model can be used to evaluate private firm credit risk and the credit spread of a risk corporate bond over the corresponding Treasure rate. The credit spread is given by

$$C(V_t;\theta) = -\frac{1}{T-\tau_t}\ln\left(\frac{V_t}{F}\Phi(-d_{1t}) + e^{-r(T-\tau_t)}\Phi(d_{2t})\right) - r. \tag{15.24}$$

The default probability is given by

$$P(V_t;\theta) = \Phi\left(\frac{\ln(F/V_t) - (\mu - \sigma^2/2)(T-\tau_t)}{\sigma\sqrt{T-\tau_t}}\right). \tag{15.25}$$

At a reasonably high frequency, S_t may be observed with errors due to the presence of various market microstructure effects. This observation motivates Duan and Fulop (2009) to consider the following generalization to Merton's model:

$$\ln S_t = \ln S(V_t;\sigma) + \delta v_t,\ v_t \sim N(0,1). \tag{15.26}$$

In a state-space framework, (15.26) is an observation equation and (15.22) is a state equation. Unfortunately, the Kalman filter is not applicable here since the observation equation is nonlinear.

Let $\mathbf{X} = (\ln S_1, \cdots, \ln S_n)'$, $\mathbf{V} = (\ln V_1, \cdots, \ln V_n)'$, and $\theta = (\mu, \sigma, \delta)'$. The likelihood function of (15.26) is given by

$$p(\mathbf{X};\theta) = \int p(\mathbf{X},\mathbf{V};\theta)d\mathbf{V} = \int p(\mathbf{X}|\mathbf{V};\mu)p(\mathbf{V};\theta)d\mathbf{V}. \tag{15.27}$$

In general this is a high-dimensional integral which does not have closed form expression due to the non-linear dependence of $\ln S_t$ on $\ln V_t$. Although in this section, our main focus is the Bayesian MCMC methods, SML is possible. Indeed all the SML methods discussed in Sect. 15.3 are applicable here. However, we will discuss a new set of SML methods – particle filters.

15.5.1 SML via Particle Filter

It is known that Kalman filter is an optimal recursive data processing algorithm for processing series of measurements generated from a linear dynamic system. It is applicable any linear Gaussian state-space model where all relevant conditional distributions are linear Gaussians. Particle filters, also known as sequential Monte Carlo methods, extend the Kalman filter to nonlinear and non-Gaussian state space models.

In a state space model, two equations have to be specified in the fully parametric manner. First, the state equation describes the evolution of the state with time. Second, the measurement equation relates the noisy measurements to the state. A recursive filtering approach means that received data can be processed sequentially rather than as a batch so that it is not necessary to store the complete data set nor to reprocess existing data if a new measurement becomes available. Such a filter consists of essentially two stages: prediction and updating. The prediction stage uses the system model to predict the state density forward from one measurement time to the next. Since the state is usually subject to unknown disturbances, prediction generally translates, deforms, and spreads the state density. The updating operation uses the latest measurement to modify the prediction density. This is achieved using Bayes theorem, which is the mechanism for updating knowledge about the target state in the light of extra information from new data. When the model is linear and Gaussian, the density in both stages is Gaussian and Kalman filter gives analytical expressions to the mean and the co-variance. As a byproduct, the full conditional distribution of measurements is available, facilitating the calculation of the likelihood.

For nonlinear and non-Gaussain state space models, the density in neither stage is not Gaussian any more and the optimal filter is not available analytically. Particle filter is a technique for implementing a recursive filter by Monte Carlo simulations. The key idea is to represent the required density in connection to prediction and updating by a set of random samples (known as "particles") with associated weights and to compute estimates based on these samples and weights. As the number of samples becomes very large, this simulation-based empirical distribution is equivalent the true distribution.

To fix the idea, assume that the nonlinear non-Gaussian state space model is of the form,

$$\begin{cases} Y_t = H(X_t, e_t) \\ X_t = F(X_{t-1}, u_t), \end{cases} \tag{15.28}$$

where X_t is a k-dimensional state vector, u_t is a l-dimensional white noise sequence with density $q(u)$, v_t is a l-dimensional white noise sequence with density $r(v)$ and assumed uncorrelated with $\{u_s\}_{s=1}^t$, H and F are possibly nonlinear functions. Let $v_t = G(Y_t, X_t)$ and G' is the derivative of G as a function of Y_t. The density of the initial state vector is assumed to be $p_0(x)$. Denote $Y_{1:k} = \{Y_1, \cdots, Y_k\}$. The objective of the prediction is to obtain $p(X_t|Y_{1:t})$. It can be seen that

$$p(X_t|Y_{1:t-1}) = \int p(X_t|X_{t-1})p(X_{t-1}|Y_{1:t-1})dX_{t-1}. \tag{15.29}$$

At time step t, when a new measurement Y_t becomes available, it may be used to update the predictive density $p(X_t|Y_{1:t-1})$ via Bayes rule in the updating stage,

$$p(X_t|Y_{1:t}) = \frac{p(Y_t|X_t)p(X_t|Y_{1:t-1})}{p(Y_t|Y_{1:t-1})}. \tag{15.30}$$

Unfortunately, for the nonlinear non-Gaussian state-space model, the recursive propagation in both stages is only a conceptual solution and cannot be determined analytically. To deal with this problem, particle filtering algorithm consists of recursive propagation of the weights and support points when each measurement is received sequentially so that the true densities can be approximated by the corresponding empirical density.

Various versions of particle filters have been proposed in the literature. In this chapter we only summarize all the steps involved in Kitagawa's algorithm (Kitagawa 1996):

1. Generate M l-dimensional particles from $p_0(x)$, $f_0^{(j)}$ for $j = 1, \ldots, M$.
2. Repeat the following steps for $t = 1, \ldots, n$.
   (a) Generate M l-dimensional particles from $q(u)$, $u_t^{(j)}$ for $j = 1, \ldots, M$.
   (b) Compute $p_t^{(j)} = F(f_{t-1}^{(j)}, u_t^{(j)})$ for $j = 1, \ldots, M$.
   (c) Compute $\alpha_t^{(j)} = r(G(Y_t, p_t^{(j)}))$ for $j = 1, \ldots, M$.
   (d) Re-sample $\{p_t^{(j)}\}_{j=1}^M$ to get $\{f_t^{(j)}\}_{j=1}^M$ with probabilities proportional to $\{r(G(Y_t, p_t^{(j)})) \times |G'(Y_t, p_t^{(j)})|\}_{j=1}^M$.

Other particle filtering algorithms include sampling importance resampling filter of Gordon et al. (1993), auxiliary sampling importance resampling filter of Pitt and Shephard (1999a), and regularized particle filter (Musso et al. 2001).

To estimate the Merton's model via ML, Duan and Fulop employed the *particle filtering* method of Pitt (2002). Unlike the method proposed by Kitagawa (1995) which samples a point $X_t^{(m)}$ when the system is advanced, Duan and Fulop sampled a pair $(V_t^{(m)}, V_{t+1}^{(m)})$ at once when the system is advanced. Since the resulting likelihood function is not smooth with respect to the parameters, to ensure a smooth surface for the likelihood function, Duan and Fulop used the smooth bootstrap procedure for resampling of Pitt (2002).

Because the log-likelihood function can be obtained as a by-product of the filtering algorithm, it can be maximized numerically over the parameter space to obtain the SMLE. If $M \to \infty$, the log-likelihood value obtained from simulations should converge to the true likelihood value. As a result, it is expected that for a sufficiently large number of particles, the estimates that maximize the approximated log-likelihood function are sufficiently close to the true ML estimates.

15.5.2 Bayesian MCMC Methods

The structure in the state-space model ensures the pivotal role played by Bayes theorem in the recursive propagation. Not surprisingly, the requirement for the updating of information on receipt of new measurements are ideally suited for the Bayesian approach for statistical inference. In this chapter, we will show that Bayesian methods provide a rigorous general approach to the dynamic state estimation problem. Since many models in financial econometrics have a state-space representation, Bayesian methods have received more and more attentions in statistical analysis of financial time series.

The general idea of the Bayesian approach is to perform posterior computations, given the likelihood function and the prior distribution. MCMC is a class of algorthims which enables one to obtain a correlated sample from a Morkov chain whose stationary transition density is the same as the posterior distribution. There are certain advantages in the Bayesian MCMC method. First, as a likelihood-based method, MCMC matches the efficiency of ML. Second, as a by-product of parameter estimation, MCMC provides smoothed estimates of latent variables because it augments the parameter space by including the latent variables. Third, unlike the frequentist's methods whose inference is almost always based on asymptotic arguments, inferences via MCMC are based on the exact posterior distribution. This advantage is especially important when the standard asymptotic theory is difficult to derive or the asymptotic distribution does not provide satisfactory approximation to the finite sample distribution. As a trade-off, one has to specify the prior distribution. In addition, with MCMC it is straightforward to obtain the exact posterior distribution of any transformation (linear or nonlinear) of model parameters and latent variables, such as the credit spread and the default probability. Therefore, the exact finite sample inference can easily be made in MCMC, whereas the ML method necessitates the delta method to obtain the asymptotic distribution. When the asymptotic distribution of the original parameters does not work well, it is expected that the asymptotic distribution yielded by the delta method may not work well. Fourth, numerical optimization is not needed in MCMC. This advantage is of practical importance when the likelihood function is difficult to optimize numerically. Finally, the proposed method lends itself easily to dealing with flexible specifications.

There are three disadvantages of the MCMC method. First, in order to obtain the filtered estimate of the latent variable, a separate method is required. This is in contrast with the ML method of Duan and Fulop (2009) where the filtered estimate of the latent variable is obtained as a by-product. Second, with the MCMC method the model has to be fully specified whereas the MLE remains consistent even when the microstructure noise is nonparametrically specified, and in this case, ML becomes quasi-ML. However, in recent years, semiparametric MCMC methods have appeared in the literature. For example, the flexibility of the error distribution may be accommodated by using a Dirichelt process mixture (DPM) prior (see Ferguson (1973) for the detailed account of DMP, and Jensen and Maheu (2008) for an application of DMP to volatility modeling). Finally, prior distributions have to be specified. In some cases, prior distributions may have important influences on the posterior analysis but it is not so obvious to specify the prior distributions.

From the Bayesian viewpoint, we understand the specification of the structural credit risk model as a hierarchical structure of conditional distributions. The hierarchy is specified by a sequence of three distributions, the conditional distribution of $\ln S_t | \ln V_t, \delta$, the conditional distribution of $\ln V_t | \ln V_{t-1}, \mu, \sigma$, and the prior distribution of θ. Hence, our Bayesian model consists of the joint prior distribution of all unobservables, here the three parameters, μ, σ, δ, and the unknown states, $\mathbf{V}$, and the joint distribution of the observables, here the sequence of contaminated log-equity prices $\mathbf{X}$. The treatment of the latent state variables $\mathbf{V}$ as the additional unknown parameters is the well known data-augmentation technique originally proposed by Tanner and Wong (1987) in the context of MCMC. Bayesian inference is then based on the posterior distribution of the unobservables given the data. In the sequel, we will denote the probability density function of a random variable θ by $p(\theta)$. By successive conditioning, the joint prior density is

$$p(\mu, \sigma, \delta, \mathbf{V}) = p(\mu, \sigma, \delta) p(\ln V_0) \prod_{t=1}^{n} p(\ln V_t | \ln V_{t-1}, \mu, \sigma). \tag{15.31}$$

We assume prior independence of the parameters μ, δ and σ. Clearly $p(\ln V_t | \ln V_{t-1}, \mu, \sigma)$ is defined through the state equations (15.22). The likelihood $p(\mathbf{X}|\mu, \sigma, \delta, \mathbf{V})$ is specified by the observation equations (15.26) and the conditional independence assumption:

$$p(\mathbf{X}|\mu, \sigma, \delta, \mathbf{V}) = \prod_{t=1}^{n} p(\ln S_t | \ln V_t, \delta). \tag{15.32}$$

Then, by Bayes' theorem, the joint posterior distribution of the unobservables given the data is proportional to the prior times likelihood, i.e.

$$p(\mu, \sigma, \delta, \mathbf{V}|\mathbf{X}) \propto p(\mu) p(\sigma) p(\delta) p(\ln V_0) \prod_{t=1}^{n} p(\ln V_t | \ln V_{t-1}, \mu, \sigma) \prod_{t=1}^{n} p(\ln S_t | \ln V_t, \delta). \tag{15.33}$$

Without data augmentation, we need to deal with the intractable likelihood function $p(\mathbf{X}|\theta)$ which makes the direct analysis of the posterior density $p(\theta|\mathbf{V})$ difficult. The particle filtering algorithm of Duan and Fulop (2009) can be used to overcome the problem. With data augmentation, we focus on the new posterior density $p(\theta, \mathbf{V}|\mathbf{X})$ given in (15.33). Note that the new likelihood function is $p(\mathbf{X}|\theta, \mathbf{V})$ which is readily available analytically once the distribution of ϵ_t is specified. Another advantage of using the data-augmentation technique is that the latent state variables $\mathbf{V}$ are the additional unknown parameters and hence we can make statistical inference about them.

The idea behind the MCMC methods is to repeatedly sample from a Markov chain whose stationary (multivariate) distribution is the (multivariate) posterior density. Once the chain converges, the sample is regarded as a correlated sample from the posterior density. By the ergodic theorem for Markov chains, the posterior moments and marginal densities can be estimated by averaging the corresponding functions over the sample. For example, one can estimate the posterior mean by the sample mean, and obtain the credible interval from the marginal density. When the simulation size is very large, the marginal densities can be regarded to be exact, enabling exact finite sample inferences. Since the latent state variables are in the parameter space, MCMC also provides the exact solution to the smoothing problem of inferring about the unobserved equity value.

While there are a number of MCMC algorithms available in the literature, we only use the Gibbs sampler which samples each variate, one at a time, from the full conditional distributions defined by (15.33). When all the variates are sampled in a cycle, we have one sweep. The algorithm is then repeated for many sweeps with the variates being updated with the most recent samples. With regularity conditions, the draws from the samplers converge to draw from the posterior distribution at a geometric rate. For further information about MCMC and its applications in econometrics, see Chib (2001) and Johannes and Polson (2003).

Defining $\ln V_{-t}$ by $\ln V_1, \ldots, \ln V_{t-1}, \ln V_{t+1}, \ldots, \ln V_n$, the Gibbs sampler is summarized as:

1. Initialize θ and $\mathbf{V}$.
2. Sample $\ln V_t$ from $\ln V_t | \ln V_{-t}, \mathbf{X}$.
3. Sample $\sigma | \mathbf{X}, \mathbf{V}, \mu, \delta$.
4. Sample $\delta | \mathbf{X}, \mathbf{V}, \mu, \sigma$.
5. Sample $\mu | \mathbf{X}, \mathbf{V}, \sigma, \delta$.

Steps 2–5 forms one cycle. Repeating steps 2–5 for many thousands of times yields the MCMC output. To mitigate the effect of initialization and to ensure the full convergence of the chains, we discard the so-call burn-in samples. The remaining samples are used to make inference.

It is easy to implement the Gibbs sampling for the credit risk model defined above. One can make use of the all purpose Bayesian software package WinBUGS. As shown in Meyer and Yu (2000) and Yu et al. (2006), WinBUGS provides an idea framework to perform the Bayesian MCMC computation when the model has a state-space form, whether it is nonlinear or non-Gaussian or both. As the Gibbs sampler updates only one variable at a time, it is referred as a single-move algorithm.

In the stochastic volatility literature, the single-move algorithm has been criticized by Kim et al. (1998) for lacking simulation efficiency because the components of state variables are highly correlated. More efficient MCMC algorithms, such as multi-move algorithms, can be developed for estimating credit risk models. In fact, Shephard and Pitt (1997), Kim et al. (1998), Chib et al. (2002), Liesenfeld and Richard (2006) and Omori et al. (2007) have developed various multi-move algorithms to estimate univariate and multivariate SV models. The idea of the multimover algorithms is to sample the latent vector $\mathbf{V}$ in a single block.

Table 15.3 MCMC and SML estimates of the credit risk model

	μ		σ		$\delta \times 100$	
	Mean	Std err	Mean	Std err	Mean	Std err
Bayesian	0.3154	0.1689	0.1686	0.0125	0.5673	0.1225
SML	0.3130	0.1640	0.1589	0.0181	0.6820	0.2082

15.5.3 An Empirical Application

For the purposes of illustration, we fit the credit risk model to daily prices of AA a company from the Dow Jones Industrial Index. The daily equity values are obtained from the CRSP database over year 2003 (the logarithmic values are contained in a file named AAlogS.txt). The initial maturity of debt is 10 years. The debt is available from the balance sheet obtained from the Compustat annual file. It is compounded for 10 years at the risk-free rate to obtain F. The risk-free rate is obtained from the US Federal Reserve. Duan and Fulop fitted the same model to the same data using SML via particle filter and approximated the variance using the Fisher information matrix. Following Huang and Yu (2010), we use the following independent prior for the three system parameters: $\mu \sim N(0.3, 4)$, $\delta \sim IG(3, 0.0001)$, and $\sigma \sim IG(2.5, 0.025)$ where IG is the inverse-gamma distribution.

WinBugs code (aa.odc) is used to implement the MCMC method based on 55,000 sweeps of which the first 5,000 sweeps are thrown away. Table 15.3 reports the estimates (the posterior means) and the standard errors (the posterior standard errors). For the purpose of comparison, the SML estimates and their asymptotic standard errors, obtained directly from Duan and Fulop (2009, Table 15.1), are also reported. While the two sets of estimates are close to each other, their standard errors are further away.

15.6 Resampling Methods and Term Structure Models

It is well known dynamic models are estimated with bias by standard estimation methods, such as least squares (LS), maximum likelihood (ML) or generalized method of moments (GMM). The bias was developed by Hurwicz (1950) for the autoregressive parameter in the context of dynamic discrete time models. The percentage bias of the corresponding parameter, i.e. the mean reversion parameter, is much more pronounced in continuous time models than their discrete time counterparts. On the other hand, estimation is fundamentally important for many practical applications. For example, it provides parameter estimators which are used directly for estimating prices of financial assets and derivatives. For another example, parameter estimation serves as an important stage for the empirical analysis of specification and comparative diagnostics. Not surprisingly, it has been found in the literature that the bias in the mean reversion estimator has important implications for the specification analysis of continuous time models (Pritsker 1998) and for pricing financial assets (Phillips and Yu 2005a, 2009b). For instance, when the true mean reversion parameter is 0.1 and 600 weekly observations (i.e. just over 10 years of data) are available to estimate a one-factor square-root term structure model (Cox et al. 1985), the bias in the ML estimator of the mean reversion parameter is 391.2% in an upwards direction. This estimation bias, together with the estimation errors and nonlinearity, produces a 60.6% downward bias in the option price of a discount bond and 2.48% downward bias in the discount bond price. The latter figures are comparable in magnitude to the estimates of bias effects discussed in Hull (2000, Chap. 21.7). The biases would be even larger when less observations are available and do not disappear even when using long spans of data that are currently available. For example, when the true mean reversion parameter is 0.1 and 600 monthly observations (i.e. 50 years of data) are available to estimate the square-root diffusion model, the bias in the ML estimator of the mean reversion parameter is 84.5% in an upwards direction. This estimation bias implies a 24.4% downward bias in the option price of a discount bond and a 1.0% downward bias in the discount bond price.

In recent years, there have been interesting advances in developing analytical formulae to approximate the bias in certain model specifications. This is typically obtained by estimating higher order terms in an asymptotic expansion of the bias. For example, in the Vasicek term structure model with a known μ,

$$dX_t = \kappa(\mu - X_t)dt + \sigma dB_t, X_0 \sim N(\mu, \sigma^2/(2\kappa))$$

Yu (2009a,b) showed that the bias in the MLE of κ can be approximated by

$$\frac{1}{2T}\left(3 + e^{2\kappa h}\right) - \frac{2(1 - e^{-2n\kappa h})}{Tn(1 - e^{-2\kappa h})}.$$

When μ has to be estimated in the Vasicek model, Tang and Chen (2009) showed that the bias in the MLE of κ can be approximated by

$$E(\widehat{\kappa}) - \kappa = \frac{1}{2T}(e^{2\kappa h} + 2e^{\kappa h} + 5).$$

Interestingly, the same bias formula applies to a QML estimate of κ, developed by Nowman (1997), under the CIR model, as shown in Tang and Chen (2009).

For more complicated models, unfortunately, the approximate bias formula is not available. To reduce this bias in parameter estimation and in pricing contingent claims, Phillips and Yu (2005a) proposed a new jackknife procedure. Phillips and Yu (2005a) show that the jackknife method always trades off the gain that may be achieved in bias reduction with a loss that arises through increased variance.

The bootstrap method of Efron (1979) is another way to reduce the bias via simulation. It was shown to be an effective method for bias correction (Hall 1992) and was illustrated in the parameter estimation in the context of continuous time model in Tang and Chen (2009). Relative to the jackknife method, it does not significantly increase the variance. Relative to the two simulation-based procedures that will be discussed below, however, bootstrap seems to use less information and hence is expected to be less efficient.

15.6.1 Indirect Inference and Median Unbiased Estimation

Resampling methods may achieve bias reduction as well as variance reduction. In this chapter, two simulation-based resampling methods are discussed, indirect inference (II) and median unbiased estimation (MUE).

II and MUE are simulation-based estimation procedures and can be understood as a generalization of the simulated method of moments approach of Duffie and Singleton (1993). MUE was first introduced by Andrews (1993). II was first introduced by Smith (1993) and coined with the term by Gouriéroux et al. (1993). II was originally proposed to deal with situations where the moments or the likelihood function of the true model are difficult to deal with (and hence traditional methods such as GMM and ML are difficult to implement), but the true model is amenable to data simulation. Because many continuous time models are easy to simulate but difficult to obtain moment and likelihood functions, the II procedure has some convenient advantages in working with continuous time models in finance.

The II and MUE procedures can have good small sample properties of parameter estimates, as shown by Andrews (1993), MacKinnon and Smith (1996), Monfort (1996), Gouriéroux et al. (2000) in the time series context and by Gouriéroux et al. (2005) in the panel context. The idea why II can remove the bias goes as follows. Whenever a bias occurs in an estimate and from whatever source, this bias will also be present in the same estimate obtained from data, which are of the same structure of the original data, simulated from the model for the same reasons. Hence, the bias can be calculated via simulations. The method therefore offers some interesting opportunities for bias correction and the improvement of finite sample properties in continuous time parameter estimation, as shown in Phillips and Yu (2009a).

To fix the idea of II/MUE for parameter estimation, consider the Vasicek model which is typically used to describe the movement of the short term interest rate. Suppose we need to estimate the parameter κ in:

$$dX(t) = \kappa(\mu - X(t))dt + \sigma(X(t))\, dW(t),$$

from observations $\{X_h, \cdots, X_{nh}\}$. An initial estimator of κ can be obtained, for example, by applying the Euler scheme to $\{X_h, \cdots, X_{nh}\}$ (call it $\hat{\kappa}_n$). Such an estimator is involved with the discretization bias (due to the use of the Euler scheme) as well as a finite sample estimation bias (due to the poor finite sample property of ML in the near-unit-root situation).

Given a parameter choice κ, we apply the Euler scheme with a much smaller step size than h (say $\delta = h/100$), which leads to

$$\tilde{X}_{t+\delta}^k = \kappa(\mu - \tilde{X}_t^k)h + \tilde{X}_t^k + \sigma(\tilde{X}_t^k)\sqrt{\delta}\varepsilon_{t+\delta},$$

where

$$t = \underbrace{0, \delta, \cdots, h(=100\delta)}, \underbrace{h+\delta, \cdots, 2h(=200\delta)}, 2h+\delta, \cdots, nh.$$

This sequence may be regarded as a nearly exact simulation from the continuous time OU model for small δ. We then choose every $(h/\delta)^{th}$ observation to form the sequence of $\{\tilde{X}_{ih}^k\}_{i=1}^n$, which can be regarded as data simulated directly from the OU model with the (observationally relevant) step size h.

Let $\{\tilde{X}_h^k, \cdots, \tilde{X}_{nh}^k\}$ be data simulated from the true model, where $k = 1, \cdots, K$ with K being the number of simulated paths. It should be emphasized that it is important to choose the number of simulated observations and the sampling interval to be the same as the number of observations and the sampling interval in the observed sequence for the purpose of the bias calibration. Another estimator of κ can be obtained by applying the Euler scheme to $\{X_h^k, \cdots, X_{nh}^k\}$ (call it $\tilde{\kappa}_n^k$). Such an estimator and hence the expected value of them across simulated paths is naturally dependent on the given parameter choice κ.

The central idea in II/MUE is to match the parameter obtained from the actual data with that obtained from the simulated data. In particular, the II estimator and median unbiased estimator of κ solve, respectively,

$$\hat{\kappa}_n = \frac{1}{K}\sum_{h=1}^{K}\tilde{\kappa}_n^k(\kappa) \text{ or } \hat{\kappa}_n = \hat{\rho}_{0.5}(\tilde{\kappa}_n^k(\kappa)), \tag{15.34}$$

where $\hat{\rho}_\tau$ is the τth sample quantile. In the case where K tends to infinity, the II estimator and median unbiased estimator solve

$$\hat{\kappa}_n = E(\tilde{\kappa}_n^k(\kappa)) \text{ or } \hat{\kappa}_n = \rho_{0.5}(\tilde{\kappa}_n^k(\kappa)), \tag{15.35}$$

where $E(\tilde{\kappa}_n^k(\kappa))$ is called the mean binding function, and $\rho_{0.5}(\tilde{\kappa}_n^k(\kappa))$ is the median binding function, i.e.

$$b_n(\kappa) = E(\tilde{\kappa}_n^k(\kappa)), \text{ or } b_N(\kappa) = \rho_{0.5}(\tilde{\kappa}_n^k(\kappa)).$$

It is a finite sample functional relating the bias to κ. In the case where b_n is invertible, the II estimator and median unbiased estimator are given by:

$$\hat{\kappa}_n^{II} = b_n^{-1}(\hat{\kappa}_n). \tag{15.36}$$

Typically, the binding functions cannot be computed analytically in either case. That is why II/MUE needs to calculate the binding functions via simulations. While often used in the literature for the binding function is the mean, the median has certain advantages over the mean. First, the median is more robust to outliers than the mean. Second, it is easier to obtain the unbiased property via the median. In particular, while the linearity of $b_n(\kappa)$ gives rise of the mean-unbiasedness in $\hat{\kappa}_n^{II}$, only monotonicity is needed for $b_n(\kappa)$ to ensure the median-unbiasedness (Phillips and Yu 2009b).

There are several advantages in the II/MUE procedure relative to the jackknife procedure. First, II is more effective on removing the bias in parameter estimates. Phillips and Yu (2009a) provided evidence to support this superiority of II. Second, the bias reduction may be achieved often without an increase in variance. In extreme cases of root near unity, the variance of II/MUE can be even smaller than that of ML (Phillips and Yu 2009a). To see this, note that (15.36) implies:

$$Var(\hat{\kappa}_n^{II}) = \left(\frac{\partial b_n}{\partial \kappa}\right)^{-1} Var(\hat{\kappa}_n^{ML}) \left(\frac{\partial b_n}{\partial \kappa'}\right)^{-1}.$$

When $\partial b_n/\partial \kappa > 1$, the II/MUE estimator has a smaller variance than MLE. Gouriéroux et al. (2000) discussed the relationship among II, MUE and bootstrap in the context of bias correction.

A disadvantage in the II/MUE procedure is the high computational cost. It is expected that with the continuing explosive growth in computing power, such a drawback is of less concern. Nevertheless, to reduce the computational cost, one can choose a fine grid of discrete points of κ and obtain the binding function on the grid. Then standard interpolation and extrapolation methods can be used to approximate the binding functions at any point.

As pointed out before, since prices of contingent-claims are always non-linear transformations of the system parameters, insertion of even unbiased estimators into the pricing formulae will not assure unbiased estimation of a contingent-claim price. The stronger the nonlinearity, the larger the bias. As a result, plugging-in the II/MUE estimates into the pricing formulae may still yield an estimate of the price with unsatisfactory finite sample performances. This feature was illustrated in a the context of various continuous time models and contingent claims in Phillips and Yu (2009d). To improve the finite sample properties of the contingent price estimate, Phillips and Yu (2009b) generalized the II/MUE procedure so that it is applied to the quantity of interest directly.

To fix the idea, suppose θ is the scalar parameter in the continuous time model on which the price of a contingent claim, $P(\theta)$, is based. Denote by $\hat{\theta}_n^{ML}$ the MLE of θ that is obtained from the actual data, and write $\widehat{P}_n^{ML} = P(\hat{\theta}_n^{ML})$ be the ML estimate of P. $\widehat{P}_n^{ML}$ involves finite sample estimation bias due to the non-linearity of the pricing function P in θ, or the use of the biased estimate $\hat{\theta}_n^{ML}$, or both these effects. The II/MUE approach involves the following steps.

Table 15.4 ML, II and median unbiased estimates of κ in the Vasicek model

	MLE	II	MUE
$\hat{\kappa}$	0.2613	0.1358	0.1642

1. Given a value for the contingent-claim price p, compute $P^{-1}(p)$ (call it $\theta(p)$), where $P^{-1}(\cdot)$ is the inverse of the pricing function $P(\theta)$.
2. Let $\widetilde{\mathbf{S}}^k(p) = \{\tilde{S}_1^k, \tilde{S}_2^k, \cdots, \tilde{S}_T^k\}$ be data simulated from the time series model (15.16) given $\theta(p)$, where $k = 1, \ldots, K$ with K being the number of simulated paths. As argued above, we choose the number of observations in $\widetilde{\mathbf{S}}^k(p)$ to be the same as the number of actual observations in $\mathbf{S}$ for the express purpose of finite sample bias calibration.
3. Obtain $\tilde{\phi}_n^{ML,k}(p)$, the MLE of θ, from the kth simulated path, and calculate $\widetilde{P}_n^{ML,k}(p) = P(\tilde{\phi}_n^{ML,k}(p))$.
4. Choose p so that the average behavior of $\widetilde{P}_n^{ML,k}(p)$ is matched with $\widehat{P}_n^{ML}$ to produce a new bias corrected estimate.

15.6.2 An Empirical Application

This empirical application compares the ML method and the simulation-based methods for estimating the mean reversion parameter in a context of Vasicek term structure model. The dataset of a short term interest rate series involves the Federal fund rate and is available from the H-15 Federal Reserve Statistical Release. It is sampled monthly and has 432 observations covering the period from January 1963 to December 1998. The same data were used in Ait-Sahalia (1999) and are contained in a file named ff.txt.

Matlab code, simVasicek.m, is used to obtain the ML, II and median unbiased estimates of κ in the Vasiecek model. Table 15.4 reports these estimates. The ML estimate is about twice as large as the II estimate. The II estimate is similar to the median unbiased estimate.

15.7 Conclusions

Simulation-based estimation of financial time series model has been ongoing in the financial econometric literature and the empirical finance literature for more than one decade. Some new developments have been made and some existing methods have been refined with the increasing complexity in models. More and more attention have been paid to the simulation-based methods in recent years. Researchers in empirical finance have sought to use these methods in practical applications in an increasing scale. we expect the need for these methods to grow further as the financial industry continues to expand and data sets become richer.

Acknowledgements I gratefully acknowledge financial support from the Ministry of Education AcRF Tier 2 fund under Grant No. T206B4301-RS. Data and program code used in this paper can be download from my website at http://www.mysmu.edu/faculty/yujun/research.html.

References

Aït-Sahalia, Y. (1999). Transition Densities for interest rate and other non-linear diffusions. *Journal of Finance*, *54*, 1361–1395.

Aït-Sahalia, Y. (2002). Maximum likelihood estimation of discretely sampled diffusion: A closed-form approximation approach. *Econometrica*, *70*, 223–262.

Aït-Sahalia, Y. (2008). Closed-form likelihood expansions for multivariate diffusions. *Annals of Statistics*, *36*, 906–937.

Aït-Sahalia, Y., & Yu, J. (2006). Saddlepoint approximations for continuous- ime markov processes. *Journal of Econometrics*, *134*, 507–551.

Andrews, D. W. K. (1993). Exactly median-unbiased estimation of first order autoregressive/unit Root models. *Econometrica*, *61*, 139–166.

Bansal, R., Gallant, A. R., Hussey, R. & Tauchen, G. (1995). Nonparametric estimation of structural models for high-frequency currency market data. *Journal of Econometrics*, *66*, 251–287.

Bauwens, L., & Galli, F. (2008). Efficient importance sampling for ML estimation of SCD models. *Computational Statistics and Data Analysis*, *53*, 1974–1992.

Bauwens, L., & Hautsch, N. (2006). Stochastic conditional intensity processes. *Journal of Financial Econometrics*, *4*, 450–493.

Black, F., & Scholes, M. (1973). The pricing of options and corporate liabilities. *Journal of Political Economy*, *81*, 637–659.

Broto, C., & Ruiz, E. (2004). Estimation methods for stochastic volatility models: a survey. *Journal of Economic Surveys*, *18*, 613–649.

Chib, S. (2001). Markov Chain Monte Carlo methods: Computation and inference. In J.J. Heckman & E. Leamer (Eds.), *Handbook of econometrics* (Vol. 5, pp. 3569–3649). Amsterdam: North-Holland

Chib, S., Nardari, F., & Shephard, N. (2002). Markov Chain Monte Carlo methods stochastic volatility models. *Journal of Econometrics*, *108*, 281–316.

Danielsson, J. (1994). Stochastic volatility in asset prices: Estimation with simulated maximum likelihood. *Journal of Econometrics*, *64*, 375–400.

Duan, J. C., & Fulop, A. (2009). Estimating the structural credit risk model when equity prices are contaminated by trading noises. *Journal of Econometrics*, *150*, 288–296.

Duffie, D., & Singleton, K. J. (1993). Simulated moments estimation of markov models of asset prices. *Econometrica*, *61*, 929–952.

Duffie, D., & Stanton, R. (2008). Evidence on simulation inference for near unit-root processes with implications for term structure estimation. *Journal of Financial Econometrics*, *6*, 108–142.

Duffie, D., Pan, J., & Singleton, K. J. (2000). Transform analysis and asset pricing for affine jump-diffusions. *Econometrica*, *68*, 1343–1376.

Durbin, J., & Koopman, S. J. (2000). Time series analysis of non-gaussian observations based on state space models from both classical and bayesian perspectives (with discussion). *Journal of the Royal Statistical Society Series B*, *62*, 3–56.

Durham, G. (2006). Monte carlo methods for estimating, moothing, and filtering one and two-factor stochastic volatility models. *Journal of Econometrics*, *133*, 273–305.

Durham, G. (2007). SV mixture models with application to S&P 500 index returns. *Journal of Financial Economics*, *85*, 822–856.

Durham, G., & Gallant, A. R. (2002). Numerical techniques for maximum likelihood estimation of continuous-time iffusion processes. *Journal of Business and Economic Statistics, 20*, 297–316.

Efron, B. (1982). *The Jackknife, the Bootstrap and other resampling method.* Philadephia: SIAM

Elerian, O. (1998). A Note on the Existence of a Closed-form Conditional Transition Density for the Milstein Scheme, Economics discussion paper 1998-W18, Nuffield College, Oxford.

Elerian, O., Chib, S., & N. Shephard (2001). Likelihood inference for discretely observed non-linear diffusions. *Econometrica, 69*, 959–993.

Ferguson, T. (1973). A bayesian analysis of some nonparametric problems. *Annals of Statistics, 1*, 209–230.

Gallant, A. R., & Long, L. R. (1997). Estimating stochastic differential equations efficiently by minimum chi-squared. *Biometrika, 84*, 125–141.

Gallant, A. R., & Tauchen, G. (1996a). Which moments to match?. *Econometric Theory, 12*, 657–681.

Gallant, A. R., & Tauchen, G. (1996b). Which moments to match?. *Econometric Theory, 12*, 657–681.

Gallant, A. R., & Tauchen, G. (1989). Seminonparametric estimation of conditionally constrained heterogeneous processes: Asset pricing applications. *Econometrica 57*, 1091–1120.

Gallant, A. R., & Tauchen, G. (1998). Reprojecting partially observed systems with application to interest rate diffusions. *Journal of the American Statistical Association, 93*, 10–24.

Gallant, A. R., & Tauchen, G. (2001a). EMM: A program for efficient method of moments estimation. User's Guide, Department of Economics, University of North Carolina.

Gallant, A. R., & Tauchen, G. (2001b). SNP: A program for nonparametric time series analysis. User's Guide, Department of Economics, University of North Carolina.

Gallant, A. R., & Tauchen, G. (2001c). Efficient method of moments. Working paper, Department of Economics, University of North Carolina.

Geweke, J. (1994). Bayesian comparison of econometric models. Working Paper, Federal Reserve Bank of Minneapolis, Minnesota.

Gordon, N. J., Salmond, D. J., & Smith, A. E. M. (1993). A novel approach to nonlinear and non-Gaussian Bayesian state estimation. *IEEE-Proceedings F, 140*, 107–133.

Gouriéroux, C., & Monfort, A. (1995). *Simulation based econometric methods.* London: Oxford University Press.

Gouriéroux, C., Monfort, A., & Renault, E. (1993). Indirect inference. *Journal of Applied Econometrics, 8*, S85–S118.

Gouriéroux, C., Renault, E. & Touzi, N. (2000). Calibration by simulation for small sample bias correction. In R.S. Mariano, T. Schuermann, & M. Weeks (Eds.), *Simulation-based inference in econometrics: Methods and applications* (pp. 328–358). London: Cambridge University Press.

Gouriéroux, C., Phillips, P. C. B., & Yu, J. (2010). Indirect inference for dynamic panel models. *Journal of Econometrics*, forthcoming.

Hall, P. (1992). *The bootstrap and edgeworth expansion.* Berlin: Springer.

Hansen, L. P. (1982). Large sample properties of generalized method of moments estimators. *Econometrica, 50*, 1029–1054.

Harvey, A. C., & Shephard, N. (1996). The estimation of an asymmetric stochastic volatility model for asset returns. *Journal of Business and Economic Statistics, 14*, 429–434.

Harvey, A. C., Ruiz, E., & Shephard, N. (1994). Multivariate stochastic variance models. *Review of Economic Studies 61*, 247–264.

He, H. (1990). Moment Approximation and Estimation of Diffusion Models of Asset Prices, Working paper, University of California at Berkeley.

Heston, S. L. (1993). A closed-form solution for options with stochastic volatility, with application to bond and currency options. *Review of Financial Studies, 6*, 327–343.

Huang, S. J., & Yu, J. (2010). Bayesian analysis of structural credit risk models with microstructure noises. *Journal of Economic Dynamics and Control, 34*, 2259–2272.

Hull, J., & White, A. (1987). The pricing of options on assets with stochastic volatilities. *Journal of Finance, 42*, 281–300.

Jacquier, E., Polson, N. G., & Rossi, P. E. (1994). Bayesian analysis of stochastic volatility models (with discussion). *Journal of Business and Economic Statistics, 12*, 371–417.

Jensen, M. J., & Maheu, J. M. (2008). Bayesian semiparametric stochastic volatility modeling, Working Paper No. 2008-15, Federal Reserve Bank of Atlanta.

Johannes, M., & Polson, N. (2009). MCMC methods for continuous time asset pricing models. In Aït-Sahalia & Hansen (Eds.), *Handbook of financial econometrics, 2*, 1–72, North-Holland.

Jungbacker, B., & Koopman, S. J. (2007). Monte Carlo estimation for nonlinear non-Gaussian state space models. *Biometrika, 94*, 827–839.

Kim, S., Shephard, N., & Chib, S. (1998). Stochastic volatility: Likelihood inference and comparison with ARCH models. *Review of Economic Studies 65*, 361–393.

Kitagawa, G. (1996). Monte Carlo filter and smoother for Gaussian nonlinear state space models. *Journal of Computational and Graphical Statistics, 5*, 1–25.

Kleppe, T., Skaug, H., & Yu, J. (2009). Simulated Maximum Likelihood estimation of continuous time stochastic volatility models. Working paper, Singapore Management University.

Koopman, S. J., Shephard, N., & Creal, D. (2009). Testing the assumptions behind importance sampling. *Journal of Econometrics, 149*, 2–11.

Lee, B. S., & Ingram, B. F. (1991). Simulation estimation of time-series models. *Journal of Econometrics, 47*, 197–205.

Lee, K. M., & Koopman, S. J. (2004). Estimating stochastic volatility models: A comparison of two importance samplers. *Studies in Nonlinear Dynamics and Econometrics, 8*, 1–15.

Liesenfeld, R., & Richard, J. F. (2003). Univariate and multivariate stochastic volatility models: Estimation and diagnostics. *Journal of Empirical Finance, 10*, 505–531.

Liesenfeld, R., & Richard, J. F. (2006). Classical and bayesian analysis of univariate and multivariate stochastic volatility models. *Econometric Reviews, 25*, 335–360.

Lo, A. W. (1988). Maximum likelihood estimation of generalized itô processes with discretely sampled data. *Econometric Theory, 4*, 231–247.

MacKinnon, J. G., & Smith, A. A. (1998). Approximate bias correction in econometrics. *Journal of Econometrics, 85*, 205–230.

McFadden, D. (1989). A method of simulated moments for estimation of discrete response models without numerical integration. *Econometrica, 57*, 995–1026.

McLeish, D. (2005). *Monte Carlo simulation and finance*. NY: Wiley.

Merton, R. C. (1974). On the pricing of corporate debt: The risk structure of interest rates. *Journal of Finance, 29*, 449–470.

Meyer, R., & Yu, J. (2000). BUGS for a Bayesian analysis of stochastic volatility models. *Econometrics Journal, 3*, 198–215.

Meyer, R., Fournier, D. A., & Berg, A. (2003). Stochastic volatility: Bayesian computation using automatic differentiation and the extended Kalman filter. *Econometrics Journal, 6*, 408–420.

Monfort, A. (1996). A reappraisal of misspecified econometric models. *Econometric Theory, 12*, 597–619.

Musso, C., Oudjane, N., & Le Gland, F. (2001). Improving regularized particle filters, In A. Doucet, N. de Freitas, & N. Gordon (Eds.), *Sequential Monte Carlo methods in practice* (pp. 247–271). New York: Springer.

Nowman, K. B. (1997). Gaussian estimation of single-factor continuous time models of the term structure of interest rates. *Journal of Finance, 52*, 1695–1703.

Omori, Y., Chib, S., Shephard, N., & Nakajima, J. (2007). Stochastic volatility with leverage: Fast and efficient likelihood inference. *Journal of Econometrics, 140*, 425–449.

Pakes, A., & Pollard, D. (1989). Simulation and the asymptotics of optimization estimators. *Econometrica, 57*(5), 1027–1057.

Pedersen, A. (1995). A new approach to maximum likelihood estimation for stochastic differential equations based on discrete observation. *Scandinavian Journal of Statistics, 22*, 55–71.

Phillips, P. C. B., & Yu, J. (2005a). Jackknifing bond option prices. *Review of Financial Studies, 18*, 707–742.

Phillips, P. C. B., & Yu, J. (2005b). Comments: A selective overview of nonparametric methods in financial econometrics. *Statistical Science, 20*, 338–343.

Phillips, P. C. B., & Yu, J. (2009a). Maximum likelihood and gaussian estimation of continuous time models in finance. In Andersen, T.G., Davis, R.A. and J.P. Kreiss (Eds.), *Handbook of financial time series*, 497–530, Springer-Verlag.

Phillips, P. C. B., & Yu, J. (2009b). Simulation-based estimation of contingent-claims prices. *Review of Financial Studies*, *22*, 3669–3705.

Pitt, M. (2002). Smooth particle filters likelihood evaluation and maximisation, Working Paper, University of Warwick.

Pitt, M., & Shephard, N. (1999a). Filtering via simulation: Auxiliary particle filter. *The Journal of the American Statistical Association*, *94*, 590–599.

Pitt, M., & Shephard, N. (1999b). Time varying covariances: A factor stochastic volatility approach. In: J. M. Bernardo, J. O. Berger, A. P. David, & A. F. M. Smith (Eds.), *Bayesian statistics 6* (pp. 547–570). Oxford: Oxford University Press.

Pritsker, M. (1998). Nonparametric density estimation and tests of continuous time interest rate models. *Review of Financial Studies*, *11*, 449–487.

Richard, J. F., & Zhang, W. (2006). Efficient high-dimensional importance. *Journal of Econometrics*, *141*, 1385–1411.

Sandmann, G., & Koopman, S. J. (1998). Estimation of stochastic volatility models via Monte Carlo maximum likelihood. *Journal of Econometrics*, *87*, 271–301.

Shephard, N., & Pitt, M. K. (1997). Likelihood analysis of non-Gaussian measurement time series. *Biometrika*, *84*, 653–667.

Skaug, H., & Yu, J. (2007). Automated Likelihood Based Inference for Stochastic Volatility Models, Working Paper, Singapore Management University.

Stern, S. (1997). Simulation-based estimation. *Journal of Economic Literature*, *35*, 2006–2039.

Tang, C. Y., & Chen, S. X. (2009). Parameter estimation and bias correction for diffusion processes. *Journal of Econometrics*, *149*, 65–81.

Taylor, S. J. (1982). Financial returns modelled by the product of two stochastic processes – a study of the daily sugar prices 1961–75. In O. D. Anderson (Ed.), *Time series analysis: Theory and practice, 1* (pp. 203–226). Amsterdam: North-Holland.

Vasicek, O. (1977). An equilibrium characterization of the term structure. *Journal of Financial Economics*, *5*, 177–186.

Yu, J. (2005). On leverage in a stochastic volatility model. *Journal of Econometrics*, 127, 165–178.

Yu, J. (2009a). Bias in the Estimation of the Mean Reversion Parameter in Continuous Time Models, Working Paper, Singapore Management University.

Yu, J. (2009b). Semiparametric Stochastic Volatility Model, Working Paper, Singapore Management University.

Yu, J., & Meyer, R. (2006). Multivariate stochastic volatility models: Bayesian estimation and model comparison. *Econometric Reviews*, *25*, 361–384.

Yu, J., Yang, Z., & Zhang, X. B. (2006). A class of nonlinear stochastic volatility models and its implications on pricing currency options, *Computational Statistics and Data Analysis*, *51*, 2218–2231.

Zehna, P., (1966). Invariance of maximum likelihood estimation. *Annals of Mathematical Statistics*, *37*, 744–744.

Part IV
Computational Methods

Chapter 16
Filtering Methods

Andras Fulop

Abstract This chapter surveys filtering methods, where the state of an unobserved dynamic model is inferred based on noisy observations. In linear and gaussian models, the Kalman Filter is applicable. We provide a brief description of the method and an example with a gaussian factor model of yields. More general models can be tackled using sequential monte carlo (SMC) techniques (also called particle filters). Here, the filtering distribution of the unobserved states is approximated by a swarm of particles and recursively update these particles using importance sampling and resampling. We give brief review of the methodology, illustrated throughout by the example of inferring asset values from noisy equity prices in a structural credit risk model. The MATLAB code implementing the examples is available.

16.1 Introduction

The methods described in this chapter are applicable to problems where a hidden dynamic Markov process needs to be filtered from the observed data containing some noise. To illustrate the problem, start with an example from fixed income taken from Diebold and Li (2006) and Diebold et al. (2006). Figure 16.1 plots 18 US treasury zero-coupon bond yields with maturities between 1 and 120 months observed in the period 1970–2000. To summarize the information in these time series, analysts often find it useful to extract a small number of dynamic factors that describe most of the variation in the yields. This parsimonious representation can help both in the interpretation of past yield curve movements and in prediction. However, in general, this low-dimensional factor structure is consistent with the observations only if the observed yields are assumed to contain some measurement

A. Fulop (✉)
ESSEC Business School, Paris, France
e-mail: fulop@essec.fr

J.-C. Duan et al. (eds.), *Handbook of Computational Finance*, Springer Handbooks of Computational Statistics, DOI 10.1007/978-3-642-17254-0_16,

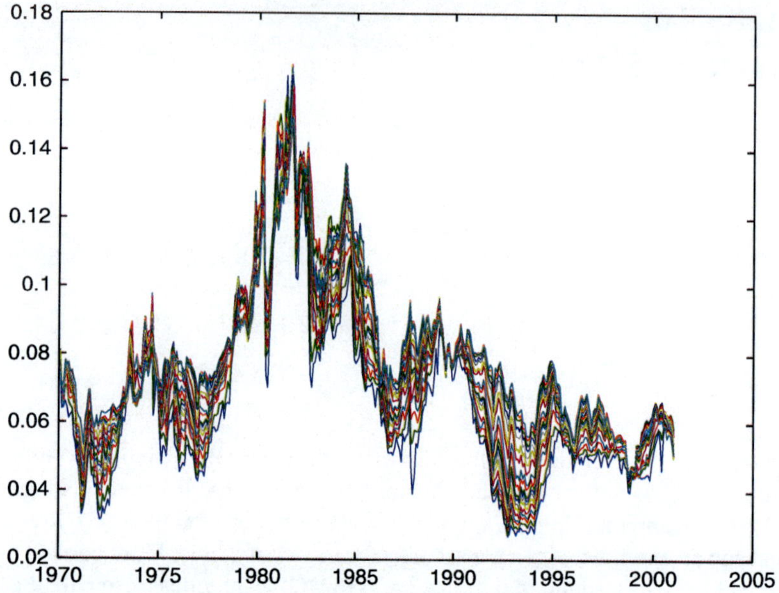

Fig. 16.1 US treasury zero-coupon-bond yields between 1970 and 2000

errors. To uncover the unobserved dynamic factors, one needs to infer them from the noisy observed yields.

Credit risk modeling is the subject of the second example. Six Flags Inc., a large operator of theme parks has been having financial difficulties in the last couple of years. On January 2, 2008 the company reported total assets of 2,945 Million USD, total liabilities of 2,912 Millions USD and preferred equities with a book value of 285 Millions, consistent with a negative −252 Millions of shareholders' equity on its balance sheets. However, in 2008, the stocks of the company were not worthless, as reported in Fig. 16.2. The main reason for the positive market value of the stock in spite of the large debt is limited liability. In case the value of the company is less than the face value of the debt when the debt is to be repaid, the stockholders can default in effect handing the company to the debtholders. As a result of this default option, both the equity and the debt can be interpreted as derivatives written on the face value of the firm. The equity holders own a long call option on the value of the firm with an exercise price equal to the face value of debt, while the debt-owners are short of this position. Unfortunately, the observed equity prices are not perfect signals on the firm value. The first order autocorrelation coefficient of the equity log-returns is equal to −0.25, showing that a considerable part of the observed price changes are due to transitory microstructure effects unrelated to permanent value innovations. Then, a methodology is needed to filter the unobserved asset value of the firm from the noisy observed equity prices.

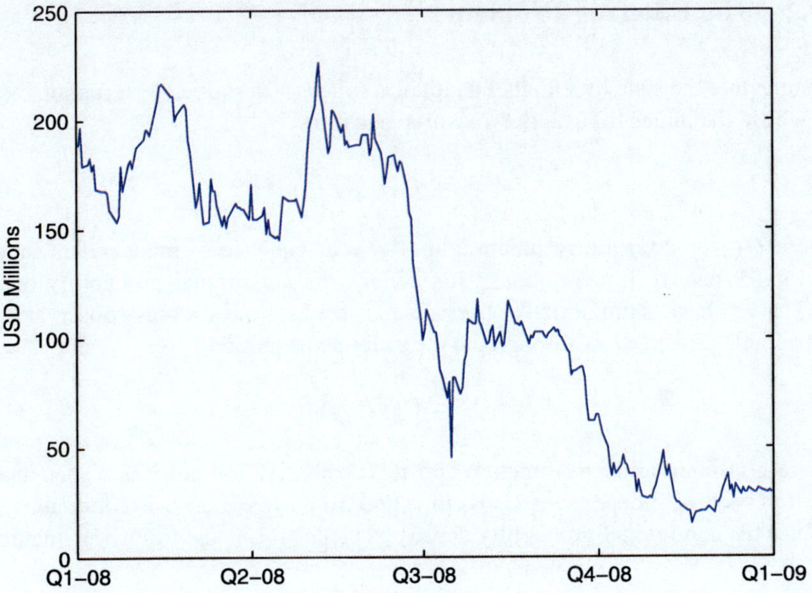

Fig. 16.2 Market capitalization of Six Flags' equity in 2008

The chapter begins in Sect. 16.2 with the description of the general filtering problem and puts the two examples in the general framework. Section 16.3 describes Kalman Filtering, applicable to linear normal systems. Here the filtering distributions are normally distributed with a mean and variance that can be recursively updated using the Kalman recursion. The method is applied to the first example on interest rate term structure. Further, some extensions of the Kalman Filter to nonlinear systems are mentioned. Section 16.4 turns to the general filtering problem, where the dynamic model is nonlinear/nongaussian. Here, the Kalman Filter is not valid any more, but analogous theoretical recursions still hold. Unfortunately, these involve integrals that need to be solved numerically. The chapter proceeds by presenting sequential Monte Carlo techniques that have been developed in last 15 year and are routinely used to solve the general filtering problem. It describes the general particle filtering algorithm, where resampling is introduced to tackle sample impoverishment, a pervasive problem in sequential importance sampling. Here and in the remaining part of the chapter, Merton's model with noisy equity observations is used to illustrate the presentation. Section 16.5 presents various strategies to produce effective proposal distributions, a crucial issue in designing an efficient importance sampling algorithm. The section ends by applying the filtering methodology on the second example, in filtering the asset value of Six Flags Inc. from its observed equity prices. Section 16.6 concludes with a brief discussion of various classical and Bayesian approaches to the estimation of the fixed model parameters in the particle filtering context.

16.2 The Filtering Problem

Assume that the state of a financial model at time k is described by a random vector x_k whose dynamics follows the transition equation

$$x_{k+1} = Q(x_k, \varepsilon_{k+1}), \tag{16.1}$$

where $Q()$ is an arbitrary function and ε_k is a sequence of independent random vectors. When x_k is continuous, this defines the conditional probability density $q(x_{k+1} \mid x_k)$. x_k is not directly observable, instead at time k a noisy observation y_k is available, linked to x_k through the measurement equation

$$y_k = G(x_k, \nu_k), \tag{16.2}$$

where $G()$ is an arbitrary function and ν_k the observation noise is a sequence of random vectors, independent across time and from ε_k. When y_k is continuous, this defines the conditional probability density $g(y_k \mid x_k)$. Use the following notation

$$x_{0:k} = (x_0, \ldots, x_k)$$
$$y_{1:k} = (y_1, \ldots, y_k)$$

Further, assume some prior distribution, $q_0(x_0)$, for the initial state variable. Then, the objective of filtering is to come up with the distribution of the hidden variable, x_k, given the observed data up to k. This quantity is the filtering distribution of x_k and is denoted by $f(x_k \mid y_{1:k})$. In the algorithms that follow these distributions are obtained sequentially, as new observations arrive.

16.2.1 Uncovering Yield Curve Factors

To tackle the first example in the introduction, this subsection describes a specific factor model of the term structure closely following Diebold and Li (2006) and Diebold et al. (2006) and shows how it fits into the general filtering framework. Denote by $y(\tau_l)_k$ the zero-coupon yield observations at time k with maturity τ_l. On each observation date k, there are 18 observed yields with maturities ranging between $\tau_1 = 1, \ldots, \tau_{18} = 120$ months. The data-set has monthly observations in the period 1975–2000. To summarize the rich cross-sectional information, the yields are assumed to depend on three common factors $(x_{1,k}, x_{2,k}, x_{3,k})$ and a yield-specific measurement noise $\nu_{l,k}$. This latter is assumed to be standard normal and independent across the yields and through time. This setup leads to the following measurement equations for $l = 1, \ldots, 18$

$$y(\tau_l)_k = x_{1,k} + x_{2,k}\left(\frac{1-e^{-\lambda\tau_l}}{\lambda\tau_l}\right) + x_{3,k}\left(\frac{1-e^{-\lambda\tau_l}}{\lambda\tau_l} - e^{-\lambda\tau_l}\right) + \sigma_\nu \nu_{l,k} \quad (16.3)$$

This factor representation is a version of the Nelson and Siegel (1987) parametric form, popular with practitioners. The interpretability of the factors is an attractive feature of this specific parameterization. First, $x_{1,k}$ has the same loading on each yield, so it can be interpreted as a level factor. Second, $x_{2,k}$ affects yields with longer maturities less, hence it is close to a slope factor. Last, $x_{3,k}$ has hump-shaped loadings and plays the role of a curvature factor. The parameter λ determines where the maximum of this hump-shaped pattern lies.

To ensure some degree of time-series consistency and to allow prediction using the model, the factors are assumed to follow independent normal AR(1) processes, resulting in the following transition equations

$$x_{i,k+1} = \mu_i + \gamma_i x_{i,k} + \sigma_{i,x}\epsilon_{i,k+1}, \quad (16.4)$$

where $\epsilon_{i,k+1}$ are independent standard normal variables. Then, if one wants to forecast the future yields, one needs to filter the last value of the unobserved factors, $x_{i,k}$ given the noisy yield observations up to k.

16.2.2 Finding the Value of the Firm in Merton's Model

Merton (1974) laid the foundation to the literature on the structural approach to credit risk modeling. The value of the firm at time t, V_t, is assumed to follow a geometric Brownian motion with respect to the physical probability law that generates the asset values

$$\frac{dV_t}{V_t} = \mu dt + \sigma dW_t$$

The risk-free rate of interest is assumed to be a constant, r. Furthermore, the firm has two classes of claims outstanding – an equity and a zero-coupon debt maturing at time T with face value F. Due to limited liability, equity is a call option on the value of the firm with payout

$$S_T = \max(V_T - F, 0) \quad (16.5)$$

Then, the equity claim in (16.5) can be priced at time $t < T$ by the standard Black-Scholes option pricing model to yield the following solution:

$$S_t \equiv S(V_t; \sigma, F, r, T-t) = V_t\Phi(d_t) - Fe^{-r(T-t)}\Phi(d_t - \sigma\sqrt{T-t}), \quad (16.6)$$

where

$$d_t = \frac{\log(\frac{V_t}{F}) + (r + \frac{\sigma^2}{2})(T - t)}{\sigma\sqrt{T - t}}$$

and $\Phi(\cdot)$ is the standard normal distribution function.

Unfortunately, the asset value of the firm, V_{τ_i} is rarely observable. In contrast, for an exchange listed firm, one can obtain a time series of equity prices denoted by $\mathcal{D}_N = \{S_{\tau_i}, i = 0, \cdots, N\}$ and try to infer the asset value using the equity prices and balance sheet information on debt. If the equity prices are not contaminated by trading noises, the asset value can be obtained by inverting the equity pricing function from (16.6) following Duan (1994). However the observed equity prices may be contaminated by microstructure noise that can be important, especially for smaller firms or firms in financial difficulties. Following Duan and Fulop (2009b) the trading noise obeys a multiplicative structure leading to the following measurement equation for the log equity price

$$\log S_{\tau_i} = \log S(V_{\tau_i}; \sigma, F, r, T - \tau_i) + \delta \nu_i, \tag{16.7}$$

where $\{\nu_i, i = 0, N\}$ are i.i.d. standard normal random variables and the nonlinear pricing function $S(V_t; \sigma, F, r, T - t)$ has been given earlier. Since the unobserved asset value process follows a geometric Brownian motion, we can derive its discrete-time form as

$$\log V_{\tau_{i+1}} = \log V_{\tau_i} + (\mu - \frac{\sigma^2}{2})h + \sigma\sqrt{h}\varepsilon_{i+1}, \tag{16.8}$$

where $\{\varepsilon_i, i = 1, N\}$ are i.i.d. standard normal random variables and $h = \tau_i - \tau_{i-1}$ is the observation frequency. Then, one needs to filter the unobserved asset price, V_{τ_i} given the noisy equity observations up to time k in the model defined by the measurement equation (16.7) and the transition equation (16.8).

16.3 Kalman Filtering

When the measurement and the transition equations are normal and linear, the filtering density is normal. Assume that the transition equation is

$$x_k = C + A x_{k-1} + \varepsilon_k, \tag{16.9}$$

where $\varepsilon_k \sim N(0, Q)$. The measurement equation is also linear and normal:

$$y_k = H x_k + \nu_k, \tag{16.10}$$

where $\nu_k \sim N(0, R)$. Introduce the following notation for conditional expectations and variances:

$$E_s(x_k) = E(x_k \mid y_{1:s})$$
$$V_s(x_k) = Var(x_k \mid y_{1:s})$$

if the initial state x_0 is distributed as $x_0 \sim N(E_0(x_0), V_0(x_0))$, the subsequent filtering distributions are also normally distributed. Further, the first two moments can be sequentially updated by first predicting the distribution of the hidden variable at k given past information up to $k-1$

$$E_{k-1}(x_k) = C + AE_{k-1}(x_{k-1})$$
$$V_{k-1}(x_k) = AV_{k-1}(x_{k-1})A' + Q$$

Then, the filtering distribution at k is obtained by including the information at k

$$K_k = V_{k-1}(x_k)H' \left(HV_{k-1}(x_k)H' + R\right)^{-1}$$
$$E_k(x_k) = E_{k-1}(x_k) + K_k\left(y_k - HE_{k-1}(x_k)\right)$$
$$V_k(x_k) = (I - K_k H)\, V_{k-1}(x_k)$$

For a general review of Kalman Filtering see Anderson and Moore (1979). For an application in yield curve modeling see Christensen et al. (2007), Diebold and Li (2006) and another in commodities see Schwartz and Smith (2000).

16.3.1 Application of Kalman Filtering: Uncovering Yield Curve Factors

It is apparent that the first example, on extracting yield curve factors, falls within the realm of Kalman Filtering. In particular both the measurement equation in (16.3) and the transition equation in (16.4) are gaussian and linear. To investigate the method on the US zero-coupon yield data-set, the parameters of the model are fitted using maximum likelihood. In-sample, the model-predicted yields have a root mean squared error of around 12 basis points, pointing towards a satisfactory fit. Figure 16.3 plots the model-implied filtered mean of the factors and the results seem to be in accord with intuition. For example, one can see that the first factor indeed acts as a level factor, with high values when the general level of interest rates is high.

Forecasting is an important application of yield-curve models. Further, investigating the out-of-sample performance of various models may be an even more important check on model validity than the in-sample fit. Hence, following Diebold and Li (2006), the out-of-sample forecasting performance of the yield curve factor model is compared with two competitors, the first being a naive random walk model while the second is an AR(1) model of the individual yields. All the models are estimated on the data up to 1993 and the quality of their forecasts is investigated on the remaining sample at the 6-months horizon. Figure 16.4 shows the RMSE of the three forecasts for all maturities and provides evidence that the discipline that the

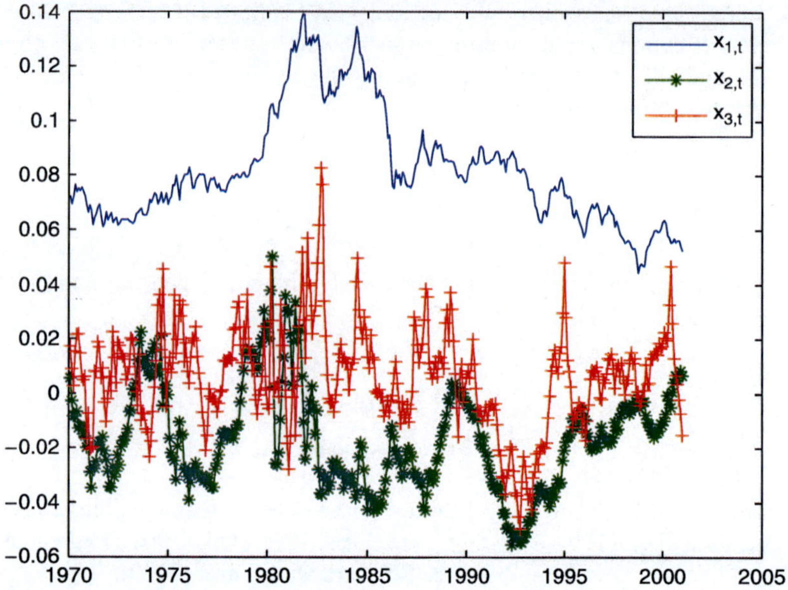

Fig. 16.3 Time series of filtered yield curve factors

factor model puts on the data considerably helps in prediction. All the results in this subsection are produced by the MATLAB script `DieboldLi_KF.m`.

16.3.2 Extensions

16.3.2.1 Extended Kalman Filter

Often, the financial model of interest is normal, but the transition and measurement equation are not linear. In particular we may have

$$x_k = Q(x_{k-1}, \varepsilon_k) \tag{16.11}$$

$$y_k = G(x_k, \nu_k), \tag{16.12}$$

where $Q()$ and $G()$ are differentiable functions and ε_k and ν_k are normally distributed. Then, the Extended Kalman Filter (EKF) approximates this system using a first-order Taylor expansion around $E_{k-1}(x_{k-1})$ and applies Kalman Filtering on the approximating linear system. In finance, this approach is often applied in term structure modeling (De Jong 2000; Duan and Simonato 1999; Duffee 2002) and in commodities modeling (Trolle and Schwartz 2008).

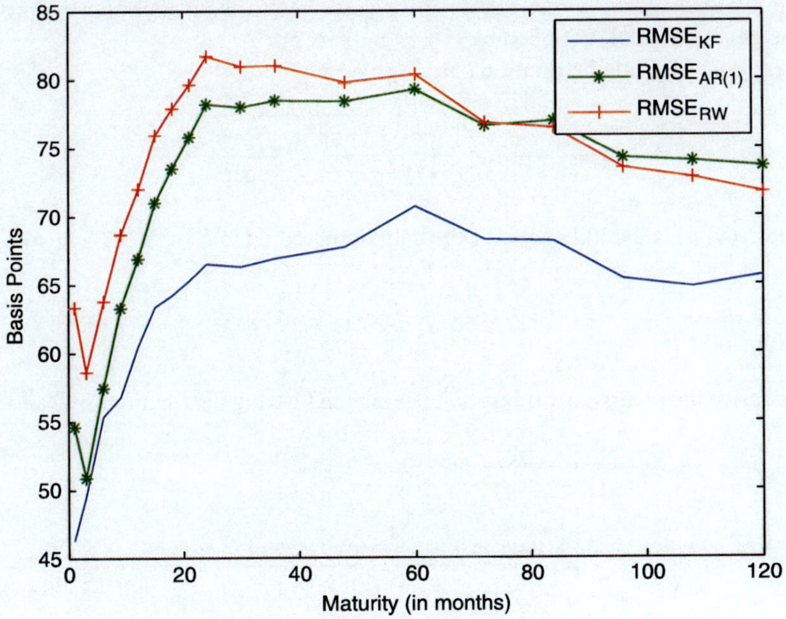

Fig. 16.4 RMSE of various forecasting methods on the 6-months horizon, 1994–2000

16.3.2.2 Unscented Kalman Filter

The EKF approximates the system only up to a first order and it can provide poor results when the nonlinearity of the measurement or transition equation is serious. An alternative approach that avoids linearization altogether is the Unscented Kalman Filter (UKF). This method approximates the normal filtering distribution using a discrete distribution that matches the mean and covariance matrix of the target gaussian random variable. Then, these points are passed through directly the nonlinear functions to obtain the quantities necessary for the Kalman recursion. In many situations the method provides a higher order approximation to the nonlinear system than the EKF. For a detailed description of the method see van der Merwe and Wan (2000). The technique has been applied to currency option pricing by Bakshi et al. (2008).

16.4 Particle Filtering

16.4.1 General Filtering Recursion

When the system is non-linear and/or non-gaussian, the filtering distribution may not be normal and the Kalman Filter is not valid any more. To appreciate the difficulty of the task, in the following we describe the sequential filtering problem in the general model described by (16.1) and (16.2).

The joint filtering distribution of $x_{0:k}$ given $y_{1:k}$ is

$$f(x_{0:k} \mid y_{1:k}) = \frac{f(x_{0:k}, y_{1:k})}{f(y_{1:k})} = \frac{f(x_{0:k}, y_{1:k})}{L(y_{1:k})},$$

where $L(y_{1:k})$ is the likelihood of the data observed up to k

$$L(y_{1:k}) = \int f(x_{0:k}, y_{1:k}) dx_{0:k}$$

Now derive the recursive formula connecting the filtering distributions at k and $k+1$

$$\begin{aligned} f(x_{0:k+1} \mid y_{1:k+1}) &= \frac{f(x_{0:k+1}, y_{1:k+1})}{L(y_{1:k+1})} \\ &= \frac{g(y_{k+1} \mid x_{k+1}) q(x_{k+1} \mid x_k) f(x_{0:k}, y_{1:k})}{L(y_{1:k})} \frac{L(y_{1:k})}{L(y_{1:k+1})} \\ &= \frac{g(y_{k+1} \mid x_{k+1}) q(x_{k+1} \mid x_k)}{f(y_{k+1} \mid y_{1:k})} f(x_{0:k} \mid y_{1:k}) \end{aligned}$$

This equation gives the recursion of the filtered distributions over the whole path space. Integrating over $x_{0:k-1}$ one gets the following relationship

$$\begin{aligned} f(x_{k:k+1} \mid y_{1:k+1}) &= \frac{g(y_{k+1} \mid x_{k+1}) q(x_{k+1} \mid x_k)}{f(y_{k+1} \mid y_{1:k})} f(x_k \mid y_{1:k}) \\ &\propto g(y_{k+1} \mid x_{k+1}) q(x_{k+1} \mid x_k) f(x_k \mid y_{1:k}) \end{aligned}$$

showing that $f(x_{0:k} \mid y_{1:k})$ is a sufficient statistic. Integrating out x_k, one arrives at the filtering distribution of x_{k+1}

$$f(x_{k+1} \mid y_{1:k+1}) \propto \int g(y_{k+1} \mid x_{k+1}) q(x_{k+1} \mid x_k) f(dx_k \mid y_{1:k})$$

The Kalman Filter is a special case where this recursion can be executed in closed-form due to the joint normality of the system. In general, the filtering distributions do not belong to a known parametric family and the integration has to be done using numerical methods. In the following a class of simulation-based methods is presented that has been extensively used in the last few years to solve the general filtering task.

16.4.2 Sequential Importance Sampling

The target is the joint filtering distribution of the hidden states

$$f(x_{0:k} \mid y_{1:k}) \propto \prod_{t=1}^{k} g(y_t \mid x_t) q(x_t \mid x_{t-1}) q_0(x_0) \tag{16.13}$$

Ideally, one would like to sample directly from the densities $g(y_t \mid x_t)q(x_t \mid x_{t-1})$, providing a straightforward recursive Monte Carlo scheme. Unfortunately, due to the complexity of these densities, this is usually not possible. Importance sampling is an approach that can be used in such cases. Here, one draws from a feasible proposal distribution $r(x_{0:k})$ instead of the target and attaches importance weights to the samples to compensate for the discrepancy between the proposal and the target. If the weighted sample is denoted by $(\xi_{0:k}^{(m)}, w_k^{(m)})$ where $m = 1, \ldots, M$, the samples and weights are obtained as

$$\begin{aligned}
\xi_{0:k}^{(m)} &\sim r(x_{0:k}) \\
w_k^{(m)} &= \frac{\prod_{t=1}^{k} g(y_t \mid \xi_t^{(m)}) q(\xi_t^{(m)} \mid \xi_{t-1}^{(m)}) q_0(\xi_0^{(m)})}{r(\xi_{0:k}^{(m)})}
\end{aligned}$$

The expectation $E\left(h(x_{0:k} \mid y_{1:k})\right)$ can be estimated by the estimator

$$\widehat{h} = \frac{\sum_{m=1}^{M} h(\xi_{0:k}^{(m)}) w_k^{(m)}}{\sum_{m=1}^{M} w_k^{(m)}}$$

Using independence of the sample the estimator is asymptotically consistent

$$\widehat{h} - E\left(h(x_{0:k} \mid y_{1:k})\right) \rightarrow_P 0 \text{ as } M \rightarrow \infty$$

and asymptotically normal

$$\sqrt{M}\left[\widehat{h} - E\left(h(x_{0:k} \mid y_{1:k})\right)\right] \rightarrow_D N\left[0, \frac{Var_r\left(h(x_{0:k}) w(x_{0:k})\right)}{\left[E_r\left(w(x_{0:k})\right)\right]^2}\right] \text{ as } M \rightarrow \infty$$

Note that the asymptotic variance can also be estimated using the simulation output, allowing inference on the reliability of the estimate.

The preceding importance sampling algorithm can be made sequential by choosing a recursive structure for the importance sampling distribution, $r(x_{0:k})$:

$$R(x_{0:k}) = \prod_{t=1}^{k} r(x_t \mid y_k, x_{t-1}) r_0(x_0)$$

Then the importance weight w_k can be written as

$$w_k = \prod_{t=1}^{k} \frac{g(y_t \mid x_t)q(x_t \mid x_{t-1})}{r(x_t \mid y_k, x_{t-1})} \frac{q_0(x_0)}{r_0(x_0)}$$

and the importance sampler can be implemented in a sequential manner

Sequential Importance Sampling

- Initial State: Draw an i.i.d. sample $\xi_0^{(m)}, m = 1, \ldots, M$ from $\xi_0^i \sim r_0(x_0)$ and set
$$w_0^{(m)} = \frac{q_0(\xi_0^{(m)})}{r_0(\xi_0^{(m)})}, m = 1, \ldots, M$$
- Recursion: For $k = 1, \ldots, N$
  1. Draw $(\xi_k^{(m)}, m = 1, \ldots, M)$ from the distribution $\xi_k^{(m)} \sim r(x_k \mid y_k, \xi_{k-1}^{(m)})$
  2. Compute the updated importance weights
$$w_k^{(m)} = w_{k-1}^{(m)} \times \frac{g(y_k \mid \xi_k^{(m)})q(\xi_k^{(m)} \mid \xi_{k-1}^{(m)})}{r(\xi_k^{(m)} \mid y_t, \xi_{k-1}^{(m)})}$$

This algorithm seems to provide a solution to the recursive filtering problem. Unfortunately after a couple of time steps the normalized weights of most points fall to zero and the weighted sample ceases to provide a reliable representation of the target distribution.

16.4.2.1 Weight Degeneracy in Merton's Model

To illustrate the phenomenon mentioned before, consider the performance of the sequential importance sampling algorithm for Merton's model with noisy equity observations. Choose the prior distribution to be a point mass assuming that the initial equity observation S_{τ_0} is observed without any error. Further, use the transition density $f(V_{\tau_{i+1}} \mid V_{\tau_i}^{(m)})$ as the proposal distribution. The procedure that results is:

Sequential Importance Sampling in Merton's Model

- Initial State: Set $V_{\tau_0}^{(m)} = S^{-1}(S_{\tau_0})$ where the function $S^{-1}(.)$ is the inverse of the equity pricing function in (16.6).

- Recursion: For $k = 1, \dots, N$
  1. Draw $V_{\tau_k}^{(m)}$ from $f(V_{\tau_k} \mid V_{\tau_{k-1}}^{(m)}, \Theta)$, which can be easily done using (16.8).
  2. Compute the updated importance weights

$$w_k^{(m)} = w_{k-1}^{(m)} f(S_{\tau_k} \mid V_{\tau_k}^{(m)}, \Theta)$$

One measure of the reliability of an importance sampler is the effective sample size, N_{eff}, defined as

$$N_{eff} = \left[\sum_{m=1}^{M} \left(\frac{w_k^{(m)}}{\sum_{m=1}^{M} w_k^{(m)}} \right)^2 \right]^{-1}$$

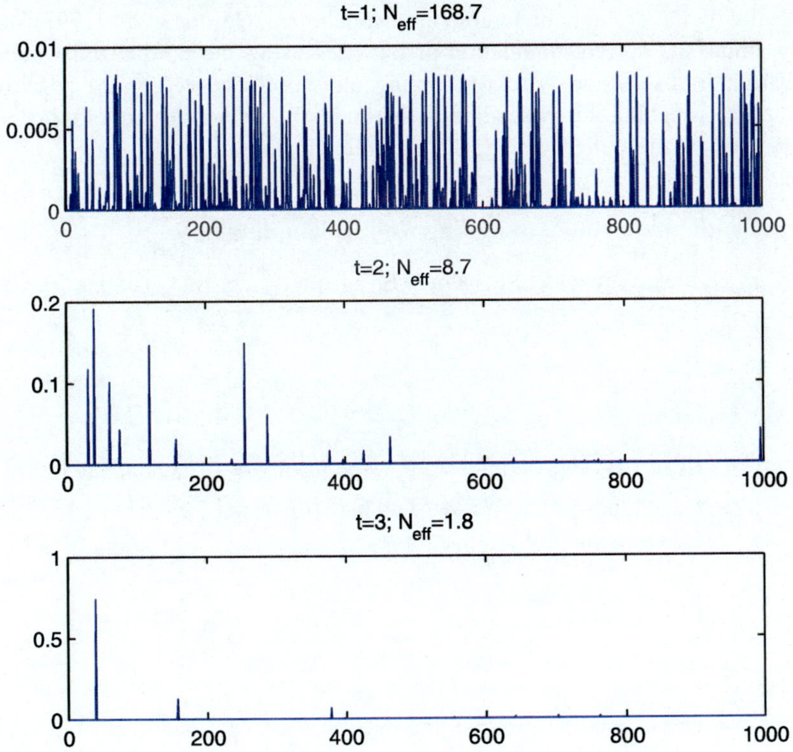

Fig. 16.5 Normalized importance weights for sequential importance sampling in Merton's model

Roughly speaking, the effective sample size measures the size of an equally-weighted Monte Carlo sample providing the same reliability as the output of the importance sampler. Figure 16.5 depicts the effective sample sizes for the first few observations in Merton's model obtained by running the SIS algorithm. The model parameters are ($\sigma = 0.2, \mu = 0.1, r = 0.05, \delta = 0.01, F = 100$). The initial asset value is 60, the initial debt maturity is 3 years, and a year of daily data is generated (i.e. $h = 1/250, n = 250$) and the sample size is $M = 1{,}000$. The MATLAB file producing this figure is `test_MertonSIS.m`. One can observe that by $t = 5$ the effective sample size collapses to one, signaling the deterioration of the filter. The underlying reason behind this phenomenon is that a fixed number of points is used to cover an increasing dimensional space.

16.4.3 Sequential Importance Sampling with Resampling (SIR or Particle Filtering)

To deal with the problem of sample impoverishment, Gordon et al. (1993) suggest to resample the current population of particles using the normalized weights as probabilities of selection. After resampling, all importance weights are reset to one. The intuition behind this procedure is that unlikely trajectories are eliminated and likely ones are multiplied. This yields the following algorithm:

Sequential Importance Sampling with Resampling

- Initial State: Draw an i.i.d. sample $\xi_0^{(m)}$ from $\xi_0^{(m)} \sim r_0(x_0)$ and set $w_0^{(m)} = \frac{q_0(\xi_0^{(m)})}{r_0(\xi_0^{(m)})}, m = 1, \ldots, M$
- For $k = 1, \ldots, N$ repeat the next steps

1. Sampling
   - Draw $(\xi_k^{(m)}, m = 1, \ldots, M)$ conditionally independently given $(\xi_{0:k-1}^{(m)}, m = 1, \ldots, M)$ from the distribution $\xi_k^{(m)} \sim r(x_k \mid y_k, \xi_{k-1}^{(m)})$
   - Compute the importance weights

$$w_k^{(m)} = \frac{g(y_k \mid \xi_k^{(m)})q(\xi_k^{(m)} \mid \xi_{k-1}^{(m)})}{r(\xi_k^{(m)} \mid y_k, \xi_{k-1}^{(m)})}$$

2. Resampling
   - Draw from the multinomial trial $(I_k^1, \ldots, I_k^M)$ with probabilities of success

$$\frac{w_k^1}{\sum_{m=1}^M w_k^{(m)}}, \ldots, \frac{w_k^M}{\sum_{m=1}^M w_k^{(m)}}$$

– Reset the importance weights $w_k^{(m)}$ to 1;

3. Trajectory update: $\xi_{0:k}^{(m)} = \xi_{0:k}^{I_k^{(m)}}, m = 1, \ldots, M$

This approach concentrates on the marginal filtering distribution $f(x_k \mid y_{0:k})$ instead of the joint one, $f(x_{0:k} \mid y_{0:k})$. Resampling helps to achieve a better characterization of the last state of the system at the expense of representing the past of the full hidden path, $x_{0:k}$.

16.4.3.1 Bootstrap Filter

In the bootstrap filter of Gordon et al. (1993) the proposal density is chosen to be equal to the transition density

$$r(x_k \mid y_k, x_{k-1}) = q(x_k \mid x_{k-1})$$

In this case the importance weights take a particularly simple form, they simply equal the measurement density

$$w_k^m = \frac{g(y_k \mid \xi_k^m) q(\xi_k^m \mid \xi_{k-1}^m)}{q(\xi_k^m \mid \xi_{k-1}^m)} = g(y_k \mid \xi_k^m)$$

16.4.3.2 Bootstrap Filter in Merton's Model

Bootstrap Filter in Merton's Model

- Initial State: Set $V_{\tau_0}^{(m)} = S^{-1}(S_{\tau_0})$ where the function $S^{-1}(.)$ is the inverse of the equity pricing function in (16.6).
- Recursion: For $k = 1, \ldots, N$

  1. Sampling

     – Draw $V_{\tau_k}^{(m)}$ from $f(V_{\tau_k} \mid V_{\tau_{k-1}}^{(m)}, \Theta)$, using equation(16.8).
     – Compute the normalized importance weights

$$\pi_k^{(m)} = \frac{w_k^{(m)}}{\sum_{m=1}^{M} w_k^{(m)}} \text{ where } w_k^{(m)} = f(S_{\tau_k} \mid V_{\tau_k}^{(m)}, \Theta)$$

2. Resample from the weighted sample $\{(V_{\tau_k}^{(m)}, \pi_k^{(m)}); m = 1, \cdots, M\}$ to obtain a new equal-weight sample of size M.

To investigate whether the resampling step successfully deals with sample depletion we repeat the simulation exercise described before on Merton's model, but now we run the bootstrap filter. Panel A of Fig. 16.6 depicts the effective sample sizes (N_{eff}) for a simulated sample path. One can see that now N_{eff} does not collapse as time progresses, so the resampling seems an effective remedy to sample depletion. Panel B reinforces this message by showing that the filter reliably tracks the unobserved asset value path. The MATLAB file producing Fig. 16.6 is `test_MertonBootstrap.m`.

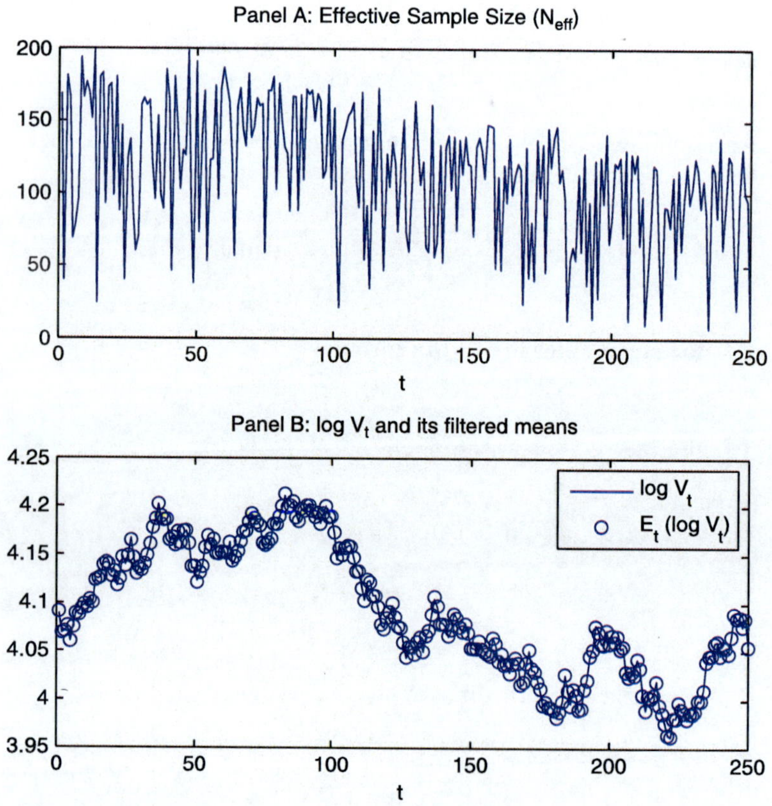

Fig. 16.6 Bootstrap filter in Merton's model

16.4.4 Theoretical Properties of Particle Filters

The filtering algorithm described above has been shown to possess attractive asymptotic properties as the number of particles, N, goes to infinity (see Crisan and Doucet 2002 for a short introduction to the theory and Del Moral 2004 for a monograph-length treatment). In particular it provides consistent estimates of any filtering quantity

$$\frac{1}{M}\sum_{m=1}^{M} h(\xi_k^m) - E\left(h(x_k \mid y_{1:k})\right) \to 0 \text{ as } M \to \infty$$

Central limit theorems has been proved for particle systems, leading to results of the type

$$\sqrt{M}\left(\frac{1}{M}\sum_{m=1}^{M} h(\xi_k^m) - E(h(x_k \mid y_{1:k}))\right) \to N(0, \sigma_k^2(h)) \text{ as } M \to \infty$$

In general the Monte-Carlo variance, $\sigma_k^2(h)$ increases with the time k. This reflects the fact that as time passes, errors accumulate in the filtering recursions. In practice this means that an ever-increasing number of particles is needed to ensure the same quality for the estimates. To rule this out and achieve uniform convergence further assumptions on the forgetting properties of the model are needed.

While these results provide the rate of convergence, $\sqrt{M}$, the constant of convergence, $\sigma_k^2(h)$ is usually not known. This means that in contrast to simple importance sampling, one cannot compute confidence intervals for the estimates.

16.5 Implementation Issues for Particle Filters

16.5.1 The Choice of Proposal in SIR

The choice of the proposal distribution is critical for the efficiency of the method. The question is how to best use the information in the next observation in sampling. The optimal choice would be the conditional distribution of the new hidden state given the past hidden state and the new observation:

$$f(x_k \mid y_k, x_{k-1}) \propto g(y_k \mid x_k) q(x_k \mid x_{k-1})$$

As direct sampling from the optimal choice is usually not feasible, approximations are needed. In the following Merton's model is used to illustrate various strategies to obtain efficient proposal distributions. The first approach uses a specific feature of Merton's model by localizing the sampler around the new observation. The second, more generic approach linearizes the model around each particle and uses the optimal sampler of the approximated model as the proposal. The third strategy adapts a parametric family of proposal and picks the best density within this family using the information in the previously sampled particles.

16.5.1.1 Localized Sampling in Merton's Model

To illustrate the importance of the proposal density, consider again Merton's model. If one uses the bootstrap filter, the importance weights are

$$f(S_{\tau_i+1} \mid V_{\tau_i+1}, \Theta) \propto \frac{1}{\delta}\phi\left(\frac{\log S_{\tau_i+1} - \log S(V_{\tau_i+1}, \Theta)}{\delta}\right)$$

When the microstructure noise δ is small, this density function is peaked, resulting in high variance of the particle weights and a poor representation of the filtering distribution. Intuitively, when the microstructure noise is relatively small, the new observation is very informative on the hidden asset value. This makes the bootstrap sampler that ignores the new observation, a poor choice for the proposal.

However, if the observation is so important, why not totally base the sampler on the observation, forgetting the past? This idea is used in Duan and Fulop (2009b) to propose an efficient sampler, localized around the new observed equity price. In particular, Duan and Fulop (2009b) suggest to draw from the microstructure noise, ν_k and to use the asset value implied by the noise and the new observation as the sampler. This results in the following algorithm:

Localized Sampling in Merton's Model

- Initial State: Set $V_{\tau_0}^{(m)} = S^{-1}(S_{\tau_0})$ where the function $S^{-1}(.)$ is the inverse of the equity pricing function in (16.6).
- Recursion: For $k = 1, \ldots, N$

  1. Sampling

     – Draw a standard normal $\nu_k^{(m)}$ and compute $V_{\tau_k}^{(m)} = V_{\tau_k}^*(S_{\tau_k}, \nu_k^{(m)})$ to obtain the pair $(V_{\tau_i}^{(m)}, V_{\tau_k}^{(m)})$, where

     $$V_{\tau_k}^*(S_{\tau_k}, \nu_k) = S^{-1}(S_{\tau_k} e^{-\delta\nu_k}; \sigma, F, r, T - \tau_k)$$

     – Compute the importance weights

     $$w_k^{(m)} = \frac{f(V_{\tau_k}^{(m)} \mid V_{\tau_{k-1}}^{(m)}, \Theta)}{\Phi(d_{\tau_k}^{*(m)}) e^{\delta \nu_k^{(m)}}}$$

     – Normalize the importance weights

$$\pi_k^{(m)} = \frac{w_k^{(m)}}{\sum_{m=1}^{M} w_k^{(m)}} \text{ where } w_k^{(m)} = f(S_{\tau_k} \mid V_{\tau_k}^{(m)}, \Theta)$$

2. Resample from the weighted sample $\{(V_{\tau_k}^{(m)}, \pi_k^{(m)}); m = 1, \cdots, M\}$ to obtain a new equal-weight sample of size M.

Here, using a change of variables formula, the density function of the sampler is

$$g(V_{\tau_k}^{(m)} \mid S_{\tau_k}, V_{\tau_{k-1}}^{(m)}) = f(V_{\tau_k}^{*}(S_{\tau_k}, v_k^{(m)}) \mid S_{\tau_k}) = \frac{\phi(v_k^{(m)})\Phi(d_{\tau_k}^{*(m)})e^{\delta v_k^{(m)}}}{\delta S_{\tau_k}}$$

Then, the expression for the importance weights can be derived as

$$\begin{aligned} w_k^{(m)} &= \frac{f(S_{\tau_k} \mid V_{\tau_k}^{(m)}, \Theta) f(V_{\tau_k}^{(m)} \mid V_{\tau_{k-1}}^{(m)}, \Theta)}{g(V_{\tau_k}^{(m)} \mid S_{\tau_k}, V_{\tau_{k-1}}^{(m)})} \\ &= \frac{f(S_{\tau_k} \mid V_{\tau_k}^{(m)}, \Theta)\delta S_{\tau_k} f(V_{\tau_k}^{(m)} \mid V_{\tau_{k-1}}^{(m)}, \Theta)}{\phi(v_k^{(m)})\Phi(d_{\tau_k}^{*(m)})e^{\delta v_k^{(m)}}} \\ &= \frac{f(V_{\tau_k}^{(m)} \mid V_{\tau_{k-1}}^{(m)}, \Theta)}{\Phi(d_{\tau_k}^{*(m)})e^{\delta v_k^{(m)}}} \end{aligned}$$

Table 16.1 shows the efficient sample sizes for the bootstrap filter and the localized sampler for different values of the measurement noise standard deviation, δ. The values are averages taken through time and across 20 simulations, run at different random seeds. The sample size $M = 1{,}000$ and all the other simulation parameters are as described before. The MATLAB file producing the table is `test_MertonLocalized.m`. Overall, the localized sampler seems to perform much better than the bootstrap filter reflected in the much higher effective sample sizes. Further, as δ decreases, the performance of the bootstrap filter deteriorates while that of the localized filter actually gets better. The reason for this phenomenon is that for smaller values of δ, the relative importance of the new observation is higher in determining the location of the new unobserved asset value. Then, the localized sampler that ignores the past overperforms the bootstrap filter that ignores the new observation.

Table 16.1 Effective sample size for the localized sampler and the bootstrap filter in Merton's model

	$\delta = 0.0005$	$\delta = 0.005$	$\delta = 0.01$	$\delta = 0.02$
N_{eff} (Localized)	999.9	993.0	974.1	916.9
N_{eff} (Bootstrap)	6.4	61.4	121.1	230.4

16.5.1.2 Using Local Linearization to Generate the Proposal in Merton's Model

The localized sampler described in the previous section completely ignores the past. An alternative approach is to follow the advice of Doucet et al. (2000) and use a local linear approximation of the model to generate a proposal. Here, both the past and the new observation is used to come up with a proposal distribution at the price of the bias due to the linearization. In Merton' model, the only non-linearity is in the measurement equation (16.7). Linearizing this equation around the conditional expected value yields the approximate measurement equation:

$$\log S_{\tau_k} \sim A\left(\log V^{*(m)}\right) + B\left(\log V^{*(m)}\right) \times \left(\log V_{\tau_k} - \log V^{*(m)}\right),$$

where

$$\begin{aligned}
\log V^{*(m)} &= (\mu - \frac{\sigma^2}{2})h + \log V_{\tau_{k-1}}^{(m)} \\
A\left(\log V^{*(m)}\right) &= \log S\left(e^{\log V^{*(m)}}; \sigma, F, r, T - \tau_k\right) \\
B\left(\log V^{*(m)}\right) &= \frac{V^* \Phi(d^{*(m)})}{S\left(e^{\log V^{*(m)}}; \sigma, F, r, T - \tau_k\right)}
\end{aligned}$$

By local normality of this system, the conditional distribution of $\log V_{\tau_k}^{(m)}$ given $\log S_{\tau_k}$ is

$$\log V_{\tau_k}^{(m)} \sim N\left(\mu(\log V_{\tau_{k-1}}^{(m)}), \sigma^2(\log V_{\tau_{k-1}}^{(m)})\right),$$

where

$$\begin{aligned}
\mu(\log V_{\tau_{k-1}}^{(m)}) &= \log V^{*(m)} + \frac{B\left(\log V^{*(m)}\right)\sigma^2 h}{B^2\left(\log V^{*(m)}\right)\sigma^2 h + \delta^2} \times \left(\log S_{\tau_k} - A\left(\log V^{*(m)}\right)\right) \\
\sigma^2(\log V_{\tau_{k-1}}^{(m)}) &= \sigma^2 h - \frac{B^2\left(\log V^{*(m)}\right)\sigma^4 h^2}{B^2\left(\log V^{*(m)}\right)\sigma^2 h + \delta^2}
\end{aligned}$$

The expression of the importance weights is

$$\begin{aligned}
w_k^{(m)} &= \frac{f(S_{\tau_k} \mid V_{\tau_k}^{(m)}, \Theta) f(V_{\tau_k}^{(m)} \mid V_{\tau_{k-1}}^{(m)}, \Theta)}{g(V_{\tau_k}^{(m)} \mid S_{\tau_k}, V_{\tau_{k-1}}^{(m)})} \\
&= \frac{f(S_{\tau_k} \mid V_{\tau_k}^{(m)}, \Theta) f(V_{\tau_k}^{(m)} \mid V_{\tau_{k-1}}^{(m)}, \Theta)}{\phi\left(\frac{\log V_{\tau_k}^{(m)} - \mu(\log V_{\tau_{k-1}}^{(m)})}{\sigma(\log V_{\tau_{k-1}}^{(m)})}\right) / \sigma(\log V_{\tau_{k-1}}^{(m)})}
\end{aligned}$$

Table 16.2 Effective sample size for the localized sampler and the bootstrap filter in Merton's model

	$\delta = 0.0005$	$\delta = 0.005$	$\delta = 0.01$	$\delta = 0.02$
N_{eff} (Linearized)	607.5	966.7	979.1	955.0
N_{eff} (Bootstrap)	6.4	61.4	121.1	230.4

Table 16.2 compares this linearized proposal with the bootstrap filter. The MATLAB file producing the table is `test_MertonLinearized.m`. The linearized filter performs much better, with results that are comparable to the localized sampler described before. Instead of using local linearization, van der Merwe et al. (2000) suggests using the unscented Kalman Filter for proposal generation.

16.5.1.3 Using Adaptation to Tune the Proposal in Merton's Model

Another generic approach to improve the efficiency of the particle filtering algorithm is to use the filter output to adapt the proposal. Cornebise et al. (2008) suggests to implement this idea by choosing a parametric family of proposal distribution and then optimize the parameters using the particles from the filter. To illustrate this method, consider the adaptation of the bootstrap filter in Merton's model. In particular, assume that the following family of proposals is chosen:

$$\log V_{\tau_k}^{(m)} \sim N\left((\mu - \frac{\sigma^2}{2})h + \log V_{\tau_{k-1}}^{(m)} + \gamma_{1,k}, \sigma^2 h \gamma_{2,k}\right) \tag{16.14}$$

Setting $\gamma_{1,k} = 0$ and $\gamma_{2,k} = 1$ one obtains bootstrap filter. In general $\gamma_{1,k}$ and $\gamma_{2,k}$ can be varied in order to find a proposal that is as close as possible to the target distribution, $f(\log V_{\tau_k}, \log V_{\tau_{k-1}} \mid \mathcal{D}_k)$. One appropriate metric to measure closeness between probability distributions is the Kullback-Leibler (K-L) distance. In the present context, if $r(\log V_{\tau_k} \mid \gamma, \log V_{\tau_{k-1}})$ is the parametric proposal conditional on $\log V_{\tau_{k-1}}$, then the overall proposal over the pair $(\log V_{\tau_k}, \log V_{\tau_{k-1}})$ is $r(\log V_{\tau_k} \mid \gamma, \log V_{\tau_{k-1}}) f(\log V_{\tau_{k-1}} \mid \mathcal{D}_{k-1})$. The K-L distance of this proposal from the target is defined as

$$\begin{aligned} D_{KL} &= \left(f(\log V_{\tau_k}, \log V_{\tau_{k-1}} \mid \mathcal{D}_k) \| r(\log V_{\tau_k} \mid \gamma, \log V_{\tau_{k-1}}) f(\log V_{\tau_{k-1}} \mid \mathcal{D}_{k-1})\right) \\ &= \int_{\{\log V_{\tau_k}, \log V_{\tau_{k-1}}\}} f(\log V_{\tau_k}, \log V_{\tau_{k-1}} \mid \mathcal{D}_k) \\ &\quad \times \log\left(\frac{f(\log V_{\tau_k}, \log V_{\tau_{k-1}} \mid \mathcal{D}_k)}{r(\log V_{\tau_k} \mid \gamma, \log V_{\tau_{k-1}}) f(\log V_{\tau_{k-1}} \mid \mathcal{D}_{k-1})}\right) \end{aligned}$$

Then, the "best" proposal within the parametric family is the one that minimizes the K-L distance to $f(\log V_{\tau_k}, \log V_{\tau_{k-1}} \mid \mathcal{D}_k)$. This is achieved by γ_k^* solving

$$\gamma_k^* = arg \max_{\gamma} \int_{\{\log V_{\tau_k}, \log V_{\tau_{k-1}}\}} f(\log V_{\tau_k}, \log V_{\tau_{k-1}} \mid \mathcal{D}_k) \times \log r(\log V_{\tau_k} \mid \gamma, \log V_{\tau_{k-1}}) \quad (16.15)$$

This optimization problem is unfeasible as the integral is not known in closed form. However, if one has a normalized weighted sample $(\pi_k^{(m)}, \log V_{\tau_k}^{(m)}, \log V_{\tau_{k-1}}^{(m)}, m = 1, \ldots, M)$ representing $f(\log V_{\tau_k}, \log V_{\tau_{k-1}} \mid \mathcal{D}_k)$ from a prior run of a particle filter, the problem can be approximated by

$$\gamma_k^* = arg \max_{\gamma} \sum_{i=1}^{M} \pi_k^{(m)} \log r(\log V_{\tau_k}^{(m)} \mid \gamma, \log V_{\tau_{k-1}}^{(m)}) \quad (16.16)$$

In the example in Merton's model with the choice of the proposal family as in (16.14), the optimization problem becomes

$$(\gamma_{1,k}^*, \gamma_{2,k}^*) = arg \max_{\gamma_{1,k}, \gamma_{2,k}} \sum_{i=1}^{M} \pi_k^{(m)} \Bigg(-\frac{(\log V_{\tau_k}^{(m)} - (\mu - \frac{\sigma^2}{2})h - \log V_{\tau_{k-1}}^{(m)} - \gamma_{1,k})^2}{2\gamma_{1,k}, \sigma^2 h \gamma_{2,k}} - \frac{\log(\gamma_{1,k}, \sigma^2 h \gamma_{2,k})}{2} \Bigg)$$

This can be solved in one step, yielding

$$\gamma_{1,k}^* = \sum_{i=1}^{M} \pi_k^{(m)} \left(\log V_{\tau_k}^{(m)} - (\mu - \frac{\sigma^2}{2})h - \log V_{\tau_{k-1}}^{(m)} \right) \quad (16.17)$$

$$\gamma_{2,k}^* = \frac{\sum_{i=1}^{M} \pi_k^{(m)} \left(\log V_{\tau_k}^{(m)} - (\mu - \frac{\sigma^2}{2})h - \log V_{\tau_{k-1}}^{(m)} - \gamma_{1,k}^* \right)^2}{\sigma^2 h} \quad (16.18)$$

The algorithm is initialized by running the bootstrap filter (setting $(\gamma_{1,k}^{(0)} = 0, \gamma_{2,k}^{(0)} = 1)$) and then the filter is adapted by the procedure described above.

Adapted Bootstrap Filter in Merton's Model

- Initial State: Set $V_{\tau_0}^{(m)} = S^{-1}(S_{\tau_0})$ where the function $S^{-1}(.)$ is the inverse of the equity pricing function in (16.6).
- Run the bootstrap filter, providing $(\gamma_{1,k}^{(1)}, \gamma_{2,k}^{(1)})$ using (16.17) and (16.18)
- Adapt the filter: For $j = 1, \ldots, N_{iter}$

– Recursion: For $k = 1, \ldots, N$

1. Sampling

· Draw $\log V_{\tau_k}^{(m)}$ from $r(V_{\tau_k}^{(m)} \mid \gamma_{1,k}^{(j)}, \gamma_{2,k}^{(j)})$

$$\log V_{\tau_k}^{(m)} \sim N\left((\mu - \frac{\sigma^2}{2})h + \log V_{\tau_{k-1}}^{(m)} + \gamma_{1,k}^{(j)}, \sigma^2 h \gamma_{2,k}^{(j)}\right)$$

· Compute the normalized importance weights

$$\pi_k^{(m)} = \frac{w_k^{(m)}}{\sum_{m=1}^{M} w_k^{(m)}}$$

where

$$w_k^{(m)} = \frac{f(S_{\tau_k} \mid V_{\tau_k}^{(m)}, \Theta) f(V_{\tau_k}^{(m)} \mid V_{\tau_{k-1}}^{(m)}, \Theta)}{r(V_{\tau_k}^{(m)} \mid \gamma_{1,k}^{(j)}, \gamma_{2,k}^{(j)}, V_{\tau_{k-1}}^{(m)})}$$

2. Compute the new value of the adaptation parameters: $(\gamma_{1,k}^{(j+1)}, \gamma_{2,k}^{(j+1)})$ using the new weighted sample and (16.17-16.18). To avoid spurious results due to a poor particle set, $\gamma_{2,k}$ is updated only when $N_{eff}(k) >= 5$.
3. Resample from the weighted sample $\{(V_{\tau_k}^{(m)}, \pi_k^{(m)}); m = 1, \cdots, M\}$ to obtain a new equal-weight sample of size M.

As $M \to \infty$, the approximating optimization problem in (16.16) converges to the true problem in (16.15). Thus if M is large enough, setting $N_{iter} = 1$ would already achieve the optimal parameters. However for finite M, the initial particle approximation may be poor and running a couple more iterations can yield further improvements.

Table 16.3 reports the results of this algorithm with $N_{iter} = 4$ and $M = 1{,}000$, with all the other simulation parameters set as in the examples before. The MATLAB file producing the table is `test_MertonAdapt.m`. Adaptation yields great improvements in the algorithm, providing acceptable results even when δ is small and the likelihood function is very peaked. In accordance with theory, most of the improvement takes place in the first iteration. Substantial further improvements are achieved only when the initial sampler is very poor, the case of small δ. In more complicated problems, wider parametric families could be used for adaptation. In particular, using the adaptive D-kernel method of Cappe et al. (2007, 2008) would allow the use of general mixture classes.

Table 16.3 Effective sample size for the adapted bootstrap filter in Merton's model

	$\delta = 0.0005$	$\delta = 0.005$	$\delta = 0.01$	$\delta = 0.02$
N_{eff} (Iteration 0)	6.4	61.4	121.1	230.4
N_{eff} (Iteration 1)	252.6	520.3	537.4	557.8
N_{eff} (Iteration 2)	457.4	542.0	546.5	557.9
N_{eff} (Iteration 3)	506.7	543.54	545.8	559.7
N_{eff} (Iteration 4)	523.9	544.3	547.51	557.6

16.5.2 *Other Variations on the Filtering Algorithm*

When the future observation is very informative on the present state, it may be better to resample the present particles before propagating them forward. This idea is used in the Auxiliary Particle Filter by Pitt and Shephard (1999) and investigated theoretically in Doucet and Johansen (2008). More sophisticated resampling routines have been proposed to reduce the variance of multinomial resampling. Some examples are residual resampling (Liu, 2001) or stratified resampling (Kitagawa, 1996).

16.5.3 *Application of Particle Filtering: Obtaining the Asset and Debt Value of Six Flags*

In the second example described before the objective is to obtain the unobserved asset value of Six Flags in 2008 using the noisy time series of equity. The application of Merton's model necessitates some assumptions on the inputs of the model. The face value of debt is chosen to be the sum of total liabilities and preferred equity (as this latter is more senior than common equity) yielding $F = 3,197.9$ (the unit is Million USD). The maturity of debt is chosen to be 1 years, while the risk-free rate is set to 2.7%, the 1-year zero-coupon yield on treasuries at the beginning of 2008. Last, to run the filter one needs estimates of the model parameters (μ, σ, δ). The estimation of the drift is unreliable using 1 year of data, so the drift is simply set equal to the riskfree rate. The other two parameters σ and δ are estimated in a Bayesian procedure using importance sampling and a flat prior. The posterior means are used as point estimates yielding $\sigma = 0.075$ and $\delta = 0.0117$. Panel A of Fig. 16.7 reports the filtered asset values while Panel B the filtered yield spread on the debt (liabilities+preferred equity) of the firm. The localized filter with $M = 1{,}000$ particles was used to produce the results. One can see that in the second half of 2008 when the asset value of the company decreased, the spread becomes more sensitive to changes in the asset value. This can be explained by the fact that by this stage, the equity buffer that protects the debt-holders is more depleted. To understand the uncertainty created by the noise in the equity prices, Fig. 16.8 plots the 90% confidence interval of the yield spread of Six Value, ranging from 7 basis

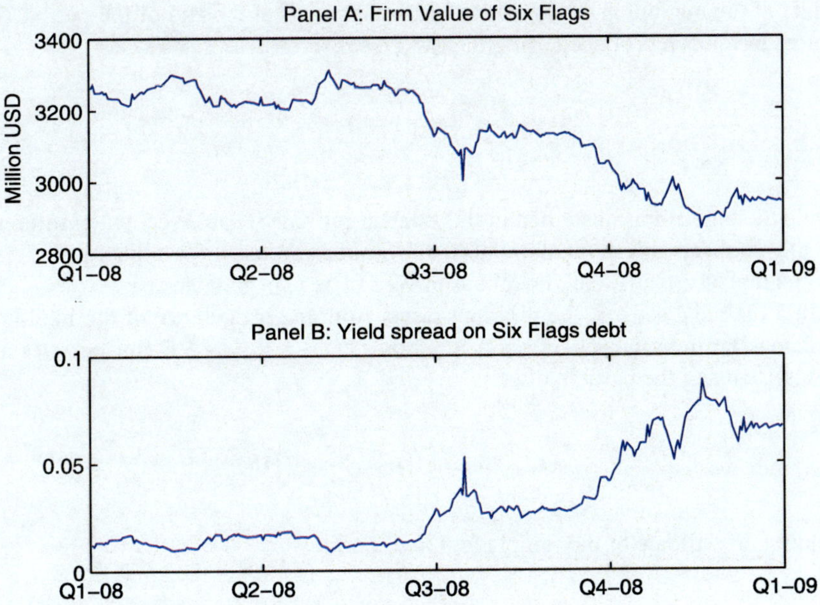

Fig. 16.7 Filtered asset value and yield spread on Six Flags' debt

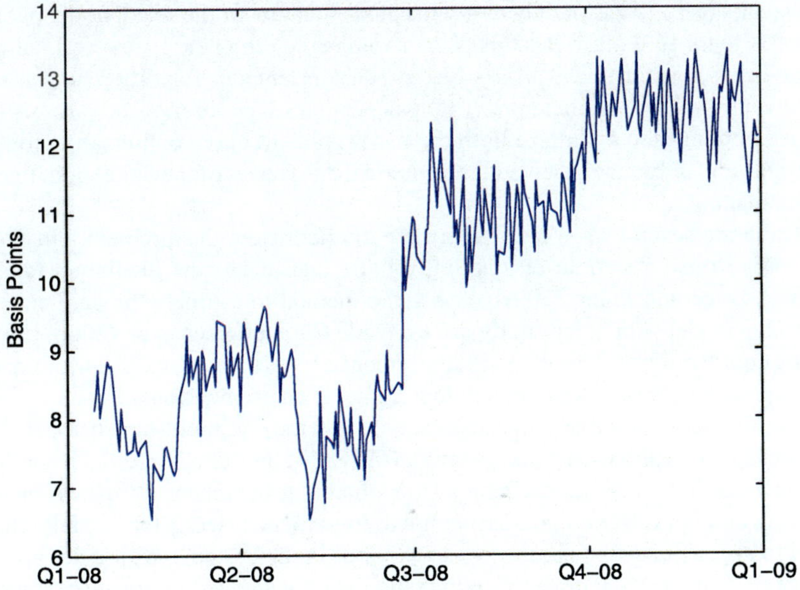

Fig. 16.8 Ninety percent confidence interval on the yield spread of Six Flag

points at the beginning of the year to 12 basis points at the end of the period. The figures have been produced using the MATLAB script `SixFlags.m`.

16.6 Outlook

While the algorithms described in this chapter provide reliable sequential inference on the unobservable dynamic states, the important task of estimating the fixed parameters of the financial model has proved to be a formidable task.

In a classical setting, the problem stems from the irregularity of the likelihood surface. The individual likelihood function, $f(y_k \mid y_{1:k-1}, \theta)$, can be estimated pointwise using the particle filter as

$$\hat{f}(y_k \mid y_{1:k-1}, \theta) \approx \sum_{i=1}^{M} w_k^{(m)}(\theta)$$

yielding an estimate of the sample loglikelihood:

$$\hat{l}(y_{1:N} \mid \theta) = \sum_{k=1}^{N} \log \hat{f}(y_k \mid y_{1:k-1}, \theta)$$

However, $\hat{l}(y_{1:N} \mid \theta)$ is an inherently irregular function of the fixed model parameters, θ. Figure 16.9 illustrates this phenomenon by plotting the estimated likelihood function of a simulated data sample in Merton's model for different values of the asset volatility parameter, σ. The local wiggles one observes here result from the resampling step and make both the usual gradient-based optimization routines unusable and inference based on the numerical derivatives of the likelihood function problematic.

There are several ways in the literature to circumvent this problem. Pitt (2002) proposes to use a smooth resampling routine that makes the likelihood function regular. Duan and Fulop (2009b) apply the method to estimate the parameters of Merton's model with noisy equity prices, while Christoffersen et al. (2008) use it in fitting equity option prices with different stochastic volatility models. Unfortunately, the approach only works when the hidden state is one-dimensional.

An alternative approach that works even when x_k is multi-dimensional is the Monte-Carlo Expectation-Maximization (MCEM) algorithm. Here the irregularity of the filter becomes inconsequential for obtaining parameter estimates, because filtering and optimization are disentangled. In particular, while the particle filter is used to approximate the necessary expectations in the E-step, the particles are kept unchanged in the M-step where optimization is implemented. Further, Olsson et al. (2008) show that it is sufficient to use fixed-lag smoothing in the E-step with a relatively small number of lags. This is important because in particle filtering the inference on the recent past is more reliable than the representation of the distant

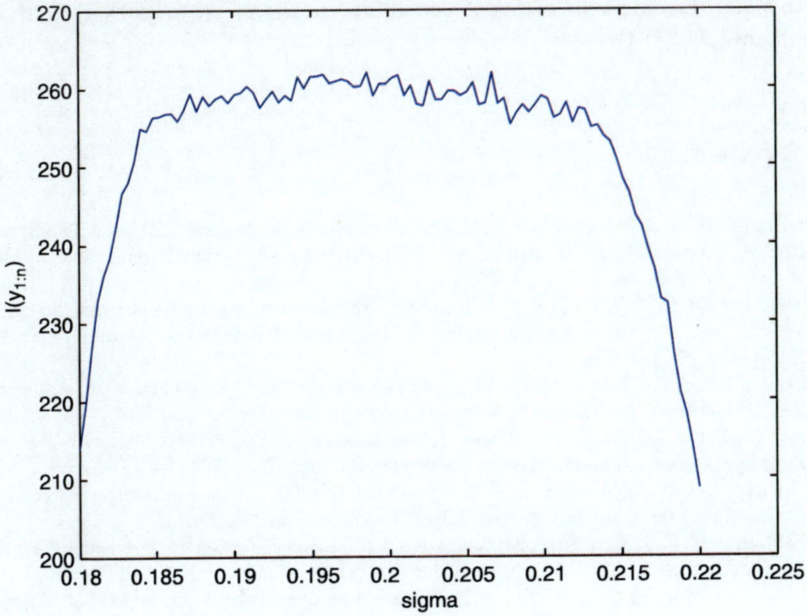

Fig. 16.9 Irregularity of the likelihood function in Merton's model

past, a result of the repeated use of the resampling step. To deal with the problem of inference, Duan and Fulop (2009) proposes the use the sample cross-products of the individual smoothed scores and a Newey-West correction. Duan and Fulop (2007) apply the MCEM algorithm to the estimation of a jump-diffusion model with high-frequency data and microstructure noise, while Fulop and Lescourret (2008) uses it to estimate intra-daily patterns of transaction costs and volatilities on the credit default swap market.

In a Bayesian setting, one could simply try to include the fixed parameters in the state-space and perform particle filtering on the extended state-space. This is very attractive as it would allow joint sequential inference on the states and the fixed parameters. Unfortunately it is well-known that this algorithm is unreliable. The underlying reason is that the extended dynamic system is not forgetting its past due to the inclusion of the fixed parameters, thus the Monte-Carlo errors committed in each stage quickly accumulate. Extending the work of Storvik (2002), Johannes and Polson (2006) suggest tracking the filtering distribution of some sufficient statistics to perform sequential inference on the parameters. Johannes et al. (2008) apply this approach to examine the predictability of the stock market and optimal portfolio allocation. The key limitation is that the method can only be applied to models that admit a finite-dimensional sufficient statistic structure for the fixed parameters. Instead of attempting sequential inference, Andrieu et al. (2010) suggest inserting particle filters into an MCMC algorithm as a proposal-generating mechanism. The method is illustrated on different economic and financial models by Flury and Shephard (2008).

References

Anderson, B. D. O. & Moore, J. B. (1979). *Optimal filtering*. Englewood Cliffs, N.J: Prentice-Hall.

Andrieu, C., Doucet, A., & Holenstein, A. (2010). Particle markov chain monte carlo. *Journal of Royal Statistical Society B*, *72*, 1–33.

Bakshi, G., Carr, P., & Wu, L. (2008). Stochastic risk premiums, stochastic skewness in currency options, and stochastic discount factors in international economies. *Journal of Financial Economics*, *87*, 132–156.

Cappe, O., Douc, R., Gullin, A., Marin, J. M., & Robert, C. P. (2007). Convergence of adaptive mixtures of importance sampling schemes. *Annals of Statistics*, *35*, 420–448.

Cappe, O., Douc, R., Gullin, A., Marin, J. M., & Robert, C. P. (2008). Adaptive importance sampling in general mixture classes. *Statistics and Computing, 18*, 447–459.

Christensen, J. H. E., Diebold, F. X., & Rudebusch, G. D. (2007). The affine arbitrage-free class of Nelson-Siegel term structure models. NBER Working Paper No. 13611.

Christoffersen, P. F., Jacobs, K., & Mimouni, K. (2008). Models for S&P 500 dynamics: Evidence from realized volatility, daily returns, and option prices. Manuscript, McGill University.

Cornebise, J., Moulines, E., & Olsson, J. (2008). Adaptive methods for sequential importance sampling with application to state space models. *Statistics and Computing, 18*, 461–480.

Crisan, D. & Doucet, A. (2002). A survey of convergence results on particle filtering methods for practitioners. *IEEE Transactions on Signal Processing*, *50*, 736–746.

De Jong, F. (2000). Time series and cross-section information in affine term-structure models. *Journal of Business & Economic Statistics*, *18*, 300–314.

Del Moral, P. (2004). *Feynman-Kac formulae genealogical and interacting particle systems with applications*. New York: Springer.

Diebold, F. X. & Li, C. (2006). Forecasting the term structure of government bond yields. *Journal of Econometrics*, *130*, 337–364.

Diebold, F. X., Rudebusch, G. D., Aruoba, B. (2006). The macroeconomy and the yield curve: A dynamic latent factor approach. *Journal of Econometrics*, *131*, 309–338.

Doucet, A. & Johansen, A. M. (2008). A note on auxiliary particle filters. *Statistics and Probability Letters, 78*, 1498–1504.

Doucet, A., Godsill, S. J., & Andrieu, C. (2000). On sequential monte carlo sampling methods for bayesian filtering. *Statistics and Computing*, *10*, 197–208.

Duan, J. C. (1994). Maximum likelihood estimation using price data of the derivative contract. *Mathematical Finance*, *4*, 155–167.

Duan, J. C. & Fulop, A. (2007). How frequently does the stock price jump? – an analysis of high-frequency data with microstructure noises. Working Paper.

Duan, J. C. & Fulop, A. (2009). A stable estimator for the information matrix under EM. *Statistics and Computing, 21*, 83–91.

Duan, J. C. & Fulop, A. (2009b). Estimating the structural credit risk model when equity prices are contaminated by trading noises. *Journal of Econometrics*, *150*, 288–296.

Duan, J. C. & Simonato, J. G. (1999). Estimating and testing exponential-affine term structure models by Kalman filter. *Review of Quantitative Finance and Accounting*, *13*, 111–135.

Duffee, G. E. (2002). Term premia and interest rate forecasts in affine models. *Journal of Finance*, *57*, 405–443.

Flury, T. & Shephard, N. (2008). Bayesian inference based only on simulated likelihood: Particle filter analysis of dynamic economic models. Technical report, Nuffield College, Oxford University.

Fulop, A. & Lescourret, L. (2008). Intra-daily variations in volatility and transaction costs in the Credit Default Swap market. Working Paper, 2009.

Gordon, N. J., Salmond, D. J., & Smith, A. F. M. (1993). A novel approach to non-linear and non-gaussian bayesian state estimation. *IEEE Proceedings F*, *140*, 107–113.

Johannes, M. & Polson, N. (2006). Exact particle filtering and parameter learning. Working Paper.

Johannes, M., Korteweg, A. G., & Polson, N. (2008). Sequential learning, predictive regressions, and optimal portfolio returns. Working Paper.

Kitagawa, G. (1996). Monte carlo filter and smoother for non-gaussian nonlinear state space models. *Journal of Computational and Graphical Statistics*, *5*, 1–25.

Liu, J. S. (2001). *Monte Carlo strategies in scientific computing*. New York: Springer.

Merton, R. C. (1974). On the pricing of corporate debt: The risk structure of interest rates. *Journal of Finance*, *29*, 449–470.

Nelson, C. R. & Siegel, A. F. (1987). Parsimonious modeling of yield curve. *Journal of Business*, *60*, 473–489.

Olsson, J., Cappe, R., Douc, R., & Moulines, E. (2008). Sequential monte carlo smoothing with application to parameter estimation in non-linear state space models. *Bernoulli*, *14*, 155–179.

Pitt, M. (2002). Smooth particle filters likelihood evaluation and maximisation. Working Paper, University of Warwick.

Pitt, M. & Shephard, N. (1999). Filtering via simulation: Auxiliary particle filter. *Journal of the American Statistical Association*, *94*, 590–599.

Schwartz, E. & Smith, J. E. (2000). Short-term variations and long-term dynamics in commodity prices. *Management Science*, *46*, 893–911.

Storvik, G. (2002). Particle filters in state space models with the presence of unknown static parameters. *IEEE Transactions on Signal Processing*, *50*, 281–289.

Trolle, A. B. & Schwartz, E. (2008). Unspanned stochastic volatility and the pricing of commodity derivatives. Working Paper.

van der Merwe, R. & Wan, E. (2000). The unscented kalman filter for nonlinear estimation. *Proceedings of Symposium 2000 on Adaptive Systems for Signal Processing, Communication and Control.*

van der Merwe, R., Doucet, A., De Freitas, N., & Wan, E. (2000). *The unscented particle filter.*

Chapter 17
Fitting High-Dimensional Copulae to Data

Ostap Okhrin

Abstract This paper make an overview of the copula theory from a practical side. We consider different methods of copula estimation and different Goodness-of-Fit tests for model selection. In the GoF section we apply Kolmogorov-Smirnov and Cramer-von-Mises type tests and calculate power of these tests under different assumptions. Novating in this paper is that all the procedures are done in dimensions higher than two, and in comparison to other papers we consider not only simple Archimedean and Gaussian copulae but also Hierarchical Archimedean Copulae. Afterwards we provide an empirical part to support the theory.

17.1 Introduction

Many practical problems arise from modelling high dimensional distributions. Precise modelling is important in fitting of asset returns, insurance payments, overflows from a dam and so on. Often practitioners stay ahead of potential problems by using assets backed up in huge portfolios, payments spatially distributed over land, and dams located on rivers where there are already other hydrological stations. This means that univariate problems are extended to multivariate ones in which all the univariate ones are dependent on each other. Until the late 1990s elliptical distribution, in particular the multivariate normal one, was the most desired distribution in practical applications. However the normal distribution does not, in practice, meet most applications. Some studies (see e.g Fama 1965; Mandelbrot 1965, etc.) show that daily returns are not normally distributed but follow stable distributions. This means that on one hand one cannot take the distribution in which

O. Okhrin (✉)
Ladislaus von Bortkiewitcz Chair of Statistics,
C.A.S.E. – Center of Applied Statistics and Economics,
Humboldt-Universität zu Berlin, 10178 Berlin, Germany
e-mail: :ostap.okhrin@wiwi.hu-berlin.de

J.-C. Duan et al. (eds.), *Handbook of Computational Finance*, Springer Handbooks of Computational Statistics, DOI 10.1007/978-3-642-17254-0_17,

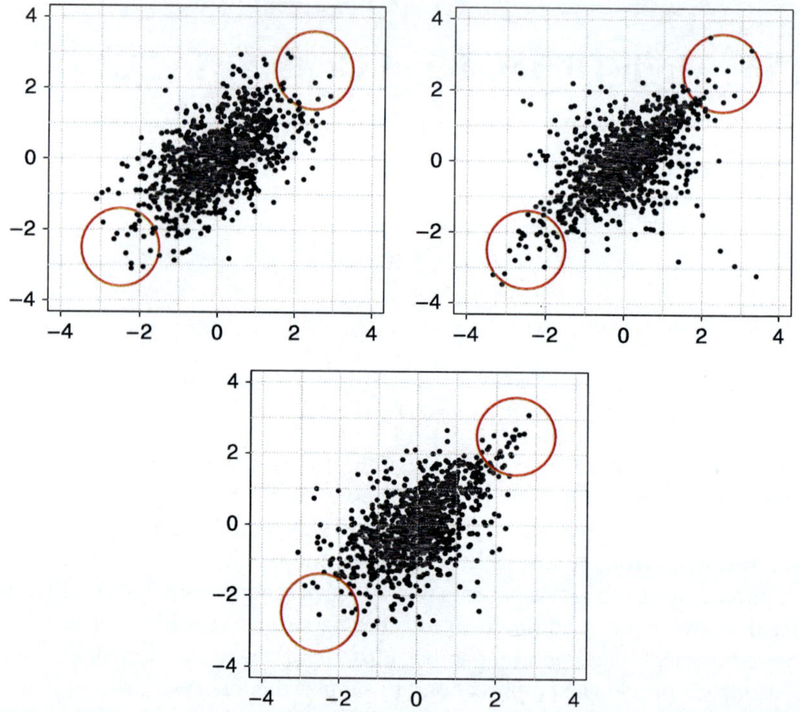

Fig. 17.1 Scatter plots of bivariate samples with different dependency structures

margins are normal, and on the other hand, stable multivariate distributions are difficult to implement. In the hydrological problem, margins arise from extreme value distribution, while one is interested in the maximal value of the water collected after the winter season over a number of years, this value arises from the family of extreme distributions. As in the previous example, the multivariate extreme value distribution family is also somewhat restrictive.

Two further problems are illustrated in Fig. 17.1. The scatter plot in the first figure shows the realisations of two Gaussian random variables. The points are symmetric and no extreme outliers can be observed. In contrary, the second picture exhibits numerous outliers. The outliers in the first and third quadrants show that extreme values often occur simultaneously for both variables. Such behaviour is observed in crisis periods, when strong negative movements on financial markets occur simultaneously. On the third figure we observe that the dependency between the negative values is different compared to the positive values. This type of non-symmetric dependency cannot be modeled by elliptical distributions, because they impose a very specific radially symmetric dependency structure.

Following these examples we need a solution to easily separate the modelling of the dependency structure and the margins. This is one of the tasks of copulae; to enable the modelling of marginals separately from the dependency. The above problem concerning assets could be solved by taking margins from the stable distribution and the dependency, as in the multivariate one. Similar solutions could be found for other problems. In finance, copulae are applied in different fields such as credit portfolio modelling and risk management.

Over the last 40 years, copula has only been attractive from a mathematical perspective, and only as late as 1999 were the different complicated properties of copula, such as the distribution (which made it more flexible), settled and solved. Nowadays dependency plays a key role in many financial models, starting from the basic portfolio theory of Markowitz. Recent developments strongly support the joint non-Gaussianity of asset returns and exploit numerous alternative approaches to model the underlying distribution. The key role of dependency can be best illustrated by the famous quote "Given the prices of single-bet financial contracts, what is the price of multiple-bet contracts? There is no unique answer to that question...". The first application of copulae to financial data was carried out by Embrechts et al. (1999). In this paper copulae were used in risk management framework which stimulated a series of ground breaking applied papers. Breymann et al. (2003) model the dependencies of high-frequency data. An application to risk management is discussed in Junker and May (2005). Portfolio selection problems were considered in Hennessy and Lapan (2002) and in Patton (2004). Theoretical foundations of copula-based GARCH models and its application were proposed by Chen and Fan (2005). Lee and Long (2009), Giacomini et al. (2009) and Härdle et al. (2010) consider time varying copulae.

The new fields of application show the need for further theoretical developments. Each proposed model should be estimated with either parametric, semi- or nonparametric methods. The semiparametric estimation of the copula-based distribution, which is based on the nonparametric estimation of margins and estimation of the parameter for the fixed copula function, is discussed in Chen and Fan (2006), Chen et al. (2006), Genest et al. (1995), Joe (2005) and Wang and Wells (2000). Fully nonparametric estimation is discussed in Fermanian and Scaillet (2003), Chen and Huang (2007) and Lejeune and Sarda (1992). To measure how well a copula-based statistical model fits the data, several goodness-of-fit tests were developed and discussed in the papers by Chen and Fan (2005), Chen et al. (2004), Fermanian (2005), Genest et al. (2006), Genest and Rémillard (2008) and Breymann et al. (2003). In-depth discussion of simulation methodologies for Archimedean copulae can be found in Whelan (2004) and McNeil (2008). A detailed review and discussion of copula theory is given in Joe (1997) and Nelsen (2006).

In this chapter we describe the attractive features of copulae from the statistical perspective, with examples and applications in real data. We consider the most important copula classes with different methods of estimation and goodness-of-fit tests. We compare different goodness-of-fit tests by their rejection rates, for which a profound simulation study has been devised. In the empirical part of the chapter we apply different copula models to the normalised residuals and test the quality of the fit by discussed goodness-of-fit tests. We found that for the selected datasets hierarchical Archimedean copula outperform the simple Archimedean copula and the Gaussian copula by all goodness-of-fit tests.

17.2 Theoretical Background

From the early days of the multivariate probability theory it is well known, that given the d-variate distribution function $F : \mathbb{R} \rightarrow [0;1]$ of a d-variate random vector $(X_1, \dots, X_d)$ the distribution function, called marginal distribution function of each of the d components $X_1, \dots, X_d$ is easily computed:

$$\begin{aligned} F_1(x) &= F(x, +\infty, \dots, +\infty), \\ F_2(x) &= F(+\infty, x, +\infty, \dots, +\infty), \\ &\dots \\ F_d(x) &= F(+\infty, \dots, +\infty, x). \end{aligned}$$

The converse problem was studied by Fréchet (1951), Hoeffding (1940) and Hoeffding (1941), where having the distribution functions $F_1, \dots, F_d$ of d random variables $X_1, \dots, X_d$ defined on the same probability space $(\Omega, \mathcal{F}, \mathcal{P})$ they wanted to make a conclusions about the set $\Gamma(F_1 \dots, F_d)$ of the d-variate distribution functions whose marginals are $F_1, \dots, F_d$

$$F \in \Gamma(F_1, \dots, F_d) \Leftrightarrow \begin{cases} F_1(x) = F(x, +\infty, \dots, +\infty), \\ F_2(x) = F(+\infty, x, +\infty, \dots, +\infty), \\ \dots \\ F_d(x) = F(+\infty, \dots, +\infty, x). \end{cases}$$

Nowadays the set $\Gamma(F_1, \dots, F_d)$ is called the Fréchet class of $F_1, \dots, F_d$. Γ is not empty, because it always contains the independence case in which $F(x_1, \dots, x_d) = F_1(x_1) \cdot \dots \cdot F_d(x_d)$, $\forall x_1, \dots, x_d \in \mathbb{R}$. Dealing with Fréchet classes, one often interests in the bounds and members of the Γ. Dall'Aglio (1972) studies conditions under which there is only one distribution function which belongs to $\Gamma(F_1, \dots, F_d)$. A nice and short review of the Fréchet classes can be found in Joe (1997).

In 1959 Sklar found the partial solution to the above mentioned problem by introducing copulae. Because there are a variety of copula definitions we will first look at the most general one. For this we will need to define the C-volume with the d-box that is a cartesian product $[\mathbf{a}, \mathbf{b}] = \prod_{j=1}^{d} [a_j, b_j]$, where, for every index $j \in \{1, 2, \dots, d\}$, $0 \le a_j \le b_j \le 1$.

Definition 1. For a function $C : [0;1]^d \rightarrow [0;1]$, the C-volume V_c of the box $[\mathbf{a}, \mathbf{b}]$ is defined via

$$V_c([\mathbf{a}, \mathbf{b}]) \stackrel{def}{=} \sum_{\mathbf{v}} sign(\mathbf{v}) C(\mathbf{v}),$$

where the sum is carried over all the 2^d vertices $\mathbf{v}$ of the box $[\mathbf{a}, \mathbf{b}]$. Here also

$$sing(\mathbf{v}) = \begin{cases} 1, & \text{if } v_j = a_j \text{ for an even number of vertices,} \\ -1, & \text{if } v_j = a_j \text{ for an odd number of vertices.} \end{cases}$$

Here is the definition of a copula, see Härdle and Simar (2007):

Definition 2. A function $C : [0,1]^d \to [0,1]$ is a d-dimensional copula if:

1. $C(x_1, \ldots, x_d) = 0$, when $x_j = 0$ for at least one index $j \in \{1, \ldots, d\}$.
2. $C(1, 1, \ldots, x_j, 1 \ldots, 1) = x_j$.
3. The V_c-volume of every d-box $[\mathbf{a}, \mathbf{b}]$ is positive: $V_c([\mathbf{a}, \mathbf{b}]) \geq 0$.

The set of all the d-dimensional copulae ($d \geq 3$) in the rest of the chapter is denoted as $\mathcal{C}_d$, while the set of all bivariate ($d = 2$) copulae is denoted by $\mathcal{C}$. As already mentioned above, this simple family of functions has been extremely popular because of its property given in the Sklar (1959) theorem

Theorem 1. *Given a d-dimensional distribution function F, a copula $C \in \mathcal{C}_d$ exists such that for all $(x_1, \ldots, x_d) \in \overline{\mathbb{R}}^d$:*

$$F(x_1, \ldots, x_d) = C\{F_1(x_1), \ldots, F_d(x_d)\}. \tag{17.1}$$

The copula C is uniquely defined on $\prod_{j=1}^d F_j(\overline{\mathbb{R}})$ and therefore unique if all margins are continuous, thus

$$C(u_1, \ldots, u_d) = F\{F_1^{-1}(u_1), \ldots, F_d^{-1}(u_d)\}. \tag{17.2}$$

Conversely, if $F_1, \ldots, F_d$ are d one-dimensional distribution functions, then the function F defined in (17.1) is a d-dimensional distribution function.

Sklar's theorem also answers the question of the uniqueness of the copula C. However, if, for example, in the two dimensional case at least one of the two distribution functions has a discrete component, there may be more than one copula extending C from $F_1(\overline{\mathbb{R}}) \times F_2(\overline{\mathbb{R}})$ to the whole unit square $[0,1]^2$. This is due to a fact that C is uniquely defined only on the product of the ranges $F_1(\overline{\mathbb{R}}) \times F_2(\overline{\mathbb{R}})$. In this case it is good to have a procedure of bilinear interpolation in order to single out a unique copula. In the variety of papers where copulae are applied in different fields, authors usually do not consider the assumption that the random variables are continuous. This assumption is necessary to avoid problems with non-uniqueness. The second part of the Sklar's theorem is based on the construction of the multivariate distribution from the margins and the copula function. It is extremely popular in practice, where, for example, in risk management, analysts may have a better idea about the marginal behaviour of individual risk factors, than about their dependency structure. This approach allows them to combine marginal models and to investigate the sensitivity of risk to the dependence specification.

New multivariate distributions are created in two steps. At first, all univariate random variables $X_1, \ldots, X_d$ are separately described by their marginal distributions

$F_1, \dots, F_d$. Then secondly, the copula $C \in \mathcal{C}_d$ which contains all the information about the relationship between the original variables $X_1, \dots, X_d$ – not taking into account the information provided by $F_1, \dots, F_d$ – is introduced.

Being armed with the remarks written above, one can write the following copula definition

Definition 3. A d-dimensional copula is a cumulative distribution function on $[0,1]^d$ with standard uniform marginal cumulative distribution functions.

As in the case of the multivariate distribution, mentioned at the beginning, setting all of the arguments equal to $+\infty$ one gets an univariate marginal distribution. A univariate marginal of copula C is obtained by setting some of its arguments equal to 1. Similarly the m-marginal of C, $m < d$ is given by setting all $d - m$ arguments equal to 1, from the simple combinatoric problem, we see that there are $\binom{d}{m}$ different m-margins of the copula C.

A copula C satisfies a set of different important conditions, one of which is the Lipschitz condition which says that:

$$|C(u_1, \dots, u_d) - C(v_1, \dots, v_d)| \le \sum_{j=1}^{d} |v_j - u_j|.$$

Another property says, that $\forall j \in \{1, \dots, d\}$, $\{u_1, \dots, u_{j-1}, t, u_{j+1}, \dots, u_d\}$, $\forall t \in [0,1]$, the functions $t \mapsto C(u_1, \dots, u_{j-1}, t, u_{j+1}, \dots, u_d)$ are increasing as functions of t.

To get a better impression of what a copula is from a definition, let us consider a special bivariate case. Explicitly, a bivariate copula is a function $C : [0,1]^2 \to [0,1]$ such that:

1. $\forall u \in [0,1] \quad C(u,0) = C(0,u) = 0.$
2. $\forall u \in [0,1] \quad C(u,1) = C(1,u) = u.$
3. $\forall u, u', v, v' \in [0,1]$ with $u \le u'$ and $v \le v'$

$$C(u',v') - C(u',v) - C(u,v') + C(u,v) \ge 0.$$

The last inequality is referred to as the rectangular inequality and the function that satisfies it is said to be 2-increasing. The bivariate copula is always of special interest, because of the properties that are difficult to derive in higher dimensions.

The property of increasingness with respect to each argument could be profound for the bivariate copula in the following way. As we know from above, if C is a bivariate copula, then functions $[0,1] \ni t \mapsto C(t,v)$ and $[0,1] \ni t \mapsto C(v,t)$ are increasing with respect to t. The increasingness with respect to each argument means that derivatives with respect to Lebegue measure exist almost everywhere, and those derivatives are positive where they exist. From the Lipschitz conditions they are also bound above

$$0 \leq \frac{\partial C(s,t)}{\partial s} \leq 1, \quad 0 \leq \frac{\partial C(s,t)}{\partial t} \leq 1.$$

Every copula can be expressed in the form of the sum of absolutely continuous and singular part and an absolutely continuous copula C has a density c such that

$$C(u_1,\dots,u_d) = \int_{[0,1]^d} c(s_1,\dots,s_d)ds_1\dots ds_d = \int_0^1 ds_1 \dots \int_0^1 c(s_1,\dots,s_d)ds_d$$

from which the copula density is found by differentiation

$$c(u_1,\dots,u_d) = \frac{\partial^d C(u_1,\dots,u_d)}{\partial u_1 \dots \partial u_d}.$$

Following the Sklar theorem, the multivariate distribution F with margins $F_1,\dots,F_d$ has multivariate density f with marginal densities $f_1,\dots,f_d$ respectively. If, from the Sklar theorem copula C exists such that $F(x_1,\dots,x_d) = C\{F_1(x_1),\dots,F_d(x_d)\}$ then the d-variate density is

$$f(x_1,\dots,x_d) = c\{F_1(x_1),\dots,F_d(x_d)\} \cdot f_1(x_1)\dots f_d(x_d). \tag{17.3}$$

Notice, however, that, as a consequence of the Lipschitz condition, for every bivariate copula C and for every $v \in [0,1]$, both functions $t \mapsto C(t,v)$ and $t \mapsto C(v,t)$ are absolutely continuous so that

$$C(t,v) = \int_0^t c_{1v}(s)ds \text{ and } C(v,t) = \int_0^t c_{2v}(s)ds.$$

Unfortunately, this representation has no application so far.

17.3 Copula Classes

Naturally, there are an infinite number of different copula functions satisfying the assumptions of definition. In this section we discuss in details three important classes of simple, elliptical and Archimedean copulae.

17.3.1 Simple Copulae

Often we are interested in some extreme, special cases, like independence and perfect positive or negative dependence. If d-random variables $X_1,\dots,X_d$ are stochastically independent from the Sklar Theorem the structure of such a relationship is given by the product (independence) copula defined as

$$\Pi(u_1,\ldots,u_d) = \prod_{j=1}^{d} u_j,\; u_1,\ldots,u_d \in [0,1].$$

Another two extremes are the lower and upper Fréchet–Hoeffding bounds. They represent the perfect negative and positive dependencies respectively

$$W(u_1,\ldots,u_d) = \max\Big(0, \sum_{j=1}^{d} u_j + 1 - d\Big),$$
$$M(u_1,\ldots,u_d) = \min(u_1,\ldots,u_d), \quad u_1,\ldots,u_d \in [0,1].$$

If, in a two dimensional case $C = W$ and $(X_1, X_2) \sim C(F_1, F_2)$ then X_2 is a decreasing function of X_1. Similarly, if $C = M$, then X_2 is an increasing function of X_1. In other words both M and W are singular, where M uniformly spreads the probability mass on the diagonal $X_1 = X_2$ and W uniformly spreads the probability mass on the opposite diagonal $X_1 = -X_2$. In general we can argue that an arbitrary copula which represents some dependency structure lies between these two bounds, i.e.

$$W(u_1,\ldots,u_d) \leq C(u_1,\ldots,u_d) \leq M(u_1,\ldots,u_d).$$

The bounds serve as benchmarks for the evaluation of the dependency magnitude. Note, however, that the lower Fréchet–Hoeffding bound is not a proper copula function for $d > 2$ but is a proper quasi-copula. Both upper and lower bounds are sharp, because there are copulae, that are either equal, at some points, to one of the two bounds.

The simple copulae for the two dimensional case are plotted in Fig. 17.2.

17.3.2 Elliptical Copulae

Due to the popularity of Gaussian and t-distributions in financial applications, elliptical copulae also play an important role. For example, in the modelling of collateralized debt obligations, where the assumption of the Gaussian one-factor dependency between joint default of the obligors, proposed by Li (2000), is seen as a standard approach. The construction of this type of copulae is based directly on the Sklar Theorem. The Gaussian copula and its copula density are given by:

$$C_N(u_1,\ldots,u_d,\boldsymbol{\Sigma}) = \boldsymbol{\Phi}_{\boldsymbol{\Sigma}}\{\Phi^{-1}(u_1),\ldots,\Phi^{-1}(u_d)\},$$
$$c_N(u_1,\ldots,u_d,\boldsymbol{\Sigma})$$
$$= |\boldsymbol{\Sigma}|^{-1/2}\exp\left\{-\frac{[\Phi^{-1}(u_1),\ldots,\Phi^{-1}(u_d)]'(\boldsymbol{\Sigma}^{-1}-\boldsymbol{I})[\boldsymbol{\Phi}^{-1}(u_1),\ldots,\boldsymbol{\Phi}^{-1}(u_d)]}{2}\right\},$$

for all $u_1,\ldots,u_d \in [0,1]$,

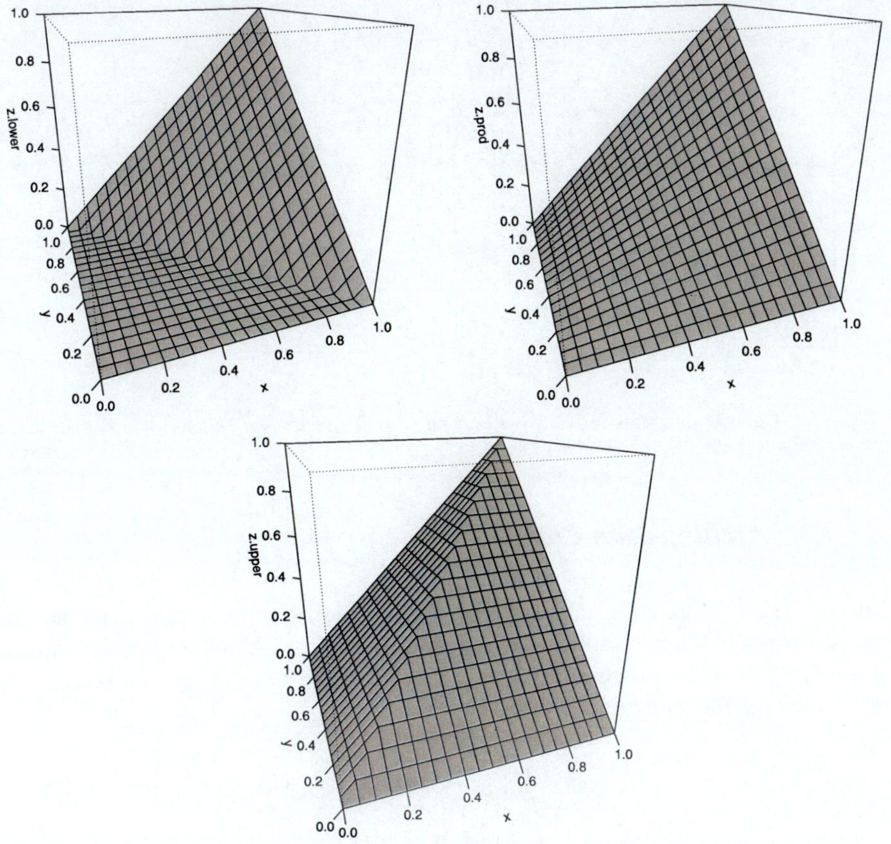

Fig. 17.2 Lower Frechet–Hoeffdings bound, Product copula and upper Frechet–Hoeffdings bound in two-dimensional case (*from left to right*)

where Φ_Σ is a d-dimensional normal distribution with a zero mean and the correlation matrix Σ. The variances of the variables are imposed by the marginal distributions. Note, that in the multivariate case the implementation of elliptical copulae is very involved due to technical difficulties with multivariate cdf's. The level plots of the two-dimensional respective densities with different margins are given in Fig. 17.3.

Using (17.2) one can derive the copula function for an arbitrary elliptical distribution. The problem is, however, that such copulae depend on the inverse distribution functions and these are rarely available in an explicit form. Therefore, the next class of copulae with its generalisations provides an important flexible and rich family of alternatives to the elliptical copulae.

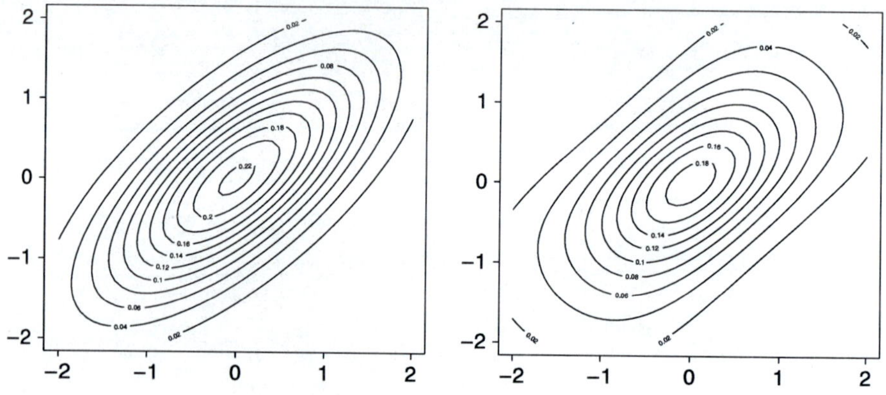

Fig. 17.3 Contour diagrams for Gaussian copula with Gaussian (*left column*) and t_3 distributed (*right column*) margins

17.3.3 Archimedean Copulae

In contrast to elliptical copulae, Archimedean copulae have a special method of construction which does not use (17.2), but fulfills all the conditions of the copula. Having M as an univariate distribution function of the positive random variable let ϕ be the Laplace transform of M, $\phi = \mathcal{LS}(M)$

$$\phi(s) = \int_0^\infty e^{-sw}\, dM(w), \quad s \geq 0. \tag{17.4}$$

Thus, M is said to be the inverse Laplace transform of ϕ, $M = \mathcal{LS}^{-1}(\phi)$. We denote as $\mathfrak{L}$ the class of Laplace transforms which contain strictly decreasing differentiable functions, see Joe (1997):

$$\mathfrak{L} = \{\phi : [0;\infty) \to [0,1] \,|\, \phi(0) = 1,\ \phi(\infty) = 0;\ (-1)^j \phi^{(j)} \geq 0;\ j = 1,\dots,\infty\}.$$

It is known, that for an arbitrary univariate distribution function F, a unique distribution function G exists such that

$$F(x) = \int_0^\infty G^\alpha(x)\, dM(\alpha) = \phi\{-\log G(x)\}.$$

This leads to $G = \exp\{-\phi^{[-1]}(F)\}$, where $\phi^{[-1]}$ is the generalised inverse

$$\phi^{[-1]}(x) = \begin{cases} \phi^{-1}(x) & \text{for } 0 \leq x < \phi(0); \\ 0 & \text{else.} \end{cases}$$

Taking d univariate distributions $F_1, \dots, F_d$, a simple extension leads to the multivariate distribution function that belongs to $\Gamma(F_1, \dots, F_d)$

$$F = \int G_1^{\alpha} \dots G_d^{\alpha} \, dM(\alpha) = \phi(-\log G_1 - \dots - \log G_d) = \phi\left\{\sum_{j=1}^{d} \phi^{[-1]}(F_j)\right\},$$

with Archimedean copula given by

$$C(u_1, \dots, u_d) = \phi\left\{\sum_{j=1}^{d} \phi^{[-1]}(u_j)\right\}. \quad (17.5)$$

The function ϕ is called the *generator* of the Archimedean copula. Throughout the chapter the notation ϕ^{-1} is understood as the generalised inverse $\phi^{[-1]}$. Usually generator function depends on the parameter θ which is set to be the parameter of the copula. It is easy to see, that Archimedean copulae are exchangeable. In two-dimensional cases they are symmetric in the sense that $C(u, v) = C(v, u), \; \forall u, v \in [0, 1]$. Joe (1997) and Nelsen (2006) provide a classified list of the typical Archimedean generators. Here we discuss the three most commonly used ones in financial applications, Archimedean copulae.

The first, widely used (in practice) copula is the Gumbel (1960) copula, which gained its popularity from the extreme value theory. The multivariate distribution based on the Gumbel copula with univariate extreme value marginal distributions is the only extreme value distribution based on an Archimedean copula, see Genest and Rivest (1989). Moreover, all distributions based on Archimedean copulae belong to its domain of attraction under common regularity conditions. Direct and inverse generators of the Gumbel copula with the copula function are given by

$$\phi(x, \theta) = \exp\{-x^{1/\theta}\}, \quad 1 \le \theta < \infty, \; x \in [0, \infty),$$
$$\phi^{-1}(x, \theta) = (-\log x)^{\theta}, \quad 1 \le \theta < \infty, \; x \in [0, 1],$$
$$C_\theta(u_1, \dots, u_d) = \exp\left[-\left\{\sum_{j=1}^{d} (-\log u_j)^{\theta}\right\}^{\theta^{-1}}\right], \; u_1, \dots, u_d \in [0, 1].$$

The Gumbel copula leads to asymmetric contour diagrams and shows stronger linkage between positive values, however, is also shows more variability and more mass in the negative tail.

For $\theta = 1$, the Gumbel copula reduces to the product copula and for $\theta \to \infty$ we obtain the Fréchet–Hoeffding upper bound. This copula does not have an extension to the negative dependence. The Gumbel copula is one of a few Archimedean copulae for which we have an explicit form of the distribution function M from (17.4). In the case of Gumbel copula M is the stable distribution, see Renyi (1970). This information is very useful in the simulation techniques, especially for the Marshall and Olkin (1988) method, see Sect. 17.4.

Another example is the Clayton (1978) copula which, in contrary to the Gumbel, has more mass on the lower tail, and less on the upper. This copula is often used in the modelling of the losses, which is of interest, for example, in insurance and finance. The necessary functions for this example are

$$\phi(x,\theta) = (\theta x + 1)^{-\frac{1}{\theta}}, \quad -1/(d-1) \le \theta < \infty,\, \theta \ne 0,\, x \in [0,\infty),$$

$$\phi^{-1}(x,\theta) = \frac{1}{\theta}(u^{-\theta} - 1),\, -1/(d-1) \le \theta < \infty,\, \theta \ne 0,\, x \in [0,1],$$

$$C_\theta(u_1,\ldots,u_d) = \left\{\left(\sum_{j=1}^{d} u_j^{-\theta}\right) - d + 1\right\}^{-\theta^{-1}}, \; u_1,\ldots,u_d \in [0,1].$$

The Clayton copula is one of few copulae that has a truncation property and has a simple explicit form of density for any dimension

$$c_\theta(u_1,\ldots,u_d) = \prod_{j=1}^{d}\{1 + (j-1)\theta\}u_j^{-(\theta+1)}\left(\sum_{j=1}^{d} u_j^{-\theta} - d + 1\right)^{-(\theta^{-1}+d)}.$$

As the parameter θ tends to infinity, dependence becomes maximal and the copula gives the upper Frechet–Hoeffding bound. As θ tends to zero, we have independence. As $\theta \to -1/(d-1)$, the distribution tends to the lower Fréchet bound.

Another interesting Archimedean copula is the so called Frank (1979) copula, which, in the bivariate case, is the only elliptical Archimedean copula in the sense that $C(u,v) = u + v - 1 + C(1-u, 1-v) = \overline{C}(u,v)$, where $\overline{C}(u,v)$ is called the survival or associative copula. $\overline{C}(u,v)$ is also a copula for a survival bivariate distribution. Direct and inverse generator of the Frank copula with the copula functions are

$$\phi(x,\theta) = -\frac{1}{\theta}\log\{1 + e^{u}(e^{-\theta} - 1)\}, \quad 0 \le \theta < \infty,\, x \in [0,\infty),$$

$$\phi^{-1}(x,\theta) = \log\left\{\frac{e^{-\theta x} - 1}{e^{-\theta} - 1}\right\}, \; 0 \le \theta < \infty,\, x \in [0,1],$$

$$C_\theta(u_1,\ldots,u_d) = -\frac{1}{\theta}\log\left[1 + \frac{\prod_{j=1}^{d}\{\exp(-\theta u_j) - 1\}}{\{\exp(-\theta) - 1\}^{d-1}}\right], \; u_1,\ldots,u_d \in [0,1].$$

The dependence becomes maximal when θ tends to infinity and independence is achieved when $\theta = 0$.

The level plots of the bivariate copula-based densities with t_3 and normal margins are given in Fig. 17.4.

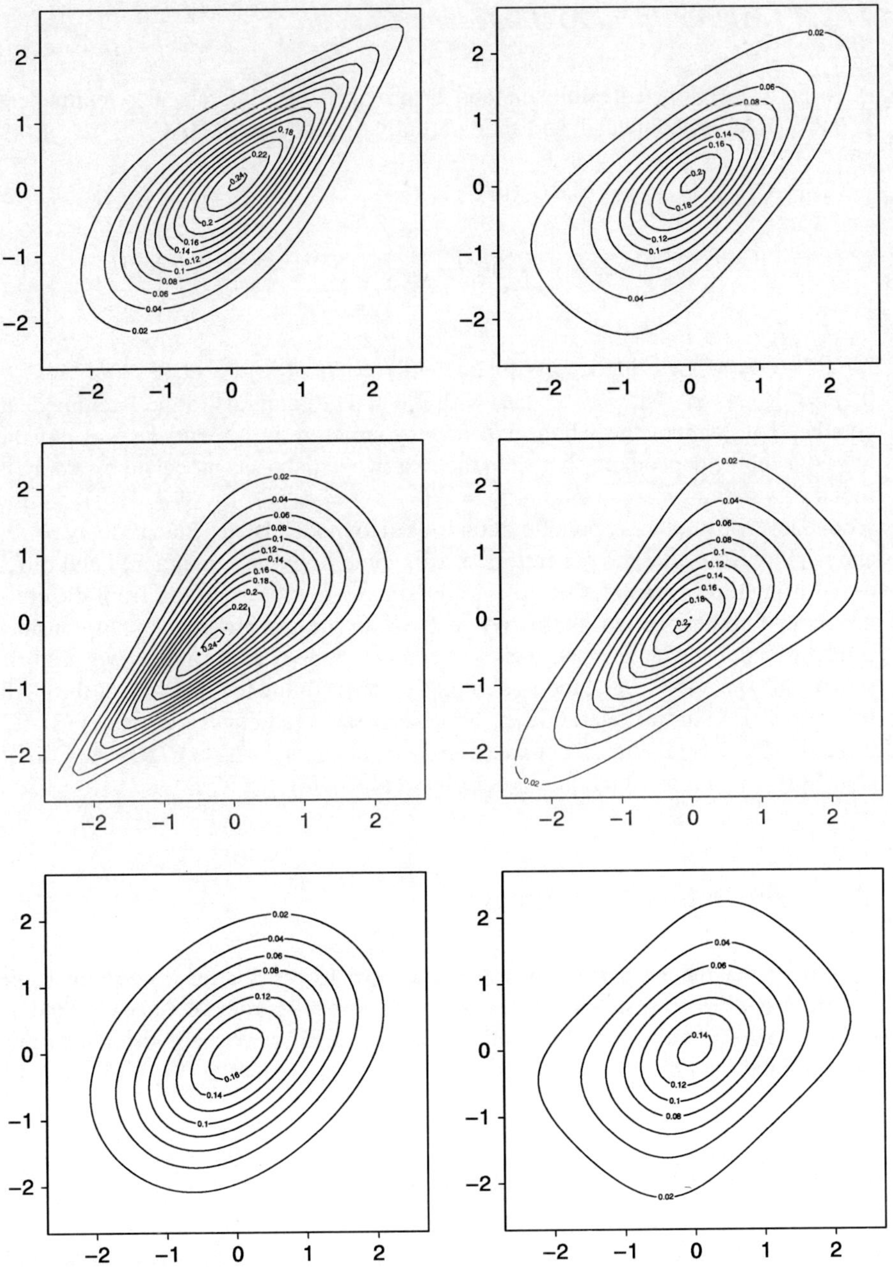

Fig. 17.4 Contour diagrams for (*from top to bottom*) Gumbel, Clayton and Frank copula with normal (*left column*) and t_3 distributed (*right column*) margins

17.3.4 Hierarchical Archimedean Copulae

A recently developed flexible method is provided by hierarchical Archimedean copulae (HAC). The special, so called partially nested, case of HAC:

$$C(u_1,\ldots,u_d) = C_0\{C_1(u_1,\ldots,u_{k_1}),\ldots,C_m(u_{k_{m-1}+1},\ldots,u_d)\} \tag{17.6}$$

$$= \phi_0\left[\sum_{p=1}^{m}\phi_0^{-1}\circ\phi_i\left\{\sum_{j=k_{p-1}+1}^{k_p}\phi_p^{-1}(u_j)\right\}\right]$$

for $\phi_0^{-1}\circ\phi_p \in \{w : [0;\infty) \to [0;\infty)|w(0) = 0;\, w(\infty) = \infty;\, (-1)^{j-1}w^{(j)} \geq 0;\, j = 1,\ldots,\infty\}$, $p = 1,\ldots,m$, with $k_0 = 1$. In contrast to the Archimedean copula, HAC defines the whole dependency structure in a recursive way. At the lowest level the dependency between the first two variables is modelled by a copula function with the generator ϕ_1, i.e. $z_1 = C(u_1,u_2) = \phi_1\{\phi_1^{-1}(u_1)+\phi_1^{-1}(u_2)\}$. At the second level an another copula function is used to model the dependency between z_1 and u_3, etc. Note, that the generators ϕ_i can come from the same family and differ only through the parameter or, to introduce more flexibility, come from different generator families. As an alternative to the fully nested model, we can consider copula functions, with arbitrarily chosen combinations at each copula level. Okhrin et al. (2008) provide several methodologies of determining the structure of the HAC from the data, Okhrin et al. (2009) provide necessary theoretical properties of HAC, there are also several empirical papers on the application HAC to CDO (see Choros et al. 2009) and to weather data (see Filler et al. 2010).

17.4 Simulation Techniques

To investigate the properties of some multivariate distributions, one needs the algorithms of the simulations because many of those properties are to be checked by Monte Carlo techniques. In this section we provide different methods of sampling from copula.

17.4.1 Conditional Inverse Method

The conditional inverse method is a general approach for the simulation of random variables from an arbitrary multivariate distribution. This method can be also used to simulate from copulae. The idea is to generate random variables recursively from the conditional distributions. To sample $U_1,\ldots,U_d$ from copula C we proceed with the following steps:

1. Sample $V_1, \dots, V_d$ from $U(0, 1)$.
2. $U_1 = V_1$.
3. $U_j = C_j^{-1}(V_j|U_1, \dots, U_{j-1})$ for $j = 2, \dots, d$ where the conditional distribution of U_j is given by

$$C_j(u_j|u_1, \dots, u_{j-1}) = P(U_j \le u_j|U_1 = u_1 \dots U_{j-1} = u_{j-1})$$
$$= \frac{\frac{\partial^{j-1} C_j(u_1,\dots,u_j)}{\partial u_1 \dots \partial u_{j-1}}}{\frac{\partial^{j-1} C_{j-1}(u_1,\dots,u_{j-1})}{\partial u_1 \dots \partial u_{j-1}}} \tag{17.7}$$

with $C_j = C(u_1, \dots, u_j, 1, \dots, 1) = C(u_1, \dots, u_j)$.

The approach is numerically expensive, due to high order derivatives of C and the calculation of the inverse of the conditional distribution function.

17.4.2 Marshall and Olkin (1988) Method

To simulate from Archimedean copulae a simpler method was introduced in Marshall and Olkin (1988). The idea of the method is based on the fact that Archimedean copulae are derived from Laplace transforms (17.4). Following Marshall and Olkin (1988) we proceed with the following three steps procedure:

1. Sample U from $M = \mathcal{LS}^{-1}(\phi)$.
2. Sample independent $(V_1, \dots, V_d) \sim U[0, 1]$.
3. $U_j = \phi\{-\ln(V_j)/U\}$ for $j = 1, \dots, d$.

This method works much faster than the classic conditional inverse technique. The drawback is that the distribution M can only be determined explicitly for a few generator functions ϕ. For example for Gumbel copula $M(\theta) = St(1/\theta, 1, [cos\{\pi/(2\theta)\}]^\theta)$ and for Clayton copula $M(\theta) = \Gamma(1/\theta, 1)$.

17.4.3 McNeil (2008) Method

Methods of simulation from the different HAC structures were proposed in McNeil (2008); this is an extension of the Marshall and Olkin (1988) method. Below is the algorithm for partially nested copulae (17.6):

1. Sample U from $M = LS^{-1}(\phi_0)$.
2. For $i = 1, \dots, m$ sample

$$V_{k_{p-1}+1}, \dots, V_{k_p} \text{ from } C[u_{k_{p-1}+1}, \dots, u_{k_p}; \exp\{-U\phi_0^{-1} \circ \phi_p(\cdot)\}]$$

using Marshall and Olkin (1988) method where

$$C[u_{k_{p-1}+1},\dots,u_{k_p};\exp\{-U\phi_0^{-1}\circ\phi_p(\cdot)\}]$$

is the simple Archimedean copula with the generator function given by $\exp\{-U\phi_0^{-1}\circ\phi_p(\cdot)\}$.

3. $(U_{k_{p-1}+1},\dots,U_{k_p})^\top = \phi_0[-\log\{(V_{k_{p-1}+1},\dots,V_{k_p})^\top\}/U],\ p=1,\dots,m.$

This method, however also has some drawbacks because the inverse Laplace transform of the composition of the generator function does not always have an explicit form. Nevertheless, McNeil (2008) provides a list of combinations, which enable this.

17.5 Estimation

For a given data-set one needs to find an appropriate model, and to estimate the parameter when the model is fixed. In this section we describe different methods of the estimation of the copula from the data. All methods are similar and are based on (17.2). Having the sample $X_{ij},\ i=1,\dots,n,\ j=1,\dots,d$ one needs to estimate the copula. To estimate the marginal distributions $\hat{F}_j(\cdot),\ j=1,\dots,d$ at least three possible methods are available. The most simple one is to use the empirical distribution function

$$\hat{F}_j(x)=\frac{1}{n+1}\sum_{i=1}^{n}\mathbf{I}\{X_{ij}\le x\}.$$

The change of the fraction before the sum from the classical $\frac{1}{n}$ to $\frac{1}{n+1}$ is made to bound the empirical distribution from 1; otherwise this causes problems in the maximum likelihood (ML) calculation. The inverse function of $\hat{F}_j(x)$ is then an empirical quantile. Instead of this simplest empirical estimation one can smooth the distribution function by using a kernel method, see Härdle and Linton (1994). Using kernel function $\kappa:\mathbb{R}\to\mathbb{R},\ \int\kappa=1$ with the bandwidth $h>0$ one gets following estimator

$$\tilde{F}_j(x)=\frac{1}{n+1}\sum_{i=1}^{n}K\left(\frac{x-X_{ij}}{h}\right),$$

with $K(x)=\int_{-\infty}^{x}\kappa(t)dt$. Apart from nonparametric methods, there is also a parametric method that is based on the assumption of a parametric form of the marginal distribution $F_j(x,\hat{\alpha}_j)$, where α_j is the parameter of the distribution, and $\hat{\alpha}_j$ is its estimator based on the ML method or method of moments. The last case considers the full knowledge of the true marginal distribution $F_j(x)$, which is rare in practice.

In the same way, there are four possible choices of the copula function. Let us first determine general margins $\breve{F}_j(x)$ that could be one of $\hat{F}_j(x)$, $\tilde{F}_j(x)$, $F_j(x,\hat{\alpha})$ or $F_j(x)$. The empirical copula is then defined as

$$\widehat{C}(u_1,\ldots,u_d) = \frac{1}{n}\sum_{i=1}^{n}\prod_{j=1}^{d}\mathbf{I}\{\breve{F}_j(X_{ij}) \le u_j\}. \tag{17.8}$$

Let $K_j,\ j=1,\ldots,d$ be the same symmetric kernel for each direction as in the estimation of marginal distributions, and let $h_j,\ j=1,\ldots,d$ be the set of bandwidths, then the kernel based copula estimation considered in Fermanian and Scaillet (2003) is

$$\widetilde{C}(u_1,\ldots,u_d) = \frac{1}{n}\sum_{i=1}^{n}\prod_{j=1}^{d}K_j\left\{\frac{u_j - \breve{F}_j(X_{ij})}{h_j}\right\}. \tag{17.9}$$

In the bivariate case ($d = 2$) to avoid boundary bias, one uses Chen and Huang (2007) local linear kernel to smooth at $u \in [0,1]$

$$K_{uh} = \frac{K(x)\{a_2(u,h) - a_1(u,h)x\}}{a_0(u,h)a_2(u,h) - a_1^2(u,h)},$$

with $a_\ell(u,h) = \int_{(u-1)/h}^{u/h} t^\ell K(t)dt,\ \ell = 0,1,2$ and $h > 0$ (see Lejeune and Sarda 1992 ; Jones 1993). Let $G_{uh}(t) = \int_{-\infty}^{t} K_{uh}(x)dx$ and $T_{uh} = G_{uh}\{(u-1)/h\}$, then an unbiased kernel based estimator of the bivariate copula is given by

$$\widetilde{C}(u_1,u_2) = \frac{1}{n}G_{u_1h}\left\{\frac{u_1 - \breve{F}_1(X_{i1})}{h}\right\} G_{u_2h}\left\{\frac{u_2 - \breve{F}_2(X_{i2})}{h}\right\} \\ -(u_1T_{u_2h} + u_2T_{u_1h} + T_{u_1h}T_{u_2h}). \tag{17.10}$$

The last situation is the parametric copula $C(\mathbf{u},\theta)$, where the copula comes from some fixed family. In this case the parameter of the copula function is estimated using the ML method. From (17.3) the likelihood function for the case $\breve{F}_j(x) = F_j(x,\alpha_j),\ j = 1,\ldots,d$ is

$$L(\theta,\alpha_1,\ldots,\alpha_d) = \prod_{i=1}^{n} f(X_{i1},\ldots,X_{id};\alpha_1,\ldots,\alpha_d,\theta)$$

and the log-likelihood function is given by

$$\ell(\theta, \alpha_1, \ldots, \alpha_d) = \sum_{i=1}^{n} \log c\{F_1(X_{i1}; \alpha_1), \ldots, F_d(X_{id}; \alpha_d); \theta\}$$
$$+ \sum_{i=1}^{n} \sum_{j=1}^{d} \log f_j(X_{ij}; \alpha_j),$$

where $f_j(\cdot)$ are marginal densities. All parameters $\{\theta, \alpha_1, \ldots, \alpha_d\}$ can be estimated in one or two steps. For practical applications, however, a two step estimation procedure is more efficient. A one step procedure, also called *full maximum likelihood*, is carried out by maximising likelihood function simultaneously over all parameters, thus by solving

$$(\partial\ell/\partial\alpha_1, \ldots, \partial\ell/\partial\alpha_d, \partial\ell/\partial\theta) = \mathbf{0},$$

with respect to $(\theta, \alpha_1, \ldots, \alpha_d)$. Following the standard theory on ML estimation estimators are efficient and asymptotically normal. However, it is often computationally demanding to solve the system simultaneously.

The two step procedure can be done for any kind of marginal distribution $\breve{F}_j(x) \in \{\widehat{F}_j(x), \widetilde{F}_j(x), F_j(x, \hat{\alpha})\}$. Firstly, we estimate the marginal distribution by using any of the above methods and secondly, we estimate the copula parameter by the pseudo log-likelihood function

$$\ell_p(\theta) = \sum_{i=1}^{n} \log c\{\breve{F}_1(X_{i1}), \ldots, \breve{F}_d(X_{id}); \theta\}.$$

The solution is then

$$\hat{\theta} = \arg\max_{\theta} \ell_p(\theta).$$

If the marginal distributions are from parametric families $\breve{F}_j(x) = F_j(x, \hat{\alpha}_j)$, $j = 1, \ldots, d$, then the method is called *inference for margins*. Otherwise, if margins, are nonparametrically estimated $\breve{F}_j(x) \in \{\widehat{F}_j(x), \widetilde{F}_j(x)\}$, $j = 1, \ldots, d$, then the method is called *canonical maximum likelihood* method.

17.6 Goodness-of-Fit (GoF) Tests

After the copula is estimated, one needs to test how well the estimated copula describes the sample. Nonparametric copula is certainly the best choice for this, and is usually considered the benchmark in many tests. With the GoF tests one checks whether the underlying copula belongs to any copula family. The test problem could be written as a composite null hypothesis

$$H_0: \ C \in \mathcal{C}_0, \quad \text{against} \quad H_1: \ C \notin \mathcal{C}_0,$$

where $\mathcal{C}_0 = \{C_\theta : \theta \in \Theta\}$ is a known parametric family of copulae. In some cases we restrict ourselves to the one element family $\mathcal{C}_0 = C_0$, thus the hypothesis in this case in the simple one. The test problem is, in general, equivalent to the GoF tests for multivariate distributions. However, since the margins are estimated we cannot apply the standard test procedures directly.

Here we consider several methodologies recently introduced in the literature. We can categorised them into three classes: tests based on the empirical copula, tests based on the Kendall's process and tests based on Rosenblatt's transform.

17.6.1 Tests Based on the Empirical Copula

These tests are based directly on the distance between C and C_0. Naturally, as C is unknown one takes the empirical copula which is fully nonparametric $\hat{C}$ or $\tilde{C}$ instead. The estimated copula C_0, that should be tested, is the parametric one $C(\cdot, \hat{\theta})$. Two statistics considered in the literature (see e.g Fermanian 2005; Genest and Rémillard 2008, etc.) are similar to Crámer-von Mises and Kolmogorov–Smirnov test statistics

$$S = n \int_{[0,1]^d} \{\widehat{C}(u_1, \dots, u_d) - C(u_1, \dots, u_d, \widehat{\theta})\}^2 \, d\widehat{C}(u_1, \dots, u_d),$$

$$T = \sup_{u_1,\dots,u_d \in [0,1]} \sqrt{n}|\widehat{C}(u_1, \dots, u_d) - C(u_1, \dots, u_d, \widehat{\theta})|.$$

Genest and Rémillard (2008) show the convergence of $\sqrt{n}\{\widehat{C}(u_1, \dots, u_d) - C(u_1, \dots, u_d, \widehat{\theta})\}$ in distribution, they also show that tests based on S and T are consistent. In actual fact, the p-values of the test statistics depends on this limiting distribution and in practice p-values are calculated using the bootstrap methods described in Genest and Rémillard (2008). This is quite expensive numerically, but leads to proper results.

17.6.2 Tests Based on Kendall's Process

Genest and Rivest (1993), Wang and Wells (2000) and Barbe et al. (1996) consider a test based on the true and empirical distributions of the pseudo random variable $V = C(U_1, \dots, U_d) \ \sim \ K$. The expectation of v is the transformation of the multivariate extension of Kendall's τ, hence the deviation of the true K and empirical $\hat{K}$ as a univariate function is called Kendall's process. The most natural empirical estimation of K is

$$\hat{K}(v) = \frac{1}{n}\sum_{i=1}^{n} \mathbf{I}\{V_i \leq v\}.$$

The theoretical form of the K was discussed in Barbe et al. (1996) and Okhrin et al. (2009) for different copula functions. In the bivariate case of the Archimedean copulae it is related to the generator function as

$$K(v,\theta) = v - \frac{\phi_\theta^{-1}(v)}{\{\phi_\theta^{-1}(v)\}'}.$$

As in the tests based on the empirical copulae Wang and Wells (2000) and Genest et al. (2006) propose to compute a Kolmogorov–Smirnov and Crámer-von-Mises statistics for the K

$$S_K = n\int_0^1 \{\hat{K}(v) - K(v,\theta)\}^2\,dv,$$

$$T_K = \sup_{v\in[0,1]} |\hat{K}(v) - K(v,\theta)|,$$

where $\hat{K}(v)$ and $K(v,\theta)$ are empirical and theoretical K-distributions of the variable $v = C(u_1,\dots,u_d)$. However, as in the previous tests, exact p-values for this statistic cannot be computed explicitly. Savu and Trede (2004) propose a χ^2-test based on the K-distribution. Unfortunately, in most cases the distribution of the test statistic does not follow a standard distribution and either a bootstrap or another computationally intensive methods should be used.

17.6.3 Tests Based on Rosenblatt's Process

An alternative global approach is based on the probability integral transform introduced in Rosenblatt (1952) and applied in Breymann et al. (2003), Chen et al. (2004) and Dobrić and Schmid (2007). The idea of the transformation is to construct the variables

$$\begin{aligned} Y_{i1} &= \check{F}_1(X_{i1}), \\ Y_{ij} &= C\{\check{F}_j(X_{ij})|\check{F}_1(X_{i1}),\dots,\check{F}_{j-1}(X_{i,j-1})\}, \quad \text{for} \quad j = 2,\dots,d \end{aligned} \tag{17.11}$$

where the conditional copula is defined in (17.7). Under H_0 the variables Y_{ij}, for $j = 1,\dots,d$ are independently and uniformly distributed on the interval [0, 1].

Here we discuss the second test based on Y_{ij} proposed in Chen et al. (2004). Consider the variable $W_i = \sum_{j=1}^{d}[\Phi^{-1}(Y_{ij})]^2$. Under H_0 it holds that $W_i \sim \chi^2_d$. Breymann et al. (2003) assume that estimating margins and copula parameters does not significantly affect the distribution of $\hat{W}_i$ and apply a standard χ^2 test directly to the pseudo-observations. Chen et al. (2004) developed a kernel-based test for the distribution of W and, thus, an account for estimation errors. Let $\tilde{g}_W(w)$ denote the kernel estimator of the density of W. Under H_0 the density $g_W(w)$ is equal to one, as the density of the uniform distribution. As a measure of divergency Chen et al. (2004) used $\hat{J}_n = \int_0^1 \{\tilde{g}_W(w) - 1\}^2 dw$. Assuming non-parametric estimator of the marginal distributions Chen et al. (2004) prove under regularity conditions that

$$T_n = (n\sqrt{h}\hat{J}_n - c_n)/\sigma \to N(0,1),$$

where the normalisation parameters h, c_n and σ are defined in Chen et al. (2004). The proof of this statement does not depend explicitly on the type of the non-parametric estimator of the marginals $\check{F}_j$, but uses the order of $\check{F}_j(X_{ij}) - F_j(X_{ij})$ as a function of n. It can be shown that if the parametric families of marginal distributions are correctly specified and their parameters are consistently estimated, then the statement also holds if we use parametric estimators for marginal distributions.

17.7 Simulation Study

A Monte Carlo experiment has been provided to discuss the finite sample properties of the goodness-of-fit tests based on the empirical copula and different estimation techniques on the simulated data. We restrict ourselves to the three dimensional case of three copula families, namely Gaussian, simple AC with Gumbel generator and HAC with Gumbel generator. For the simulation from the AC we use the Marshall and Olkin (1988) method and for simulation from HAC the McNeil (2008) method. To simulate from the Gaussian copula we simulate first from normal distribution and then apply the Sklar's theorem (1).

The main characteristic of interest in this study is to see whether the tests are able to maintain their nominal level fixed at $\alpha = 0.1$ and to see the power of the tests under the variety of alternatives. This is the only study that discusses the power of goodness-of-fit tests for copula in dimensions higher that $d = 2$. We consider all possible copulae with parameters $\tau \in \{0.25, 0.5, 0.75\}$. This means that under consideration were three AC: $C_{\theta(0.25)}(\cdot)$, $C_{\theta(0.5)}(\cdot)$, $C_{\theta(0.75)}(\cdot)$, three HAC: $C_{\theta(0.25)}\{C_{\theta(0.50)}(u_1,u_2),u_3\}$, $C_{\theta(0.25)}\{C_{\theta(0.75)}(u_1,u_2),u_3\}$, $C_{\theta(0.75)}\{C_{\theta(0.50)}(u_1,u_2),u_3\}$, and 15 Gaussian copulae with all possible positive definite correlation matrices containing values $\rho \in \{0.25, 0.5, 0.75\}$. Here $\theta(\tau)$ converts Kendall's τ correlation coefficient into a corresponding copula parameter.

The results are provided in Table 17.1 for AC, in Table 17.2 for HAC and in Table 17.3 for Gaussian copulae. To save the workspace we provide results for

Table 17.1 Non-rejection rate of the different models, where the sample is drawn from the simple AC

θ		AC							
		$n=50$				$n=150$			
		T		S		T		S	
		emp.	par.	emp.	par.	emp.	par.	emp.	par.
	HAC	0.88	0.51	0.83	0.38	0.93	0.36	0.90	0.35
$\theta(0.25)$	AC	0.88	0.51	0.89	0.50	0.95	0.32	0.90	0.34
	Gauss	0.71	0.29	0.56	0.22	0.69	0.11	0.43	0.08
	HAC	0.90	0.38	0.94	0.30	0.87	0.35	0.88	0.27
$\theta(0.5)$	AC	0.96	0.55	0.95	0.45	0.90	0.45	0.92	0.35
	Gauss	0.76	0.30	0.65	0.19	0.47	0.13	0.31	0.02
	HAC	0.93	0.29	0.93	0.15	0.89	0.27	0.89	0.10
$\theta(0.75)$	AC	0.93	0.29	0.93	0.22	0.90	0.25	0.91	0.13
	Gauss	0.77	0.19	0.65	0.10	0.57	0.11	0.24	0.05

Table 17.2 Non-rejection rate of the different models, where the sample is drawn from the HAC

$\boldsymbol{\theta}$		HAC							
		$n=50$				$n=150$			
		T		S		T		S	
		emp.	par.	emp.	par.	emp.	par.	emp.	par.
	HAC	0.88	0.29	0.90	0.24	0.96	0.31	0.92	0.26
$\theta(0.25, 0.5)$	AC	0.91	0.26	0.93	0.36	0.54	0.13	0.53	0.07
	Gauss	0.82	0.20	0.69	0.19	0.57	0.14	0.37	0.04
	HAC	0.93	0.21	0.92	0.13	0.88	0.18	0.88	0.09
$\theta(0.25, 0.75)$	AC	0.46	0.14	0.54	0.07	0.00	0.00	0.00	0.00
	Gauss	0.84	0.19	0.71	0.13	0.52	0.10	0.42	0.01
	HAC	0.86	0.31	0.87	0.18	0.91	0.20	0.94	0.08
$\theta(0.5, 0.75)$	AC	0.89	0.36	0.92	0.28	0.44	0.04	0.47	0.02
	Gauss	0.70	0.19	0.55	0.12	0.50	0.11	0.30	0.05

only 3 Gaussian copulae out of 15 with the largest difference between parameters. For HAC, a vector function $\boldsymbol{\theta}(\tau_1, \tau_2)$ converts two Kendall's τ into HAC copula parameters. If $\tau_1 < \tau_2$ then copula $C_{\theta(\tau_1)}\{C_{\theta(\tau_2)}(u_1, u_2), u_3\}$ is considered. For Gaussian copula

$$\boldsymbol{\Sigma}(\tau_1, \tau_2, \tau_3) = \begin{pmatrix} 1 & \tau_1 & \tau_2 \\ \tau_1 & 1 & \tau_3 \\ \tau_2 & \tau_3 & 1 \end{pmatrix}.$$

From each copula we simulate a sample of $n = 50$ or $n = 150$ observations with standard normal margins. The margins are then estimated parametrically (normal distribution with estimated mean and variance) or nonparametrically. Respective columns in the tables are marked by "par." and "emp.". For each sample we estimate the AC using inference for the margins method, HAC using Okhrin et al. (2008) and

Table 17.3 Non-rejection rate of the different models, where the sample is drawn from the Gaussian copula

		Gauss							
		$n = 50$				$n = 150$			
Σ		T		S		T		S	
		emp.	par.	emp.	par.	emp.	par.	emp.	par.
	HAC	0.89	0.20	0.93	0.11	0.78	0.08	0.81	0.02
$\Sigma(0.25, 0.25, 0.75)$	AC	0.43	0.13	0.47	0.09	0.00	0.00	0.00	0.00
	Gauss	0.88	0.22	0.89	0.12	0.87	0.11	0.86	0.03
	HAC	0.92	0.20	0.91	0.14	0.76	0.07	0.69	0.04
$\Sigma(0.25, 0.75, 0.25)$	AC	0.39	0.12	0.39	0.04	0.00	0.00	0.00	0.00
	Gauss	0.90	0.18	0.87	0.13	0.92	0.12	0.94	0.10
	HAC	0.89	0.30	0.93	0.16	0.78	0.10	0.75	0.04
$\Sigma(0.75, 0.25, 0.25)$	AC	0.51	0.16	0.46	0.07	0.00	0.00	0.00	0.00
	Gauss	0.91	0.28	0.90	0.17	0.88	0.13	0.86	0.06

the Gaussian copula using the generalised method of moments. Then we test how good these distributions fit the sample. The empirical copula for both tests has been calculated as in (17.8). Number of bootstrap steps provided for the tests is equal to $N = 1000$. To sum up the simulation procedure, we used:

1. F : two methods of estimation of margins (parametric and nonparametric).
2. C_0 : hypothesised copula models under H_0 (three models).
3. C : copula model from which the data were generated (three models with 3, 3 and 15 levels of dependence respectively).
4. n : size of each sample drawn from C (two possibilities, $n = 50$ and $n = 150$).

Thus, for all these $2\times3\times(3+3+15)\times2 = 252$ situations we perform 100 repetitions in order to calculate the power of both tests. This study is hardly comparable to other similar studies, because, as far as we know, this is the only one that considers the three dimensional case, and the only one that considers a hierarchical Archimedean copulae.

To understand the numbers in the tables more deeply let us consider first the value in Table 17.1. The number 0.88 says, that testing using Kolmogorov–Smirnov type statistic T_n for the AC with $\tau = 0.25$ from the sample of a size $n = 50$, with nonparametrically estimated margins, rejects the null hypotheses H_0, assuming that the data are from HAC, in $100\% - 88\% = 12\%$ of chances. It is very natural that the rejection rate for the AC, that have HAC under H_0, is very close to the case, where AC is under H_0. In general AC is a special case of HAC. If the true distribution is AC, the rejection rates should be equal, or close to each other, and the difference based only on the estimation error.

Figure 17.6 represents the level of both goodness-of-fit tests for different sizes in terms of three quartiles; the outliers are marked with closed dots. In general,

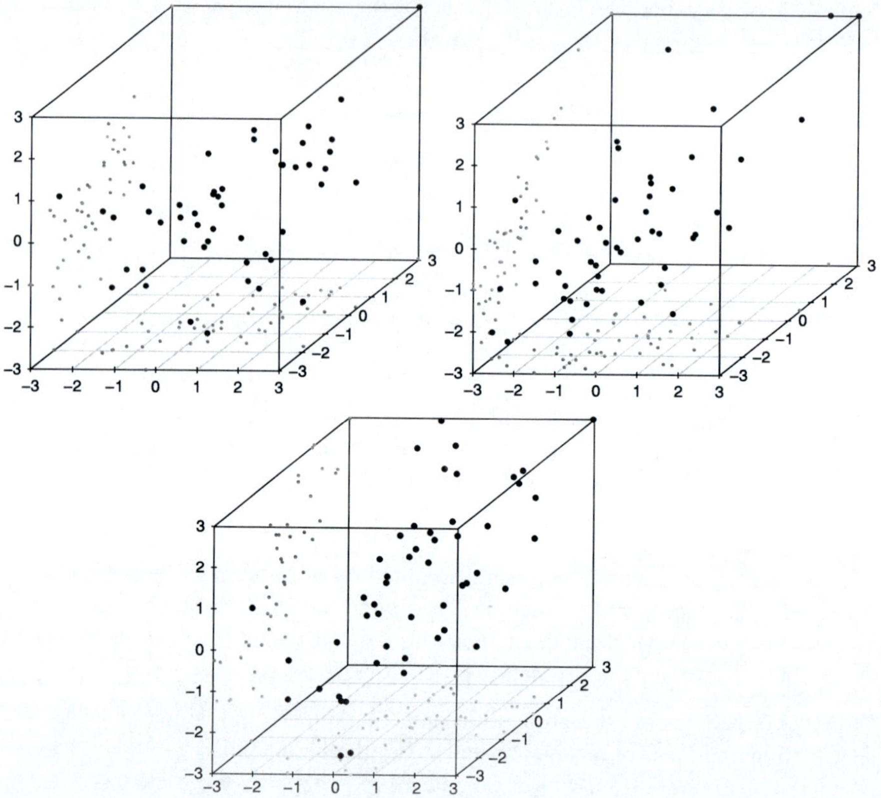

Fig. 17.5 Samples of size $n = 50$ from $C_{0.25}(\cdot)$, $C_{\theta(0.75)}\{C_{\theta(0.50)}(u_1, u_2), u_3\}$ and Gaussian copula with upper diagonal elements of the correlation matrix given by $\rho = (0.25, 0.25, 0.75)^\top$

values lies below 0.1, which implies that the bootstrap performs well. Increasing the number of runs improves this graph. We see that if the sample size has enlarged three times, then the tests have approximately doubled their power in S statistics, and a slightly smaller coefficient is given for the T statistics. In general, small size samples from different models look very similar (see Fig. 17.5), this makes detection of the model that best fits the data hardly applicable, this also explains a lot of outliers in Fig. 17.6.

From the tables we see, that S_n performs, on average, better than T_n statistics, this can be also seen from the Fig. 17.6. In the tables, rejection rates for S_n under false H_0 are in general higher, than for T_n statistics. We can also conclude that the larger the difference between parameters of the model is the faster AC is rejected. This can be expressed by the only parameter in AC that does not covers the whole dependency.

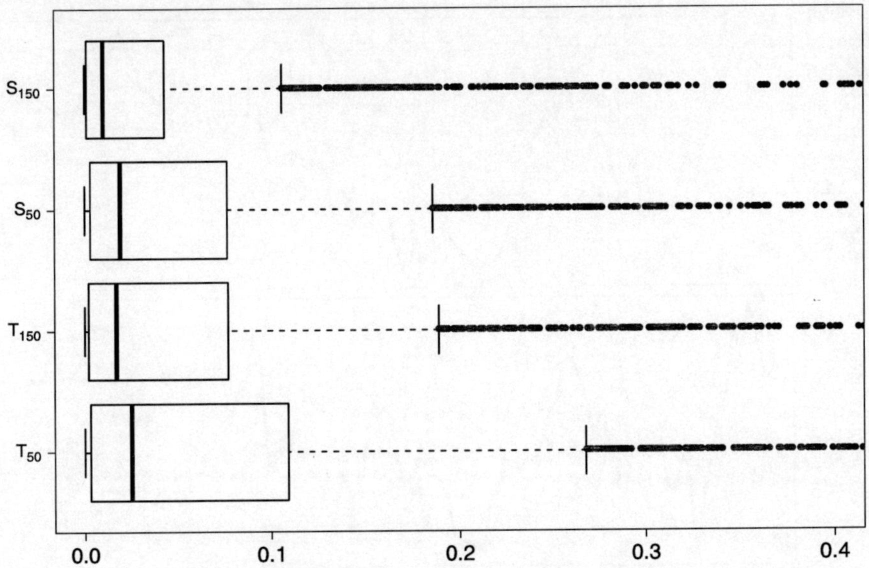

Fig. 17.6 Levels of goodness-of-fit tests for different sample size, for parametric margins

17.8 Empirical Results

The empirical part of this study is based on the calculation of the Value-at-Risk for the Profit and Loss function of the portfolio containing three assets. Asset returns follow some GARCH-type process with residuals from copula based models. We consider the daily stock prices of three American banks, namely Bank of America, Citigroup and Santander from 29.09.2000 to 16.02.2001. This results in $T = 100$ observations being consistent with the simulation study provided above. We take this time interval because several U.S. banks have recorded strong earnings in the fourth quarter of 2000. Rising profits were reported by U.S. industry leaders, namely Citigroup and Bank of America. At the same time bad forecasts for technology companies were reported; these influence the financial sector as well. Prices $\{X_{tj}\}$, $j = 1, 2, 3$ behave (over the chosen period) as in Fig. 17.7. Assuming the log-returns $R_{tj} = \log(X_{tj}/X_{t-1,j})$, $j = 1, 2, 3$, $t = 1, \ldots, T$ (see Fig. 17.8) follow an ARMA(1,1)-GARCH(1,1) process, we have

$$R_{tj} = \mu_j + \gamma_j R_{t-1,j} + \zeta_j \sigma_{t-1,j} \varepsilon_{t-1,j} + \sigma_{tj} \varepsilon_{tj},$$

where

$$\sigma_{tj}^2 = \omega_j + \alpha_j \sigma_{t-1,j}^2 + \beta_j \sigma_{t-1,j}^2 \varepsilon_{t-1,j}^2$$

and $\omega > 0, \alpha_j \geq 0, \beta_j \geq 0, \alpha_j + \beta_j < 1, |\zeta| < 1$.

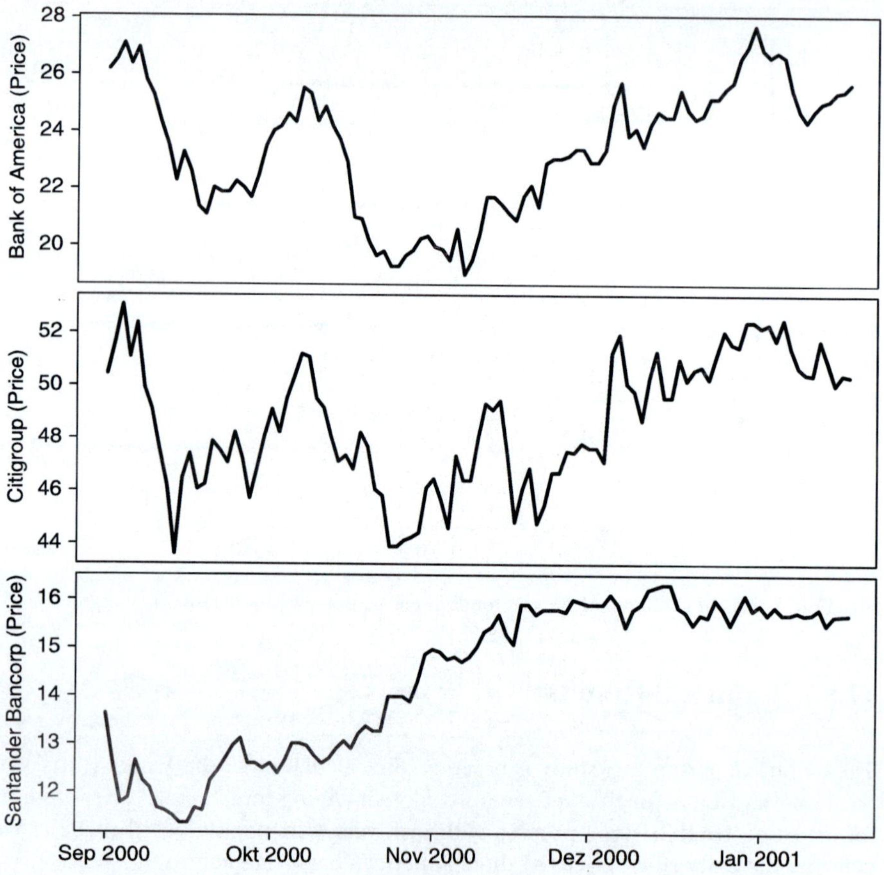

Fig. 17.7 Stock prices for Bank of America, Citigroup and Santander (*from top to bottom*)

The fit of an ARMA(1,1)-GARCH(1,1) model to the log returns $\mathbf{R}_t = (R_{t1}, R_{t2}, R_{t3})^\top$, $T = 100$, gives the estimates $\hat{\omega}_j, \hat{\alpha}_j, \hat{\beta}_j, \hat{\xi}_j$ and $\hat{\gamma}_j$, as in Table 17.4. Empirical residuals $\{\hat{\varepsilon}_t\}_{t=1}^T$, where $\hat{\varepsilon}_t = (\hat{\varepsilon}_{t1}, \hat{\varepsilon}_{t2}, \hat{\varepsilon}_{t3})^\top$ are assumed to be normally distributed; this is not rejected by the Kolmogorov–Smirnov test at the high level of significance for all three banks. Residuals are also assumed to be independent, because of the Box–Ljung autocorrelation test with lag 12. Thus, in the estimation of copula we use an inference for margins method, where margins are normal, thus, estimated parametrically.

Upper diagonal cells of Fig. 17.9 represent pair wise scatterplots of ARMA-GARCH residuals. In the lower diagonal cells of the same figure we show the scatterplots of the residuals mapped on the unit square by the estimated marginal cdf, $\hat{F}(\hat{\varepsilon})$.

We estimated three different models, namely simple AC, HAC and Gaussian copula. Afterwards two tests, used in the simulation study, were applied to see how

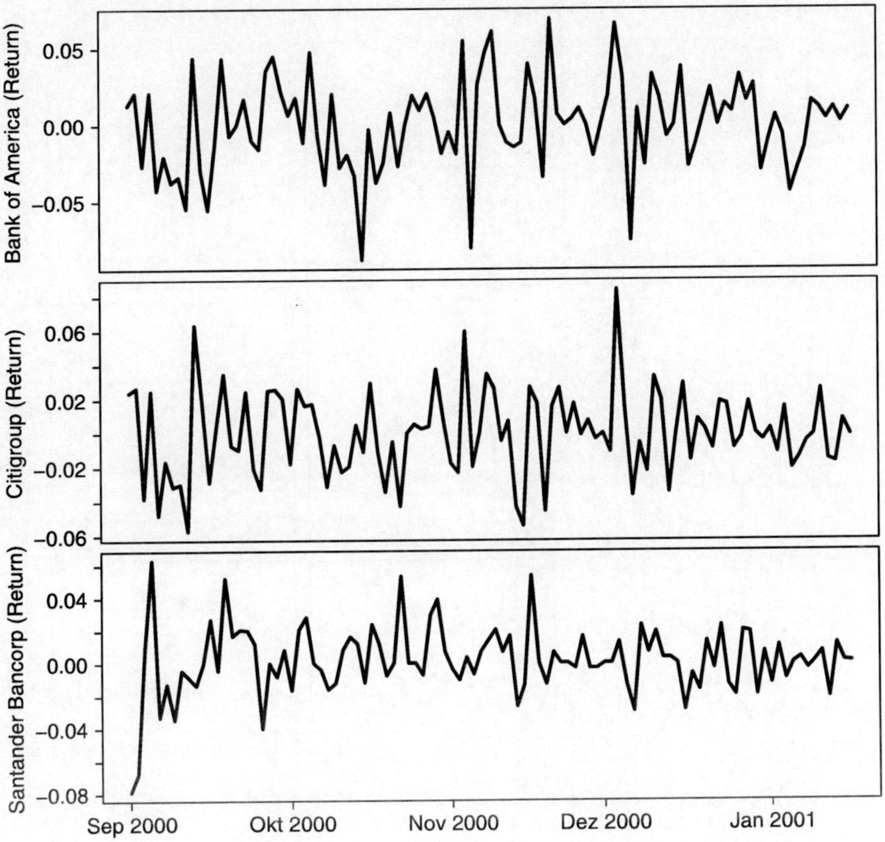

Fig. 17.8 Log-returns for Bank of America, Citigroup and Santander (*from top to bottom*)

Table 17.4 Fitting of univariate ARMA(1,1)-GARCH(1,1) to asset returns. The standard deviation of the parameters, which are quiet big because of the small sample size, are given in parentheses. The last two columns provide the p-values of the Box–Ljung test (BL) for autocorrelations and Kolmogorov–Smirnov test (KS) for testing of normality of the residuals

	$\hat{\mu}_j$	$\hat{\gamma}_j$	$\hat{\zeta}_j$	$\hat{\omega}_j$	$\hat{\alpha}_j$	$\hat{\beta}_j$	BL	KS
Bank of	1.879e−03	0.226	−0.232	3.465e−04	0.551	0.170	0.567	0.829
America	(2.598e−03)	(0.642)	(0.654)	(1.369e−04)	(0.284)	(0.155)		
Citigroup	0.116e−03	0.305	−0.455	2.669e−04	0.096	0.471	0.569	0.786
	(1.487e−03)	(0.296)	(0.288)	(5.533e−04)	(0.165)	(1.008)		
Santander	1.359e−03	0.430	−0.566	4.512e−10	0.012	0.979	0.914	0.781
	(0.908e−03)	(0.149)	(0.174)	(1.376e−05)	(0.018)	(0.049)		

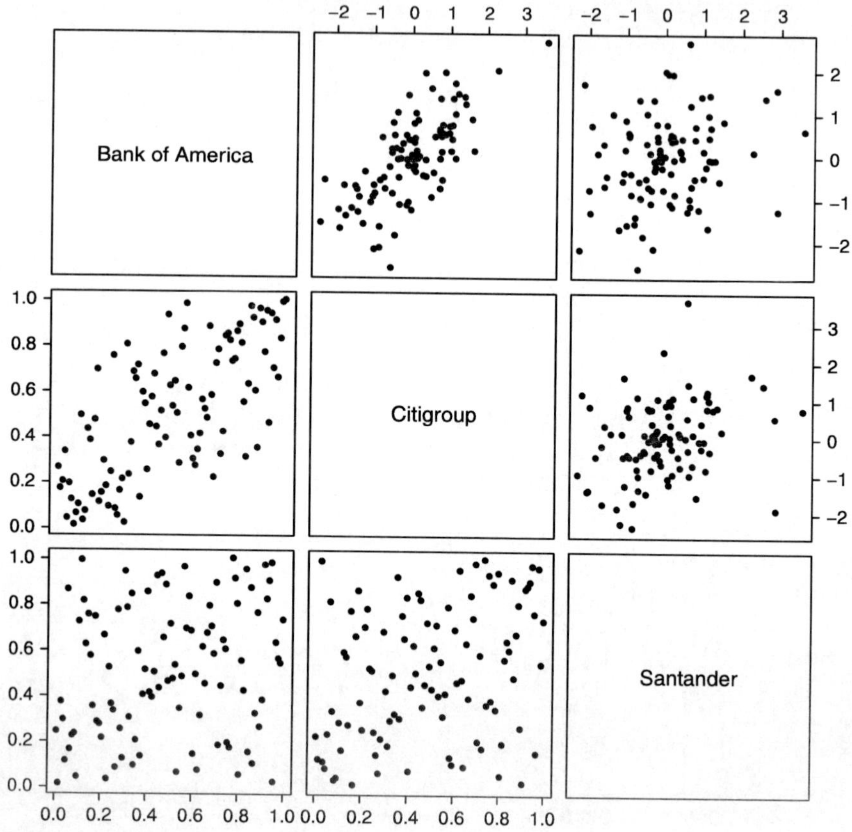

Fig. 17.9 Scatterplots from ARMA-GARCH residuals (*upper triangular*) and from residuals mapped on unit square by the cdf (*lower triangular*)

good these models describe data. In this case the number of bootstrap runs has been increased to $N = 10000$ to make the test results more precise. Estimated models, and p-values are represented in Table 17.5. We see that parameters in the HAC model deviate from each other, we may conclude therefore, that a simple AC is not a proper model that fits the data. On the other hand, from Fig. 17.9 we see that the points are not elliptical; this convinces us to expect a low p-value of the test where the Gaussian copula is under H_0. In the first two columns of Table 17.5 we put p-values for all tests. We conclude that HAC is the most appropriate model for this particular dataset, because it has the largest p-value. Based on two tests only HAC can not be rejected under significance level $\alpha = 0.05$. This means that our data may not be described by the simple three-dimensional normal distribution, but the margins are still normal.

To see if knowledge of preferable distribution is worth knowing in a financial problem, we estimate the Vale-at-Risk from a Profit and Loss of a linear portfolio

Table 17.5 p-values of both GoFs and estimates of the models under different H_0 hypotheses

	T_{100}	S_{100}	Estimates
HAC	0.3191	0.1237	$C\{C(u_1, u_2; 1.996), u_3; 1.256\}$
AC	0.0012	0.0002	$C(u_1, u_2, u_3; 1.276)$
Gauss	0.0160	0.0078	$C_N\{u_1, u_2, u_3; \Sigma(0.697, 0.215, 0.312)\}$

using copulae. The portfolio is composed of the stocks discussed above. We also perform an evaluation of the estimators through backtesting. Let w be the portfolio, which is represented by the number of assets for a specified stock in the portfolio, $w = \{w_1, \dots, w_d\}$, $w_i \in \mathbb{Z}$. The value V_t of the portfolio w is given by

$$V_t = \sum_{j=1}^{d} w_j X_{tj} \tag{17.12}$$

and the random variable defined as the absolute change in the portfolio

$$L_{t+1} = (V_{t+1} - V_t) = \sum_{j=1}^{d} w_j X_{tj} \left\{\exp(R_{t+1,j}) - 1\right\} \tag{17.13}$$

also called profit and loss (P&L) function, expresses the absolute change in the portfolio value in one period. The distribution function of L, dropping the time index, is given by

$$F_L(x) = P(L \leq x). \tag{17.14}$$

As usual the Value-at-Risk at level α from a portfolio w is defined as the α-quantile from F_L:

$$\mathrm{VaR}(\alpha) = F_L^{-1}(\alpha). \tag{17.15}$$

It follows from (17.14) that F_L depends on the d-dimensional distribution of log-returns F_X. In general, the loss distribution F_L depends on a random process representing the risk factors influencing the P&L from a portfolio. In the present case log-returns modelled by an ARMA(1,1)-GARCH(1,1) model are a suitable risk factor choice. Thus, modelling their distribution is essential to obtain the quantiles from F_L. To estimate the VaR we simulate samples of residuals ε_t from HAC, AC and Gaussian copula with normal margins, then apply simulated residuals to the estimated ARMA(1,1)-GARCH(1,1) model and calculate it based on the values of the Profit and Loss $\hat{L}$ with $w = (1, 1, 1)^\top$. The $\widehat{VaR}(\alpha)$ is then an empirical α-quantile from the $\hat{L}$. In Fig. 17.10 we represent the series of estimated Value-at-Risk with $\alpha = 0.1$ and the P&L function. Afterwards backtesting is used to evaluate

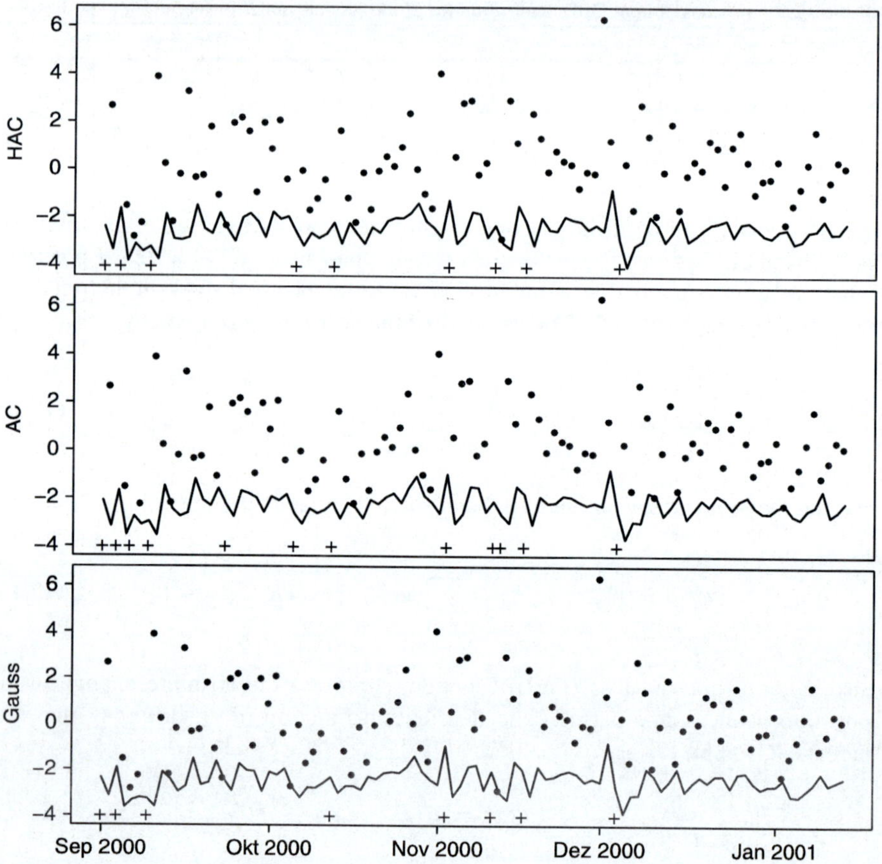

Fig. 17.10 $\widehat{VaR}(\alpha)$, P&L (*dots*) and exceedances (*crosses*), estimated with 3-dimensional HAC with Gumbel generator (*top*), simple Gumbel copula (*middle*) and Gaussian copula (*bottom*) with $\alpha = 0.1$

the performance of the specified copula family $\mathcal{C}$. The estimated values for the VaR are compared with the true realisations $\{L_t\}$ of the P&L function, an exceedance occuring for each L_t smaller than $\widehat{VaR}_t(\alpha)$. The ratio of the number of exceedances to the number of observations gives the exceedances ratio $\hat{\alpha}$:

$$\hat{\alpha} = \frac{1}{T}\sum_{t=1}^{T} \mathbf{I}\{L_t < \widehat{VaR}_t(\alpha)\}.$$

The backtesting results are provided in Table 17.6. From them we see that the Gaussian copula usually underestimates the VaR. This is natural because this copula does not have nor upper nor a lower tail dependence. The simple Archimedean copula overestimates the VaR. Results provided by HAC are the closest to the true

Table 17.6 Backtesting for the estimation of VaR under different alternatives

α	$\hat{\alpha}_{HAC}$	$\hat{\alpha}_{AC}$	$\hat{\alpha}_{Gauss}$
0.10	0.091	0.122	0.081
0.05	0.040	0.061	0.031
0.01	0.000	0.010	0.000

ones, but this copula underestimates the true VaR in all levels of significance. This is also natural because Gumbel copula describes wins rather than losses best. In general these results were expected due to the fact, that HAC is the only copula that was accepted by both tests under a high level of significance.

17.9 Conclusions

In this chapter we gave a short survey on copulae. We discussed different copula classes, methods of simulation and estimation and several goodness-of-fit tests. We provided an extensive simulation study in which two goodness-of-fit tests and two estimation techniques were considered. Afterwards, copulae were applied to de-GARCHed real world time-series. From the empirical study we conclude that, in some cases, even if margins are normal, the dependency is certainly not linearly normal, and more flexible dependency models are asked for.

Acknowledgements The financial support from the Deutsche Forschungsgemeinschaft through SFB 649 "Ökonomisches Risiko", Humboldt-Universität zu Berlin is gratefully acknowledged.

References

Barbe, P., Genest, C., Ghoudi, K., & Rémillard, B. (1996). On Kendalls's process. *Journal of Multivariate Analysis*, *58*, 197–229.

Breymann, W., Dias, A., & Embrechts, P. (2003). Dependence structures for multivariate high-frequency data in finance. *Quantitative Finance*, *1*, 1–14.

Chen, S. X., & Huang, T. (2007). Nonparametric estimation of copula functions for dependence modeling. *The Canadian Journal of Statistics*, *35*, 265–282.

Chen, X., & Fan, Y. (2005). Pseudo-likelihood ratio tests for model selection in semiparametric multivariate copula models. *The Canadian Journal of Statistics*, *33*(2), 389–414.

Chen, X., & Fan, Y. (2006). Estimation and model selection of semiparametric copula-based multivariate dynamic models under copula misspesification. *Journal of Econometrics*, *135*, 125–154.

Chen, X., Fan, Y., & Patton, A. (2004). Simple tests for models of dependence between multiple financial time series, with applications to U.S. equity returns and exchange rates. Discussion paper 483, Financial Markets Group, London School of Economics.

Chen, X., Fan, Y., & Tsyrennikov, V. (2006). Efficient estimation of semiparametric multivariate copula models. *Journal of the American Statistical Association*, *101*(475), 1228–1240.

Choros, B., Härdle, W., & Okhrin, O. (2009). CDO and HAC. SFB 649 Discussion Paper 2009-038, Sonderforschungsbereich 649, Humboldt Universität zu Berlin, Germany. Available at http://sfb649.wiwi.hu-berlin.de/papers/pdf/SFB649DP2009-038.pdf.

Clayton, D. G. (1978). A model for association in bivariate life tables and its application in epidemiological studies of familial tendency in chronic disease incidence. *Biometrika, 65*, 141–151.

Dall'Aglio, G. (1972). Fréchet classes and compatibility of distribution functions. *Symp. Math., 9*, 131–150.

Dobrić, J., & Schmid, F. (2007). A goodness of fit test for copulas based on Rosenblatt's transformation. *Computational Statistics and Data Analysis, 51*, 4633 – 4642.

Embrechts, P., McNeil, A. J., & Straumann, D. (1999). Correlation and dependence in risk management: Properties and pitfalls. *RISK, 12*(5), 69–71.

Fama, E. F. (1965). The behavior of stock market prices. *Journal of Business, 38*(1), 34–105.

Fermanian, J.-D. (2005). Goodness-of-fit tests for copulas. *Journal of Multivariate Analysis, 95*(1), 119–152.

Fermanian, J.-D., & Scaillet, O. (2003). Nonparametric estimation of copulas for time series. *Journal of Risk, 5*, 25–54.

Filler, G., Odening, M., Okhrin, O., & Xu, W. (2010). On the systemic nature of weather risk. *Agricultural Finance Review. 70*(2), 267–284.

Frank, M. J. (1979). On the simultaneous associativity of $f(x, y)$ and $x+y-f(x, y)$. *Aequationes Mathematicae, 19*, 194–226.

Fréchet, M. (1951). Sur les tableaux de corrélation dont les marges sont donnés. *Annales de l 'Université de Lyon, 4*, 53–84.

Genest, C., Ghoudi, K., & Rivest, L.-P. (1995). A semi-parametric estimation procedure of dependence parameters in multivariate families of distributions. *Biometrika, 82*, 543–552.

Genest, C., Quessy, J.-F., & Rémillard, B. (2006). Goodness-of-fit procedures for copula models based on the probability integral transformation. *Scandinavian Journal of Statistics, 33*, 337–366.

Genest, C., & Rémillard, B. (2008). Validity of the parametric bootstrap for goodness-of-fit testing in semiparametric models. *Annales de l'Institut Henri Poincaré: Probabilités et Statistiques, 44*, 1096–1127.

Genest, C., & Rivest, L.-P. (1989). A characterization of Gumbel family of extreme value distributions. *Statistics and Probability Letters, 8*, 207–211.

Genest, C., & Rivest, L.-P. (1993). Statistical inference procedures for bivariate Archimedean copulas. *Journal of the American Statistical Association, 88*, 1034–1043.

Giacomini, E., Härdle, W. K., & Spokoiny, V. (2009). Inhomogeneous dependence modeling with time-varying copulae. *Journal of Business and Economic Statistics, 27*(2), 224–234.

Gumbel, E. J. (1960). Distributions des valeurs extrêmes en plusieurs dimensions. *Publ. Inst. Statist. Univ. Paris, 9*, 171–173.

Härdle, W., Okhrin, O., & Okhrin, Y. (2010). Time varying hierarchical Archimedean copulae. *submitted for publication.*

Härdle, W., & Simar, L. (2007). *Applied Multivariate Statistical Analysis* (2 ed.). Heidelberg: Springer.

Härdle, W. K., & Linton, O. (1994). Applied nonparametric methods. In R. Engle, & D. McFadden (Eds.) *Handbook of Econometrics*. North-Holland: Elsevier.

Hennessy, D. A., & Lapan, H. E. (2002). The use of Archimedean copulas to model portfolio allocations. *Mathematical Finance, 12*, 143–154.

Hoeffding, W. (1940). Masstabinvariante Korrelationstheorie. *Schriften des Mathematischen Instituts und des Instituts für Angewandte Mathematik der Universität Berlin, 5*(3), 179–233.

Hoeffding, W. (1941). Masstabinvariante Korrelationsmasse für diskontinuierliche Verteilungen. *Arkiv für matematischen Wirtschaften und Sozial forschung, 7*, 49–70.

Joe, H. (1997). *Multivariate Models and Dependence Concepts*. London: Chapman & Hall.

Joe, H. (2005). Asymptotic efficiency of the two-stage estimation method for copula-based models. *Journal of Multivariate Analisys, 94*, 401–419.

Jones, M. C. (1993). Simple boundary corrections for kernel density estimation. *Statistic Computing*, *3*, 135–146.

Junker, M., & May, A. (2005). Measurement of aggregate risk with copulas. *Econometrics Journal*, *8*, 428–454.

Lee, T.-H., & Long, X. (2009). Copula-based multivariate garch model with uncorrelated dependent errors. *Journal of Econometrics*, *150*(2), 207–218.

Lejeune, M., & Sarda, P. (1992). Smooth estimators of distribution and density functions. *Computational Statistics & Data Analysis*, *14*, 457–471.

Li, D. X. (2000). On default correlation: A copula function approach. *The Journal of Fixed Income*, *6*, 43–54.

Mandelbrot, B. (1965). The variation of certain speculative prices. *Journal of Business*, *36*(4), 34–105.

Marshall, A. W., & Olkin, J. (1988). Families of multivariate distributions. *Journal of the American Statistical Association*, *83*, 834–841.

McNeil, A. J. (2008). Sampling nested Archimedean copulas. *Journal Statistical Computation and Simulation*, *78*(6), 567–581.

Nelsen, R. B. (2006). *An Introduction to Copulas*. New York: Springer.

Okhrin, O., Okhrin, Y., & Schmid, W. (2008). On the structure and estimation of hierarchical Archimedean copulas. *under revision*.

Okhrin, O., Okhrin, Y., & Schmid, W. (2009). Properties of Hierarchical Archimedean Copulas. SFB 649 Discussion Paper 2009-014, Sonderforschungsbereich 649, Humboldt-Universität zu Berlin, Germany. Available at http://sfb649.wiwi.hu-berlin.de/papers/pdf/SFB649DP2009-014.pdf.

Patton, A. J. (2004). On the out-of-sample importance of skewness and asymmetric dependence for asset allocation. *Journal of Financial Econometrics*, *2*, 130–168.

Renyi, A. (1970). *Probability Theory*. Amsterdam: North-Holland.

Rosenblatt, M. (1952). Remarks on a multivariate transformation. *The Annals of Mathematical Statistics*, *23*, 470–472.

Savu, C., & Trede, M. (2004). Goodness-of-fit tests for parametric families of Archimedean copulas. Discussion paper, University of Muenster.

Sklar, A. (1959). Fonctions dé repartition á n dimension et leurs marges. *Publ. Inst. Stat. Univ. Paris*, *8*, 299–231.

Wang, W., & Wells, M. (2000). Model selection and semiparametric inference for bivariate failure-time data. *Journal of the American Statistical Association*, *95*, 62–76.

Whelan, N. (2004). Sampling from Archimedean copulas. *Quantitative Finance*, *4*, 339–352.

Chapter 18
Numerical Methods for Nonlinear PDEs in Finance

Peter A. Forsyth and Kenneth R. Vetzal

Abstract Several examples of nonlinear Hamilton Jacobi Bellman (HJB) partial differential equations are given which arise in financial applications. The concept of a visocisity solution is introduced. Sufficient conditions which ensure that a numerical scheme converges to the viscosity solution are discussed. Numerical examples based on an uncertain volatility model are presented which show that seemingly reasonable discretization methods (which do not satisfy the sufficient conditions for convergence) fail to converge to the viscosity solution.

18.1 Introduction

Many problems in finance can be posed in terms of an optimal stochastic control. Some well-known examples include transaction cost/uncertain volatility models (Leland 1985; Avellaneda et al. 1995; Pooley et al. 2003), passport options (Andersen et al. 1998; Shreve and Vecer 2000), unequal borrowing/lending costs in option pricing (Bergman 1995), risk control in reinsurance (Mnif and Sulem 2001), optimal withdrawals in variable annuities (Dai et al. 2008), optimal execution of trades (Lorenz and Almgren 2007; Lorenz 2008), and asset allocation (Zhou and Li 2000; Li and Ng 2000). A recent survey on the theoretical aspects of this topic is given in Pham (2005).

These optimal stochastic control problems can be formulated as nonlinear Hamilton-Jacobi-Bellman (HJB) partial differential equations (PDEs). In general,

P.A. Forsyth (✉)
Cheriton School of Computer Science, University of Waterloo, Waterloo, ON, Canada N2L 3G1
e-mail: paforsyt@uwaterloo.ca

K.R. Vetzal
School of Accounting and Finance, University of Waterloo, Waterloo, ON, Canada N2L 3G1
e-mail: kvetzal@uwaterloo.ca

J.-C. Duan et al. (eds.), *Handbook of Computational Finance*, Springer Handbooks of Computational Statistics, DOI 10.1007/978-3-642-17254-0_18,

especially in realistic situations where the controls are constrained (e.g. in the case of asset allocation, we may require that trading must cease upon insolvency, that short positions are not allowed, or that position limits are imposed), there are no analytical solutions to the HJB PDEs. At first glance, it would appear to be a formidable task to develop a numerical method for solving such complex PDEs. In addition, there may be no smooth classical solutions to the HJB equations. In this case, we must seek the viscosity solution (Crandall et al. 1992) of these equations.

However, using the powerful theory developed in Barles and Souganidis (1991), Barles et al. (1995) and Barles (1997) we can devise a general approach for numerically solving these HJB PDEs. This approach ensures convergence to the viscosity solution.

The contributions of this article are as follows:

- We discuss several examples of optimal stochastic control in finance.
- We give an intuitive description of the concept of a viscosity solution.
- We present a general approach for discretizing the HJB PDEs. This technique ensures that the discrete solutions converge to the viscosity solution (Barles and Souganidis 1991; Barles et al. 1995; Barles 1997). The method uses fully implicit time stepping. Consequently, there are no time step restrictions due to stability considerations, an advantage over the Markov chain approach (Kushner and Dupuis 1991).
- We also discuss some techniques for the solution of the nonlinear discretized algebraic equations and an important property of the discrete solutions (i.e. preservation of arbitrage inequalities).
- Finally, we present a numerical example, illustrating that seemingly reasonable discretization methods, which do not satisfy the conditions in Barles and Souganidis (1991) can converge to incorrect (i.e. non-viscosity) solutions, and even solutions which embed arbitrage opportunities.

18.2 Examples

18.2.1 Uncertain Volatility

Let $V(S,t)$ be the value of a contingent claim written on an asset which has a price S that evolves according to the stochastic process

$$dS = \mu S \, dt + \sigma S \, dZ, \tag{18.1}$$

where μ is the drift rate, σ is volatility, and dZ is the increment of a Wiener process. There are a number of situations where $V(S,t)$ must be determined by solving an optimal control problem.

Consider for example, the uncertain volatility model developed in Avellaneda et al. (1995) and Lyons (1995). This provides a pricing mechanism for cases where volatility is uncertain, but lies within a band, $\sigma \in [\sigma_{\min}, \sigma_{\max}]$. In this case, the PDE which is used to determine the value of a contingent claim is determined by the two extremal volatilities. Let the expiry time of the claim be T, and let $\tau = T - t$. For a short position the optimal control problem is given by

$$V_\tau = \sup_{Q \in \hat{Q}} \left\{ \frac{Q^2 S^2}{2} V_{SS} + S V_S - rV \right\} = 0, \tag{18.2}$$

where $\hat{Q} = \{\sigma_{\min}, \sigma_{\max}\}$ and r is the borrowing/lending rate. Replacing the sup by an inf gives the corresponding pricing equation for a long position. It should also be pointed out that a PDE of precisely the same form as (18.2) arises in the completely different context of option valuation under transaction costs (Leland 1985).

18.2.2 Continuous Time Mean-Variance Asset Allocation

We suppose that an investor may divide his wealth W into a fraction p in a risky asset, the price of which follows process (18.1), and a fraction $(1 - p)$ in a risk-free bond, the value of which follows

$$\frac{dB}{dt} = rB, \tag{18.3}$$

where r is the risk-free rate. If α is the number of units of S owned, then $W = \alpha S + B$, and the process followed by W is

$$dW = [p\mu + (1-p)r]\, W\, dt + p\sigma W\, dZ. \tag{18.4}$$

We suppose that the investor follows an asset allocation strategy $p(t)$ for time $t \in [0, T]$. If W_T is the wealth at the terminal time T, then the optimal strategy may be posed as finding the $p(t)$ that maximizes the expected return less a penalty for risk (as measured by variance), i.e.

$$\sup_{p(t) \in z} \left\{ E^{t=0}[W_T] - \lambda \mathrm{var}^{t=0}[W_T] \right\}, \tag{18.5}$$

where

$$\begin{aligned} E^{t=0}[\cdot] &\text{ is the expectation as seen at } t = 0 \\ \mathrm{var}^{t=0}[\cdot] &\text{ is the variance as seen at } t = 0 \\ z &\text{ is the set of admissible controls, and} \\ \lambda &\text{ is the risk aversion parameter.} \end{aligned}$$

Varying λ allows us to generate a set of points $\left(\sqrt{var^{t=0}[W_T]}, E^{t=0}[W_T]\right)$ on the mean-variance efficient frontier.

Problem (18.5) is the *pre-commitment* version of the mean-variance trade-off (Basak and Chabakauri 2007). There is no direct dynamic programming formulation of problem (18.5). However, we can solve a different problem which has the same optimal control $p(t)$ and which is easier to solve.

We would like to use dynamic programming to determine the efficient frontier, given by (18.5). However, the presence of the variance term causes some difficulty. This can be avoided with the help of the results in Li and Ng (2000) and Zhou and Li (2000):

Theorem 1 (Equivalent Linear Quadratic (LQ) problem). *If $p^*(t)$ is the optimal control of problem (18.5), then $p^*(t)$ is also the optimal control of problem*

$$\sup_{p(t)\in z} \left\{E^{t=0}[\mu W_T - \lambda W_T^2]\right\}, \tag{18.6}$$

where

$$\mu = 1 + 2\lambda E_{p^*}^{t=0}[W_T], \tag{18.7}$$

with p^* being the optimal control of problem (18.6).

The notation $E_{p^*}^{t=0}[\cdot]$ refers to the expected value given the strategy $p^*(t)$. This result seems at first sight to be not very useful, since the parameter μ is a function of the optimal control p^*, which is not known until the problem is solved. However, we can write (18.6) in the form

$$-\lambda \inf_{p(t)\in z} E^{t=0}[W_T^2 - \gamma W_T] \tag{18.8}$$

with $\gamma = \mu/\lambda$, since $\lambda > 0$. Consequently, for fixed γ, an optimal control of problem (18.8) is an optimal control of

$$\inf_{p(t)\in z} \left\{E^{t=0}\left[(W_T - \frac{\gamma}{2})^2\right]\right\}. \tag{18.9}$$

As a result, for fixed γ, we can determine the optimal control $p(t)$ of problem (18.5) as follows. Let

$$V(W,\tau) = E^{T-\tau}\left[(W_T - \gamma)^2\right]. \tag{18.10}$$

Then, V is given from the solution to

$$V_\tau = \inf_{p\in z} \{(p\mu + (1-p)r)W V_W + (p\sigma)^2 W^2 V_{WW}\} \tag{18.11}$$

$$V(W, \tau = 0) = (W - \gamma/2)^2. \tag{18.12}$$

Having solved (18.12), we then have the optimal control $p^*(W, t)$. This can be used to determine a pair $\left(\sqrt{\text{var}^{t=0}[W_T]}, E^{t=0}[W_T]\right)$. Varying γ allows us to trace out an efficient frontier.

18.2.3 Guaranteed Minimum Withdrawal Benefit Variable Annuity

Guaranteed Minimum Withdrawal Benefit (GMWB) variable annuities are discussed at length in Milevsky and Salisbury (2006), Dai et al. (2008) and Chen and Forsyth (2008). We briefly review the final equations here. Let $W \equiv W(t)$ be the stochastic process of the personal variable annuity account and $A \equiv A(t)$ be the stochastic process of the account balance of the guarantee. We assume that the reference portfolio $S \equiv S(t)$, which underlies the variable annuity policy before the deduction of any proportional fees, follows a geometric Brownian motion under the risk-neutral measure with a volatility of σ and a risk-free interest rate of r:

$$dS = rS\,dt + \sigma S\,dZ. \tag{18.13}$$

The major feature of the GMWB is the guarantee on the return of the entire premium via withdrawal. The insurance company charges the policy holder a proportional annual insurance fee η for this guarantee. Therefore we have the following stochastic differential equation for W:

$$dW = \begin{cases} (r - \eta)W dt + \sigma W dZ + dA & \text{if } W > 0, \\ 0 & \text{if } W = 0. \end{cases} \tag{18.14}$$

Let $\gamma \equiv \gamma(t)$ denote the withdrawal rate at time t and assume $0 \le \gamma \le \lambda$ (λ is the maximum possible withdrawal rate). The policy guarantees that the accumulated sum of withdrawals throughout the policy's life is equal to the premium paid up front, which is denoted by ω_0. Consequently, we have $A(0) = \omega_0$, and

$$A(t) = \omega_0 - \int_0^t \gamma(u)\,du. \tag{18.15}$$

In addition, almost all policies with GMWB put a cap on the maximum allowed withdrawal rate without penalty. Let G be such a contractual withdrawal rate, and κ be the proportional penalty charge applied on the portion of withdrawal exceeding G. The net withdrawal rate $f(\gamma)$ received by the policy holder is then

$$f(\gamma) = \begin{cases} \gamma & 0 \le \gamma \le G, \\ G + (1-\kappa)(\gamma - G) & G < \gamma \le \lambda. \end{cases} \tag{18.16}$$

The no-arbitrage value $V(W, A, t)$ of the variable annuity with GMWB therefore is given by

$$V(W, A, t) = \max_{\gamma \in [0,\lambda]} E^t \Bigg[e^{-r(T-t)} \max\left(W(T), (1-\kappa)A(T)\right) + \int_t^T e^{-r(u-t)} f\left(\gamma(u)\right) du \Bigg], \tag{18.17}$$

where T is the policy maturity time and the expectation E_t is taken under the risk-neutral measure. The withdrawal rate γ is the control variable chosen to maximize the value of $V(W, A, t)$.

Define

$$\mathcal{L}V = \frac{\sigma^2}{2} W^2 V_{WW} + (r - \eta) W V_W - rV, \tag{18.18}$$

and

$$\mathcal{F}V = 1 - V_W - V_A. \tag{18.19}$$

If we let the maximum possible withdrawal rate $\lambda \to \infty$ (withdrawing instantaneously a finite amount), then we obtain the singular control problem (Dai et al. 2008)

$$\min\left[V_\tau - \mathcal{L}V - G\max(\mathcal{F}V, 0), \kappa - \mathcal{F}V\right] = 0. \tag{18.20}$$

18.3 Viscosity Solutions

The highly nonlinear PDEs ((18.2), (18.12), and (18.20)) do not have smooth (i.e. differentiable) solutions in general. In this case, it is not obvious what we mean by the solution to a differential equation. To clarify, it is useful to give an intuitive description of the concept of a *viscosity solution*. For sake of illustration, consider (18.2).

We can write our PDE as

$$g(V, V_S, V_{SS}, V_\tau) = V_\tau - \sup_{Q \in \hat{Q}} \left\{ \frac{Q^2 S^2}{2} V_{SS} + S V_S - rV \right\} = 0. \tag{18.21}$$

We assume that $g(x, y, z, w)$ $(x = V, y = V_S, z = V_{SS}, w = V_\tau)$ satisfies the ellipticity condition

$$g(x, y, z + \epsilon, w) \leq g(x, y, z, w) \quad \forall \epsilon \geq 0, \tag{18.22}$$

which in our case usually means that the coefficient of the V_{SS} term in $\mathcal{L}V$ is non-negative. Suppose for the moment that smooth solutions to (18.21) exist, i.e. $V \in C^{2,1}$, where $C^{2,1}$ refers to a continuous function $V = V(S, \tau)$ having continuous first and second derivatives in S, and a continuous first derivative in τ. Let ϕ be a set of $C^{2,1}$ test functions. Suppose $V - \phi \leq 0$, and that $\phi(S_0, \tau_0) = V(S_0, \tau_0)$ at the single point (S_0, τ_0). Then the single point (S_0, τ_0) is a global maximum of $(V - \phi)$,

$$\begin{aligned} V - \phi &\leq 0, \\ \max(V - \phi) &= V(S_0, \tau_0) - \phi(S_0, \tau_0) = 0. \end{aligned} \tag{18.23}$$

Consequently, at (S_0, τ_0)

$$\begin{aligned} \phi_\tau &= V_\tau \\ \phi_S &= V_S \\ (V - \phi)_{SS} \leq 0 \quad &\Rightarrow \phi_{SS} \geq V_{SS}. \end{aligned} \tag{18.24}$$

Hence, from (18.22,18.24), we have

$$\begin{aligned} &g\left(V(S_0, \tau_0), \phi_S(S_0, \tau_0), \phi_{SS}(S_0, \tau_0), \phi_\tau(S_0, \tau_0)\right) \\ &\quad = g\left(V(S_0, \tau_0), V_S(S_0, \tau_0), \phi_{SS}(S_0, \tau_0), V_\tau(S_0, \tau_0)\right) \\ &\quad \leq g\left(V(S_0, \tau_0), V_S(S_0, \tau_0), V_{SS}(S_0, \tau_0), V_\tau(S_0, \tau_0)\right) = 0, \end{aligned} \tag{18.25}$$

or, to summarize,

$$\begin{aligned} g\left(V(S_0, \tau_0), \phi_S(S_0, \tau_0), \phi_{SS}(S_0, \tau_0), \phi_\tau(S_0, \tau_0)\right) &\leq 0 \\ V - \phi &\leq 0 \\ \max(V - \phi) = V(S_0, \tau_0) - \phi(S_0, \tau_0) &= 0. \end{aligned} \tag{18.26}$$

If this is true for any test function ϕ, then we say that V is a *viscosity subsolution* of (18.21).

Now, suppose that χ is a $C^{2,1}$ test function, with $V - \chi \geq 0$, and $V(S_0, \tau_0) = \chi(S_0, \tau_0)$ at the single point (S_0, τ_0). Then, (S_0, τ_0) is the global minimum of $V - \chi$,

$$\begin{aligned} V - \chi &\geq 0 \\ \min(V - \chi) &= V(S_0, \tau_0) - \chi(S_0, \tau_0) = 0. \end{aligned} \tag{18.27}$$

Consequently, at (S_0, τ_0)

$$\begin{aligned}\chi_\tau &= V_\tau \\ \chi_S &= V_S \\ (V-\chi)_{SS} \geq 0 &\quad \Rightarrow \chi_{SS} \leq V_{SS}.\end{aligned} \tag{18.28}$$

Hence, from (18.27,18.28), we have

$$\begin{aligned}&g\left(V(S_0,\tau_0), \chi_S(S_0,\tau_0), \chi_{SS}(S_0,\tau_0), \chi_\tau(S_0,\tau_0)\right) \\ &\quad = g\left(V(S_0,\tau_0), V_S(S_0,\tau_0), \chi_{SS}(S_0,\tau_0), V_\tau(S_0,\tau_0)\right) \\ &\quad \geq g\left(V(S_0,\tau_0), V_S(S_0,\tau_0), V_{SS}(S_0,\tau_0), V_\tau(S_0,\tau_0)\right) = 0.\end{aligned} \tag{18.29}$$

Summarizing,

$$\begin{aligned}g\left(V(S_0,\tau_0), \chi_S(S_0,\tau_0), \chi_{SS}(S_0,\tau_0), \chi_\tau(S_0,\tau_0)\right) &\geq 0 \\ V-\chi &\geq 0 \\ \min(V-\chi) = V(S_0,\tau_0) - \chi(S_0,\tau_0) &= 0.\end{aligned} \tag{18.30}$$

If this is true for any test function χ, we say that V is a *viscosity supersolution* of (18.21). A solution which is both a viscosity subsolution and a viscosity supersolution is a viscosity solution.

Now, suppose that V is continuous but not smooth. This means that we cannot define V as the solution to $g(V, V_S, V_{SS}, V_\tau) = 0$. However, we can still use conditions (18.26) and (18.30) to define a viscosity solution to (18.21), since all derivatives are applied to smooth test functions. Informally, a viscosity solution V to (18.21) is defined such that:

- For any $C^{2,1}$ test function ϕ, such that

$$V - \phi \leq 0; \quad \phi(S_0,\tau_0) = V(S_0,\tau_0), \tag{18.31}$$

  (ϕ touches V at the single point (S_0,τ_0)), then

$$g\left(V(S_0,\tau_0), \phi_S(S_0,\tau_0), \phi_{SS}(S_0,\tau_0), \phi_\tau(S_0,\tau_0)\right) \leq 0. \tag{18.32}$$

- As well, for any $C^{2,1}$ test function χ such that

$$V - \chi \geq 0; \quad V(S_0,\tau_0) = \chi(S_0,\tau_0), \tag{18.33}$$

  (χ touches V at the single point (S_0,τ_0)), then

$$g\left(V(S_0,\tau_0), \chi_S(S_0,\tau_0), \chi_{SS}(S_0,\tau_0), \chi_\tau(S_0,\tau_0)\right) \geq 0. \tag{18.34}$$

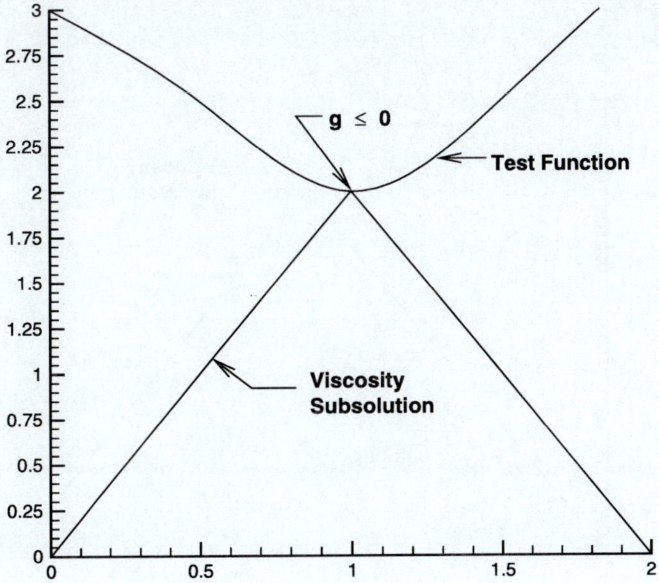

Fig. 18.1 Illustration of viscosity subsolution definition

An example of a subsolution and a typical test function is shown in Fig. 18.1. Similarly, the supersolution case is shown in Fig. 18.2.

Note that there may be some points where a smooth test function can touch the viscosity solution only from above or below, but not both. The kink at $S = 1$ in Fig. 18.2 is an example of such a situation. It is not possible for a smooth $C^{2,1}$ test function χ satisfying $V - \chi \geq 0$, $\chi(1, \tau_0) = V(1, \tau_0)$ to exist.

There may also be some points where a smooth $C^{2,1}$ test function cannot touch the solution from either above or below. As a pathological example, consider the function

$$f(x) = \begin{cases} \sqrt{x} & x \geq 0, \\ -\sqrt{-x} & x < 0. \end{cases} \tag{18.35}$$

This function cannot be touched at the origin from below (or above) by any smooth function with bounded derivatives. Note that the definition of a viscosity solution only specifies what happens when the test function touches the viscosity solution at a single point (from either above or below). The definition is silent about cases where this cannot happen.

18.4 General Form for the Example Problems

We can treat many control problems in finance using a similar approach. Even singular control problems, as in (18.20), can be solved using the methods described here, if we use the penalty technique described in Dai et al. (2008).

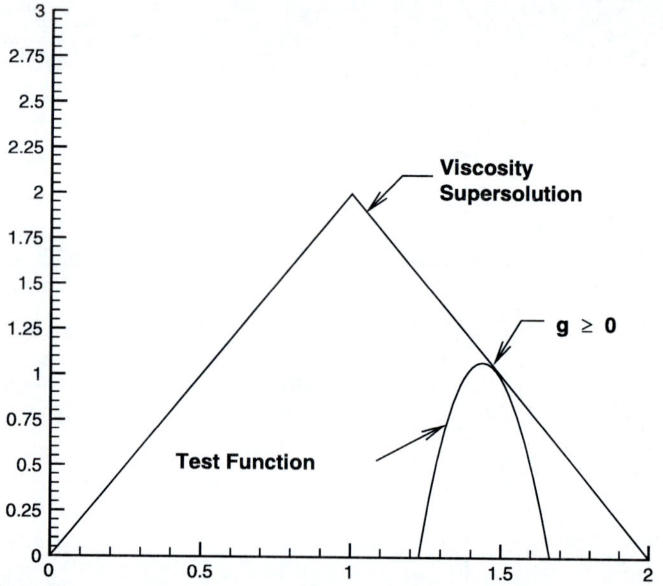

Fig. 18.2 Illustration of viscosity supersolution definition

For ease of exposition, we will focus on single factor optimal control problems. We give a brief overview of the methods here – see Forsyth and Labahn (2008) for more details. Let the value function be denoted by $V = V(S, \tau)$, where $\tau = T - t$, with T being the expiry time of the contract or claim being considered. Set

$$\mathcal{L}^Q V \equiv a(S, \tau, Q)V_{SS} + b(S, \tau, Q)V_S - c(S, \tau, Q)V, \tag{18.36}$$

where Q is a control parameter. We write our problem in the general form

$$V_\tau = \sup_{Q \in \hat{Q}} \left\{ \mathcal{L}^Q V + d(S, \tau, Q) \right\}, \tag{18.37}$$

$\hat{Q}$ being a compact set of feasible controls. Note that we can replace the sup in (18.37) by an inf and all the methods remain essentially the same.

We will assume in the following that $a(S, \tau, Q) \geq 0$ and $c(S, \tau, Q) \geq 0$. In a financial context this corresponds to non-negative interest rates and volatilities.

18.4.1 Boundary Conditions

We will assume that the problem is posed on a bounded domain $[S_{\min}, S_{\max}]$. In many cases, the original problem is posed on an unbounded domain. We assume that the problem has been *localized* for computational purposes. We will assume that the boundary conditions at $[S_{\min}, S_{\max}]$ are either the limit of the PDE as $S \to S_{\min}, S_{\max}$ or some type of given Dirichlet condition.

18.4.2 Strong Comparison Result

We assume that the HJB PDE (18.37) along with appropriate boundary conditions satisfies the *strong comparison property* (Crandall et al. 1992), which then implies that there exists a unique, continuous viscosity solution to (18.37).

18.5 Discretization

Define a grid $\{S_0, S_1, \ldots, S_p\}$ with $S_p = S_{\max}$, and let V_i^n be a discrete approximation to $V(S_i, \tau^n)$. Let $V^n = [V_0^n, \ldots, V_p^n]'$, and let $(\mathcal{L}_h^Q V^n)_i$ denote the discrete form of the differential operator (18.36) at node (S_i, τ^n). The operator (18.36) can be discretized using forward, backward or central differencing in the S direction to give

$$(\mathcal{L}_h^Q V^{n+1})_i = \alpha_i^{n+1}(Q) V_{i-1}^{n+1} + \beta_i^{n+1}(Q) V_{i+1}^{n+1} - (\alpha_i^{n+1}(Q) + \beta_i^{n+1}(Q) + c_i^{n+1}(Q)) V_i^{n+1}. \tag{18.38}$$

It is important that central, forward or backward discretizations be used to ensure that (18.40) is a positive coefficient discretization. To be more precise, this condition is

Condition 3. Positive Coefficient Condition

$$\alpha_i^{n+1}(Q) \geq 0,\ \beta_i^{n+1}(Q) \geq 0,\ c_i^{n+1}(Q) \geq 0,\ i = 0, \ldots, p-1,\ \forall Q \in \hat{Q}. \tag{18.39}$$

We will assume that all models have $c_i^{n+1}(Q) \geq 0$. Consequently, we choose central, forward or backward differencing at each node so as to ensure that $\alpha_i^{n+1}(Q), \beta_i^{n+1}(Q) \geq 0$. Appendix A provides details concerning forward, backward and central differencing. Note that different nodes can have different discretization schemes. If we use forward and backward differencing, then (18.57) in Appendix A guarantees a positive coefficient method. However, since this discretization is only first order correct, it is desirable to use central differencing as much as possible (and yet still obtain a positive coefficient method). This issue is discussed in detail in Wang and Forsyth (2007).

Equation (18.37) can now be discretized using fully implicit time stepping together with the discretization (18.38) to give

$$\frac{V_i^{n+1} - V_i^n}{\Delta\tau} = \sup_{Q^{n+1} \in \hat{Q}} \left\{ (\mathcal{L}_h^{Q^{n+1}} V^{n+1})_i + d_i^{n+1} \right\}. \tag{18.40}$$

Of course, an explicit method would involve evaluating the terms on the right hand side of (18.40) at the old time level n instead of $n+1$. A Crank–Nicolson scheme would be an equally-weighted average of the fully implicit scheme (18.40) and an explicit scheme.

18.5.1 Matrix Form of the Discrete Equations

Set $V^{n+1} = [V_0^{n+1}, V_1^{n+1}, \ldots, V_p^{n+1}]'$ and $Q = [Q_0, Q_1, \ldots, Q_p]'$. We can write the discrete operator $(\mathcal{L}_h^Q V^n)_i$ as

$$\begin{aligned}(\mathcal{L}_h^Q V^n)_i &= [A(Q)V^n]_i \\ &= \left[\alpha_i^n(Q)V_{i-1}^n + \beta_i^n(Q)V_{i+1}^n - (\alpha_i^n(Q) + \beta_i^n(Q) + c_i^n(Q))V_i^n\right],\ i < p.\end{aligned} \tag{18.41}$$

The first and last rows of A are modified as needed to handle the boundary conditions. Let F^{n+1} be a vector which encodes boundary conditions (i.e. $F_i^{n+1} = 0$ except possibly at $i = 0, p$).

Let $D^n(Q)$ be the vector with entries

$$[D(Q)]_i^n = \begin{cases} d_i^n(Q) & \text{for } i < p \rightarrow i \text{ is not a Dirichlet node} \\ 0 & \text{for } i = p \rightarrow i \text{ is a Dirichlet node} \end{cases}.$$

Remark 1 (Matrix Supremum Notational Convention). In the following, we will denote

$$\sup_{Q \in \hat{Q}} \left\{ \left[A^{n+1}(Q)V^{n+1} + D^{n+1}(Q)\right]_i \right\}$$

by

$$A^{n+1}(Q^{n+1})V^{n+1} + D^{n+1}(Q^{n+1}),$$

where the optimal control at time level $n+1$ for node i is

$$Q_i^{n+1} \in \arg\sup_{Q \in \hat{Q}} \left\{ \left[A^{n+1}(Q)V^{n+1} + D^{n+1}(Q)\right]_i \right\}.$$

If the local objective function is a continuous function of Q, then the supremum is simply the maximum value (since $\hat{Q}$ is compact), and Q^{n+1} is the point where a maximum is reached. Alternatively, if the local objective function is discontinuous, $A^{n+1}(Q^{n+1})$ is interpreted as the appropriate limiting value of $[A^{n+1}(Q)]_i$ which generates the supremum at the limit point Q^{n+1}. An example of an algorithm for computing this limit point is given in Wang and Forsyth (2007) for the case of maximizing the usage of central weighting. Note that Q^{n+1} is not necessarily unique.

The discrete equations (18.40) can be written as

$$\left[I - \Delta\tau A^{n+1}(Q^{n+1})\right] V^{n+1} = V^n + \Delta\tau D^{n+1}(Q^{n+1}) + (F^{n+1} - F^n), \quad (18.42)$$

where

$$Q_i^{n+1} \in \arg\sup_{Q \in \hat{Q}} \left\{\left[A^{n+1}(Q)V^{n+1} + D^{n+1}(Q)\right]_i\right\}.$$

For convenience, define

$$(\Delta\tau)_{\max} = \max_n(\tau^{n+1} - \tau^n) \text{ and } (\Delta\tau)_{\min} = \min_n(\tau^{n+1} - \tau^n),$$

where we assume that there are mesh size/time step parameters $h_{\min}, h_{\max}$ such that

$$(\Delta S)_{\max} = C_1 h_{\max}, \quad (\Delta\tau)_{\max} = C_2 h_{\max},$$
$$(\Delta S)_{\min} = C_3 h_{\min}, \quad (\Delta\tau)_{\min} = C_4 h_{\min},$$

with C_1, C_2, C_3, C_4 being positive constants independent of h.

We can then write the discrete equations (18.40) or (18.42) at each node in the form

$$G_i^{n+1}(h_{\max}, V_i^{n+1}, V_{i+1}^{n+1}, V_{i-1}^{n+1}, V_i^n, V_{i+1}^n, V_{i-1}^n) = 0,$$

where

$$G_i^{n+1} \equiv \frac{V_i^{n+1} - V_i^n}{\Delta\tau} - \sup_{Q^{n+1} \in \hat{Q}} \left\{\left(A^{n+1}(Q^{n+1})V^{n+1} + D^{n+1}(Q^{n+1})\right)_i\right\}$$
$$- \frac{F_i^{n+1} - F_i^n}{\Delta\tau}. \quad (18.43)$$

For notational brevity, we shall occasionally write

$$G_i^{n+1}(h_{\max}, V_i^{n+1}, \{V_j^{n+1}\}_{j \neq i}, V_i^n) \equiv G_i^{n+1}(h_{\max}, V_i^{n+1}, V_{i+1}^{n+1}, V_{i-1}^{n+1}, V_i^n), \quad (18.44)$$

where $\{V_j^{n+1}\}_{j \neq i}$ is the set of values V_j^{n+1}, for $j = 1, \dots, p$, with $j \neq i$.

18.6 Convergence to the Viscosity Solution

In Pooley et al. (2003), examples were given in which seemingly reasonable discretizations of nonlinear option pricing PDEs were either unstable or converged to the incorrect solution. It is important to ensure that we can generate discretizations which are guaranteed to converge to the viscosity solution (Barles 1997; Crandall et al. 1992). Assuming that (18.37) satisfies the strong comparison property (Barles and Burdeau 1995; Barles and Rouy 1998; Chaumont 2004), then, from Barles and Souganidis (1991) and Barles (1997), a numerical scheme converges to the viscosity solution if the method is (1) consistent, (2) stable (in the l_∞ norm), and (3) monotone. To be precise, we define these terms.

Definition 1 (Stability). Discretization (18.43) is stable if

$$\|V^{n+1}\|_\infty \leq C_5,$$

for $0 \leq n \leq N$, $T = N\Delta\tau$, for $(\Delta\tau)_{\min} \to 0$, $(\Delta S)_{\min} \to 0$, where C_5 is independent of $(\Delta\tau)_{\min}$, $(\Delta S)_{\min}$.

Lemma 1 (Stability). *If the discretization (18.43) satisfies the positive coefficient condition (18.39), then the scheme is l_∞ stable.*

Proof. This is easily shown using a maximum analysis as in Forsyth and Labahn (2008). □

For ease of exposition, we consider the simple case where we restrict attention to interior nodes. This allows us to use the following definition of consistency.

Definition 2 (Consistency). Let ϕ denote any smooth function with $\phi_i^n = \phi(S_i, \tau^n)$, and let

$$\Phi = \left(\phi_\tau - \sup_{Q \in \hat{Q}} \{\mathcal{L}^Q \phi + d\}\right)_i^{n+1} - G_i^{n+1}\left(h_{\max}, \phi_i^{n+1}, \phi_{i+1}^{n+1}, \phi_{i-1}^{n+1}, \phi_i^n, \phi_{i+1}^n, \phi_{i-1}^n\right).$$

Scheme (18.43) is consistent if

$$\lim_{h_{\max} \to 0} |\Phi| = 0. \tag{18.45}$$

Remark 2. For the general case where the HJB PDE degenerates at the boundary, a more complicated definition of consistency is required in order to handle boundary data (Barles 1997). We refer the reader to Barles (1997) for this definition, and to Chen and Forsyth (2008) for a specific application of this more complex definition.

Remark 3. Note that Definition 2 is given in terms of smooth test functions ϕ, and does not require differentiability of the actual solution.

Lemma 2 (Consistency). *If the discrete equation coefficients are as given in Appendix A, then the discrete scheme (18.43) is consistent as defined in Definition 2.*

Proof. This follows from a Taylor series argument. □

Definition 3 (Monotonicity). The discrete scheme (18.43) is monotone if for all $\epsilon_j^l \geq 0$ and i

$$G_i^{n+1}\left(h_{\max}, V_i^{n+1}, \{V_j^{n+1} + \epsilon_j^{n+1}\}_{j \neq i}, \{V_j^n + \epsilon_j^n\}\right)$$
$$\leq G_i^{n+1}\left(h_{\max}, V_i^{n+1}, \{V_j^{n+1}\}_{j \neq i}, \{V_j^n\}\right). \tag{18.46}$$

Lemma 3 (Monotonicity). *If the discretization (18.43) satisfies the positive coefficient condition (18.39), then it is monotone as defined in Definition 3.*

Proof. We write (18.43) out in component form (at the interior nodes so that $F_i = 0$)

$$G_i^{n+1}\left(h, V_i^{n+1}, V_{i+1}^{n+1}, V_{i-1}^{n+1}, V_i^n\right)$$
$$= \frac{V_i^{n+1} - V_i^n}{\Delta\tau} + \inf_{Q^{n+1} \in \hat{Q}} \Big\{\left(\alpha_i^{n+1}(Q) + \beta_i^{n+1}(Q) + c_i^{n+1}(Q)\right) V_i^{n+1}$$
$$-\alpha_i^{n+1}(Q) V_{i-1}^{n+1} - \beta_i^{n+1}(Q) V_{i+1}^{n+1} - d_i^{n+1}(Q)\Big\}. \tag{18.47}$$

Note that, given two functions $X(x), Y(x)$,

$$\inf_x X(x) - \inf_y Y(y) \leq \sup_x (X(x) - Y(x)).$$

Then, for $\epsilon \geq 0$, we have

$$G_i^{n+1}\left(h, V_i^{n+1}, V_{i+1}^{n+1} + \epsilon, V_{i-1}^{n+1}, V_i^n\right) - G_i^{n+1}\left(h, V_i^{n+1}, V_{i+1}^{n+1}, V_{i-1}^{n+1}, V_i^n\right)$$
$$= \inf_{Q \in \hat{Q}} \Big\{\left(\alpha_i^{n+1}(Q) + \beta_i^{n+1}(Q) + c_i^{n+1}(Q)\right) V_i^{n+1}$$
$$- \alpha_i^{n+1}(Q) V_{i-1}^{n+1} - \beta_i^{n+1}(Q) V_{i+1}^{n+1} - \beta_i^{n+1}(Q)\epsilon - d_i^{n+1}(Q)\Big\}$$
$$- \inf_{Q^* \in \hat{Q}} \Big\{\left(\alpha_i^{n+1}(Q^*) + \beta_i^{n+1}(Q^*) + c_i^{n+1}(Q^*)\right) V_i^{n+1}$$
$$- \alpha_i^{n+1}(Q^*) V_{i-1}^{n+1} - \beta_i^{n+1}(Q^*) V_{i+1}^{n+1} - d_i^{n+1}(Q^*)\Big\}$$
$$\leq \sup_{Q \in \hat{Q}} \Big\{-\beta_i^{n+1}(Q)\epsilon\Big\} = -\epsilon \inf_{Q \in \hat{Q}} \Big\{\beta_i^{n+1}(Q)\Big\} \leq 0. \tag{18.48}$$

This follows from the fact that $\beta_i^{n+1}(Q) \geq 0$. Similarly,

$$G_i^{n+1}\left(h, V_i^{n+1}, V_{i+1}^{n+1}, V_{i-1}^{n+1}+\epsilon, V_i^n\right) - G_i^{n+1}\left(h, V_i^{n+1}, V_{i+1}^{n+1}, V_{i-1}^{n+1}, V_i^n\right) \leq 0. \tag{18.49}$$

Finally, it is obvious from (18.47) that

$$G_i^{n+1}\left(h, V_i^{n+1}, V_{i+1}^{n+1}, V_{i-1}^{n+1}, V_i^n+\epsilon\right) - G_i^{n+1}\left(h, V_i^{n+1}, V_{i+1}^{n+1}, V_{i-1}^{n+1}, V_i^n\right) \leq 0, \tag{18.50}$$

concluding the proof. □

Theorem 2 (Convergence to the Viscosity Solution). *Provided that the original HJB PDE satisfies the strong comparison property, and discretization (18.42) satisfies all the conditions required for Lemmas 1–3, then scheme (18.42) converges to the viscosity solution of (18.37).*

Proof. This follows directly from the results in Barles and Souganidis (1991) and Barles (1997). □

18.7 Solution of the Nonlinear Discretized Equations

Note that an implicit time stepping method requires the solution of highly nonlinear algebraic equations at each time step. We use a Newton-like form of policy iteration to solve the discrete equations:

Policy Iteration

Let $(V^{n+1})^0 = V^n$

Let $\hat{V}^k = (V^{n+1})^k$

For $k = 0, 1, 2, \ldots$ until convergence

Solve $\left[I - (1-\theta)\Delta\tau A^{n+1}(Q^k)\right]\hat{V}^{k+1} =$

$$\begin{aligned}&[I + \theta\Delta\tau A^n(Q^n)]\,V^n + (F^{n+1} - F^n)\\ &+ (1-\theta)\Delta\tau D^{n+1}(Q^k) + \theta\Delta\tau D^n\end{aligned}$$

$$Q_i^k \in \arg\sup_{Q\in\hat{Q}}\left\{\left[A^{n+1}(Q)\hat{V}^k + D^{n+1}(Q)\right]_i\right\} \tag{18.51}$$

If $k > 0$ and

$$\left(\max_i \frac{\left|\hat{V}_i^{k+1} - \hat{V}_i^k\right|}{\max\left(scale, \left|\hat{V}_i^{k+1}\right|\right)} < tolerance\right)$$

then quit

EndFor

The term *scale* in scheme (18.51) is used to preclude unrealistic levels of accuracy when the value is very small. Typically, $scale = 1$ for values expressed in dollars.

Theorem 3 (Convergence of the Policy Iteration). *Provided that the discretization (18.43) satisfies the positive coefficient condition (18.39), then the policy iteration (18.51) converges to the unique solution of (18.42) for any initial iterate $\hat{V}^0$. Moreover, the iterates converge monotonically.*

Proof. See Forsyth and Labahn (2008). □

The most fundamental principle of valuation in finance is the absence of arbitrage (i.e. there are no free lunches). One way of stating this is as follows. Imagine that we have two contingent claims with the same expiry time that are written on the same underlying asset, which has a price of S. Denote these two claims by $V(S,\tau)$ and $W(S,\tau)$. No-arbitrage implies that if the terminal payoff for V is always at least as high as that for W, then V must be worth at least as much as W at any time prior to expiry. More succinctly,

$$V(S,0) \geq W(S,0) \Rightarrow V(S,\tau) \geq W(S,\tau). \tag{18.52}$$

Let V^n and W^n denote discrete solutions to (18.42). We would like to ensure that these solutions are arbitrage-free, i.e.

$$V^n \geq W^n \Rightarrow V^{n+1} \geq W^{n+1}. \tag{18.53}$$

It can be shown that this property is satisfied under certain conditions, which we state in the following theorem:

Theorem 4 (Discrete no-arbitrage principle). *Assume that:*

(i) Discretization (18.43) satisfies the positive coefficient condition (18.39);
(ii) Fully implicit time stepping is used; and
(iii) Appropriate boundary conditions are imposed at the end-points of the discrete grid (see Forsyth and Labahn 2008 for details).

Then the discrete no-arbitrage condition (18.53) holds.

Proof. See Forsyth and Labahn (2008). □

18.8 Numerical Example: Uncertain Volatility

As a simple illustration of the methods outlined above, we will consider the case of pricing an option contract in an uncertain volatility model, as described in Avellaneda et al. (1995) and Lyons (1995) and outlined above in Sect. 18.2.1. Recall that we are interested in valuing an option under the assumption that the volatility σ lies between two bounds, $\sigma_{\min}$ and $\sigma_{\max}$, but is otherwise unknown. From the standpoint of the option writer, the best case is found by solving (18.2), reproduced here for convenience:

$$V_\tau = \sup_{Q \in \hat{Q}} \left\{ \frac{Q^2 S^2}{2} V_{SS} + S V_S - rV \right\} = 0, \tag{18.54}$$

with $\hat{Q} = \{\sigma_{\min}, \sigma_{\max}\}$. Of course, from the perspective of the purchaser of the option, this would represent the worst possible case. Conversely, the worst case for the writer (found by replacing the sup by an inf in the equation above) corresponds to the best situation for the purchaser. At first glance this problem might appear to be trivial, since option values are increasing in volatility. However, while this is the case for a plain vanilla European option, it is not true in general provided that the option "gamma" V_{SS} can change sign. This can happen, for example, in the case of barrier options. Consider the case of an up-and-out call option, which is just like a regular call option unless the underlying asset price S moves above some barrier H during the contract's life, in which case the payoff becomes zero. The gamma of this contract can be positive for some values of S and negative for others, as noted, for example, in Derman and Kani (1996).

Another example arises in the context of a portfolio of plain vanilla European options, and it is this case that we will consider here. Note that this highlights the nonlinear nature of the problem, in that the problem is trivial for each of the options in the portfolio, but not for the linear combination that forms the portfolio. Suppose that an investor purchases a butterfly spread from a financial institution. This involves taking a long position in a low strike (K_1) option, a short position in two middle strike (K_2) options, and a long position in a high strike (K_3) option, all with identical maturities. Assume that the strikes are evenly spaced, and that all options are calls. Our test case uses the input parameters provided in Table 18.1.

The payoff function at maturity is plotted in Fig. 18.3. The sharp peak around the middle strike $K_2 = 100$ will generate rapid changes with S in the solution value as we solve over time. This can be expected to cause problems with numerical methods unless we are careful.

Our numerical experiment uses a discrete grid ranging from $S_{\min} = 0$ to $S_{\max} = 500$. The coarsest grid has 94 unevenly spaced nodes (a finer spacing is placed near the strikes), and uses 100 (constant-sized) time steps. Successive grid refinements

Table 18.1 Input parameters for test case

Parameter	Value
r	0.04
T	0.5
K_1	95
K_2	100
K_3	105
$\sigma_{\min}$	0.30
$\sigma_{\max}$	0.45

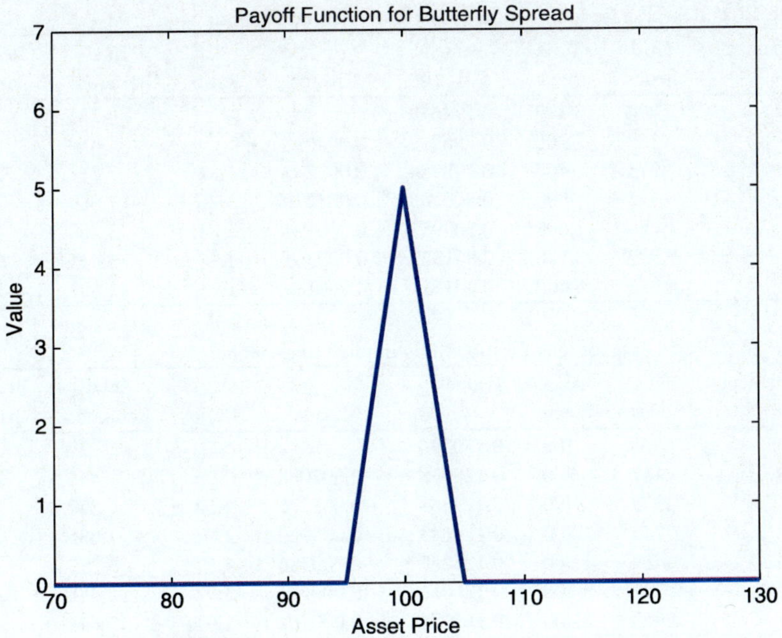

Fig. 18.3 Payoff function for butterfly spread

involve doubling the number of time steps and inserting new grid points midway between previously existing nodes.

We begin by considering the results for the best case for a long position with fully implicit time stepping. Results are provided in Table 18.2. In this table, the column labelled "Change" is the difference in the computed solution from the previous grid refinement level, and the column labelled "Ratio" is the change for the current refinement level divided by that for the previous level. Values of "Ratio" around two indicate approximate first order convergence. Approximate second order convergence would be shown by values of "Ratio" of about four. As can be seen from the table, fully implicit time stepping leads asymptotically to approximate first order convergence. The last two columns of the table show the total number of nonlinear iterations taken during the solution, and the average number of nonlinear iterations per time step. For this particular case, about two iterations are required for each time step.

Table 18.3 repeats the analysis, but for the worst case for a long position. Clearly, the value at $S = 100$ is much lower, but we again see that the algorithm exhibits approximate linear convergence and that around two iterations are needed per time step. Figure 18.4 plots the solution profile obtained for the best and worst cases for a long position using fully implicit time steps.

Tables 18.4 and 18.5 document the serious problems which can occur when we use numerical methods which are not guaranteed to converge to the viscosity

Table 18.2 Best case for long position, fully implicit time stepping

Refinement level	Grid nodes	Time steps	Value at $S = 100$	Change	Ratio	Total iterations	Iterations per step
0	94	100	0.792639			227	2.27
1	187	200	0.796737	0.004098		450	2.25
2	373	400	0.798984	0.002247	1.82	871	2.18
3	745	800	0.800263	0.001279	1.76	1,689	2.11
4	1,489	1,600	0.800957	0.000694	1.84	3,260	2.04
5	2,977	3,200	0.801322	0.000365	1.90	6,445	2.01
6	5,953	6,400	0.801511	0.000189	1.93	12,802	2.00

Table 18.3 Worst case for long position, fully implicit time stepping

Refinement level	Grid nodes	Time steps	Value at $S = 100$	Change	Ratio	Total iterations	Iterations per step
0	94	100	0.130726			227	2.27
1	187	200	0.128638	−0.002088		443	2.22
2	373	400	0.127363	−0.001275	1.64	870	2.18
3	745	800	0.126643	−0.000720	1.77	1,685	2.11
4	1,489	1,600	0.126257	−0.000386	1.87	3,297	2.06
5	2,977	3,200	0.126056	−0.000201	1.92	6,488	2.03
6	5,953	6,400	0.125954	−0.000102	1.97	12,844	2.01

solution and are not necessarily arbitrage-free. The only difference here compared to Tables 18.2 and 18.3 is the switch from fully implicit time stepping to Crank–Nicolson. The key results from Table 18.4 are as follows. Although Crank–Nicolson is in theory second order accurate in time, the convergence rate here is actually less than first order. More importantly, the scheme is converging to a different answer than that obtained in Table 18.2. Since the fully implicit scheme used in Table 18.2 is guaranteed to converge to the viscosity solution, the implication here is that the Crank–Nicolson approach is converging to some other (i.e. non-viscosity) solution. Comparing Tables 18.2 and 18.3, we can also see that the Crank–Nicolson approach requires more than twice as many nonlinear iterations.

The same general conclusions apply to Table 18.5: the Crank–Nicolson scheme converges at a rate which is slower than first order, it requires more than twice as many iterations than does the fully implicit approach, and it is converging to an answer which is not the viscosity solution. In fact, the Crank–Nicolson method converges here to a negative value. This represents an obvious arbitrage opportunity and is clearly an absurd result. Cases like this are in a sense reassuring, since it is obvious that the answer makes no sense. From this perspective, the Crank–Nicolson results for the best case long position are possibly of greater concern. Without calculating the correct answer via the fully implicit approach, it is not immediately clear that the Crank–Nicolson answer is incorrect. Figure 18.5 plots the solution profile obtained for the best and worst cases for a long position using the Crank–Nicolson scheme.

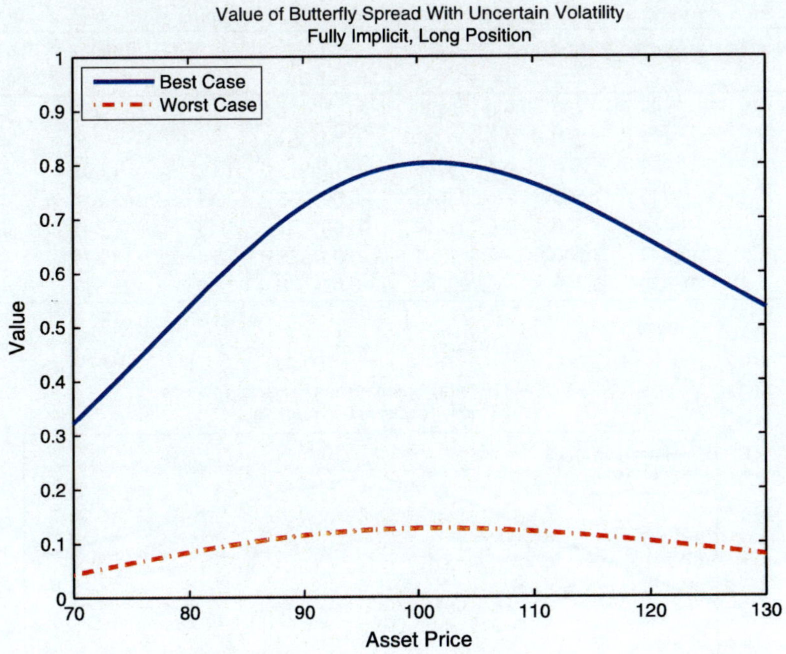

Fig. 18.4 Value of butterfly spread with uncertain volatility. Fully implicit time stepping, long position

Table 18.4 Best case for long position, Crank–Nicolson time stepping

Refinement level	Grid nodes	Time steps	Value at $S = 100$	Change	Ratio	Total iterations	Iterations per step
0	94	100	4.410778			428	4.28
1	187	200	4.571876	0.161098		897	4.49
2	373	400	4.687534	0.115658	1.39	1,780	4.45
3	745	800	4.765390	0.077856	1.49	3,539	4.42
4	1,489	1,600	4.816438	0.051048	1.53	7,161	4.48
5	2,977	3,200	4.849302	0.032864	1.55	13,995	4.37
6	5,953	6,400	4.870269	0.020967	1.57	27,529	4.30

In addition to calculating the value of the position, we are often interested in hedging parameters such as delta and gamma. Figures 18.6 and 18.7 plot the delta and gamma respectively for the best case for a long position with fully implicit time steps. The corresponding plots for the Crank–Nicolson case for delta and gamma are given in Figs. 18.8 and 18.9 respectively. Comparing Figs. 18.6 and 18.8, we see that the plot for delta is much smoother for the fully implicit case (in addition to being far smaller in magnitude). In fact, there appears to be a discontinuity in the delta at $S = 100$ for the Crank–Nicolson case. Figure 18.7 shows a smooth profile for the option gamma using fully implicit time steps. On the other hand, Fig. 18.9

Table 18.5 Worst case for long position, Crank–Nicolson time stepping

Refinement level	Grid nodes	Time steps	Value at $S = 100$	Change	Ratio	Total iterations	Iterations per step
0	94	100	−6.178730			457	4.57
1	187	200	−6.399983	−0.221253		926	4.63
2	373	400	−6.545795	−0.145812	1.52	1,901	4.75
3	745	800	−6.643648	−0.097853	1.49	3,815	4.77
4	1,489	1,600	−6.709119	−0.065471	1.49	7,341	4.59
5	2,977	3,200	−6.751707	−0.042588	1.54	14,379	4.49
6	5,953	6,400	−6.778385	−0.026678	1.60	28,317	4.42

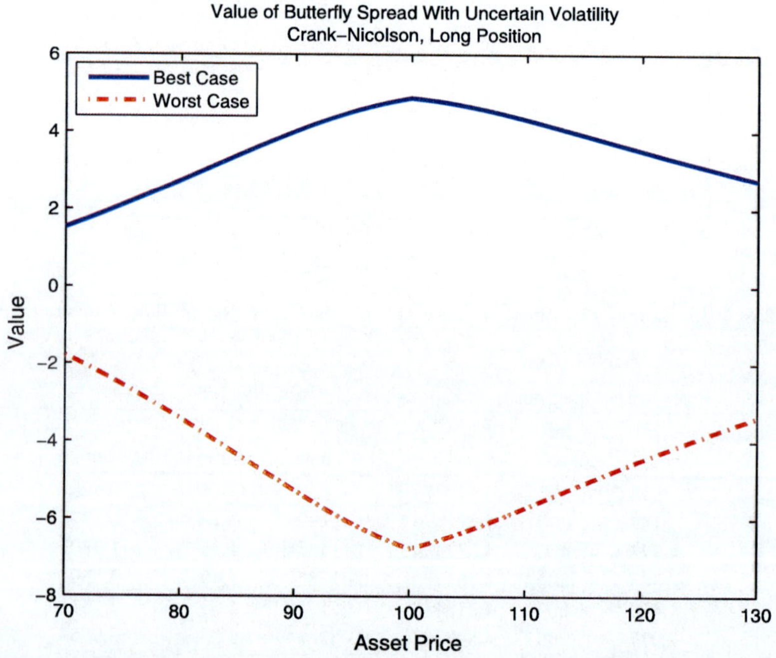

Fig. 18.5 Value of butterfly spread with uncertain volatility. Crank–Nicolson time stepping, long position

shows severe oscillations around values of $S = 100$. Taken collectively, these plots again provide a strong warning against the naïve use of Crank–Nicolson methods in that the calculation of important hedging parameters is prone to serious errors. This is not surprising – if the solution itself is not accurate, we should expect the estimates of its derivatives to be even worse.

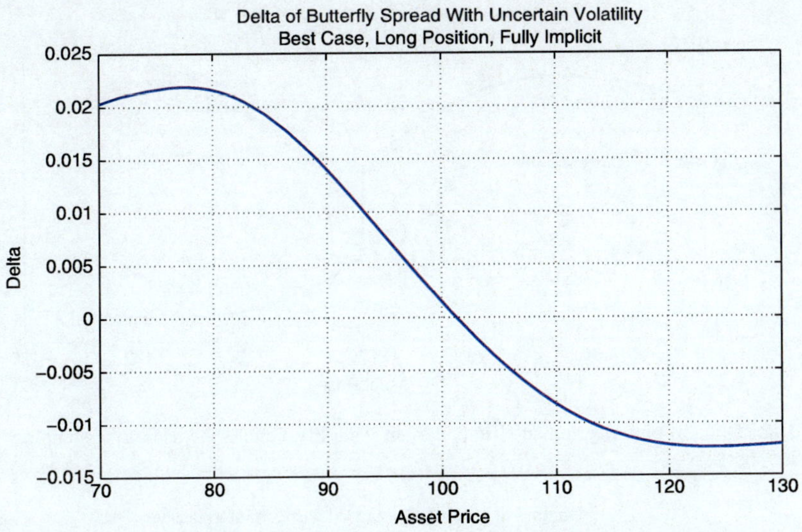

Fig. 18.6 Delta of butterfly spread with uncertain volatility. Fully implicit time stepping, long position, best case

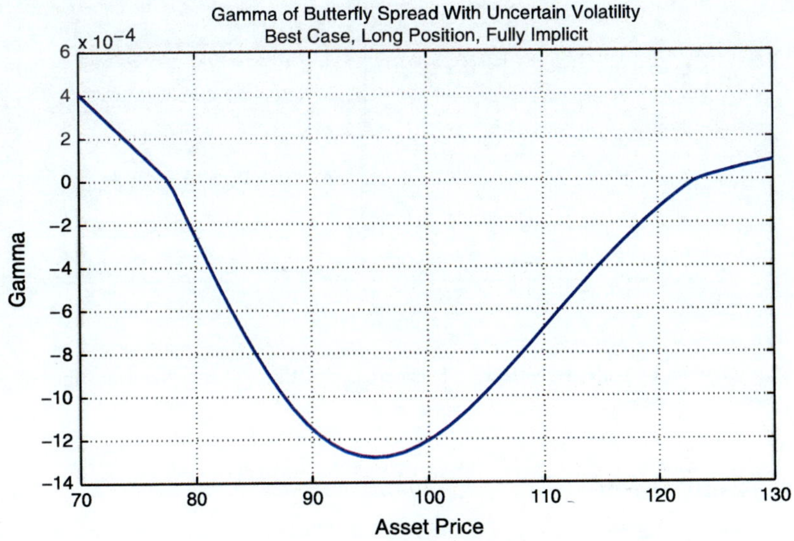

Fig. 18.7 Gamma of butterfly spread with uncertain volatility. Fully implicit time stepping, long position, best case

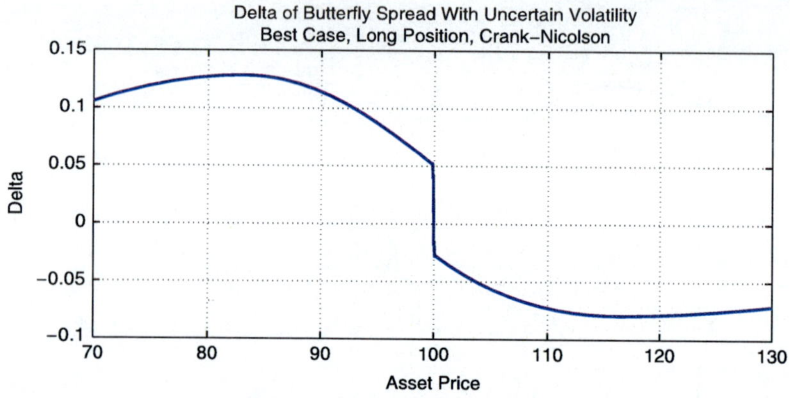

Fig. 18.8 Delta of butterfly spread with uncertain volatility. Crank–Nicolson time stepping, long position, best case

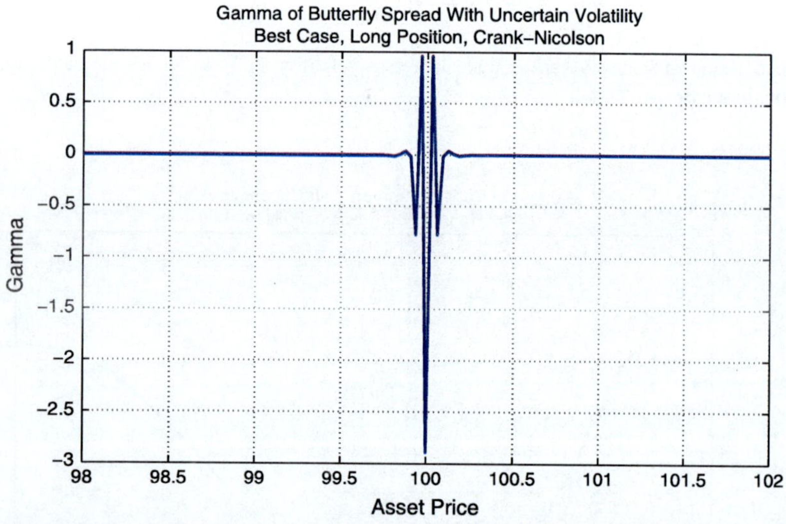

Fig. 18.9 Gamma of butterfly spread with uncertain volatility. Crank–Nicolson time stepping, long position, best case

18.9 Conclusions

Many problems of practical interest in finance can be cast as stochastic optimal control problems. These problems are generally nonlinear and require numerical solution. This article has described some of these problems, along with a general approach that can be taken to solve them numerically. This approach stresses the importance of using a positive coefficient discretization and fully implicit time stepping. This guarantees convergence to the viscosity solution, and has the important feature that the discrete solutions are arbitrage-free. Apparently reasonable discretizations such as Crank–Nicolson methods are not guaranteed to converge to the viscosity solution, nor can we be sure that they do not lead to free lunches. Moreover, the use of such methods can lead to serious errors in the estimation of hedging parameters.

Appendix A: Discrete Equation Coefficients

Let Q_i^n denote the optimal control at node i and time level n, and set

$$a_i^{n+1} = a(S_i, \tau^n, Q_i^n), \quad b_i^{n+1} = b(S_i, \tau^n, Q_i^n), \quad c_i^{n+1} = c(S_i, \tau^n, Q_i^n). \tag{18.55}$$

Then we can use central, forward or backward differencing at any node. For central differencing:

$$\begin{aligned} \alpha_{i,central}^n &= \left[\frac{2a_i^n}{(S_i - S_{i-1})(S_{i+1} - S_{i-1})} - \frac{b_i^n}{S_{i+1} - S_{i-1}} \right] \\ \beta_{i,central}^n &= \left[\frac{2a_i^n}{(S_{i+1} - S_i)(S_{i+1} - S_{i-1})} + \frac{b_i^n}{S_{i+1} - S_{i-1}} \right]. \end{aligned} \tag{18.56}$$

For forward/backward differencing: $(b_i^n > 0/b_i^n < 0)$

$$\begin{aligned} \alpha_{i,forward/backward}^n &= \left[\frac{2a_i^n}{(S_i - S_{i-1})(S_{i+1} - S_{i-1})} + \max\left(0, \frac{-b_i^n}{S_i - S_{i-1}}\right) \right] \\ \beta_{i,forward/backward}^n &= \left[\frac{2a_i^n}{(S_{i+1} - S_i)(S_{i+1} - S_{i-1})} + \max\left(0, \frac{b_i^n}{S_{i+1} - S_i}\right) \right]. \end{aligned} \tag{18.57}$$

References

Andersen, L., Andreasen, J., & Brotherton-Ratcliffe, R. (1998). The passport option. *Journal of Computational Finance*, *1*(3), 15–36.

Avellaneda, M., Levy, A., & Parás, A. (1995). Pricing and hedging derivative securities in markets with uncertain volatilities. *Applied Mathematical Finance*, 2, 73–88.

Barles, G. (1997). Convergence of numerical schemes for degenerate parabolic equations arising in finance. In L. C. G. Rogers & D. Talay (Eds.), *Numerical methods in finance* (pp. 1–21). Cambridge: Cambridge University Press.

Barles, G., & Burdeau, J. (1995). The Dirichlet problem for semilinear second-order degenerate elliptic equations and applications to stochastic exit time control problems. *Communications in Partial Differential Equations*, *20*, 129–178.

Barles, G., & Rouy, E. (1998). A strong comparison result for the Bellman equation arising in stochastic exit time control problems and applications. *Communications in Partial Differential Equations*, *23*, 1995–2033.

Barles, G., & Souganidis, P. E. (1991). Convergence of approximation schemes for fully nonlinear equations. *Asymptotic Analysis*, *4*, 271–283.

Barles, G., Daher, C. H., & Romano, M. (1995). Convergence of numerical shemes for parabolic eqations arising in finance theory. *Mathematical Models and Methods in Applied Sciences*, *5*, 125–143.

Basak, S., & Chabakauri, G. (2007). Dynamic mean-variance asset allocation. Working Paper, London Business School.

Bergman, Y. (1995). Option pricing with differential interest rates. *Review of Financial Studies*, *8*, 475–500.

Chaumont, S. (2004). A strong comparison result for viscosity solutions to Hamilton-Jacobi-Bellman equations with Dirichlet conditions on a non-smooth boundary. Working paper, Institute Elie Cartan, Université Nancy I.

Chen, Z., & Forsyth, P. A. (2008). A numerical scheme for the impulse control formulation for pricing variable annuities with a guaranteed minimum withdrawal benefit (GMWB). *Numerische Mathematik*, *109*, 535–569.

Crandall, M. G., Ishii, H., & Lions, P. L. (1992). User's guide to viscosity solutions of second order partial differential equations. *Bulletin of the American Mathematical Society*, *27*, 1–67.

Dai, M., Kwok, Y. K., & Zong, J. (2008). Guaranteed minimum withdrawal benefit in variable annuities. *Mathematical Finance*, *18*, 595–611.

Derman, E., & Kani, I. (1996). The ins and outs of barrier options: Part 1. *Derivatives Quarterly*, *3*(Winter), 55–67.

Forsyth, P. A., & Labahn, G. (2008). Numerical methods for controlled Hamilton-Jacobi-Bellman PDEs in finance. *Journal of Computational Finance*, *11*, 1–44.

Kushner, H. J., & Dupuis, P. G. (1991). *Numerical methods for stochastic control problems in continuous time*. New York: Springer.

Leland, H. E. (1985). Option pricing and replication with transaction costs. *Journal of Finance*, *40*, 1283–1301.

Li, D., & Ng, W.-L. (2000). Optimal dynamic portfolio selection: Multiperiod mean variance formulation. *Mathematical Finance*, *10*, 387–406.

Lorenz, J. (2008). Optimal trading algorithms: Portfolio transactions, mulitperiod portfolio selection, and competitive online search. PhD Thesis, ETH Zurich.

Lorenz, J., & Almgren, R. (2007). Adaptive arrival price. In B. R. Bruce (Ed.), *Algorithmic trading III: Precision, control, execution*. New York: Institutional Investor Journals.

Lyons, T. (1995). Uncertain volatility and the risk free synthesis of derivatives. *Applied Mathematical Finance*, *2*, 117–133.

Milevsky, M. A., & Salisbury, T. S. (2006). Financial valuation of guaranteed minimum withdrawal benefits. *Insurance: Mathematics and Economics*, *38*, 21–38.

Mnif, M., & Sulem, A. (2001). Optimal risk control under excess of loss reinsurance. Working paper, Université Paris 6.

Pham, H. (2005). On some recent aspects of stochastic control and their applications. *Probability Surveys*, *2*, 506–549.

Pooley, D. M., Forsyth, P. A., & Vetzal, K. R. (2003). Numerical convergence properties of option pricing PDEs with uncertain volatility. *IMA Journal of Numerical Analysis*, *23*, 241–267.

Shreve, S., & Vecer, J. (2000). Options on a traded account: Vacation calls, vacation puts, and passport options. *Finance and Stochastics*, *4*, 255–274.

Wang, J., & Forsyth, P. A. (2007). Maximal use of central differencing for Hamilton-Jacobi-Bellman PDEs in finance. *SIAM Journal on Numerical Analysis*, *46*, 1580–1601.

Zhou, X. Y., & Li, D. (2000). Continuous time mean variance portfolio selection: A stochastic LQ framework. *Applied Mathematics and Optimization*, *42*, 19–33.

Chapter 19
Numerical Solution of Stochastic Differential Equations in Finance

Timothy Sauer

Abstract This chapter is an introduction and survey of numerical solution methods for stochastic differential equations. The solutions will be continuous stochastic processes that represent diffusive dynamics, a common modeling assumption for financial systems. We include a review of fundamental concepts, a description of elementary numerical methods and the concepts of convergence and order for stochastic differential equation solvers.

In the remainder of the chapter we describe applications of SDE solvers to Monte-Carlo sampling for financial pricing of derivatives. Monte-Carlo simulation can be computationally inefficient in its basic form, and so we explore some common methods for fostering efficiency by variance reduction and the use of quasi-random numbers. In addition, we briefly discuss the extension of SDE solvers to coupled systems driven by correlated noise, which is applicable to multiple asset markets.

19.1 Stochastic Differential Equations

Stochastic differential equations (SDEs) have become standard models for financial quantities such as asset prices, interest rates, and their derivatives. Unlike deterministic models such as ordinary differential equations, which have a unique solution for each appropriate initial condition, SDEs have solutions that are continuous-time stochastic processes. Methods for the computational solution of stochastic differential equations are based on similar techniques for ordinary differential equations, but generalized to provide support for stochastic dynamics.

T. Sauer (✉)
Department of Mathematics, George Mason University, Fairfax, VA 22030, USA
e-mail: tsauer@gmu.edu

J.-C. Duan et al. (eds.), *Handbook of Computational Finance*, Springer Handbooks of Computational Statistics, DOI 10.1007/978-3-642-17254-0_19,

We will begin with a quick survey of the most fundamental concepts from stochastic calculus that are needed to proceed with our description of numerical methods. For full details, the reader may consult Klebaner (1998), Oksendal (1998) and Steele (2001).

A set of random variables X_t indexed by real numbers $t \geq 0$ is called a *continuous-time stochastic process*. Each instance, or *realization* of the stochastic process is a choice from the random variable X_t for each t, and is therefore a function of t.

Any (deterministic) function $f(t)$ can be trivially considered as a stochastic process, with variance $V(f(t)) = 0$. An archetypal example that is ubiquitous in models from physics, chemistry, and finance is the *Wiener process* W_t, a continuous-time stochastic process with the following three properties:

Property 1. For each t, the random variable W_t is normally distributed with mean 0 and variance t.

Property 2. For each $t_1 < t_2$, the normal random variable $W_{t_2} - W_{t_1}$ is independent of the random variable W_{t_1}, and in fact independent of all $W_t, 0 \leq t \leq t_1$.

Property 3. The Wiener process W_t can be represented by continuous paths.

The Wiener process, named after Norbert Wiener, is a mathematical construct that formalizes random behavior characterized by the botanist Robert Brown in 1827, commonly called Brownian motion. It can be rigorously defined as the scaling limit of random walks as the step size and time interval between steps both go to zero. Brownian motion is crucial in the modeling of stochastic processes since it represents the integral of idealized noise that is independent of frequency, called white noise. Often, the Wiener process is called upon to represent random, external influences on an otherwise deterministic system, or more generally, dynamics that for a variety of reasons cannot be deterministically modeled.

A typical *diffusion process* in finance is modeled as a differential equation involving deterministic, or *drift* terms, and stochastic, or *diffusion* terms, the latter represented by a Wiener process, as in the equation

$$dX = a(t, X)\, dt + b(t, X)\, dW_t \tag{19.1}$$

Notice that the SDE (19.1) is given in differential form, unlike the derivative form of an ODE. That is because many interesting stochastic processes, like Brownian motion, are continuous but not differentiable. Therefore the meaning of the SDE (19.1) is, by definition, the integral equation

$$X(t) = X(0) + \int_0^t a(s, y)\, ds + \int_0^t b(s, y)\, dW_s,$$

where the meaning of the last integral, called an Ito integral, will be defined next.

Let $c = t_0 < t_1 < \ldots < t_{n-1} < t_n = d$ be a grid of points on the interval $[c, d]$. The Riemann integral is defined as a limit

$$\int_c^d f(x)\,dx = \lim_{\Delta t \to 0} \sum_{i=1}^{n} f(t_i')\Delta t_i,$$

where $\Delta t_i = t_i - t_{i-1}$ and $t_{i-1} \le t_i' \le t_i$. Similarly, the *Ito integral* is the limit

$$\int_c^d f(t)\,dW_t = \lim_{\Delta t \to 0} \sum_{i=1}^{n} f(t_{i-1})\Delta W_i$$

where $\Delta W_i = W_{t_i} - W_{t_{i-1}}$, a step of Brownian motion across the interval. Note a major difference: while the t_i' in the Riemann integral may be chosen at any point in the interval (t_{i-1}, t_i), the corresponding point for the Ito integral is required to be the left endpoint of that interval.

Because f and W_t are random variables, so is the Ito integral $I = \int_c^d f(t)\,dW_t$. The *differential* dI is a notational convenience; thus

$$I = \int_c^d f\,dW_t$$

is expressed in differential form as

$$dI = f\,dW_t.$$

The differential dW_t of Brownian motion W_t is called *white noise*. A typical solution is a combination of drift and the diffusion of Brownian motion.

To solve SDEs analytically, we need to introduce the chain rule for stochastic differentials, called the *Ito formula*:
If $Y = f(t, X)$, then

$$dY = \frac{\partial f}{\partial t}(t, X)\,dt + \frac{\partial f}{\partial x}(t, X)\,dx + \frac{1}{2}\frac{\partial^2 f}{\partial x^2}(t, X)\,dx\,dx \tag{19.2}$$

where the $dx\,dx$ term is interpreted by using the identities

$$\begin{aligned} dt\,dt &= 0 \\ dt\,dW_t &= dW_t\,dt = 0 \\ dW_t\,dW_t &= dt \end{aligned} \tag{19.3}$$

The Ito formula is the stochastic analogue to the chain rule of conventional calculus. Although it is expressed in differential form for ease of understanding, its meaning is precisely the equality of the Ito integral of both sides of the equation. It is proved under rather weak hypotheses by referring the equation back to the definition of Ito integral (Oksendal, 1998).

Some of the important features of typical stochastic differential equations can be illustrated using the following historically-pivotal example from finance, often called the Black–Scholes diffusion equation:

$$\begin{cases} dX = \mu X\, dt + \sigma X\, dW_t \\ X(0) = X_0 \end{cases} \tag{19.4}$$

with constants μ and σ. Although the equation is comparatively simple, the fact that it can be exactly solved led to its central importance, by making a closed-form formula available for the pricing of simple options (Black and Scholes, 1973).

The solution of the Black–Scholes stochastic differential equation is geometric Brownian motion

$$X(t) = X_0 e^{(\mu - \frac{1}{2}\sigma^2)t + \sigma W_t}. \tag{19.5}$$

To check this, write $X = f(t, Y) = X_0 e^Y$, where $Y = (\mu - \frac{1}{2}\sigma^2)t + \sigma W_t$. By the Ito formula,

$$dX = X_0 e^Y\, dY + \tfrac{1}{2} e^Y\, dY\, dY$$

where $dY = (\mu - \frac{1}{2}\sigma^2)\, dt + \sigma\, dW_t$. Using the differential identities from the Ito formula,

$$dY\, dY = \sigma^2\, dt,$$

and therefore

$$\begin{aligned} dX &= X_0 e^Y (r - \tfrac{1}{2}\sigma^2)\, dt + X_0 e^Y \sigma\, dW_t + \tfrac{1}{2}\sigma^2 e^Y\, dt \\ &= X_0 e^Y \mu\, dt + X_0 e^Y \sigma\, dW_t \\ &= \mu X\, dt + \sigma X\, dW_t \end{aligned}$$

as claimed.

Figure 19.1 shows a realization of geometric Brownian motion with constant drift coefficient μ and diffusion coefficient σ. Similar to the case of ordinary differential equations, relatively few stochastic differential equations have closed-form solutions. It is often necessary to use numerical approximation techniques.

19.2 Numerical Methods for SDEs

The simplest effective computational method for the approximation of ordinary differential equations is Euler's method (Sauer, 2006). The Euler-Maruyama method (Maruyama, 1955) is the analogue of the Euler method for ordinary differential equations. To develop an approximate solution on the interval $[c, d]$, assign a grid of points

$$c = t_0 < t_1 < t_2 < \ldots < t_n = d.$$

Approximate x values

$$w_0 < w_1 < w_2 < \ldots < w_n$$

will be determined at the respective t points. Given the SDE initial value problem

$$\begin{cases} dX(t) = a(t, X)dt + b(t, X)dW_t \\ X(c) = X_c \end{cases} \tag{19.6}$$

we compute the approximate solution as follows:

Euler-Maruyama Method

$$\begin{aligned} w_0 &= X_0 \\ w_{i+1} &= w_i + a(t_i, w_i)\Delta t_{i+1} + b(t_i, w_i)\Delta W_{i+1} \end{aligned} \tag{19.7}$$

where

$$\begin{aligned} \Delta t_{i+1} &= t_{i+1} - t_i \\ \Delta W_{i+1} &= W(t_{i+1}) - W(t_i). \end{aligned} \tag{19.8}$$

The crucial question is how to model the Brownian motion ΔW_i. Define $N(0,1)$ to be the standard random variable that is normally distributed with mean 0 and standard deviation 1. Each random number ΔW_i is computed as

$$\Delta W_i = z_i \sqrt{\Delta t_i} \tag{19.9}$$

where z_i is chosen from $N(0,1)$. Note the departure from the deterministic ordinary differential equation case. Each set of $\{w_0, \ldots, w_n\}$ produced by the Euler-Maruyama method is an approximate realization of the solution stochastic process $X(t)$ which depends on the random numbers z_i that were chosen. Since W_t is a stochastic process, each realization will be different and so will our approximations.

As a first example, we show how to apply the Euler-Maruyama method to the Black–Scholes SDE (19.4). The Euler-Maruyama equations (19.7) have form

$$w_0 = X_0 \tag{19.10}$$

$$w_{i+1} = w_i + \mu w_i \Delta t_i + \sigma w_i \Delta W_i.$$

We will use the drift coefficient $\mu = 0.75$ and diffusion coefficient $\sigma = 0.30$, which are values inferred from the series of market close share prices of Google, Inc. (NYSE ticker symbol GOOG) during the 250 trading days in 2009. To calculate the values μ and σ^2, the mean and variance, respectively, of the daily stock price returns were converted to an annual basis, assuming independence of the daily returns.

An exact realization, generated from the solution (19.5), along with the corresponding Euler-Maruyama approximation, are shown in Fig. 19.1. By

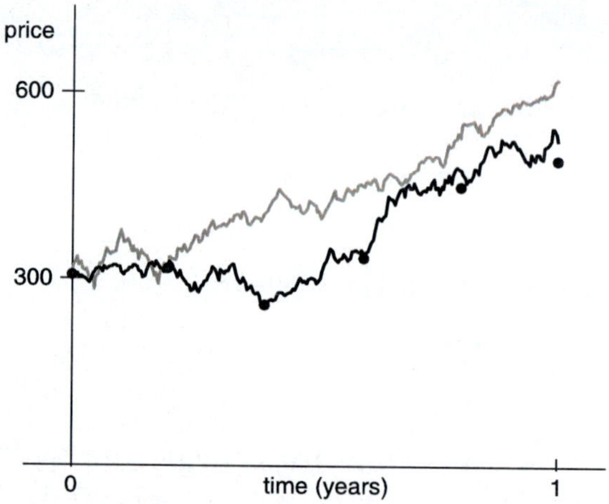

Fig. 19.1 Solution to the Black–Scholes stochastic differential equation (19.4). The exact solution (19.5) is plotted as a *black curve*. The Euler-Maruyama approximation with time step $\Delta t = 0.2$ is plotted as *circles*. The drift and diffusion parameters are set to $\mu = 0.75$ and $\sigma = 0.30$, respectively. Shown in *grey* is the actual stock price series, from which μ and σ were inferred

corresponding, we mean that the approximation used the same Brownian motion realization as the true solution. Note the close agreement between the solution and the approximating points, plotted as small circles every 0.2 time units. In addition, the original time series of Google share prices is shown for comparison. Both the original time series (grey curve) and the simulation from (19.5) (black curve) should be considered as realizations from the same diffusion process, with identical μ, σ and initial price $X_0 = 307.65$.

As another example, consider the *Langevin equation*

$$dX(t) = -\mu X(t)\, dt + \sigma\, dW_t \tag{19.11}$$

where μ and σ are positive constants. In this case, it is not possible to analytically derive the solution to this equation in terms of simple processes. The solution of the Langevin equation is a stochastic process called the *Ornstein-Uhlenbeck process*. Figure 19.2 shows one realization of the approximate solution. It was generated from an Euler-Maruyama approximation, using the steps

$$\begin{aligned} w_0 &= X_0 \\ w_{i+1} &= w_i - \mu w_i \Delta t_i + \sigma \Delta W_i \end{aligned} \tag{19.12}$$

for $i = 1, \ldots, n$. This stochastic differential equation is used to model systems that tend to revert to a particular state, in this case the state $X = 0$, in the presence of a noisy background. Interest-rate models, in particular, often contain mean-reversion assumptions.

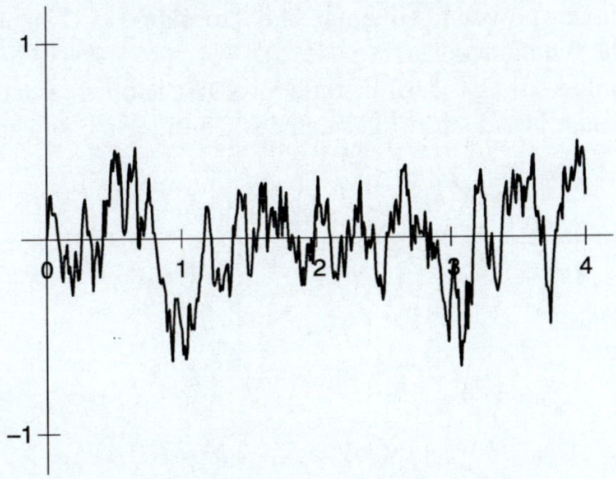

Fig. 19.2 Solution to Langevin equation (19.11). The upper path is the solution approximation for parameters $\mu = 10, \sigma = 1$, computed by the Euler-Maruyama method

19.3 Strong Convergence of SDE Solvers

The definition of convergence is similar to the concept for ordinary differential equation solvers, aside from the differences caused by the fact that a solution to an SDE is a stochastic process, and each computed trajectory is only one realization of that process. Each computed solution path $w(t)$, using Euler-Maruyama for example, gives a random value at T, so that $w(T)$ is a random variable as well. The difference between the values at time T, $e(T) = X(T) - w(T)$, is therefore a random variable.

A discrete-time approximation is said to *converge strongly* to the solution $X(t)$ at time T if

$$\lim_{\Delta t \to 0} E\{|X(T) - w_{\Delta t}(T)|\} = 0$$

where $w_{\Delta t}$ is the approximate solution computed with constant stepsize Δt, and E denotes expected value (Platen 1999). For strongly convergent approximations, we further quantify the rate of convergence by the concept of order. An SDE solver *converges strongly with order m* if the expected value of the error is of mth order in the stepsize, i.e. if for any time T,

$$E\{|X(T) - w_{\Delta t}(T)|\} = O((\Delta t)^m)$$

for sufficiently small stepsize Δt. This definition generalizes the standard convergence criterion for ordinary differential equations, reducing to the usual definition when the stochastic part of the equation goes to zero (Higham 2001, Higham and Kloeden 2005).

Although the Euler method for ordinary differential equations has order 1, the strong order for the Euler-Maruyama method for stochastic differential equations is $1/2$. This fact was proved in Gikhman and Skorokhod (1972), under appropriate conditions on the functions a and b in (19.6).

In order to build a strong order 1 method for SDEs, another term in the "stochastic Taylor series" must be added to the method. Consider the stochastic differential equation

$$\begin{cases} dX(t) = a(X,t)dt + b(X,t)dW_t \\ X(0) = X_0. \end{cases} \tag{19.13}$$

Milstein Method

$$\begin{aligned} w_0 &= X_0 \\ w_{i+1} &= w_i + a(w_i, t_i)\Delta t_i + b(w_i, t_i)\Delta W_i \\ &\quad + \frac{1}{2} b(w_i, t_i) \frac{\partial b}{\partial x}(w_i, t_i)(\Delta W_i^2 - \Delta t_i) \end{aligned} \tag{19.14}$$

The Milstein Method has order one (Milstein 1985, 1995, 1997, 2004, 2005). Note that the Milstein Method is identical to the Euler-Maruyama Method if there is no X term in the diffusion part $b(X,t)$ of the equation. In case there is, Milstein will in general converge to the correct stochastic solution process more quickly than Euler-Maruyama as the step size Δt_i goes to zero.

For comparison of the Euler-Maruyama and Milstein methods, we apply them to the Black–Scholes stochastic differential equation

$$dX = \mu X \, dt + \sigma X \, dW_t. \tag{19.15}$$

We discussed the Euler-Maruyama approximation above. The Milstein Method becomes

$$\begin{aligned} w_0 &= X_0 \\ w_{i+1} &= w_i + \mu w_i \Delta t_i + \sigma w_i \Delta W_i + \tfrac{1}{2}\sigma(\Delta W_i^2 - \Delta t_i) \end{aligned} \tag{19.16}$$

Applying the Euler-Maruyama Method and the Milstein Method with decreasing stepsizes Δt results in successively improved approximations, as Table 19.1 shows:

The two columns represent the average, over 100 realizations, of the error $|w(T) - X(T)|$ at $T = 8$. The orders $1/2$ for Euler-Maruyama and 1 for Milstein are clearly visible in the table. Cutting the stepsize by a factor of 4 is required to reduce the error by a factor of 2 with the Euler-Maruyama method. For the Milstein method, cutting the stepsize by a factor of 2 achieves the same result. The data in the table is plotted on a log-log scale in Fig. 19.3.

The Milstein method is a Taylor method, meaning that it is derived from a truncation of the stochastic Taylor expansion of the solution. This is in many cases a disadvantage, since the partial derivative appears in the approximation method, and must be provided explicitly by the user. This is analogous to Taylor methods for

Table 19.1 Average error at $T = 8$ of approximate solutions of (19.4). The error scales as $\Delta t^{1/2}$ for Euler-Maruyama and Δt for Milstein

Δt	Euler-Maruyama	Milstein
2^{-1}	0.169369	0.063864
2^{-2}	0.136665	0.035890
2^{-3}	0.086185	0.017960
2^{-4}	0.060615	0.008360
2^{-5}	0.048823	0.004158
2^{-6}	0.035690	0.002058
2^{-7}	0.024277	0.000981
2^{-8}	0.016399	0.000471
2^{-9}	0.011897	0.000242
2^{-10}	0.007913	0.000122

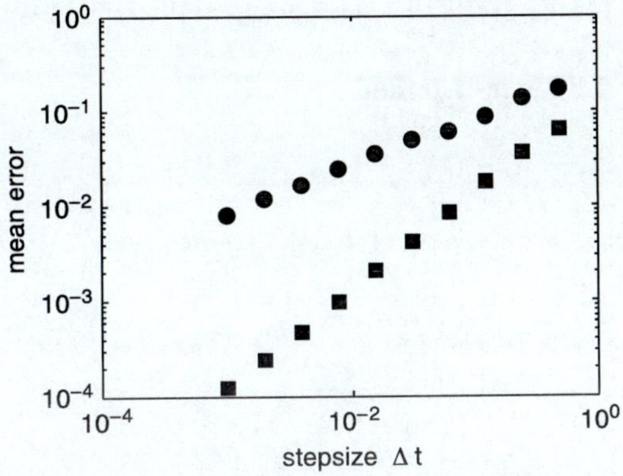

Fig. 19.3 Error in the Euler-Maruyama and Milstein methods. Solution paths are computed for the geometric Brownian motion equation (19.15) and are compared to the correct $X(T)$ given by (19.5). The absolute difference is plotted versus stepsize h for the two different methods. The Euler-Maruyama errors are plotted as *circles* and the Milstein error as *squares*. Note the slopes $1/2$ and 1, respectively, on the log-log plot

solving ordinary differential equations, which are seldom used in practice for that reason. To counter this problem, Runge–Kutta methods were developed for ODEs, which trade these extra partial derivatives in the Taylor expansion for extra function evaluations from the underlying equation.

In the stochastic differential equation context, the same trade can be made with the Milstein method, resulting in a strong order 1 method that requires evaluation of $b(X)$ at two places on each step. A heuristic derivation can be carried out by making the replacement

$$b_x(w_i) \approx \frac{b(w_i + b(w_i)\sqrt{\Delta t_i}) - b(w_i)}{b(w_i)\sqrt{\Delta t_i}}$$

in the Milstein formula (19.14), which leads to the following method (Rumelin, 1982):

Strong Order 1.0 Runge–Kutta Method

$$
\begin{aligned}
w_0 &= X_0 \\
w_{i+1} &= w_i + a(w_i)\Delta t_i + b(w_i)\Delta W_i \\
&\quad + \frac{1}{2}[b(w_i + b(w_i)\sqrt{\Delta t_i}) - b(w_i)](\Delta W_i^2 - \Delta t_i)/\sqrt{\Delta t_i}
\end{aligned}
$$

The orders of the methods introduced here for SDEs, $1/2$ for Euler-Maruyama and 1 for Milstein and the Runge–Kutta counterpart, would be considered low by ODE standards. Higher-order methods can be developed for SDEs, but become much more complicated as the order grows (Saito and Mitsui 1996, Burrage et al. 2000, Burrage et al. 2004, Higham et al. 2002). As an example, consider the strong order 1.5 scheme for the SDE (19.13) proposed in Platen and Wagner (1982):

Strong Order 1.5 Taylor Method

$$
\begin{aligned}
w_0 &= X_0 \\
w_{i+1} &= w_i + a\Delta t_i + b\Delta W_i + \frac{1}{2}bb_x\left(\Delta W_i^2 - \Delta t_i\right) \\
&\quad + a_y\sigma\Delta Z_i + \frac{1}{2}\left(aa_x + \frac{1}{2}b^2a_{xx}\right)\Delta t_i^2 \\
&\quad + \left(ab_x + \frac{1}{2}b^2b_{xx}\right)(\Delta W_i\Delta t_i - \Delta Z_i) \\
&\quad + \frac{1}{2}b\left(bb_{xx} + b_x^2\right)\left(\frac{1}{3}\Delta W_i^2 - \Delta t_i\right)\Delta W_i
\end{aligned}
\tag{19.17}
$$

where partial derivatives are denoted by subscripts, and where the additional random variable ΔZ_i is normally distributed with mean 0, variance $E(\Delta Z_i^2) = \frac{1}{3}\Delta t_i^3$ and correlated with ΔW_i with covariance $E(\Delta Z_i \Delta W_i) = \frac{1}{2}\Delta t_i^2$. Note that ΔZ_i can be generated as

$$\Delta Z_i = \tfrac{1}{2}\Delta t_i(\Delta W_i + \Delta V_i/\sqrt{3})$$

where ΔV_i is chosen independently from $\sqrt{\Delta t_i}N(0,1)$.

Whether higher-order methods are needed in a given application depends on how the resulting approximate solutions are to be used. In the ordinary differential equation case, the usual assumption is that the initial condition and the equation are known with accuracy. Then it makes sense to calculate the solution as closely as possible to the same accuracy, and higher-order methods are called for. In the context of stochastic differential equations, in particular if the initial conditions are chosen from a probability distribution, the advantages of higher-order solvers are often less compelling, and if they come with added computational expense, may not be warranted.

19.4 Weak Convergence of SDE Solvers

Strong convergence allows accurate approximations to be computed on an individual realization basis. For some applications, such detailed pathwise information is required. In other cases, the goal is to ascertain the probability distribution of the solution $X(T)$, and single realizations are not of primary interest.

Weak solvers seek to fill this need. They can be simpler than corresponding strong methods, since their goal is to replicate the probability distribution only. The following additional definition is useful.

A discrete-time approximation $w_{\Delta t}$ with step-size Δt is said to *converge weakly* to the solution $X(T)$ if

$$\lim_{\Delta t \to 0} E\{f(w_{\Delta t}(T))\} = E\{f(X(T))\}$$

for all polynomials $f(x)$. According to this definition, all moments converge as $\Delta t \to 0$. If the stochastic part of the equation is zero and the initial value is deterministic, the definition agrees with the strong convergence definition, and the usual ordinary differential equation definition.

Weakly convergent methods can also be assigned an order of convergence. We say that a the solver *converges weakly with order* m if the error in the moments is of mth order in the stepsize, or

$$|E\{f(X(T))\} - E\{f(w_{\Delta t}(T))\}| = O((\Delta t)^m)$$

for sufficiently small stepsize Δt.

In general, the rates of weak and strong convergence do not agree. Unlike the case of ordinary differential equations, where the Euler method has order 1, the Euler-Maruyama method for SDEs has strong order $m = 1/2$. However, Euler-Maruyama is guaranteed to converge weakly with order 1.

Higher order weak methods can be much simpler than corresponding strong methods, and are available in several different forms. The most direct approach is to exploit the Ito-Taylor expansion (Kloeden and Platen, 1992), the Ito calculus analogue of the Taylor expansion of deterministic functions. An example SDE solver that converges weakly with order 2 is the following:

Weak Order 2 Taylor Method

$$\begin{aligned} w_0 &= X_0 \\ w_{i+1} &= w_i + a\Delta t_i + b\Delta W_i + \tfrac{1}{2}bb_x(\Delta W_i^2 - \Delta t_i) \\ &\quad + a_x b\Delta Z_i + \tfrac{1}{2}(aa_x + \tfrac{1}{2}a_{xx}b^2)\Delta t^2 \\ &\quad + (ab_x + \tfrac{1}{2}b_{xx}b^2)(\Delta W_i \Delta t_i - \Delta Z_i) \end{aligned} \tag{19.18}$$

where ΔW_i is chosen from $\sqrt{\Delta t_i}N(0,1)$ and ΔZ_i is distributed as in the above Strong Order 1.5 Method.

A second approach is to mimic the idea of Runge–Kutta solvers for ordinary differential equations. These solvers replace the explicit higher derivatives in the Ito-Taylor solvers with extra function evaluations at interior points of the current solution interval. Platen (1987) proposed the following weak order 2 solver of Runge–Kutta type:

Weak Order 2 Runge–Kutta Method

$$\begin{aligned} w_0 &= X_0 \\ w_{i+1} &= w_i + \tfrac{1}{2}[a(u) + a(w_i)]\Delta t_i \\ &\quad + \frac{1}{4}[b(u_+) + b(u_-) + 2b(w_i)]\Delta W_i \\ &\quad + \frac{1}{4}[b(u_+) - b(u_-)](\Delta W_i^2 - \Delta t)/\sqrt{\Delta t_i} \end{aligned} \tag{19.19}$$

where

$$\begin{aligned} u &= w_i + a\Delta t_i + b\Delta W_i \\ u_+ &= w_i + a\Delta t_i + b\sqrt{\Delta t_i} \\ u_- &= w_i + a\Delta t_i - b\sqrt{\Delta t_i}. \end{aligned}$$

Figure 19.4 compares the Euler-Maruyama method, which converges with order 1 in the weak sense, to the Weak Order 2 Runge–Kutta-Type Method. Note the difference between strong and weak convergence. In the previous Fig. 19.3, which considers strong convergence, the mean error of the estimate of a point $X(T)$ on the solution curve was plotted. In Fig. 19.4, on the other hand, the mean error of the estimate of the expected value $E[X(T)]$ is plotted, since we are comparing weak convergence of the methods. The weak orders are clearly revealed by the log-log plot.

Several other higher-order weak solvers can be found in Kloeden and Platen (1992). Weak Taylor methods of any order can be constructed, as well as Runge–Kutta analogues that reduce or eliminate the derivative calculations (Talay and Tubaro 1990, Tocino and Ardanuy 2002, Jentzen et al. 2008). In addition, standard Richardson extrapolation techniques (Sauer, 2006) can be used to bootstrap weak method approximations of a given order to the next order. See (Kloeden and Platen 1992) or (Kloeden et al. 1994) for details.

Weak solvers are often an appropriate choice for financial models, when the goal is to investigate the probability distribution of an asset price or interest rate, or when Monte-Carlo sampling is used to price a complicated derivative. In such cases it is typical to be primarily interested in one of the statistical moments

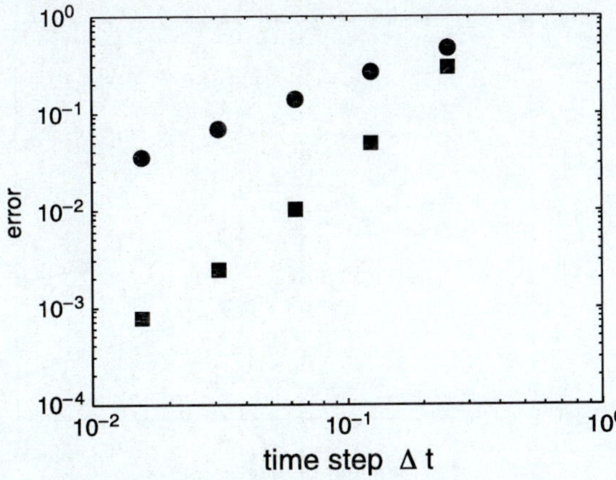

Fig. 19.4 The mean error of the estimation of $E(X(T))$ for SDE (19.15). The plot compares the Euler-Maruyama method (*circles*) which has weak order 1, and the weak order 2 Runge–Kutta type method (*squares*) given in (19.19). Parameter used were $X(0) = 10, T = 1, \mu = -3, \sigma = 0.2$

of a stochastically-defined quantity, and weak methods may be simpler and still sufficient for the sampling purpose. In the next section we explore some of the most common ways SDE solvers are used to carry out Monte-Carlo simulations for derivative pricing.

19.5 Monte-Carlo Sampling of SDE Paths for Option Pricing

As an illustrative example of the use of SDE solvers for option pricing, consider the European call, whose value at expiration time T is $\max\{X(T) - K, 0\}$, where $X(t)$ is the price of the underlying stock, K is the strike price (Hull 2002). The no-arbitrage assumptions of Black–Scholes theory imply that the present value of such an option is

$$C(X_0, T) = e^{-rT} E(\max\{X(T) - K, 0\}) \tag{19.20}$$

where r is the fixed prevailing interest rate during the time interval $[0, T]$, and where the underlying stock price $X(t)$ satisfies the stochastic differential equation

$$dX = rX\,dt + \sigma X\,dW_t.$$

The value of the call option can be determined by calculating the expected value (19.20) explicitly. Using the Euler-Maruyama method for following solutions to the Black–Scholes SDE, the value $X(T)$ at the expiration time T can be determined for each path, or realization of the stochastic process. For a given n realizations, the average $\langle \max\{X(T) - K, 0\}\rangle$ can be used as an approximation to the expected

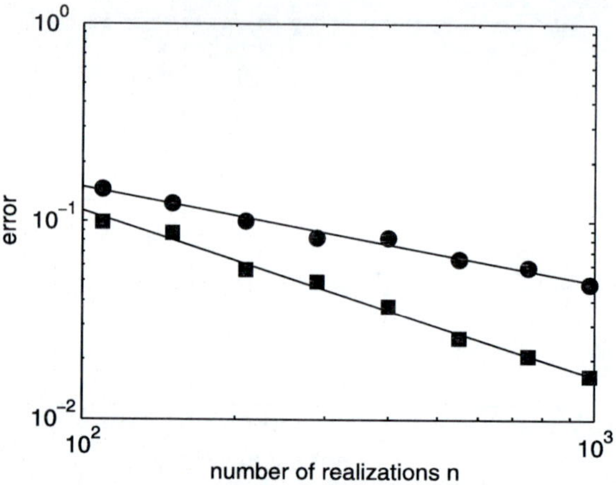

Fig. 19.5 Option pricing comparison between pseudo-random and quasi-random numbers. *Circles* (*squares*) represent error in Monte-Carlo estimation of European call by following SDE paths using pseudo-random (quasi-random) numbers to generate increments. Settings were $X(0) = 10, K = 12, r = 0.05, \sigma = 0.5$, expiration time $T = 0.5$. The number of Wiener increments per trajectory was $m = 8$

value in (19.20). Carrying this out and comparing with the exact solution from the Black–Scholes formula

$$C(X,T) = XN(d_1) - Ke^{-rT}N(d_2) \tag{19.21}$$

where

$$d_1 = \frac{\log(X/K) + (r + \frac{1}{2}\sigma^2)T}{\sigma\sqrt{T}}, \quad d_2 = \frac{\log(X/K) + (r - \frac{1}{2}\sigma^2)T}{\sigma\sqrt{T}},$$

yields the errors plotted as circles in Fig. 19.5.

The results above were attained using pseudo-random numbers (Park and Miller, 1988, Hellekalek 1998, Marsaglia and Zaman 1991, Marsaglia and Tsang 2000) to generate the Wiener increments ΔW in the Euler-Maruyama method. An improvement in accuracy can be achieved by using *quasi-random* numbers instead.

By definition, standard normal pseudo-random numbers are created to be independent and identically-distributed, where the distribution is the standard normal distribution. For many Monte-Carlo sampling problems, the independence is not crucial to the computation (Rubinstein, 1981, Fishman 1996, Gentle 2003, Glasserman 2004). If that assumption can be discarded, then there are more efficient ways to sample, using what are called *low-discrepancy* sequences. Such quasi-random sequences are identically-distributed but not independent. Their advantage is that they are better at self-avoidance than pseudo-random numbers, and by essentially reducing redundancy they can deliver Monte-Carlo approximations of significantly reduced variance with the same number of realizations.

Consider the problem of estimating an expected value like (19.20) by calculating many realizations. By Property 2 of the Wiener process, the m increments $\Delta W_1, \ldots, \Delta W_m$ of each realization must be independent. Therefore along the trajectories, independence must be preserved. This is accomplished by using m different low-discepancy sequences along the trajectory. For example, the base-p low discrepancy sequences due to Halton (1960) for m different prime numbers p can be used along the trajectory, while the sequences themselves run across different realizations.

Figure 19.5 shows a comparison of errors for the Monte-Carlo pricing of the European call, using this approach to create quasi-random numbers. The low-discrepancy sequences produce nonindependent uniform random numbers, and must be run through the Box-Muller method (Box and Muller, 1958) to produce normal quasi-random numbers. The pseudo-random sequences show error proportional to $n^{-0.5}$, while the quasi-random appear to follow approximately $n^{-0.7}$.

More sophisticated low-discrepancy sequences, due to Faure, Niederreiter, Xing, and others, have been developed and can be shown to be more efficient than the Halton sequences (Niederreiter 1992). The chapter in this volume by Niederreiter (Niederreiter, 2010) describes the state of the art in generating such sequences.

The quasi-random approach can become too cumbersome if the number of steps m along each SDE trajectory becomes large. As an example, consider a *barrier* option, whose value is a function of the entire trajectory. For a *down-and-out* barrier call, the payout is canceled if the underlying stock drops belong a certain level during the life of the option. Therefore, at time T the payoff is $\max(X(T) - K, 0)$ if $X(t) > L$ for $0 < t < T$, and 0 otherwise. For such an option, accurate pricing

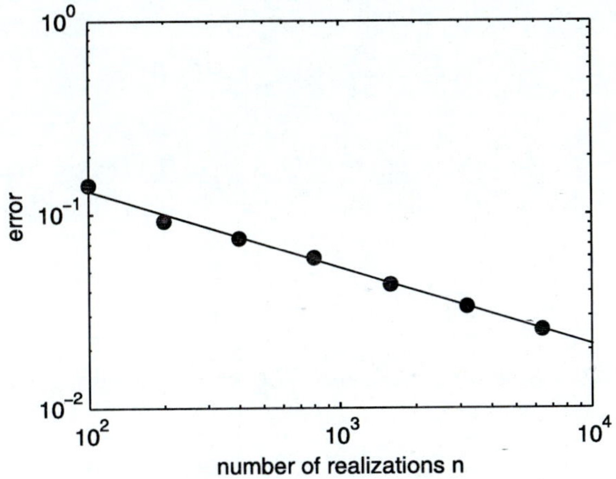

Fig. 19.6 Pricing error for barrier down-and-out call option. Error is proportional to the square root of the number of Monte-Carlo realizations

is dependent on using a relatively large number of steps m per trajectory. Results of a Monte-Carlo simulation of this modified call option are shown in Fig. 19.6, where the error was computed by comparison with the exact price

$$V(X,T) = C(X,T) - \left(\frac{X}{L}\right)^{1-\frac{2r}{\sigma^2}} C(L^2/X,T)$$

where $C(X,t)$ is the standard European call value with strike price K. The trajectories were generated with Euler-Maruyama approximations with pseudo-random number increments, where $m = 1000$ steps were used.

Other approaches to making Monte-Carlo sampling of trajectories more efficient fall under the umbrella of variance reduction. The idea is to calculate the expected value more accurately with fewer calls to the random number generator. The concept of *antithetic variates* is to follow SDE solutions in pairs, using the Wiener increment in one solutions and its negative in the other solution at each step. Due to the symmetry of the Wiener process, the solutions are equally likely. For the same number of random numbers generated, the standard error is decreased by a factor of $\sqrt{2}$.

A stronger version of variance reduction in computing averages from SDE trajectories can be achieved with *control variates*. We outline one such approach, known as variance reduction by delta-hedging. In this method the quantity that is being estimated by Monte-Carlo is replaced with an equivalent quantity of smaller variance. For example, instead of approximating the expected value of (19.20), the cash portion of the replicating portfolio of the European call can be targeted, since it must equal the option price at expiration.

Let C_0 be the option value at time $t = 0$, which is the goal of the calculation. At the time $t = 0$, the seller of the option hedges by purchasing $\Delta = \frac{\partial C}{\partial X}$ shares of the underlying asset. Thus the cash account, valued forward to time T, holds

$$\left[C_0 - \frac{\partial C}{\partial X}(t_0)X_{t_0}\right]e^{r(T-t_0)}.$$

At time step $t = t_1$, the seller needs to hold $\Delta = \frac{\partial C}{\partial X}(t_1)$ shares, so after purchasing $\frac{\partial C}{\partial X}(t_1) - \frac{\partial C}{\partial X}(t_0)$ shares, the cash account (valued forward) drops by

$$\left[\frac{\partial C}{\partial X}(t_1) - \frac{\partial C}{\partial X}(t_0)]X_{t_1}\right]e^{r(T-t_1)}.$$

Continuing in this way, the cash account of the replicating portfolio at time T, which must be C_T, equals

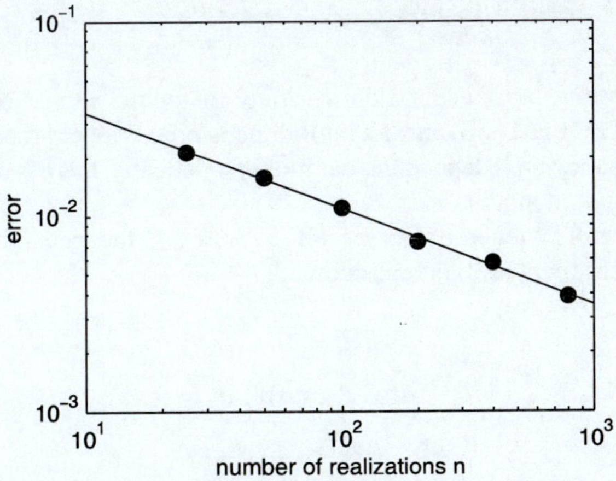

Fig. 19.7 Estimation errors for European call using control variates. Error is proportional to the square root of the number of Monte-Carlo realizations. Compare absolute levels of error with Fig. 19.5

$$C_0 e^{r(T-t_0)} - \sum_{k=0}^{N} \left[\frac{\partial C}{\partial X}(t_k) - \frac{\partial C}{\partial X}(t_{k-1}) \right] X_{t_k} e^{r(T-t_k)}$$

$$= C_0 e^{r(T-t_0)} + \sum_{k=0}^{N-1} \frac{\partial C}{\partial X}(t_k)(X_{t_{k+1}} - X_{t_k} e^{r\Delta t}) e^{r(T-t_{k+1})}$$

and so

$$C_0 = e^{-r(T-t_0)} \left[C_T - \sum_{k=0}^{N-1} \frac{\partial C}{\partial X}(t_k)(X_{t_{k+1}} - X_{t_k} e^{r\Delta t}) e^{r(T-t_{k+1})} \right]$$

$$= e^{-r(T-t_0)} [C_T - \mathrm{cv}]$$

where cv denotes the control variate. Estimating the expected value of this expression yields fast convergence, as demonstrated in Fig. 19.7. Compared to Fig. 19.5, the errors in pricing of the European call are lower by an order of magnitude for a similar number of realizations. However, the calculation of the control variate adds significantly to the computational load, and depending on the form of the derivative, may add more overhead than is gained from the reduced variance in some cases.

19.6 Multifactor Models

Financial derivatives that depend on a variety of factors should be modeled as a stochastic process that is driven by a multidimensional Wiener process. The various random factors may be independent, but more realistically, there is often correlation between the random inputs.

For multifactor Wiener processes $(W_t^1, \ldots, W_t^k)$, the generalization of Ito's Formula requires that (19.3) is replaced with

$$\begin{aligned} dt\, dt &= 0 \\ dt\, dW_t^i &= dW_t^i\, dt = 0 \\ dW_t^i\, dW_t^j &= \rho_{ij}\, dt \end{aligned} \tag{19.22}$$

where ρ_{ij} represents the statistical correlation between W_t^i and W_t^j. As usual, correlation ρ of two random variables X_1 and X_2 is defined as

$$\rho(X_1, X_2) = \frac{\text{cov}(X_1, X_2)}{\sqrt{V(X_1)}\sqrt{V(X_2)}}.$$

Note that $\rho(X_1, X_1) = 1$, and X_1 and X_2 are uncorrelated if $\rho(X_1, X_2) = 0$.

To construct discretized correlated Wiener processes for use in SDE solvers, we begin with a desired correlation matrix

$$R = \begin{bmatrix} \rho_{11} & \cdots & \rho_{1k} \\ \vdots & & \vdots \\ \rho_{k1} & \cdots & \rho_{kk} \end{bmatrix}$$

that we would like to specify for Wiener processes $W^1, \ldots, W^k$. The matrix R is symmetric with units on the main diagonal. A straightforward way to create noise processes with a specified correlation is through the singular value decomposition (SVD) (see Sauer 2006 for a description). The SVD of R is

$$R = \Gamma \Lambda \Gamma^\top$$

where Γ is an orthogonal matrix ($\Gamma^{-1} = \Gamma^\top$), and Λ is a diagonal matrix with nonzero entries on the main diagonal.

Begin with k independent, uncorrelated Wiener processes $Z_1, \ldots, Z_k$, satisfying $dZ_i dZ_i = dt, dZ_i dZ_j = 0$ for $i \neq j$. Define the column vector $\mathbf{dW} = \Gamma \Lambda^{1/2} \mathbf{dZ}$, and check that the covariance matrix, and therefore the correlation matrix, of $\mathbf{dW}$ is

$$\begin{aligned}\mathbf{dWdW}^\top &= \Gamma\Lambda^{1/2}\mathbf{dZ}(\Gamma\Lambda^{1/2}\mathbf{dZ})^\top \\ &= \Gamma\Lambda^{1/2}\mathbf{dZdZ}^\top\Lambda^{1/2}\Gamma^\top \\ &= \Gamma\Lambda\Gamma^\top dt = R\,dt\end{aligned}$$

For example, a two-asset market has correlation matrix

$$R = \begin{bmatrix} 1 & \rho \\ \rho & 1 \end{bmatrix} = \begin{bmatrix} \mathrm{corr}(W^1, W^1) & \mathrm{corr}(W^1, W^2) \\ \mathrm{corr}(W^2, W^1) & \mathrm{corr}(W^2, W^2) \end{bmatrix}.$$

Since the SVD of this 2×2 correlation matrix is

$$\begin{bmatrix} 1 & \rho \\ \rho & 1 \end{bmatrix} = \begin{bmatrix} \frac{1}{\sqrt{2}} & \frac{1}{\sqrt{2}} \\ \frac{1}{\sqrt{2}} & -\frac{1}{\sqrt{2}} \end{bmatrix} \begin{bmatrix} 1+\rho & 0 \\ 0 & 1-\rho \end{bmatrix} \begin{bmatrix} \frac{1}{\sqrt{2}} & \frac{1}{\sqrt{2}} \\ \frac{1}{\sqrt{2}} & -\frac{1}{\sqrt{2}} \end{bmatrix},$$

we calculate

$$\begin{aligned} dW^1 &= \frac{\sqrt{1+\rho}}{\sqrt{2}}\,dZ^1 + \frac{\sqrt{1-\rho}}{\sqrt{2}}\,dZ^2 \\ dW^2 &= \frac{\sqrt{1+\rho}}{\sqrt{2}}\,dZ^1 - \frac{\sqrt{1-\rho}}{\sqrt{2}}\,dZ^2. \end{aligned} \tag{19.23}$$

With a change of variables, the correlation ρ can be generated alternatively as

$$\begin{aligned} dW^1 &= dZ^1 \\ dW^2 &= \rho\,dZ^1 + \sqrt{1-\rho^2}\,dZ^2. \end{aligned} \tag{19.24}$$

As a simple example, we calculate the value of a European spread call using Monte-Carlo estimation of noise-coupled stochastic differential equations using a two-factor model. Assume there are two assets X_1 and X_2 satisfying arbitrage-free SDE's of form

$$\begin{aligned} dX_1 &= rX_1\,dt + \sigma_1 X_1\,dW^1 \\ dX_1 &= rX_2\,dt + \sigma_2 X_3\,dW^2 \end{aligned} \tag{19.25}$$

where $dW^1 dW^2 = \rho\,dt$, and that the payout at expiration time T is $\max\{X_1(T) - X_2(T) - K, 0\}$ for a strike price K. The Monte-Carlo approach means estimating the expected value

$$E(e^{-rT}\max\{X_1(T) - X_2(T) - K, 0\}).$$

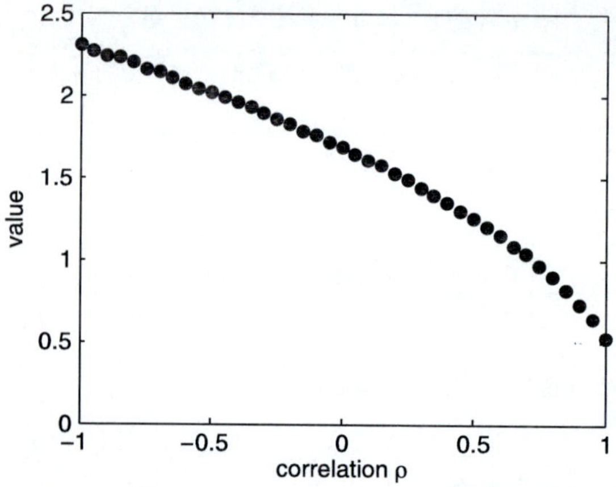

Fig. 19.8 European spread call value as a function of correlation. The Euler-Maruyama solver was used with multifactor correlated Wiener processes. The initial values of the underlying assets were $X_1(0) = 10, X_2(0) = 8$, the interest rate was $r = 0.05$, strike price $K = 2$, and expiration time $T = 0.5$

Using either form (19.23) or (19.24) for the coupled Wiener increments in the Euler-Maruyama paths, the correct price can be calculated. Figure 19.8 shows the dependence of the price on the two-market correlation ρ. As can be expected, the more the assets move in an anticorrelated fashion, the more probable the spread call will land in the money.

19.7 Summary

Numerical methods for the solution of stochastic differential equations are essential for the analysis of random phenomena. Strong solvers are necessary when exploring characteristics of systems that depend on trajectory-level properties. Several approaches exist for strong solvers, in particular Taylor and Runge–Kutta type methods, although both increase greatly in complication for orders greater than one. We have restricted our discussion to fixed stepsize methods; consult Romisch and Winkler (2006) and Lamba et al. (2007) for extensions to adaptive stepsize selection.

In many financial applications, major emphasis is placed on the probability distribution of solutions, and in particular mean and variance of the distribution. In such cases, weak solvers may suffice, and have the advantage of comparatively less computational overhead, which may be crucial in the context of Monte-Carlo simulation.

Independent of the choice of stochastic differential equation solver, methods of variance reduction exist that may increase computational efficiency. The replacement of pseudorandom numbers with quasirandom analogues from low-discrepancy sequences is applicable as long as statistical independence along trajectories is maintained. In addition, control variates offer an alternate means of variance reduction and increases in efficiency in Monte-Carlo simulation of SDE trajectories.

References

Black, F., & Scholes, M. (1973). The pricing of options and corporate liabilities. *Journal of Political Economy*, *81*, 637–654.

Box, G. E. P., & Muller, M. (1958). A note on the generation of random normal deviates. *Annals of Mathematical Statistics*, *29*, 610–611.

Burrage, K., Burrage, P. M., & Mitsui, T. (2000). Numerical solutions of stochastic differential equations - implementation and stability issues. *Journal of Computational and Applied Mathematics*, *125*, 171–182.

Burrage, K., Burrage, P. M., & Tian, T. (2004). Numerical methods for strong solutions of stochastic differential equations: an overview. *Proceedings of the Royal Society of London A*, *460*, 373–402.

Fishman, G. S. (1996). *Monte Carlo: concepts, algorithms, and applications*. Berlin: Springer.

Gentle, J. E. (2003). *Random number generation and Monte Carlo methods* (2nd ed.). Berlin: Springer.

Gikhman, I., & Skorokhod, A. (1972). *Stochastic differential equations*. Berlin: Springer.

Glasserman, P. (2004). *Monte Carlo methods in financial engineering*. New York: Springer.

Halton, J. H. (1960). On the efficiency of certain quasi-random sequences of points in evaluating multi-dimensional integrals. *Numerische Mathematik*, 2, 84–90.

Hellekalek, P. (1998). Good random number generators are (not so) easy to find. *Mathematics and Computers in Simulation*, *46*, 485–505.

Higham, D. J. (2001). An algorithmic introduction to numerical simulation of stochastic differential equations. *SIAM Review*, *43*, 525–546.

Higham, D. J., & Kloeden, P. (2005). Numerical methods for nonlinear stochastic differential equations with jumps. *Numerische Mathematik*, *101*, 101–119.

Higham, D. J., Mao, X., & Stuart, A. (2002). Strong convergence of Euler-type methods for nonlinear stochastic differential equations. *SIAM Journal on Numerical Analysis*, *40*, 1041–1063.

Hull, J. C. (2002). *Options, futures, and other derivatives* (5th ed.). Upper Saddle River, NJ: Prentice Hall.

Jentzen, A., Kloeden, P., & Neuenkirch, A. (2008). Pathwise approximation of stochastic differential equations on domains: higher order convergence rates without global Lipschitz coefficients. *Numerische Mathematik*, *112*, 41–64.

Klebaner, F. (1998). *Introduction to stochastic calculus with applications*. London: Imperial College Press.

Kloeden, P., & Platen, E. (1992). *Numerical solution of stochastic differential equations*. Berlin: Springer.

Kloeden, P., Platen, E., & Schurz, H. (1994). *Numerical solution of SDE through computer experiments*. Berlin: Springer.

Lamba, H., Mattingly, J. C., & Stuart, A. (2007). An adaptive Euler-Maruyama scheme for SDEs: convergence and stability. *IMA Journal of Numerical Analysis*, *27*, 479–506.

Marsaglia, G., & Zaman, A. (1991). A new class of random number generators. *Annals of Applied Probability*, *1*, 462–480.

Marsaglia, G., & Tsang, W. W. (2000). The ziggurat method for generating random variables. *Journal of Statistical Software*, *5*, 1–7.

Maruyama, G. (1955). Continuous Markov processes and stochastic equations. *Rendiconti del Circolo Matematico di Palermo*, *4*, 48–90.

Milstein, G. (1988). A theorem on the order of convergence of mean-square approximations of solutions of stochastic differential equations. *Theory of Probability and Its Applications*, *32*, 738–741.

Milstein, G. (1995). *Numerical integration of stochastic differential equations*. Dordrecht: Kluwer.

Milstein, G., & Tretyakov, M. (1997). Mean-square numerical methods for stochastic differential equations with small noises. *SIAM Journal on Scientific Computing*, *18*, 1067–1087.

Milstein, G., & Tretyakov, M. (2004). *Stochastic numerics for mathematical physics*. Berlin: Springer.

Milstein, G., & Tretyakov, M. (2005). Numerical integration of stochastic differential equations with nonglobally Lipschitz coefficients. *SIAM Journal on Numerical Analysis*, *43*, 1139–1154.

Niederreiter, H. (1992). *Random number generation and quasi-Monte Carlo methods*. Philadelphia: SIAM Publications.

Niederreiter, H. (2010). Low-discrepancy simulation. In H. Niederreiter (Ed.), Handbook of Computational Finance 2011 (pp. 715–741). Berlin: Springer.

Oksendal, B. (1998). *Stochastic differential equations: an introduction with applications* (5th ed.). Berlin: Springer.

Park, S., & Miller, K. (1988). Random number generators: good ones are hard to find. *Communications of the ACM*, *31*, 1192–1201.

Platen, E. (1987). Derivative-free numerical methods for stochastic differential equations. *Lecture Notes in Control and Information Sciences*, *96*, 187–193.

Platen, E. (1999). An introduction to numerical methods for stochastic differential equations. *Acta Numerica*, *8*, 197–246.

Platen, E., & Wagner, W. (1982). On a Taylor formula for a class of Ito processes. *Probability and Mathematical Statistics*, *3*, 37–51.

Romisch, W., & Winkler, R. (2006). Stepsize control for mean-square numerical methods for stochastic differential equations with small noise. *SIAM Journal on Scientific Computing*, *28*, 604–625.

Rubinstein, R. Y. (1981). *Simulation and the Monte Carlo method*. New York: Wiley.

Rumelin, W. (1982). Numerical treatment of stochastic differential equations. *SIAM Journal of Numerical Analysis*, *19*, 604–613.

Saito, Y., & Mitsui, T. (1996). Stability analysis of numerical schemes for stochastic differential equations. *SIAM Journal of Numerical Analysis*, *33*, 2254–2267.

Sauer, T. (2006). *Numerical analysis*. Boston: Pearson.

Steele, J. M. (2001). *Stochastic calculus and financial applications*. New York: Springer.

Talay, D., & Tubaro, L. (1990). Expansion of the global error for numerical schemes solving stochastic differential equations. *Stochastic Analysis and Applications*, *8*, 483–509.

Tocino, A., & Ardanuy, R. (2002). Runge-Kutta methods for numerical solution of stochastic differential equations. *Journal of Computational and Applied Mathematics*, *138*, 219–241.

Chapter 20
Lattice Approach and Implied Trees

Rüdiger U. Seydel

Abstract Lattice methods or tree methods have become standard tools for pricing many types of options, since they are robust and easy to implement. The basic method is built on a binomial tree and assumes constant volatility and constant relative node spacing. The tree grows from the initial spot price, until maturity is reached. There the payoff is evaluated, and a subsequent backward recursion yields the value of the option. The resulting discrete-time approach is consistent with the continuous Black-Scholes model. This basic lattice approach has been extended to cope with a variable local volatility. Here the lattice nodes are determined based on market data of European-style options. In this way an "implied tree" is created matching the volatility smile. This chapter introduces into tree methods.

Lattice methods or tree methods play an important role in option pricing. They are robust, and relatively easy to implement. The first lattice method for option pricing is attributed to Cox et al. (1979). It is based on a binomial tree in the (S,t)-plane, where S denotes the price of the underlying and t is time. The recombining tree grows from the initial point of the current spot price $(S_0, 0)$, branching at equidistantly spaced time instances with grid spacing Δt, until the maturity is reached. Following the Black–Scholes model, the original lattice framework assumes a constant volatility σ. This enables a uniform mesh generation with a constant relative node spacing, and a minimum number of parameters. The discrete-time approach is consistent with the continuous Black–Scholes model, converging to the continuous value when the time step Δt goes to zero. The resulting *binomial method* is widely applicable. This basic version has been extended to handle payment of dividends, and exotic options.

R.U. Seydel (✉)
Mathematisches Institut der Universität zu Köln, Weyertal 86, 50931 Köln, Germany
e-mail: seydel@math.uni-koeln.de; www.compfin.de

J.-C. Duan et al. (eds.), *Handbook of Computational Finance*, Springer Handbooks of Computational Statistics, DOI 10.1007/978-3-642-17254-0_20,

As is well-known, the Black–Scholes model endures some shortcomings. One remedy is to tune the parameter σ such that it is not solely calibrated to the process S_t, but rather linked to option prices as seen through the eyes of the Black–Scholes formula. The adjusted parameter is the *implied volatility*. While this approach improves the ability to price exotic options under the Black–Scholes model, it is not fully satisfying because of the *smile*, which means the variation of σ with strike K and maturity T. In about 1994, Derman and Kani, Dupire, Rubinstein and others in a series of papers suggested lattice approaches that overcome the assumption of constant volatility. The new methods cope with a variable local volatility $\sigma(S,t)$. The S-values of the lattice nodes are calibrated on market data of European-style options. In this way a somewhat irregular grid is created, which allows to match the volatility smile. Resulting trees are called *implied trees*. The greater flexibility of the tree goes along with an increased number of parameters.

Today many different variants of lattice methods are in use. This chapter starts with an introduction into the basic idea of a tree method. Then variants will be reviewed and discussed, including a basic implied-tree approach.

20.1 Preparations

Figure 20.1 illustrates the geometrical setting, here for a vanilla put. The surface of the value function $V(S,t)$ of the option's price is shown, with the asset S-axis with strike K, and the time t-axis with maturity T. As solution of the Black–Scholes partial differential equation, the entire surface $V(S,t)$ for the rectangular half strip $S > 0$, $0 \leq t \leq T$ may be of interest, see Seydel (2009). This half strip is the *domain* of the value function. In practice one is often interested in the one value $V(S_0, 0)$ of an option at the current spot price S_0. Then it can be unnecessarily costly to calculate the surface $V(S,t)$ for the entire domain to extract the required information $V(S_0, 0)$. The relatively small task of calculating $V(S_0, 0)$

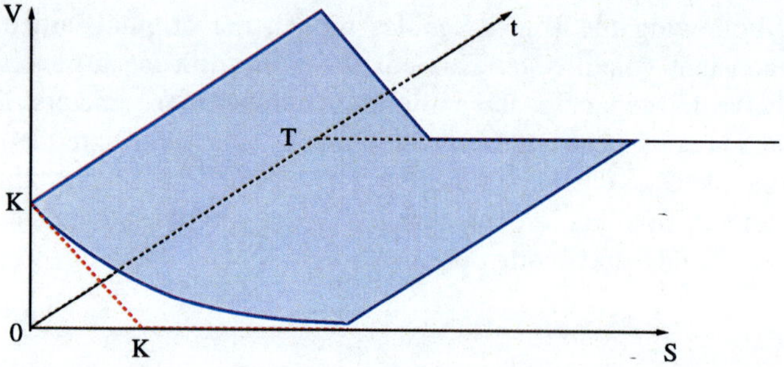

Fig. 20.1 Vanilla put option, schematically: the geometry of the value-function surface $V(S,t)$, and the main variables. At maturity $t = T$, the surface equals the payoff, which is redrawn for $t = 0$ (*dashed*)

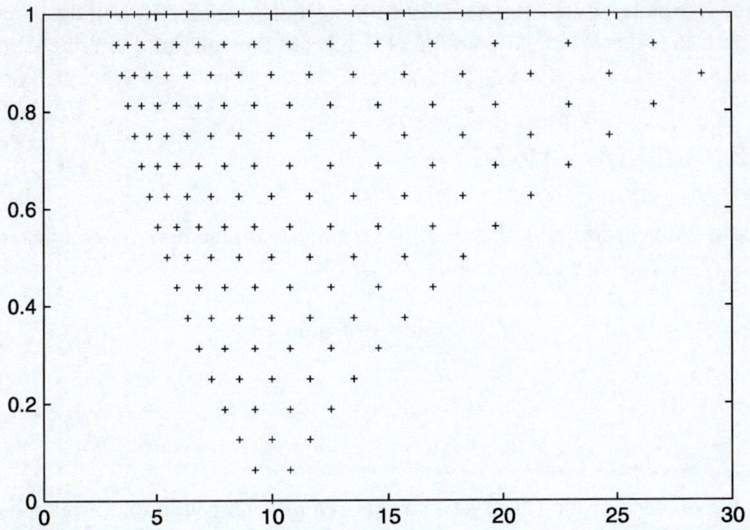

Fig. 20.2 The domain of the half strip $S > 0$, $0 \leq t \leq T$ with $T = 1$; nodes of a binomial tree with $M = 16$ layers of time steps; horizontally: S-axis cut to $S \leq 30$, with $S_0 = 10$; vertically: t-axis

can be comfortably solved using the binomial method. This method is based on a tree-type grid applying appropriate binary rules at each grid point. The grid is not predefined but is constructed by the method. For illustration see the grid nodes in Fig. 20.2.

20.1.1 The Continuous Problem

In this chapter, put or call options are to be valued. For vanilla options, the final payoff $\Psi(S_T)$ at maturity T is

$$\Psi(S_T) \stackrel{\text{def}}{=} \begin{cases} (S_T - K)^+ & \text{for a call,} \\ (K - S_T)^+ & \text{for a put,} \end{cases} \tag{20.1}$$

where $(f(S))^+ \stackrel{\text{def}}{=} \max\{f(S), 0\}$. For vanilla options there are market prices available. Market models help to valuate or analyze options. The famous model due to Black, Merton, and Scholes is a continuous-time model based on the assumption of a geometrical Brownian motion (GBM) for the asset S_t

$$\mathrm{d}S_t = (\mu - \delta)\, S_t \,\mathrm{d}t + \sigma\, S_t \,\mathrm{d}W_t \tag{20.2}$$

where W_t denotes a standard Wiener process. For the scenario of Sects. 20.1, 20.2, 20.3, the growth rate μ, the rate δ of a dividend yield, and the volatility σ are assumed constant. Later (in Sect. 20.4) we shall consider a nonlinear volatility function $\sigma(S,t)$. In Sect. 20.1 and Sect. 20.2, for the sake of a simple exposition, we assume $\delta = 0$.

20.1.2 A Discrete Model

We begin with discretizing the continuous time t, replacing t by equidistant time instances t_i. Let us use the notations

$$\begin{aligned} M &= \text{number of time steps} \\ \Delta t &= \frac{T}{M} \\ t_i &= i \cdot \Delta t, \quad i = 0, \ldots, M \\ S_i &= S(t_i) \end{aligned} \tag{20.3}$$

So far the domain of the (S,t) half strip is *semidiscretized* in that it is replaced by parallel straight lines with distance Δt apart, leading to a discrete-time model. For later reference, we list expectation and variance of GBM for constant μ, σ for this discrete time setting with time step Δt. Expectations of the continuous model, under the risk-free probability with constant risk-free interest rate r, are

$$\mathsf{E}(S_{i+1}) = S_i \, \mathrm{e}^{r\Delta t} \tag{20.4}$$

$$\mathsf{E}(S_{i+1}^2) = S_i^2 \, \mathrm{e}^{(2r+\sigma^2)\Delta t} \tag{20.5}$$

from which the variance follows.

The next step of discretization will replace the continuous values S_i along the line $t = t_i$ by discrete values $S_{j,i}$, for all i and appropriate j. A tree structure will emerge, for example, as illustrated by its nodes in Fig. 20.2. The root of the tree is the current asset price S_0 for $t = 0$. For a binomial tree, the tree will branch into two branches at each node (Figs. 20.2, 20.3, 20.4). The resulting tree serves a the grid on which the computation and valuation of an option will be performed.

What is needed now are rules how such a binomial tree should evolve. In the following Sect. 20.2, we describe the classical tree valuation method introduced by Cox et al. (1979), see also Rendleman and Bartter (1979), and Hull and White (1988). We label the binomial tree method with CRR. The CRR tree matches the Black–Merton–Scholes model. For a more general tree suitable for local volatility functions, we refer to Sect. 20.4.

20.2 The Basic CRR Binomial Method

For a better understanding of the S-discretization consult Fig. 20.3. This figure shows a mesh of the CRR grid, namely, the transition from t to $t + \Delta t$, or from t_i to t_{i+1}.

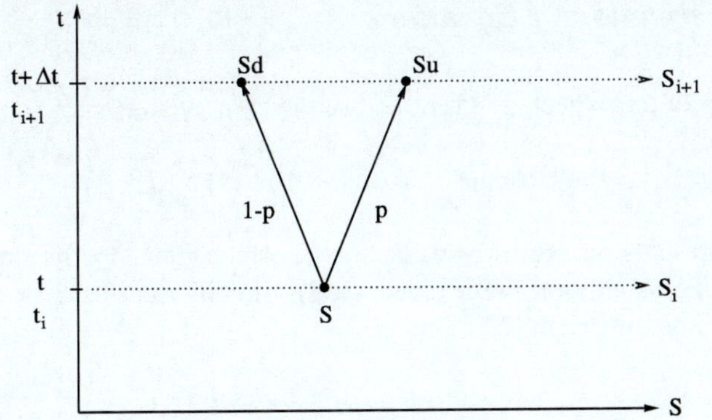

Fig. 20.3 The principle setup of the CRR binomial method

20.2.1 Assumptions

We assume (Bi1), (Bi2), and (Bi3) below.

(Bi1) The price S over each period of time Δt can only have two possible outcomes: An initial value S either evolves up to Su or down to Sd with $0 < d < u$. Here u is the factor of an upward movement and d is the factor of a downward movement.

(Bi2) The probability of an up movement is p, $\mathsf{P}(\text{up}) = p$.

The rules (Bi1) and (Bi2) represent the framework of a binomial process. Such a process behaves like tossing a biased coin where the outcome "head" (up) occurs with probability p. At this stage of the modeling, the values of the three parameters u, d and p are undetermined. They are fixed in a way such that the model is consistent with the continuous model in case $\Delta t \to 0$. This aim leads to further assumptions. The basic idea of the approach is to equate the expectation and the variance of the discrete model with the corresponding values of the continuous model. This amounts to require

(Bi3) Expectation and variance of S refer to their continuous counterparts, evaluated for the risk-free interest rate r.

This assumption leads to two equations for the parameters u, d, p. The resulting probability P of (Bi2) does not reflect the expectations of an individual in the market. Rather P is an artificial risk-neutral probability that matches (Bi3). The expectation E in (20.4) refers to this probability, which is sometimes written E_{P}. As noted above, we assume that no dividend is paid within the time period of interest. This assumption simplifies the derivation of the method and can be removed later.

20.2.2 *Derivation of Equations*

By definition of the expectation for the discrete case, we have

$$\mathsf{E}(S_{i+1}) = p\, S_i\, u + (1-p)\, S_i\, d.$$

Here S_i represents an arbitrary value for t_i, which develops randomly to S_{i+1}, following the assumptions (Bi1) and (Bi2). In this sense, E is a conditional expectation. Equating with (20.4) gives

$$e^{r\Delta t} = pu + (1-p)d \tag{20.6}$$

This is the first of three equations required to fix u, d, p. Solved for the risk-neutral probability p we obtain

$$p = \frac{e^{r\Delta t} - d}{u - d}.$$

To be a valid model of probability, $0 \leq p \leq 1$ must hold. This is equivalent to

$$d \leq e^{r\Delta t} \leq u.$$

These inequalities relate the upward and downward movements of the asset price to the riskless interest rate r. The inequalities are no new assumption but follow from the no-arbitrage principle. The assumption $0 < d < u$ remains sustained.

Next we equate variances. Via the variance the volatility σ enters the model. Recall that the variance satisfies $\mathsf{Var}(S) = \mathsf{E}(S^2) - (\mathsf{E}(S))^2$. Equations (20.4) and (20.5) combine to

$$\mathsf{Var}(S_{i+1}) = S_i^2 e^{2r\Delta t}(e^{\sigma^2\Delta t} - 1).$$

On the other hand the discrete model satisfies

$$\begin{aligned}\mathsf{Var}(S_{i+1}) &= \mathsf{E}(S_{i+1}^2) - (\mathsf{E}(S_{i+1}))^2 \\ &= p(S_i u)^2 + (1-p)(S_i d)^2 - S_i^2(pu + (1-p)d)^2.\end{aligned}$$

Equating variances of the continuous and the discrete model, and applying (20.6) leads to

$$e^{2r\Delta t + \sigma^2\Delta t} = pu^2 + (1-p)d^2 \tag{20.7}$$

Equations (20.6) and (20.7) constitute two relations for the three unknowns u, d, p.

20.2.2.1 Anchoring the Equations

Because there is one degree of freedom in (20.6) and (20.7) we are free to impose an arbitrary third equation. One class of examples is defined by the assumption

$$u \cdot d = \gamma, \tag{20.8}$$

for a suitable constant γ. The simple and plausible choice $\gamma = 1$ reflects a symmetry between upward and downward movement of the asset price. Alternatively to the choice $ud = 1$ in (20.8), the choice $p = \frac{1}{2}$ has been suggested, see Rendleman and Bartter (1979), Hull (2000), §16.5, or Wilmott et al. (1996).

When the strike K is not well grasped by the tree and its grid points, the error depending on M may oscillate. To facilitate extrapolation, it is advisable to have the strike value K on the medium grid point, $S_T = K$, no matter what (even) value of M is chosen. The error can be smoothed by special choices of u and d. To anchor the grid such that at the final line (for $t = T$) the center grid point always equals the strike K, one proceeds as follows. On the final line the grid points are

$$S_{j,M} = S_0 u^j d^{M-j}$$

for $j = 0, \ldots, M$. For even M, the center grid point has index $j = M/2$, and S-value

$$S_0 u^{M/2} d^{M/2}.$$

That is, for even M, set

$$S_0 u^{M/2} d^{M/2} = K$$

and the tree is centered at the strike. A straightforward calculation with

$$(ud)^{M/2} = \frac{K}{S_0} \quad \Rightarrow \quad \frac{2}{M} \log \frac{K}{S_0} = \log (ud)$$

gives the proper constant γ:

$$\gamma = ud = \exp\left[\frac{2}{M} \log \frac{K}{S_0}\right] \tag{20.9}$$

Now the parameters u, d and p are fixed by (20.6), (20.7), (20.8). They depend on r, σ and Δt. So does the grid, which is analyzed next (Fig. 20.4).

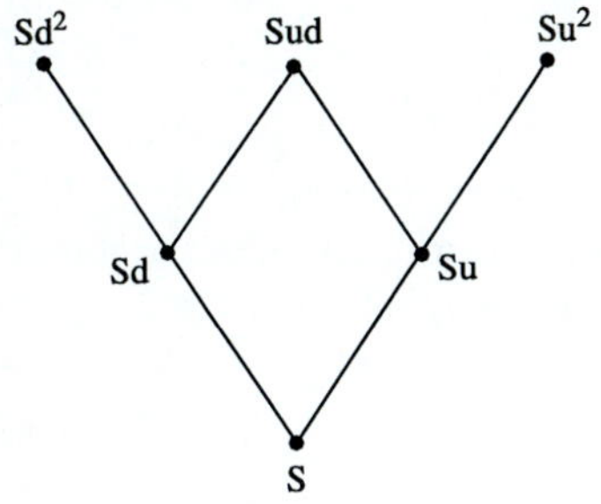

Fig. 20.4 Sequence of several meshes (schematically)

20.2.2.2 The Shape of the Grid

The above rules are applied to each grid line $i = 0, \ldots, M$, starting at $t_0 = 0$ with the specific value $S = S_0$. Attaching meshes of the kind depicted in Fig. 20.3 for subsequent values of t_i builds a tree with values $Su^j d^k$ and $j + k = i$. In this way, specific discrete values $S_{j,i}$ of S_i and the nodes of the tree are defined. Since the same constant factors u and d underlie all meshes and since $Sud = Sdu$ holds, after the time period $2\Delta t$ the asset price can only take three values rather than four: The tree is *recombining*. It does not matter which of the two possible paths we take to reach Sud. This property extends to more than two time periods. Consequently the binomial process defined by Assumptions (Bi1)–(Bi3) is *path independent*. Accordingly at expiration time $T = M\Delta t$ the price S can take only the $(M + 1)$ discrete values $Su^j d^{M-j}$, $j = 0, 1, \ldots, M$. For $ud = 1$ these are the values $Su^{2j-M} = S_{j,M}$. The number of nodes in the tree grows quadratically in M.

The symmetry of the choice $ud = 1$ becomes apparent in that after two time steps the asset value S repeats. (Compare also Fig. 20.2.) In the (t, S)-plane the tree can be interpreted as a grid of exponential-like curves. The binomial approach defined by (Bi1) with the proportionality between S_i and S_{i+1} reflects exponential growth or decay of S. So all grid points have the desirable property $S > 0$.

20.2.3 The Algorithm

Next we give a solution to the equations and set up the classical CRR algorithm.

20.2.3.1 Solution of the Equations

Using the abbreviations

$$\alpha = e^{r\Delta t}, \quad \beta = \frac{1}{2}\left(\frac{\gamma}{\alpha} + \alpha e^{\sigma^2 \Delta t}\right),$$

we obtain by elimination the quadratic equation

$$0 = u^2 - u\left(\frac{\gamma}{\alpha} + \alpha e^{\sigma^2 \Delta t}\right) + \gamma = u^2 - 2\beta u + \gamma,$$

with solutions $u = \beta \pm \sqrt{\beta^2 - \gamma}$. By virtue of $ud = \gamma$ and Vieta's Theorem, d is the solution with the minus sign. In summary, the three parameters u, d, p are given by

$$\begin{aligned}
\beta &= \tfrac{1}{2}(\gamma e^{-r\Delta t} + e^{(r+\sigma^2)\Delta t}) \\
u &= \beta + \sqrt{\beta^2 - \gamma} \\
d &= \gamma / u = \beta - \sqrt{\beta^2 - \gamma} \\
p &= \frac{e^{r\Delta t} - d}{u - d}
\end{aligned} \tag{20.10}$$

A consequence of this approach is that for $\gamma = 1$ the relation $u = e^{\sigma\sqrt{\Delta t}}$ holds up to terms of higher order. Therefore the extension of the tree in S-direction matches the volatility of the asset. So the tree is well-scaled and covers a relevant range of S-values. The original CRR choice is

$$u = e^{\sigma\sqrt{\Delta t}}, \quad d = e^{-\sigma\sqrt{\Delta t}}, \quad \tilde{p} \stackrel{\text{def}}{=} \frac{1}{2}\left(1 + \frac{r}{\sigma}\sqrt{\Delta t}\right),$$

where $\tilde{p}$ is a first-order approximation to the p of (20.10). The choice $p = 1/2$ leads to the parameters

$$u = e^{r\Delta t}\,(1 + \sqrt{e^{\sigma^2\Delta t} - 1}), \quad d = e^{r\Delta t}\,(1 - \sqrt{e^{\sigma^2\Delta t} - 1}).$$

In what follows, we stick to (20.10) with $\gamma = 1$.

20.2.3.2 Forward Phase: Initializing the Tree

Now the parameters u and d can be considered known, and the discrete node values of S for each t_i for all $i \leq M$ can be calculated. (To adapt the matrix-like notation to the two-dimensional grid of the tree, the initial price and root of the tree will be also denoted $S_{0,0}$.) Each initial price S_0 leads to another tree of node values $S_{j,i}$.

$$\text{For } i = 1, 2, \ldots, M \text{ calculate :}$$
$$S_{j,i} = S_0 u^j d^{i-j}, \quad j = 0, 1, \ldots, i$$

Now the grid points $(t_i, S_{j,i})$ are fixed, on which approximations to the option values $V_{j,i} \stackrel{\text{def}}{=} V(S_{j,i}, t_i)$ are to be calculated.

20.2.3.3 Calculating the Option Value, Valuation on the Tree

For t_M the values $V(S, t_M)$ are known from the payoff (20.1). This payoff is valid for each S, including $S_{j,M} = S u^j d^{M-j}$, $j = 0, \ldots, M$, and defines the values $V_{j,M} = \Psi(S_{j,m})$:

Call:

$$V_{j,M} = (S_{j,M} - K)^+ \tag{20.11}$$

Put:

$$V_{j,M} = (K - S_{j,M})^+ \tag{20.12}$$

The **backward phase** calculates recursively for $t_{M-1}, t_{M-2}, \ldots$ the option values V for all t_i, starting from $V_{j,M}$. Recall that based on Assumption (Bi3) the equation that corresponds to (20.6) with double index leads to

$$S_{j,i}\,\mathrm{e}^{r\Delta t} = p\,S_{j,i}\,u + (1-p)\,S_{j,i}\,d,$$

and

$$S_{j,i} = \mathrm{e}^{-r\Delta t}\,(pS_{j+1,i+1} + (1-p)S_{j,i+1}).$$

This manifestation of risk neutrality is valid also for V, $V_i = \mathrm{e}^{-r\Delta t}\mathsf{E}(V_{i+1})$. In double-index notation the recursion is

$$V_{j,i} = \mathrm{e}^{-r\Delta t}\,\left(pV_{j+1,i+1} + (1-p)V_{j,i+1}\right). \tag{20.13}$$

This recursion for $V_{j,i}$ is no further assumption, but a consequence of the no-arbitrage principle and the risk-neutral valuation. For **European options**, (20.13) is a recursion for $i = M-1,\ldots,0$, starting from (20.11), (20.12), and terminating with $V_{0,0}$. The obtained value $V_{0,0}$ is an approximation to the value $V(S_0,0)$ of the continuous model, which results in the limit $M\to\infty$ ($\Delta t\to 0$). The accuracy of the approximation $V_{0,0}$ depends on M. This is reflected by writing $V_0^{(M)}$. The basic idea of the approach implies that the limit of $V_0^{(M)}$ for $M\to\infty$ is the Black–Scholes value $V(S_0,0)$, see below Sect. 2.5.

For **American options**, the above recursion must be modified by adding a test whether early exercise is to be preferred. To this end the value of (20.13) is compared with the value of the payoff Ψ. In this context, the value (20.13) is the *continuation value*, denoted $V_{j,i}^{\text{cont}}$. And at any time t_i the holder optimizes the position and decides which of the two choices

$$\{\text{exercise, hold}\}$$

is preferable. So the holder chooses the maximum

$$\max\{\,\Psi(S_{j,i}),\ V_{j,i}^{\text{cont}}\,\}.$$

This amounts to a *dynamic programming* procedure. In summary, the dynamic-programming principle, based on the (20.11), (20.12) for i rather than M, combined with (20.13), reads as follows:

Call:

$$V_{j,i} = \max\left\{(S_{j,i}-K)^+,\ \mathrm{e}^{-r\Delta t}\cdot(pV_{j+1,i+1} + (1-p)V_{j,i+1})\right\} \tag{20.14}$$

Put:

$$V_{j,i} = \max\left\{(K-S_{j,i})^+,\ \mathrm{e}^{-r\Delta t}\cdot(pV_{j+1,i+1} + (1-p)V_{j,i+1})\right\} \tag{20.15}$$

The resulting algorithm is

Algorithm (CRR binomial method)

input: $r,\ \sigma,\ S = S_0,\ T,\ K,$ choice of put or call, European or American, M

calculate: $\Delta t = T/M,\ u,\ d,\ p$ from (10)

$S_{0,0} = S_0$

$S_{j,M} = S_{0,0} u^j d^{M-j},\ j = 0, 1, \ldots, M$

(for American options, also $S_{j,i} = S_{0,0} u^j d^{i-j}$ for $0 < i < M,\ j = 0, 1, \ldots, i$)

valuation: $V_{j,M}$ from (11) or (12)

$V_{j,i}$ for $i < M$ $\begin{cases} \text{from (13) for European options} \\ \text{from (14) or (15) for American options} \end{cases}$

output: $V_{0,0}$ is the approximation $V_0^{(M)}$ to $V(S_0, 0)$

Note that this algorithm is a basic version of a binomial method. Several improvements are possible, see the Remarks below.

20.2.4 *Practical Experience and Improvements*

The above CRR algorithm is easy to implement and highly robust. Figure 20.5 illustrates the result of the algorithm for an American put. In two examples, we present numbers for comparison. For anchoring, the classical $ud = 1$ is used.

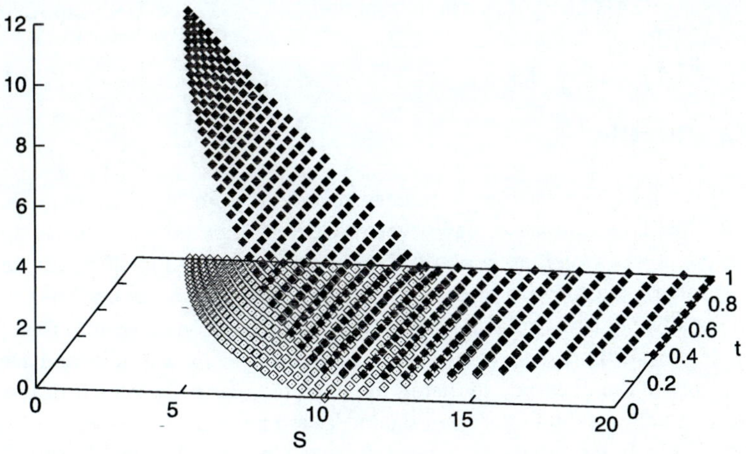

Fig. 20.5 Tree in the (S, t)-plane with nodes (*empty squares*) and (S, t, V)-points (*full squares*) for $M = 32$ (American put with $r = 0.06, \sigma = 0.30, K = 10, T = 1, S_0 = 10$)

Table 20.1 Results of example 1

M	$V^{(M)}(5,0)$
8	4.42507
16	4.42925
32	4.429855
64	4.429923
128	4.430047
256	4.430390
2,048	4.430451
Black–Scholes	4.43046477621

Example 1 (European put): We choose the parameters $K = 10$, $S = 5$, $r = 0.06$, $\sigma = 0.3$, $T = 1$.

The Table 20.1 lists approximations $V^{(M)}$ to $V(5,0)$. The convergence towards the Black–Scholes value $V(S,0)$ is visible; the latter was calculated by evaluating the Black–Scholes formula, see Seydel (2009). The number of printed decimals illustrates at best the attainable accuracy and does not reflect economic practice. The convergence rate is reflected by the results in Table 20.1.

The convergence rate is linear, $\mathcal{O}(\Delta t) = \mathcal{O}(M^{-1})$, which may be seen by plotting $V^{(M)}$ over M^{-1}. In such a plot, the values of $V^{(M)}$ roughly lie close to a straight line, which reflects the linear error decay. The reader may wish to investigate more closely how the error decays with M. It turns out that for the described version of the binomial method the convergence in M is not monotonic. It will not be recommendable to extrapolate the $V^{(M)}$-data to the limit $M \rightarrow \infty$, at least not the data of Table 20.1.

Example 2 (American put): For the parameters $K = 50$, $S = 50$, $r = 0.1$, $\sigma = 0.4$, $T = 0.41666\ldots$ ($\frac{5}{12}$ for 5 months), $M = 32$, the CRR approximation to V_0 is 4.2719.

20.2.4.1 Remarks

Table 20.1 might suggest that it is easy to obtain high accuracy with binomial methods. This is not the case; flaws were observed in particular close to the early-exercise curve, see Coleman et al. (2002). As illustrated by Fig. 20.2, the described standard version wastes many nodes $S_{j,i}$ close to zero and far away from the strike region. For advanced binomial methods and speeding up convergence, consult Breen (1991), Figlewski and Gao (1999), and Klassen (2001). Broadie and Detemple (1996) improve the accuracy by using the analytic Black–Scholes formula for the continuation value at the first step of the backward phase $i = M - 1$. For a detailed account of the binomial method consult also Cox and Rubinstein (1985). The approximation of the Greeks delta, gamma, and theta exploit the calculated V-values at the nodes in an elegant way, see Pelsser and Vorst (1994). Finite differences are used for a slightly extended tree, which starts already at $t = -2\Delta t$ so that a tree node hits the point $(S_0, 0)$. Honore and Poulsen (2002) explain how to implement the binomial method in spreadsheets. Many applications of binomial trees are found in Lyuu (2002). In case of barrier options, the nodes of the tree should be placed with care to maintain high accuracy. Dai and Lyuu (2010) suggest an initial trinomial step, tuned so that the following CRR tree has layers coinciding with barriers.

20.2.5 Convergence to the Black–Scholes Formula

Consider a European call in the CRR binomial model. Suppose the calculated value is $V_0^{(M)}$. In the limit $M \to \infty$ the sequence $V_0^{(M)}$ converges to the value $V_{\text{call}}(S_0, 0)$ of the continuous Black–Scholes model. In what follows, we sketch the proof, again for the case of no dividend payment, $\delta = 0$. For later reference we state the famous Black–Scholes formula:

$$V_{\text{call}}(S_0, 0) = S_0 F(d_1) - \mathrm{e}^{-rT} K \cdot F(d_2) \tag{20.16}$$

with

$$d_1 = \frac{\log \frac{S}{K} + (r + \frac{\sigma^2}{2})T}{\sigma\sqrt{T}}, \quad d_2 = \frac{\log \frac{S}{K} + (r - \frac{\sigma^2}{2})T}{\sigma\sqrt{T}}, \tag{20.17}$$

and $F(a)$ denotes the standard normal distribution

$$F(a) = \frac{1}{\sqrt{2\pi}} \int_{-\infty}^{a} \mathrm{e}^{-z^2/2}\, \mathrm{d}z.$$

Let $X = S_{j,M}$ be the final value at t_M of a path that traverses the tree starting at S_0. The index j reflects the number of "up's" after the M decisions at the nodes of the tree. For the binomial distribution the probability P that the path arrives at node $S_{j,M}$ is

$$\mathsf{P}(X = S_{j,M}) = \binom{M}{j} p^j (1-p)^{M-j} \tag{20.18}$$

(See Fig. 20.6 for an illustration of this probability.) Hence the value of the CRR approximation of the European call at $t = 0$ is

$$V_0^{(M)} = \mathrm{e}^{-rT} \sum_{j=0}^{M} \binom{M}{j} p^j (1-p)^{M-j} \left(S_0 u^j d^{M-j} - K\right)^+. \tag{20.19}$$

Let J be the smallest index j with $S_{j,M} \geq K$. This J is determined by the parameters, as seen from the equivalent statements

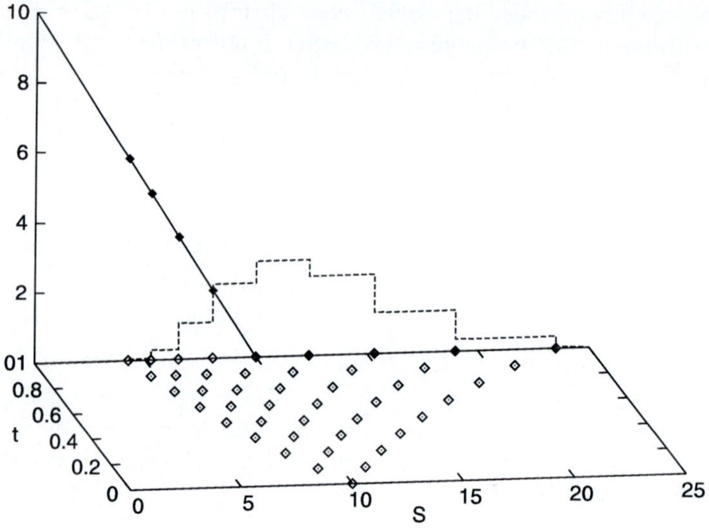

Fig. 20.6 Illustration of a binomial tree and payoff, here for a put, (S,t)-grid points for $M = 8$, $K = S_0 = 10$. The binomial density (*dashed line*) of the risk-free probability is shown, scaled with factor 10

$$S_0 u^J d^M d^{-J} - K \geq 0$$

$$\left(\frac{u}{d}\right)^J \geq \frac{K}{S_0} d^{-M}$$

$$J \log\left(\frac{u}{d}\right) \geq \log \frac{K d^{-M}}{S_0}$$

$$J \geq \alpha \stackrel{\text{def}}{=} \frac{\log \frac{K}{S_0} - M \log d}{\log u - \log d} = -\frac{\log \frac{S_0}{K} + M \log d}{\log u - \log d} \tag{20.20}$$

With this well-defined $\alpha = \alpha(M)$ we have J as the smallest index $\geq \alpha$. Now the zero part of the payoff in (20.19) can be split off, and

$$\begin{aligned} V_0^{(M)} &= \mathrm{e}^{-rT} S_0 \sum_{j=J}^{M} \binom{M}{j} p^j (1-p)^{M-j} u^j d^{M-j} \\ &\quad - \mathrm{e}^{-rT} K \sum_{j=J}^{M} \binom{M}{j} p^j (1-p)^{M-j}. \end{aligned} \tag{20.21}$$

We make use of

$$\mathrm{e}^{-rT} = \mathrm{e}^{-rM\Delta t} = (\mathrm{e}^{-r\Delta t})^M$$

and rewrite the first sum in (20.21)

$$\sum_{j=J}^{M} \binom{M}{j} \left(\frac{pu}{e^{r\Delta t}}\right)^j \left(\frac{(1-p)d}{e^{r\Delta t}}\right)^{M-j}.$$

Note that

$$\frac{pu}{e^{r\Delta t}} + \frac{(1-p)d}{e^{r\Delta t}} = \frac{(pu-pd)+d}{e^{r\Delta t}} = \frac{(e^{r\Delta t}-d)+d}{e^{r\Delta t}} = 1.$$

With the notation $\tilde{p} = \frac{pu}{e^{r\Delta t}}$ the first sum is equal to

$$\sum_{j=J}^{M} \binom{M}{j} \tilde{p}^j (1-\tilde{p})^{M-j},$$

the same type as the second sum in (20.21). Now we can express $V_0^{(M)}$ by means of the binomial probability P with (complementary) distribution function $B_{M,p}(J)$,

$$\mathsf{P}(j > J) = B_{M,p}(J) = \sum_{k=J}^{M} \binom{M}{k} p^k (1-p)^{M-k},$$

as

$$V_0^{(M)} = S_0 B_{M,\tilde{p}}(J) - e^{-rT} K \cdot B_{M,p}(J). \tag{20.22}$$

Recall the central limit theorem,

$$\lim_{M\to\infty} \mathsf{P}\left(\frac{j-Mp}{\sqrt{Mp(1-p)}} \le a\right) = F(a),$$

where Mp is the expectation of j and $Mp(1-p)$ its variance. Hence,

$$\lim_{M\to\infty} \mathsf{P}(Y > a) = 1 - \lim_{M\to\infty} \mathsf{P}(Y \le a) = 1 - F(a) = F(-a). \tag{20.23}$$

The observation

$$\mathsf{P}(j > J) = \mathsf{P}\left(\frac{j-Mp}{\sqrt{Mp(1-p)}} > \frac{J-Mp}{\sqrt{Mp(1-p)}}\right)$$

reveals the a in (20.23),

$$a = \frac{J-Mp}{\sqrt{Mp(1-p)}}.$$

Since $J \to \alpha$ for $M \to \infty$ it remains to show

$$\lim_{M\to\infty} \frac{M\tilde{p}-\alpha}{\sqrt{M\tilde{p}(1-\tilde{p})}} = d_1 \quad \text{and} \quad \lim_{M\to\infty} \frac{Mp-\alpha}{\sqrt{Mp(1-p)}} = d_2.$$

To this end one substitutes the p, u, d by their expressions from (20.10). We leave this to the reader. A reference is Kwok (1998).

20.3 Extensions

Lattice approaches can be adjusted to actual market data. For example, the terminal probabilities can be corrected appropriately, see Rubinstein (1994). In that respect, implied trees are basic means, and will be explained in some detail in Sect. 20.4. Tree methods can be applied to value exotic options as well, see Hull (2000) or Lyuu (2002). In this Sect. 20.3 we briefly comment on dividends and trinomial models.

20.3.1 Dividends

Discrete paying of dividends can be incorporated into the binomial algorithm. If a dividend is paid at a specific time t_k, the price of the asset drops by the same amount. To take this jump into account, the tree is cut at t_k and the S-values are reduced appropriately, see Hull (2000), §16.3, or Wilmott et al. (1996). Note that when the stock pays an amount D, then the part of the tree for $t \geq t_k$ is no longer recombining. As is easily seen from adjusting node values $S_0 u^j d^{k-j}$ to $S_0 u^j d^{k-j} - D$, the nodes on the next time level differ by $D(u-d)$, and the number of nodes doubles. Hull (2000) discusses this matter and suggests ways how to fix the problem. For a constant dividend yield rate δ, the formulas of the preceeding section are easily adapted. For example, in (20.4), (20.5), (20.6), (20.7), (20.10) the rate r must be replaced by $r - \delta$, but the discount factor in (20.13) remains unchanged. This more general case will be considered in Sect. 20.4.

20.3.2 Trinomial Model

Another extension of the binomial method is the *trinomial method*. Here each mesh offers three outcomes, with probabilities p_1, p_2, p_3 and $p_1 + p_2 + p_3 = 1$, see the illustration of Fig. 20.7. One possible set of parameters is

$$u = e^{\sigma\sqrt{3\Delta t}}$$

$$d = \frac{1}{u}$$

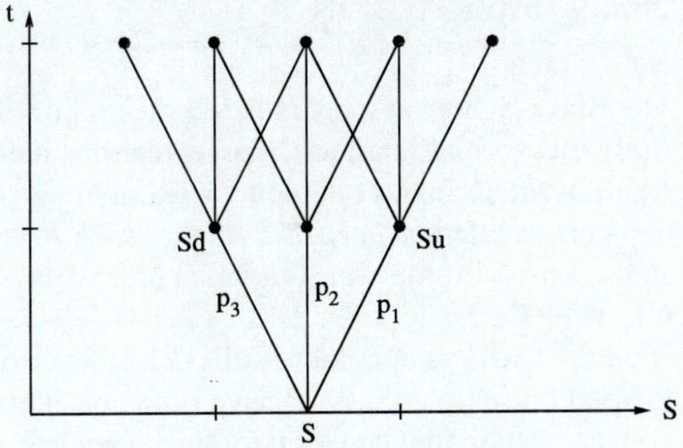

Fig. 20.7 First meshes of a trinomial tree

$$p_1 = \sqrt{\frac{\Delta t}{12\sigma^2}} \left(r - \frac{\sigma^2}{2}\right) + \frac{1}{6}$$

$$p_2 = \frac{2}{3}$$

$$p_3 = -\sqrt{\frac{\Delta t}{12\sigma^2}} \left(r - \frac{\sigma^2}{2}\right) + \frac{1}{6}$$

see Hull (2000). Lyuu (2002) suggests another set of parameters. The trinomial model is more flexible and allows for higher accuracy. Figlewski and Gao (1999) work with patches of finer meshes to improve accuracy, in particular, close to $(S, t) = (K, T)$.

20.3.3 Trees in Higher Dimension

Boyle et al. (1989) generalized the binomial method canonically to multivariate contingent claims with n assets. But already for $n = 2$ the recombining standard tree with M time levels requires $\frac{1}{3}M^3 + \mathcal{O}(M^2)$ nodes, and for $n = 3$ the number of nodes is of the order $\mathcal{O}(M^4)$. Tree methods also suffer from the curse of dimension. But obviously not all of the nodes of the canonical binomial approach are needed. The ultimate aim is to approximate the lognormal distribution, and this can be done with fewer nodes. Nodes in $\mathbb{R}^n$ should be constructed in such a way that the number of nodes grows comparably slower than the quality of the approximation of the distribution function. Lyuu (2002) presents an example of a two-dimensional approach. Generalizing the trinomial approach to higher dimensions is not recommendable because of storage requirements. Instead, other geometrical structures as icosahedral volumes can be applied. McCarthy and Webber (2001/02) discuss such approaches. For a convergence analysis of tree methods, and for an extension to Lévy processes, see Forsyth et al. (2002), and Maller et al. (2006).

20.4 Implied Trees

The Black-Scholes model is based on simplifying assumptions that are not necessarily met by real markets. In particular this holds for the assumption of a constant volatility σ. In market data of traded options, one observes the volatility smile, a non-constant dependence of σ on the strike K and on the time for maturity T. This smile consists of the *skew* (variation of σ with K), and the *term structure* (variation of σ with T).

In the classical approaches, the dependence of the value $V(S,t;\ K,T)$ has been focused on (S,t), and K,T have been considered constant. Dupire (1994) derived a partial differential equation for the dependence of V on K,T. From relevant data it is possible to approximate a *local volatility* $\sigma(S,t)$. Inserting the local volatility into the Black-Scholes approach allows to improve its pricing ability.

Such an additional flexibility of the Black-Scholes approach can be adapted also by tree methods. Such methods were suggested, for example, by Derman and Kani (1994), and by Rubinstein (1994). Here we describe in some detail the implied tree of Derman and Kani (1994). Implied trees take advantage of market prices and are calibrated right away from option data. Note the contrast to the CRR tree, which is calibrated to the underlying process S_t independently of the option.

20.4.1 Arrow-Debreu Prices

An essential tool for the derivation is the Arrow-Debreu price.

Definition Arrow-Debreu price $\lambda_{j,i}$: $\lambda_{j,i}$ is the sum of the products of all riskless-discounted transition probabilities, with summation over all paths leading from the root $(0,0)$ to the node (j,i).

For example,

$$\lambda_{1,2} = \mathrm{e}^{-r2\Delta t}\,[p_{0,1}(1-p_{0,0}) + (1-p_{1,1})p_{0,0}],$$

compare Fig. 20.8. As is easily seen, there is a recursion for these prices. Fixing $\lambda_{0,0} = 1$,

$$\lambda_{0,1} = \mathrm{e}^{-r\Delta t}\,\lambda_{0,0}(1-p_{0,0}), \quad \lambda_{1,1} = \mathrm{e}^{-r\Delta t}\,\lambda_{0,0}\,p_{0,0}$$

holds. The general recursion for the interior nodes is

$$\lambda_{j+1,i+1} = \mathrm{e}^{-r\Delta t}\,[\lambda_{j,i}\,p_{j,i} + \lambda_{j+1,i}\,(1-p_{j+1,i})] \quad \text{for } 0 \le j \le i-1, \tag{20.24}$$

because two entries exist for each of the nodes. And for the two boundary paths, each node has only one entry, hence

$$\begin{aligned} \lambda_{i+1,i+1} &= \mathrm{e}^{-r\Delta t}\,\lambda_{i,i}\,p_{i,i} \\ \lambda_{0,i+1} &= \mathrm{e}^{-r\Delta t}\,\lambda_{0,i}\,(1-p_{0,i}) \end{aligned} \tag{20.25}$$

completes the recursion.

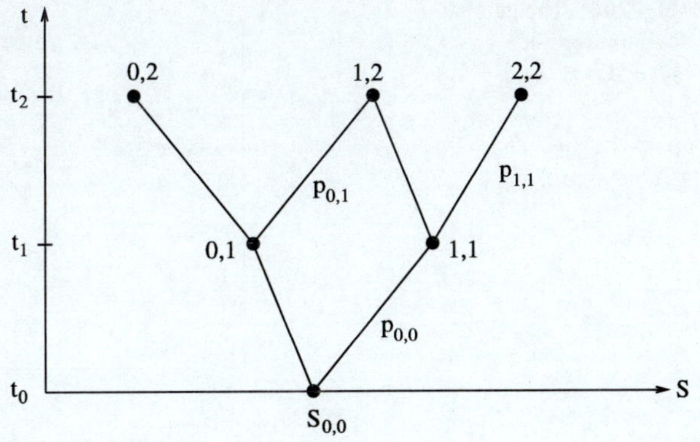

Fig. 20.8 Nodes (j, i) of a general binomial grid, with probabilities $p_{j,i}$ and variable positions of thc nodes; initial part of the tree

Recall that in the special case of the classical CRR tree, with $p_{j,i} = p$ for all j, i, the Bernoulli experiment results in the probabilities (20.18)

$$\binom{M}{j} p^j (1-p)^{M-j},$$

describing the probability that the node (j, M) is hit. Since the Arrow-Debreu prices distribute the discounting over the time slices,

$$\lambda_{j,M} = \mathrm{e}^{-rT} \binom{M}{j} p^j (1-p)^{M-j}$$

holds, and the expectation of a European vanilla option can be written

$$V(S_{0,0}, 0) = \sum_{j=0}^{M} \lambda_{j,M} \Psi(S_{j,M}). \tag{20.26}$$

The same pricing formula (20.26) holds true for a general binomial path with probabilities $p_{j,i}$. The final probability to hit the node (j, M) is $e^{rT} \lambda_{j,M}$.

20.4.2 *Derman & Kani Tree*

The method of Derman and Kani sets up reasonable probabilities $p_{j,i}$ and positions $(S_{j,i}, t_i)$ of the nodes (j, i). The grid is designed such that it matches market data. Assume that a bunch of market prices of options are known. These option data are subjected to a suitable smoothing algorithm as described by Fengler (2005), and by Glaser and Heider (2010). Based on this cumbersome preparatory work, the data

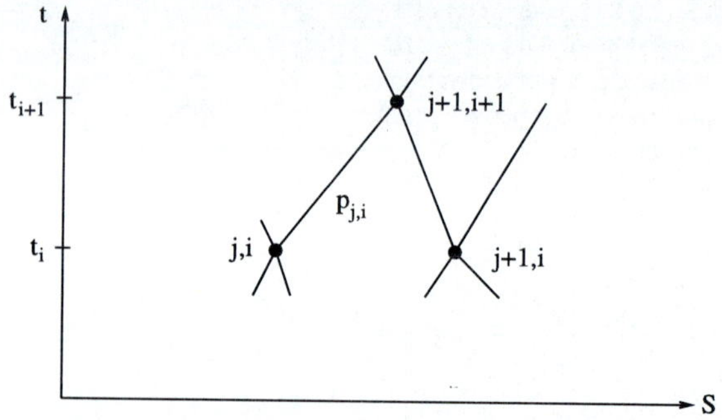

Fig. 20.9 The general buildup for node $(j+1, i+1)$

can be interpolated or approximated such that "market data" are given for *any* value of strike and maturity.

Suppose all nodes are placed and all probabilities are fixed for the time level t_i. That is, the $2i+2$ numbers

$$S_{0,i}, S_{1,i}, \ldots, S_{i,i},$$
$$\lambda_{0,i}, \lambda_{1,i}, \ldots, \lambda_{i,i}$$

are available. For the next time level t_{i+1} the $2i+4$ numbers

$$S_{0,i+1}, S_{1,i+1}, \ldots, S_{i+1,i+1},$$
$$\lambda_{0,i+1}, \lambda_{1,i+1}, \ldots, \lambda_{i+1,i+1}$$

are to be calculated. This requires $2i+3$ equations, because the recursion (20.24), (20.25) for the Arrow-Debreu prices requires only $i+1$ probabilities

$$p_{0,i}, \ldots, p_{i,i}$$

see Figs. 20.8, 20.9, 20.10. $i+1$ of the equations are easily set up, requesting as in CRR that the expectation over the time step Δt matches that of the continuous model (20.4). With the *forward price* $F_{j,i}$

$$F_{j,i} \stackrel{\text{def}}{=} S_{j,i}\, e^{(r-\delta)\Delta t}$$

this can be written

$$p_{j,i}\, S_{j+1,i+1} + (1 - p_{j,i})\, S_{j,i+1} = F_{j,i} \tag{20.27}$$

for $0 \le j \le i$. This sets up $i+1$ equations for the probabilities,

$$p_{j,i} = \frac{F_{j,i} - S_{j,i+1}}{S_{j+1,i+1} - S_{j,i+1}}, \tag{20.28}$$

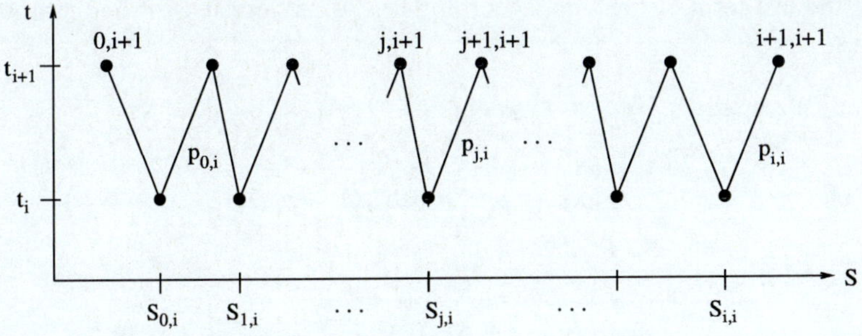

Fig. 20.10 All nodes of lines t_i and t_{i+1}; transition from line t_i to t_{i+1}

which in turn fix the Arrow-Debreu values via (20.24), (20.25). It remains to set up $i+2$ equations for the unknown grid coordinates $S_{j,i+1}$ for $0 \leq j \leq i+1$.

At this stage, the market data enter. According to the assumption, (approximate) vanilla put and call prices are available also for the specific choices of the maturity t_{i+1} and the $i+1$ strikes $S_{0,i}, \ldots, S_{i,i}$. For ease of notation, we denote the market values

$$\begin{aligned} C_{j,i} &= V_{\text{call}}^{\text{market}}(S_{0,0}, 0;\ S_{j,i}, t_{i+1}) \\ P_{j,i} &= V_{\text{put}}^{\text{market}}(S_{0,0}, 0;\ S_{j,i}, t_{i+1}) \end{aligned} \tag{20.29}$$

for $0 \leq j \leq i$. In (20.29), there are only $i+1$ independent option values, because put and call are related through the put-call parity.

20.4.2.1 Recursion Based on Call Data

Next we discuss how the call values $C_{j,i}$ enter; the put values $P_{j,i}$ will enter analogously. For the strike $S_{j,i}$, we apply (20.26), where M is replaced by $i+1$. Then, by (20.29), the grid values $S_{k,i+1}$ are to be chosen such that

$$C_{j,i} = \sum_{k=0}^{i+1} \lambda_{k,i+1}(S_{k,i+1} - S_{j,i})^+ ,$$

which for $S_{j,i+1} < S_{j,i} < S_{j+1,i+1}$ can be written

$$C_{j,i} = \sum_{k=j+1}^{i+1} \lambda_{k,i+1}(S_{k,i+1} - S_{j,i}). \tag{20.30}$$

Substituting the recursion (20.24), (20.25), this is

$$C_{j,i} = \mathrm{e}^{-r\Delta t} \sum_{k \geq j+1} [\lambda_{k-1,i}\, p_{k-1,i} + \lambda_{k,i}(1 - p_{k,i})] \cdot (S_{k,i+1} - S_{j,i}) \tag{20.31}$$

where the last term of the sum, according to (20.25), consists of one term only. It follows

$$
\begin{aligned}
e^{r\Delta t} C_{j,i} &= [\lambda_{j,i}\, p_{j,i} + \lambda_{j+1,i}\,(1 - p_{j+1,i})]\,(S_{j+1,i+1} - S_{j,i}) \\
&\quad + [\lambda_{j+1,i}\, p_{j+1,i} + \lambda_{j+2,i}(1 - p_{j+2,i})]\,(S_{j+2,i+1} - S_{j,i}) \\
&\quad + \ldots + [\lambda_{i-1,i}\, p_{i-1,i} + \lambda_{i,i}(1 - p_{i,i})]\,(S_{i,i+1} - S_{j,i}) \\
&\quad + [\lambda_{i,i}\, p_{i,i}](S_{i+1,i+1} - S_{j,i}) \\
&= \lambda_{j,i}\, p_{j,i}\,(S_{j+1,i+1} - S_{j,i}) \\
&\quad + \lambda_{j+1,i}(F_{j+1,i} - S_{j,i}) + \ldots + \lambda_{i,i}(F_{i,i} - S_{j,i}) \\
&= \lambda_{j,i}\, p_{j,i}\,(S_{j+1,i+1} - S_{j,i}) + \sum_{k=j+1}^{i} \lambda_{k,i}(F_{k,i} - S_{j,i})
\end{aligned}
$$

Note that the sum in this expression is completely known from the previous line t_i. The known summation term is combined with the data of the smile into the known numbers

$$
A_{j,i} \stackrel{\text{def}}{=} e^{r\Delta t} C_{j,i} - \sum_{k=j+1}^{i} \lambda_{k,i}(F_{k,i} - S_{j,i}).
$$

This gives the relation

$$
A_{j,i} = \lambda_{j,i}\, p_{j,i}\,(S_{j+1,i+1} - S_{j,i}) \tag{20.32}
$$

which involves *only two* unknowns $p_{j,i}$ and $S_{j+1,i+1}$. We substitute $p_{j,i}$ from (20.28) into (20.32), and solve for $S_{j+1,i+1}$. The result

$$
S_{j+1,i+1} = \frac{S_{j,i+1}(A_{j,i} + \lambda_{j,i} S_{j,i}) - \lambda_{j,i} S_{j,i} F_{j,i}}{S_{j,i+1}\lambda_{j,i} + A_{j,i} - \lambda_{j,i} F_{j,i}} \tag{20.33}
$$

is a recursion $S_{j+1,i+1} = f(S_{j,i+1})$ along the line t_{j+1}, fixing a new node $S_{j+1,i+1}$ after the previous node $S_{j,i+1}$ was set. The probabilities are then given by (20.28) and (20.24).

20.4.2.2 Starting the Recursion

This raises a new question: Where should the recursion (20.33) start? Recall that we have $i + 2$ unknown nodes on line t_{j+1}, but only $i + 1$ independent option values in (20.29). That is, there is one degree of freedom. For example, one node can be set freely. Following Derman and Kani (1994), we make the center of the tree coincide with the center of the standard CRR tree. This requires to discuss two situations: Either the line t_{1+1} has an even number of nodes, or an odd number.

The simple situation is the odd number of nodes at line t_{i+1}. Then we set artificially for the center node with j-index m,

$$(i \text{ odd}) \quad m \stackrel{\text{def}}{=} \frac{i+1}{2}, \qquad S_{m,i+1} \stackrel{\text{def}}{=} S_{0,0}. \tag{20.34}$$

For the upper part of the line ($j > m$) the recursion (20.33) defines all nodes, starting from $S_{m,i+1}$. For the lower part of line ($j < m$) the corresponding recursion based on put values $P_{j,i}$ will be applied, see (20.40) below.

In the other situation, when the number of nodes at line t_{i+1} and i are even, the center of the tree is straddled by the two nodes with j-indices $m = i/2$ and $m+1$. Recall from CRR with $d = 1/u$ that its logarithmic spacing for the scenario of Fig. 20.3 amounts to

$$S_{j,i+1} = S_{j,i}^2 / S_{j+1,i+1} \tag{20.35}$$

for any j, i. We assume this spacing for the center nodes, and substitute (20.35) into (20.33). This gives a quadratic equation for the $S_{j+1,i+1}$ at the center position. One of the two solutions ($S_{j,i}$) is meaningless, because nodes are not separated. The other node is

$$S_{j+1,i+1} = \frac{S_{j,i}(\lambda_{j,i} S_{j,i} + A_{j,i})}{\lambda_{j,i} F_{j,i} - A_{j,i}}.$$

Note from the previous line, where i is odd, we have $S_{m,i} = S_{0,0}$. So

$$(i \text{ even}) \quad m \stackrel{\text{def}}{=} \frac{i}{2}, \qquad S_{m+1,i+1} \stackrel{\text{def}}{=} \frac{S_{0,0}(\lambda_{m,i} S_{0,0} + A_{m,i})}{\lambda_{m,i} F_{m,i} - A_{m,i}}. \tag{20.36}$$

This defines the starting point for the recursion (20.33) [or for (20.40) below].

20.4.2.3 Recursion Based on Put Data

For the put, the recursion is derived in a similar way as done above for the call. We demand for the strike $S_{j,i}$ and the put data $P_{j,i}$

$$P_{j,i} = \sum_{k=0}^{i+1} \lambda_{k,i+1} \, (S_{j,i} - S_{k,i+1})^+.$$

Then for an ordered grid with $S_{j,i+1} < S_{j,i} < S_{j+1,i+1}$

$$P_{j,i} = \sum_{k=0}^{j} \lambda_{k,i+1} \, (S_{j,i} - S_{k,i+1}). \tag{20.37}$$

Hence,

$$\begin{aligned} e^{r\Delta t} P_{j,i} &= \lambda_{0,i}(1 - p_{0,i}) \, (S_{j,i} - S_{0,i+1}) \\ &\quad + [\lambda_{0,i} \, p_{0,i} + \lambda_{1,i} \, (1 - p_{1,i})] \, (S_{j,i} - S_{1,i+1}) \end{aligned}$$

$$
\begin{aligned}
&+ [\lambda_{1,i}\, p_{1,i} + \lambda_{2,i}\,(1 - p_{2,i})]\,(S_{j,i} - S_{2,i+1}) \\
&+ \ldots \\
&+ [\lambda_{j-1,i}\, p_{j-1,i} + \lambda_{j,i}\,(1 - p_{j,i})]\,(S_{j,i} - S_{j,i+1}) \\
&= \lambda_{j,i}(1 - p_{j,i})(S_{j,i} - S_{j,i+1}) + \sum_{k=0}^{j-1} \lambda_{k,i}(F_{k,i} - S_{j,i})
\end{aligned}
$$

With the well-defined numbers

$$B_{j,i} \stackrel{\text{def}}{=} \mathrm{e}^{r\Delta t} P_{j,i} - \sum_{k=0}^{j-1} \lambda_{k,i}(F_{k,i} - S_{j,i}) \tag{20.38}$$

we arrive at

$$B_{j,i} = \lambda_{j,i}(1 - p_{j,i})(S_{j,i} - S_{j,i+1}). \tag{20.39}$$

After substituting $p_{j,i}$ the final recursion based on put data is

$$S_{j,i+1} = \frac{B_{j,i} S_{j+1,i+1} + \lambda_{j,i} S_{j,i}(F_{j,i} - S_{j+1,i+1})}{B_{j,i} + \lambda_{j,i}(F_{j,i} - S_{j+1,i+1})} \tag{20.40}$$

This is the recursion for the lower half of the nodes on line t_{i+1}. The starting point is again provided by (20.34), or (20.36).

The pricing of an option works in a backward loop analogously as in the CRR algorithm.

20.4.2.4 Adjustment of Node Spacing

To avoid arbitrage, it is crucial that the probabilities $p_{j,i}$ must lie between zero and one. From (20.28) we see what a violation $p_{j,i} < 0$ or $p_{j,i} > 1$ would mean. The latter is equivalent to $F_{j,i} > S_{j+1,i+1}$, the former depends on whether $S_{j+1,i+1} > S_{j,i+1}$ is guaranteed. Sufficient for $0 \le p_{j,i} \le 1$ is to demand for all j, i

$$F_{j,i} < S_{j+1,i+1} < F_{j+1,i}. \tag{20.41}$$

In addition to (20.41), the ordered-grid condition $S_{j,i} < S_{j+1,i+1} < S_{j+1,i}$ anticipated by Fig. 20.9 must hold. In case these requirements are not satisfied by the node values provided by (20.33) or (20.40), the values of $S_{j+1,i+1}$ must be adjusted accordingly. Derman and Kani (1994) suggest to escape to the logarithmic spacing.

The practical work with the above implied grid is not without problems. When the values of $S_{j,i+1}$ provided by (20.33) and (20.40) are frequently overridden in order to maintain an ordered grid that satisfies (20.41), then the influence of some of the market data $C_{j,i}$, $P_{j,i}$ is cut off. The loss of information caused by many such repairs deteriorates the quality of the approximation. This happens often for fine grids. Also, the denominators in (20.33) and (20.40) may take unfavorable values (say, close to zero), which can lead to unplausible values of the nodes.

20.4.3 Local Volatility

Estimates of the local volatility can be obtained from the implied grid. To this end, the return R is investigated at each node j,i. For a binomial tree, we have two samples for $R_{j,i}$. Taking the return of the underlying process S in the sense $R = \log(S_{\text{new}}/S_{\text{old}})$ as in Seydel (2009), the expectation and variance is

$$\mathsf{E}(R_{j,i}) = p_{j,i} \log \frac{S_{j+1,i+1}}{S_{j,i}} + (1 - p_{j,i}) \log \frac{S_{j,i+1}}{S_{j,i}}$$
$$\mathsf{Var}(R_{j,i}) = p_{j,i} \left[\log \frac{S_{j+1,i+1}}{S_{j,i}} - \mathsf{E}(R_{j,i})\right]^2 + (1 - p_{j,i}) \left[\log \frac{S_{j,i+1}}{S_{j,i}} - \mathsf{E}(R_{j,i})\right]^2$$

For the model (20.2), the scaling is $\mathsf{Var}(R_{j,i}) = \sigma_{j,i}^2 \Delta t$, which defines the local volatility $\sigma_{j,i}$ at node j,i. A short calculation shows

$$\sigma_{j,i} = \sqrt{\frac{p_{j,i}(1 - p_{j,i})}{\Delta t}} \log \frac{S_{j+1,i+1}}{S_{j,i}} \tag{20.42}$$

This allows to estimate the local volatility σ from the values of the grid.

A worked example can be found in Derman and Kani (1994), and in Fengler (2005). For the test, for example, an artificial implied volatility function $\hat{\sigma}(K)$ is set up, and corresponding Black–Scholes values are calculated. These in turn serve as the data $C_{j,i}$, $P_{j,i}$ for the computational experiment. The same test example is also used to illustrate the trinomial tree. For the handling of actual market data, consult Fengler (2005), and Glaser and Heider (2010), who discuss the calculation of reasonable values of $C_{j,i}$, $P_{j,i}$. Exotic options of the European style are priced using the Arrow-Debreu prices calculated by an implied tree.

20.4.4 Alternative Approaches

Barle and Cakici (1998) modify the Derman and Kani algorithm by setting the strikes equal to the forward prices $F_{j,i}$. The central node is set such that the tree bends along with the interest rate. Rubinstein (1994) implements the tree by a backward recursion. An improvement was suggested by Jackwerth (1977); see also the discussion in Fengler (2005). An implied trinomial tree is constructed by Derman et al. (1996). The trinomial tree gives more flexibility because it involves more parameters. But also for the trinomial tree and for the Barle and Cakici variant the "probabilities" p must be checked, and nodes must be overridden in case $p < 0$ or $p > 1$. Derman et al. (1996) discuss also variations in the time step Δt. A comparison of the Barle and Cakici approach with the Derman and Kani approach is found in Hardle and Myvsivckova (2009).

References

Barle, S., Cakici. N. (1998). How to grow a smiling tree. *Journal of Financial Engineering, 7*, 127–146.

Boyle, P.P., Evnine, J., Gibbs, S. (1989). Numerical evaluation of multivariate contingent claims. *Review Financial Studies 2*, 241–250.

Breen, R. (1991). The accelerated binomial option pricing model. *Journal of Financial and Quantitative Analysis, 26*, 153–164.

Broadie, M., Detemple, J. (1996). American option valuation: new bounds, approximations, and a comparison of existing methods. *Review Financial Studies, 9*, 1211–1250.

Coleman, T.F., Li, Y., Verma, Y. (2002). A Newton method for American option pricing. *Journal of Computational Finance, 5*(3), 51–78.

Cox, J.C., Ross, S., Rubinstein, M. (1979). Option pricing: A simplified approach. *Journal of Financial Economics, 7*, 229–264.

Cox, J.C., Rubinstein, M. (1985). Options Markets. Prentice Hall, Englewood Cliffs.

Dai, T.S., Lyuu, Y.D. (2010). The binomial tree: A simple model for efficient and accurate option pricing. to appear: J. Derivatives

Derman, E., Kani, I. (1994). Riding on a smile. *Risk, 7*(2), 32–39.

Derman, E., Kani, I., Chriss, N. (1996). Implied trinomial trees of the volatility smile. *Journal of Derivatives, 3*, 7–22.

Dupire, B. (1994). Pricing with a smile. *Risk*, **7**, 18–20.

Fengler, M.R. (2005). Semiparametric Modeling of Implied Volatility. Springer, Berlin.

Figlewski, S., Gao, B. (1999). The adaptive mesh model: A new approach to efficient option pricing. *Journal of Financial Economics, 53*, 313–351.

Forsyth, P.A., Vetzal, K.R., Zvan, R. (2002). Convergence of numerical methods for valuing path-dependent options using interpolation. *Review of Derivatives Research, 5*, 273–314.

Glaser, J., Heider, P. (2010). Arbitrage-free approximation of call price surfaces and input data risk. *Quantitative Finance*, DOI: 10.1080/14697688.2010.514005

Härdle, W., Myšičková, A. (2009). Numerics of implied binomial trees. In: Härdle, W., Hautsch, N., Overbeck, L. (eds.) Applied Quantitative Finance. Springer, Berlin.

Honoré, P., Poulsen, R. (2002). Option pricing with EXCEL. In: Nielsen, S. (eds.) Programming Languages and Systems in Computational Economics and Finance. Kluwer, Amsterdam.

Hull, J., White, A. (1988). The use of the control variate technique in option pricing. *Journal of Financial Quantitative Analysis, 23*, 237–251.

Hull, J.C. (2000). Options, Futures, and Other Derivatives. Fourth Edition. Prentice Hall International Editions, Upper Saddle River.

Jackwerth, J.C. (1977). Generalized binomial trees. *Journal of Derivatives, 5*, 7–17.

Klassen, T.R. (2001). Simple, fast and flexible pricing of Asian options. *Journal of Computational Finance, 4*(3), 89–124.

Kwok, Y.K. (1998). Mathematical Models of Financial Derivatives. Springer, Singapore.

Lyuu, Y.D. (2002). Financial Engineering and Computation. Principles, Mathematics, Algorithms. Cambridge University Press, Cambridge.

Maller, R.A., Solomon, D.H., Szimayer, A. (2006). A multinomial approximation for American option prices in Lévy process models. *Mathematical Finance, 16*, 613–633.

McCarthy, L.A., Webber, N.J. (2001/02). Pricing in three-factor models using icosahedral lattices. *Journal of Computational Finance*, *5*(2), 1–33.

Pelsser, A., Vorst, T. (1994). The binomial model and the Greeks. *Journal of Derivatives*, *1*, 45–49.

Rendleman, R.J., Bartter, B.J. (1979). Two-state option pricing. *Journal of Finance*, *34*, 1093–1110.

Rubinstein, M. (1994). Implied binomial trees. *Journal of Finance*, *69*, 771–818.

Seydel, R.U. (2009). Tools for Computational Finance. 4th Edition. Springer, Berlin.

Wilmott, P., Dewynne, J., Howison, S. (1996). Option Pricing. Mathematical Models and Computation. Oxford Financial Press, Oxford.

Chapter 21
Efficient Options Pricing Using the Fast Fourier Transform

Yue Kuen Kwok, Kwai Sun Leung, and Hoi Ying Wong

Abstract We review the commonly used numerical algorithms for option pricing under Levy process via Fast Fourier transform (FFT) calculations. By treating option price analogous to a probability density function, option prices across the whole spectrum of strikes can be obtained via FFT calculations. We also show how the property of the Fourier transform of a convolution product can be used to value various types of option pricing models. In particular, we show how one can price the Bermudan style options under Levy processes using FFT techniques in an efficient manner by reformulating the risk neutral valuation formulation as a convolution. By extending the finite state Markov chain approach in option pricing, we illustrate an innovative FFT-based network tree approach for option pricing under Levy process. Similar to the forward shooting grid technique in the usual lattice tree algorithms, the approach can be adapted to valuation of options with exotic path dependence. We also show how to apply the Fourier space time stepping techniques that solve the partial differential-integral equation for option pricing under Levy process. This versatile approach can handle various forms of path dependence of the asset price process and embedded features in the option models. Sampling errors and truncation errors in numerical implementation of the FFT calculations in option pricing are also discussed.

Y.K. Kwok (✉)
Hong Kong University of Science and Technology, Clear Water Bay, Kowloon, Hong Kong
e-mail: maykwok@ust.hk

K.S. Leung · H.Y. Wong
The Chinese University of Hong Kong, Shatin, NT, Hong Kong
e-mail: ksleung@se.cuhk.edu.hk; hywong@cuhk.edu.hk

J.-C. Duan et al. (eds.), *Handbook of Computational Finance*, Springer Handbooks of Computational Statistics, DOI 10.1007/978-3-642-17254-0_21,

21.1 Introduction

The earliest option pricing models originated by Black and Scholes (1973) and Merton (1973) use the Geometric Brownian process to model the underlying asset price process. However, it is well known among market practitioners that the lognormal assumption of asset price returns suffers from serious deficiencies that give rise to inconsistencies as exhibited by smiles (skewness) and term structures in observed implied volatilities. The earlier remedy to resolve these deficiencies is the assumption of state and time dependence of the volatility of the asset price process (see Derman and Kani 1998; Dupire 1994). On the other hand, some researchers recognize the volatility of asset returns as a hidden stochastic process, which may also undergo regime change. Examples of these pioneering works on stochastic volatility models are reported by Stein and Stein (1991), Heston (1993), and Naik (2000). Starting from the seminar paper by Merton (1976), jumps are introduced into the asset price processes in option pricing. More recently, researchers focus on option pricing models whose underlying asset price processes are the Levy processes (see Cont and Tankov 2004; Jackson et al. 2008).

In general, the nice analytic tractability in option pricing as exhibited by Black-Scholes-Merton's Geometric Brownian process assumption cannot be carried over to pricing models that assume stochastic volatility and Levy processes for the asset returns. Stein and Stein (1991) and Heston (1993) manage to obtain an analytic representation of the European option price function in the Fourier domain. Duffie et al. (2000) propose transform methods for pricing European options under the affine jump-diffusion processes. Fourier transform methods are shown to be an effective approach to pricing an option whose underlying asset price process is a Levy process. Instead of applying the direct discounted expectation approach of computing the expectation integral that involves the product of the terminal payoff and the density function of the Levy process, it may be easier to compute the integral of their Fourier transform since the characteristic function (Fourier transform of the density function) of the Levy process is easier to be handled than the density function itself. Actually, one may choose a Levy process by specifying the characteristic function since the Levy-Khinchine formula allows a Levy process to be fully described by the characteristic function.

In this chapter, we demonstrate the effective use of the Fourier transform approach as an effective tool in pricing options. Together with the Fast Fourier transform (FFT) algorithms, real time option pricing can be delivered. The underlying asset price process as modeled by a Levy process can allow for more general realistic structure of asset returns, say, excess kurtosis and stochastic volatility. With the characteristic function of the risk neutral density being known analytically, the analytic expression for the Fourier transform of the option value can be derived. Option prices across the whole spectrum of strikes can be obtained by performing Fourier inversion transform via the efficient FFT algorithms.

This chapter is organized as follows. In the next section, the mathematical formulations for building the bridge that links the Fourier methods with option pricing are discussed. We first provide a brief discussion on Fourier transform and FFT algorithms. Some of the important properties of Fourier transform, like the Parseval relation, are presented. We also present the definition of a Lévy process and the statement of the Lévy-Khintchine formula. In Sect. 21.3, we derive the Fourier representation of the European call option price function. The Fourier inversion integrals in the option price formula can be associated with cumulative distribution functions, similar to the Black-Scholes type representation. However, due to the presence of a singularity arising from non-differentiability in the option payoff function, the Fourier inversion integrals cannot be evaluated by applying the FFT algorithms. We then present various modifications of the Fourier integral representation of the option price using the damped option price method and time value method (see Carr and Madan 1999). Details of the FFT implementation of performing the Fourier inversion in option valuation are illustrated. In Sect. 21.4, we consider the extension of the FFT techniques for pricing multi-asset options. Unlike the finite difference approach or the lattice tree methods, the FFT approach does not suffer from the curse of dimensionality of the option models with regard to an increase in the number of risk factors in defining the asset return distribution (see Dempster and Hong 2000; Hurd and Zhou 2009). In Sect. 21.5, we show how one can price Bermudan style options under Lévy processes using the FFT techniques by reformulating the risk neutral valuation formulation as a convolution. We show how the property of the Fourier transform of a convolution product can be effectively applied in pricing a Bermudan option (see Lord et al. 2008). In Sect. 21.6, we illustrate an innovative FFT-based network approach for pricing options under Lévy processes by extending the finite state Markov chain approach in option pricing. Similar to the forward shooting grid technique in the usual lattice tree algorithms, the approach can be adapted to valuation of options with exotic path dependence (see Wong and Guan 2009). In Sect. 21.7, we derive the partial integral-differential equation formulation that governs option prices under the Lévy process assumption of asset returns. We then show how to apply the Fourier space time stepping techniques that solve the partial differential-integral equation for option pricing under Lévy processes. This versatile approach can handle various forms of path dependence of the asset price process and features/constraints in the option models (see Jackson et al. 2008). We present summary and conclusive remarks in the last section.

21.2 Mathematical Preliminaries on Fourier Transform Methods and Lévy Processes

Fourier transform methods have been widely used to solve problems in mathematics and physical sciences. In recent years, we have witnessed the continual interests in developing the FFT techniques as one of the vital tools in option pricing. In fact, the Fourier transform methods become the natural mathematical tools when we consider option pricing under Lévy models. This is because a Lévy process X_t can be fully described by its characteristic function $\phi_X(u)$, which is defined as the Fourier transform of the density function of X_t.

21.2.1 Fourier Transform and Its Properties

First, we present the definition of the Fourier transform of a function and review some of its properties. Let $f(x)$ be a piecewise continuous real function over $(-\infty, \infty)$ which satisfies the integrability condition:

$$\int_{-\infty}^{\infty} |f(x)|\, dx < \infty.$$

The Fourier transform of $f(x)$ is defined by

$$\mathcal{F}_f(u) = \int_{-\infty}^{\infty} e^{iuy} f(y)\, dy. \tag{21.1}$$

Given $\mathcal{F}_f(u)$, the function f can be recovered by the following Fourier inversion formula:

$$f(x) = \frac{1}{2\pi}\int_{-\infty}^{\infty} e^{-iux} \mathcal{F}_f(u)\, du. \tag{21.2}$$

The validity of the above inversion formula can be established easily via the following integral representation of the Dirac function $\delta(y-x)$, where

$$\delta(y-x) = \frac{1}{2\pi}\int_{-\infty}^{\infty} e^{iu(y-x)}\, du.$$

Applying the defining property of the Dirac function

$$f(x) = \int_{-\infty}^{\infty} f(y)\delta(y-x)\, dy$$

and using the above integral representation of $\delta(y-x)$, we obtain

$$\begin{aligned} f(x) &= \int_{-\infty}^{\infty} f(y)\frac{1}{2\pi}\int_{-\infty}^{\infty} e^{iu(y-x)}\, du dy \\ &= \frac{1}{2\pi}\int_{-\infty}^{\infty} e^{-iux}\left(\int_{-\infty}^{\infty} f(y)e^{iuy}\, dy\right) du. \end{aligned}$$

This gives the Fourier inversion formula (21.2).

Sometimes it may be necessary to take u to be complex, with $\operatorname{Im} u \neq 0$. In this case, $\mathcal{F}_f(u)$ is called the generalized Fourier transform of f. The corresponding

Fourier inversion formula becomes

$$f(x) = \frac{1}{2\pi} \int_{i\,\mathrm{Im}u-\infty}^{i\,\mathrm{Im}u+\infty} e^{-iux} \mathcal{F}_f(u)\, du.$$

Suppose the stochastic process X_t has the density function p, then the Fourier transform of p

$$\mathcal{F}_p(u) = \int_{-\infty}^{\infty} e^{iux} p(x)\, dx = E\left[e^{iuX}\right] \tag{21.3}$$

is called the characteristic function of X_t.

The following mathematical properties of $\mathcal{F}_f$ are useful in our later discussion.

1. *Differentiation*
$$\mathcal{F}_{f'}(u) = -iu\mathcal{F}_f(u).$$
2. *Modulation*
$$\mathcal{F}_{e^{\lambda x} f}(u) = \mathcal{F}_f(u - i\lambda), \qquad \lambda \text{ is real.}$$
3. *Convolution*
Define the convolution between two integrable functions $f(x)$ and $g(x)$ by
$$h(x) = f * g(x) = \int_{-\infty}^{\infty} f(y)g(x-y)\, dy,$$
then
$$\mathcal{F}_h = \mathcal{F}_f \mathcal{F}_g.$$
4. *Parseval relation*
Define the inner product of two complex-valued square-integrable functions f and g by
$$< f, g > = \int_{-\infty}^{\infty} f(x)\bar{g}(x)\, dx,$$
then
$$< f, g > = \frac{1}{2\pi} < \mathcal{F}_f(u), \mathcal{F}_g(u) > .$$

We would like to illustrate an application of the Parseval relation in option pricing. Following the usual discounted expectation approach, we formally write the option price V with terminal payoff $V_T(x)$ and risk neutral density function $p(x)$ as

$$V = e^{-rT} \int_{-\infty}^{\infty} V_T(x) p(x)\, dx = e^{-rT} < V_T(x), p(x) > .$$

By the Parseval relation, we obtain

$$V = \frac{e^{-rT}}{2\pi} < \mathcal{F}_p(u), \mathcal{F}_{V_T}(u) > . \qquad (21.4)$$

The option price can be expressed in terms of the inner product of the characteristic function of the underlying process $\mathcal{F}_p(u)$ and the Fourier transform of the terminal payoff $\mathcal{F}_{V_T}(u)$. More applications of the Parseval relation in deriving the Fourier inversion formulas in option pricing and insurance can be found in Dufresne et al. (2009).

21.2.2 Discrete Fourier Transform

Given a sequence $\{x_k\}$, $k = 0, 1, \cdots, N-1$, the discrete Fourier transform of $\{x_k\}$ is another sequence $\{y_j\}$, $j = 0, 1, \cdots, N-1$, as defined by

$$y_j = \sum_{k=0}^{N-1} e^{\frac{2\pi ijk}{N}} x_k, \qquad j = 0, 1, \cdots, N-1. \qquad (21.5)$$

If we write the N-dimensional vectors

$$\mathbf{x} = (x_0\ x_1\ \cdots\ x_{N-1})^{\mathbf{T}} \quad \text{and} \quad \mathbf{y} = (y_0\ y_1\ \cdots\ y_{N-1})^{\mathbf{T}},$$

and define a $N \times N$ matrix F^N whose (j,k)th entry is

$$F^N_{j,k} = e^{\frac{2\pi ijk}{N}}, \qquad 1 \le j,k \le N,$$

then $\mathbf{x}$ and $\mathbf{y}$ are related by

$$\mathbf{y} = F^N \mathbf{x}. \qquad (21.6)$$

The computation to find $\mathbf{y}$ requires N^2 steps.

However, if N is chosen to be some power of 2, say, $N = 2^L$, the computation using the FFT techniques would require only $\frac{1}{2}NL = \frac{N}{2}\log_2 N$ steps. The idea behind the FFT algorithm is to take advantage of the periodicity property of the Nth root of unity. Let $M = \frac{N}{2}$, and we split vector $\mathbf{x}$ into two half-sized vectors as defined by

$$\mathbf{x}' = (x_0\ x_2\ \cdots\ x_{N-2})^{\mathbf{T}} \quad \text{and} \quad \mathbf{x}'' = (x_1\ x_3\ \cdots\ x_{N-1})^{\mathbf{T}}.$$

We form the M-dimensional vectors

$$\mathbf{y}' = F^M \mathbf{x}' \quad \text{and} \quad \mathbf{y}'' = F^M \mathbf{x}'',$$

where the (j,k)th entry in the $M \times M$ matrix F^M is

$$F^M_{j,k} = e^{\frac{2\pi ijk}{M}}, \qquad 1 \le j,k \le M.$$

It can be shown that the first M and the last M components of $\mathbf{y}$ are given by

$$\begin{aligned} y_j &= y'_j + e^{\frac{2\pi ij}{N}} y''_j, \qquad j = 0,1,\cdots,M-1, \\ y_{j+M} &= y'_j - e^{\frac{2\pi ij}{N}} y''_j, \qquad j = 0,1,\cdots,M-1. \end{aligned} \tag{21.7}$$

Instead of performing the matrix-vector multiplication $F^N\mathbf{x}$, we now reduce the number of operations by two matrix-vector multiplications $F^M\mathbf{x}'$ and $F^M\mathbf{x}''$. The number of operations is reduced from N^2 to $2\left(\frac{N}{2}\right)^2 = \frac{N^2}{2}$. The same procedure of reducing the length of the sequence by half can be applied repeatedly. Using this FFT algorithm, the total number of operations is reduced from $O(N^2)$ to $O(N \log_2 N)$.

21.2.3 Lévy Processes

An adapted real-valued stochastic process X_t, with $X_0 = 0$, is called a Lévy process if it observes the following properties:

1. *Independent increments*
   For every increasing sequence of times $t_0, t_1, \cdots, t_n$, the random variables $X_{t_0}, X_{t_1} - X_{t_0}, \cdots, X_{t_n} - X_{t_{n-1}}$ are independent.
2. *Time-homogeneous*
   The distribution of $\{X_{t+s} - X_s; t \ge 0\}$ does not depend on s.
3. *Stochastically continuous*
   For any $\epsilon > 0$, $P\left[|X_{t+h} - X_t| \ge \epsilon\right] \to 0$ as $h \to 0$.
4. *Cadlag process*
   It is right continuous with left limits as a function of t.

Lévy processes are a combination of a linear drift, a Brownian process, and a jump process. When the Lévy process X_t jumps, its jump magnitude is non-zero. The Lévy measure w of X_t defined on $\mathbb{R} \setminus \{0\}$ dictates how the jump occurs. In the finite-activity models, we have $\int_{\mathbb{R}} w(dx) < \infty$. In the infinite-activity models, we observe $\int_{\mathbb{R}} w(dx) = \infty$ and the Poisson intensity cannot be defined. Loosely speaking, the Lévy measure $w(dx)$ gives the arrival rate of jumps of size $(x, x + dx)$. The characteristic function of a Lévy process can be described by the Lévy-Khinchine representation

Table 21.1 Characteristic functions of some parametric Levy processes

Lévy process X_t	Characteristic function $\phi_X(u)$
Finite-activity models	
Geometric Brownian motion	$\exp\left(iu\mu t - \frac{1}{2}\sigma^2 tu^2\right)$
Lognormal jump diffusion	$\exp\left(iu\mu t - \frac{1}{2}\sigma^2 tu^2 + \lambda t(e^{iu\mu_J - \frac{1}{2}\sigma_J^2 u^2} - 1)\right)$
Double exponential jump diffusion	$\exp\left(iu\mu t - \frac{1}{2}\sigma^2 tu^2 + \lambda t\left(\frac{1-\eta^2}{1+u^2\eta^2}e^{iu\kappa} - 1\right)\right)$
Infinite-activity models	
Variance gamma	$\exp(iu\mu t)(1 - iu\nu\theta + \frac{1}{2}\sigma^2\nu u^2)^{\frac{t}{\nu}}$
Normal inverse Gaussian	$\exp\left(iu\mu t + \delta t\sqrt{\alpha^2-\beta^2} - \sqrt{\alpha^2-(\beta+iu)^2}\right)$
Generalized hyperbolic	$\exp(iu\mu t)\left(\frac{\alpha^2-\beta^2}{\alpha^2-(\beta+iu)^2}\right)^{\frac{\lambda t}{2}}\left(\frac{K_\lambda\left(\delta\sqrt{\alpha^2-(\beta+iu)^2}\right)}{K_\lambda\left(\delta\sqrt{\alpha^2-\beta^2}\right)}\right)^t,$
	where $K_\lambda(z) = \frac{\pi}{2}\frac{I_\nu(z) - I_{-\nu}(z)}{\sin(\nu\pi)},$
	$I_\nu(z) = \left(\frac{z}{2}\right)^\nu \sum_{k=0}^{\infty}\frac{(z^2/4)^k}{k!\Gamma(\nu+k+1)}$
Finite-moment stable	$\exp\left(iu\mu t - t(iu\sigma)^\alpha \sec\frac{\pi\alpha}{2}\right)$
CGMY	$\exp(C\Gamma(-Y))[(M-iu)^Y - M^Y + (G+iu)^Y - G^Y],$
	where $C, G, M > 0$ and $Y > 2$

$$
\begin{aligned}
\phi_X(u) &= E[e^{iuX_t}] \\
&= \exp\left(aitu - \frac{\sigma^2}{2}tu^2 + t\int_{\mathbb{R}\setminus\{0\}}\left(e^{iux} - 1 - iux\mathbf{1}_{|x|\le 1}\right)w(dx)\right) \\
&= \exp(t\psi_X(u)),
\end{aligned}
\tag{21.8}
$$

where $\int_{\mathbb{R}}\min(1,x^2)\,w(dx) < \infty$, $a \in \mathbb{R}$, $\sigma^2 \ge 0$. We identify a as the drift rate and σ as the volatility of the diffusion process. Here, $\psi_X(u)$ is called the characteristic exponent of X_t. Actually, $X_t \stackrel{d}{=} tX_1$. All moments of X_t can be derived from the characteristic function since it generalizes the moment-generating function to the complex domain. Indeed, a Lévy process X_t is fully specified by its characteristic function ϕ_X. In Table 21.1, we present a list of Lévy processes commonly used in finance applications together with their characteristic functions.

21.3 FFT Algorithms for Pricing European Vanilla Options

The renowned discounted expectation approach of evaluating a European option requires the knowledge of the density function of the asset returns under the risk neutral measure. Since the analytic representation of the characteristic function rather than the density function is more readily available for Lévy processes, we prefer to express the expectation integrals in terms of the characteristic function. First, we derive the formal analytic representation of a European option price as cumulative distribution functions, like the Black-Scholes type price formula. We then examine the inherent difficulties in the direct numerical evaluation of the Fourier integrals in the price formula.

Under the risk neutral measure Q, suppose the underlying asset price process assumes the form

$$S_t = S_0 \exp(-rt + X_t), \quad t > 0,$$

where X_t is a Lévy process and r is the riskless interest rate. We write $Y = \log S_0 + rT$ and let $\mathcal{F}_{V_T}$ denote the Fourier transform of the terminal payoff function $V_T(x)$, where $x = \log S_T$. By applying the discounted expectation valuation formula and the Fourier inversion formula (21.2), the European option value can be expressed as (see Lewis 2001)

$$\begin{aligned} V(S_t, t) &= e^{-r(T-t)} E_Q[V_T(x)] \\ &= \frac{e^{-r(T-t)}}{2\pi} E_Q\left[\int_{i\mu-\infty}^{i\mu+\infty} e^{-izx}\mathcal{F}_{V_T}(z)\,dz\right] \\ &= \frac{e^{-r(T-t)}}{2\pi} \int_{i\mu-\infty}^{i\mu+\infty} e^{-izx}\phi_{X_T}(-z)\mathcal{F}_{V_T}(z)\,dz, \end{aligned}$$

where $\mu = \operatorname{Im} z$ and $\Phi_{X_T}(z)$ is the characteristic function of X_T. The above formula agrees with (21.4) derived using the Parseval relation.

In our subsequent discussion, we set the current time to be zero and write the current stock price as S. For the T-maturity European call option with terminal payoff $(S_T - K)^+$, its value is given by (see Lewis 2001)

$$\begin{aligned} C(S,T;K) &= \frac{-Ke^{-rT}}{2\pi} \int_{i\mu-\infty}^{i\mu+\infty} \frac{e^{-iz\kappa}\phi_{X_T}(-z)}{z^2 - iz}\,dz \\ &= \frac{-Ke^{-rT}}{2\pi}\left[\int_{i\mu-\infty}^{i\mu+\infty} e^{-iz\kappa}\phi_{X_T}(-z)\frac{i}{z}\,dz \right.\\ &\quad \left. - \int_{i\mu-\infty}^{i\mu+\infty} e^{-iz\kappa}\phi_{X_T}(-z)\frac{i}{z-i}\,dz\right] \\ &= S\left[\frac{1}{2} + \frac{1}{\pi}\int_0^\infty \operatorname{Re}\left(\frac{e^{iu\log\kappa}\phi_{X_T}(u-i)}{iu\phi_{X_T}(-i)}\right)du\right] \\ &\quad - Ke^{-rT}\left[\frac{1}{2} + \frac{1}{\pi}\int_0^\infty \operatorname{Re}\left(\frac{e^{iu\log\kappa}\phi_{X_T}(u)}{iu}\right)du\right], \end{aligned} \tag{21.9}$$

where $\kappa = \log\frac{S}{K} + rT$. This representation of the call price resembles the Black-Scholes type price formula. However, due to the presence of the singularity at $u = 0$ in the integrand function, we cannot apply the FFT to evaluate the integrals. If we expand the integrals as Taylor series in u, the leading term in the expansion for both integral is $O\left(\frac{1}{u}\right)$. This is the source of the divergence, which arises from the discontinuity of the payoff function at $S_T = K$. As a consequence, the Fourier transform of the payoff function has large high frequency terms. Carr and Madan (1999) propose to dampen the high frequency terms by multiplying the payoff by an exponential decay function.

21.3.1 Carr–Madan Formulation

As an alternative formulation of European option pricing that takes advantage of the analytic expression of the characteristic function of the underlying asset price process, Carr and Madan (1999) consider the Fourier transform of the European call price (considered as a function of log strike) and compute the corresponding Fourier inversion to recover the call price using the FFT. Let $k = \log K$, the Fourier transform of the call price $C(k)$ does not exist since $C(k)$ is not square integrable. This is because $C(k)$ tends to S as k tends to $-\infty$.

21.3.1.1 Modified Call Price Method

To obtain a square-integrable function, Carr and Madan (1999) propose to consider the Fourier transform of the damped call price $c(k)$, where

$$c(k) = e^{\alpha k} C(k),$$

for $\alpha > 0$. Positive values of α are seen to improve the integrability of the modified call value over the negative k-axis. Carr and Madan (1999) show that a sufficient condition for square-integrability of $c(k)$ is given by

$$E_Q\left[S_T^{\alpha+1}\right] < \infty.$$

We write $\psi_T(u)$ as the Fourier transform of $c(k)$, $p_T(s)$ as the density function of the underlying asset price process, where $s = \log S_T$, and $\phi_T(u)$ as the characteristic function (Fourier transform) of $p_T(s)$. We obtain

$$\begin{aligned}\psi_T(u) &= \int_{-\infty}^{\infty} e^{iuk} c(k)\, dk \\ &= \int_{-\infty}^{\infty} e^{-rT} p_T(s) \int_{-\infty}^{s} \left[e^{s+\alpha k} - e^{(1+\alpha)k}\right] e^{iuk}\, dkds \\ &= \frac{e^{-rT}\phi_T(u-(\alpha+1)i)}{\alpha^2+\alpha-u^2+i(2\alpha+1)u}. \qquad (21.10)\end{aligned}$$

The call price $C(k)$ can be recovered by taking the Fourier inversion transform, where

$$C(k) = \frac{e^{-\alpha k}}{2\pi} \int_{-\infty}^{\infty} e^{-iuk} \psi_T(u)\, du$$
$$= \frac{e^{-\alpha k}}{\pi} \int_{0}^{\infty} e^{-iuk} \psi_T(u)\, du, \tag{21.11}$$

by virtue of the properties that $\psi_T(u)$ is odd in its imaginary part and even in its real part [since $C(k)$ is real]. The above integral can be computed using FFT, the details of which will be discussed next. From previous numerical experience, usually $\alpha = 3$ works well for most models of asset price dynamics. It is important to observe that α has to be chosen such that the denominator has only imaginary roots in u since integration is performed along real value of u.

21.3.1.2 FFT Implementation

The integral in (21.11) with a semi-infinite integration interval is evaluated by numerical approximation using the trapezoidal rule and FFT. We start with the choice on the number of intervals N and the stepwidth Δu. A numerical approximation for $C(k)$ is given by

$$C(k) \approx \frac{e^{-\alpha k}}{\pi} \sum_{j=1}^{N} e^{-iu_j k} \psi_T(u_j) \Delta u, \tag{21.12}$$

where $u_j = (j-1)\Delta u$, $j = 1, \cdots, N$. The semi-infinite integration domain $[0, \infty)$ in the integral in (21.11) is approximated by a finite integration domain, where the upper limit for u in the numerical integration is $N\Delta u$. The error introduced is called the *truncation error*. Also, the Fourier variable u is now sampled at discrete points instead of continuous sampling. The associated error is called the *sampling error*. Discussion on the controls on various forms of errors in the numerical approximation procedures can be found in Lee (2004).

Recall that the FFT is an efficient numerical algorithm that computes the sum

$$y(k) = \sum_{j=1}^{N} e^{-i\frac{2\pi}{N}(j-1)(k-1)} x(j), \qquad k = 1, 2, \cdots, N. \tag{21.13}$$

In the current context, we would like to compute around-the-money call option prices with k taking discrete values: $k_m = -b + (m-1)\Delta k$, $m = 1, 2, \cdots, N$. From one set of the FFT calculations, we are able to obtain call option prices for a range of strike prices. This facilitates the market practitioners to capture the price sensitivity of a European call with varying values of strike prices. To effect the FFT calculations, we note from (21.13) that it is necessary to choose Δu and Δk such that

$$\Delta u \Delta k = \frac{2\pi}{N}. \tag{21.14}$$

A compromise between the choices of Δu and Δk in the FFT calculations is called for here. For fixed N, the choice of a finer grid Δu in numerical integration leads to a larger spacing Δk on the log strike.

The call price multiplied by an appropriate damping exponential factor becomes a square-integrable function and the Fourier transform of the modified call price becomes an analytic function of the characteristic function of the log price. However, at short maturities, the call value tends to the non-differentiable terminal call option payoff causing the integrand in the Fourier inversion to become highly oscillatory. As shown in the numerical experiments performed by Carr and Madan (1999), this causes significant numerical errors. To circumvent the potential numerical pricing difficulties when dealing with short-maturity options, an alternative approach that considers the time value of a European option is shown to exhibit smaller pricing errors for all range of strike prices and maturities.

21.3.1.3 Modified Time Value Method

For notational convenience, we set the current stock price S to be unity and define

$$z_T(k) = e^{-rT} \int_{-\infty}^{\infty} \left[(e^k - e^s)\mathbf{1}_{\{s<k,k<0\}} + (e^s - e^k)\mathbf{1}_{\{s>k,k<0\}} \right] p_T(s)\, ds, \tag{21.15}$$

which is seen to be equal to the T-maturity call price when $K > S$ and the T-maturity put price when $K < S$. Therefore, once $z_T(k)$ is known, we can obtain the price of the call or put that is currently out-of-money while the call-put parity relation can be used to obtain the price of the other option that is in-the-money.

The Fourier transform $\zeta_T(u)$ of $z_T(k)$ is found to be

$$\begin{aligned} \zeta_T(u) &= \int_{-\infty}^{\infty} e^{iuk} z_T(k)\, dk \\ &= e^{-rT} \left[\frac{1}{1+iu} - \frac{e^{rT}}{iu} - \frac{\phi_T(u-i)}{u^2 - iu} \right]. \end{aligned} \tag{21.16}$$

The time value function $z_T(k)$ tends to a Dirac function at small maturity and around-the-money, so the Fourier transform $\zeta_T(u)$ may become highly oscillatory. Here, a similar damping technique is employed by considering the Fourier transform of $\sinh(\alpha k) z_T(k)$ (note that $\sinh \alpha k$ vanishes at $k = 0$). Now, we consider

$$\begin{aligned} \gamma_T(u) &= \int_{-\infty}^{\infty} e^{iuk} \sinh(\alpha k) z_T(k)\, dk \\ &= \frac{\zeta_T(u - i\alpha) - \zeta_T(u + i\alpha)}{2}, \end{aligned}$$

and the time value can be recovered by applying the Fourier inversion transform:

$$z_T(k) = \frac{1}{\sinh(\alpha k)} \frac{1}{2\pi} \int_{-\infty}^{\infty} e^{-iuk} \gamma_T(u) \, du. \tag{21.17}$$

Analogous FFT calculations can be performed to compute the numerical approximation for $z_T(k_m)$, where

$$z_T(k_m) \approx \frac{1}{\pi \sinh(\alpha k_m)} \sum_{j=1}^{N} e^{-i\frac{2\pi}{N}(j-1)(m-1)} e^{ibu_j} \gamma_T(u_j) \Delta u, \tag{21.18}$$

$$m = 1, 2, \cdots, N, \quad \text{and} \quad k_m = -b + (m-1)\Delta k.$$

21.4 Pricing of European Multi-Asset Options

Apparently, the extension of the Carr–Madan formulation to pricing European multi-asset options would be quite straightforward. However, depending on the nature of the terminal payoff function of the multi-asset option, the implementation of the FFT algorithm may require some special considerations.

The most direct extension of the Carr–Madan formulation to the multi-asset models can be exemplified through pricing of the correlation option, the terminal payoff of which is defined by

$$V(S_1, S_2, T) = (S_1(T) - K_1)^+ (S_2(T) - K_2)^+. \tag{21.19}$$

We define $s_i = \log S_i$, $k_i = \log K_i$, $i = 1, 2$, and write $p_T(s_1, s_2)$ as the joint density of $s_1(T)$ and $s_2(T)$ under the risk neutral measure Q. The characteristic function of this joint density is defined by the following two-dimensional Fourier transform:

$$\phi(u_1, u_2) = \int_{-\infty}^{\infty} \int_{-\infty}^{\infty} e^{i(u_1 s_1 + u_2 s_2)} p_T(s_1, s_2) \, ds_1 ds_2. \tag{21.20}$$

Following the Carr–Madan formulation, we consider the Fourier transform $\psi_T(u_1, u_2)$ of the damped option price $e^{\alpha_1 k_1 + \alpha_2 k_2} V_T(k_1, k_2)$ with respect to the log strike prices k_1, k_2, where $\alpha_1 > 0$ and $\alpha_2 > 0$ are chosen such that the damped option price is square-integrable for negative values of k_1 and k_2. The Fourier transform $\psi_T(u_1, u_2)$ is related to $\phi(u_1, u_2)$ as follows:

$$\psi_T(u_1, u_2) = \frac{e^{-rT} \phi(u_1 - (\alpha_1 + 1)i, u_2 - (\alpha_2 + 1)i)}{(\alpha_1 + iu_1)(\alpha_1 + 1 + iu_1)(\alpha_2 + iu_2)(\alpha_2 + 1 + iu_2)}. \tag{21.21}$$

To recover $C_T(k_1, k_2)$, we apply the Fourier inversion on $\psi_T(u_1, u_2)$. Following analogous procedures as in the single-asset European option, we approximate the two-dimensional Fourier inversion integral by

$$C_T(k_1,k_2) \approx \frac{e^{-\alpha_1 k_1 - \alpha_2 k_2}}{(2\pi)^2} \sum_{m=0}^{N-1}\sum_{n=0}^{N-1} e^{-i(u_m^1 k_1 + u_n^2 k_2)} \psi_T(u_m^1, u_n^2)\Delta_1\Delta_2, \qquad (21.22)$$

where $u_m^1 = \left(m - \frac{N}{2}\right)\Delta_1$ and $u_n^2 = \left(n - \frac{N}{2}\right)\Delta_2$. Here, Δ_1 and Δ_2 are the stepwidths, and N is the number of intervals. In the two-dimensional form of the FFT algorithm, we define

$$k_p^1 = \left(p - \frac{N}{2}\right)\Delta_1 \quad \text{and} \quad k_q^1 = \left(q - \frac{N}{2}\right)\Delta_2,$$

where λ_1 and λ_2 observe

$$\lambda_1\Delta_1 = \lambda_2\Delta_2 = \frac{2\pi}{N}.$$

Dempster and Hong (2000) show that the numerical approximation to the option price at different log strike values is given by

$$C_T(k_p^1, k_q^2) \approx \frac{e^{-\alpha_1 k_p^1 - \alpha_2 k_q^2}}{(2\pi)^2} \Gamma(k_p^1, k_q^2)\Delta_1\Delta_2, \qquad 0 \le p, q \le N, \qquad (21.23)$$

where

$$\Gamma(k_p^1, k_q^2) = (-1)^{p+q} \sum_{m=0}^{N-1}\sum_{n=0}^{N-1} e^{-\frac{2\pi i}{N}(mp+nq)} \left[(-1)^{m+n}\psi_T(u_m^1, u_n^2)\right].$$

The nice tractability in deriving the FFT pricing algorithm for the correlation option stems from the rectangular shape of the exercise region Ω of the option. Provided that the boundaries of Ω are made up of straight edges, the above procedure of deriving the FFT pricing algorithm still works. This is because one can always take an affine change of variables in the Fourier integrals to effect the numerical evaluation. What would be the classes of option payoff functions that allow the application of the above approach? Lee (2004) lists four types of terminal payoff functions that admit analytic representation of the Fourier transform of the damped option price. Another class of multi-asset options that possess similar analytic tractability are options whose payoff depends on taking the maximum or minimum value among the terminal values of a basket of stocks (see Eberlein et al. 2009). However, the exercise region of the spread option with terminal payoff

$$V_T(S_1, S_2) = (S_1(T) - S_2(T) - K)^+ \qquad (21.24)$$

is shown to consist of a non-linear edge. To derive the FFT algorithm of similar nature, it is necessary to approximate the exercise region by a combination of rectangular strips. The details of the derivation of the corresponding FFT pricing algorithm are presented by Dempster and Hong (2000).

Hurd and Zhou (2009) propose an alternative approach to pricing the European spread option under Lévy model. Their method relies on an elegant formula of the Fourier transform of the spread option payoff function. Let $P(s_1, s_2)$ denote the terminal spread option payoff with unit strike, where

$$P(s_1, s_2) = (e^{s_1} - e^{s_2} - 1)^+.$$

For any real numbers ϵ_1 and ϵ_2 with $\epsilon_2 > 0$ and $\epsilon_1 + \epsilon_2 < -1$, they establish the following Fourier representation of the terminal spread option payoff function:

$$P(s_1, s_2) = \frac{1}{(2\pi)^2} \int_{-\infty+i\epsilon_2}^{\infty+i\epsilon_2} \int_{-\infty+i\epsilon_1}^{\infty+i\epsilon_1} e^{i(u_1 s_1 + u_2 s_2)} \hat{P}(u_1, u_2)\, du_1 du_2, \quad (21.25)$$

where

$$\hat{P}(u_1, u_2) = \frac{\Gamma(i(u_1 + u_2) - 1)\Gamma(-iu_2)}{\Gamma(iu_1 + 1)}.$$

Here, $\Gamma(z)$ is the complex gamma function defined for $\text{Re}(z) > 0$, where

$$\Gamma(z) = \int_0^\infty e^{-t} t^{z-1}\, dt.$$

To establish the Fourier representation in (21.25), we consider

$$\hat{P}(u_1, u_2) = \int_{-\infty}^{\infty} \int_{-\infty}^{\infty} e^{-i(u_1 s_1 + u_2 s_2)} P(s_1, s_2)\, ds_2 ds_1.$$

By restricting to $s_1 > 0$ and $e^{s_2} < e^{s_1} - 1$, we have

$$\begin{aligned}
\hat{P}(u_1, u_2) &= \int_0^\infty e^{-iu_1 s_1} \int_{-\infty}^{\log(e^{s_1}-1)} e^{-iu_2 s_2}(e^{s_1} - e^{s_2} - 1)\, ds_2 ds_1 \\
&= \int_0^\infty e^{-iu_1 s_1} (e^{s_1} - 1)^{1-iu_2} \left(\frac{1}{-iu_2} - \frac{1}{1 - iu_2} \right) ds_1 \\
&= \frac{1}{(1 - iu_2)(-iu_2)} \int_0^1 z^{iu_1} \left(\frac{1-z}{z} \right)^{1-iu_2} \frac{dz}{z},
\end{aligned}$$

where $z = e^{-s_1}$. The last integral can be identified with the beta function:

$$\beta(a,b) = \frac{\Gamma(a)\Gamma(b)}{\Gamma(a+b)} = \int_0^1 z^{a-1}(1-z)^{b-1}\,dz,$$

so we obtain the result in (21.25). Once the Fourier representation of the terminal payoff is known, by virtue of the Parseval relation, the option price can be expressed as a two-dimensional Fourier inversion integral with integrand that involves the product of $\hat{P}(u_1, u_2)$ and the characteristic function of the joint process of s_1 and s_2. The evaluation of the Fourier inversion integral can be affected by the usual FFT calculations (see Hurd and Zhou 2009). This approach does not require the analytic approximation of the two-dimensional exercise region of the spread option with a non-linear edge, so it is considered to be more computationally efficient.

The pricing of European multi-asset options using the FFT approach requires availability of the analytic representation of the characteristic function of the joint price process of the basket of underlying assets. One may incorporate a wide range of stochastic structures in the volatility and correlation. Once the analytic forms in the integrand of the multi-dimensional Fourier inversion integral are known, the numerical evaluation involves nested summations in the FFT calculations whose dimension is the same as the number of underlying assets in the multi-asset option. This contrasts with the usual finite difference/lattice tree methods where the dimension of the scheme increases with the number of risk factors in the prescription of the joint process of the underlying assets. This is a desirable property over other numerical methods since the FFT pricing of the multi-asset options is not subject to this curse of dimensionality with regard to the number of risk factors in the dynamics of the asset returns.

21.5 Convolution Approach and Pricing of Bermudan Style Options

We consider the extension of the FFT technique to pricing of options that allow early exercise prior to the maturity date T. Recall that a Bermudan option can only be exercised at a pre-specified set of time points, say $\mathcal{T} = \{t_1, t_2, \cdots, t_M\}$, where $t_M = T$. On the other hand, an American option can be exercised at any time prior to T. By taking the number of time points of early exercise to be infinite, we can extrapolate a Bermudan option to become an American option. In this section, we would like to illustrate how the convolution property of Fourier transform can be used to price a Bermudan option effectively (see Lord et al. 2008).

Let $F(S(t_m), t_m)$ denote the exercise payoff of a Bermudan option at time t_m, $m = 1, 2, \cdots, M$. Let $V(S(t_m), t_m)$ denote the time-t_m value of the Bermudan option with exercise point set $\mathcal{T}$; and we write $\Delta t_m = t_{m+1} - t_m, m = 1, 2, \cdots, M-1$. The Bermudan option can be evaluated via the following backward induction procedure:

terminal payoff: $V(S(t_M), t_M) = F(S(t_M), t_M)$
For $m = M-1, M-2, \cdots, 1$, compute

$$C(S(t_m), t_m) = e^{-r\Delta t_m} \int_{-\infty}^{\infty} V(y, t_{m+1}) p(y|S(t_m))\, dy$$
$$V(S(t_m), t_m) = \max\{C(S(t_m), t_m), F(S(t_m), t_m)\}.$$

Here, $p(y|S(t_m))$ represents the probability density that relates the transition from the price level $S(t_m)$ at t_m to the new price level y at t_{m+1}. By virtue of the early exercise right, the Bermudan option value at t_m is obtained by taking the maximum value between the time-t_m continuation value $C(S(t_m), t_m)$ and the time-t_m exercise payoff $F(S(t_m), t_m)$.

The evaluation of $C(S(t_m), t_m)$ is equivalent to the computation of the time-t_m value of a t_{m+1}-maturity European option. Suppose the asset price process is a monotone function of a Lévy process (which observes the independent increments property), then the transition density $p(y|x)$ has the following property:

$$p(y|x) = p(y - x). \tag{21.26}$$

If we write $z = y - x$, then the continuation value can be expressed as a convolution integral as follows:

$$C(x, t_m) = e^{-r\Delta t_m} \int_{-\infty}^{\infty} V(x + z, t_{m+1}) p(z)\, dz. \tag{21.27}$$

Following a similar damping procedure as proposed by Carr and Madan (1999), we define

$$c(x, t_m) = e^{\alpha x + r\Delta t_m} C(x, t_m)$$

to be the damped continuation value with the damping factor $\alpha > 0$. Applying the property of the Fourier transform of a convolution integral, we obtain

$$\mathcal{F}_x\{c(x, t_m)\}(u) = \mathcal{F}_y\{v(y, t_{m+1})\}(u)\phi(-(u - i\alpha)), \tag{21.28}$$

and $\phi(u)$ is the characteristic function of the random variable z.

Lord et al. (2008) propose an effective FFT algorithm to calculate the following convolution:

$$c(x, t_m) = \frac{1}{2\pi} \int_{-\infty}^{\infty} e^{-iux} \hat{v}(u)\phi(-(u - i\alpha))\, du, \tag{21.29}$$

where $\hat{v}(u) = \mathcal{F}\{v(y, t_m)\}$. The FFT calculations start with the prescription of uniform grids for u, x and y:

$$u_j = u_0 + j\Delta u, \quad x_j = x_0 + j\Delta x, \quad y_j = y_0 + j\Delta y, \qquad j = 0, 1, \cdots, N-1.$$

The mesh sizes Δx and Δy are taken to be equal, and Δu and Δy are chosen to satisfy the Nyquist condition:

$$\Delta u \Delta y = \frac{2\pi}{N}.$$

The convolution integral is discretized as follows:

$$c(x_p) \approx \frac{e^{-iu_0(x_0+p\Delta y)}}{2\pi} \Delta u \sum_{j=0}^{N-1} e^{-ijp\frac{2\pi}{N}} e^{ij(y_0-x_0)\Delta u} \phi(-(u_j - i\alpha))\hat{v}(u_j), \quad (21.30)$$

where

$$\hat{v}(u_j) \approx e^{iu_0 y_0} \Delta y \sum_{n=0}^{N-1} e^{ijn2\pi/N} e^{inu_0\Delta y} w_n v(y_n),$$

$$w_0 = w_{N-1} = \frac{1}{2}, \quad w_n = 1 \quad \text{for } n = 1, 2, \cdots, N-2.$$

For a sequence x_p, $p = 0, 1, \cdots, N-1$, its discrete Fourier transform and the corresponding inverse are given by

$$\mathcal{D}_j\{x_n\} = \sum_{n=0}^{N-1} e^{ijn2\pi/N} x_n, \quad \mathcal{D}_n^{-1}\{x_j\} = \frac{1}{N}\sum_{j=0}^{N-1} e^{-ijn2\pi/N} x_j.$$

By setting $u_0 = -\frac{N}{2}\Delta u$ so that $e^{inu_0\Delta y} = (-1)^n$, we obtain

$$c(x_p) \approx e^{iu_0(y_0-x_0)}(-1)^p \mathcal{D}_p^{-1}\{e^{ij(y_0-x_0)\Delta u}\phi(-(u_j - i\alpha))\mathcal{D}_j\{(-1)^n w_n v(y_n)\}\}. \quad (21.31)$$

In summary, by virtue of the convolution property of Fourier transform, we compute the discrete Fourier inversion of the product of the discrete characteristic function of the asset returns $\phi(-(u_j - i\alpha))$ and the discrete Fourier transform of option prices $\mathcal{D}_j\{(-1)^n w_n v(y_n)\}$. It is seen to be more efficient when compared to the direct approach of recovering the density function by taking the Fourier inversion of the characteristic function and finding the option prices by discounted expectation calculations (see Zhylyevsky 2010).

21.6 FFT-Based Network Method

As an extension to the usual lattice tree method, an FFT-based network approach to option pricing under Lévy models has been proposed by Wong and Guan (2009). The network method somewhat resembles Duan-Simonato's Markov chain approximation method (Duan and Simonato 2001). This new approach is developed for option pricing for which the characteristic function of the log-asset value is known. Like the lattice tree method, the network method can be generalized to valuation of path dependent options by adopting the forward shooting grid technique (see Kwok 2010).

First, we start with the construction of the network. We perform the space-time discretization by constructing a pre-specified system of grids of time and state: $t_0 < t_1 < \cdots < t_M$, where t_M is the maturity date of the option, and $x_0 < x_1 < \cdots < x_N$, where $\mathcal{X} = \{x_j | j = 0, 1, \cdots, N\}$ represents the set of all possible values of log-asset prices. For simplicity, we assume uniform grid sizes, where $\Delta x = x_{j+1} - x_j$ for all j and $\Delta t = t_{i+1} - t_i$ for all i. Unlike the binomial tree where the number of states increases with the number of time steps, the number of states is fixed in advance and remains unchanged at all time points. In this sense, the network resembles the finite difference grid layout. The network approach approximates the Lévy process by a finite state Markov chain, like that proposed by Duan and Simonato (2001). We allow for a finite probability that the log-asset value moves from one state to any possible state in the next time step. This contrasts with the usual finite difference schemes where the linkage of nodal points between successive time steps is limited to either one state up, one state down or remains at the same state. The Markov chain model allows greater flexibility to approximate the asset price dynamics that exhibits finite jumps under Lévy model with enhanced accuracy. A schematic diagram of a network with seven states and three time steps is illustrated in Fig. 21.1.

After the construction of the network, the next step is to compute the transition probabilities that the asset price goes from one state x_i to another state x_j under the Markov chain model, $0 \leq i, j \leq N$. The corresponding transition probability is defined as follows:

$$p_{ij} = P[X_{t+\Delta t} = x_j | X_t = x_i], \tag{21.32}$$

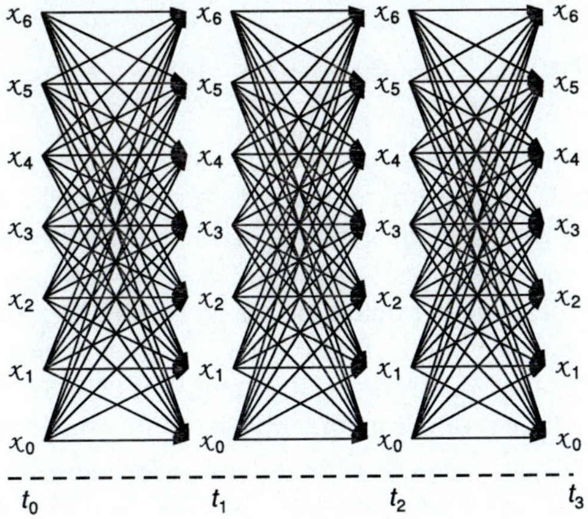

Fig. 21.1 A network model with three time steps and seven states

which is independent of t due to the time homogeneity of the underlying Lévy process. We define the corresponding characteristic function by

$$\phi_i(u) = \int_{-\infty}^{\infty} e^{iuz} f_i(z|x_i)\, dz,$$

where $f_i(z|x_i)$ is the probability density function of the increment $X_{t+\Delta t} - X_t$ conditional on $X_t = x_i$. The conditional probability density function can be recovered by Fourier inversion:

$$f_i(x_j|x_i) = \mathcal{F}_u^{-1}\{\phi_i(u)\}(x_j). \tag{21.33}$$

If we take the number of Markov chain states to be $N+1 = 2^L$ for some integer L, then the above Fourier inversion can be carried out using the FFT techniques. The FFT calculations produce approximate values for $f_i(x_j|x_i)$ for all i and j. We write these approximate conditional probability values obtained from the FFT calculations as $\tilde{f}_i(x_j|x_i)$. The transition probabilities among the Markov chain states are then approximated by

$$\tilde{p}_{ij} \approx \frac{\tilde{f}_i(x_j|x_i)}{\sum_{i=0}^{N} \tilde{f}_i(x_j|x_i)}, \qquad 0 \le i, j \le N. \tag{21.34}$$

Once the transition probabilities are known, we can perform option valuation using the usual discounted expectation approach. The incorporation of various path dependent features can be performed as in usual lattice tree calculations. Wong and Guan (2009) illustrate how to compute the Asian and lookback option prices under Lévy models using the FFT-based network approach. Their numerical schemes are augmented with the forward shooting grid technique (see Kwok 2010) for capturing the asset price dependency associated with the Asian and lookback features.

21.7 Fourier Space Time Stepping Method

When we consider option pricing under Lévy models, the option price function is governed by a partial integral-differential equation (PIDE) where the integral terms in the equation arise from the jump components in the underlying Lévy process. In this section, we present the Fourier space time stepping (FST) method that is based on the solution in the Fourier domain of the governing PIDE (see Jackson et al. 2008). This is in contrast with the usual finite difference schemes which solve the PIDE in the real domain. We discuss the robustness of the FST method with regard to its symmetric treatment of the jump terms and diffusion terms in the PIDE and the ease of incorporation of various forms of path dependence in the option models. Unlike the usual finite difference schemes, the FST method does not require time stepping calculations between successive monitoring dates in pricing Bermudan options and discretely monitored barrier options. In the numerical implementation procedures, the FST method does not require the analytic expression for the Fourier transform of the terminal payoff of the option so it can deal easier with more exotic forms of the payoff functions. The FST method can be easily extended to multi-asset option models with exotic payoff structures and pricing models that allow regime switching in the underlying asset returns.

First, we follow the approach by Jackson et al. (2008) to derive the governing PIDE of option pricing under Lévy models and consider the Fourier transform of the PIDE. We consider the model formulation under the general multi-asset setting. Let $\mathbf{S}(t)$ denote a d-dimensional price index vector of the underlying assets in a multi-asset option model whose T-maturity payoff is denoted by $V_T(\mathbf{S}(T))$. Suppose the underlying price index follows an exponential Lévy process, where

$$\mathbf{S}(t) = \mathbf{S}(0)e^{\mathbf{X}(t)},$$

and $\mathbf{X}(t)$ is a Lévy process. Let the characteristic component of $\mathbf{X}(t)$ be the triplet $(\boldsymbol{\mu}, M, \boldsymbol{\nu})$, where $\boldsymbol{\mu}$ is the non-adjusted drift vector, M is the covariance matrix of the diffusion components, and $\boldsymbol{\nu}$ is the d-dimensional Lévy density. The Lévy process $\mathbf{X}(t)$ can be decomposed into its diffusion and jump components as follows:

$$\mathbf{X}(t) = \boldsymbol{\mu}(t) + M\mathbf{W}(t) + \mathbf{J}^l(t) + \lim_{\epsilon \to 0} \mathbf{J}^\epsilon(t), \tag{21.35}$$

where the large and small components are

$$\mathbf{J}^l(t) = \int_0^t \int_{|\mathbf{y}| \geq 1} \mathbf{y}\, m(d\mathbf{y} \times ds)$$

$$\mathbf{J}^\epsilon(t) = \int_0^t \int_{\epsilon \leq |\mathbf{y}| < 1} \mathbf{y}\, [m(d\mathbf{y} \times ds) - \nu(d\mathbf{y} \times ds)],$$

respectively. Here, $\mathbf{W}(t)$ is the vector of standard Brownian processes, $m(d\mathbf{y} \times ds)$ is a Poisson random measure counting the number of jumps of size $\mathbf{y}$ occurring at time s, and $\nu(d\mathbf{y} \times ds)$ is the corresponding compensator. Once the volatility and Lévy density are specified, the risk neutral drift can be determined by enforcing the risk neutral condition:

$$E_0[e^{\mathbf{X}(1)}] = e^r,$$

where r is the riskfree interest rate. The governing partial integral-differential equation (PIDE) of the option price function $V(\mathbf{X}(t), t)$ is given by

$$\frac{\partial V}{\partial t} + \mathcal{L}V = 0 \tag{21.36}$$

with terminal condition: $V(\mathbf{X}(T), T) = V_T(\mathbf{S}(0), e^{\mathbf{X}(T)})$, where $\mathcal{L}$ is the infinitesimal generator of the Lévy process operating on a twice differentiable function $f(\mathbf{x})$ as follows:

$$\mathcal{L}f(\mathbf{x}) = \left(\boldsymbol{\mu}^{\mathbf{T}}\frac{\partial}{\partial \mathbf{x}} + \frac{\partial}{\partial \mathbf{x}}^{\mathbf{T}} M \frac{\partial}{\partial \mathbf{x}}\right) f(\mathbf{x})$$
$$+ \int_{\mathbb{R}^n\setminus\{\mathbf{0}\}} \{[f(\mathbf{x}+\mathbf{y}) - f(\mathbf{x})] - \mathbf{y}^{\mathbf{T}}\frac{\partial}{\partial \mathbf{x}} f(\mathbf{x})\mathbf{1}_{|\mathbf{y}|<1}\}\, \nu(d\mathbf{y}). \quad (21.37)$$

By the Lévy-Khintchine formula, the characteristic component of the Lévy process is given by

$$\psi_{\mathbf{X}}(\mathbf{u}) = i\boldsymbol{\mu}^{\mathbf{T}}\mathbf{u} - \frac{1}{2}\mathbf{u}^{\mathbf{T}}M\mathbf{u} + \int_{\mathbb{R}^n} \left(e^{i\mathbf{u}^{\mathbf{T}}\mathbf{y}} - 1 - i\mathbf{u}^{\mathbf{T}}\mathbf{y}\mathbf{1}_{|\mathbf{y}|<1}\right) \nu(d\mathbf{y}). \quad (21.38)$$

Several numerical schemes have been proposed in the literature that solve the PIDE (21.36) in the real domain. Jackson et al. (2008) propose to solve the PIDE directly in the Fourier domain so as to avoid the numerical difficulties in association with the valuation of the integral terms and diffusion terms. An account on the deficiencies in earlier numerical schemes in treating the discretization of the integral terms can be found in Jackson et al. (2008).

By taking the Fourier transform on both sides of the PIDE, the PIDE is reduced to a system of ordinary differential equations parametrized by the d-dimensional frequency vector $\mathbf{u}$. When we apply the Fourier transform to the infinitesimal generator $\mathcal{L}$ of the process $\mathbf{X}(t)$, the Fourier transform can be visualized as a linear operator that maps spatial differentiation into multiplication by the factor $i\mathbf{u}$. We define the multi-dimensional Fourier transform as follows (a slip of sign in the exponent of the Fourier kernel is adopted here for notational convenience):

$$\mathcal{F}[f](\mathbf{u}) = \int_{-\infty}^{\infty} f(\mathbf{x})e^{-i\mathbf{u}^{\mathbf{T}}\mathbf{x}}\, d\mathbf{x}$$

so that

$$\mathcal{F}^{-1}[\mathcal{F}_f](\mathbf{u}) = \frac{1}{2\pi}\int_{-\infty}^{\infty} \mathcal{F}_f e^{i\mathbf{u}^{\mathbf{T}}\mathbf{x}}\, d\mathbf{u}.$$

We observe

$$\mathcal{F}\left[\frac{\partial}{\partial \mathbf{x}} f\right] = i\mathbf{u}\mathcal{F}[f] \qquad \text{and} \qquad \mathcal{F}\left[\frac{\partial^2}{\partial \mathbf{x}^2} f\right] = i\mathbf{u}\,\mathcal{F}[f]\,i\mathbf{u}^{\mathbf{T}}$$

so that

$$\mathcal{F}[\mathcal{L}V](\mathbf{u}, t) = \psi_{\mathbf{X}}(\mathbf{u})\mathcal{F}[V](\mathbf{u}, t). \quad (21.39)$$

The Fourier transform of $\mathcal{L}V$ is elegantly given by multiplying the Fourier transform of V by the characteristic component $\psi_{\mathbf{X}}(\mathbf{u})$ of the Lévy process $\mathbf{X}(t)$. In the Fourier domain, $\mathcal{F}[V]$ is governed by the following system of ordinary differential equations:

$$\frac{\partial}{\partial t}\mathcal{F}[V](\mathbf{u},t) + \psi_{\mathbf{X}}(\mathbf{u})\mathcal{F}[V](\mathbf{u},t) = 0 \tag{21.40}$$

with terminal condition: $\mathcal{F}[V](\mathbf{u},T) = \mathcal{F}_{V_T}(\mathbf{u},T)$.

If there is no embedded optionality feature like the knock-out feature or early exercise feature between t and T, then the above differential equation can be integrated in a single time step. By solving the PIDE in the Fourier domain and performing Fourier inversion afterwards, the price function of a European vanilla option with terminal payoff V_T can be formally represented by

$$V(\mathbf{x},t) = \mathcal{F}^{-1}\left\{\mathcal{F}[V_T](\mathbf{u},T)e^{\psi_{\mathbf{X}}(\mathbf{u})(T-t)}\right\}(\mathbf{x},t). \tag{21.41}$$

In the numerical implementation procedure, the continuous Fourier transform and inversion are approximated by some appropriate discrete Fourier transform and inversion, which are then effected by FFT calculations. Let $\mathbf{v}_T$ and $\mathbf{v}_t$ denote the d-dimensional vector of option values at maturity T and time t, respectively, that are sampled at discrete spatial points in the real domain. The numerical evaluation of $\mathbf{v}_t$ via the discrete Fourier transform and inversion can be formally represented by

$$\mathbf{v}_t = \mathcal{FFT}^{-1}[\mathcal{FFT}[\mathbf{v}_T]e^{\psi_{\mathbf{X}}(T-t)}], \tag{21.42}$$

where $\mathcal{FFT}$ denotes the multi-dimensional FFT transform. In this numerical FFT implementation of finding European option values, it is not necessary to know the analytic representation of the Fourier transform of the terminal payoff function. This new formulation provides a straightforward implementation of numerical pricing of European spread options without resort to elaborate design of FFT algorithms as proposed by Dempster and Hong (2000) and Hurd and Zhou (2009) (see Sect. 21.4).

Suppose we specify a set of preset discrete time points $\mathcal{X} = \{t_1, t_2, \cdots, t_N\}$, where the option may be knocked out (barrier feature) or early exercised (Bermudan feature) prior to maturity T (take $t_{N+1} = T$ for notational convenience). At these time points, we either impose constraints or perform optimization based on the contractual specification of the option. Consider the pricing of a discretely monitored barrier option where the knock-out feature is activated at the set of discrete time points $\mathcal{X}$. Between times t_n and t_{n+1}, $n = 1, 2, \cdots, N$, the barrier option behaves like a European vanilla option so that the single step integration can be performed from t_n to t_{n+1}. At time t_n, we impose the contractual specification of the knock-out feature. Say, the option is knocked out when S stays above the up-and-out barrier B. Let R denote the rebate paid upon the occurrence of knock-out, and $\mathbf{v}^n$ be the vector of option values at discrete spatial points. The time stepping algorithm can be succinctly represented by

$$\mathbf{v}^n = H_B(\mathcal{FFT}^{-1}[\mathcal{FFT}[\mathbf{v}^{n+1}]e^{\psi_{\mathbf{X}}(t_{n+1}-t_n)}]),$$

where the knock-out feature is imposed by defining H_B to be (see Jackson et al. 2008)

$$H_B(\mathbf{v}) = \mathbf{v}\mathbf{1}_{\left\{x<\log \frac{B}{S(0)}\right\}} + R\mathbf{1}_{\left\{x\geq\log \frac{B}{S(0)}\right\}}.$$

No time stepping is required between two successive monitoring dates.

21.8 Summary and Conclusions

The Fourier transform methods provide the valuable and indispensable tools for option pricing under Lévy processes since the analytic representation of the characteristic function of the underlying asset return is more readily available than that of the density function itself. When used together with the FFT algorithms, real time pricing of a wide range of option models under Lévy processes can be delivered using the Fourier transform approach with high accuracy, efficiency and reliability. In particular, option prices across the whole spectrum of strikes can be obtained in one set of FFT calculations.

In this chapter, we review the most commonly used option pricing algorithms via FFT calculations. When the European option price function is expressed in terms of Fourier inversion integrals, option pricing can be delivered by finding the numerical approximation of the Fourier integrals via FFT techniques. Several modifications of the European option pricing formulation in the Fourier domain, like the damped option price method and time value method, have been developed so as to avoid the singularity associated with non-differentiability of the terminal payoff function. Alternatively, the pricing formulation in the form of a convolution product is used to price Bermudan options where early exercise is allowed at discrete time points. Depending on the structures of the payoff functions, the extension of FFT pricing to multi-asset models may require some ingeneous formulation of the corresponding option model. The order of complexity in the FFT calculations for pricing multi-asset options generally increases with the number of underlying assets rather than the total number of risk factors in the joint dynamics of the underlying asset returns. When one considers pricing of path dependent options whose analytic form of the option price function in terms of Fourier integrals is not readily available, it becomes natural to explore various extensions of the lattice tree schemes and finite difference approach. The FFT-based network method and the Fourier space time stepping techniques are numerical approaches that allow greater flexibility in the construction of the numerical algorithms to handle various form of path dependence of the underlying asset price processes through the incorporation of the auxiliary conditions that arise from modeling the embedded optionality features. The larger number of branches in the FFT-based network approach can provide better accuracy to approximate the Lévy process with jumps when compared to the usual trinomial tree approach. The Fourier space time stepping method solves the governing partial integral-differential equation of option pricing under Lévy model in the Fourier domain. Unlike usual finite difference schemes, no time stepping procedures are required between successive monitoring instants in option models with discretely monitored features.

In summary, a rich set of numerical algorithms via FFT calculations have been developed in the literature to perform pricing of most types of option models under Lévy processes. For future research topics, one may consider the pricing of volatility derivatives under Lévy models where payoff function depends on the realized variance or volatility of the underlying price process. Also, more theoretical works should be directed to error estimation methods and controls with regard to sampling errors and truncation errors in the approximation of the Fourier integrals and other numerical Fourier transform calculations.

Acknowledgements This work was supported by the Hong Kong Research Grants Council under Project 642110 of the General Research Funds.

References

Black, F., & Scholes, M. (1973). The pricing of option and corporate liabilities. *Journal of Political Economy, 81*, 637–659.

Carr, P., & Madan, D. (1999). Option valuation using the fast Fourier transform. *Journal of Computational Finance*, *2*(4), 61–73.

Cont, R., & Tankov, P. (2004). *Financial modelling with jump processes*. Boca Raton: Chapman and Hall.

Dempster, M. A. H., & Hong, S. S. G. (2000). *Spread option valuation and the fast Fourier transform*. Technical report WP 26/2000. Cambridge: The Judge Institute of Management Studies, University of Cambridge.

Derman, E., & Kani, T. (1998). Stochastic implied trees: Arbitrage pricing with stochastic term and strike structure of volatility. *International Journal of Theoretical and Applied Finance, 1*, 61–110.

Duan, J., & Simonato, J. G. (2001). American option pricing under GARCH by a Markov chain approximation. *Journal of Economic Dynamics and Control, 25*, 1689–1718.

Duffie, D., Pan, J., & Singleton, K. (2000). Transform analysis and asset pricing for affine jump-diffusion. *Econometrica, 68*(6), 1343–1376.

Dufresne, D., Garrido, J., & Morales, M. (2009). Fourier inversion formulas in option pricing and insurance. *Methodology and Computing in Applied Probability, 11*, 359–383.

Dupire, B. (1994). Pricing with smile. *Risk*, *7*(1), 18–20.

Eberlein, E., Glau, K., & Papapantoleon, A. (2009). Analysis of Fourier transform valuation formulas and applications. Working paper of Freiburg University.

Heston, S. (1993). A closed-form solution for options with stochastic volatility with applications to bond and currency options. *Review of Financial Studies, 6*, 327–343.

Hurd, T. R., & Zhou, Z. (2009). A Fourier transform method for spread option pricing. Working paper of McMaster University.

Jackson, K. R., Jaimungal, S., & Surkov, V. (2008). Fourier space time-stepping for option pricing with Levy models. *Journal of Computational Finance, 12*(2), 1–29.

Kwok, Y. K. (2010). Lattice tree methods for strongly path dependent options. *Encyclopedia of Quantitative Finance*, Cont R. (ed.), John Wiley and Sons Ltd, 1022–1027.

Lee, R. (2004). Option pricing by transform methods: Extensions, unification, and error control. *Journal of Computational Finance*, *7*(3), 51–86.

Lewis, A. L. (2001). A simple option formula for general jump-diffusion and other exponential Levy processes. Working paper of Envision Financial Systems and OptionsCity.net, Newport Beach, California.

Lord, R., Fang, F., Bervoets, F., & Oosterlee, C. W. (2008). A fast and accurate FFT-based method for pricing early-exercise options under Lévy processes. *SIAM Journal on Scientific Computing*, *30*, 1678–1705.

Merton, R. (1973). Theory of rational option pricing. *Bell Journal of Economics and Management Sciences*, *4*, 141–183.

Merton, R. (1976). Option pricing when the underlying stock returns are discontinuous. *Journal of Financial Economics*, *3*, 125–144.

Naik, V. (2000). Option pricing with stochastic volatility models. *Decisions in Economics and Finance*, *23*(2), 75–99.

Stein, E., & Stein, J. (1991). Stock price distribution with stochastic volatility: An analytic approach. *Review of Financial Studies*, *4*, 727–752.

Wong, H. Y., & Guan, P. (2009). An FFT network for Lévy option pricing. Working Paper of The Chinese University of Hong Kong.

Zhylyevsky, O. (2010). A fast Fourier transform technique for pricing American options under stochastic volatility. *Review of Derivatives Research*, *13*, 1–24.

Chapter 22
Dynamic Programming and Hedging Strategies in Discrete Time

Shih-Feng Huang and Meihui Guo

Abstract In this chapter, we introduce four hedging strategies for path-independent contingent claims in discrete time – superhedging, local expected shortfall-hedging, local quadratic risk-minimizing and local quadratic risk-adjusted-minimizing strategies. The corresponding dynamic programming algorithms of each trading strategy are introduced for making adjustment at each rebalancing time. The hedging performances of these discrete time trading strategies are discussed in the trinomial, Black-Scholes and GARCH models. Moreover, the hedging strategies of path-dependent contingent claims are introduced in the last section, and the hedges of barrier options are illustrated as examples.

22.1 Introduction

A hedge is an important financial strategy used to reduce the risk of adverse price movements in an asset by buying or selling others. Recently, hedging becomes a more important issue consequent on the catastrophe for the global financial system caused by the bankruptcy of major financial-services firm such as the Lehman Brothers Holdings Inc. In practice, practitioners are impossible to adjust their hedging positions continuously, such as in the Black-Scholes framework, and need to reduce the rebalancing times to lower down their transaction costs. Thus how to set up hedging portfolios in discrete time is of more practical importance. Herein, we introduce four hedging strategies for path-independent contingent claims in

S.-F. Huang (✉)
Department of Applied Mathematics, National University of Kaohsiung, Kaohsiung, Taiwan
e-mail: huangsf@nuk.edu.tw

M. Guo
Department of Applied Mathematics, National Sun Yat-sen University, Kaohsiung, Taiwan
e-mail: guomh@math.nsysu.edu.tw

J.-C. Duan et al. (eds.), *Handbook of Computational Finance*, Springer Handbooks of Computational Statistics, DOI 10.1007/978-3-642-17254-0_22,

discrete time – superhedging, local expected shortfall-hedging, local quadratic risk-minimizing and local quadratic risk-adjusted-minimizing strategies. The related dynamic programmings are introduced for making adjustment of the hedging portfolio at each rebalancing time.

Normally, a hedge consists of taking an offsetting position in a related security, such as a derivative. In complete financial markets, contingent claims can be replicated by self-financing strategies, and the costs of replication define the prices of the claims. In incomplete financial markets, one can still eliminate the risk completely by using a "superhedging" strategy (or called the perfect-hedging). However, from a practical point of view the cost of superhedging is often too expensive. Therefore investors turn to hedging strategies with less capitals by considering risk minimization criteria. Different hedging strategies are proposed from different economic perspectives such as minimizing the quadratic hedging risks or the expected shortfall risks. To simplify the illustration of the hedging strategies, we employ a trinomial model, which is a discrete time and discrete state incomplete market model, to introduce the construction of these hedging strategies. In addition, we will compare the hedging performances of different discrete time hedging strategies in the Black-Scholes and GARCH models. In the last section, we discuss the problem of hedging path-dependent contingent claims and introduce the hedging strategies of barrier options.

22.2 Discrete Time Hedging Strategies and Dynamic Programmings

In this section, several discrete time hedging strategies in incomplete market models are introduced. We illustrate four hedging strategies in a trinomial model, which is a discretized description of geometric Brownian motion often used to describe asset behavior. One can extend the results to multinomial market model analogously. In a trinomial tree the asset price at each node moves in three possible ways, up movement, down movement and jump movement. The general form of a one period trinomial tree is as shown in Fig. 22.1a. Given the stock price S_{t-1} at time $t-1$, where $t = 1, 2, \cdots$, suppose that there are three possible stock prices at time t, $S_t = uS_{t-1}$, $S_t = dS_{t-1}$ and $S_t = jS_{t-1}$, with probability p_1, p_2 and p_3, respectively, where $u > d > j$, p_i's are positive and $p_1 + p_2 + p_3 = 1$. If a practitioner shorts a European call option with expiration date T and strike price K at the initial time, how could she set up a hedging portfolio to hedge her short position?

22.2.1 *Superhedging Strategy*

First of all, we discuss the superhedging strategy, which was introduced by Bensaid et al. (1992) in discrete time. They concluded that for a contingent claim with

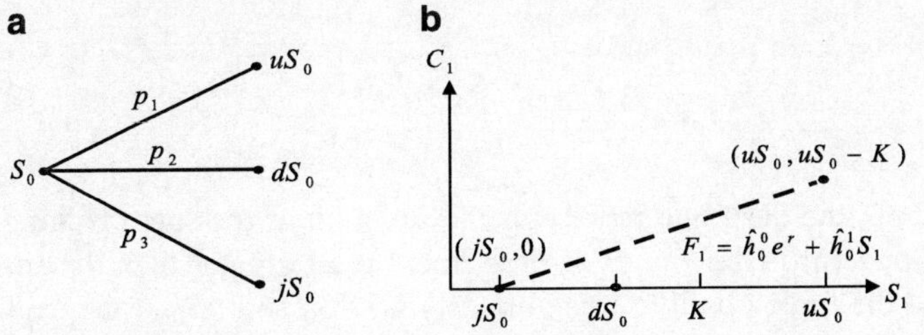

Fig. 22.1 (**a**) One period trinomial model; (**b**) Superhedging of the one period trinomial model

expiration date T and payoff V_T the initial hedging capital of the corresponding superhedging strategy is identical to

$$\sup_{Q \in \mathcal{Q}} E^Q(e^{-rT} V_T),$$

where $\mathcal{Q}$ is the set containing all the risk-neutral probability measures Q, and r is the continuously compounded riskless interest rate. In order to construct the superhedging strategy in the trinomial model, a comprehensive method is introduced in the following. We employ a one-period trinomial model for illustration. Denote the initial hedging capital by F_0 and let

$$F_0 = h_0^0 + h_0^1 S_0,$$

where h_0^0 and h_0^1 are the holding units of the riskless bond and the stock at the initial time, respectively. At the expiration date, the value of this hedging portfolio becomes $F_1 = h_0^0 e^r + h_0^1 S_1$. Our aim is to search a hedging strategy $H^{SH} = (\hat{h}_0^0, \hat{h}_0^1)$ such that

$$\begin{aligned} F_0(H^{SH}) = \min{}_H\{F_0(H) : F_1(H) \geq C_1 = (S_1 - K)^+, \\ \text{for } S_1 = uS_0,\ dS_0, \text{ and } jS_0\}, \end{aligned} \tag{22.1}$$

where $C_1 = (S_1 - K)^+$ is the payoff function of a European call option with strike price K. Note that F_1 is a linear function of S_1 and C_1 is convex in S_1. Thus the linear function passing through the two terminal points $(uS_0, C_1(uS_0))$ and $(jS_0, C_1(jS_0))$ is the optimal solution of (22.1). Hence, in the trinomial model with assuming $K < uS_0$, the superhedging strategy H^{SH} is defined as

$$\begin{cases} \hat{h}_0^0 = \dfrac{K - uS_0}{u - j} j e^{-r} \\ \hat{h}_0^1 = \dfrac{uS_0 - K}{(u - j)S_0} \end{cases},$$

and thus

$$F_0(H^{SH}) = \frac{(uS_0 - K)(1 - je^{-r})}{u - j}.$$

In Fig. 22.1b, the dash line represents the function of the superhedging strategy at time 1. Apparently, the values of dash line are all greater than the corresponding payoffs of the three possible stock prices jS_0, dS_0 and uS_0 at the expiration date. In other words, practitioners can eliminate all the risk of the short position of the European call option by setting the superhedging strategy H^{SH} with initial capital $F_0(H^{SH})$.

Example 1. If $u = 1.1$, $d = 0.9$, $j = 0.8$, $r = 0$, $S_0 = 100$ and $K = 100$, then $F_0(H^{SH}) = \frac{20}{3}$. Furthermore, let q_1, q_2 and q_3 denote the risk-neutral probability measures of the events $S_1 = uS_0$, $S_1 = dS_0$ and $S_1 = jS_0$, respectively. Using the constraints of $S_0 = e^{-r}E^Q(S_1)$, $q_1 + q_2 + q_3 = 1$, and q_i's are positive, we have $\frac{1}{2} < q_1 < \frac{2}{3}$, $q_2 = 2 - 3q_1$ and $q_3 = 2q_1 - 1$. Hence, the no-arbitrage price of the European call option, $C_0 = e^{-r}E^Q(C_1)$, is between 5 and $\frac{20}{3}$, where the upper bound is exactly the same as the hedging capital of the superhedge. This result is consistent with the conclusion in Bensaid et al. (1992).

Although the superhedging can always keep investors staying on the safe side, it is often too expensive. Therefore, practitioners are unwilling to put up the initial amount of capital required by a superhedging and are ready to accept some risk with some risk minimizing criteria. In the following sections, we introduce several different risk minimizing criteria.

22.2.2 *Local Expected Shortfall-Hedging and the Related Dynamic Programming*

The expected shortfall of a self-financing hedging strategy H of a contingent claim with payoff V_T is defined as

$$E\{(V_T - F_T(H))^+\},$$

where $F_T(H)$ is the terminal wealth of the self-financing hedging strategy H at the expiration date T. Practitioners want to know whether there exists an optimal hedging strategy, denoted by H^{ES}, such that the expected shortfall risk is minimized with a pre-fixed initial hedging capital V_0, that is,

$$H^{ES} = \arg\min_{H \in \mathcal{S}} E\{(V_T - F_T(H))^+\},$$

where

$$\mathsf{S} = \{H \mid H \text{ is a self-financing hedging strategy with initial hedging capital } V_0\}.$$

Cvitanić and Karatzas (1999) and Föllmer and Leukert (2000) pioneered the expected shortfall-hedging approach and showed the existence of this hedging strategy. Schulmerich and Trautmann (2003) proposed a searching algorithm to construct a hedging strategy which minimizes the expected shortfall risk in complete and incomplete discrete markets. But the searching algorithm often spends large of computation time. In order to overcome this time-consuming problem, Schulmerich and Trautmann (2003) further proposed a local expected shortfall-hedging strategy. The idea of the local expected shortfall-hedging strategy is introduced in the following.

The first step is to find an optimal modified contingent claim X^*, which is a contingent claim that belongs to the set χ of all modified contingent claims, where

$$\chi \equiv \{X \mid X < V_T \text{ and } E^Q(X/B^T) \leq V_0$$
$$\text{for all risk-netral probability measure } Q\},$$

and

$$X^* = \arg\min_{X \in \chi} E(V_T - X). \tag{22.2}$$

The above definition implies that the superhedging cost of any modified contingent claim is lower or equal than the initial hedging capital V_0. By Proposition 2 of Schulmerich and Trautmann (2003), the superhedging cost of the optimal modifined contingent claim X^* is identical to the shortfall risk of the hedging strategy H^{ES}, that is,

$$E\{(V_T - F_T(H^{ES}))^+\} = E(V_T - X^*).$$

Therefore, one can determine the desired hedging strategy H^{ES} by the following two steps:

[Dynamic programming of expected shortfall-hedging]

1. Find an optimal modified contingent claim $X^* \in \chi$ with criterion (22.2).
2. Construct a superhedging strategy for X^*.

Since Step-2 can be accomplished by the method introduced in the previous section, the main concern is the first step. In complete markets, the optimal modified contingent claim X^* is a direct consequence of a slight modification of the Neyman-Pearson lemma (see Föllmer and Leukert 2000; Schulmerich and Trautmann 2003). The solution of the optimal modified contingent claim is given in Proposition 4 of Schulmerich and Trautmann (2003), that is,

$$X^*(\omega) = V_T(\omega) I_{(P(\omega)/Q(\omega) > c)} + \gamma I_{(P(\omega)/Q(\omega) = c)} \tag{22.3}$$

with $c_{ES} = \min_\omega\{P(\omega)/Q(\omega)\}$ and

$$\gamma = \{V_0 B_T - V_T E^Q(I_{(P(\omega)/Q(\omega)>c)})\}\Big/E^Q(I_{(P(\omega)/Q(\omega)=c)}).$$

If the market is incomplete, the construction of the optimal expected shortfall hedging strategy is much more complicated than that in complete markets due to the fact that the risk-neutral probability measures are not unique. For continuous time models, Föllmer and Leukert (2000) showed that an optimal hedging strategy exists but didn't provide an explicit algorithm to calculate it. As for the discrete models, an algorithm of the optimal expected shortfall-hedging is given in Proposition 5 of Schulmerich and Trautmann (2003). The basic idea of this algorithm is still based on (22.3). The main difficulty is to deal with the non-uniqueness of the equivalent martingale measures. Let $\overline{Q}$ denote the smallest polyhedron containing all the martingale measures. Since $\overline{Q}$ is convex, there exists a finite number of extreme points of the convex polyhedron $\overline{Q}$, denoted by $Q_1, \cdots, Q_L$, and thus the criterion of choosing optimal modified contingent claim X^*, $\max_{Q\in\overline{Q}} E^Q(X^*/B_T) \leq V_0$, could be simplified by

$$\max_{i=1,\cdots,L} E^{Q_i}(X^*/B_T) \leq V_0.$$

However, it consumes a lot of computational effort to check this condition. Therefore, Schulmerich and Trautmann (2003) further proposed the following local expected shortfall-hedging strategy, denoted by H^{LES}:

[Dynamic programming of local expected shortfall-hedging]

Let V_T be a European type contingent claim and F_t^{SH} be the corresponding superhedging values at time $t = 1, \cdots, T$. Then find sequentially a self-financing strategy $H^{LES} = (H_1^{LES}, \cdots, H_T^{LES})$ with H_t^{LES} minimizing the local expected shortfall

$$E_{t-1}\{(F_t^{SH} - F_t(H))^+\},$$

for $t = 1, \cdots, T$, where $E_{t-1}(\cdot)$ denotes the conditional expectation under the dynamic probability measure given the information up to time $t-1$.

In the following, we give two examples to illustrate the construction of H_t^{LES}. Example 2 gives a one-period trinomial case and Example 3 considers a two-period situation.

Example 2. Consider the same one-period trinomial model as in Example 1. Let ω_1, ω_2 and ω_3 denote the states of $S_1 = uS_0$, dS_0 and jS_0, respectively, and P denote the dynamic probability measure with $P(\omega_1) = 0.55$, $P(\omega_2) = 0.40$ and $P(\omega_3) = 0.05$. As shown in Example 1, the set $\mathcal{Q}$ of risk-neutral probability measures can be expressed as

$$\mathcal{Q} = \{(q_1, q_2, q_3) : \frac{1}{2} < q_1 < \frac{2}{3},\ q_2 = 2 - 3q_1 > 0 \text{ and } q_3 = 2q_1 - 1 > 0\}.$$

Let $\overline{\mathcal{Q}}$ be the smallest polyhedron containing $\mathcal{Q}$, that is,

$$\overline{\mathcal{Q}} = \{(q_1, q_2, q_3) : \frac{1}{2} \leq q_1 \leq \frac{2}{3}, q_2 = 2 - 3q_1 \geq 0 \text{ and } q_3 = 2q_1 - 1 \geq 0\}.$$

Then $Q_1(\omega_1, \omega_2, \omega_3) = (\frac{1}{2}, \frac{1}{2}, 0)$ and $Q_2(\omega_1, \omega_2, \omega_3) = (\frac{2}{3}, 0, \frac{1}{3})$ be two extreme points of this convex polyhedron $\overline{\mathcal{Q}}$.

A practitioner is willing to set her initial hedging capital to be 6, which is less than the initial capital required by the superhedging strategy $\frac{20}{3}$. Our aim is to determine a trading strategy minimizing the expected shortfall with the initial hedging capital. By Proposition 5 of Schulmerich and Trautmann (2003), since $E^{Q_2}(V_1) > E^{Q_1}(V_1)$, we consider Q_2 first. In order to determine the modified contingent claim, one can apply Proposition 4 of Schulmerich and Trautmann (2003). However, Proposition 4 of Schulmerich and Trautmann (2003) can not be implemented directly to the trinomial model since trinomial model is not a complete market model. Nevertheless, due to the fact that $Q_2(\omega_2) = 0$, we can ignore the state ω_2 temporarily and only determine the modified contingent claim by the states ω_1 and ω_3:

$$X(\omega) = V_1(\omega) I_{(P(\omega)/Q_2(\omega) > c)} + \gamma I_{(P(\omega)/Q_2(\omega) = c)},$$

where $c = \min\{P(\omega)/Q_2(\omega) : \omega = \omega_i, i = 1, 3\}$ and γ is chosen to ensure $E^{Q_2}(X) \leq 6$. By straightforward computation, we have $X(\omega_1) = 10$ and $X(\omega_3) = -2$.

Next, construct the superhedging strategy for $X(\omega_i)$, $i = 1, 3$. By the same way introduced in Sect. 22.2.1, one can obtain the hedging portfolio $H_0^{LES} = (\tilde{h}_0^0, \tilde{h}_0^1)$ to be

$$\begin{cases} \tilde{h}_0^0 = -34 \\ \tilde{h}_0^1 = 0.4. \end{cases},$$

which satisfies $H_1^{LES}(\omega_1) = X(\omega_1) = 10$ and $H_1^{LES}(\omega_3) = X(\omega_3) = -1$. Finally, we defined the value of the modified contingent claim of state ω_2 by $X(\omega_2) = H_1^{LES}(\omega_2) = 2$. Note that for this modified contingent claim, we have $E^{Q_2}(X) = E^{Q_1}(X) = 6$ and since any risk-neutral probability measure can be expressed by $Q = aQ_1 + (1-a)Q_2$, $0 < a < 1$, thus for all risk-neutral probability measure $Q \in \mathcal{Q}$ we conclude that $E^Q(X) = 6$, and the corresponding minimal expected shortfall is

$$E\{V_1 - F_1(H_0^{LES})\}^+ = E(V_1 - X)^+ = 0.1.$$

Example 3. In this example, we extend the one-period trinomial model discussed in previous examples to two period. In each period, given the stock price S_t, $t = 0, 1$, let the stock prices at the next time point be $S_{t+1} = uS_t$, dS_t and jS_t, with dynamic probability 0.55, 0.40 and 0.05, respectively. In Fig. 22.3, the values of S_0, S_1 and S_2 are set with $S_0 = 100$, $u = 1.1$, $d = 0.9$ and $j = 0.8$. The payoff at time-2 is defined by $V_2 = (S_2 - 100)^+$, which is the payoff of a European call option with

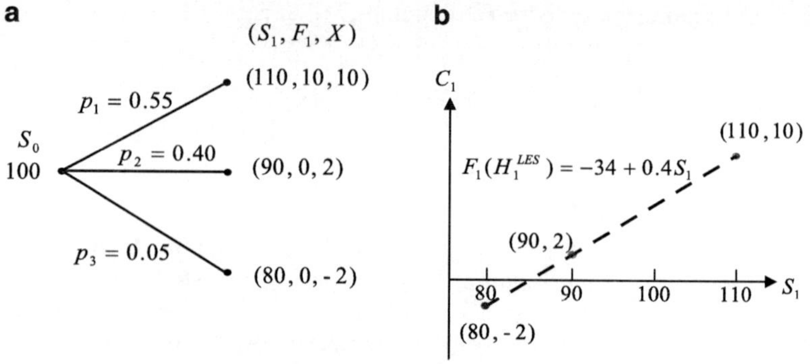

Fig. 22.2 (**a**) One period trinomial model in Example 2; (**b**) Local expected shortfall-hedging strategy of the one period trinomial model

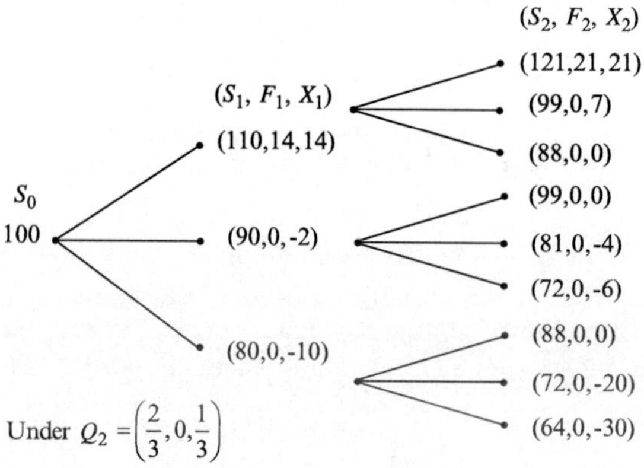

Fig. 22.3 Local expected shortfall-hedging strategy of the two period trinomial model in Example 3

expiration date $T = 2$ and strike price $K = 100$. As in Example 2, the probability measures $Q_1(\omega_1, \omega_2, \omega_3) = (\frac{1}{2}, \frac{1}{2}, 0)$ and $Q_2(\omega_1, \omega_2, \omega_3) = (\frac{2}{3}, 0, \frac{1}{3})$ are the two extreme points of the convex polyhedron $\overline{\mathcal{Q}}$. Also assume that an investor's initial hedging capital is set to be 6, which is still less than the initial capital required by the superhedging strategy

$$\max_{Q \in \overline{\mathcal{Q}}} E^{Q}(e^{-2r} V_2) = E^{Q_2}(e^{-2r} V_2) = \frac{28}{3},$$

where the riskless interest rate r is set to be 0. In the following, we illustrate how to obtain the modified contingent claim X_t at each time point $t = 1, 2$, and then construct the local expected shortfall-hedging strategy.

Since $E^{Q_2}(e^{-2r} F_2) > E^{Q_1}(e^{-2r} F_2)$ where $F_2 = V_2$, thus under the probability measure Q_2, compute the time-1 payoff F_1 by the conditional expectation,

$E^{Q_2}(e^{-r}F_2|S_1)$, given the stock price S_1. The first step is to find the one period optimal expected shortfall-hedging with initial hedging capital 6 and payoff F_1 in the one period trinomial model. This step can be solve by similar way as in Example 2. Hence, we obtain the modified contingent claim X_1 and the corresponding hedging strategy.

For the second period, given any stock price S_1, the problem can be treated as another one period trinomial hedging task, that is, find the one period optimal expected shortfall-hedging with initial hedging capital $X_1(S_1)$ and payoff F_2. Therefore, we can still adopt similar way as in Example 2 to obtain the modified contingent claim X_2 and the corresponding hedging strategy. The values of the modified contingent claim X_i, $i = 1, 2$, are given in Fig. 22.3. Note that for the modified contingent claim X_2, we have $E^{Q_2}(X_2) = E^{Q_1}(X_2) = 6$ and since any risk-neutral probability measure can be expressed by $Q = aQ_1 + (1-a)Q_2$, $0 < a < 1$, thus for all risk-neutral probability measure $Q \in \mathcal{Q}$ we conclude that $E^{Q}(X) = 6$, and the corresponding expected shortfall is $E[(F_2 - X_2)^+] = 1.235$.

22.2.3 *Local Quadratic Risk-Minimizing Hedging Strategy and Its Dynamic Programming*

In the following two sections, we introduce two different local quadratic risk-minimizing hedging strategies. Both are allowed to be non-self-financing trading strategies. That is, practitioners are allowed to put or withdraw money at each rebalancing time point.

Consider a contingent claim, underlying the risky stock S_T, that pays the value V_T at expiration date T. Practitioners interested in hedging this claim could attempt to set up a hedging scheme by a dynamic trading strategy in the underlying assets. Let F_{t-1} be the value of the hedging portfolio consisting of riskless bond and the underlying stock at time $t-1$,

$$F_{t-1} = h_{t-1}^0 B_{t-1} + h_{t-1}^1 S_{t-1}, \tag{22.4}$$

where h_{t-1}^0 and h_{t-1}^1 are the holding units of riskless bond B_{t-1} and the stock S_{t-1} at time $t-1$, respectively. Retain h_{t-1}^0 and h_{t-1}^1 constant till time t and the value of the hedging portfolio becomes $h_{t-1}^0 B_t + h_{t-1}^1 S_t$ before relocating the hedging positions at time t. Denote the difference between before and after relocating the hedging portfolio by δ_t, which is called the additional capital at time t and is defined as follows,

$$\delta_t(S_t) = F_t(S_t) - (h_{t-1}^0 B_t + h_{t-1}^1 S_t), \tag{22.5}$$

for $t = 1, \cdots, T$. Note that if $\delta_t(S_t) = 0$ for all $t = 1, \cdots, T$, then the trading strategy is called self-financing. Here we release this restriction and consider to construct a trading strategy which is capable to reduce the risk caused by the additional capital in some risk-minimizing sense. In order to achieve this objective, let the holding positions at the expiration date be $h_T^0 = V_T/B_T$ and $h_T^1 = 0$, and thus $F_T = V_T$, which means that the hedging portfolio is set to replicate the payoff of the claim after relocating the hedging positions at time T. The holding positions at rebalancing time $t = 1, \cdots, T-1$, are then determined by a backward scheme with a specific risk-minimizing criterion.

In this section, we first introduce the local quadratic risk-minimizing criterion. Based on this criterion, the holding units are determined by

$$\min_{h_{t-1}^0, h_{t-1}^1} E_{t-1}\left(\{\delta_t(S_t)/B_t\}^2\right), \tag{22.6}$$

and the closed-form expression of h_{t-1}^0 and h_{t-1}^1 for $t = 1, \cdots, T$ can be obtained by solving

$$\frac{\partial}{\partial h_{t-1}^0} E_{t-1}\left(\{\delta_t(S_t)/B_t\}^2\right) = 0 \text{ and } \frac{\partial}{\partial h_{t-1}^1} E_{t-1}\left(\{\delta_t(S_t)/B_t\}^2\right) = 0.$$

The dynamic programming of the local quadratic risk-minimizing hedging, abbreviated by LQR-hedging, is summarized as follows.

[Dynamic programming of LQR-hedging]

$$\begin{cases} (\hat{h}_T^0, \hat{h}_T^1) = (\tilde{V}_T, 0) \\ \hat{h}_{t-1}^1 = \dfrac{\mathrm{Cov}_{t-1}(\tilde{F}_t, \tilde{S}_t)}{\mathrm{Var}_{t-1}(\tilde{S}_t)} \\ \hat{h}_{t-1}^0 = E_{t-1}(\tilde{F}_t) - \hat{h}_{t-1}^1 E(\tilde{S}_t) = E_{t-1}\{\hat{h}_t^0 + (\hat{h}_t^1 - \hat{h}_{t-1}^1)\tilde{S}_t\} \end{cases}, \tag{22.7}$$

where $\tilde{V}_T = V_T/B_T$, $\tilde{F}_t = F_t/B_t$ and $\tilde{S}_t = S_t/B_t$, for $t = 1, \cdots, T$.

In the following, we give an example to illustrate the local expected squared risk minimizing hedging strategy in a trinomial model.

Example 4. Consider the same one-period trinomial model as in Example 2. By the dynamic programming (22.7), the holding units of riskless bonds and the risky security are given by $(\hat{h}_0^0, \hat{h}_0^1) = (-40.26, 0.4553)$. The initial hedging capital is

$$F_0 = \hat{h}_0^0 + \hat{h}_0^1 S_0 = 5.5,$$

which lies in the no-arbitrage region $(5, \frac{20}{3})$. That is, practitioners can use an initial hedging capital of 5.5, which is less than the initial capital required by the superhedging strategy, to construct a hedging portfolio which minimizes the quadratic risk.

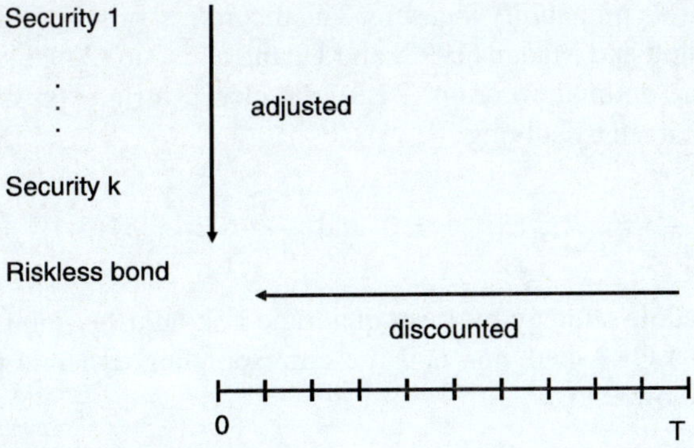

Fig. 22.4 Risk-adjusted and discounted values

22.2.4 *Local Quadratic Risk-Adjusted-Minimizing Hedging and Its Dynamic Programming*

In this section, we introduce another type of quadratic risk-minimizing hedging under consideration of risk-adjusted. Define the one-step-ahead risk-adjusted hedging cost as

$$\delta_t^*(S_t) \equiv \delta(S_t e^{-\lambda_t}),$$

where $\lambda_t \equiv \log[E_{t-1}\{(S_t/B_t)/(S_{t-1}/B_{t-1})\}]$ is the risk premium. Figure 22.4 illustrates the concepts of risk-adjusted with k risky securities and discounted values. Instead of the risk-minimizing criterion (22.6) mentioned in Sect. 22.2.3, we consider to construct a trading strategy which minimizes the one-step-ahead quadratic discounted risk-adjusted hedging costs, that is,

$$\min_{h_{t-1}^0, h_{t-1}^1} E_{t-1}\{\tilde{\delta}_t^*(S_t)\}^2, \tag{22.8}$$

where $\tilde{\delta}_t^*(S_t) = \delta_t^*(S_t)/B_t = \tilde{F}_t^* - (h_{t-1}^0 + h_{t-1}^1 \tilde{S}_t^*)$, and $\tilde{F}_t^* = F_t(S_t e^{-\lambda_t})/B_t$ and $\tilde{S}_t^* = S_t e^{-\lambda_t}/B_t$ denote the discounted adjusted values of F_t and S_t, respectively.

Statistically speaking, the criterion (22.8) is equivalent to find a best linear approximation of $\tilde{F}_t^*$ with the shortest L^2-distance under the physical measure P. Hence, this best linear approximation will pass through the point $(E_{t-1}(\tilde{S}_t^*), E_{t-1}(\tilde{F}_t^*))$. By the definition of λ_t, we have $E_{t-1}(\tilde{S}_t^*) = \tilde{S}_{t-1}$. Therefore, the amount $E_{t-1}(\tilde{F}_t^*)$ is treated as the discounted hedging capital for a given discounted stock price $\tilde{S}_{t-1}$ at time $t-1$ and thus

$$E_{t-1}(\tilde{F}_t^*) = \tilde{F}_{t-1},$$

under the dynamic probability measure. The theoretical proof of this equality could be found in Elliott and Madan (1998) and Huang and Guo (2009c).

Based on the optimal criterion (22.8), the closed-form expression of h_{t-1}^0 and h_{t-1}^1 can be obtained by solving

$$\frac{\partial}{\partial h_{t-1}^0} E_{t-1}\{\tilde{\delta}_t^*(S_t)\}^2 = 0 \text{ and } \frac{\partial}{\partial h_{t-1}^1} E_{t-1}\{\tilde{\delta}_t^*(S_t)\}^2 = 0.$$

We call this trading strategy by local quadratic risk-adjusted-minimizing hedging, abbreviated by LQRA-hedging, and the corresponding dynamic programming is described as follows.

[Dynamic programming of LQRA-hedging]

$$\begin{cases} (\hat{h}_T^0, \hat{h}_T^1) = (\tilde{V}_T, 0) \\ \hat{h}_{t-1}^1 = \dfrac{\mathrm{Cov}_{t-1}(\tilde{F}_t^*, \tilde{S}_t^*)}{\mathrm{Var}_{t-1}(\tilde{S}_t^*)} \\ \hat{h}_{t-1}^0 = E_{t-1}(\tilde{F}_t^*) - \hat{h}_{t-1}^1 E(\tilde{S}_t^*) = E_{t-1}\{\hat{h}_t^0 + (\hat{h}_t^1 - \hat{h}_{t-1}^1)\tilde{S}_t^*\} \end{cases} \quad (22.9)$$

In the following, we give an example to illustrate the LQRA-hedging in a trinomial model.

Example 5. Consider the same one-period trinomial model as in Example 2. First, we compute the risk premium

$$\lambda = \log\left\{E\left(e^{-r}\frac{S_t}{S_{t-1}}\right)\right\} = e^{-r}\log(up_1 + dp_2 + jp_3) = \log(1.0005).$$

The discounted risk-adjusted stock prices at time 1 are

$$\tilde{S}_1^*(\omega_1) = S_1(\omega_1)e^{-r-\lambda} = \frac{22000}{201} > K = 100,$$
$$\tilde{S}_1^*(\omega_2) = S_1(\omega_2)e^{-r-\lambda} = \frac{1800}{201} < K,$$

and

$$\tilde{S}_1^*(\omega_3) = S_1(\omega_3)e^{-r-\lambda} = \frac{16000}{201} < K.$$

Hence, the corresponding discounted risk-adjusted option values are $\tilde{V}_1^*(\omega_1) = \frac{1900}{201}$, and $\tilde{V}_1^*(\omega_2) = \tilde{V}_1^*(\omega_3) = 0$. By the dynamic programming (22.9), the holding units of riskless bonds and the security are given by $(\hat{h}_0^0, \hat{h}_0^1) = (-38.06, 0.4326)$. Thus the initial hedging capital is

$$F_0 = \hat{h}_0^0 + \hat{h}_0^1 S_0 = 5.199,$$

which also lies in the interval of no-arbitrage prices, $(5, \frac{20}{3})$. In other words, the criterion of quadratic risk-adjusted-minimizing not only provides a hedging strategy, but also a no-arbitrage price of the European call option in this incomplete market.

If the market model is discrete time and continuous state type, such as the GARCH model (Bollerslev 1986), Elliott and Madan (1998) showed that

$$E_{t-1}(\tilde{F}_t^*) = E_{t-1}^Q(\tilde{F}_t) = \tilde{F}_{t-1},$$

where the measure Q is the martingale measure derived by the extended Girsanov change of measure. In particular, Huang and Guo (2009c) showed that if the innovation is assumed to be Gaussian distributed, then the GARCH martingale measure derived by the extended Girsanov principle is identical to that obtained by Duan (1995). Moreover, the formula of the optimal $(\hat{h}_{t-1}^0, \hat{h}_{t-1}^1)$ obtained in (22.9) can be expressed as

$$\begin{cases} \hat{h}_{t-1}^0 = \dfrac{\tilde{F}_{t-1} E_{t-1}^Q(\tilde{S}_t^2) - \tilde{S}_{t-1} E_{t-1}^Q(\tilde{F}_t \tilde{S}_t)}{\text{Var}_{t-1}^Q(\tilde{S}_t)} \\ \hat{h}_{t-1}^1 = \dfrac{\text{Cov}_{t-1}^Q(\tilde{F}_t, \tilde{S}_t)}{\text{Var}_{t-1}^Q(\tilde{S}_t)} \end{cases} \tag{22.10}$$

under the risk-neutral measure Q, for $t = 1, \cdots, T$, where Cov_{t-1}^Q and Var_{t-1}^Q are the conditional covariance and variance given $\mathcal{F}_{t-1}$ computed under the risk-neutral measure Q, respectively.

Both (22.9) and (22.10) provide recursive formulae for building the LQRA-hedging backward from the expiration date. In practical implementation, practitioners may want to keep the holding units of the hedging portfolio constant for ℓ units of time due to the impact of the transaction costs. If we denote the discounted hedging capital with hedging period ℓ at time t by

$$\tilde{F}_{t,\ell} = \hat{h}_{t,\ell}^0 + \hat{h}_{t,\ell}^1 \tilde{S}_t, \tag{22.11}$$

where $\hat{h}_{t,\ell}^0$ and $\hat{h}_{t,\ell}^1$ are the holding units of riskless bonds and the underlying asset, respectively, and are determined instead by the following optimal criterion

$$\min_{h_{t,\ell}^0, h_{t,\ell}^1} E_t^Q\{\tilde{\delta}_{t+\ell}(S_{t+\ell})\}^2. \tag{22.12}$$

Note that the optimal holding units $(\hat{h}_{t,\ell}^0, \hat{h}_{t,\ell}^1)$ are functions of the hedging period ℓ. By similar argument as (22.10), $\hat{h}_{t,\ell}^0$ and $\hat{h}_{t,\ell}^1$ can be represented as

$$\begin{cases} \hat{h}_{t,\ell}^1 = \text{Cov}_t^Q(\tilde{F}_{t+\ell}, \tilde{S}_{t+\ell}) / \text{Var}_t^Q(\tilde{S}_{t+\ell}) \\ \hat{h}_{t,\ell}^0 = E_t^Q(\tilde{F}_{t+\ell}) - \hat{\eta}_{t,\ell}^1 \tilde{S}_{t+\ell} \end{cases}. \tag{22.13}$$

Equation (22.13) is really handy in building LQRA-hedging. For example, suppose that a practitioner writes a European call option with strike price K and expiration date T. She wants to set up a ℓ-period hedging portfolio consisting of the underlying stock and the riskless bonds at time t with the hedging capital F_t to hedge her short position, and the hedging portfolio remains constant till time $t+\ell$, $0<\ell\le T-t$. Huang and Guo (2009c) proved that the hedging capital of the ℓ-period LQRA-hedging is independent of the hedging period ℓ and is equal to the no-arbitrage price derived by the extended Girsanov principle. A dynamic programming of the ℓ-period LQRA-hedging for practical implementation is also proposed by Huang and Guo (2009c). The algorithm is summarized as follows.

[Dynamic programming of ℓ-period LQRA-hedging]

1. For a given stock price S_t at time t, generate n stock prices $\{S_{t+\ell,j}\}_{j=1}^n$, at time $t+\ell$ conditional on S_t from the risk-neutral model.
2. For each $S_{t+\ell,j}$, derive the corresponding European call option prices, $F_{t+\ell}(S_{t+\ell,j})$, by either the dynamic semiparametric approach (DSA) (Huang and Guo 2009a,b) or empirical martingale simulation (EMS) method (Duan and Simonato 1998) for $t+\ell<T$. If $t+\ell=T$, then $F_T(S_{T,j})=(S_{T,j}-K)^+$.
3. Regress $\tilde{F}_{t+\ell}(S_{t+\ell,j})$ on $\tilde{S}_{t+\ell,j}$, $j=1,\cdots,n$. Then $(\hat{h}^0_{t,\ell},\hat{h}^1_{t,\ell})$ are the corresponding regression coefficients.

Since the above trading strategy employs the simple linear regression to determine the hedging positions, it is easy to be implemented and computed in a personal computer. The main computational burden might comes from Step-2 of the above algorithm, where we have to compute the European call option values by the DSA or EMS method. Herein, we give a brief introduction of this method. The DSA is proposed by Huang and Guo (2009a) for solving the multi-step conditional expectation problems where the multi-step conditional density doesn't have closed-form representation. It is an iterative procedure which uses nonparametric regression to approximate derivative values and parametric asset models to derive the one-step conditional expectations. The convergence order of the DSA is derived under continuity assumption on the transition densities of the underlying asset models. For illustration, suppose we want to compute the multi-step conditional expectation $E_0(F_T)$. We transform the problem into $E_0[E_1[\cdots[E_{T-1}(F_T)]\cdots]]$. and then compute the one-step backward conditional expectation. Denote $F_t=E_t(F_{t+1})$, $t=0,\cdots,T-1$. In general, F_t is a nonlinear function of the underlying asset for $t=1,\cdots,T-1$, and the conditional expectation $E_{t-1}(F_t)$ does not have closed-form representation, which makes the multi-step conditional expectation complexity. Huang and Guo (2009a) adopted piecewise regression function to approximate the derivative value function F_t at each discrete time point t, denoted by $\hat{F}_t$, and then compute the conditional expectation of $\hat{F}_t$, that is, $\tilde{F}_t=E_t(\hat{F}_{t+1})$ and treated $\tilde{F}_t$ as an approximation of F_t. The procedure keeps iterating till the initial time to obtain the derivative price. A flow chart of the DSA is given in Fig. 22.5.

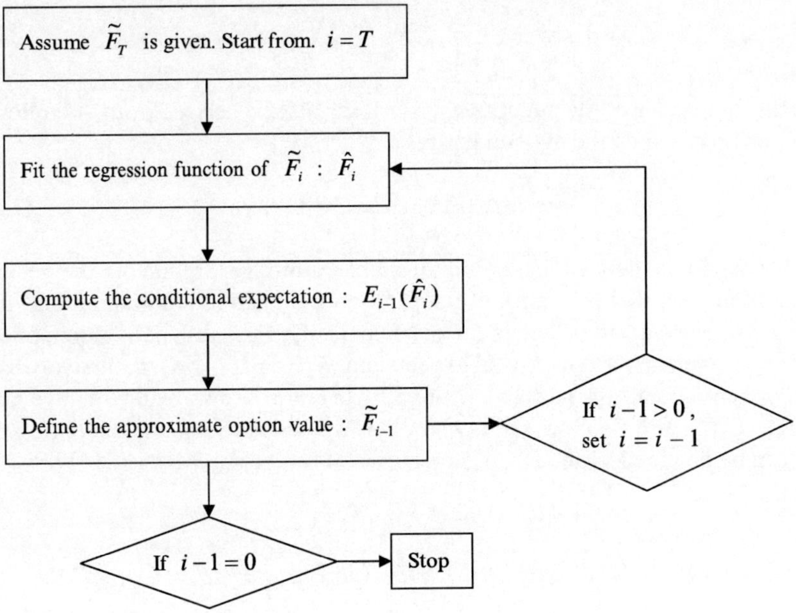

Fig. 22.5 Flow chart of the DSA

22.3 The Comparison of the Discrete Time Hedging Strategies

In this section, we are interested in comparing commonly used delta-hedging strategy with the discrete time hedging strategies introduced in Sect. 22.2. The delta of a derivative is referred to as the rate of change in the price of a derivative security relative to the price of the underlying asset. Mathematically, the delta value Δ_t at time t is defined as the partial derivative of the price of the derivative with respect to the price of the underlying, that is, $\Delta_t = \partial V_t / \partial S_t$, where V_t and S_t are the prices of the derivative and the underlying asset at time t, respectively.

For example, considering a European call option with expiration date T and strike price K, the no-arbitrage option value at time t is $C_t = e^{-r(T-t)} E_t^Q\{(S_T - K)^+\}$, where r is the riskless interest rate. After simplifying the partial derivative $\partial C_t / \partial S_t$ and exploiting the following property,

$$\lim_{h\to 0} \frac{1}{h} \int_{\log(K/(S_t+h))}^{\log(K/S_t)} \{(S_t + h)e^y - K\} dG_t(y) = 0,$$

where $G_t(y)$ is the conditional distribution of $\log(S_T/S_t)$ given $\mathcal{F}_t$ under the martingale measure Q, one can show that the delta of the European call option can be expressed as

$$\Delta_t(c) = \frac{\partial C_t}{\partial S_t} = e^{-r(T-t)} E_t^Q\left(\frac{S_T}{S_t} I_{(S_T \geq K)}\right). \tag{22.14}$$

And the delta value of the put option, $\Delta_t(p)$, can also be derived from the following relationship based on the put-call parity:

$$\Delta_t(c) - \Delta_t(p) = 1. \tag{22.15}$$

Since (22.15) is derived based on a simple arbitrage argument, the result is distribution-free, that is it does not depend on the distribution assumption of the underlying security. To calculate the delta value of a European call option, one can either approximate the conditional expectation, $E_t^Q(S_T I_{\{S_T \geq K\}})$, recursively by the DSA or approximate the partial derivative, $\Delta_t(c) = \partial C_t / \partial S_t$, by the relative rate of change $\{C_t(S_t + h) - C_t(S_t)\}/h$, where h is a small constant and the option price C_t's can be obtained by the DSA.

22.3.1 *LQRA-Hedging and Delta-Hedging Under Complete Markets*

In a complete market every contingent claim is marketable, and the risk neutral probability measure is unique. There exists a self-financing trading strategy and the holding units of the stocks and bonds in the replicating portfolio are uniquely determined. This trading strategy is called the perfect hedging which attains the lower bound of the criteria (22.6) and (22.8). Thus we expect both the LQR- and LQRA-hedging strategies will coincide with the delta-hedging under the complete market models. In the following, we show directly the holding units of the stocks in an LQRA-hedging is the same as in the delta-hedging for the two complete market models – the binomial tree and the Black-Scholes models (Black and Scholes 1973). For simplicity, let the bond price $B_t = e^{rt}$ where r represents a constant riskless interest rate. First, consider a binomial tree model. Assumes at each step that the underlying instrument will move up or down by a specific factor (u or d) per step of the tree, where (u, d) satisfies $0 < d < e^r < u$. For example, if $S_{t-1} = s$, then S_t will go up to $s_u = us$ or down to $s_d = ds$ at time t, with the risk neutral probability

$$q = P(S_t = s_u | S_{t-1} = s) = \frac{e^r - d}{u - d} = 1 - P(S_t = s_d | S_{t-1} = s).$$

By straightforward computation, we have

$$\mathrm{Cov}_{t-1}^Q(F_t, S_t) = q(1-q)(s_u - s_d)\{F_t(s_u) - F_t(s_d)\}$$

and

$$\mathrm{Var}_{t-1}^Q(S_t) = q(1-q)(s_u - s_d)^2.$$

Thus by (22.10) the holding units of the stock in the η-hedging is

$$\hat{h}_{t-1}^1 = \frac{\mathrm{Cov}_{t-1}^Q(\tilde{F}_t, \tilde{S}_t)}{\mathrm{Var}_{t-1}^Q(\tilde{S}_t)} = \frac{F_t(s_u) - F_t(s_d)}{s_u - s_d},$$

which is consistent with the delta-hedging of the binomial tree model.

Next, consider the Black-Scholes model,

$$dS_t = rS_t dt + \sigma S_t dW_t, \tag{22.16}$$

where r and σ are constants and W_t is the Wiener process. For a European call option with strike price K and expiration date T, the holding units of the stock in the delta-hedging of the Black-Scholes model is $\Delta_t = \Phi(d_1(S_t))$ at time t, where

$$d_1(S_t) = \frac{\log(S_t/K) + (r + 0.5\sigma^2)(T-t)}{\sigma\sqrt{T-t}}$$

and $\Phi(\cdot)$ is the cumulative distribution function of the standard normal random variable. We claim in the following that

$$h_t^1 \to \Phi(d_1)$$

as $dt \to 0$, where dt denotes the length of the time period $[t, t+dt]$. Denote the discounted stock price and option value at time t by $\tilde{S}_t$ and $\tilde{F}_t$, respectively. Note that

$$\begin{aligned}
\mathrm{Cov}_t^Q(\tilde{F}_{t+dt}, \tilde{S}_{t+dt}) &= E_t^Q(\tilde{F}_{t+dt}\tilde{S}_{t+dt}) - \tilde{F}_t\tilde{S}_t \\
&= E_t^Q\Big(e^{-r(t+dt)}\Big\{S_{t+dt}\Phi(d_1(S_{t+dt})) \\
&\quad -Ke^{-r(T-t-dt)}\Phi(d_2(S_{t+dt}))\Big\}\tilde{S}_{t+dt}\Big) - \tilde{F}_t\tilde{S}_t \\
&\approx E_t^Q(\tilde{S}_{t+dt}^2)\Phi(d_1(S_t)) - \tilde{S}_t K e^{-rT}\Phi(d_2(S_t)) \\
&\quad -e^{-rt}\Big\{S_t\Phi(d_1(S_t)) - Ke^{-r(T-t)}\Phi(d_2(S_t))\Big\}\tilde{S}_t \\
&= \Phi(d_1(S_t))\mathrm{Var}_t^Q(\tilde{S}_{t+dt}),
\end{aligned}$$

where $d_2(S_t) = d_1(S_t) - \sigma\sqrt{T-t}$ and the approximation ($\approx$) is due to

$$E_t^Q\{\tilde{S}_{t+dt}^k\Phi(d_i(S_{t+dt}))\} \approx E_t^Q(\tilde{S}_{t+dt}^k)\Phi(d_i(S_t))$$

for small dt, $i = 1, 2$ and $k = 1, 2$. Therefore, by (22.10) we have $h_t^1 \to \Phi(d_1)$ as $dt \to 0$. This result indicates that if practitioners are allowed to adjust the hedging portfolio continuously, then LQRA-hedging coincides with delta-hedging.

However, we should be aware that practitioners are not allowed to rebalance the hedging portfolio continuously and may want to reduce the number of the rebalancing time as less as possible due to the impact of transaction costs in practice.

22.3.2 LQRA-Hedging and Delta-Hedging Under Incomplete Markets

In this section, we consider that the log-return of the underlying assets follows a GARCH model such as

$$\begin{cases} R_t = r - \frac{1}{2}\sigma_t^2 + \lambda\sigma_t + \sigma_t\varepsilon_t, & \varepsilon_t \sim D(0,1) \\ \sigma_t^2 = \alpha_0 + \alpha_1\sigma_{t-1}^2\varepsilon_{t-1}^2 + \alpha_2\sigma_{t-1}^2 \end{cases}, \tag{22.17}$$

where the parameters are set the same as in Duan (1995)

$$\begin{aligned} &\lambda = 0.007452,\ \alpha_0 = 0.00001524,\ \alpha_1 = 0.1883,\ \alpha_2 = 0.7162,\\ &\sigma_d = \sqrt{\tfrac{\alpha_0}{1-\alpha_1-\beta_1}} = 0.01263 \text{ (per day, i.e. 0.2413 per annum)},\\ &K = 40,\ r = 0, \end{aligned}$$

and the innovation ε_t is assumed to be normal or double exponential distributed with zero mean and unit variance. Suppose that a practitioner writes a European call option with strike price K and expiration date T, and set up a delta-hedging portfolio at the initial time, with the hedging capital F_0, that is,

$$F_0 = h_0 + \Delta_0 S_0,$$

where F_0 denotes the risk-neutral price derived by the extended Girsanov principle and thus the cash position h_0 can be obtained by $F_0 - \Delta_0 S_0$. Similarly, we can construct the LQRA-hedging portfolio by

$$F_0 = h_0^{\eta} + \eta_0 S_0.$$

We simulate $n = 10{,}000$ random paths to generate the stock price, $\{S_{T,i}\}_{i=1}^{n}$, under the physical model (22.17), and then compute the ratio of the average variations of the delta hedging and LQRA-hedging portfolios

$$G_T = \frac{\sum_{i=1}^{n}\{h_0 e^{rT} + \Delta_0 S_{T,i} - (S_{T,i} - K)^+\}^2}{\sum_{i=1}^{n}\{h_0^{\eta} e^{rT} + \eta_0 S_{T,i} - (S_{T,i} - K)^+\}^2},$$

for $T = 5, 10, 30$ (days). Table 22.1 shows the simulation results of G_T, $T = 5, 10, 30$, of the GARCH-normal and GARCH-dexp models with $K = 35, 40, 45$ and several different parameter settings.

Table 22.1 The relative values of the average squared hedging costs of delta-hedging and LQRA-hedging in the GARCH(1,1) log-return model

	GARCH-normal				GARCH-dexp			
	kur.	$K = 35$	$K = 40$	$K = 45$	kur.	$K = 35$	$K = 40$	$K = 45$
Case 1: $\alpha_0 = 0.00001524, \alpha_1 = 0.1883, \alpha_2 = 0.7162, \lambda = 0.007452$								
G_5	4.10	1.01	1.00	1.01	6.67	1.02	1.01	1.03
G_{10}	4.33	1.01	1.00	1.02	7.39	1.03	1.01	1.07
G_{30}	4.29	1.01	1.01	1.05	9.24	1.04	1.03	1.16
Case 2: $\alpha_0 = 0.00002, \alpha_1 = 0.1, \alpha_2 = 0.8, \lambda = 0.01$								
G_5	3.50	1.00	1.01	1.01	4.91	1.02	1.01	1.05
G_{10}	3.53	1.01	1.01	1.03	4.76	1.02	1.01	1.07
G_{30}	3.42	1.01	1.03	1.04	4.28	1.00	1.03	1.08
Case 3: $\alpha_0 = 0.00002, \alpha_1 = 0.2, \alpha_2 = 0.7, \lambda = 0.01$								
G_5	4.18	1.01	1.01	1.03	6.95	1.04	1.01	1.09
G_{10}	4.42	1.01	1.01	1.06	8.34	1.05	1.02	1.16
G_{30}	4.39	1.00	1.03	1.08	9.21	1.01	1.06	1.21
Case 4: $\alpha_0 = 0.00002, \alpha_1 = 0.3, \alpha_2 = 0.6, \lambda = 0.01$								
G_5	5.09	1.02	1.01	1.06	10.52	1.06	1.02	1.15
G_{10}	6.06	1.04	1.02	1.11	20.50	1.08	1.04	1.27
G_{30}	8.87	1.01	1.06	1.20	53.67	1.04	1.25	1.79

Note that the values of G_T's in Table 22.1 are all greater than 1, which means the average variation of the LQRA-hedging is smaller than the delta-hedging. Under the same parameter setting in both GARCH-normal and GARCH-dexp models, the kurtosis of the GARCH-dexp models is greater than the GARCH-normal model. The results shows that G_T tends to increase in the kurtosis of the log-returns, especially when the option is out-of-the-money.

In Fig. 22.6, we plot the hedging strategies of delta- and LQRA-hedging for one period case, where $\tilde{F}_t$ is the discounted option value function at time t and the point $(\tilde{S}_t, \tilde{F}_t)$ denotes the time-t discounted stock price and discounted hedging capital. In the left-hand panel, the dash-line, $\Delta(t-1, t)$, denotes the delta-hedging values, which is the tangent of the curve $\tilde{F}_t$ at the point $(\tilde{S}_t, \tilde{F}_t)$. In the right-hand panel, the dot-line, $\eta(t-1, t)$, represents the LQRA-hedging, which is regression line of $\tilde{F}_{t+1}$ under the risk-neutral probability measure derived by the extended Girsanov principle (see Elliott and Madan 1998; Huang and Guo 2009d).

If the hedging period increases to ℓ, $\ell > 1$, then the delta-hedging $\Delta(t, t+\ell)$ remains the same, that is, $\Delta(t, t+\ell) = \Delta(t, t+1)$, (see the left-hand panel of Fig. 22.7). However, the LQRA-hedging, $\eta(t, t+\ell)$ (see the red line in the right-hand panel of Fig. 22.7), would be different from $\eta(t, t+1)$ since the hedging target is changed from $\tilde{F}_t$ to $\tilde{F}_{t+\ell}$. This phenomenon states that the LQRA-hedging is capable of making adjustment to the hedging period ℓ, which makes it more suitable for various time period hedging. This phenomenon also provides an explanation

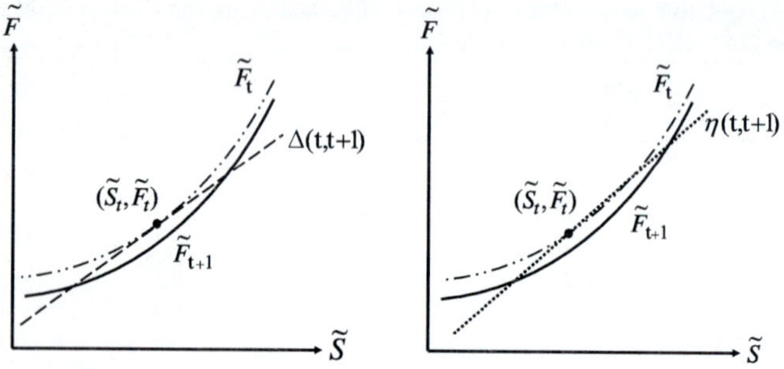

Fig. 22.6 One period delta- and LQRA-hedging

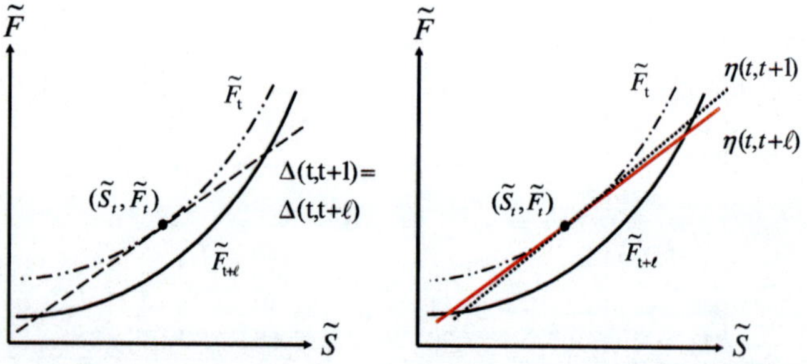

Fig. 22.7 ℓ-period delta- and LQRA-hedging

of why the G_T's in most cases of Table 22.1 tends to increase in the hedging period or in the kurtosis of the log-returns. Because when the hedging period or the kurtosis of the log-returns increases, the LQRA-hedging would reduce more variability between the hedging portfolio and the hedging target than delta-hedging.

22.4 Hedging Strategies of Barrier Options

In previous sections, we introduce several discrete time hedging strategies and illustrate the corresponding dynamic programming for vanilla options. In this section, we consider the hedging strategies of barrier options. A barrier option is one type of exotic option and the corresponding payoff depends on the underlying reaching or crossing a given barrier. Since barrier options are cheaper than a similar option without barrier, thus traders might like to buy a barrier option instead of the vanilla option in some situations. For example, if a trader wants to buy the IBM stock with a certain price in the future and believes that the stock price will go up next few months, but won't go above 100, then she can buy an up-and-out call option with a certain barrier and pay less premium than the vanilla call option. In Korea, one type of barrier option, called Knock-In-Knock-Out (KIKO) option, was very popular among smaller firms in early 2008, when the Korean won was stronger against the US dollar. The KIKO contracts allow subscribers to hedge against currency rate fluctuations with a preferential exchange rate as long as the currency stays within the knock-in and knock-out barriers. Many local small and midsized exporters signed KIKO contracts with local banks to hedge against moderate currency swings. Unfortunately, the currency exchange rate dropped unexpectedly in the summer of 2008, and local firms that bought KIKO contracts were forced to take massive losses. Some of the firms later sued the banks that sold them the derivatives. From this event, we realize that the potential risk of the financial derivative might cause huge disaster to both the issuer and the subscriber of the contract. Therefore, it is of practical importance to construct the hedging strategy of the exotic options. In the following, we focus on the hedging of barrier option and introduce a trading strategy under consideration of hedging performance and transaction costs.

Barrier option is a path-dependent contingent claim, that is, the payoff of a barrier option depends on the underlying asset values during the time interval $[0, T]$, where 0 and T stand for the initial and expiration dates, respectively. In general, the evaluation of path-dependent contingent claims is more complicated than path-independent ones since the randomness comes from not only the underlying asset value at maturity but also those before the expiration date. For example, the payoff of a down-and-in call option is defined as

$$DIC_T = (S_T - K)^+ I_{(\min_{t \in [0,T]} S_t \leq B)},$$

where S_T is the underlying asset value at time T, K is the strike price, B denotes the barrier and $I_{(\cdot)}$ is an indicator function. By the definition of DIC_T, one can see that the payoff not only depends on S_T but also depends on the minimum value of the underlying asset, $\min_{t \in [0,T]} S_t$, in the time interval $[0, T]$. Once the value of the underlying asset reaches or bellows the barrier B prior to maturity, the option is active immediately and the payoff is then identical to the European call option with the same strike price and expiration date.

Suppose that a practitioner shorts a down-and-in call option and wants to set up a hedging portfolio to hedge her short position. Under consideration of transaction costs, the trading strategies introduced in the previous sections may not be optimal since the increasing frequency of portfolio rebalancing costs more transaction fees. Due to the trade-off between risk reduction and transaction costs, Huang and Huang (2009) proposed a hedging strategy which rebalances the hedging positions only once during the duration of the barrier option. In the following, we illustrate the hedging strategy for down-and-in call options when the underlying asset follows a geometric Brownian motion process.

Assume that the underlying asset follows Model (22.16). We have the following results:

1. If $B \geq K$, then the no-arbitrage price of the down-and-in call option is

$$
\begin{aligned}
DIC_0(B,K) &= P_0(K) - K \cdot DIB_0(B) \\
&\quad + B\left\{\left(\frac{B}{S_0}\right)^{\frac{2r}{\sigma^2}} \Phi\left(\frac{\log\frac{B}{S_0}+\left(r+\frac{\sigma^2}{2}\right)T}{\sigma\sqrt{T}}\right) + \frac{S_0}{B}\Phi\left(\frac{\log\frac{B}{S_0}-\left(r+\frac{\sigma^2}{2}\right)T}{\sigma\sqrt{T}}\right)\right\};
\end{aligned}
\tag{22.18}
$$

2. If $B < K$, then the no-arbitrage price of the down-and-in call option is

$$
\begin{aligned}
DIC_0(B,K) &= \left(\frac{B}{S_0}\right)^{\frac{2r}{\sigma^2}} B\Phi\left(\frac{\log\frac{B^2}{KS_0}+(r+\frac{\sigma^2}{2})T}{\sigma\sqrt{T}}\right) \\
&\quad - \left(\frac{B}{S_0}\right)^{\frac{2r}{\sigma^2}} \frac{KS_0}{B}e^{-rT}\Phi\left(\frac{\log\frac{B^2}{KS_0}+(r-\frac{\sigma^2}{2})T}{\sigma\sqrt{T}}\right),
\end{aligned}
\tag{22.19}
$$

where $P_0(K)$ is the European put option price with the same strike price and maturity as the down-and-in call option, $\Phi(\cdot)$ is the distribution function of a standard normal random variable, and $DIB_0(B)$ denotes the no-arbitrage price of a down-and-in bond, which is defined as

$$
DIB_T(B) = I_{(\min_{0\le t\le T} S_t \le B)},
$$

and can be evaluated by

$$
DIB_0(B) = e^{-rT}\left\{\Phi\left(\frac{\log\frac{B}{S_0}-(r-\frac{\sigma^2}{2})T}{\sigma\sqrt{T}}\right) + \left(\frac{B}{S_0}\right)^{(\frac{2r}{\sigma^2}-1)}\Phi\left(\frac{\log\frac{B}{S_0}+(r-\frac{\sigma^2}{2})T}{\sigma\sqrt{T}}\right)\right\}.
\tag{22.20}
$$

In particular, if the riskless interest rate equals zero, $r = 0$, then the above results can be simplified as:

(i′) If $B \ge K$, $DIC_0(B,K) = P_0(K) + (B-K)DIB_0(B)$;
(ii′) If $B < K$, $DIC_0(B,K) = \frac{K}{B}P_0(\frac{B^2}{K})$.

By (i′) and (ii′), we can construct a perfect hedging strategy of a down-and-in call option. If $B \ge K$, since the right-hand side of (i′) is a linear combination of a European put option and a down-and-in bond, thus the practitioner who shorts a down-and-in call option can hedge her short position via the following two steps:

1. At time 0, long a European put option with strike price K and expiration date T and also long $(B-K)$ shares of a down-and-in bond with barrier B and expiration date T. Notice that the cost of this portfolio is exactly the same as the no-arbitrage price of the down-and-in call option.
2. Let τ denote the first hitting time when the underlying asset value reaches the barrier price, that is,

$$
\tau = \inf\{t, S_t = B, 0 \le t \le T\},
\tag{22.21}
$$

and let $\tau = \infty$ if $\min_{t\in[0,T]} S_t > B$. If $\tau = \infty$, then the down-and-in call option and the hedging portfolio set up in Step-1 are both worthless. If $\tau \le T$, by the fact that $DIC_\tau(B,K) = C_\tau(K)$ and put-call parity, $C_\tau(K) = P_\tau(K) + B - K$, one can then hedge the down-and-in call option perfectly via shorting the hedging portfolio set up in Step-1 and longing a European call option with strike price K and expiration date T at time τ.

Similarly, if $B < K$, from (ii′), the trader can hedge her short position by longing K/B shares of a European put option with strike price B^2/K and expiration date T at time 0. And then by put-call symmetry (see Carr and Lee 2009), $C_\tau(K)I_{(\tau\le T)} = (K/B)P\big(B^2/K\big)I_{(\tau\le T)}$, the trader can short the European put option and long a European call option with strike price K and expiration date T without putting in or withdrawing any capital to hedge the down-and-in call option perfectly at time τ $(< T)$. Moreover, if $\tau = \infty$, then the down-and-in call option and the hedging portfolio set up at time 0 are both worthless. Therefore, in the case of $r = 0$, a down-and-in call option can be perfectly hedged by plain vanilla options and down-and-in bond. This result is also obtained by Bowie and Carr (1994).

If the riskless interest is greater than zero, then the hedging strategy of barrier options mentioned above would not be the perfect-hedging. Details are as follows. If $B \le K$ and at time τ $(< T)$, one can set up a perfect-hedging by longing a European call option with the same strike price and expiration date as the down-and-in call option, that is, $DIC_\tau(B,K)I_{(\tau\le T)} = C_\tau(K)I_{(\tau\le T)}$. Then by put-call parity, $C_\tau(K)I_{(\tau\le T)} = (P_\tau(K) + B - Ke^{-r(T-\tau)})I_{(\tau\le T)}$, one can have the desired European call option by shorting a European put option and $B - Ke^{-r(T-\tau)}$ shares of down-and-in bond at time τ. Therefore, at time 0, the hedging portfolio comprises a European put option with strike price K and expiration date T and $Be^{-r\tau} - Ke^{-rT}$ shares of down-and-in bond. Since τ is a random variable, thus the above hedging portfolio can not be set up at time 0 in practice. On the other hand, if $B > K$ and at time τ $(< T)$, by put-call symmetry, we have

$$DIC_\tau(B,\ K)I_{(\tau\le T)} = C_\tau(K)I_{(\tau\le T)} = \frac{K}{Be^{r(T-\tau)}}P\left(\frac{B^2e^{2r(T-\tau)}}{K}\right)I_{(\tau\le T)}.$$

In this case, one can long Ke^{-rT}/B shares of European put option with strike price $B^2e^{2r(T-\tau)}/K$ and expiration date T to construct a perfect-hedging portfolio at time 0. However, since the strike price of the European put option is now a random variable, this hedging strategy can't be obtained in practice as well. In order to overcome this problem, Huang and Huang (2009) proposed the following method.

If $B \le K$, by using the inequalities $e^{-rT}I_{(\tau\le T)} \le e^{-r\tau}I_{(\tau\le T)} \le I_{(\tau\le T)}$, we have the upper and lower bounds of the down-and-in call option value: $L \le DIC_0(B,K) \le U$, where $L = P_0(K) + (B-K)DIB_0(B)$ and $U = P_0(K) + (Be^{rT} - K)DIB_0(B)$ are both linear combinations of European put option and down-and-in bond. Next, we adopt a linear combination of L and U to construct a hedging portfolio, that is, $V_0 = (1-\alpha)L + \alpha U$, where V_0 denotes the initial hedging capital and $0 < \alpha < 1$. Further let V_0 be identical to the price of the down-and-in call option and we then have $\alpha = (DIC_0 - L)/(U - L)$ and

$$V_0 = DIC_0(B, K) = P_0(K) + \beta DIB_0(B), \tag{22.22}$$

where $\beta = \{DIC_0(B, K) - P_0(K)\}/DIB_0(B)$. The hedging strategy proposed by Huang and Huang (2009) is to set up the portfolio (22.22), which comprises a European put option and β shares of down-and-in bond, at time 0 and hold the portfolio till time $\min(\tau, T)$. If $\tau \le T$, then short the portfolio and long a European call option with strike price K and expiration date T. Notice that at time τ $(\le T)$ the value of the hedging portfolio (22.22) becomes $P_\tau(K) + \beta$, which may not be identical to the European call option price, $C_\tau(K)$. Therefore, the trader has to put in some additional capital for portfolio rebalancing and the total payoff of this hedging strategy is

$$\{V_T - DIC_T(B, K)\}I_{(\tau\le T)} = \{\beta - e^{r(T-\tau)}(B - Ke^{-r(T-\tau)})\}I_{(\tau\le T)}, \tag{22.23}$$

at expiration date, where the equality holds by the put-call parity, $C_\tau(K) = P_\tau(K) + B - Ke^{-r(T-\tau)}$ and the profit β comes from the down-and-in bond. On the other hand, if the underlying asset prices are never equal to or less than the barrier price, then the down-and-in call option and the hedging portfolio (22.22) are both worthless at expiration date.

Similarly, if $B > K$, Huang and Huang (2009) adopted the inequalities,

$$L^* \le DIC_0(B, K) \le U^*,$$

where $L^* = (K/B)P_0(B^2/K)$ and $U^* = (K/Be^{rT})P_0(B^2e^{2rT}/K)$ are both European put options. And then set up the hedging portfolio by a linear combination of L^* and U^*, $V_0 = (1-\alpha^*)L^* + \alpha^* U^*$. Further let $V_0 = DIC_0$ and hence $\alpha^* = (DIC_0 - L^*)/(U^* - L^*)$ and the initial hedging portfolio is

$$V_0 = \frac{(1-\alpha^*)K}{B} P_0\left(\frac{B^2}{K}\right) + \frac{\alpha^* K}{Be^{rT}} P_0\left(\frac{B^2 e^{2rT}}{K}\right), \tag{22.24}$$

which comprises $(1-\alpha^*)K/B$ shares of European put option with strike price B^2/K and expiration date T and $\alpha^* K/(Be^{rT})$ shares of European put option with strike price B^2e^{2rT}/K and expiration date T. Hold the hedging portfolio till time $\min(\tau, T)$, and if $\tau < T$, then short the portfolio and long a European call option with strike price K and expiration date T. As in the case of $B \le K$, the value of the hedging portfolio (22.24) may not be identical to the European call option price at time τ and the trader needs some additional capital for the portfolio rebalancing. The total payoff of this hedging strategy is

$$\begin{aligned}&\{V_T - DIC_T(B, K)\}I_{(\tau\le T)} \\ &= e^{r(T-\tau)}\left\{\frac{(1-\alpha^*)K}{B}P_\tau\left(\frac{B^2}{K}\right) + \frac{\alpha^* K}{Be^{rT}}P_\tau\left(\frac{B^2e^{2rT}}{K}\right) - C_\tau(K)\right\}I_{(\tau\le T)}, \end{aligned}\tag{22.25}$$

and

$$\{V_T - DIC_T(B,K)\}I_{(\tau>T)} = \frac{\alpha^* K}{Be^{rT}} P_T\left(\frac{B^2 e^{2rT}}{K}\right) I_{(\tau>T)},$$

at maturity.

The trading strategies of barrier option proposed in (22.22) and (22.24) are handy in practice since both are comprised of the derivatives traded in the financial market such as the vanilla options and down-and-in bond. Also, the corresponding strike prices and expiration dates of the components in the hedging portfolio are all deterministic functions of those of the original barrier option. Therefore, it is easy to be implemented in practice and the traders who adopt this trading strategy only need to determine which derivative should be bought or sold in the market. In addition, the strike prices and the units of the derivatives could be computed directly by the formulae through (22.22)–(22.25). Next, we are interested in investigating the hedging performance of the proposed strategy in the real market. However, barrier options are usually traded over-the-counter (OTC) and thus the option data is not publicly available. In the following, we employ a simulation study instead to compare the hedging performance of the proposed hedging portfolio with non-hedging strategy for down-and-in call options, where the non-hedging strategy means that the practitioner didn't set up any hedging portfolio but put the money, obtained from shorting the down-and-in call option, in a bank. By generating N simulation paths, let

$$\overline{D}_0 = \frac{1}{N}\sum_{j=1}^{N}(V_0 e^{rT} - DIC_{T,j})/DIC_0$$

and

$$\overline{D}_1 = \frac{1}{N}\sum_{j=1}^{N}(V_{T,j} - DIC_{T,j})/DIC_0$$

denote the average payoff of non-hedging strategy divided by DIC_0 and the proposed hedging strategy divided by DIC_0, respectively, where $DIC_{T,j}$ and $V_{T,j}$ are the values of the down-and-in call option and the proposed hedging strategy, respectively, obtained from the jth simulated path at maturity. Further let q_α^0 and q_α^1 are the α-quantiles derived from the empirical distributions of $(V_0 e^{rT} - DIC_{T,j})/DIC_0$ and $(V_{T,j} - DIC_{T,j})/DIC_0$, $j = 1, \cdots, N$, respectively. Table 22.2 gives the simulation results with parameters $r = 0.05$, $\mu = 0.10$, $\sigma = 0.20$ and $S_0 = 100$. In Table 22.2, one can see that $\overline{D}_1 < \overline{D}_0$ in most cases, which means that the average loss of the proposed hedging strategy is less than that of non-hedging strategy. Moreover, $q_{0.005}^0$ and $q_{0.005}^1$ are adopted to measure the range of the risk of the hedging strategies. And the ratio $q_{0.005}^1/q_{0.005}^0$ is used to compare the hedging efficiency of the proposed hedging strategy with respect to the non-hedging strategy. In Table 22.2, the values of $q_{0.005}^1/q_{0.005}^0$ decrease as the barrier price B increases if T is fixed, that is, the hedging performance of the proposed strategy becomes better with large barrier prices. Furthermore, the values

Table 22.2 The simulation results of hedging strategies (22.22) and (22.24) with parameters $r = 0.05$, $\mu = 0.10$, $\sigma = 0.20$, $S_0 = 100$ and $N = 10{,}000$

T(days)		$(B/S_0, K/S_0)$	$\bar{D}_0$	$\bar{D}_1$	$q^0_{0.005}$	$q^1_{0.005}$	$\frac{q^1_{0.005}}{q^0_{0.005}}$
90	(22.22)	(0.90,0.90)	0.0506	−0.0004	−20.4710	−0.3172	0.0155
		(0.925,0.90)	0.0584	−0.0001	−10.9191	−0.1220	0.0112
		(0.95,0.95)	−0.0723	−0.0002[a]	−10.7767	−0.0973	0.0090
		(0.975,0.95)	−0.0692	−0.0001[a]	−5.9533	−0.0290	0.0049
	(22.24)	(0.925,0.95)	0.0045	0.0251[a]	−20.2215	−0.1753	0.0087
		(0.95,0.975)	−0.0809	0.0134[a]	−14.1331	−0.0990	0.0070
180	(22.22)	(0.90,0.90)	−0.0536	−0.0018[a]	−15.4581	−0.2480	0.0160
		(0.925,0.90)	−0.0085	−0.0003[a]	−9.6246	−0.1155	0.0120
		(0.95,0.95)	−0.1320	−0.0005[a]	−9.1158	−0.0839	0.0092
		(0.975,0.95)	−0.1405	−0.0002[a]	−6.1975	−0.0295	0.0048
	(22.24)	(0.925,0.95)	−0.0645	0.0032[a]	−13.7629	−0.1827	0.0133
		(0.95,0.975)	−0.1322	0.0013[a]	−11.6451	−0.1039	0.0089

[a]Denotes $\bar{D}_1 > \bar{D}_0$

of $q^1_{0.005}/q^0_{0.005}$ are all less than 2%, which means that the proposed hedging strategy is able to hedge over 98% risk of the non-hedging strategy. The simulation study shows that the hedging performance of the proposed hedging strategy is much more better than the non-hedging one. Therefore, comparing with saving the money in the bank, practitioner is suggested to adopt the proposed hedging strategy to hedge her short position of a down-and-in call option.

References

Bensaid, B., Lesne, J. P., Pagès, H., & Scheinkman, J. (1992). Derivatives asset pricing with trasaction costs. *Mathematical Finance*, *2*, 63–86.

Black, F., & Scholes, M. (1973). The pricing of options and corporate liabilities. *Journal of Political Economy*, *81*, 637–654.

Bollerslev, T. (1986). Generalized autoregressive conditional heteroskedasticity. *Journal of econometrics*, *31*, 307–327.

Bowie, J., & Carr, P. (1994). Static simplicity. *Risk*, *7*, 45–49.

Carr, P., & Lee, R. (2009). Put-call symmetry: Extensions and applications. *Mathematical Finance*, *19*, 23–560.

Cvitanić, J., & Karatzas, I. (1999). On dynamic measures of Risk. *Finance and Stochastics*, *3*, 451–482.

Duan, J. C. (1995). The GARCH option pricing model. *Mathematical Finance*, *5*, 13–32.

Duan, J. C., & Simonato, J. G. (1998). Empirical martingale simulation for asset prices. *Management Science*, *44*, 1218–1233.

Elliott, R. J., & Madan, D. B. (1998). A discrete time equivalent martingale measure. *Mathematical Finance*, *8*, 127–152.

Föllmer, H., & Leukert, P. (2000). Efficient hedging: Cost versus shortfall risk. *Finance and Stochastics*, *4*, 117–146.

Huang, S. F., Guo, M. H. (2009a). Financial derivative valuation - a dynamic semiparametric approach. *Statistica Sinica, 19*, 1037–1054.

Huang, S. F., & Guo, M. H. (2009b). Valuation of multidimensional Bermudan options. In W. Härdle (Ed.), *Applied quantitative finance* (2nd ed.). Berlin: Springer.

Huang, S. F., & Guo, M. H. (2009c). *Multi-period hedging strategy minimizing squared risk adjusted costs*. Working paper.

Huang, S. F., & Guo, M. H. (2009d). *Model risk of implied conditional heteroscedastic models with Gaussian innovation in option pricing and hedging*. Working paper.

Huang, S. F., & Huang, J. Y. (2009). Hedging strategies against path-dependent contingent claims. *Journal of Chinese Statistical Association, 47*, 194–218.

Schulmerich, M., & Trautmann, S. (2003). Local expected shortfall-hedging in discrete time. *European Finance Review, 7*, 75–102.

Chapter 23
Approximation of Dynamic Programs

Michèle Breton and Javier de Frutos

Abstract Under some standard market assumptions, evaluating a derivative implies computing the discounted expected value of its future cash flows and can be written as a stochastic Dynamic Program (DP), where the state variable corresponds to the underlying assets' observable characteristics. Approximation procedures are needed to discretize the state space and to reduce the computational burden of the DP algorithm. One possible approach consists in interpolating the function representing the value of the derivative using polynomial basis functions. This chapter presents basic interpolation approaches used in DP algorithms for the evaluation of financial options, in the simple setting of a Bermudian put option.

23.1 Introduction

A derivative security is a financial instrument whose value depends on the value of other basic underlying assets. A stock option, for example, is a derivative whose value depends on the price of a stock. Derivative securities include forward and future contracts, swaps, and options of various kinds. They are characterized by their payoff function, maturity, and cash-flow schedule. Options are further characterized by their "optionality" and by their exercise schedule; holders have the right, but not the obligation, to exercise options at previously defined dates. European options can be exercised only at the maturity date; Bermudian options can be exercised at a finite number of predefined dates; and, American options can be exercised at any time during their life.

Michèle Breton (✉)
GERAD, HEC Montréal, 3000, chemin de la Côte-Sainte-Catherine, Montréal, QC, Canada H3T 2A7, michele.breton@hec.ca

Javier de Frutos
GERAD and Dpto de Matemática Aplicada, Universidad de Valladolid, Valladolid, Spain, frutos@mac.uva.es

J.-C. Duan et al. (eds.), *Handbook of Computational Finance*, Springer Handbooks of Computational Statistics, DOI 10.1007/978-3-642-17254-0_23,

Under some standard market assumptions, evaluating a derivative implies computing the discounted expected value of its future cash flows. In that context, the evaluation of a financial option can be written as a stochastic Dynamic Program, where the state variable corresponds to the underlying assets' observable characteristics (e.g. prices), the stages correspond to the dates where the option can be exercised, the decision at a given stage amounts to a choice between exercising the option or not, and the value function represents the value of the financial option as a function of the date and of the underlying asset's observable characteristics.

The discrete-stage Dynamic Programming (DP) algorithm over a finite horizon requires the evaluation of the value function at each decision stage for all possible states, by comparing expected values of actions over all possible state transitions. Approximation procedures are used to reduce the computational burden of the DP algorithm.

In most financial models, the state variable is continuous. One way to approximate the dynamic program is to divide the state space in convex sections, and to suppose that the value function and strategies are constant on a given section. This discretization approach leads to a Markov Decision Problem with finite state and action spaces, which can be solved with the standard DP recursion.

Another interesting approach consists in interpolating the value function, as a linear combination of suitably chosen basis functions, in order to replicate the properties of the financial derivative (e.g. matching, continuity, convexity), and such that the approximant is easy to use inside the DP algorithm. Two classical interpolation schemes are spectral methods and finite element methods.

Spectral methods use basis functions that are non-zero over the entire domain (most commonly families of polynomial functions). They usually require few interpolation nodes, and relatively high degree polynomials. On the other hand, finite element methods use basis functions with support on small subsets of the domain (most commonly spline functions). They usually are carried-out on a large number of interpolation nodes and use polynomials of low degree over sub-intervals.

A third approximation approach is to use prospective DP, which consists in evaluating the value function on a subset of the state space, for example by generating possible trajectories. Monte-Carlo simulation is often used in that context to approximate the value function. One salient characteristic of the DP model for the evaluation of financial derivatives is the fact that the possible trajectories do not depend on the decisions taken by the option holder, which simplifies considerably the state space exploration strategies of prospective DP. A drawback of such methods is the introduction of statistical error, but this may be the only possible way when the state space is too large to be explored exhaustively.

This chapter presents basic interpolation approaches used in DP algorithms for the evaluation of financial options. The approaches are presented in the simple setting of a Bermudian put option, and references to specific applications are provided. Section 23.2 presents the DP model and notation, as well as the Markov Decision Problem (MDP) resulting from the state space discretization. Section 23.3 introduces interpolation approaches, and Sects. 23.4 and 23.5 present respectively finite element and spectral interpolation of the value function. Section 23.6 proposes a hybrid algorithm combining elements from both methods. Section 23.7 is a conclusion.

23.2 Model and Notation

Consider a Bermudian put option written on the underlying asset, with maturity T. This option gives its holder the right to sell the underlying asset for a pre-determined price K, called the strike price, at any date in a given set $\mathcal{T} = \{t_n\}$ in $[0, T]$. To simplify, and without loss of generality, assume that exercise dates are equally spaced, with $n = 1, \ldots, N$, where t_0 is the contract inception date and $t_N = T$ is the maturity of the contract. Bermudian options admit European options as a special case (when the set $\mathcal{T}$ contains a single date T) and American options as a limiting case (when the time between two exercise dates becomes arbitrarily small).

Let the price of the underlying asset $\{S\}$ be a Markov process that verifies the fundamental no-arbitrage property. The value of the option at any date t_n when the price of the underlying asset is s is given by

$$v_n(s) = \begin{cases} v_0^h(s) & \text{for } n = 0 \\ \max\left(v^e(s), v_n^h(s)\right) & \text{for } n = 1, \ldots, N-1, \\ v^e(s) & \text{for } n = N \end{cases} \tag{23.1}$$

where v^e denotes the *exercise value* of the option:

$$v^e(s) = \max\{K - s; 0\}, \tag{23.2}$$

and v_n^h denotes the *holding value* of the option at t_n. Under standard no-arbitrage assumptions, the discounted price of the underlying asset is a martingale with respect to some probability measure Q, and the expected value of the future potentialities of the option contract is given by

$$v_n^h(s) = \beta E[v_{n+1}(S_{t_{n+1}}) \mid S_{t_n} = s], \quad \text{for } n = 0, \ldots, N-1, \tag{23.3}$$

where β is the discount factor and $E[\cdot]$ denotes the expectation with respect to measure Q.

One way to price the option is to solve the discrete-time stochastic dynamic program (23.1)–(23.3), by backward induction from the known function $v_N = v^e$. Even for the most simple cases, the value function cannot be expressed in closed-form and the option value must be obtained numerically.

In most market models, the underlying asset price is assumed to take values in $[0, +\infty)$. Since the state space of the dynamic program is continuous, the first step is to partition it into a collection of convex subsets, and obtain a corresponding finite set of grid points where the option value is to be evaluated.

Define a partition of the positive real line into $(p+1)$ intervals

$$[a_i, a_{i+1}) \quad \text{for } i = 0, \ldots, p, \tag{23.4}$$

where $0 = a_0 \leq a_1 < \ldots < a_p < a_{p+1} = +\infty$ and grid $\mathcal{G} = \{a_i\}_{i=1,\ldots,p}$. Here, to simplify notation, we assume that the partitions and grids are identical for $n = 1, \ldots, N$, but stage-specific grids $\mathcal{G}^n = \{a_i^n\}_{i=1,\ldots,p_n}$ can also be used. The standard projection of the dynamic program (23.1)–(23.3) into a finite-state Markov Decision Program (MDP) is given, for $i = 1, \ldots, p$, by

$$\widetilde{v}_n(i) = \begin{cases} \widetilde{v}_0^h(i) & \text{for } n = 0 \\ \max\left(\widetilde{v}^e(i), \widetilde{v}^h(i)\right) & \text{for } n = 1, \ldots, N-1 \\ \widetilde{v}^e(i) & \text{for } n = N \end{cases} \tag{23.5}$$

$$\widetilde{v}^e(i) = \max\{K - a_i; 0\} \tag{23.6}$$

$$\widetilde{v}_n^h(i) = \beta \sum_{j=0}^{p} p_{ij}^n \widetilde{v}_{n+1}(j) \quad \text{for } n = 1, \ldots, N-1, \tag{23.7}$$

where each state i corresponds to a grid point a_i, and where the transition probabilities p_{ij}^n are obtained from the Markov price process $\{S\}$ under measure Q so that:

$$p_{ij}^n = Q\left[S_{t_{n+1}} \in [a_j, a_{j+1}) | S_{t_n} = a_i\right].$$

In many applications, the transition probabilities p_{ij}^n are independent of n if the discretization grids are identical at all stages and they can be pre-computed in $O(p^2)$ operations. Solution of the MDP (23.5)–(23.7) is straightforward by backward induction and yields, in $O(N \times p^2)$ operations, the value of the option and the optimal strategy (exercise or not) for all decision dates and all asset prices on the grid. The option value and exercise strategy in the discrete problem can then be extended to an approximate option value and sub-optimal strategy in the original continuous problem through some form of interpolation; for instance, one may use linear interpolation for the option value, and constant strategies over the intervals $[a_i, a_{i+1})$, $i = 0, \ldots, p$. Typically, this approximation scheme will converge to the solution of the original problem, as the discretization becomes finer and finer, if there is a sufficient amount of continuity in the original problem (see Whitt 1978, 1979). In addition, since the state space is unbounded, one usually has to show that the approximation error outside the localization interval $[a_1, a_p]$ becomes negligible when this interval is large enough.

This approach was used in Duan and Simonato (2001) to price American options in a GARCH setting. In their model, the option price depends on two state variables (asset price and volatility), and the state space is projected on a two-dimensional grid. The numerical algorithm proposed by the authors, termed Markov chain approximation, relies on the approximation of the underlying GARCH asset price process by a finite-state Markov chain. The authors show that the MDP converges to the option value as the grid becomes finer while the localization interval becomes wider. The advantage of Markov chain approximation lies in the fact that the price of an option can be computed by simple matrix operations, making use of matrix representation and computation available in high-level programming languages.

23.3 Interpolation of the Value Function

An interpolation function $\widehat{v}$ is generally defined as a linear combination of p basis functions, denoted φ_j, $j = 1, \ldots, p$:

$$\widehat{v}(s) = \sum_{j=1}^{p} c_j \varphi_j(s). \tag{23.8}$$

Interpolation achieves two interesting purposes in option evaluation; First, it allows to approximate a complicated function, which cannot be expressed in closed-form, by a simpler function which can be used efficiently in computations; Second, it allows to describe and store the continuous function approximating the option value using a finite set of coefficients.

Assume that, at stage $n+1 \leq N$, the option value as a function of the underlying state price is described by a (continuous) interpolation function $\widehat{v}_{n+1}(s)$. At stage n, the approximate DP algorithm consists in evaluating, on the set of grid points $\mathcal{G}$,

$$\widetilde{v}_n^h(i) = \beta E[\widehat{v}_{n+1}(S_{t_{n+1}}) \mid S_{t_n} = a_i], \tag{23.9}$$

$$\widetilde{v}_n(i) = \begin{cases} \widetilde{v}_0^h(i) & \text{for } n = 0 \\ \max\left(v^e(a_i), \widetilde{v}_n^h(i)\right) & \text{for } n = 1, \ldots, N-1. \end{cases} \tag{23.10}$$

An interpolation function $\widehat{v}_n(s)$ is then obtained, using the values $\widetilde{v}_n(i)$, $i = 1, \ldots p$, which are called interpolation nodes. Interpolation consists in computing the value of coefficients c_j in (23.8) by specifying conditions that need to be satisfied by the interpolation function. The simplest and most usual property is that the interpolation function coincide with the the option value at the p interpolation nodes, yielding a system of p linear equations

$$\sum_{j=1}^{p} c_j^n \varphi_j(a_i) = \widetilde{v}_n(i), i = 1, \ldots p, \tag{23.11}$$

where c_j^n, $j = 1, \ldots p$, denote the coefficients obtained at stage n.

The choice of an interpolation scheme consists in selecting a family of basis functions φ_j and a set of interpolation nodes $\mathcal{G}$ that are unisolvent, that is, for any data set $\widetilde{v}_n(i)$, $i = 1, \ldots p$, the system (23.11) has a unique solution. Three fundamental principles motivate these choices: convergence of the interpolation function to the value function, efficiency and accuracy in the computation of the interpolation coefficients, and efficiency in the evaluation of the expected value in (23.9). In the two following sections, we describe two families of interpolation methods.

23.4 Finite Element Interpolation

In finite element interpolation, basis functions are non-zero on sub-intervals of the domain. The most common schemes are spline interpolation, and piecewise polynomial interpolation. Both consist in using polynomials of low degree over each sub-interval. Piecewise polynomial interpolation yields continuous interpolation functions. Spline interpolation of order k yields continuous interpolation functions with continuous derivatives of order $k-1$ or less. When $k = 1$, spline interpolation and polynomial interpolation coincide, yielding a piecewise linear continuous approximation. Here, we will focus on spline approximation of degree 1, which is easy to implement and results in an efficient approximation of the option value. The approach is easy to extend to higher degree polynomials or splines. Figure 23.1 gives a graphical representation of a piecewise linear approximation of the value of a Bermudian put option, using equally spaced grid points.

Consider the grid $\mathcal{G} = \{a_i\}_{i=1,...,p}$ and the corresponding partition (23.4). Define p piecewise linear basis functions such that, for $j = 1, ..., p$

$$\varphi_j(a_i) = \begin{cases} 1 \text{ if } i = j \\ 0 \text{ if } i \neq j \end{cases}, i = 1, ..., p$$

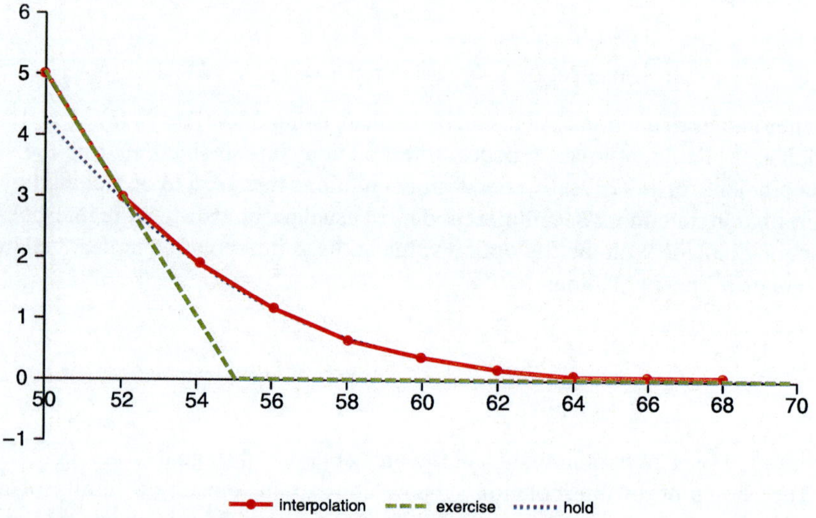

Fig. 23.1 Value and finite-element interpolation of a Bermudean put option in the Black–Scholes model, one period before maturity. Parameters are $K = 55, \sigma = 0.1, r = 0.0285$

and where the φ_j are continuous on $[a_0, a_p]$, linear on each $[a_i, a_{i+1}]$ and null on (a_0, a_1) and $(a_p, +\infty)$. At stage n, given the values $\widetilde{v}_n(i)$, $i = 1, ...p$ computed by (23.9) and (23.10), the system of conditions (23.11) satisfied by the coefficients reduces to

$$c_i^n = \widetilde{v}_n(i), i = 1, ...p$$

and the interpolation function is given by

$$\widehat{v}_n(s) = \sum_{j=1}^{p} \widetilde{v}_n(j)\, \varphi_j(s). \tag{23.12}$$

Given a function v and a grid $\mathcal{G} = \{a_i\}$, denote $E_{in}[v(S)] \equiv E[v(S_{t_{n+1}}) \mid S_{t_n} = a_i]$. Using (23.12), the expectation in (23.9) of an interpolation function $\widehat{v}_{n+1}$ is written

$$\begin{aligned} E_{in}[\widehat{v}_{n+1}(S)] &= E_{in}\left[\sum_{j=1}^{p} \widetilde{v}_{n+1}(j)\, \varphi_j(S)\right] \\ &= \sum_{j=1}^{p} \widetilde{v}_{n+1}(j)\, E_{in}[\varphi_j(S)] \\ &= \sum_{j=1}^{p} \widetilde{v}_{n+1}(j)\, A_{ij}^n, \end{aligned}$$

where the transition parameters A_{ij}^n are given by

$$A_{ij}^n = E_{in}[\varphi_j(S)] = \begin{cases} E_{in}\left[\frac{a_2 - S}{a_2 - a_1}\mathcal{I}_1^n\right] & \text{for } i = 1 \\ E_{in}\left[\frac{S - a_{i-1}}{a_i - a_{i-1}}\mathcal{I}_{i-1}^n + \frac{a_{i+1} - S}{a_{i+1} - a_i}\mathcal{I}_i^n\right] & \text{for } 1 < i < p \\ E_{in}\left[\frac{S - a_{i-1}}{a_i - a_{i-1}}\mathcal{I}_{i-1}^n\right] & \text{for } i = p \end{cases}$$

and where $\mathcal{I}_i^n$ denotes the indicator function of the event $\{a_i \leq S_{t_n+1} \leq a_{i+1}\}$. Notice that the transition parameters can be obtained in closed-form for a large class of models for the underlying asset price process; otherwise, numerical integration may be used. In many applications, if the interpolation grids are identical at all stages, these transition parameters are independent of time and can be pre-computed in $O(p^2)$ operations. The option value can then be readily obtained in $O(N \times p^2)$ operations by the dynamic program

$$\widetilde{v}_n^h(i) = \beta \sum_{j=1}^{p} \widetilde{v}_{n+1}(j)\, A_{ij}^n, \quad i = 1, ...p$$

$$\widetilde{v}_n(i) = \begin{cases} \widetilde{v}_0^h(i) & \text{for } n = 0 \\ \max\left(v^e(a_i), \widetilde{v}_n^h(i)\right) & \text{for } n = 1, \ldots, N-1 \\ v^e(a_i) & \text{for } n = N \end{cases}, \quad i = 1, ...p.$$

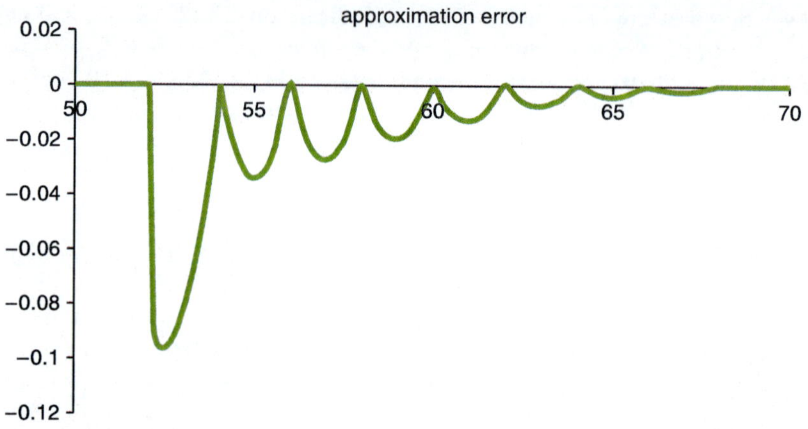

Fig. 23.2 Interpolation error of a Bermudean put option in the Black–Scholes model, one period before maturity. Parameters are $K = 55, \sigma = 0.1, r = 0.0285$. Localization interval is [50,68] and $p = 10$

Notice that a linear spline approximation does not require computation of interpolation coefficients, and that the expected value in (23.9) is obtained by simple matrix operations. Ideally, the density of grid points should be higher in the regions where the value function has higher curvature, and around the exercise frontier where the curvature changes abruptly – however there is often a clear advantage in keeping the grid constant over time. Finally, notice that linear extrapolation can also be used outside the localization interval, for instance in the exercise region where the value function is linear. For convex value functions, piecewise linear interpolation provides an upper bound on the option value if the approximation error outside the localization interval is small enough. Figure 23.2 shows the approximation error from the interpolation of the Bermudian put illustrated in Fig. 23.1.

A higher convergence rate can be achieved when interpolating with higher degree polynomials on each sub-interval, but this increases the computational and storage burden for both the transition parameters and the computation of the interpolation coefficients.

Finite element interpolation was used in Ben-Ameur et al. (2002) to price Asian options in a Black–Scholes framework; in this case, the option price depends on two state variables (current price and arithmetic average). They use a linear interpolation with respect to the price, and a quadratic interpolation with respect to the average, and the grid is defined from the quantiles of the log-normal distribution. It was used in Ben-Ameur et al. (2006) to price installment options, in Ben-Ameur et al. (2007) to price call and put options embedded in bonds, and in Ben Abdallah et al. (2009) to price the delivery options of the CBOT T-Bond futures. In the last two cases, the underlying asset is the spot interest rate. Ben-Ameur et al. (2009) use finite element interpolation with time-varying, equally spaced grids to price options in a GARCH framework, using asset price and asset volatility as state variables.

To conclude, we point out that Markov chain approximation can be shown to be equivalent to a special case of finite element interpolation, where the basis functions are piecewise constant, such that, for $j = 1, ..., p$

$$\varphi_j(s) = \begin{cases} 1 \text{ if } s \in (a_j, a_{j+1}] \\ 0 \quad \text{otherwise.} \end{cases}$$

Finite element interpolation using linear splines is usually more efficient than Markov chain approximation, as it does not require more computational work, while it produces a continuous approximation to the option value rather than a discontinuous one, and converges at a faster rate.

23.5 Spectral Interpolation

In spectral interpolation, basis functions are non-zero over the entire domain. The most common scheme is polynomial interpolation. It yields a polynomial of degree $p - 1$ matching the value function on p interpolation nodes. Figure 23.3 gives a graphical representation of the interpolation of a Bermudian put option with a polynomial of degree 9, using Chebyshev interpolation nodes.

Define p polynomial basis functions such that φ_j is a polynomial of degree $j-1$, for $j = 1, ..., p$. Consider the grid $\mathcal{G} = \{a_i\}_{i=1,...,p}$. At stage n, denote $\widetilde{v}_n$ the

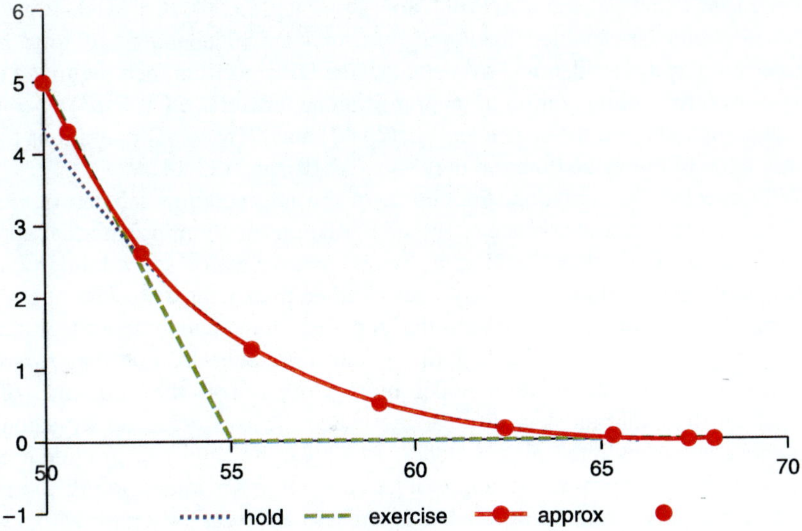

Fig. 23.3 Value and spectral interpolation of a Bermudean put option in the Black–Scholes model, one period before maturity. Parameters are $K = 55, \sigma = 0.1, r = 0.0285$

column vector $[\widetilde{v}_n(i)]$, $i = 1, \ldots p$ computed by (23.9) and (23.10). The system (23.11) satisfied by the coefficients is written in matrix form

$$c^n = \Phi^{-1}\widetilde{v}_n,$$

where Φ is a square matrix of dimension p, with $\Phi_{ij} = \varphi_j(a_i)$, and c^n is the column vector of coefficients $\left[c_j^n\right]$, $j = 1, \ldots, p$. The interpolation function is a polynomial of degree $p - 1$ given by

$$\widehat{v}_n(s) = \sum_{j=1}^{p} c_j^n \varphi_j(s). \tag{23.13}$$

Using (23.13), the expectation in (23.9) of an interpolation function $\widehat{v}_{n+1}$ is written

$$\begin{aligned} E_{in}[\widehat{v}_{n+1}(S)] &= E_{in}\left[\sum_{j=1}^{p} c_j^{n+1}\varphi_j(s)\right] \\ &= \sum_{j=1}^{p} c_j^{n+1} E_{in}[\varphi_j(S)] \\ &= \sum_{j=1}^{p} c_j^{n+1} B_{ij}^n, \end{aligned} \tag{23.14}$$

where the transition parameters B_{ij}^n are expectations of polynomial functions. As in the case of finite-element interpolation, for a large class of models, transition parameters can be obtained in closed-form and are independent of time if the interpolation grids are identical over stages. The DP algorithm then requires $O(p^2)$ operations for the computation of the transition parameters, $O(N \times p^2)$ operations for the computation of the coefficients in (23.13) and $O(N \times p^2)$ operations for the computation of the value function in (23.9), (23.10), and (23.14).

With spectral interpolation, the choice of the interpolation scheme is crucial. With respect to the convergence of the interpolation function, polynomial approximation tends to produce oscillating errors; with equally spaced interpolation nodes, the interpolation error may grow, rather than decrease, with increasing number of interpolation nodes (this is the so-called Runge phenomenon), producing large errors near the boundary of the localization interval. For that reason, it is better to space interpolation nodes more closely near the endpoints of the localization interval, and less so near the center. This is achieved by using, for instance, Chebyshev interpolation nodes, which can be shown to provide a better approximation. Moreover, the approximation error from interpolating a smooth function by a polynomial using Chebyshev nodes converges very rapidly to 0 when p increases. Chebyshev nodes over the interval $\left[a_1, a_p\right]$ are given by:

$$a_i = \frac{a_1 + a_p}{2} - \frac{a_p - a_1}{2} \cos\left(\frac{i-1}{p-1}\pi\right), \; i = 1, ..p.$$

Figure 23.4 shows the approximation error from the interpolation of the Bermudian put illustrated in Fig. 23.3 using Chebyshev interpolation nodes, compared to the approximation error obtained with equally spaced nodes. Notice that because of the early exercise opportunity, the option value is not smooth at the exercise barrier, thus requiring a high degree polynomial for a precise interpolation.

A second concern is the accuracy of the computation of the interpolation coefficients. Indeed, choosing the power functions $\varphi_j(s) = s^{j-1}$ as a basis yields an ill-conditioned interpolation matrix (the so-called Vandermonde matrix), which becomes increasingly difficult to invert as p increases, and produces large numerical errors, even for moderate p. A better choice is the Chebyshev polynomial basis. Chebyshev polynomials are defined recursively as:

$$\widetilde{\varphi}_1(z) = 1$$
$$\widetilde{\varphi}_2(z) = z$$
$$\widetilde{\varphi}_j(z) = 2z\widetilde{\varphi}_{j-1}(z) - \widetilde{\varphi}_{j-2}(z),$$

where, for $s \in [a_1, a_p]$, $z = 2\frac{s-a_1}{a_p-a_1} - 1 \in [-1, 1]$ and $\left|\widetilde{\varphi}_j(z)\right| \leq 1$. Chebyshev polynomials evaluated at Chebyshev interpolation nodes yield a well-conditioned interpolation matrix, which can be solved accurately even when p is large. Moreover, in that case coefficients can be obtained efficiently by Fast Fourier Transforms, which reduces the computational burden to $O(p \log p)$ operations (see Canuto et al. 2006).

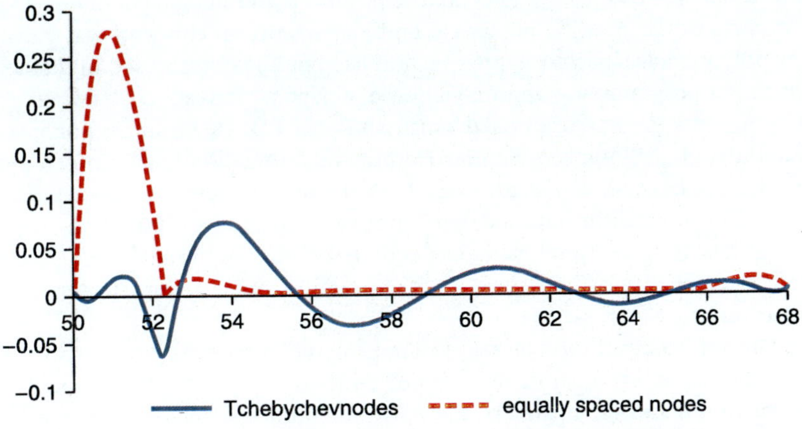

Fig. 23.4 Interpolation error for a spectral interpolation of a Bermudean put option in the Black–Scholes model, one period before maturity. Parameters are $K = 55, \sigma = 0.1, r = 0.0285$

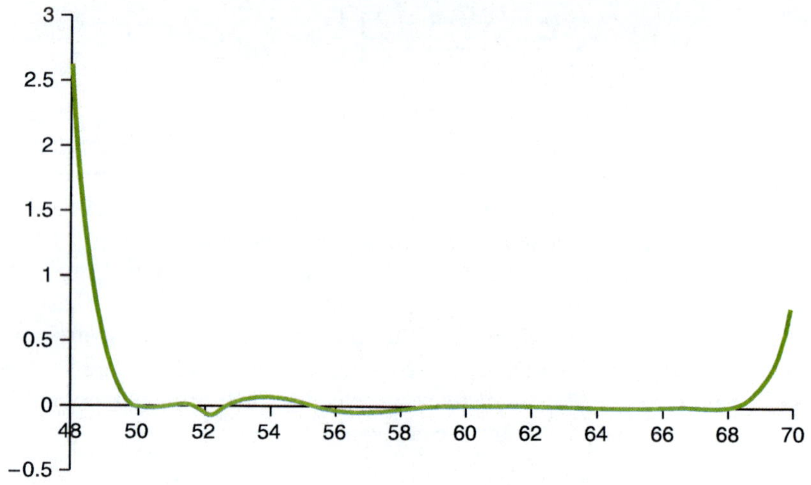

Fig. 23.5 Approximation error outside the localization interval [50,68]

Finally, even when transition parameters B_{ij}^n are known is closed-form, the evaluation of the expected value in (23.9) can be problematic with spectral approximation. A first important remark is that the polynomial function $\sum_{j=1}^{p} c_j^n \varphi_j(s)$ is generally very far from the value function outside the localization interval. Theoretical convergence of the DP approximation is obtained if the probability of leaving the localization interval converges to 0 faster than a polynomial, which is the case in most market models. However, even in that case, using the expectation of the interpolation function in (23.14) may still cause significant errors if the localization interval is not large enough. To illustrate, Fig. 23.5 represents the approximation error outside the localization interval [50,68] for the Bermudian put in Fig. 23.3.

Because of the possible numerical problems involved with working with high-order polynomials, spectral approximation has not been much used in the context of dynamic programming approximation of financial derivatives. However, spectral approximation has been used successfully in PDE algorithms. For instance, Chiarella et al. (1999) use a Fourier-Hermite representation for American options while Chiarella et al. (2008a,b) extend this method for the evaluation of barrier options and jump-diffusion models respectively, de Frutos (2008) uses Laguerre polynomials to price options embedded in bonds and Breton and de Frutos (2010) use a Fourier-Chebyshev approximation to price options under GARCH specifications.

Inside the localization interval, spectral interpolation exhibits spectral convergence, that is, an exponential decrease of the error with respect to the number of interpolation nodes (or degree of the polynomial interpolant). As a consequence, for smooth functions, a very good precision can be reached with few interpolation nodes. This means that the value function can be represented and stored using a relatively small number of coefficients, which is a definite advantage for options defined over a multi-dimensional state space. In the following section, we propose an hybrid approach maintaining spectral convergence, while minimizing the approximation error outside the localization interval and around the exercise barrier.

23.6 Hybrid Approach

A natural way to reduce the approximation error outside the localization interval $[a_1, a_p]$ consists in dividing the state space in three distinct regions, where the spectral interpolation is used in $[a_1, a_p]$, while a suitable extrapolation is used in $[0, a_1)$ and $(a_p, +\infty)$. For instance, Fig. 23.6 shows the approximation error in the case of the put option in Fig. 23.3 when the exercise value is used over $[0, a_1)$ while the null function is used over $(a_p, +\infty)$.

Notice that an even better fit can be obtained if a_1 coincides with the exercise barrier, so that the function to be interpolated is smooth in the interval $[a_1, a_p]$, as illustrated in Fig. 23.7.

However, the computation of conditional expectations of high-order polynomial functions is often numerically unstable. To illustrate, in the Black–Scholes model, the transition parameter B^n_{ij} is given by

$$B^n_{ij} = \frac{1}{\sqrt{2\pi}} \int_{-\infty}^{\infty} \varphi_j \left(a_i \exp\left(r - \frac{\sigma^2}{2} + \sigma\varepsilon \right) \right) \exp\left(\frac{-\varepsilon^2}{2} \right) d\varepsilon,$$

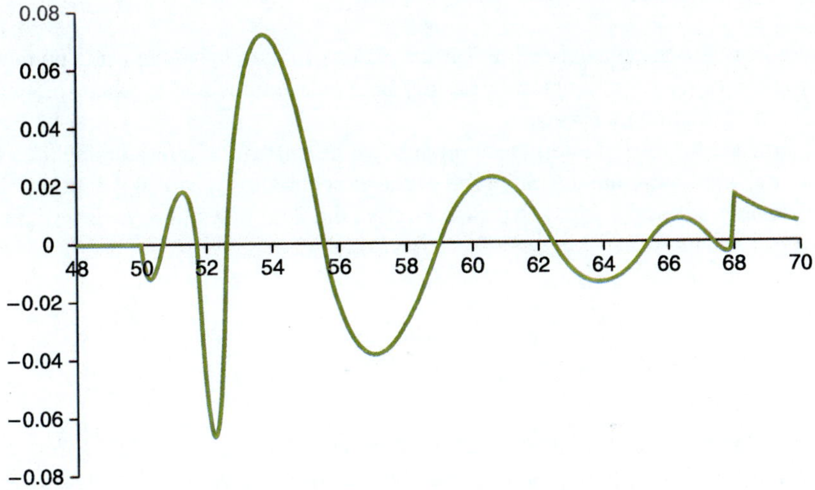

Fig. 23.6 Approximation error when exercise value is used to the left and null function is used to the right of the localization interval [50,68]

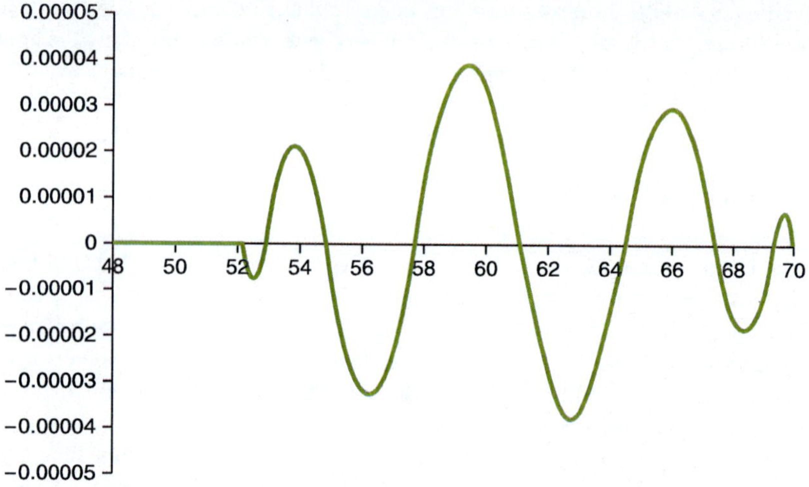

Fig. 23.7 Interpolation error when localization interval is [52.22, 70] and exercise value is used to the left of the exercise frontier

where r is the risk-less interest rate and σ the volatility of the underlying asset between two exercise dates. When φ_j is a polynomial of degree q, restricting the expectation to the interval for ε corresponding to $a_i \exp\left(r - \frac{\sigma^2}{2} + \sigma\varepsilon\right) \in [a_1, a_p]$ involves the numerical evaluation of expressions of the form

$$a_i^q \exp(qb),$$

where b is a constant depending on the values of r, σ, a_i, a_1 and a_p; numerical evaluation of such an expression rapidly becomes impracticable when the degree of the interpolation increases.

Here, we propose an alternate approach, where, instead of interpolating the value function, we interpolate the integrand of the expectation $E_{in}[v_{n+1}(\cdot)]$ at grid points.

Consider a grid $\mathcal{G} = \{a_i\}$. At stage $n < N$, the holding value at a_i is written

$$\begin{aligned} v_n^h(a_i) &= \beta E_{in}[v_{n+1}(S)] \\ &= \beta E_{in}[v_{n+1}(S)\mathcal{I}_{[0,a_1)}^n] \\ &\quad + \beta E_{in}[v_{n+1}(S)\mathcal{I}_{[a_1,a_p]}^n] + \beta E_{in}[v_{n+1}(S)\mathcal{I}_{(a_p,\infty)}^n], \end{aligned}$$

where $\mathcal{I}_I^n$ denotes the indicator function of the event $\{S_{t_n+1} \in I\}$. For appropriately chosen localization interval $[a_1, a_p]$, the holding value may be approximated by

$$\begin{aligned} \widetilde{v}_n^h(i) &= \beta\left(E_{in}[v^e(S)\mathcal{I}_{[0,a_1)}^n] + E_{in}[v_{n+1}(S)\mathcal{I}_{[a_1,a_p]}^n]\right) \\ &= \beta E_{in}[(K - S)\mathcal{I}_{[0,a_1)}^n] + \beta E_{in}[v_{n+1}(S)\mathcal{I}_{[a_1,a_p]}^n] \end{aligned}$$

$$= \beta \int_0^{a_1} (K - S)\, f_{in}(S) dS + \beta \int_{a_1}^{a_p} v_{n+1}(S)\, f_{in}(S) dS$$

$$= \beta C_i^n + \beta \int_{a_1}^{a_p} v_{n+1}(S)\, f_{in}(S) dS, \tag{23.15}$$

where $f_{in}(S)$ denotes the probability density of $[S_{t_{n+1}} \mid S_{t_n} = a_i]$. The transition parameters C_i^n can be readily obtained in closed-form for a large class of models, and in many cases can be pre-computed as they do not depend on n if the interpolation grids are identical at all stages.

Define the function $w_{n+1}^i(s) \equiv v_{n+1}(s)\, f_{in}(s)$ and assume that an approximation $\widetilde{v}_{n+1}$ of v_{n+1} is known on $\mathcal{G}$. The spectral interpolation of $w_{n+1}^i(s)$ is then

$$\widehat{w}_{n+1}^i(s) = \sum_{j=1}^{p} c_{ij}^{n+1} \varphi_j(s),$$

where the coefficients c_{ij}^{n+1} satisfy, for $i = 1, ..., p$, the linear system

$$\sum_{j=1}^{p} c_{ij}^{n+1} \varphi_j(a_k) = \widetilde{v}_{n+1}(a_k)\, f_{in}(a_k), k = 1, ...p.$$

Replacing in (23.15), we finally obtain

$$\widetilde{v}_n^h(i) = \beta C_i^n + \beta \sum_{j=1}^{p} c_{ij}^{n+1} \int_{a_1}^{a_p} \varphi_j(S) dS.$$

Notice that if $\widetilde{\varphi}_j$ is the Chebyshev polynomial of degree $j - 1$ defined over the interval $[-1, 1]$, it satisfies

$$\int_{-1}^{1} \widetilde{\varphi}_j(u) du = \begin{cases} 0 & \text{if } j \text{ is even} \\ \frac{2}{j(2-j)} & \text{if } j \text{ is odd;} \end{cases}$$

Using $\varphi_j(s) = \widetilde{\varphi}_j(z)$ with $z = 2\frac{s-a_1}{a_p-a_1} - 1$, a simple change of variable then yields

$$\widetilde{v}_n^h(i) = \beta C_i^n + \beta \sum_{j=1}^{p} c_{ij}^{n+1} \frac{a_p - a_1}{j(2-j)} \mathcal{I}_{\{j \text{ odd}\}}.$$

The dynamic program yielding the option value is then the following:

$$\widetilde{v}_N(i) = v^e(a_i),\ i = 1, ..., p.$$

For $n = N-1, \ldots, 0$

$$w^i_{n+1}(a_k) = \widetilde{v}_{n+1}(a_k)\, f_{in}(a_k),\ i = 1, \ldots, p,\ k = 1, \ldots, p \tag{23.16}$$

$$c^{n+1}_i = \Phi^{-1} w^i_{n+1},\ i = 1, \ldots p \tag{23.17}$$

$$C^n_i = \int_{-\infty}^{a_1} (K-S)\, f_{in}(S) dS,\ i = 1, \ldots, p$$

$$\widetilde{v}^h_n(i) = \beta C^n_i + \beta \sum_{j=1}^{p} c^{n+1}_{ij} \frac{a_p - a_1}{j\,(2-j)} \mathcal{I}_{\{j \text{ odd}\}},\ i = 1, \ldots, p \tag{23.18}$$

$$\widetilde{v}_n(i) = \max\left\{v^e(a_i), \widetilde{v}^h_n(i)\right\},\ i = 1, \ldots, p,$$

where $c^{n+1}_i = \left[c^{n+1}_{ij}\right]$ is the vector of coefficients, $w^i_{n+1} = \left[w^i_{n+1}(a_k)\right]$ and $\Phi = \left[\varphi_j(a_k)\right]$.

The DP algorithm requires $O(N \times p)$ operations for the computation of the parameters C^n_i and the value function, $O(N \times p^2)$ operations for the computation of the function in (23.16) and the holding value in (23.15), and $O(N \times p^2 \log p)$ operations for the computation of the coefficients in (23.17) using Fast Fourier Transform techniques.

Higher precision – or conversely less grid points – can be achieved by selecting the lower bound of the localization interval to coincide with the exercise barrier, at the expense of performing a search for the exercise barrier at each stage. Breton et al. (2010) recently proposed such a hybrid approach to price options in the GARCH framework, using a tridimensional state variable. They interpolate the value function by Chebyshev polynomials in both the asset price and volatility spaces.

23.7 Conclusion

This chapter presented basic interpolation approaches for the approximation of financial derivatives by dynamic programs. Finite element using linear spline interpolation is easy to implement and numerically robust, but requires a relatively large number of interpolation nodes to attain high precision. Spectral interpolation converges exponentially fast, and very good precision can be attained with a relatively small number of interpolation nodes when the function to be approximated is smooth and defined over a bounded domain. However, early exercise opportunities introduce discontinuities in the derivative at the exercise barrier. On the other hand, spectral interpolation using high degree polynomials may cause numerical instability. In this chapter, we propose a novel hybrid approach, allowing to obtain spectral convergence, while avoiding localization errors and numerical instability.

Acknowledgements Research supported by IFM2 and by NSERC (Canada) to the first author, and Spanish MICINN, grant MTM2007-60528 (co-financed by FEDER funds) to the second author.

References

Ben Abdallah, R., Ben-Ameur, H., & Breton, M. (2009). An analysis of the true notional bond system applied to the CBOT T-Bond futures. *Journal of Banking and Finance*, *33*, 534–545.

Ben-Ameur, H., Breton, M., & l'Écuyer, P. (2002). A dynamic programming procedure for pricing American-style Asian options. *Management Science*, *48*, 625–643.

Ben-Ameur, H., Breton, M., & François, P. (2006). A dynamic programming approach to price installment options. *European Journal of Operational Research*, *169*(2), 667–676.

Ben-Ameur, H., Breton, M., Karoui, L., & L'Écuyer, P. (2007). A dynamic programming approach for pricing options embedded in bonds. *Journal of Economic Dynamics and Control*, *31*, 2212–2233.

Ben-Ameur, H., Breton, M., & Martinez, J. (2009). Dynamic programming approach for valuing options in the GARCH model. *Management Science*, *55*(2), 252–266.

Breton, M., & de Frutos, J. (2010). Option pricing under GARCH processes by PDE methods. *Operations Research*, *58*, 1148–1157.

Breton, M., de Frutos, J., & Serghini-Idrissi, S. (2010). Pricing options under GARCH in a dynamic programming spectral approximation framework, GERAD working paper.

Canuto, C., Hussaini, M. Y., Quarteroni, A., & Zang, T. A. (2006). Spectral methods. *Fundamentals in single domains*. Heidelberg: Springer.

Chiarella, C., El-Hassan, N., & Kucera, A. (1999). Evaluation of American option prices in a path integral framework using Fourier-Hermite series expansion. *Journal of Economic Dynamics and Control*, *23*(9–10), 1387–1424.

Chiarella, C., El-Hassan, N., & Kucera, A. (2008a). The evaluation of discrete barrier options in a path integral framework. In E. Kontoghiorghes, B. Rustem & P. Winker (Eds.), *Computational methods in financial engineering: essays in honour of Manfred Gilli* (pp. 117–144). Heidelberg: Springer.

Chiarella, C., Meyer, G., & Ziogas, A. (2008b). Pricing american options under stochastic volatility and jump-diffusion dynamics. In K. Muller & U. Steffens (Eds.), *Die Zukunft der Finanzdienstleistungs-industrie in Deutschland* (pp. 213–236). Germany: Frankfurt School.

Duan, J. C., & Simonato, J. G. (2001). American option pricing under GARCH by a Markov chain approximation. *Journal of Economic Dynamics and Control*, *25*, 1689–1718.

de Frutos, J. (2008). A spectral method for bonds. *Computers and Operations Research*, *35*, 64–75.

Whitt, W. (1978). Approximation of dynamic programs I. *Mathematics of Operations Research*, *3*, 231–243.

Whitt, W. (1979). Approximation of dynamic programs II. *Mathematics of Operations Research*, *4*, 179–185.

Chapter 24
Computational Issues in Stress Testing

Ludger Overbeck

Abstract Stress testing should be an integral part of any risk management approach for financial institutions. It can be basically viewed as an analysis of a portfolio of transaction under severe but still reasonable scenarios. Those scenarios might be based on a sensitivity analysis with respect to the model parameter, like a large shift in spread curves, or an increase in default probabilities. Then the corresponding transaction and portfolios are revalued at those stressed parameters. This does not increase the computational effort compared to the revaluation under the assumed normal statistical scenario. However a second class of stress testing approaches relies on the factor model usually underlying most portfolio risk models. In credit risk this might be an asset-value model or a macro-economic model for the default rates. In market risk the factor model are interest rates, spread indices and equity indices. The stress can then be formulated in terms of severe shocks on those factors. Technically this is based on the restricting the sample space of factors. If one wants now to assess the risk of a portfolio under those factor stress scenarios, again the worst case losses should be considered from this sub-sample. In a plain Monte-Carlo-based sample a huge number of simulations are necessary. In the contributions we will show how this problem is solved with importance sampling techniques. Usually the Monte-Carlo sample of the underlying factors is shifted to the regions of interest, i.e. much more stress scenarios are generated than in the original scenario generation. This is in particular successful for portfolios, like in credit, which are mostly long the risk.

L. Overbeck (✉)
Institute of Mathematics, University of Giessen, Giessen, Germany
e-mail: ludger.overbeck@math.uni-giessen.de

J.-C. Duan et al. (eds.), *Handbook of Computational Finance*, Springer Handbooks of Computational Statistics, DOI 10.1007/978-3-642-17254-0_24,

24.1 Introduction

Since the financial crisis starting in 2007 the risk management community has revisited the concept of stress testing. The importance of stress testing can be seen in particular in the large literatur on stress testing in the regulatory environment (BIS 2000, 2001, 2005b; Blaschke et al. 2001; Cherubini and Della Lunga 1999; Cihak 2004, 2007; DeBandt and Oung 2004; Elsinger et al. 2006; Gray and walsh 2008; Lopez 2005; Peura and Jokivuolle 2004). It is also an integral part of the so-called second Pillar in the new basel accord (BIS 2005a).

Stress testing means to measure the impact of severe changes in the economic and financial environment to the risk of financial institutions. Basically this can be viewed to expose the portfolio of a financial institution to some downturn scenarios. Conceptually it is related to the general theory of coherent risk measures, since those can also be represented as the supremum of the value of financial positions under some generalized scenarios (Artzner et al. 1997, 1999). Mathematically however those scenarios are described by absolutely continuous measures with respect to a single reference measures. Stress tests are in many cases, from this conceptual point of view, usually point (Dirac-) measures on very specific point in the set of future states of the world. Some literatur in that direction can be found in Berkowitz (1999), Kupiec (1998), Schachter (2001) and Longin (2000), where the last two articles use extreme value theory, cf. Embrechts et al. (1997) and McNeil et al. (2005).

Portfolio based stress testing which is the main technique we are going discuss in this article, however, considers more generally new probability measures on the probability space spanned by the risk factors underlying the risk models. Details on this approach can be found in the Sect. 24.2, Stress testing in credit risk, which has grown out of Bonti et al. (2006) and Kalkbrener and Overbeck (2010). Similar portfolio related stress testing is also considered in Breuer et al. (2007), Breuer and Krenn (2000), Kim and Finger (2000), Elsinger et al. (2006), Glassermann and Li (2005) and Cihak (2004). All approaches, as part of risk measurement systems, rely on the same basic structure, how to measure risk: Risks comes from the uncertainty of future states of the world. The best we can get is a distribution of future states. Stress tests are based on some subjective distributional assumptions on those states, sometimes even Dirac, e.g. deterministic, measures. More intuitively, the probability of a stress scenario is ignored or just set to 1. But still a functional relationship between the underlying factors and the impact on the portfolio is necessary as described in the following paragraph.

24.1.1 General Structure of Risk Models

In general risk models consists of two components. Risk in financial institutions results from the uncertainty about the future state of the factors underlying the economy and the financial markets. Hence for each financial position which we identify with its value V^i there is a set of factors f_i and a function V_i such that

$$V^i = V_i(f_i).$$

For the two main risk types, market and credit risk, we will first comment on the risk factors used and then on the valuation topic.

24.1.1.1 Risk Factors

The definition and identification of risk factors is crucial for risk measurement and even more for risk management. In the subprime crisis many risk systems failed since the ignored the subprime delinquency rate as a risk factors. Those risk systems used just the quoted market spread on subprime assets without analyzing the underlying primary risk, default rates. Usually systematic and idiosyncratic (non-systematic) factors are considered. In market risk the factors are usually of systematic character, like interest rates, equity indices, commodity prices. In between systematic and idiosyncratic risk there are spreads and equity time series, since they are influenced by general market movements and by firm specific issues. The most idiosyncratic risk factors are default events or, depending on the model, the idiosyncratic part of the asset-value model or the idiosyncratic part of the spread. In any risk system the states of the world are fully specified by values or changes these risk factors will take. For stress testing purposes some of the interesting states of the world might a priori not be given by values of the risk factors. Additional statistical or modeling must be done to associated a stress scenario, which is for example formulated in terms of GDP or unemployment rate, values of the underlying risk factors. This is a great challenge in stress testing, but not a specific computational issue. For our more computational oriented analysis the large number of risk factors poses the main problem concerning risk factors.

Market Risk

In market risk the identification of risk factors is a very straightforward tasks. Usually one relies on the factors constituting the financial market. The problem there is the large number of factors, since there might be several thousands of them. Below find a table of some (Table 24.1):

Credit Risk

In credit risk modelling the question of risk factors depends usually on the model which is used. An overview on credit risk models can be found in Bluhm et al. (2002). There are different classification schemes. If we consider the one which separates reduced form models from structural models, the consequence for the underlying risk factors are as follows. In reduced form models the defaults rates

Table 24.1 List of risk factors

Factor	Dimensions
Spreads	
Per rating	10
× industrie	50
× region	20
× maturity	5
Equity indices	10
Per industrie	50
× region	20
Volatility matrices	25
Exchange rate	20
Interest rates	
Per currency	20
× maturity	10

are the basic prime variables and therefore the default rates can be viewed as the underlying risk factors. Default rates might be historically calibrated as in Credit Risk+ and Credit Portfolio View or they might be derived from market information, like in spread based models. They later are of course very similar to market risk models. Structural models are conceptually based on asset-value variables.

24.1.1.2 Re-Valuation

The second component of a risk systems models how the new values of the risk factors will impact the value of the transactions in the portfolio of a financial institution. Of course to determine the value of a transaction is in general a very complex problem – even outside the context of stress testing. In addition to this general problem under normal market and economic condition it is of course questionable whether in stress situation the same valuation formulas can be applied. This can be already described in the context of parameter of a valuation function. As an example consider the standard valuation formula for synthetic CDO before the breakout of the crisis. It was the base correlation approach with a constant recovery parameter of 40%. In the crisis however the quoted prices were not compatible with this 40% recovery. One had either to change the model assumption of constant 40% leading to more complex valuation function or assume another fixed recovery of e.g. 10%. However then the valuation with single name Credit Default Swaps would be inconsistent. Another way out of this and this might nowadays – after the crisis – followed by most banks is to replace the simple base correlation by more complex valuation routine, e.g. based on different copulas, since the base correlation uses the simple one factor Gaussian copula. More accurate copulas, like hierachial archimedean copula, are often computational much more involved. This is of course a computational issue, but not really specific to stress testing, since more accurate valuation formulas are also important in non-stressed situations. However in normal economic environment also simple valuation might be sufficient.

Credit Risk

In credit risk the classical revaluation. sometimes called book-value, is still the most widespread valuation approach in the banking book. There are then basically to potential values in the future states of each loan, namely the book value, usually 100, if no default has accured or if default happened the recovery value R of the transaction. In most implementation the recovery R is even a deterministic number and not a random variable. Then the only risk – and hence the only uncertainty – is the default event. However, since in the recent financial crisis many losses of the banks where not actually defaults but revaluations even of their banking book positions – mainly accounting wise displayed as provisions or write-downs, it became obvious, that additional risks should be included in credit risk, namely migration risk and spread risk. Migration risk is defined as the change in value coming from changes in the rating of the counterparty or transaction, i.e. the transaction will have a new value if the underlying counterparty has migrated to a different rating class. Here it is important to observe that the migration of a loan to a different rating has only an effect on valuation if we go beyond default only or more precisely book value method. The valuation of a loan has then to depend on the credit quality of the loan. This can be achieved by a kind of fair-value approach. The most credit like approach would be based on expected loss. To put it simple the loan value is 100-Expected Loss (EL). The EL then depends on the rating an hence there is a different discount for expected loss in each rating class. More wide spread however is the discounting with different spreads for each rating class. Usually this spread might depend also on industry and region. For most counterparty a single name spread curve is probably not available. One therefore restores on generic spread curve for a rating, (industry/region) bucket. In this migration step the spread is fixed and known, i.e. not stochastic. As a last component in credit risk spread volatility is included. This mean after migrating in a new rating class also the spread in this rating class might have changed, due to stochastic fluctuation. Of course, since spread include in addition to the market assessment of "historical or statistical risk" also the risk aversion of the market. The risk aversion measures somehow how much more than the expected loss is necessary to pay in order to convince the investor to take this risk. In reality the investor is not exposed to expected loss but to realized losses which in credit in each single case is far away (positive = no loss, negative = 1-recovery) from expected loss. Another line of arguments would see the spread above the expected loss as the unexpected loss component of the market which also changes with risk aversion. For the purpose of stress testing the assumption how spread, i.e. risk aversion, will change in crisis is very important. Unfortunately not many studies exits around that topic. We will show in the example on portfolio level stress testing in credit risk, how a shift in model implied unexpected loss can be derived. Another important feature of spread risk comes from its bridging function to market risk. Since it describes the market risk aversion it has many components in common with market which we will describe in the next paragraph. Before let us mention the asset-value approach to loan valuation which comprises, default, migration and spread risk in a consistent model (Merton 1974; Gupton et al. 1997; Crosby and Bohn 2003).

Market Risk

There are different way to value a financial product as a function of underlying risk variables. The full pricing approach would be to take the valuation function V to be the market standard or for Sometimes in market risk management systems the Delta or Delta-Gamma approach is applied in which ΔV is approximated by the first and second derivative (∇, Γ) of V with respect to the risk factors

$$\Delta V = V_i(f_i^{\text{new}}) - V_i(f_i) \tag{24.1}$$

$$\sim (f_i^{\text{new}} - f_i)^T \nabla(V_i(f_i)) + 1/2 * (f_i^{\text{new}} - f_i)^T \Gamma(f(i)) f_i^{\text{new}} - f_i \tag{24.2}$$

Of course in the case of largest differences $(f_i^{\text{new}} - f_i)$ and non-linear products this is not a good approximation anymore and we could not use it in stress testing. Therefore we arrive at a second computational issue in the stress testing, namely the efficient and fast calculation of value functions. Usually, for large market moves one has to carry out a full revaluation, which is usually computational very costly. However we do not deal with this in the current overview paper.

24.1.2 Stress Testing

There are several approaches to stress testing. It can be basically viewed as an analysis of a portfolio of transaction under severe but still reasonable scenarios. Those scenarios might be based on a sensitivity analysis with respect to the model parameter, like a large shift in spread curves, or an increase in default probabilities. Then the corresponding transaction and portfolios are revalued at those stressed parameters. This does not increase the computational effort compared to the revaluation under the assumed normal statistical scenario. This stress testing as mentioned above is based on a Dirac-measure and several specification of those point scenarios can be found in BIS (2001) and BIS (2005b). However a second class of stress testing approaches relies on the factor model usually underlying most portfolio risk models. This approach is the main topic discussed in the article. In credit risk this might be an asset-value model or a macro-economic model for the default rates. In market risk the factor model are interest rates, spread indices and equity indices. The stress can then be formulated in terms of severe shocks on those factors. Technically this is based on the restricting the sample space of factors. If one wants now to assess the risk of a portfolio under those factor stress scenarios, again the worst case losses should be considered from this sub-sample. But also the expected loss or the mean of the P&L distribution is already very informative in such severe scenarios.

24.1.3 Calibration of Stress Scenarios

Many stress scenarios – especially in more regulatory driven approaches – are formulated in terms of macroeconomic variables, like GDP. Then a first task in the calibration of stress tests is the translation of those scenarios into scenarios for the risk factors. As it is well known it is difficult to explain market moves by some macro-economic variables, i.e. a regression has often small explanatory power. Attempts like this can be found in Pesaran et al. (2004, 2005, 2006) in the context of asset-value models. This is an important building block in the stress testing approach and requires surely more academic input as well. As a practical way around this weak statistical and econometric link between macro-economic variables and risk factors might be to formulate the stress directly in the risk factors world. The mapping can then be done by a probability mapping. For example, if an increase in oil price by 100% has a probability of 0.1%, then we might translate this event to the 0.1%-qunatile of the first principal component of the risk factors system, since an oil price shock is global, cf. Kalkbrener and Overbeck (2010). From a computational point of view however it does not add to more complexity.

24.1.4 Overview of Paper

As we will concentrated now on portfolio level stress testing we will explain this approach first in more detail in the context of credit risk in the following section. This part is based on the papers (Bonti et al. 2006; Kalkbrener and Overbeck 2010). Then we will discuss some importance sampling issues in this context which are adopted from approaches developed in Kalkbrener et al. (2004), cf. also Kalkbrener et al. (2007) and Egloff et al. (2005). In Sect. 24.3 we discuss some issues related to market risk. First the potential of carry out a similar portfolio level stress testing as in market risk and secondly the use of importance sampling technique for the joint calculation of several (point-measure) stress testing.

24.2 Stress Testing in Credit Risk

To be specific we will first construct the portfolio model which we will then expose to stress.

24.2.1 The Credit Portfolio

For the sake of simplicity we describe the default-only structural model with deterministic exposure-at-default. Extension to incorporate migration risk or volatile exposures and loss given defaults are straight-forward. The credit portfolio P

consists of n loans. With each loan we associate an "Ability-to-Pay" variable $A_i : \mathbb{R}^{m+1} \to \mathbb{R}$, which is a linear combination of the m systematic factors $f_1, \dots, f_m$ and a specific variable z_i:

$$A_i(x_1, \dots, x_m, z_i) := \sum_{j=1}^{m} \phi_{ij} f_j + \sqrt{1 - R_i^2} z_i \tag{24.3}$$

with $0 \le R_i^2 \le 1$ and weight vector $(\phi_{i1}, \dots, \phi_{im})$. The m systematic $f_i, i = 1, \dots, m$ will be entry point for the portfolio level stress testing.

The loan loss $L_i : \mathbb{R}^{m+1} \to \mathbb{R}$ and the portfolio loss function $L : \mathbb{R}^{m+n} \to \mathbb{R}$ are defined by

$$L_i := l_i \cdot \mathbf{1}_{\{A_i \le D_i\}}, \quad L := \sum_{i=1}^{n} L_i,$$

where $0 < l_i$ and $D_i \in \mathbb{R}$ are the exposure-at-default and the default threshold respectively. As probability measure $\mathbb{P}$ on $\mathbb{R}^{m+n}$ we use the product measure

$$\mathbb{P} := N_{\mathbf{0},\Gamma} \times \prod_{i=1}^{n} N_{0,1},$$

where $N_{0,1}$ is the standardized one-dimensional normal distribution and $N_{\mathbf{0},\Gamma}$ the m-dimensional normal distribution with mean $\mathbf{0} = (0, \dots, 0) \in \mathbb{R}^m$ and non-singular covariance matrix $C \in \mathbb{R}^m_m$. Note that each f_i, z_i and A_i is a centered and normally distributed random variable under $\mathbb{P}$. We assume that the weight vector $(\phi_{i1}, \dots, \phi_{im})$ has been normalized in such a way that the variance of A_i is 1. Hence, the default probability p_i of the ith loan equals

$$p_i := \mathbb{P}(A_i \le D_i) = N(D_i),$$

where N denotes the standardized one-dimensional normal distribution function. This relation is used to determine the default threshold from empirical default probabilities.

24.2.2 *Coherent Risk Measurement and Capital Allocation*

The risk characteristics under stress should include at least the Expected Loss and Expected Shortfall both under the stressed measure $\tilde{\mathbb{P}}$ and the unstressed measure $\mathbb{P}$:

$$\tilde{EL} = \tilde{E}[L]$$

The expected shortfall of L at level α is defined by

$$\mathrm{ES}_\alpha(L) := (1-\alpha)^{-1} \int_\alpha^1 \mathrm{VaR}_u(L) du,$$

where the value-at-risk $\mathrm{VaR}_\alpha(L)$ of L at level $\alpha \in (0, 1)$ is simply an α-quantile of L once under $\mathbb{P}$ and then under $\tilde{\mathbb{P}}$. For most practical applications the average of all losses above the α-quantile is a good approximation of $\mathrm{ES}_\alpha(L)$: for $c := \mathrm{VaR}_\alpha(L)$ we have

$$\mathrm{ES}_\alpha(L) \approx \tilde{\mathrm{E}}(L|L > c) = (1-\alpha)^{-1} \int L \cdot \mathbf{1}_{\{L>c\}}\, d\tilde{\mathbb{P}}. \tag{24.4}$$

24.2.3 *Portfolio Level Stress Testing*

Often stress scenarios are formulated in a macro-economic manner. Hence, in order to translate such a given stress scenario into model constraints, a precise meaning has to be given to the systematic factors of the portfolio model. Recall that each ability-to-pay variable

$$A_j = \sum_{i=1}^{m} \phi_{ji} f_i + \sqrt{1 - R_j^2} z_j$$

is a weighted sum of m systematic factors $f_1, \ldots, f_m$ and one specific factor ε_j. The systematic factors often correspond either to geographic regions or industries. The systematic weights ϕ_{ji} are chosen according to the relative importance of the corresponding factors for the given counterparty, e.g. the automobile company BMW might have the following (unscaled) representation:

$$\begin{aligned}\text{BMW assets} &= 0.8 \times \text{German factor} + 0.2 \times \text{US factor}\\ &\quad + 0.9 \times \text{Automotive factor} + 0.1 \times \text{Finance factor}\\ &\quad + \text{BMW's non-systematic risk.}\end{aligned}$$

The specific factor is assumed independent of the systematic factors. Its role is to model the remaining (non-systematic) risk of the counterparty.

The economic interpretation of the systematic factors is essential for implementing stress scenarios in the model. The actual translation of a scenario into model constraints is done in two steps:

1. Identification of the appropriate risk factors based on their economic interpretation
2. Truncation of their distributions by specifying upper bounds that determine the severity of the stress scenario

Using the credit portfolio model introduced in Sect. 24.2.1 as quantitative framework, the specification of the model constraints is formalized as follows. A subset $S \subseteq \{1, \ldots, m\}$ is defined, which identifies the stressed factors f_i, $i \in S$. For each of these factors a cap $C_i \in \mathbb{R}$ is specified. The purpose of the thresholds C_i, $i \in S$, is to restrict the sample space of the model. If $i \notin S$ we set $C_i = \infty$, unrestricted. More formally, the restricted sample space $\tilde{\Omega} \subseteq \Omega$ is defined by

$$\tilde{\Omega} := \{\omega \in \Omega \mid f_i(\omega) \leq C_i \text{ for all } i\}. \tag{24.5}$$

The probability measure describing this stress is then restrictions of the σ-algebra $\mathcal{A}$ and the probability measure $\mathbb{P}$ to $\bar{\Omega}$ are denoted by $\bar{\mathcal{A}}$ and

$$\tilde{\mathbb{P}} = \mathbb{P}[\cdot|\tilde{\Omega}] = \mathbb{P}[\cdot|f < C].$$

24.2.4 *Importance Sampling Unstressed*

The portfolio level stress testing results can be naively obtained from a subsampling approach, i.e. run a huge Monte Carlo simulation for the actual calculation of risk capital at portfolio and transaction level and then select those Monte-Carlo-Scenarios which satisfy the constraint of the main practical problem in applying expected shortfall to realistic credit portfolios is the computation of numerically stable MC estimates. In this section we adapt importance sampling to our credit portfolio model and show that this technique significantly reduces the variance of Monte Carlo simulation. The efficient computation of expected shortfall (24.4) is a challenging task for realistic portfolios, even in the unstressed case. Straightforward Monte Carlo simulation does not work well . As an example, assume that we want to compute expected shortfall with respect to the $\alpha = 99.9\%$ quantile and compute $\nu = 100{,}000$ MC samples $s_1 \geq s_2 \geq \ldots \geq s_\nu$ of the portfolio loss L. Then $\mathrm{ES}_\alpha(L)$ becomes

$$(1-\alpha)^{-1}\mathrm{E}(L \cdot \mathbf{1}_{\{L>c\}}) = (1-\alpha)^{-1} \int L \cdot \mathbf{1}_{\{L>c\}}\, d\mathbb{P} = \sum_{i=1}^{100} s_i/100, \tag{24.6}$$

where $c := \mathrm{VaR}_\alpha(L)$. Since the computation of $\mathrm{ES}_\alpha(L)$ is only based on 100 samples it is subject to large statistical fluctuations numerically unstable. A significantly higher number of samples has to be computed which makes straightforward MC simulation impracticable for large credit portfolios. For this purpose we will present the basic variance reduction idea first for the unstressed probability measure and then later give some ideas how to modify it for stress scenarios.

24.2.4.1 Monte Carlo Simulation Based on Importance Sampling

Importance sampling is a technique for reducing the variance of MC simulations and – as a consequence – the number of samples required for stable results. It has been successfully applied to problems in market risk (Glasserman et al. 1999). In our setting, the integral in (24.6) is replaced by the equivalent integral

$$\int L \cdot \mathbf{1}_{\{L>c\}} \cdot \phi \, d\bar{\mathbb{P}}, \tag{24.7}$$

where $\mathbb{P}$ is continuous with respect to the probability measure $\bar{\mathbb{P}}$ and has density ϕ. The objective is to choose $\bar{\mathbb{P}}$ in such a way that the variance of the Monte-Carlo estimate for the integral (24.7) is minimal under $\bar{\mathbb{P}}$. This MC estimate is

$$\mathrm{ES}_\alpha(L)_{\nu,\bar{\mathbb{P}}} := \frac{1}{\nu}\sum_{i=1}^{\nu} L_{\bar{\mathbb{P}}}(i) \cdot \mathbf{1}_{\{L_{\bar{\mathbb{P}}}(i)>c\}}\phi(i), \tag{24.8}$$

where $L_{\bar{\mathbb{P}}}(i)$ is a realization of the portfolio loss L and $\phi(i)$ a realization of the density ϕ under the probability measure $\bar{\mathbb{P}}$. Under suitable conditions as $\nu \to \infty$, $\mathrm{ES}_\alpha(L)_{\nu,\bar{\mathbb{P}}}$ converges to (24.7) and the sampling error converges as

$$\sqrt{\nu} \cdot (\mathrm{ES}_\alpha(L)_{\nu,\bar{\mathbb{P}}} - \mathrm{ES}_\alpha(L)) \xrightarrow{d} N(0, \sigma_{\mathrm{ES}_\alpha(L)}(\bar{\mathbb{P}})), \tag{24.9}$$

where $\sigma^2_{\mathrm{ES}_\alpha(L)}(\bar{\mathbb{P}})$ is the variance of $L \cdot \mathbf{1}_{\{L>c\}} \cdot \phi$ under $\bar{\mathbb{P}}$, i.e.

$$\sigma^2_{\mathrm{ES}_\alpha(L)}(\bar{\mathbb{P}}) = \int \left(L \cdot \mathbf{1}_{\{L>c\}} \cdot \phi\right)^2 d\bar{\mathbb{P}} - \left(\int L \cdot \mathbf{1}_{\{L>c\}} \cdot d\mathbb{P}\right)^2. \tag{24.10}$$

In the following we restrict the set of probability measures $\bar{\mathbb{P}}$, which we consider to determine a minimum of (24.10): for every $M = (M_1, \ldots, M_m) \in \mathbb{R}^m$ define the probability measure $\mathbb{P}_M$ by

$$\mathbb{P}_M := N_{M,\Gamma} \times \prod_{i=1}^{n} N_{0,1}, \tag{24.11}$$

where $N_{M,\Gamma}$ is the m-dimensional normal distribution with mean M and covariance matrix Γ. In other words, those probability measures are considered which only change the mean of the systematic components $x_1, \ldots, x_m$ in the definition of the "Ability-to-Pay" variables $A_1, \ldots, A_n$. This choice is motivated by the nature of the problem. The MC estimate of integral (24.7) can be improved by increasing the number of scenarios which lead to high portfolio losses, i.e. portfolio losses above threshold c. This can be realized by generating a sufficiently large number of defaults in each sample. Since defaults occur when "Ability-to-Pay" variables fall below default thresholds we can enforce a high number of defaults by adding a negative mean to the systematic components.

Having thus restricted importance sampling to measures of the form (24.11) we consider $\sigma^2_{\mathrm{ES}_\alpha(L)}(\mathbb{P}_M)$ as a function from $\mathbb{R}^m$ to $\mathbb{R}$ and rephrase

The Variance Reduction Problem

Compute a minimum $M = (M_1, \ldots, M_m)$ of the variance

$$\sigma^2_{\mathrm{ES}_\alpha(L)}(\mathbb{P}_M) = \int \left(L \cdot \mathbf{1}_{L>c} \cdot \frac{n_{0,\Gamma}}{n_{M,\Gamma}} \right)^2 d\mathbb{P}_M - \left(\int L \cdot \mathbf{1}_{L>c} \, d\mathbb{P} \right)^2 \quad (24.12)$$

in $\mathbb{R}^m$, where $n_{\mathbf{0},\Gamma}$ and $n_{M,\Gamma}$ denote the probability density functions of $N_{\mathbf{0},\Gamma}$ and $N_{M,\Gamma}$ respectively.

We can formulate the minimization condition as

$$\partial_{M_i} \sigma^2_{\mathrm{ES}_\alpha(L)}(\mathbb{P}_M) = 0, \quad \forall\, i = 1, \ldots, m. \quad (24.13)$$

Using the representation in (24.7) and the specification of the portfolio model this leads to the system of m equations

$$2 \sum_{j=1}^{m} C^{-1}_{ij} M_j$$

$$= -\partial_{M_i} \log \left(\int L(x - M, z)^2 \cdot \mathbf{1}_{\{L(x-M,z)>c\}} \, dN_{\mathbf{0},\Gamma}(x) \prod_{i=1}^{n} dN_{0,1}(z_i) \right).$$

and the explicit representation of the portfolio loss reads

$$L(x, z) = \sum_{i=1}^{n} l_j \cdot \mathbf{1}_{\{N^{-1}(p_i) > \sum_{k=1}^{m} \phi_{ik} x_k + \sqrt{1-R_i^2} z_i\}}. \quad (24.14)$$

For realistic portfolios with thousands of loans this system is analytically and numerically intractable.

24.2.4.2 Approximation by a Homogeneous Portfolio

To progress we therefore approximate the original portfolio P by a homogeneous and infinitely granular portfolio $\bar{P}$. This means that the losses, default probabilities and "Ability-to-Pay" variables of all loans in $\bar{P}$ are identical and that the number of loans n is infinite with fixed total exposure. We emphasize that this approximation technique is only used for determining a mean vector M for importance sampling. The actual calculations of expected shortfall and expected shortfall contributions are based on Monte Carlo simulation of the full portfolio model as specified in Sect. 24.2.1. There is no unique procedure to establish the homogeneous portfolio, which is closest to a given portfolio. We propose the following technique as in Kalkbrener et al. (2004) for determining the parameters of the homogeneous portfolio $\bar{P}$, i.e. exposure-at-default l, default probability p, R^2 and the set of factor weights ρ_j, $(j = 1, \ldots, m)$:

Loss and Default Probability

The homogeneous loss l is the average of the individual losses l_i and the homogeneous default probability p is the exposure-at-default weighted default probability of all loans in the portfolio:

$$l := \frac{\sum_{i=1}^{n} l_i}{n}, \qquad p := \frac{\sum_{i=1}^{n} p_i l_i}{\sum_{i=1}^{n} l_i}.$$

Weight Vector

The homogeneous weight vector is the normalized, weighted sum of the weight vectors of the individual loans: in this paper the positive weights $g_1, \ldots, g_n \in \mathbb{R}$ are given by $g_i := p_i l_i$, i.e. the ith weight equals the ith expected loss, and the homogeneous weight vector $\rho = (\rho_1, \ldots, \rho_m)$ is defined by

$$\rho := \psi/s \quad \text{with} \quad \psi = (\psi_1, \ldots, \psi_m) := \sum_{i=1}^{n} g_i \cdot (\phi_{i1}, \ldots, \phi_{im}).$$

The scaling factor $s \in \mathbb{R}$ is chosen such that

$$R^2 = \sum_{i,j=1}^{m} \rho_i \cdot \rho_j \cdot \mathrm{Cov}(x_i, x_j) \tag{24.15}$$

holds, where R^2 is defined in (24.16).

R^2

The specification of the homogeneous R^2 is based on the condition that the weighted sum of "Ability-to-Pay" covariances is identical in the original and the homogeneous portfolio. More precisely, define

$$R^2 := \frac{\sum_{k,l=1}^{m} \psi_k \psi_l \mathrm{Cov}(x_k, x_l) - \sum_{i=1}^{n} g_i^2 R_i^2}{(\sum_{i=1}^{n} g_i)^2 - \sum_{i=1}^{n} g_i^2} \tag{24.16}$$

and the ith homogeneous "Ability-to-Pay" variable by

$$\bar{A}_i := \sum_{j=1}^{m} \rho_j x_j + \sqrt{1 - R^2} z_i.$$

Proposition 1. *The following equality holds for the weighted sum of "Ability-to-Pay" covariances of the original and the homogeneous portfolio:*

$$\sum_{i,j=1}^{n} g_i g_j \mathrm{Cov}(A_i, A_j) = \sum_{i,j=1}^{n} g_i g_j \mathrm{Cov}(\bar{A}_i, \bar{A}_j). \tag{24.17}$$

24.2.4.3 Analytic Loss Distributions of Infinite Homogeneous Portfolios

In this section we approximate the loss function of the homogeneous portfolio by its infinite limit $n \to \infty$. The approximation technique is based on the law of large numbers and works well for large portfolios as already developed by Vasicek (1991) and used in the Basel II framework (Gordy (2003)).

Proposition 2. *Let the loss function $\bar{L}_i$ of the ith facility in the homogeneous portfolio $\bar{P}$ be defined by*

$$\bar{L}_i := l \cdot \mathbf{1}_{\bar{A}_i \leq N^{-1}(p)}.$$

Then

$$\lim_{n\to\infty} (1/n) \cdot \sum_{i=1}^{n} \bar{L}_i = l \cdot N\left(\frac{N^{-1}(p) - \sum_{j=1}^{m} \rho_j x_j}{\sqrt{1-R^2}}\right)$$

holds almost surely on Ω.

Based on the above result we define the function $L^\infty : \mathbb{R} \to \mathbb{R}$ by

$$L^\infty(x) := n \cdot l \cdot N\left(\frac{N^{-1}(p) - x}{\sqrt{1-R^2}}\right) \tag{24.18}$$

and approximate the portfolio loss function $L(x_1, \ldots, x_m, z_1, \ldots, z_n)$ of the original portfolio P by the loss function

$$L_m^\infty(x_1, \ldots, x_m) := L^\infty\left(\sum_{j=1}^{m} \rho_j x_j\right) \tag{24.19}$$

of the infinite homogeneous portfolio. The threshold $c^\infty := \mathrm{VaR}_\alpha(L_m^\infty)$ is defined as the α-quantile of L_m^∞ with respect to the m-dimensional Gaussian measure $N_{\mathbf{0},\Gamma}$. By approximating the finite inhomogeneous portfolio P by an infinite homogeneous portfolio we have transformed the variance reduction problem (24.12) to

The Variance Reduction Problem for Infinite Homogeneous Portfolios: compute a minimum $M = (M_1, \ldots, M_m)$ of

$$\sigma^2_{\mathrm{ES}_\alpha(L_m^\infty)}(M) = \int \left(L_m^\infty \cdot \mathbf{1}_{L_m^\infty > c^\infty} \cdot \frac{n_{\mathbf{0},\Gamma}}{n_{M,\Gamma}}\right)^2 dN_{M,C} \tag{24.20}$$

in $\mathbb{R}^m$.

Note that we have achieved a significant reduction of complexity: the dimension of the underlying probability space has been reduced from $m + n$ to m and the loss function L_m^∞ is not a large sum but has a concise analytic form. In the next section we will present as in Kalkbrener et al. (2004) simple and efficient algorithm which solves the variance reduction problem for infinite homogeneous portfolios.

24.2.4.4 Optimal Mean for Infinite Homogeneous Portfolios

The computation of the minimum of (24.20) is done in two steps:

One-factor model

Instead of m systematic factors $x_1, \ldots, x_m$ we consider the corresponding one-factor model and compute the minimum $\mu^{(1)} \in \mathbb{R}$ of (24.20) in the case $m = 1$. This $\mu^{(1)}$ is the minimum of

$$\int_{-\infty}^{N^{-1}(1-\alpha)} \frac{(L_1^\infty \cdot n_{0,1})^2}{n_{M,1}} \, dx.$$

Multi-factor model

The one-dimensional minimum $\mu^{(1)}$ can be lifted to the m-dimensional minimum $\mu^{(m)} = (\mu_1^{(m)}, \ldots, \mu_m^{(m)})$ of (24.20) by

$$\mu_i^{(m)} := \frac{\mu^{(1)} \cdot \sum_{j=1}^m \mathrm{Cov}(x_i, x_j) \cdot \rho_j}{\sqrt{R^2}}. \qquad (24.21)$$

24.2.5 *Importance Sampling for Stress Testing*

The procedure above gives as a very successfull approach for a very efficient variance reduction technique based on shifting the underlying factor model. Since the expected shortfall can be viewed as an expected loss under a severe, but very specific scenario, for the portfolio, namely that the loss exceeds a specific quantile, this approach is good starting point for importance sampling techniques in stress testing.

24.2.5.1 Subsampling

As a first remark, one can say that the shifted factor model produces also more losses in those factor scenarios which finally hurts the portfolio most. Hence many stress testing procedures which are also hot spots of the portfolio might be found by the simple subsampling technique. That is choose now those scenarios under the shifted measures, where also the constraint of the stress test are fullfilled. I.e. for each functional T of the loss variable L we compute $\tilde{E}[T(L)]$ by

$$\tilde{E}[T(L)] = E[T(L)|X < C] = (P[X < C])^{-1} E[T(L)1_{\{X<C\}}] \quad (24.22)$$

$$E[T(L)1_{\{X<C\}}] = \int \left(T(L)1_{\{X<C\}} \frac{n_{0,\Gamma}}{n_{M,\Gamma}} \right) d\mathbb{P}_M \quad (24.23)$$

$$\sim \frac{1}{\nu} \sum_{i=1}^{\nu} T(L(x_i, z_i))\mathbf{1}_{x_i<C} \frac{n_{0,\Gamma}}{n_{M,\Gamma}}(x_i), \quad (24.24)$$

where $(x_i, z_i), i = 1, \ldots, \nu$ are k simulations of the factor model and the idiosyncratic asset risk vectors. Γ is the original covariance matrix of the factor model and M is the optimal drift for the Expected Shortfall calculation (unstressed) as above.

Remark

There might be functional T of L which give in combination with the restriction a very fast calculation by the above approach. For example if $T(x) = x$ and we restrict only the most sensitive factor for the portfolio. Like for a bank lending mainly to European customer we only restrict the European factor. For other pairs of functionals T and restriction vector C this might be not efficient. E.g. calculation of the Expected Loss in normal times $C_i = \infty$ all i and $T(x) = x$ importance sampling with the optimal expected shortfall shift is not efficient.

24.2.5.2 Stress Specific Shifts

Lead by the successfull application of the importance sampling scheme to Expected Shortfall calculation and the observation, that expected shortfall can also be viewed as an Expected Loss under a very specific downturn scenario, namely the scenario "loss is larger than quantile", we will now propose some specific importance sampling schemes for stress testing. First the optimisation equation for the general portfolio level stress testing defined by the restriction vector C and a functional T, whose expectation should be related to risk measures is

The Variance Reduction Problem for Stress Testing: compute a minimum $M = (M_1, \ldots, M_m)$ of the variance

$$\sigma_T^2(\mathbb{P}_M) = \int \left(T(L) \cdot \mathbf{1}_{X<C} \cdot \frac{n_{0,\Gamma}}{n_{M,\Gamma}} \right)^2 d\mathbb{P}_M - \left(\int T(L) \cdot \mathbf{1}_{X<C} \, d\mathbb{P} \right)^2 \quad (24.25)$$

Also this minimization problem is not feasible and we propose several approaches to improve the efficiency:

Infinite Granular Approximation

We can proceed as in the section on unstressed importance sampling by taking the infinite granular approximation as in the first part leading to Proposition 1. Then we have to minimize over the drift vector M

$$\int \left(T(L_m^\infty)) \cdot \mathbf{1}_{X<C} \cdot \frac{n_{\mathbf{0},\Gamma}}{n_{M,\Gamma}} \right)^2 dN_{M,\Gamma} \tag{24.26}$$

If we consider now a matrix A with $A \cdot A^T = \Gamma$ we can re-formulate this

$$\int_{\mathbb{R}^m} \left(T(L_m^\infty(Ax+M)) \cdot \mathbf{1}_{Ax+M<C} \cdot \frac{n_{\mathbf{0},\Gamma}}{n_{M,\Gamma}}(Ax+M) \right)^2 \Pi_{i=1}^m n_{0,1}(x_i)dx \tag{24.27}$$

If one wants to solve the normal integral by a Monte-Carlo simulation one just has to generate ν vectors $x^{(j)} = (x_1^{(j)}, \ldots, x_m^{(j)})^T$ of $m-$ independent standard normal random numbers $x_i^{(j)}$ and minimize

$$\sum_{j=1}^{\nu} \left(T(L_m^\infty(Ax^{(j)}+M)) \cdot \mathbf{1}_{Ax^{(j)}+M<C} \cdot \frac{n_{\mathbf{0},\Gamma}}{n_{M,\Gamma}}(Ax+M) \right)^2 . \tag{24.28}$$

This is a feasible optimization problem. Of course a reduction to a one-factor optimization as for the calculation of unstressed expected shortfall is not always straight forward and subject to future research.

Remark

(i) In this approach we have avoided to work with the probability measure $\tilde{\mathbb{P}}$, but did all the analysis under the original $\mathbb{P}$. The reason was that the restricted probability measure is not normal anymore and we expect less analytic tractability of the optimization required in importance sampling approach. In the next section we will therefore replace the restricted probability measure by a shifted one.

(ii) We have so far assumed that the risk characteristic of interest can be written as an expectation of a functional T of L. Also in the unstressed case we have approximated the expected shortfall by $(1-\alpha)^{-1}E[L|L > c]$ with c the somewhere known or approximated quantil of L.

For the derivation of the quantil of the unstressed loss distribution can be also carried out under the transformation to P_M: Generate the Monte-Carlo sample $L(i), i =$

$1, \dots, \nu$ under P_M. Then calculate the sum $\sum_{i=1}^{n}([i])\frac{n_{0,\Gamma}}{n_{M,\Gamma}}(x([i]))$, where $[i]$ is the index of the $[i]$-largest loss, until this sum equals $1-\alpha$. If we denote the index of last summand by n_α then $L(n_\alpha)$ corresponds to the quantile under $\mathbb{P}$. In stress testing we face the additional problem that we want to know such type of risk characteristics, in particular Value-at-Risk and Expected Shortfall, which can not be written as an integral of a function of the loss distribution, under $\tilde{\mathbb{P}}$.

24.2.6 Shifted Factor Model

Let us assume we want to calculate the α- quantile of the loss distribution under the measure $\tilde{\mathbb{P}}$. This means we want to find the smallest $0 \leq y$ such that

$$\alpha \cdot \mathbb{P}[X < C] = \int \mathbf{1}_{X<C}\mathbf{1}_{L<y}d\mathbb{P} \tag{24.29}$$

$$= \int \mathbf{1}_{X<C}\mathbf{1}_{L<y}\frac{n_{0,\Gamma}}{n_{M,\Gamma}}(X)d\mathbb{P}_M \tag{24.30}$$

With the techniques presented so far we can only find a reasonable drift $M = M(y)$ for each y. To find the quantile we therefore restore to a plausible, perhaps not optimal, drift transformation. We set $\tilde{M}_i = C_i$ if $C < \infty$ and 0 else. Another possibility would be to set $\tilde{M}_i = C_i - \kappa \cdot \sqrt{\gamma_{ii}}$ with some multiplier κ for the volatility $\sqrt{\gamma_{ii}}$ of the i-th factor.

Second Drift Transformation

If we are interested – as it was the main motivation for the importance sampling approach of the unstressed model – in the calculation of risk contributions, like contributions to expected shortfall, we can then proceed as before with a second drift transformation. Let us assume we have now calculated the quantile $\tilde{q}$ with the help of $\mathbb{P}_{\tilde{M}}$, suitable $\tilde{M}$. Then we have two possibilities:

- We replace in the unstressed case $\mathbb{P}$ by $\mathbb{P}_{\tilde{M}}$ and c by $\tilde{q}$ and proceed in exactly the same way as in the unstressed case until we derive at the optimal drift for the expected shortfall calculation $\tilde{M}_{opt}$. In ordèr to obtain the stressed expected shortfall we have then to apply the subsampling technique, i.e. we have to generate $f_1, \dots, f_\nu$ Monte-carlo samples of f under $\mathbb{P}_{\tilde{M}_{opt}}$ and sample z_i of the idiosyncratic risk and then calculate

$$((1-\alpha) \cdot P[X < C])^{-1}\frac{1}{\nu}\sum_{i=1}^{\nu} L_j(f_i, z_i)\mathbf{1}_{L>\tilde{q}, f_i<C}\frac{n_{0,\Gamma}}{n_{\tilde{M}_{opt},\Gamma}}(f_i) \tag{24.31}$$

for the risk contribution of counterparty j with L_j the loss function of counterparty j.

- Define in (24.28), (24.27) or (24.26) the transformation $T(l) = \mathbf{1}_{l>\tilde{q}}$ and try to find the optimal drift.

24.2.7 *Reformulation of the Scenarios*

Instead of the implementation of portfolio level stress testing scenarios in terms of the conditional distribution $\tilde{\mathbb{P}}$ the conditional distribution that a given set of factor will stay below some level C_i one can simplify the implementation approach, if the scenarios are specified in terms of the probability that a certain level is not reached. To fix the idea assume that we have a single factor scenario, on factor f_1, of the type f_1 will drop below C_1 with a probability of $s\%$. This gives directly a reasonable new drift $M_1 = C_1 - N_{(0,1)}^{-1}(s\%)$. More generally we now consider stress tests directly expressed in terms of a shifted normal distribution or even a new normal distribution on the factors with density $n_{\tilde{M},\tilde{\Gamma}}$. Then we can proceed exactly as in the unstressed case to obtain the optimal importance sampling shift for each single scenario.

24.2.7.1 Simultaneous Stress Test Implementation

The relative straight forward importance sampling technique for a given stress scenario based on new normal distribution of the factor model might let to the attempt to do a simultaneous optimisation of the drift. For that let us assume we have k new stress scenarios associated with the pairs $(M_i, \Gamma_i), i = 1, \ldots, k$ then we have to find an optimal pair (M, Γ) minimizing

$$\sum_{i=1}^{k} \int \left(T_i(L) \cdot \frac{n_{M_i,\Gamma_i}}{n_{M,\Gamma}} \right)^2 d\mathbb{P}_M \tag{24.32}$$

If this optimial drift transformation leads also to a reasonable sampling variance for the single scenarios, all scenarios can be calculated in one run of the portfolio model, namely under the measure $\mathbb{P}_M$. This would reduce the computation time for multiple portfolio stress tests considerably.

Since in credit risk most portfolio are "long only" most scenarios which will hurt the bank have similar drift. Hence a simultaneous minimization might be efficient. If the risk of the portfolio is more homogenous distributed and up-wards and downwards trends of the underlying factor might both hurt the portfolio, simultaneous minimization might be not efficient. Unfortunately this "dispersion" property of potential risky investments and risky scenarios is one feature of trading books which are more exposed to market risk.

24.3 Stress Testing in Market Risk

As mentioned in the introduction in Market Risk analysis many risk factors have to be considered. As reported in BIS (2001) and BIS (2005b) most stress tests in market risk are actually based in single point measures scenarios. Portfolio dependent stress tests are rarely handled in the literatur. In general the approach should be similar to credit risk. Once a certain intuitive scenario is specified, usually only by the specification of values of a single or only a few risk factors, one considered a kind of truncated or conditional distribution of the factor model taking into account this specification. The factor model is considered under this shifted or constraint distribution and the re-valuation takes place in the same way as in the unstressed model. However as mentioned in the introduction, it might also be sensible to use stressed valuation models instead of the one in normal market situation For example it is well-known that the base correlation model for CDOs did not work in the credit crisis, e.g. Krekel (2008).

A more challenging feature of market risk is the absence of a simple one factor approximation as in credit risk for the importance sampling and the portfolio stress testing approach. Since trading books might be long and short w.r.t. to any type of risk factors, it is usually not straightforward to apply a general importance sampling scheme for market risk models. It depends usually heavily on the specific portfolio which are the factors such that a severe shift in the mean of the factor (the importance sampling shift) will actually lead to large losses. It might e.g. be that a fall in equity indices leads to large losses, but that the corresponding (i.e. correlated) decrease in bond prices might actually give a profit for the financial institutions since the are protection buyer in the corresponding default swap market. This means the will profit from more expensive protection. This means that the optimal shift is actually a multidimensional problem and can not easily attacked. Usually a good starting point of the determination of the mean shift is given by the vector of sensitivities. If $\mathbf{f} = (f_1, \ldots, f_n)$ represents the vector of risk factors then the sensitivities are given by the vector

$$\left(\frac{\partial V_p}{\partial f_1}(\mathbf{f}), \ldots, \frac{\partial V_p}{\partial f_n}(\mathbf{f}) \right).$$

The single factor with weights proportional to this sensitivities might be a good starting point to find reasonable stress tests for the traded portfolio and as a consequence also for the implementation of the importance sampling scheme. A main disadvantage of the sensitivity based approach is the omission of non-linearity. An improvement would be to consider the Delta-Gamma-Approach in valuation in which the second derivatives

$$\Gamma_{ij} = \frac{\partial^2 V_p}{\partial f_i \partial f_j}$$

are also considered. Some ideas for applying importance sampling techniques in the unstressed case can be found in Chap. 9 of Glasserman (2004).

Stress Tests by Restriction

Now we consider the subsampling based stress tests in analogy with Sect. 24.2.5.1. If we define the stressed probability distribution again by the conditional distribution that the factors are restricted to a multivariate interval $[\mathbf{x}, \mathbf{y}] = [x_1, y_1] \times \cdots \times [x_m, y_m]$ where m is the number of risk factors. Then we want to minimize the following expression with respect to the new drift $\tilde{M}$.

$$\int \left(V_p(f) \cdot \frac{n_{M,\Sigma}}{n_{\tilde{M},\Sigma}} \right)^2 \mathbf{1}_{f \in [\mathbf{x},\mathbf{y}]} n_{\tilde{M},\Sigma}(df). \tag{24.33}$$

Here we assume that the factors have a joint normal distribution, Σ is the covariance matrix of the factor vector f, M is the unstressed mean of the factor distribution, and $V_p(f)$ is the valuation function, more precisely the change in portfolio value when f is the change in factor values. In general this problem seems to be hard to solve. We therefore recommend in a first step to formulate the stress scenarios in terms of a new shifted distribution of the factor which we then denote by $\tilde{\mathbb{P}}$. In order then to calculate risk characteristics T of the portfolio value V_p we proceed as in any standard, non-stressed, Monte-Carlo technique. Again, some procedures are described in Glasserman (2004).

Revaluation Under Stress

Let us go back to the frequently used Dirac- measures as a stress scenario, but assume now that we have to calculate the portfolio for a series of such point measures. From a valuation point of view V_i the value of a transaction i is the expected discounted cashflows $V_i = E_Q[C_i]$, $C_i =$ discounted cashflows, under the "risk neutral" measure Q. The measure Q is always parameterised by the current values of the risk factors, meaning $Q = Q(f)$. Each new value will give a new valuation measure Q. This means, identifying each scenario with a new value of f we have to carry out many revaluation

$$V_i(f_j) = E_{Q(f_j}[C_i], j = 1, .., K,$$

where K is the number of scenarios under consideration. In many cases we might have the situation where the $Q(f_j)$ have densities with respect to $Q(f_0)$ the current pricing measure. For example if we assume that all cashflows are again functions of a multivariate normal distribution driven by the factors. Then we can generate ν samples of $C_i(\omega_k), k = 1, \ldots, \nu$ under the measure $Q(f_0)$ and the also values of the density $dQ(f_j)/dQ(f_0)(\omega_k)$ and obtain the an estimation of $V_i(f_j)$ by

$$\frac{1}{\nu} \sum_{i=1}^{\nu} C_i(\omega_k) dQ(f_j)/dQ(f_0)(\omega_k).$$

If the computation of the density is considerably faster than a re-run of the scenario generation of ω_k under all the different measure, this procedure might improve the calculation involved with stress testing in market risk. This might in particular helpful if there is no analytic form for V_i and therefore if the calculation of the value V_i has to use Monte-Carlo-Simulation anyway.

24.4 Summary

For credit risk we gave a detailed survey of potential combination of well developed importance sampling techniques with some portfolio level stress testing as in the papers (Kalkbrener and Overbeck 2010; Bonti et al. 2006) about stress testing and (Kalkbrener et al. 2004) about importance sampling. For market risk we gave a short overview of potential computational techniques – in particular importance sampling – which might be possible to apply in several levels of stress testing, like portfolio level stress testing and revaluation under specific single scenarios.

References

Artzner, P., Delbaen, F., Eber, J.-M., & Heath, D. (1997). Thinking coherently. *RISK*, *10*, 68–71.

Artzner, P., Delbaen, F., Eber, J.-M., & Heath, D. (1999). Coherent measures of risk. *Mathematical Finance*, *9*, 203–228.

Berkowitz, J. (1999). A coherent framework for stress-testing. *Journal of Risk*, *2*(2), 1–11.

BIS (2000). Stress testing in credit portfolio models: Current practice and aggregation issues. Committee on the global financial system, Basel.

BIS (2001). A survey of stress tests and current practice at major financial institutions. Committee on the global financial system, Basel.

BIS (2005a). International convergence of capital measurement and capital standards. A revised framework. Basel Committee on Banking Supervision, Basel.

BIS (2005b). Stress testing at major financial institutions: Survey results and practice. Committee on the Global Financial System, Basel.

Blaschke, W., Jones, M. T., Majnoni, G., & Martinez Peria, S. (2001). Stress testing of financial systems: An overview of issues, methodologies, and FSAP experiences. IMF Working Paper, International Monetary Fund, Washington DC.

Bluhm, C., Overbeck, L., & Wagner, C. K. J.(2002). An introduction to credit risk modeling. *Financial mathematics series*. London: Chapman & Hall.

Bonti, G., Kalkbrener, M., Lotz, C., & Stahl, G. (2006). Credit risk concentrations under stress. *Journal of Credit Risk*, *2*(3), 115–136.

Breuer, T. & Krenn, G. (2000). Identifying stress test scenarios. Working Paper, Fachhochschule Vorarlberg, Dornbirn.

Breuer, T., Jandaika, M., Rheinberger, K., & Summer, M. (2007). Macro stress and worst case analysis of loan portfolios. Working Paper, Fachhochschule Vorarlberg, Dornbirn.

Cherubini, U. & Della Lunga, G. (1999). Stress testing techniques and value at risk measures: A unified approach. Working Paper, Banca Commerciale Italiana, Milano.

Cihak, M. (2004). Stress testing: A review of key concepts. Czech National Bank Research Policy Note 2/2004.

Cihak, M. (2007). Introduction to applied stress testing. IMF Working Paper No 07/59, International Monetary Fund, Washington DC.

Crosby, P. J. & Bohn, J. R. (2003). Modeling default risk. Manuscript, Moody's KMV LLC. http://www.moodyskmv.com/research/whitepaper/ModelingDefaultRisk.pdf.

DeBandt, O., & Oung, V. (2004). Assessment of stress tests conducted on the french banking system. Banque de France, Financial Stability Review No 5, November 2004.

Egloff, D., Leippold, M., Jöhri, S., & Dalbert, C. (2005). Optimal importance sampling for credit portfolios with stochastic approximations. Working paper, Zürcher Kantonalbank, Zurich.

Elsinger, H., Lehar, A., & Summer, M. (2006). Using market information for banking system risk assessment. *International Journal of Central Banking*, *2*(1), 137–165.

Embrechts, P., Klüppelberg, C., & Mikosch, T. (1997). *Modelling extremal events*. Berlin: Springer.

Glasserman, P. (2004). *Monte Carlo methods in financial engineering*. New York: Springer.

Glasserman, P. & Li, J. (2005). Importance sampling for portfolio credit risk. *Management Science*, *51*, 1643–1656.

Gordy, M. B. (2003). A risk-factor model foundation for ratings-based bank capital rules. *Journal of Financial Intermediation*, *12*(3), 199–232.

Gray, D. & Walsh, J. P. (2008). Factor model for stress-testing with a contingent claims model of the chilean banking system. IMF Working Paper, International Monetary Fund, Washington DC.

Gupton, G. M., Finger, C. C., & Bhatia, M. (1997). CreditMetrics – Technical document. J. P. Morgan.

Kalkbrener, M. & Overbeck, L. (2010). Stress testing in credit portfolio models.

Kalkbrener, M., Lotter, H., & Overbeck, L. (2004). Sensible and efficient capital allocation for credit portfolios. *RISK*, 17 (2004), 19–24.

Kalkbrener, M., Kennedy, A., & Popp, M. (2007). Efficient calculation of expected shortfall contributions on large credit portfolios. *Journal of Computational Finance*, *11*, 1–33.

Kim, J. & Finger, C. C. (2000). A stress test to incorporate correlation breakdown. *Journal of Risk* *2*(3): 5–19.

Krekel, M. (2008). Pricing distressed CDOs with base correlation and stochastic recovery. HypoVereinsbank – Quantitative research.

Kupiec, P. (1998). Stress testing in a value at risk framework. *Journal of Derivatives*, *24*, 7–24.

Longin, F. (2000). From value at risk to stress testing: The extreme value approach. *Journal of Banking and Finance*, *24*, 1097–1130.

Lopez, J. A. (2005). Stress tests: Useful complements to financial risk models. FRBSF Economic Letter 2005–14, Federal Reserve Bank of San Francisco.

McNeil, A.J, Frey, R., & Embrechts, P. (2005). *Quantitative risk management*. Princeton: Princeton University Press.

Merton, R. C.(1974). On the pricing of corporate debt: The risk structure of interest rates. *Journal of Finance*, *29*, 449–470.

Pesaran, M. H., Schuermann, T., & Weiner, S. M. (2004). Modeling regional interdependencies using a global error-correcting macroeconomic model. *Journal of Business and Economic Statistics*, *22*, 129–162.

Pesaran, M. H., Schuermann, T., & Treutler, B. J. (2005). Global business cycles and credit risks. NBER Working Paper #11493.

Pesaran, M. H., Schuermann, T., Treutler, B. J., & Weiner, S. M. (2006). Macroeconomic dynamics and credit risk: A global perspective. *Journal of Money, Credit and Banking*, *38*(5), 1211–1262.

Peura, S. & Jokivuolle, E. (2004). Simulation-based stress testing of banks' regulatory capital adequacy. *Journal of Banking and Finance*, *28*, 1801–1824.

Schachter, B. (2001). How well can stress tests complement VaR. Derivatives risk management service (January 2001).

Vasicek, O. A. (1991). The loan loss distribution. Technical report, KMV Corporation.

Chapter 25
Portfolio Optimization

Jérôme Detemple and Marcel Rindisbacher

Abstract This paper reviews a Monte Carlo method for consumption-portfolio decision problems in models with complete markets and diffusion processes. It starts with a review of various characterizations of optimal policies. It then focuses on characterizations amenable to simulation and discusses the Monte Carlo Malliavin Derivative Method (MCMD). Various aspects of the method are examined. Numerical schemes for solutions of SDEs are reviewed and compared. An error analysis is carried out. Explicit formulas for convergence rates and asymptotic error distributions are given. An illustration for HARA utility and multiple risky assets is provided.

25.1 Introduction

A question of long-standing interest in finance pertains to the optimal allocation of funds among various financial assets available, in order to sustain lifetime consumption and bequest. The answer to this question is important for practical purposes, both from an institutional and an individual point of view. Mutual funds, pension funds, hedge funds and other institutions managing large portfolios are routinely confronted with this type of decision. Individuals planning for retirement are also concerned about the implications of their choices. Quantitative portfolio models help to address various issues of relevance to the parties involved.

Mean-variance analysis, introduced by Markowitz (1952), has long been a popular approach to determine the structure and composition of an optimal portfolio. This type of analysis, unfortunately, suffers from several shortcomings. It suggests, in particular, optimal portfolios, that are independent of an investor's wealth

J. Detemple (✉) · M. Rindisbacher
Boston University School of Management, Boston University, 595 Commonwealth Avenue, Boston, MA 02215, USA
e-mail: detemple@bu.edu; rindisbm@bu.edu

J.-C. Duan et al. (eds.), *Handbook of Computational Finance*, Springer Handbooks of Computational Statistics, DOI 10.1007/978-3-642-17254-0_25,

and horizon. A rigorous dynamic analysis of the consumption-portfolio choice problem, as originally carried out by Merton (1969, 1971), reveals some of the missing ingredients. It shows that optimal portfolios should include, in addition to mean-variance terms, dynamic hedging components designed to insure against fluctuations in the opportunity set. Merton's analysis highlights the restrictive nature of mean-variance portfolios. Policies of this type are only optimal under extreme circumstances, namely for investors with logarithmic utility (who display myopic behavior) or when opportunity sets are deterministic (means and variances of asset returns do not vary stochastically). It also shows that dynamic hedging terms depend on an investor's horizon and modulate the portfolio composition as the individual ages.

Merton's portfolio formula is based on a partial differential equation (PDE) characterization of the value function associated with the consumption-portfolio choice problem. This type of characterization, while leading to interesting economic insights, presents challenges for implementation. PDEs are indeed notoriously difficult (if not impossible) to solve numerically in the case of high-dimensional problems. This precludes implementations for large scale investment models with many assets and state variables, and for investors with wealth-dependent relative risk aversion (Brennan et al. (1997) provide numerical results for a class of low dimentional problems when utilities are constant relative risk averse).

An alternative characterization of optimal portfolios is obtained by using probabilistic concepts and methods, introduced with the advent of the martingale approach. Major contributions, by Pliska (1986), Karatzas et al. (1987) and Cox and Huang (1989), lead to the identification of explicit solutions for optimal consumption and bequest. Optimal portfolio formulas are derived by Ocone and Karatzas (1991) for Ito processes and Detemple et al. (2003) for diffusions. These formulas take the form of conditional expectations of random variables that are explicitly identified and involve auxiliary factors solving stochastic differential equations (SDEs). For implementation, Monte Carlo simulation is naturally suggested by the structure of these expressions.

This paper reviews the different characterizations of optimal consumption-portfolio policies derived in the literature. It then focuses more specifically on the formulas that can be implemented by Monte Carlo simulation. The particular approach to optimal portfolios which is highlighted is the *Monte Carlo Malliavin Derivatives* method (MCMD). Various aspects of MCMD are discussed. Numerical schemes for simulation of SDEs are reviewed and compared. An extensive asymptotic error analysis is also provided. Convergence rates and asymptotic distributions are reviewed. An example illustrating the power and flexibility of the MCMD method is presented. Finally, the paper briefly discusses alternative simulation-based approaches that have been proposed in the literature.

Section 25.2 presents the elements of the consumption-portfolio choice problem. Section 25.3 describes optimal policies that are obtained using various approaches to the problem. A Monte Carlo method for the computation of optimal policies is reviewed in Sect. 25.4. Asymptotic properties of discretization errors and of MCMD portfolio estimators are described. An illustrative example appears in Sect. 25.5.

Alternative simulation approaches for optimal portfolio choice are briefly described in Sect. 25.6.

25.2 The Consumption-Portfolio Choice Problem

The canonical continuous time consumption-portfolio choice model was introduced by Merton (1969, 1971). In his framework, uncertainty is generated by a d-dimensional Brownian motion W and prices/state variables follow a vector diffusion process. The investor has a finite horizon $[0, T]$. The presentation, throughout this review, focuses on the special case of complete financial markets.

25.2.1 *Financial Market*

The financial market consists of d risky assets and a riskless asset. The riskless asset is a money market account that pays interest at the rate $r(t, Y_t)$, where Y is a d_y-dimensional vector of state variables. Risky assets are dividend-paying stocks, with returns evolving according to

$$\begin{cases} dR_t = (r(t,Y_t)\mathbf{1} - \delta(t,Y_t))\,dt + \sigma(t,Y_t)(\theta(t,Y_t)\,dt + dW_t), & S_0 \text{ given} \\ dY_t = \mu^Y(t,Y_t)\,dt + \sigma^Y(t,Y_t)\,dW_t, & Y_0 \text{ given.} \end{cases} \tag{25.1}$$

The vector R is the $d \times 1$ vector of cumulative stock returns, $\mathbf{1} \equiv (1, ..., 1)'$ is the $d \times 1$ vector of ones, $\delta(t, Y_t)$ is the $d \times 1$ vector of dividend yields and $\sigma(t, Y_t)$ the $d \times d$ matrix of return volatility coefficients. The volatility matrix is assumed to be invertible, ensuring that all risks are hedgeable (the market is complete). The quantity $\theta(t, Y_t)$ is the *market price of Brownian motion risk*, given by $\theta(t, Y_t) \equiv \sigma(t, Y_t)^{-1}(\mu(t, Y_t) - r(t, Y_t)\mathbf{1})$ where $\mu(t, Y_t)$ is the vector of instantaneous expected stock returns. All the coefficients of the return process depend the vector of state variables Y, that satisfies the stochastic differential equation described on the second line of (25.1). The coefficients of this equation, $\mu^Y(t, Y_t), \sigma^Y(t, Y_t)$, are assumed to satisfy standard conditions for the existence of a unique strong solution (see Karatzas and Shreve 1991, p. 338).

The state price density (SPD) implied by the return process (25.1) is

$$\xi_t = \exp\left(-\int_0^t r(s,Y_s)\,ds - \int_0^t \theta(s,Y_s)'\,dW_s - \frac{1}{2}\int_0^t \theta(s,Y_s)'\theta(s,Y_s)\,ds\right). \tag{25.2}$$

The SPD ξ_t represents the stochastic discount factor that can be used for valuation at date 0 of cash flows received at the future date t.

The conditional state price density (CSPD) is defined as $\xi_{t,v} \equiv \xi_v/\xi_t$. It represents the stochastic discount factor for valuation at t of random cash flows received at $v \geq t$.

25.2.2 Choices and Preferences

An investor operating in the market above will consume, invest and leave a bequest at the terminal date. A consumption policy c is a nonnegative stochastic process, adapted to the Brownian filtration. A bequest policy X_T is a measurable nonnegative random variable at the terminal date. A portfolio policy π is a d-dimensional adapted stochastic process, representing the fractions of wealth invested in the risky stocks. Portfolio components are allowed to take negative values (short sales are permitted).

A consumption-bequest-portfolio policy (c, X, π) generates the wealth process X given by

$$dX_t = (X_t r(t, Y_t) - c_t)\, dt + X_t \pi_t' \sigma(t, Y_t)(\theta(t, Y_t)\, dt + dW_t) \qquad (25.3)$$

subject to the initial condition $X_0 = x$, where x is initial wealth.

Investor preferences are defined over consumption-bequest policies. Preferences are assumed to have the von Neumann-Morgenstern (expected utility) representation

$$\mathbf{E}\left[\int_0^T u(c_v, v)\, dv + U(X_T, T)\right], \qquad (25.4)$$

where $u(c_v, v)$ is the instantaneous utility of consumption at date v and $U(X_T, T)$ is the utility of terminal bequest. Utility functions $u : [A_u, \infty) \times [0, T] \to \mathbb{R}$ and $U : [A_U, \infty) \to \mathbb{R}$, are assumed to be twice continuously differentiable, strictly increasing and strictly concave. Marginal utilities are zero at infinity. They are assumed to be infinite at A_u, A_U. If $A_u, A_U > 0$, the utility functions are extended over the entire positive domain by setting $u(c, v) = -\infty$, $U(X, T) = -\infty$ for $c \in [0, A_u)$, $X \in [0, A_U)$.

A standard example of utility function is the Hyperbolic Absolute Risk Aversion (HARA) specification

$$u(c, t) = \frac{1}{1 - R}(c - A_u)^{1-R},$$

where $R > 0$ and A_u is a constant (A_u can be positive of negative).

The inverses $I : \mathbb{R}_+ \times [0, T] \to [A_u, \infty)$ and $J : \mathbb{R}_+ \to [A_U, \infty)$ of the marginal utility functions $u'(c, t)$ and $U'(X, T)$ play a fundamental role. Given the assumptions above, these inverses exist and are unique. They are also strictly decreasing with limiting values $\lim_{y\to 0} I(y, t) = \lim_{y\to 0} J(y, T) = \infty$ and $\lim_{y\to\infty} I(y, t) = A_u$,$\lim_{y\to\infty} J(y, T) = A_U$.

Throughout the paper it will be assumed that initial wealth is sufficient to finance the minimum consumption level. This condition is $x \geq \mathbf{E}\left[\int_0^T \xi_v A_u^+ dv + \xi_T A_U^+\right]$, where $A^+ \equiv \max(0, A)$.

25.2.3 The Dynamic Choice Problem

The investor maximizes preferences over consumption, bequest and portfolio policies. The *dynamic consumption-portfolio choice* problem is

$$\max_{(c,X_T,\pi)} \mathbf{E}\left[\int_0^T u(c_v, v)dv + U(X_T, T)\right] \tag{25.5}$$

subject to the constraints

$$dX_t = (X_t r(t, Y_t) - c_t)\,dt + X_t \pi_t' \sigma(t, Y_t)(\theta(t, Y_t)\,dt + dW_t);\ X_0 = x \tag{25.6}$$

$$c_t \geq 0, \quad X_t \geq 0 \tag{25.7}$$

for all $t \in [0, T]$. Equation (25.6) is the dynamic evolution of wealth. The first constraint in (25.7) is the nonnegativity restriction on consumption. The second (25.7) is a no-default condition, imposed to ensure that wealth is nonnegative at all times, including the bequest time.

25.2.4 The Static Choice Problem

Pliska (1986), Karatzas et al. (1987) and Cox and Huang (1989) show that the dynamic problem is equivalent to the following *static consumption-portfolio choice* problem

$$\max_{(c,X)} \mathbf{E}\left[\int_0^T u(c_v, v)\,dv + U(X_T, T)\right] \tag{25.8}$$

subject to the static budget constraint

$$\mathbf{E}\left[\int_0^T \xi_s c_s + \xi_T X_T\right] \leq x \tag{25.9}$$

and the nonnegativity constraints $c \geq 0$ and $X_T \geq 0$. Equation (25.9) is a budget constraint. It mandates that the present value of consumption and bequest be less than or equal to initial wealth. The objective in (25.8) is to maximize lifetime utility with respect to consumption and bequest, which satisfy the usual nonnegativity restrictions.

In the static problem (25.8) and (25.9) there is no reference to the portfolio, which is treated as a residual decision. The reason for this is because of market completeness. Once a consumption-bequest policy has been identified, there exists a replicating portfolio that finances it.

25.3 Optimal Policies

There are two approaches for characterizing optimal policies. The first is the one followed by Merton (1969, 1971). If focuses on the dynamic problem and relies on dynamic programming principles for resolution. Optimal policies are characterized in terms of a value function solving a nonlinear Partial Differential Equation (PDE). The second approach was introduced by Pliska (1986), Karatzas et al. (1987) and Cox and Huang (1989), and is based on probabilistic methods. This approach, often called the *Martingale approach*, identifies optimal consumption and bequest as the explicit solutions of the static optimization problem. The optimal portfolio is the replicating strategy that synthesizes consumption and bequest.

25.3.1 A PDE Characterization

Merton's classic approach to the dynamic consumption-portfolio problem is based on dynamic programming. Optimal policies are expressed in terms of the derivatives of the value function $V(t, X_t, Y_t)$, associated with the optimization problem (25.5)–(25.7).

Theorem 1 (Merton 1971). *Optimal consumption and bequest are*

$$c_t^* = I\left(V_x\left(t, X_t^*, Y_t\right), t\right)\right)^+, \quad X_T^* = J\left(V_x\left(T, X_T^*, Y_T\right), T\right)^+, \tag{25.10}$$

where $x^+ \equiv \max(0, x)$. *The optimal portfolio has two components, a mean-variance term* π_t^m *and a dynamic hedging term* π_t^y. *Thus,* $X_t^* \pi_t^* = X_t^* \pi_t^m + X_t^* \pi_t^y$ *with*

$$X_t^* \pi_t^m = -\frac{V_x\left(t, X_t^*, Y_t\right)}{V_{xx}\left(t, X_t^*, Y_t\right)} \left(\sigma(t, Y_t)'\right)^{-1} \theta(t, Y_t) \tag{25.11}$$

$$X_t^* \pi_t^y = -\left(\sigma(t, Y_t)'\right)^{-1} \sigma^Y(t, Y_t)' \frac{V_{yx}\left(t, X_t^*, Y_t\right)}{V_{xx}\left(t, X_t^*, Y_t\right)}, \tag{25.12}$$

where V_x, V_{xx}, V_{yx} *are partial first and second derivatives of the value function. The value function solves the partial differential equation*

$$\begin{aligned} 0 = {} & u\left(I\left(V_x, t\right)^+, t\right) + V_x\left(r(t, Y_t) X_t - I\left(V_x, t\right)^+\right) + V_t + V_y \mu^Y(t, Y_t) \\ & + \frac{1}{2} trace\left\{V_{yy} \sigma^Y(t, Y_t)\left(\sigma^Y(t, Y_t)\right)'\right\} - \frac{1}{2} V_{xx} \left\|\psi\left(t, X_t^*, Y_t\right)\right\|^2 \end{aligned} \tag{25.13}$$

with

$$\psi\left(t, X_t^*, Y_t\right) \equiv -\left(\left(\frac{V_x\left(t, X_t^*, Y_t\right)}{V_{xx}\left(t, X_t^*, Y_t\right)}\right) \theta(t, Y_t) + \sigma^Y(t, Y_t)' \left(\frac{V_{yx}\left(t, X_t^*, Y_t\right)}{V_{xx}\left(t, X_t^*, Y_t\right)}\right)\right) \tag{25.14}$$

and subject to the boundary conditions $V(T,x,y) = U(x,T)$ *and* $V(t,0,y) = \int_t^T u(0,v)\,dv + U(0,T)$.

Consumption and bequest depend both on the marginal value of wealth $V_x\left(t, X_t^*, Y_t\right)$, i.e., the derivative of the value function with respect to wealth. The latter measures the opportunity cost of wealth. In states where this marginal value is high, the cost of consumption is high. It is then optimal to consume little.

As stated in the proposition, the optimal portfolio has two components. The first, π_t^m, is a static mean-variance term capturing the desire to diversify. It depends on the instantaneous risk-return trade-off, reflected in $\left(\sigma(t,Y_t)'\right)^{-1}\theta(t,Y_t)$. The second, π_t^y, is the dynamic hedging component first identified by Merton (1971). It reflects the investor's desire to protect against stochastic fluctuations in the opportunity set, i.e., fluctuations in $(r(t,Y_t), \theta(t,Y_t))$. If the market price of risk and the interest rate are independent of the state variables Y, the value function solving (25.13) is also independent of Y. The dynamic hedging component vanishes.

Another situation in which the dynamic hedging term vanishes is when utility functions are logarithmic (unit relative risk aversion). In this instance, the solution of (25.13) is additively separable ($V_x\left(t, X_t^*, Y_t\right) = V_x\left(t, X_t^*\right) + G(t,Y_t)$) even if $(r(t,Y_t), \theta(t,Y_t))$ are stochastic. The log investor displays myopia, in the sense of not caring about stochastic variations in the state variables. These variations determine future market prices of risk and interest rates.

In rare instances the PDE (25.13) can be solved explicitly. In most cases, numerical resolution methods are required. Lattice-based methods have been extensively used for that purpose. Unfortunately, lattice methods suffer from a curse of dimensionality (the computational complexity grows exponentially with the number of state variables). As a result, only low-dimensional problems can be tackled with this type of numerical approach.

25.3.2 A Probabilistic Representation for Complete Markets

Pliska (1986), Karatzas et al. (1987) and Cox and Huang (1989) approach the problem from the static point of view. Optimal consumption and bequest policies for general utilities are derived in the latter two references. Formulas for the financing portfolio, in settings with Ito processes, were first derived by Ocone and Karatzas (1991) using the Clark-Ocone formula. Diffusion models were considered in Detemple et al. (2003). The next theorem is a variation of their results which emphasizes the role of risk tolerance.

Theorem 2 (Detemple et al. 2003). *Optimal consumption and bequest are*

$$c_t^* = I\left(y^*\xi_v, v\right)^+, \quad X_T^* = J\left(y^*\xi_T, T\right)^+, \tag{25.15}$$

where ξ *the state price density in (25.2) and* y^* *is the unique solution of the nonlinear equation*

$$\mathbf{E}\left[\int_0^T \xi_v I\left(y^*\xi_v, v\right)^+ dv + \xi_T J\left(y^*\xi_T, T\right)^+\right] = x. \tag{25.16}$$

The optimal portfolio has the decomposition $X_t^*\pi_t^* = X_t^*\pi_t^m + X_t^*\pi_t^h$ *with*

$$X_t^*\pi_t^m = \mathbf{E}_t\left[\int_t^T \xi_{t,v}\Gamma_v^* 1_{\{I_v^*\geq 0\}} dv + \xi_{t,T}\Gamma_T^* 1_{\{J_T^*\geq 0\}}\right]\left(\sigma\left(t, Y_t\right)'\right)^{-1}\theta\left(t, Y_t\right) \tag{25.17}$$

$$\begin{aligned} X_t^*\pi_t^h = &-\left(\sigma\left(t, Y_t\right)'\right)^{-1}\mathbf{E}_t\left[\int_t^T \xi_{t,v}\left(c_v^* - \Gamma_v^*\mathbf{1}_{\{I_v^*\geq 0\}}\right) H_{t,v} dv\right] \\ &-\left(\sigma\left(t, Y_t\right)'\right)^{-1}\mathbf{E}_t\left[\xi_{t,T}\left(X_T^* - \Gamma_T^*\mathbf{1}_{\{J_T^*\geq 0\}}\right) H_{t,T}\right], \end{aligned} \tag{25.18}$$

where $I_v^* \equiv I\left(y^*\xi_v, v\right)$, $J_T^* \equiv J\left(y^*\xi_T, T\right)$ *and* Γ_v^*, Γ_T^* *are the absolute risk tolerance measures* $\Gamma^u(c, v) \equiv -u_x\left(x, v\right)/u_{xx}\left(x, v\right)$ *and* $\Gamma^U\left(x\right) \equiv -U_x\left(x\right)/U_{xx}\left(x\right)$ *evaluated at optimal consumption* $c_v^* = \left(I_v^*\right)^+$ *and bequest* $X_T^* = \left(J_T^*\right)^+$. *Furthermore*

$$H_{t,v}' = \int_t^v \left(\partial r\left(s, Y_s\right) + \theta\left(s, Y_s\right)'\partial\theta\left(s, Y_s\right)\right)\mathcal{D}_t Y_s ds + \int_t^v dW_s'\partial\theta\left(s, Y_s\right)\mathcal{D}_t Y_s, \tag{25.19}$$

where $\mathcal{D}_t Y_s$ *is the Malliavin derivative process that satisfies the linear SDE*

$$d\mathcal{D}_t Y_s = \left[\partial\mu^Y\left(s, Y_s\right) ds + \sum_{j=1}^d \partial\sigma_j^Y\left(s, Y_s\right) dW_s^j\right]\mathcal{D}_t Y_s; \qquad \mathcal{D}_t Y_t = \sigma\left(t, Y_t\right) \tag{25.20}$$

and $\partial r\left(s, Y_s\right), \partial\theta\left(s, Y_s\right), \partial\mu^Y\left(s, Y_s\right), \partial\sigma_j^Y\left(s, Y_s\right)$ *are gradients with respect to* Y.

The probabilistic formulas in Theorem 2 provide further insights about portfolio structure. Expression (25.17) shows that the size of the position in the mean-variance portfolio depends on the cost of optimal risk tolerance. Expression (25.18) shows the determinants of the hedging demand. Note in particular that the Malliavin derivative $\mathcal{D}_t Y_v$ measures the impact of an infinitesimal perturbation of the Brownian motion W_t at time t on the position of the state variable at the future time v. If the investment opportunity set is deterministic, then $\partial r\left(s, Y_s\right) = \partial\theta\left(s, Y_s\right) = H_{t,v} = 0$ and $V_{xy} = 0$. The dynamic hedging demand vanishes. Similarly, if the investor has unit relative risk aversion, then $c_v^* = \Gamma_v^*$, $X_T^* = \Gamma_T^*$ and $V_{xy}\left(t, X_t^*, Y_t\right) = 0$. Moreover, in this case, the cost of optimal risk tolerance becomes $-V_x(t, X_t^*, Y_t)/V_{xx}(t, X_t^*, Y_t) = X_t^*$.

The formulas is Theorem 2 permit implementations based on Monte Carlo simulation. Indeed, the formulas express the portfolio components in terms of expected values of random variables that are completely identified and can be calculated by simulation. In particular, Malliavin derivatives appearing in these expressions solve the linear SDE (25.20) and can be computed by simulation.

The Monte Carlo simulation approach to portfolio choice based on the formulas in Theorem 2 is called the *Monte Carlo Malliavin Derivative* (MCMD) method. MCMD is extremely flexible. It permits implementations for large numbers of assets and state variables. It also permits arbitrary utility functions, up to the regularity conditions imposed. With simulation, the computational complexity grows only linearly with the number of risky assets and state variables.

A comparison of the optimal policies in Theorems 1 and 2 gives $y^* \equiv V_x(0, x, Y_0)$,

$$\frac{V_x(t, X_t^*, Y_t)}{V_x(0, x, Y_0)} = \xi_t \tag{25.21}$$

$$-\frac{V_x(t, X_t^*, Y_t)}{V_{xx}(t, X_t^*, Y_t)} = \mathbf{E}_t\left[\int_t^T \xi_{t,v}\Gamma_v^* 1_{\{I_v^* \geq 0\}} dv + \xi_{t,T}\Gamma_T^* 1_{\{J_T^* \geq 0\}}\right] \tag{25.22}$$

and

$$-\frac{V_{xy}(t, X_t^*, Y_t)}{V_{xx}(t, X_t^*, Y_t)} = -\mathbf{E}_t\left[\int_t^T \xi_{t,v}\left(c_v^* - \Gamma_v^* 1_{\{I_v^* \geq 0\}}\right) H_{t,v} dv\right]$$

$$-\mathbf{E}_t\left[\xi_{t,T}\left(X_T^* - \Gamma_T^* 1_{\{J_T^* \geq 0\}}\right) H_{t,T}\right]. \tag{25.23}$$

Malliavin calculus and the martingale approach identify the probabilistic representations (25.22) and (25.23) for the derivatives of the value function, in the same way as the Feynman-Kac formula represents solutions of PDEs. The Feynman-Kac formula has led to the development of Monte Carlo methods for the computation of derivative securities prices characterized by PDEs. These methods have proven particularly useful in the case of derivatives written on large baskets of financial assets. The probabilistic portfolio formula in Theorem 2 performs a similar role. It connects derivatives of the value function from the dynamic programming approach to conditional expectations of functionals of Brownian motion. With these relations, optimal portfolios from dynamic programming can be calculated using forward Monte Carlo simulation methods.

25.3.3 *Measure Change and Portfolio Representation*

An alternative formula for the optimal portfolio can be derived by using long term bonds as numeraires. Let $B_t^v \equiv \mathbf{E}_t[\xi_{t,v}]$ be the price of a pure discount bond with maturity date v. Expressing the conditional SPD $\xi_{t,v}$ in terms of this bond numeraire defines the density

$$Z_{t,v} \equiv \frac{\xi_{t,v}}{\mathbf{E}_t\left[\xi_{t,v}\right]} = \frac{\xi_{t,v}}{B_t^v}$$
$$= \exp\left(\int_t^v \sigma^z(s,T)'\,dW_s - \frac{1}{2}\int_t^v \sigma^z(s,T)'\sigma^z(s,T)\,ds\right), \quad (25.24)$$

where $\sigma^z(s,v) \equiv \sigma^B(s,v) - \theta_s$ is the volatility of the martingale $\eta_s^z \equiv \mathbf{E}_s\left[Z_{t,v}\right]$ and $\sigma^B(s,v)' \equiv \mathcal{D}_s \log B_s^v$ is the bond return volatility (see Detemple and Rindisbacher 2010). The random variable $Z_{t,v}$ is the density of the forward-v measure. This measure, introduced by Geman (1989) and Jamshidian (1989), permits calculations of present values directly in the bond numeraire. The volatility $\sigma^z(s,v)$ is a $1 \times d$ vector representing the negative of the market price of risk evaluated in the bond numeraire.

Expressing optimal policies in terms of bond numeraires leads to the following formulas

Theorem 3 (Detemple and Rindisbacher 2010). *Optimal consumption and bequest are* $c_v^* = J\left(y^*\xi_t B_t^v Z_{t,v}, v\right)^+$ *and* $X_T^* = I\left(y^*\xi_t B_t^T Z_{t,T}\right)^+$. *Intermediate wealth is*

$$X_t^* = \int_t^T B_t^v \mathbf{E}_t^v\left[c_v^*\right]dv + B_t^T \mathbf{E}_t^T\left[X_T^*\right], \quad (25.25)$$

where $B_t^v = \mathbf{E}_t[\xi_{t,v}]$ *is the price of a pure discount bond with maturity date* $v \in [0,T]$. *Define the random variables* $J_v^* \equiv J\left(y^*\xi_t B_t^v Z_{t,v}, v\right)$ *and* $I_T^* \equiv I\left(y^*\xi_t B_t^T Z_{t,T}\right)$. *The optimal portfolio has three components, a mean-variance term* π_t^m, *a static bond hedge* π_t^b *and a forward density hedge* π_t^z. *It writes* $X_t^*\pi_t^* = X_t^*\pi_t^m + X_t^*\pi_t^b + X_t^*\pi_t^z$, *where*

$$X_t^*\pi_t^m = \int_t^T \mathbf{E}_t^v\left[\Gamma_v^* 1_{\{J_v^* \geq 0\}}\right] B_t^v dv \left(\sigma(t,Y_t)'\right)^{-1}\theta(t,Y_t)$$
$$+\mathbf{E}_t^T\left[\Gamma_T^* 1_{\{I_T^* \geq 0\}}\right] B_t^T \left(\sigma(t,Y_t)'\right)^{-1}\theta(t,Y_t) \quad (25.26)$$

$$X_t^*\pi_t^b = \left(\sigma(t,Y_t)'\right)^{-1}\int_t^T \sigma^B(t,v)\,B_t^v \mathbf{E}_t^v\left[\left(c_v^* - \Gamma_v^*\right)1_{\{J_v^* \geq 0\}}\right]dv$$
$$+\left(\sigma(t,Y_t)'\right)^{-1}\sigma^B(t,T)\,B_t^T \mathbf{E}_t^T\left[\left(X_T^* - \Gamma_T^*\right)1_{\{I_T^* \geq 0\}}\right] \quad (25.27)$$

$$X_t^*\pi_t^z = \left(\sigma(t,Y_t)'\right)^{-1}\left(\int_t^T \mathbf{E}_t^v\left[\left(c_v^* - \Gamma_v^*\right)1_{\{J_v^* \geq 0\}}\mathcal{D}_t \log Z_{t,v}\right]B_t^v dv\right)'$$
$$+\left(\sigma(t,Y_t)'\right)^{-1}\left(\mathbf{E}_t^T\left[\left(X_T^* - \Gamma_T^*\right)1_{\{I_T^* \geq 0\}}\mathcal{D}_t \log Z_{t,T}\right]B_t^T\right)', \quad (25.28)$$

where, for $v \in [t, T]$, $Z_{t,v}$ is the density of the forward v-measure (given by (25.24)). The volatility of $Z_{t,v}$ is $\sigma^z(s, v) \equiv \sigma^B(s, v) - \theta_s$, where $\sigma^B(s, v)' \equiv \mathcal{D}_s \log B_s^v$ is the volatility of the return on the pure discount bond B_s^v. The expectation $\mathbf{E}_t^v[\cdot] \equiv \mathbf{E}_t[Z_{t,v}\cdot]$ is under the forward v-measure, $v \in [t, T]$.

The portfolio formula in Theorem 3 is in the spirit of the Heath-Jarrow-Morton (HJM) term structures models (Heath et al. 1992). To see this connection, let $f_t^v \equiv -\partial_v \log(B_t^v)$ be the continuously compounded forward rate for maturity v. As $B_t^v = \exp\left(-\int_t^v f_t^s ds\right)$ the bond return volatility is

$$\sigma^B(t, v)' = \mathcal{D}_t \log B_t^v = -\int_t^v \mathcal{D}_t f_t^s ds = -\int_t^v \sigma^f(t, s) ds, \tag{25.29}$$

where $\sigma^f(t, s)$ is the volatility of f_t^s. The optimal allocation is then given by (25.26)–(25.28), with (25.29) and the Malliavin derivative

$$\mathcal{D}_t \log Z_{t,v} = \int_t^v \left(dW_s + \left(\theta(s, Y_s) + \int_s^v \sigma^f(s, u) du\right) ds\right)' \times \left(\mathcal{D}_t \theta_s + \int_s^v \mathcal{D}_t \sigma^f(s, u) du\right). \tag{25.30}$$

This shows that the portfolio can be expressed directly in terms of the primitives in HJM model, i.e., the forward rates. Monte Carlo methods for HJM term structure models are therefore easily adapted to solve portfolio choice problems.

Theorem 3 provides further interpretation of the structure of the optimal portfolio. It effectively shows that the dynamic hedging demand decomposes into a static bond hedge and a forward market price of risk ($-\sigma^z$) hedge. If σ^z is deterministic, a three fund separation result holds: optimal portfolios of investors with arbitrary risk aversion are spanned by the money market account, the mean-variance portfolio and a portfolio hedging fluctuations in the term-structure of interest rates. In the case of a pure bequest motive (null instantaneous utility), the term-structure hedge reduces to a single pure discount bond with maturity date matching the investor's horizon.

The formulas for the portfolio components in Theorem 3 involve conditional expectations of random variables that are obtained in explicit form. As before these expressions involve auxiliary factors (Malliavin derivatives) that solve SDEs. The approach for numerical implementation suggested by these formulas is again an MCMD method. Simulation can be carried out under the forward measures.

25.3.4 Examples

If the utility functions display constant relative risk aversion, optimal portfolio weights are wealth-independent.

Corollary 1. *Suppose that the investor exhibits constant relative risk aversion $R_u = R_U = R$ and has subjective discount factor $a_t \equiv \exp(-\beta t)$ where β is constant. The optimal consumption-bequest policy is*

$$c_v^* = \left(\frac{y^* \xi_v}{a_v}\right)^{-1/R} = \left(\frac{y^* B_0^v Z_{0,v}}{a_v}\right)^{-1/R}$$

$$X_T^* = \left(\frac{y^* \xi_T}{a_T}\right)^{-1/R} = \left(\frac{y^* B_0^T Z_{0,T}}{a_T}\right)^{-1/R}.$$

The optimal portfolio is $X_t^* \pi_t^* = X_t^* \left[\pi_t^m + \pi_t^h\right]$, where

$$X_t^* \pi_t^m = \frac{X_t^*}{R} \left(\sigma(t, Y_t)'\right)^{-1} \theta(t, Y_t) \tag{25.31}$$

$$X_t^* \pi_t^h = -X_t^* \rho \left(\sigma(t, Y_t)'\right)^{-1} \frac{\mathbf{E}_t\left[\int_t^T \xi_{t,v}^{\rho} a_{t,v}^{1/R} H_{t,v} dv + \xi_{t,T}^{\rho} a_{t,T}^{1/R} H_{t,T}\right]}{\mathbf{E}_t\left[\int_t^T \xi_{t,v}^{\rho} a_{t,v}^{1/R} dv + \xi_{t,T}^{\rho} a_{t,T}^{1/R}\right]} \tag{25.32}$$

with $\rho = 1 - 1/R$ and $H_{t,v}$ defined in (25.19). Alternatively, using bonds as units of account, $X_t^* \pi_t^* = X_t^* \left[\pi_t^m + \pi_t^b + \pi_t^z\right]$ where

$$\pi_t^m = \frac{1}{R}\left(\int_t^T \Pi_t^v B_t^v dv + \Pi_t^T B_t^T\right) \left(\sigma(t, Y_t)'\right)^{-1} \theta(t, Y_t)$$

$$\pi_t^b = \rho \left(\sigma(t, Y_t)'\right)^{-1} \left(\int_t^T \sigma^B(t, v) B_t^v \Pi_t^v dv + \sigma^B(t, T) B_t^T \Pi_t^T\right)$$

$$\pi_t^z = \rho \left(\sigma(t, Y_t)'\right)^{-1} \left(\int_t^T E_t^v\left[c_v^* \mathcal{D}_t \log Z_{t,v}\right] B_t^v dv + E_t^T\left[X_T^* \mathcal{D}_t \log Z_{t,T}\right] B_t^T\right)',$$

where $\Pi_t^v = E_t^v\left[c_v^*\right]$ (resp. $\Pi_t^T = E_t^T\left[X_T^*\right]$) is the date t cost in the bond numéraire of date v consumption (resp. terminal wealth).

In the case of utilities with hyperbolic absolute risk aversion, optimal policies become

Corollary 2. *Suppose that the investor exhibits hyperbolic absolute risk aversion with utility parameters $R_u = R_U = R$, $A_u = A_U = A$ and subjective discount factor $a_t \equiv \exp(-\beta t)$ where β is constant. The optimal consumption-bequest policy is*

$$c_v^* = \left(\left(\frac{y^* \xi_v}{a_v}\right)^{-1/R} + A\right)^+ = \left(\left(\frac{y^* B_0^v Z_{0,v}}{a_v}\right)^{-1/R} + A\right)^+$$

$$X_T^* = \left(\left(\frac{y^* \xi_T}{a_T}\right)^{-1/R} + A\right)^+ = \left(\left(\frac{y^* B_0^T Z_{0,T}}{a_T}\right)^{-1/R} + A\right)^+ .$$

The portfolio components are given by (25.17) and (25.18) or (25.26)–(25.28) with

$$\Gamma^u(c,v) \equiv -\frac{u_x(c,v)}{u_{xx}(c,v)} = \frac{1}{R}(c - A_u)$$

$$\Gamma^U(X) \equiv -\frac{U_x(X)}{U_{xx}(X)} = \frac{1}{R}(X - A) .$$

25.4 Monte Carlo Methods

Optimal portfolio formulas in Theorems 2 and 3 can be calculated by Monte Carlo simulation (MCMD method). Implementation of the MCMD method is straightforward when the transition densities of the random variables appearing in these formulas are known. Conditional expectations are then estimated by averaging over i.i.d. realizations drawn from exact distributions. In this case, there is a unique source of estimation error, the *Monte Carlo error*. Asymptotic convergence properties of the Monte Carlo error are found by a central limit theorem for i.i.d. random variables. Unfortunately, functionals of diffusion processes with explicit transition densities are rare. In order to implement MCMD estimators when transition densities are unknown, the solutions of the relevant SDEs must be calculated by simulation. For this purpose a discretization scheme for the SDEs is needed. Two sources of error will then affect the numerical accuracy of portfolio estimators and determine their convergence properties. The first type of error is the Monte Carlo error associated with the approximation of conditional expectations by sample averages. The second type is the *discretization error* associated with the discretization of the diffusions. Both types of approximation errors must be controlled simultaneously in order to implement an efficient simulation scheme (see Talay and Tubaro 1990; Duffie and Glynn 1995; Bally and Talay 1996a,b).

Errors due to numerical discretization schemes for diffusions are analyzed in Sect. 25.4.1. Asymptotic properties of MCMD portfolio estimators are discussed in Sect. 25.4.2.

25.4.1 Numerical Solutions of SDEs

Several numerical schemes for SDEs are presented and their convergence properties discussed. The simplest scheme, the Euler-Maruyama scheme, is examined first. In general, Euler-Maruyama is less costly from a computational point of view, than some of the alternatives such as the Euler-Doss scheme and the Milhstein scheme. These higher order discretization procedures are discussed second.

Numerical solutions of SDEs rely on a partition of the time interval $\mathcal{P}([0,T]) = \{t_0, \ldots, t_{N-1}\}$. Increments of the Brownian motion, $\Delta W_{t_k} \equiv W_{t_{k+1}} - W_{t_k}$ for $k = 0, \ldots, N-1$, are drawn using pseudo- or quasi Monte Carlo random number generators. The random variables of interest are then obtained as finite-dimensional functionals of these innovations, using a forward simulation of the discretized diffusion.

25.4.1.1 Euler-Maruyama Scheme

Consider the process for state variables (second line of (25.1)). The Euler-Maruyama approximation of the process is given by the solution of the difference equation

$$Y_{t_{k+1}}^N = Y_{t_k}^N + \mu^Y\left(Y_{t_k}^N\right)\Delta t_k + \sum_{j=1}^{d} \sigma_j^Y\left(Y_{t_k}^N\right)\Delta W_{t_k}^j, \quad Y_{t_0} \text{ given} \tag{25.33}$$

where $\Delta t_k \equiv t_{k+1} - t_k$ and $k = 0, \ldots, N-1$. The next Theorem gives the weak limit of the scaled approximation error associated with the scheme, when the number of discretization points N goes to infinity. In order to state this result define the random variable

$$\Psi_v \equiv \mathcal{E}^R\left(\int_0^{\cdot} \partial\mu^Y(Y_s)\,ds + \sum_{j=1}^{d}\int_0^{\cdot} \partial\sigma_j^Y(Y_s)\,dW_s^j\right)_v, \tag{25.34}$$

where $\mathcal{E}^R(\cdot)$ is the right stochastic exponential (For a $d \times d$ semimartingale M, the right stochastic exponential $Z_v = \mathcal{E}^R(M)_v$ is the unique solution of the $d \times d$ matrix SDE $dZ_v = dM_v Z_v$ with $Z_0 = I_d$.) and where $\partial\mu^Y, \partial\sigma_j^Y$ are the $d_y \times d_y$ matrices of derivatives of the vectors μ^Y, σ_j^Y with respect to the elements of Y.

Theorem 4 **(Kurtz and Protter 1991).** *The approximation error $Y_T^N - Y_T$ converges weakly at the rate $1/\sqrt{N}$,*

$$\sqrt{N}\left(Y_T^N - Y_T\right) \Rightarrow -\frac{1}{\sqrt{2}}\Psi_T \int_0^T \Psi_v^{-1} \sum_{l,j=1}^{d} \left[\partial\sigma_j^Y \sigma_l^Y\right](Y_v)\,dZ_v^{l,j} \tag{25.35}$$

as $N \to \infty$, where $\left[Z^{l,j}\right]_{l,j\in\{1,\ldots,d\}}$ is a $d^2 \times 1$ standard Brownian motion independent of W.

The asymptotic distribution of the Euler-Maruyama scheme is that of a random variable centered at zero. Inspection of the expression for the weak limit reveals that the approximation error might converge at a higher rate when the volatility coefficient is deterministic (in this case $\sqrt{N}\left(Y_T^N - Y_T\right) \Rightarrow 0$ implying that there may exist an $\alpha > 1/2$ such that $N^{\alpha}\left(Y_T^N - Y_T\right) \Rightarrow U \neq 0$). These observations provide motivation for the Euler scheme with Doss transformation, or Euler-Doss scheme, presented next.

25.4.1.2 Euler-Doss Scheme

Doss (1977) introduces a transformation that eliminates stochastic fluctuations in volatility coefficients. Consider again the process for state variables. If σ^Y has full rank and if the vector field generated by the columns of the volatility matrix is Abelian, i.e., if $\partial\sigma_i^Y\sigma_j^Y = \partial\sigma_i^Y\sigma_j^Y$ (commutative noise), there exists an invertible function F such that $\partial F\left(Y_v\right)\sigma^Y\left(Y_v\right) = I_d$ where I_d is the d-dimensional identity matrix. The inverse of F, denoted by G, solves the total differential equation (See Detemple et al. (2005a) for details).

$$\partial G\left(z\right) = \sigma^Y\left(G\left(z\right)\right), \quad G(0) = 0. \tag{25.36}$$

This gives $Y_t = G\left(\hat{Y}_t\right)$, where

$$d\hat{Y}_v = \hat{\mu}^Y\left(\hat{Y}_v\right)dv + dW_v, \quad \text{with } \hat{Y}_0 = F\left(Y_0\right) \tag{25.37}$$

and

$$\hat{\mu}^Y\left(x\right) \equiv \sigma^Y\left(x\right)^{-1}\mathcal{A}_t G\left(x\right), \qquad \mathcal{A}G \equiv \mu^Y(G) - \frac{1}{2}\sum_{j=1}^{d}\partial\sigma_j^Y(G). \tag{25.38}$$

The transformed process $\hat{Y}$ has identity volatility matrix. The Euler-Maruyama approximation of $\hat{Y}$ satisfies

$$\hat{Y}_{t_{k+1}}^N = \hat{Y}_{t_k}^N + \hat{\mu}^Y\left(\hat{Y}_{t_k}^N\right)\Delta t_k + \Delta W_{t_k}$$

and an approximation of Y is $\tilde{Y}_{t_k}^N \equiv G\left(\hat{Y}_{t_k}^N\right), k = 0, \ldots, N-1$. The error distribution of this approximation is given next. Let

$$\hat{\Psi}_v \equiv \mathcal{E}^R\left(\int_0^{\cdot}\partial\hat{\mu}^Y(\hat{Y}_s)ds\right)_v \tag{25.39}$$

and denote by $\partial\hat{\mu}^Y\left(\hat{Y}_v\right) = \left[\partial_1\hat{\mu}^Y(\hat{Y}_v), \ldots, \partial_d\hat{\mu}^Y(\hat{Y}_v)\right]$ the $d_y \times d_y$ matrix with columns given by the derivatives of the vector $\hat{\mu}^Y\left(\hat{Y}_v\right)$ and by $\partial_{l,k}\hat{\mu}^Y\left(\hat{Y}_s\right)$ the $d \times 1$ vector of cross derivatives of $\hat{\mu}^Y\left(\hat{Y}_v\right)$ with respect to arguments l, k.

Theorem 5 (Detemple et al. 2006). *Suppose that σ^Y has full rank and that the commutativity conditions, $\partial\sigma_i^Y\sigma_j^Y = \partial\sigma_j^Y\sigma_i^Y$, hold for all $i, j = 1, \ldots, d_y$. The approximation error $\tilde{Y}_T^N - Y_T \equiv G\left(\hat{Y}_T^N\right) - Y_T$ converges weakly at the rate $1/N$,*

$$N\left(\tilde{Y}_T^N - Y_T\right) \Rightarrow -\partial G\left(\hat{Y}_T\right)\hat{\Psi}_T \int_0^T \hat{\Psi}_v^{-1}\partial\hat{\mu}^Y\left(\hat{Y}_v\right)\left(\frac{1}{2}d\hat{Y}_v + \frac{1}{\sqrt{12}}dZ_v\right)$$

$$-\frac{1}{2}\partial G\left(\hat{Y}_T\right)\hat{\Psi}_T\int_0^T \hat{\Psi}_v^{-1}\partial\hat{\mu}^Y\left(\hat{Y}_v\right)\sum_{k,l=1}^{d}\partial_{l,k}\hat{\mu}^Y\left(\hat{Y}_v\right)d\langle\quad (25.40)$$

as $N \to \infty$, where $\left[Z^j\right]_{j\in\{1,\ldots,d\}}$ is a $d \times 1$ standard Brownian motion independent of W and of $Z^{l,j}$ defined in Theorem 4.

Theorem 5 shows that the asymptotic error distribution is non-centered. Imposing additional uniform integrability conditions and taking expectations shows that the expected approximation error is the expected value of the random variable on the right hand side of (25.40). Relative to Euler-Maruyama, the speed of convergence increases from $1/\sqrt{N}$ to $1/N$. This increase in speed is achieved because the Doss transformation eliminates the error in the approximation of the martingale component. The commutativity condition is necessary for application of the transformation. As discussed next, this condition also plays an instrumental role for higher order discretization schemes like the Mihlstein scheme.

25.4.1.3 Mihlstein Scheme

If the drift and diffusion coefficients μ^Y and σ^Y are sufficiently smooth, higher order schemes can be derived using stochastic Taylor expansions. Under these smoothness conditions, the martingale part of the diffusion is approximated by

$$\int_{t_k}^{t_{k+1}} \sigma_j^Y(Y_v)\,dW_v^j \approx \sigma_j^Y\left(Y_{t_k}\right)\int_{t_k}^{t_{k+1}} dW_v^j$$

$$+\sum_{i=1}^{d}\left[\partial\sigma_j^Y\sigma_i^Y\right]\left(Y_{t_k}\right)\int_{t_k}^{t_{k+1}} dW_s^j\int_{t_k}^{s} dW_v^i.$$

where the first term underlies the Euler-Maruyama approximation and the second one involves second-order Wiener integrals. Derivation of this approximation uses the fact that the covariation between increments of a finite variation process and an infinite variation process is null.

The corresponding discretization scheme, known as the Mihlstein scheme, is

$$Y^N_{t_{k+1}} = Y^N_{t_k} + \mu^Y\left(Y^N_{t_k}\right)\Delta t_k + \sum_{j=1}^{d}\sigma^Y_j\left(Y^N_{t_k}\right)\Delta W^j_{t_k}$$

$$+\sum_{i,j=1}^{d}\left[\partial\sigma^Y_j\sigma^Y_i\right]\left(Y^N_{t_k}\right)\int_{t_k}^{t_{k+1}} dW^j_s\int_{t_k}^{s} dW^i_v, \tag{25.41}$$

where Y_{t_0} is given. The simulation of $\sum_{i=1}^{d}\left[\partial\sigma^Y_j\sigma^Y_i\right]\left(Y^N_{t_k}\right)\int_{t_k}^{t_{k+1}} dW^j_s\int_{t_k}^{s} dW^i_v$ typically requires an additional sub-discretization of the time-step $t_{k+1}-t_k$, which increases the computational cost. An exception is when volatility satisfies the commutativity condition $\partial\sigma^Y_j\sigma^Y_i = \partial\sigma^Y_i\sigma^Y_j$. In this case, with the help of the Ito formula applied to

$$\int_{t_k}^{t_{k+1}} dW^i_s\int_{t_k}^{t_{k+1}} dW^j_v = \int_{t_k}^{t_{k+1}} dW^i_s\int_{t_k}^{s} dW^j_v + \int_{t_k}^{t_{k+1}} dW^j_s\int_{t_k}^{s} dW^i_v + \delta_{ij}\Delta t_k,$$

where $\delta_{ij} = 1$ if $i = j$ and 0 otherwise, it holds that

$$\sum_{i,j=1}^{d}\int_{t_k}^{t_{k+1}}\int_{t_k}^{s}\left[\partial\sigma^Y_j\sigma^Y_i\right](Y_{t_k})\,dW^j_s\,dW^i_v = \frac{1}{2}\sum_{i,j=1}^{d}\left[\partial\sigma^Y_j\sigma^Y_i\right](Y_{t_k})$$
$$\times\left(\int_{t_k}^{t_{k+1}} dW^j_s\int_{t_k}^{t_{k+1}} dW^i_v - \delta_{ij}\Delta t_k\right)$$

and the Mihlstein-scheme simplifies to

$$Y^N_{t_{k+1}} = Y^N_{t_k} + \mu^Y\left(Y^N_{t_k}\right)\Delta t_k + \sum_{j=1}^{d}\sigma^Y_j\left(Y^N_{t_k}\right)\Delta W^j_{t_k} \tag{25.42}$$

$$+\frac{1}{2}\sum_{j=1}^{d}\left[\partial\sigma^Y_j\sigma^Y_j\right]\left(Y^N_{t_k}\right)\left(\left(\Delta W^j_{t_k}\right)^2 - \Delta t_k\right)$$

$$+\frac{1}{2}\sum_{\substack{i,j=1\\ i\neq j}}^{d}\left[\partial\sigma^Y_j\sigma^Y_i\right]\left(Y^N_{t_k}\right)\Delta W^j_{t_k}\Delta W^i_{t_k} \tag{25.43}$$

with Y_{t_0} given.

The commutativity condition obviates the need for sub-discretizations of the time steps, thereby reducing the computational cost of the Mihlstein scheme. This condition is always satisfied in the case of a one-dimensional SDE. The additional terms (relative to Euler-Maruyama) in (25.43) compensate for the term with slowest convergence rate $(1/\sqrt{N})$ in the error expansion of the Euler-Maruyama scheme (see Detemple et al. 2006 for details). Adding these terms therefore improves the rate of convergence.

Theorem 6 (Detemple et al. 2006). *The approximation error $\check{Y}_T^N - Y_T$ converges weakly at the rate $1/N$,*

$$N\left(\check{Y}_T^N - Y_T\right) \Rightarrow -\frac{1}{2}\Psi_T \int_0^T \Psi_s^{-1}\left(\partial\mu^Y(Y_s)\,dY_s - \sum_{j=1}^d \left[\left(\partial\sigma_j^Y\right)\left(\partial\mu^Y\right)\sigma_j^Y\right](Y_s)\,ds\right)$$
$$-\frac{1}{2}\Psi_T \int_0^T \Psi_s^{-1} \sum_{j=1}^d \left[\left(\partial\left[\left(\partial\mu^Y\right)\sigma_j^Y\right]\right)\sigma_j^Y\right](Y_s)\,ds$$
$$+\frac{1}{2}\Psi_T \int_0^T \Psi_s^{-1} \sum_{j=1}^d \left[\left(\partial\sigma_j^Y\right)\left(\partial\sigma_j^Y\right)\mu^Y\right](Y_s)\,ds$$
$$-\frac{1}{2}\Psi_T \int_0^T \Psi_s^{-1} \sum_{j=1}^d \left[\left(\partial\sigma_j^Y\right)\mu^Y\right](Y_s)\,dW_s^j$$
$$-\frac{1}{\sqrt{12}}\Psi_T \int_0^T \Psi_s^{-1} \sum_{j=1}^d \left[\left(\partial\mu^Y\right)\sigma_j^Y - \left(\partial\sigma_j^Y\right)\mu^Y\right](Y_s)\,dZ_s^j$$
$$-\frac{1}{\sqrt{6}}\Psi_T \int_0^T \Psi_s^{-1} \sum_{i,l,j=1}^d \left[\partial\sigma_i^Y \partial\sigma_l^Y \sigma_j^Y\right](Y_s)\,d\tilde{Z}_s^{l,j,i} \tag{25.44}$$

when $N \to \infty$, where $\left((Z^j)\, j \in \{1,\ldots,d\}, (\tilde{Z}^{l,j,i})_{i,l,j=1,\ldots,d}\right)$ is a $d + d^3 \times 1$ standard Brownian motion independent of W. The random variable Ψ_T is given in (25.34).

As for the Euler-Doss scheme, the asymptotic error distribution of the Mihlstein scheme is non-centered. Under additional uniform integrability assumptions, the expected approximation error is the first moment of the weak limit on the right hand side of (25.44). The expected approximation error is the second order bias of the discretization scheme. Second order biases play an important role in efficiency comparisons based on the length of asymptotic confidence intervals (see Sect. 25.4.2). Whether the Mihlstein scheme is more efficient and/or has a lower second order bias than the Euler-Doss scheme depends on the slopes of the drift and volatility coefficients. Uniform results are not available.

25.4.1.4 Numerical Example

This section presents the error densities of the three discretization schemes discussed above for the square-root process

$$dY_t = \sigma\sqrt{Y_t}\,dW_t, \quad Y_0 \text{ given.}$$

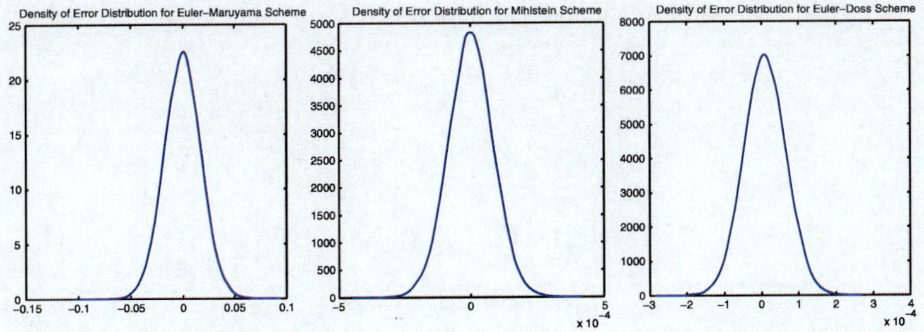

Fig. 25.1 The graphs show the densities of the error distribution for $Y_0 = 10, \sigma = 0.05, \gamma = 0.5$ and $T = 1$. Densities are calculated using a kernel estimator with discretization step $N = 500$ and $M = 50{,}000$ replications

For this SDE $\sigma^Y(y) \equiv \sigma\sqrt{y}$ and $\mu^Y(y) \equiv 0$. Limit error distributions are

$$N\left(Y_T^N - Y_t\right) \Rightarrow -\frac{1}{2}\Psi_T \int_0^T \Psi_s^{-1}\left(\partial\sigma^Y(Y_s)\right)\sigma^Y(Y_s)\,dZ_s \tag{25.45}$$

$$\begin{aligned} N\left(\tilde{Y}_T - Y_T\right) &\Rightarrow \partial G\left(\hat{Y}_T\right)\hat{\Psi}_T \int_0^T \hat{\Psi}_s^{-1}\partial\hat{\mu}^Y\left(\hat{Y}_s\right)\left(\frac{1}{2}d\hat{Y}_s + \frac{1}{\sqrt{12}}d\tilde{Z}_s\right) \\ &\quad + \frac{1}{2}\partial G\left(\hat{Y}_T\right)\hat{\Psi}_T \int_0^T \hat{\Psi}_s^{-1}\partial\hat{\mu}^Y\left(\hat{Y}_s\right)\partial^2\hat{\mu}^Y\left(\hat{Y}_s\right)ds \end{aligned} \tag{25.46}$$

$$N\left(\check{Y}_T - Y_T\right) \Rightarrow \frac{1}{\sqrt{6}}\Psi_T \int_0^T \Psi_s^{-1}\left(\partial\sigma^Y(Y_s)\right)^2\sigma^Y(Y_s)\,d\check{Z}_s, \tag{25.47}$$

where $Z, \tilde{Z}, \check{Z}$ are independent Brownian motions, $G(z) \equiv (\sigma(1-\gamma)z)^{(1/(1-\gamma))}$ and $\hat{\mu}^Y(y) \equiv -(1/2)\left(\gamma/(1-\gamma)\right)y^{-1}$ with $\gamma = 1/2$.

Figure 25.1 shows that the density of the Euler-Doss scheme is the most concentrated around zero. The range of the error distribution of the Mihlstein scheme is slightly larger, but considerably smaller than that of the Euler-Maruyama scheme. The Euler-Doss scheme is the most efficient simulation scheme in this example: confidence intervals constructed from the quantiles of the error density are the shortest.

25.4.2 Asymptotic Error Analysis

This section discusses the asymptotic error distribution of Monte Carlo portfolio estimators. It is assumed that the Lagrange multiplier $y^*$is known. To simplify notation let $\sigma_t \equiv \sigma(t, Y_t)$ and $\theta_t \equiv \theta(t, Y_t)$.

The optimal portfolio estimator can be written as

$$\begin{aligned}\widehat{X_t^* \pi_t^*}^{N,M} &= -\left(\sigma_t'\right)^{-1} \theta_t \mathbf{E}_t^M \left[g_1^{MV}\left(Z_{t,T}^N; y_t^*\right)\right] \\ &\quad - \left(\sigma_t'\right)^{-1} \theta_t \mathbf{E}_t^M \left[\int_t^T g_2^{MV}\left(Z_{t,\eta_v^N}^N; y_t^*\right) dv\right] \\ &\quad - \left(\sigma_t'\right)^{-1} \mathbf{E}_t^M \left[g_1^H\left(Z_{t,T}^N; y_t^*\right)\right] \\ &\quad - \left(\sigma_t'\right)^{-1} \mathbf{E}_t^M \left[\int_t^T g_2^H\left(Z_{t,\eta_v^N}^N; y_t^*\right) dv\right], \end{aligned} \tag{25.48}$$

where $y_t^* \equiv y^* \xi_t$ and $\left\{Z_{t,v}^N : v \in [t,T]\right\}$ is a numerical approximation of the d_z-dimensional process

$$\left\{Z_{t,v}' \equiv \left[\xi_{t,v}, H_{t,v}', vec\left(\mathcal{D}_t Y_v\right)', Y_v', v\right] : v \in [t,T]\right\}$$

solving

$$dZ_{t,v} = a\left(Z_{t,v}\right) dv + \sum_{j=1}^{d} b_j\left(Z_{t,v}\right) dW_v^j; \quad Z_{t,t} \text{ given.}$$

The operator $E_T^M[X] \equiv \frac{1}{M}\sum_{i=1}^M X^i$ is the empirical mean for i.i.d. replications X^i of the random variable X. The functions $g_1^{MV}, g_1^H, g_2^{MV}, g_2^H$ are C^3-functions that appear in the portfolio components related to terminal wealth (g_1) and intermediate consumption (g_2),

$$\begin{aligned} g_1^{MV}(z;y) &\equiv z_1 J'\left(yz_1, z_5\right); \quad g_1^H(z;y) \equiv z_1 J'\left(yz_1, z_5\right) z_2 \\ g_2^{MV}(z;y) &\equiv z_1 I'\left(yz_1, z_5\right); \quad g_2^H(z;y) \equiv z_1 I'\left(yz_1, z_5\right) z_2. \end{aligned}$$

The error components associated with the four terms in (25.48) are, with $\eta_v^N \equiv [Nv]/N$ where $[x]$ stands for the largest integer lower bound,

$$e_{1,t,T}^{MV,M,N} \equiv -\left(\mathbf{E}_t^M\left[g_1^{MV}\left(Z_{t,T}^N; y_t^*\right)\right] - \mathbf{E}_t\left[g_1^{MV}\left(Z_{t,T}; y_t^*\right)\right]\right)\left(\sigma_t'\right)^{-1}\theta_t \tag{25.49}$$

$$e_{1,t,T}^{H,M,N} \equiv -\left(\sigma_t'\right)^{-1}\left(\mathbf{E}_t^M\left[g_1^H\left(Z_{t,T}^N; y_t^*\right)\right] - \mathbf{E}_t\left[g_1^H\left(Z_{t,T}; y_t^*\right)\right]\right) \tag{25.50}$$

$$\begin{aligned} e_{2,t,T}^{MV,M,N} &\equiv -\mathbf{E}_t^M\left[\int_t^T g_2^{MV}\left(Z_{t,\eta_v^N}^N; y_t^*\right) dv\right]\left(\sigma_t'\right)^{-1}\theta_t \\ &\quad -\mathbf{E}_t\left[\int_t^T g_2^{MV}\left(Z_{t,v}; y_t^*\right) dv\right]\left(\sigma_t'\right)^{-1}\theta_t \end{aligned} \tag{25.51}$$

$$e_{2,t,T}^{H,M,N} \equiv -\left(\sigma_t'\right)^{-1}\left(\mathbf{E}_t^M\left[\int_t^T g_2^H\left(Z_{t,\eta_v^N}^N; y_t^*\right)\right] - \mathbf{E}_t\left[\int_t^T g_2^H\left(Z_{t,v}; y_t^*\right)\right]dv\right). \tag{25.52}$$

For $j \in \{1,2\}$, let $\left(e_{j,t,T}^{M,N}\right)' = \left[\left(e_{j,t,T}^{MV,M,N}\right)', \left(e_{j,t,T}^{H,M,N}\right)'\right]$ be the $1 \times 2d$ vector of approximation errors associated with the mean-variance and hedging demands for terminal wealth ($j = 1$) and intermediate consumption ($j = 2$). Finally, let $\left(e_{t,T}^{M,N}\right)' = \left[\left(e_{1,t,T}^{M,N}\right)', \left(e_{2,t,T}^{M,N}\right)'\right]$ be the $1\times 4d$ vector that incorporates all the error components. Similarly, define the $1 \times 4d$ random vector $C_{t,T}' \equiv \left[C_{1,t,T}', C_{2,t,T}'\right]$ where

$$C_{1,t,T}' \equiv \left[-g_1^{MV}\left(Z_{t,T}; y_t^*\right)\theta_t'\sigma_t^{-1}, -g_1^H\left(Z_{t,T}; y_t^*\right)'\sigma_t^{-1}\right]$$

$$C_{2,t,T}' \equiv \left[-\int_t^T g_2^{MV}\left(Z_{t,v}; y_t^*\right)dv\theta_t'\sigma_t^{-1}, -\int_t^T g_2^H\left(Z_{t,v}; y_t^*\right)' dv\sigma_t^{-1}\right]$$

are random variables in the portfolio components. The vector $C_{t,T}$ plays a critical role for the joint variance of the asymptotic error distribution.

In order to present the asymptotic convergence result, define for $v \in [t,T]$ and for a C^3-function f such that $f\left(Z_{t,v}\right) \in \mathbb{D}^{1,2}$ (The space $\mathbb{D}^{1,2}$ is the domain of the Malliavin derivative operator (see Nualart 1995)), the conditional expectations

$$K_{t,v}\left(Y_t; f\right) \equiv \frac{1}{2}\mathbf{E}_t\left[\partial f\left(Z_{t,v}\right)V_1(t,v) + V_2(t,v;f)\right] \tag{25.53}$$

$$k_{t,v}\left(Y_t; f\right) \equiv \mathbf{E}_t\left[\int_t^v\left[\partial f\left(a + \sum_{j=1}^d\left(\partial b_j\right)b_j\right) + \sum_{j=1}^d b_j'\partial^2 f b_j\right]\left(Z_{t,s}\right)ds\right] \tag{25.54}$$

and set

$$\kappa_{t,v}\left(Y_t; f\right) \equiv K_{t,v}\left(Y_t; f\right) - k_{t,v}\left(Y_t; f\right). \tag{25.55}$$

The random variables $V_1(t,v)$ and $V_2(t,v;f)$ in (25.53) are

$$\begin{aligned} V_1(t,v) \equiv & -\nabla_t Z_{t,v}\int_t^v\left(\nabla_t Z_{t,s}\right)^{-1}\partial a\left(Z_{t,s}\right)dZ_{t,s} \\ & -\nabla_t Z_{t,v}\int_t^v\left(\nabla_t Z_{t,s}\right)^{-1}\sum_{j=1}^d\left[\partial b_j a - \sum_{i=1}^d\left(\partial b_j\right)\left(\partial b_j\right)b_i\right]\left(Z_{t,s}\right)dW_s^j \\ & +\nabla_t Z_{t,v}\int_t^v\left(\nabla_t Z_{t,s}\right)^{-1}\sum_{j=1}^d\left[\partial b_j\partial b_j a\right]\left(Z_{t,s}\right)ds \\ & -\nabla_t Z_{t,v}\int_t^v\left(\nabla_t Z_{t,s}\right)^{-1}\left[\sum_{k,l=1}^d\partial_k\left(\partial_l a b_{l,j}\right)b_{k,j}\right]\left(Z_{t,s}\right)ds \end{aligned}$$

$$+\nabla_t Z_{t,v} \int_t^v (\nabla_t Z_{t,s})^{-1} \sum_{i,j=1}^{d} \left[\partial\left(\partial b_j \partial b_j b_i\right) b_i\right](Z_{t,s})\, ds$$

$$-\nabla_t Z_{t,v} \int_t^v (\nabla_t Z_{t,s})^{-1} \sum_{i,j=1}^{d} \left[\partial b_i \partial b_j \partial b_j b_i\right](Z_{t,s})\, ds, \tag{25.56}$$

where $\nabla_t Z_{t,s}$ is the tangent process of $Z_{t,s}$, i.e., the process obtained by an infinitesimal perturbation of the initial value z at time t (see Detemple et al. (2008) for a discussion and the relation with the Malliavin derivative), and

$$V_2(t, v; f) \equiv -\int_t^v \sum_{i,j=1}^{d} v_{i,j}(s, v; f)\, ds, \tag{25.57}$$

where

$$v_{i,j}(s, v; f) \equiv \left[h^{i,j} (\nabla_t Z_{t,\cdot})^{-1} \left[(\partial b_j) b_i\right](Z_{t,\cdot}), W^i\right]_s \tag{25.58}$$

$$h_t^{i,j} \equiv \mathbf{E}_t\left[\mathcal{D}_{jt}\left(\partial f(Z_{t,T}) \nabla_t Z_{t,T} e_i\right)\right] \tag{25.59}$$

and e_i the ith unit vector.

The next proposition gives the asymptotic error distribution for the Monte Carlo portfolio estimator based on the Euler scheme.

Theorem 7 (Detemple et al. 2006, 2008). *Suppose $g \in C^3(\mathbb{R}^{d_z})$ and $g(Z_{t,v}; y_t^*) \in \mathbb{D}^{1,2}$ for all $v \in [t, T]$. Also suppose that all the uniform integrability conditions of Proposition 4 in Detemple et al. (2008) are satisfied. Then, as $M \to \infty$,*

$$\sqrt{M} e_{t,T}^{M,N_M} \Rightarrow \epsilon^{md} \frac{1}{2} \begin{bmatrix} -K_{t,T}(Y_t; g_1^{MV})(\sigma_t')^{-1}\theta_t \\ -(\sigma_t')^{-1}\left[K_{t,T}(Y_t; g_{i,1}^H)\right]_{i=1,\ldots,d} \\ -\int_t^T \kappa_{t,v}(Y_t; g_2^{MV})\, dv (\sigma_t')^{-1}\theta_t \\ -(\sigma_t')^{-1} \int_t^T \left[\kappa_{t,v}(Y_t; g_{i,2}^H)\right]_{i=1,\ldots,d} dv \end{bmatrix} + L_{t,T}(Y_t; g), \tag{25.60}$$

where $N_M \to \infty$, as $M \to \infty$, $\epsilon^{md} = \lim_{M\to\infty} \sqrt{M}/N_M$ and

$$L_{t,T}(Y_t; g)' \equiv \left[L_{t,T}(Y_t; g_1^{MV})', L_{t,T}(Y_t; g_1^H)', L_{t,T}(Y_t; g_2^{MV})', L_{t,T}(Y_t; g_2^H)'\right] \tag{25.61}$$

is the terminal value of a Gaussian martingale with (deterministic) quadratic variation and conditional variance given by

$$[L, L]_{t,T}(Y_t; g) = \int_t^T \mathbf{E}_t\left[N_s (N_s)'\right] ds = \mathbf{VAR}_t\left[C_{t,T}\right] \tag{25.62}$$

$$N_s = \mathbf{E}_s\left[\mathcal{D}_s C_{t,T}\right]. \tag{25.63}$$

The mean-variance component associated with terminal wealth $g_1^{MV}(z; y_t^*)$ induces the second-order bias function $K_{t,T}\left(Y_t; g_1^{MV}\right)$. The components of the d-dimensional vector of hedging terms for terminal wealth $[g_1^H(z; y_t^*)]_i$ induce the second-order bias functions $K_{t,T}\left(Y_t; g_{i,1}^H\right)$. The mean-variance component for running consumption $g_2^{MV}(z; y_t^*)$ induces two second-order bias terms embedded in the function $\kappa_{t,v}\left(Y_t; g_2^{MV}\right)$. Similarly, the components of the d-dimensional vector of hedging terms for running consumption $\left[g_2^H(z; y_t^*)\right]_i$ induce the second-order bias functions $\kappa_{t,v}\left(Y_t; g_{i,2}^H\right)$. The functions $K_{t,T}(Y_t; \cdot)$, $\kappa_{t,v}(Y_t; \cdot)$ are defined in (25.53)–(25.55).

The asymptotic error distribution is non-centered. Estimators are therefore affected by a second-order bias. This bias, i.e. the expected value of the asymptotic error distribution, depends on the parameter ϵ^{md} and the functions (25.53)–(25.55). As indicated above, the second-order bias affects the coverage probability of confidence intervals. In the limit, as $M \to \infty$, it can be shown (see Detemple et al. 2006) that the coverage probability of a confidence interval based on the limit error distribution is smaller than the prescribed size of the confidence interval. In contrast, if $\epsilon^{md} = 0$, that is if the number of discretization points in the Euler scheme converges faster than the square root of the number of Monte Carlo replications, this size distortion disappears and the asymptotic error distribution is free of second-order biases. Unfortunately, as for given number of discretization points, N, the number of Monte Carlo replications M is restricted in size, the asymptotic error variance of the Gaussian martingale L, $\mathbf{VAR}_t\left[C_{t,T}\right]/M$ is larger. It follows that asymptotic confidence intervals are wider. This implies that the portfolio estimator is asymptotically less efficient. In the presence of a second order bias, efficient estimators are only attained if the limit convergence parameter ϵ^{md} differs from zero (see Duffie and Glynn 1995).

To summarize, efficient estimators are affected by a second order bias and there is a trade-off between asymptotic efficiency and second-order bias. Detemple et al. (2006) discuss these issues further and present second-order bias corrected estimators. They also characterize the asymptotic error distribution of estimators based on the Euler-Doss and the Mihlstein scheme. In particular, they show that the Gaussian martingale L, which emerges in the application of a central limit theorem to the Monte Carlo error, is the same. But the second-order bias terms (25.53)–(25.55) differ. Model-independent orderings of second order biases are generally not available.

It should also be noted that, from a weak convergence perspective, alternative random number generation schemes, such as quasi-Monte Carlo schemes, do not alter the asymptotic convergence behavior. They are asymptotically equivalent to the schemes examined above. Gains may nevertheless be realized in finite samples by adopting such schemes.

25.5 A Numerical Example

This section illustrates the MCMD algorithm for a setting with multiple assets/state variables and an investor with HARA utility,

$$U(x) = \frac{(x-A)^{1-R}}{1-R}.$$

Portfolio policies are described in Corollary 2.

The investment opportunity set is characterized by a constant market price of risk θ and a stochastic process for the interest rate and state variables (r, Y),

$$r_t = \bar{r} + \delta_r' (Y_t - Y_0) \tag{25.64}$$

$$dY_t = diag\left[\sigma_i^Y\right] \Sigma\, diag\left[Y_{it}^{\gamma}\right] dW_t, \quad Y_0 \text{ given.} \tag{25.65}$$

In these expressions, the $1 \times d_y$ row vector δ_r' is constant and measures the interest rate sensitivity to term-structure factors Y. Factors follow the multivariate CEV process (25.65), where $diag\,[a_i]$ is the matrix with elements a_i on the diagonal and zeroes elsewhere. The matrix Σ is lower triangular with $\Sigma_{i,j} = \varrho_{ij}$ for $j < i$, $\Sigma_{i,j} = 0$ for $j > i$, and $\Sigma_{ii} = \sqrt{1 - \sum_{j=1}^{i-1} \varrho_{ij}^2}$ for $i = 1, \ldots, d_y$. The parameter ϱ_{ij} is the correlation coefficient between increments in Y_i and Y_j. The volatility coefficient of factor i is $\sigma_i^Y Y_{it}^{\gamma}$.

Risk factors Y are local martingales. A multivariate affine process is obtained as a special case, for $\gamma = 1/2$. The short rate starts at r_0 and has a long run mean $\bar{r}$.

The asset return volatility is assumed to be constant, $\sigma = diag\left[\sigma_i^s\right] \Sigma^s$ where Σ^s is a lower-triangular matrix with elements $\Sigma_{ij}^s = \varrho_{ij}^s$ for $i < j$, $\Sigma_{ij}^s = 0$ for $j > i$, and $\Sigma_{ii} = \sqrt{1 - \sum_{j=1}^{i-1} \left(\varrho_{ij}^s\right)^2}$ for $i = 1, \ldots, d$. The coefficient ϱ_{ij}^s measures the correlation between returns of asset i and j. The volatility σ_i^s is the standard deviation of asset return i.

Table 25.1 presents portfolio weights for $d_y = d = 5$. With non-homothetic utilities, portfolio weights depend on initial wealth (see Corollary 2). With $A < 0$, marginal utility is finite at zero and the bequest constraint (the no-default condition) binds. It is of interest to note that the value function for this problem depends on state variables and on wealth in a non-multiplicative manner. In this example, lattice-based computational methods are difficult to implement and computationally costly. This is because boundary conditions with respect to wealth are not readily available and because the numerical resolution of a six-dimensional PDE is a non-trivial task. Implementation of the MCMD method, on the other hand, is straightforward.

Table 25.1 This table shows the impact of bequest constraints for HARA utility with $A = -200$, $R = 4$. Portfolio weights in five assets are shown for initial levels of wealth $x \in [10, 15, 20, 25, 30]$. The investment horizon is $T = 10$. Simulations are for $M = 100{,}000$ replications and $N = 365$ discretization points per year. Parameter values are $\bar{r} = 0.02$, $\gamma = 0.75$, $\theta' = [0.2, 0.05, 0.025, 0.075, 0.10]$, $\delta_r = [0.2, 0.15, 0.05, 0.02, 0.06]$, $Y_0' = [10, 15, 20, 25, 30]$, $\left[\sigma_i^Y\right]_{i=1,\ldots,5} = [0.15, 0.08, 0.2, 0.01, 0.05]$, $\varrho_{21} = 0.95$, $\left[\varrho_{3j}\right]_{j=1,2} = [-0.45, 0.65]$, $\left[\varrho_{4j}\right]_{j=1,2,3} = [0.2, 0.3, -0.85]$, $\left[\varrho_{5j}\right]_{j=1,2,3,4} = [-0.5, 0.75, 0.9, -0.2]$, $\left[\sigma_i^s\right]_{i=1,\ldots,5} = [0.4, 0.25, 0.20, 0.25, 0.35]$, $\varrho_{21}^s = 0.1$, $\left[\varrho_{3j}^s\right]_{j=1,2} = [-0.1, 0.5]$, $\left[\varrho_{4j}^s\right]_{j=1,2,3} = [-0.2, 0.3, 0.0]$, $\left[\varrho_{5j}^s\right]_{j=1,2,3,4} = [0.4, -0.15, -0.2, 0.1]$

		Assets					Assets				
		1	2	3	4	5	1	2	3	4	5
Wealth		With bequest constraint					Without bequest constraint				
	π	1.038	−0.035	0.129	0.981	0.888	1.667	0.110	0.090	1.468	1.333
$x = 10$	π^m	1.505	0.266	0.975	0.956	1.144	2.246	0.397	1.455	1.427	1.707
	π^h	−0.467	−0.301	−0.846	0.025	−0.256	−0.579	−0.287	−1.365	0.041	−0.374
	π	0.935	0.046	0.168	0.805	0.759	1.191	0.109	0.161	0.999	0.939
$x = 15$	π^m	1.238	0.219	0.802	0.787	0.941	1.533	0.271	0.993	0.974	1.165
	π^h	−0.303	−0.173	−0.634	0.018	−0.181	−0.342	−0.162	−0.832	0.025	−0.226
	π	0.840	0.080	0.196	0.681	0.664	0.953	0.108	0.197	0.764	0.743
$x = 20$	π^m	1.049	0.185	0.679	0.666	0.797	1.176	0.208	0.762	0.748	0.894
	π^h	−0.208	−0.106	−0.484	0.014	−0.133	−0.223	−0.100	−0.565	0.017	−0.151
	π	0.760	0.095	0.217	0.587	0.589	0.811	0.108	0.219	0.624	0.625
$x = 25$	π^m	0.906	0.160	0.587	0.576	0.689	0.963	0.170	0.624	0.612	0.732
	π^h	−0.146	−0.066	−0.370	0.011	−0.099	−0.152	−0.063	−0.405	0.012	−0.107
	π	0.693	0.101	0.232	0.514	0.530	0.716	0.107	0.233	0.530	0.546
$x = 30$	π^m	0.795	0.141	0.515	0.505	0.604	0.820	0.145	0.531	0.521	0.623
	π^h	−0.102	−0.039	−0.283	0.008	−0.074	−0.104	−0.038	−0.298	0.009	−0.077

Coding takes a few lines and computation times are low. (The Lagrange multiplier is calculated using a Monte Carlo estimator of the function $h(y) \equiv E[\xi_T I(y\xi_T)] - x$ and a robust bisection scheme for finding the root of the equation $h(y^*) = 0$.)

Table 25.1 displays the portfolio policy in two cases. The first one is when the bequest constraint is enforced, the second when negative terminal wealth is permitted. The results illustrate the impact of the constraint on the portfolio weights. Ignoring the bequest constraint is seen to have a first order effect on the optimal policy. Studies ignoring it are therefore prone to reaching misleading conclusions. The results also show that wealth effects are important, both for the mean-variance term and the dynamic hedging component. Finally, it is also worth noting that the intertemporal hedging demand is significant even though the investment horizon is of moderate length ($T = 10$).

25.6 Alternative Monte Carlo Portfolio Methods

An alternative Monte Carlo portfolio estimator has been developed by Cvitanic et al. (2003). Their method relies on an approximation of the covariation of the optimal wealth process with the Brownian innovation,

$$X_t^* (\pi_t^*)' \sigma_t = \lim_{\tau \to 0} \frac{1}{\tau} \left([X^*, W]_{t+\tau} - [X^*, W]_t \right) = \lim_{\tau \to 0} \frac{1}{\tau} \int_t^{\tau} X_s^* \pi_s' \sigma_s ds. \tag{25.66}$$

As $X_t^* = \mathbf{E}_t [\xi_T I(y\xi_T)]$, it follows that

$$[X^*, W]_{t+\tau} - [X^*, W]_t = \mathbf{E}_t [\xi_T I(y\xi_T)(W_{t+\tau} - W_t)]. \tag{25.67}$$

A Monte Carlo estimator of the portfolio, for a given Lagrange multiplier y^*, is

$$\widehat{X_t^* \pi_t^*}^{M,N} = (\sigma_t')^{-1} \mathbf{E}_t^M \left[\xi_T^N I(y^* \xi_T^N) \left(\frac{W_{t+\tau} - W_t}{\tau} \right) \right], \tag{25.68}$$

where τ is some selected time interval.

The Monte Carlo portfolio estimator (25.68) is easy to calculate. But it is based on a formula which approximates the optimal portfolio (and not the exact portfolio rule). As shown by Detemple et al. (2005c), this introduces an additional second-order bias term and reduces the speed of convergence to $M^{-1/3}$. The numerical efficiency studies in Detemple et al. (2005b,c) also show that the efficiency gains of estimators based on exact portfolio formulas (such as MCMD) can be considerable.

Similar results apply to Monte Carlo finite difference estimators (MCFD). These estimators use Monte Carlo simulation, but approximate the derivatives of the value function by finite difference perturbations of the relevant arguments. These finite difference perturbations replace the correct expressions given by expected values of functionals containing Malliavin derivatives. Again, the corresponding portfolio estimator relies on a formula that holds just in the limit, introducing an additional approximation error and reducing the convergence speed. The numerical studies in Detemple et al. (2005b,c) also illustrate the superior performance of MCMD relative to MCFD estimators.

The Monte Carlo methods based on Malliavin calculus rely only on forward simulations. In contrast, the Bellman equation naturally suggests backward simulation algorithms. In order to obtain adapted portfolio policies using backward algorithms, conditional expectations have to be calculated repeatedly. This is computationally costly, especially for high-dimensional problems. Brandt et al. (2005) consider polynomial regressions on basis functions to reduce the computational burden in backward calculations of the value function. These methods have proven useful for optimal stopping problems (Tsitsiklis and Van Roy 2001; Longstaff and Schwartz 2001; Gobet et al. 2005) where the control is binary. The optimal policy in a portfolio problem is more complex and the numerical precision depends directly on the estimates of the derivatives of the value function. A general convergence proof for these methods is unavailable at this stage (Partial results are reported in Clément et al. (2002) and Glasserman and Yu (2004)). The numerical efficiency study in Detemple et al. (2005b) shows that this approach is dominated by other MC methods.

References

Bally, V., & Talay, D. (1996a). The law of the Euler scheme for stochastic differential equations (I): Convergence rate of the distribution function, *Probability Theory and Related Fields*, *104*, 43–60.

Bally, V., & Talay, D. (1996b). The law of the Euler scheme for stochastic differential equations (II): Convergence rate of the density. *Monte Carlo Methods and its Applications*, *2*, 93–128.

Brandt, M. W., Goyal, A. Santa-Clara, P., & Stroud, J. R. (2005). A simulation approach to dynamic portfolio choice with an application to learning about return predictability. *Review of Financial Studies*, *18*, 831–873.

Brennan, M., Schwartz, E., & Lagnado, R. (1997). Strategic asset allocation. *Journal of Economic Dynamics and Control*, *21*, 1377–1403.

Clément, E., Lamberton, D., & Protter, P. (2002). An analysis of a least squares regression method for American option pricing. *Finance and Stochastics*, *5*, 449–471.

Cox, J. C., & Huang, C.-F. (1989). Optimal consumption and portfolio policies when asset prices follow a diffusion process. *Journal of Economic Theory*, *49*, 33–83.

Cvitanic, J. Goukasian, L., & Zapatero, F. (2003). Monte Carlo computation of optimal portfolio in complete markets. *Journal of Economic Dynamics and Control*, *27*, 971–986.

Detemple, J., & Rindisbacher, M. (2010). Dynamic asset allocation: Portfolio decomposition formula and applications. *Review of Financial Studies*, *23*, 25–100.

Detemple, J., Garcia, R., & Rindisbacher, M. (2005a). Representation formulas for Malliavin Derivatives of diffusion processes. *Finance and Stochastics*, *9*, 349–367.

Detemple, J., Garcia, R., & Rindisbacher, M. (2005b). Intertemporal asset allocation: A comparison of methods. *Journal of Banking and Finance*, *29*, 2821–2848.

Detemple, J., Garcia, R., & Rindisbacher, M. (2005c). Asymptotic properties of Monte Carlo estimators of derivatives of diffusion processes. *Management Science* (with online supplement *http://mansci.pubs.informs.org/ecompanion.html*), *51*, 1657–1675.

Detemple, J., Garcia, R., & Rindisbacher, M. (2006). Asymptotic properties of Monte Carlo estimators of diffusion processes. *Journal of Econometrics, 34*, 1–68.

Detemple, J., Garcia, R., & Rindisbacher, M. (2008). Simulation methods for optimal portfolios. In J.R. Birge & V. Linetsky (Eds.), *Handbooks in operations research and management science, Financial engineering* (Vol. 15, pp. 867–923). Amsterdam: Elsevier.

Detemple, J. B., Garcia, R., & Rindisbacher, M. (2003). A Monte-Carlo method for optimal portfolios. *Journal of Finance, 58*, 401–446.

Doss, H. (1977). Liens entre équations différentielles stochastiques et ordinaires. *Annales de l'Institut H. Poincaré, 13*, 99–125.

Duffie, D., & Glynn, P. (1995). Efficient Monte Carlo simulation of security prices. *Annals of Applied Probability, 5*, 897–905.

Geman, H. (1989). The importance of the forward neutral probability in a stochastic approach of interest rates. Working Paper, ESSEC.

Glasserman, P., & Yu, B. (2004). Number of paths versus number of basis functions in American option pricing. *Annals of Applied Probability, 14*, 2090–2119.

Gobet, E., Lemor, J-P., & Warin, X. (2005). A regression-based Monte-Carlo method to solve backward stochastic differential equations. *Annals of Applied Probability, 15*, 2172–2002.

Heath, D., Jarrow, R. A., & Morton, A. (1992). Bond pricing and the term structure of interest rates: A new methodology for contingent claims valuation. *Econometrica, 60*, 77–105.

Jamshidian, F. (1989). An exact bond option formula. *Journal of Finance, 44*, 205–09.

Karatzas, I., & Shreve, S. E. (1991). *Brownian Motion and Stochastic Calculus* (2nd Edn.). New York: Springer.

Karatzas, I., Lehoczky, J. P., & Shreve, S .E. (1987). Optimal portfolio and consumption decisions for a small investor on a finite horizon. *SIAM Journal of Control and Optimization, 25*, 1557–1586.

Kurtz, T. G., & Protter, P. (1991). Wong-Zakai corrections, random evolutions and numerical schemes for SDEs. In *Stochastic analysis* (pp. 331–346). New York: Academic.

Longstaff, F., & Schwartz, E. (2001). Valuing american options by simulation: A simple least-squares approach. *Review of Financial Studies, 14*, 113–147.

Markowitz, H. (1952). Portfolio selection. *Journal of Finance, 7*, 77–91.

Merton, R. C. (1969). Lifetime portfolio selection under uncertainty: The continuous time case. *Review of Economics and Statistics, 51*, 247–257.

Merton, R. C. (1971). Optimum consumption and portfolio rules in a continuous time model. *Journal of Economic Theory, 3*, 273–413.

Nualart, D. (1995). *The Malliavin calculus and related topics*. New York: Springer.

Ocone, D., & Karatzas, I. (1991). A generalized clark representation formula, with application to optimal portfolios. *Stochastics and Stochastics Reports, 34*, 187–220.

Pliska, S. (1986). A stochastic calculus model of continuous trading: Optimal portfolios. *Mathematics of Operations Research, 11*, 371–382.

Talay, D., & Tubaro, L. (1990). Expansion of the global error for numerical schemes solving stochastic differential equations. *Stochastic Analysis and its Application, 8*, 483–509.

Tsitsiklis, J., & Van Roy, B. (2001). Regression methods for pricing complex American-style options. *IEEE Transactions on Neural Networks, 12*, 694–703.

Chapter 26
Low-Discrepancy Simulation

Harald Niederreiter

Abstract This article presents a survey of low-discrepancy sequences and their applications to quasi-Monte Carlo methods for multidimensional numerical integration. Quasi-Monte Carlo methods are deterministic versions of Monte Carlo methods which outperform Monte Carlo methods for many types of integrals and have thus been found enormously useful in computational finance. First a general background on quasi-Monte Carlo methods is given. Then we describe principles for the construction of low-discrepancy sequences, with a special emphasis on the currently most powerful constructions based on the digital method and the theory of $(\mathbf{T}, s)$-sequences. Next, the important concepts of effective dimension and tractability are discussed. A synopsis of randomized quasi-Monte Carlo methods and their applications to computational finance is presented. A numerical example concludes the article.

26.1 Introduction

Many typical problems of modern computational finance, such as option pricing, can be rephrased mathematically as problems of calculating integrals with high-dimensional integration domains. Very often in such finance problems the integrand will be quite complicated, so that the integral cannot be evaluated analytically and precisely. In such cases, one has to resort to *numerical integration*, i.e., to a numerical scheme for the approximation of integrals.

A powerful approach to multidimensional numerical integration employs Monte Carlo methods. In a nutshell, a *Monte Carlo method* is a numerical method based on random sampling. A comprehensive treatment of Monte Carlo methods can be

H. Niederreiter (✉)
RICAM, Austrian Academy of Sciences, Altenbergerstr. 69, 4040 Linz, Austria
e-mail: ghnied@gmail.com

J.-C. Duan et al. (eds.), *Handbook of Computational Finance*, Springer Handbooks of Computational Statistics, DOI 10.1007/978-3-642-17254-0_26,

found in the book of Fishman (1996). The monograph of Glasserman (2004) covers Monte Carlo methods in computational finance.

Monte Carlo methods for numerical integration can be explained in a straightforward manner. In many cases, by using suitable transformations, we can assume that the integration domain is an s-dimensional unit cube $I^s := [0, 1]^s$, so this is the situation on which we will focus. We assume also that the integrand f is square integrable over I^s. Then the *Monte Carlo approximation* for the integral is

$$\int_{I^s} f(\mathbf{u})\, d\mathbf{u} \approx \frac{1}{N} \sum_{n=1}^{N} f(\mathbf{x}_n), \tag{26.1}$$

where $\mathbf{x}_1, \ldots, \mathbf{x}_N$ are independent random samples drawn from the uniform distribution on I^s. The law of large numbers guarantees that with probability 1 (i.e., for "almost all" sequences of sample points) we have

$$\lim_{N\to\infty} \frac{1}{N} \sum_{n=1}^{N} f(\mathbf{x}_n) = \int_{I^s} f(\mathbf{u})\, d\mathbf{u},$$

and so the Monte Carlo method for numerical integration converges almost surely.

We can, in fact, be more precise about the error committed in the Monte Carlo approximation (26.1). It can be verified quite easily that the square of the error in (26.1) is, on the average over all samples of size N, equal to $\sigma^2(f)N^{-1}$, where $\sigma^2(f)$ is the variance of f. Thus, with overwhelming probability we have

$$\int_{I^s} f(\mathbf{u})\, d\mathbf{u} - \frac{1}{N} \sum_{n=1}^{N} f(\mathbf{x}_n) = O(N^{-1/2}). \tag{26.2}$$

A rigorous version of this statement can be obtained from the central limit theorem (see Niederreiter 1992, p. 5). An important fact here is that the convergence rate in (26.2) is independent of the dimension s, and this makes Monte Carlo methods attractive for high-dimensional problems.

Despite the initial appeal of Monte Carlo methods for numerical integration, there are several drawbacks of these methods:

1. It is difficult to generate truly random samples.
2. Monte Carlo methods for numerical integration provide only probabilistic error bounds.
3. In many applications the convergence rate in (26.2) is considered too slow.

Note that the Monte Carlo error bound (26.2) describes the average performance of integration points (also called integration nodes) $\mathbf{x}_1, \ldots, \mathbf{x}_N$, and there should thus exist points that perform better than average. We want to focus on these points and we would like to construct them deterministically and explicitly. The criterion for choosing these points will be developed in Sect. 26.2. This ultimately leads to the concepts of low-discrepancy point sets and low-discrepancy sequences (see Sect. 26.3). The numerical integration techniques based on low-discrepancy point sets and low-discrepancy sequences are called *quasi-Monte Carlo methods*, or *QMC methods* for short.

26.2 Background on QMC Methods

Just as we did for Monte Carlo methods, we consider QMC methods in the context of numerical integration over an s-dimensional unit cube $I^s = [0,1]^s$. The approximation scheme is the same as for the Monte Carlo method, namely

$$\int_{I^s} f(\mathbf{u})\,d\mathbf{u} \approx \frac{1}{N}\sum_{n=1}^{N} f(\mathbf{x}_n),$$

but now $\mathbf{x}_1, \ldots, \mathbf{x}_N$ are deterministic points in I^s. For such a deterministic numerical integration scheme we expect a deterministic error bound, and this is indeed provided by the Koksma-Hlawka inequality (see Theorem 1 below). It depends on the star discrepancy, a measure for the deviation of the empirical distribution of a point set P (consisting of $\mathbf{x}_1, \ldots, \mathbf{x}_N \in I^s$, say) from the uniform distribution on I^s. For any Borel set $M \subseteq I^s$, let $A(M; P)$ be the number of integers n with $1 \le n \le N$ such that $\mathbf{x}_n \in M$. We put

$$R(M; P) = \frac{A(M; P)}{N} - \lambda_s(M),$$

which is the difference between the relative frequency of the points of P in M and the s-dimensional Lebesgue measure $\lambda_s(M)$ of M. If the points of P have a very uniform distribution over I^s, then the values of $R(M; P)$ will be close to 0 for a reasonable collection of Borel sets, such as for all subintervals of I^s.

Definition 1. The *star discrepancy* of the point set P is given by

$$D_N^* = D_N^*(P) = \sup_J |R(J; P)|,$$

where the supremum is extended over all intervals $J = \prod_{i=1}^{s}[0, u_i)$ with $0 < u_i \le 1$ for $1 \le i \le s$.

Theorem 1 (Koksma-Hlawka Inequality). *For any function f of bounded variation $V(f)$ on I^s in the sense of Hardy and Krause and any points $\mathbf{x}_1, \ldots, \mathbf{x}_N \in [0,1)^s$, we have*

$$\left| \int_{I^s} f(\mathbf{u})\, d\mathbf{u} - \frac{1}{N} \sum_{n=1}^{N} f(\mathbf{x}_n) \right| \le V(f) D_N^*,$$

where D_N^* *is the star discrepancy of* $\mathbf{x}_1, \ldots, \mathbf{x}_N$.

Note that $V(f)$ is a measure for the oscillation of the function f. The precise definition of the variation $V(f)$ can be found in (Niederreiter 1992, p. 19). For $f(\mathbf{u}) = f(u_1, \ldots, u_s)$, a sufficient condition for $V(f) < \infty$ is that the partial derivative $\partial^s f / \partial u_1 \cdots \partial u_s$ be continuous on I^s. A detailed proof of the Koksma-Hlawka inequality is given in the book of Kuipers and Niederreiter (1974, Sect. 2.5). There all types of variants of this inequality; see Niederreiter (1978), Niederreiter (1992, Sect. 2.2), and Hickernell (1998a).

A different kind of error bound for QMC integration was shown by Niederreiter (2003). It relies on the following concept.

Definition 2. Let $\mathcal{M}$ be a nonempty collection of Borel sets in I^s. Then a point set P of elements of I^s is called *$(\mathcal{M}, \lambda_s)$-uniform* if $R(M; P) = 0$ for all $M \in \mathcal{M}$.

Theorem 2. *Let* $\mathcal{M} = \{M_1, \ldots, M_k\}$ *be a partition of* I^s *into nonempty Borel subsets of* I^s. *For a bounded Lebesgue-integrable function* f *on* I^s *and for* $1 \le j \le k$, *we put*

$$G_j(f) = \sup_{\mathbf{u} \in M_j} f(\mathbf{u}), \quad g_j(f) = \inf_{\mathbf{u} \in M_j} f(\mathbf{u}).$$

Then for any $(\mathcal{M}, \lambda_s)$*-uniform point set consisting of* $\mathbf{x}_1, \ldots, \mathbf{x}_N \in I^s$ *we have*

$$\left| \int_{I^s} f(\mathbf{u})\, d\mathbf{u} - \frac{1}{N} \sum_{n=1}^{N} f(\mathbf{x}_n) \right| \le \sum_{j=1}^{k} \lambda_s(M_j)(G_j(f) - g_j(f)). \tag{26.3}$$

An analog of the bound (26.3) holds, in fact, for numerical integration over any probability space (see Niederreiter 2003). The primary tool for the error analysis of QMC integration is the Koksma-Hlawka inequality. The bound (26.3) is used for integrands of a very low degree of regularity which occasionally occur in computational finance. An example can be found in the work of Jiang (2007) on the pricing of European-style options and interest-rate derivatives.

26.3 Low-Discrepancy Sequences

The Koksma-Hlawka inequality leads to the conclusion that point sets with small star discrepancy guarantee small errors in QMC integration over I^s. This raises the question of how small we can make the star discrepancy of N points in I^s for fixed

N and s. For any $N \geq 2$ and $s \geq 1$, the least order of magnitude that can be achieved at present is

$$D_N^*(P) = O(N^{-1}(\log N)^{s-1}), \tag{26.4}$$

where the implied constant is independent of N. (Strictly speaking, one has to consider infinitely many values of N, i.e., an infinite collection of point sets of increasing size, for this O-bound to make sense in a rigorous fashion, but this technicality is often ignored.) A point set P achieving (26.4) is called a *low-discrepancy point set*. It is conjectured that the order of magnitude in (26.4) is best possible, that is, the star discrepancy of any $N \geq 2$ points in I^s is at least of the order of magnitude $N^{-1}(\log N)^{s-1}$. This conjecture is proved for $s = 1$ and $s = 2$ (see Kuipers and Niederreiter 1974, Sects. 2.1 and 2.2).

A very useful concept is that of a *low-discrepancy sequence*, which is an infinite sequence S of points in I^s such that for all $N \geq 2$ the star discrepancy $D_N^*(S)$ of the first N terms of S satisfies

$$D_N^*(S) = O(N^{-1}(\log N)^s) \tag{26.5}$$

with an implied constant independent of N. It is conjectured that the order of magnitude in (26.5) is best possible, but in this case the conjecture has been verified only for $s = 1$ (see Kuipers and Niederreiter 1974, Sect. 2.2).

Low-discrepancy sequences have several practical advantages. In the first place, if $\mathbf{x}_1, \mathbf{x}_2, \ldots \in I^s$ is a low-discrepancy sequence and $N \geq 2$ is an integer, then it is easily seen that the points

$$\mathbf{y}_n = \left(\frac{n-1}{N}, \mathbf{x}_n\right) \in I^{s+1}, \quad n = 1, \ldots, N,$$

form a low-discrepancy point set. Thus, if a low-discrepancy sequence has been constructed, then we immediately obtain arbitrarily large low-discrepancy point sets. Hence in the following we will concentrate on the construction of low-discrepancy sequences. Furthermore, given a low-discrepancy sequence S and a budget of N integration nodes, we can simply use the first N terms of the sequence S to get a good QMC method. If later on we want to increase N to achieve a higher accuracy, we can do so while retaining the results of the earlier computation. This is an advantage of low-discrepancy sequences over low-discrepancy point sets.

It is clear from the Koksma-Hlawka inequality and (26.5) that if we apply QMC integration with an integrand f of bounded variation on I^s in the sense of Hardy and Krause and with the first N terms $\mathbf{x}_1, \ldots, \mathbf{x}_N \in [0, 1)^s$ of a low-discrepancy sequence, then

$$\int_{I^s} f(\mathbf{u})\, d\mathbf{u} - \frac{1}{N} \sum_{n=1}^{N} f(\mathbf{x}_n) = O(N^{-1}(\log N)^s). \tag{26.6}$$

This yields a significantly faster convergence rate than the convergence rate $O(N^{-1/2})$ in (26.2). Thus, for many types of integrals the QMC method will outperform the Monte Carlo method.

26.4 Special Families of Low-Discrepancy Sequences

26.4.1 Halton Sequences

Over the years, various constructions of low-discrepancy sequences have been obtained. Historically, the first low-discrepancy sequences were designed by Halton (1960). For integers $b \geq 2$ and $n \geq 1$, let

$$n - 1 = \sum_{j=0}^{\infty} a_j(n)\, b^j, \quad a_j(n) \in \{0, 1, \ldots, b-1\},$$

be the digit expansion of $n - 1$ in base b. Then put

$$\phi_b(n) = \sum_{j=0}^{\infty} a_j(n)\, b^{-j-1}.$$

Now let $p_1 = 2, p_2 = 3, \ldots, p_s$ be the first s prime numbers. Then

$$\mathbf{x}_n = (\phi_{p_1}(n), \ldots, \phi_{p_s}(n)) \in I^s, \quad n = 1, 2, \ldots,$$

is the *Halton sequence* in the bases $p_1, \ldots, p_s$. This sequence S satisfies

$$D_N^*(S) = O(N^{-1}(\log N)^s)$$

for all $N \geq 2$, with an implied constant depending only on s (see Niederreiter 1992, Theorem 3.6). A discrepancy bound for Halton sequences with a small implied constant can be found in Atanassov (2004). The standard software implementation of Halton sequences is that of Fox (1986). More generally, one can replace $p_1, \ldots, p_s$ by pairwise coprime integers $b_1, \ldots, b_s \geq 2$, but the choice we have made yields the smallest discrepancy bound.

26.4.2 Nets

The starting point for current methods of constructing low-discrepancy sequences is the following definition which is a special case of Definition 2.

Definition 3. Let $s \geq 1$, $b \geq 2$, and $0 \leq t \leq m$ be integers and let $\mathcal{M}_{b,m,t}^{(s)}$ be the collection of all subintervals J of I^s of the form

$$J = \prod_{i=1}^{s} [a_i b^{-d_i}, (a_i + 1) b^{-d_i})$$

with integers $d_i \geq 0$ and $0 \leq a_i < b^{d_i}$ for $1 \leq i \leq s$ and with $\lambda_s(J) = b^{t-m}$. Then an $(\mathcal{M}_{b,m,t}^{(s)}, \lambda_s)$-uniform point set consisting of b^m points in I^s is called a *(t, m, s)-net in base b*.

It is important to note that the smaller the value of t for given b, m, and s, the larger the collection $\mathcal{M}_{b,m,t}^{(s)}$ of intervals in Definition 3, and so the stronger the uniform point set property in Definition 2. Thus, the primary interest is in (t, m, s)-nets in base b with a small value of t.

The standard method of constructing (t, m, s)-nets in base b proceeds as follows. Let integers $s \geq 1$, $b \geq 2$, and $m \geq 1$ be given. Let R be a finite commutative ring with identity and of order b. For $1 \leq i \leq s$ and $1 \leq j \leq m$, choose bijections $\eta_j^{(i)} : R \to \mathbb{Z}_b := \{0, 1, \ldots, b-1\}$. Furthermore, choose $m \times m$ matrices $C^{(1)}, \ldots, C^{(s)}$ over R. Now let $\mathbf{r} \in R^m$ be an m-tuple of elements of R and define

$$p_j^{(i)}(\mathbf{r}) = \eta_j^{(i)}(\mathbf{c}_j^{(i)} \cdot \mathbf{r}) \in \mathbb{Z}_b \qquad \text{for } 1 \leq i \leq s,\ 1 \leq j \leq m,$$

where $\mathbf{c}_j^{(i)}$ is the jth row of the matrix $C^{(i)}$ and $\cdot$ denotes the standard inner product. Next we put

$$p^{(i)}(\mathbf{r}) = \sum_{j=1}^{m} p_j^{(i)}(\mathbf{r}) b^{-j} \in [0, 1] \qquad \text{for } 1 \leq i \leq s$$

and

$$P(\mathbf{r}) = (p^{(1)}(\mathbf{r}), \ldots, p^{(s)}(\mathbf{r})) \in I^s.$$

By letting $\mathbf{r}$ range over all b^m possibilities in R^m, we arrive at a point set P consisting of b^m points in I^s.

In practice, the ring R is usually chosen to be a finite field $\mathbb{F}_q$ of order q, where q is a prime power. In this case, the t-value of the net constructed above can be conveniently determined by using an approach due to Niederreiter and Pirsic (2001). Let $\mathbb{F}_q^{ms}$ be the vector space of dimension ms over $\mathbb{F}_q$. We introduce a weight function W_m on $\mathbb{F}_q^{ms}$ as follows. We start by defining a weight function v on $\mathbb{F}_q^m$. We put $v(\mathbf{a}) = 0$ if $\mathbf{a} = \mathbf{0} \in \mathbb{F}_q^m$, and for $\mathbf{a} = (a_1, \ldots, a_m) \in \mathbb{F}_q^m$ with $\mathbf{a} \neq \mathbf{0}$ we set

$$v(\mathbf{a}) = \max\{j : a_j \neq 0\}.$$

Then we extend this definition to $\mathbb{F}_q^{ms}$ by writing a vector $\mathbf{A} \in \mathbb{F}_q^{ms}$ as the concatenation of s vectors of length m, that is,

$$\mathbf{A} = (\mathbf{a}^{(1)}, \ldots, \mathbf{a}^{(s)}) \in \mathbb{F}_q^{ms} \quad \text{with } \mathbf{a}^{(i)} \in \mathbb{F}_q^m \text{ for } 1 \le i \le s,$$

and putting

$$W_m(\mathbf{A}) = \sum_{i=1}^{s} v(\mathbf{a}^{(i)}).$$

Definition 4. The *minimum distance* $\delta_m(\mathcal{N})$ of a nonzero $\mathbb{F}_q$-linear subspace $\mathcal{N}$ of $\mathbb{F}_q^{ms}$ is given by

$$\delta_m(\mathcal{N}) = \min_{\mathbf{A} \in \mathcal{N} \setminus \{\mathbf{0}\}} W_m(\mathbf{A}).$$

Now let $C^{(1)}, \ldots, C^{(s)}$ be the $m \times m$ matrices over $\mathbb{F}_q$ chosen in the construction of the point set P above. Set up an $m \times ms$ matrix M as follows: for $1 \le j \le m$, the jth row of M is obtained by concatenating the jth columns of $C^{(1)}, \ldots, C^{(s)}$. Let $\mathcal{R} \subseteq \mathbb{F}_q^{ms}$ be the row space of M and let $\mathcal{R}^{\perp}$ be its dual space, that is,

$$\mathcal{R}^{\perp} = \{\mathbf{A} \in \mathbb{F}_q^{ms} : \mathbf{A} \cdot \mathbf{R} = 0 \ \text{ for all } \mathbf{R} \in \mathcal{R}\}.$$

The following result of Niederreiter and Pirsic (2001) requires $s \ge 2$, but this is no loss of generality since the case $s = 1$ is trivial in this context.

Theorem 3. *Let $m \ge 1$ and $s \ge 2$ be integers. Then, with the notation above, the point set P is a (t, m, s)-net in base q with*

$$t = m + 1 - \delta_m(\mathcal{R}^{\perp}).$$

The t-value is an important quality parameter of a net. But there are also more refined ways of measuring the quality of a net. For instance, one may also want to take into account the t-values of various lower-dimensional projections of a given net. The t-value of a lower-dimensional projection can be bounded from above by the t-value of the original net, but in some cases it can be substantially smaller. We refer to L'Ecuyer (2004) and L'Ecuyer and Lemieux (2002) for discussions of refined quality measures for nets.

26.4.3 $(\mathbf{T}, s)$-Sequences

There is an important sequence analog of Definition 3. Given a real number $x \in [0, 1]$, let

$$x = \sum_{j=1}^{\infty} z_j b^{-j}, \quad z_j \in \{0, 1, \ldots, b-1\},$$

be a b-adic expansion of x, where the case $z_j = b - 1$ for all but finitely many j is allowed. For an integer $m \geq 1$, we define the truncation

$$[x]_{b,m} = \sum_{j=1}^{m} z_j b^{-j}.$$

If $\mathbf{x} = (x^{(1)}, \ldots, x^{(s)}) \in I^s$ and the $x^{(i)}$, $1 \leq i \leq s$, are given by prescribed b-adic expansions, then we define

$$[\mathbf{x}]_{b,m} = ([x^{(1)}]_{b,m}, \ldots, [x^{(s)}]_{b,m}).$$

We write $\mathbb{N}$ for the set of positive integers and $\mathbb{N}_0$ for the set of nonnegative integers.

Definition 5. Let $s \geq 1$ and $b \geq 2$ be integers and let $\mathbf{T} : \mathbb{N} \to \mathbb{N}_0$ be a function with $\mathbf{T}(m) \leq m$ for all $m \in \mathbb{N}$. Then a sequence $\mathbf{x}_1, \mathbf{x}_2, \ldots$ of points in I^s is a *$(\mathbf{T}, s)$-sequence in base b* if for all $k \in \mathbb{N}_0$ and $m \in \mathbb{N}$, the points $[\mathbf{x}_n]_{b,m}$ with $kb^m < n \leq (k+1)b^m$ form a $(\mathbf{T}(m), m, s)$-net in base b. If for some integer $t \geq 0$ we have $\mathbf{T}(m) = m$ for $m \leq t$ and $\mathbf{T}(m) = t$ for $m > t$, then we speak of a *(t, s)-sequence in base b*.

A general theory of (t, m, s)-nets and (t, s)-sequences was developed by Niederreiter (1987). The concept of a $(\mathbf{T}, s)$-sequence was introduced by Larcher and Niederreiter (1995), with the variant in Definition 5 being due to Niederreiter and Özbudak (2007). Recent surveys of this topic are presented in Niederreiter (2005, 2008). For a (t, s)-sequence in base b we have

$$D_N^*(S) = O(b^t N^{-1} (\log N)^s) \tag{26.7}$$

for all $N \geq 2$, where the implied constant depends only on b and s. Thus, any (t, s)-sequence is a low-discrepancy sequence.

26.4.4 The Digital Method

The standard technique of constructing $(\mathbf{T}, s)$-sequences is the *digital method*. Fix a dimension $s \geq 1$ and a base $b \geq 2$. Let R again be a finite commutative ring with identity and of order b. Set up a map $\rho : R^\infty \to [0, 1]$ by selecting a bijection $\eta : R \to \mathbb{Z}_b := \{0, 1, \ldots, b-1\}$ and putting

$$\rho(r_1, r_2, \ldots) = \sum_{j=1}^{\infty} \eta(r_j) b^{-j} \quad \text{for } (r_1, r_2, \ldots) \in R^\infty.$$

Furthermore, choose $\infty \times \infty$ matrices $C^{(1)}, \dots, C^{(s)}$ over R which are called *generating matrices*. For $n = 1, 2, \dots$ let

$$n - 1 = \sum_{j=0}^{\infty} a_j(n)\, b^j, \quad a_j(n) \in \mathbb{Z}_b,$$

be the digit expansion of $n-1$ in base b. Choose a bijection $\psi : \mathbb{Z}_b \to R$ with $\psi(0) = 0$ and associate with n the sequence

$$\mathbf{n} = (\psi(a_0(n)), \psi(a_1(n)), \dots) \in R^{\infty}.$$

Then the sequence $\mathbf{x}_1, \mathbf{x}_2, \dots$ of points in I^s is defined by

$$\mathbf{x}_n = (\rho(\mathbf{n}C^{(1)}), \dots, \rho(\mathbf{n}C^{(s)})) \quad \text{for } n = 1, 2, \dots.$$

Note that the products $\mathbf{n}C^{(i)}$ are well defined since $\mathbf{n}$ contains only finitely many nonzero terms. As in Sect. 26.4.2, the ring R is usually chosen to be a finite field $\mathbb{F}_q$ of order q, where q is a prime power. The success of the digital method depends on a careful choice of the generating matrices $C^{(1)}, \dots, C^{(s)}$.

26.4.5 Sobol' Sequences and Niederreiter Sequences

The first application of the digital method occurred in the construction of *Sobol' sequences* in Sobol' (1967). This construction uses primitive polynomials over $\mathbb{F}_2$ to set up the generating matrices $C^{(1)}, \dots, C^{(s)}$ and leads to (t, s)-sequences in base 2. The wider family of irreducible polynomials was used in the construction of *Niederreiter sequences* in Niederreiter (1988), and this construction works for arbitrary prime-power bases q. Let $f_1, \dots, f_s$ be the first s monic irreducible polynomials over $\mathbb{F}_q$, ordered according to nondecreasing degrees, and put

$$T_q(s) = \sum_{i=1}^{s} (\deg(f_i) - 1). \tag{26.8}$$

The construction of Niederreiter sequences yields (t, s)-sequences in base q with $t = T_q(s)$. Let $U(s)$ denote the least value of t that is known to be achievable by Sobol' sequences for given s. Then $T_2(s) = U(s)$ for $1 \le s \le 7$ and $T_2(s) < U(s)$ for all $s \ge 8$. Thus, according to (26.7), for all dimensions $s \ge 8$ Niederreiter sequences in base 2 lead to a smaller upper bound on the star discrepancy than Sobol' sequences. Convenient software implementations of Sobol' and Niederreiter sequences are described in Bratley and Fox (1988) and Bratley et al. (1992, 1994), respectively.

In recent work, Dick and Niederreiter (2008) have shown that for given q and s, the value given in (26.8) is indeed the least t-value that is possible for s-dimensional Niederreiter sequences in base q. In the same paper, an analogous result was proved regarding $U(s)$ and Sobol' sequences for a range of dimensions of practical interest.

The potentially smallest, and thus best, t-value for any (t, s)-sequence in base b would be $t = 0$. However, according to (Niederreiter 1992, Corollary 4.24), a necessary condition for the existence of a $(0, s)$-sequence in base b is $s \leq b$. For primes p, a construction of $(0, s)$-sequences in base p for $s \leq p$ was given by Faure (1982). For prime powers q, a construction of $(0, s)$-sequences in base q for $s \leq q$ was given by Niederreiter (1987). Since $T_q(s) = 0$ for $s \leq q$ by (26.8), the Niederreiter sequences in Niederreiter (1988) also yield $(0, s)$-sequences in base q for $s \leq q$.

26.4.6 Niederreiter–Xing Sequences

An important advance in the construction of low-discrepancy sequences was made in the mid 1990s with the design of *Niederreiter–Xing sequences* which utilizes powerful tools from the theory of algebraic function fields and generalizes the construction of Niederreiter sequences. The basic papers here are Niederreiter and Xing (1996a) and Xing and Niederreiter (1995), and expository accounts of this work and further results are given in Niederreiter and Xing (1996b, 1998) and (Niederreiter and Xing 2001, Chap. 8). Niederreiter–Xing sequences are (t, s)-sequences in a prime-power base q with $t = V_q(s)$. Here $V_q(s)$ is a number determined by algebraic function fields with full constant field $\mathbb{F}_q$, and we have $V_q(s) \leq T_q(s)$ for all $s \geq 1$. In fact, much more is true. If we fix q and consider $V_q(s)$ and $T_q(s)$ as functions of s, then $V_q(s)$ is of the order of magnitude s, whereas $T_q(s)$ is of the order of magnitude $s \log s$. This yields an enormous improvement on the bound for the star discrepancy in (26.7). It is known that for any (t, s)-sequences in base b the parameter t must grow at least linearly with s for fixed b (see Niederreiter and Xing 1996b, Sect. 10), and so Niederreiter–Xing sequences yield t-values of the optimal order of magnitude as a function of s. A software implementation of Niederreiter–Xing sequences is described in Pirsic (2002) and available at

`http://math.iit.edu/~mcqmc/Software.html`

by following the appropriate links.

To illustrate the comparative quality of the above constructions of (t, s)-sequences, we consider the case of the most convenient base 2 and tabulate some values of $U(s)$ for Sobol' sequences, of $T_2(s)$ for Niederreiter sequences, and of $V_2(s)$ for Niederreiter–Xing sequences in Table 26.1. Note that the values of $V_2(s)$ in Table 26.1 for $2 \leq s \leq 7$ are the least values of t for which a (t, s)-sequence in base 2 can exist.

The approach to the construction of low-discrepancy sequences by algebraic function fields was followed up recently by Mayor and Niederreiter (2007) who gave an alternative construction of Niederreiter–Xing sequences using differentials

Table 26.1 Values of $U(s)$, $T_2(s)$, and $V_2(s)$

s	2	3	4	5	6	7	8	9	10	11	12	13	14	15	20
$U(s)$	0	1	3	5	8	11	15	19	23	27	31	35	40	45	71
$T_2(s)$	0	1	3	5	8	11	14	18	22	26	30	34	38	43	68
$V_2(s)$	0	1	1	2	3	4	5	6	8	9	10	11	13	15	21

of algebraic function fields. Niederreiter and Özbudak (2007) obtained the first improvement on Niederreiter–Xing sequences for some special pairs (q, s) of prime-power bases q and dimensions s. For instance, consider the case where q is an arbitrary prime power and $s = q + 1$. Then $T_q(q + 1) = 1$ by (26.8) and this is the least possible t-value for a $(t, q + 1)$-sequence in base q. However, the construction in Niederreiter and Özbudak (2007) yields a $(\mathbf{T}, q + 1)$-sequence in base q with $\mathbf{T}(m) = 0$ for even m and $\mathbf{T}(m) = 1$ for odd m, which is even better.

26.4.7 Additional Remarks

We remark that all constructions mentioned in Sects. 26.4.5 and 26.4.6 are based on the digital method. We note also that the extensive database at

`http://mint.sbg.ac.at`

is devoted to (t, m, s)-nets and (t, s)-sequences.

In summary, for a given prime-power base q, the currently best low-discrepancy sequences are:

1. The Faure or Niederreiter sequences (depending on whether q is prime or not) for all dimensions $s \le q$.
2. The Niederreiter–Xing sequences for all dimensions $s > q$, except for some special values of $s > q$ where the Niederreiter-Özbudak sequences are better.

We emphasize that the bound (26.7) on the star discrepancy of (t, s)-sequences is completely explicit; see Niederreiter 1992, Sect. 4.1 and a recent improvement in Kritzer (2006). For the best (t, s)-sequences, the coefficient of the leading term $N^{-1}(\log N)^s$ in the bound on the star discrepancy tends to 0 at a superexponential rate as $s \to \infty$.

We recall that the convergence rate $O(N^{-1}(\log N)^s)$ in (26.6) achieved by using low-discrepancy sequences is valid under mild regularity conditions on the integrand, namely that it be of bounded variation on I^s in the sense of Hardy and Krause. On the other hand, additional smoothness properties of the integrand are not reflected in applications of the Koksma-Hlawka inequality. For smooth integrands, there are other techniques of analyzing the integration error in suitable QMC methods in order to obtain faster convergence rates.

One approach uses a special family of sequences called Kronecker sequences. A *Kronecker sequence* is obtained from a point $(\alpha_1, \ldots, \alpha_s) \in \mathbb{R}^s$ by considering its multiples $(n\alpha_1, \ldots, n\alpha_s)$, $n = 1, 2, \ldots$, modulo 1, or in other words by considering the sequence $(\{n\alpha_1\}, \ldots, \{n\alpha_s\})$, $n = 1, 2, \ldots$, where $\{u\} = u - \lfloor u \rfloor$ denotes the fractional part of $u \in \mathbb{R}$. It was shown in (Niederreiter 1973, Sect. 9) that for special choices of $(\alpha_1, \ldots, \alpha_s)$ one obtains QMC methods that yield faster convergence rates for smooth integrands. We refer also to an account of this work in the survey article (Niederreiter 1978, Sect. 5). A convenient choice is $\alpha_i = \sqrt{p_i}$ for $1 \leq i \leq s$, where p_i is the ith prime number. An efficient implementation of these QMC methods was recently described by Vandewoestyne and Cools (2008).

Another approach to obtaining faster convergence rates for smooth integrands uses special QMC methods called lattice rules. For a given dimension $s \geq 1$, consider the factor group $\mathbb{R}^s/\mathbb{Z}^s$ which is an abelian group under addition of residue classes. Let $L/\mathbb{Z}^s$ be an arbitrary finite subgroup of $\mathbb{R}^s/\mathbb{Z}^s$ and let $\mathbf{x}_n + \mathbb{Z}^s$ with $\mathbf{x}_n \in [0, 1)^s$ for $1 \leq n \leq N$ be the distinct residue classes making up the group $L/\mathbb{Z}^s$. The points $\mathbf{x}_1, \ldots, \mathbf{x}_N$ form the integration nodes of an *N-point lattice rule*. This terminology stems from the fact that the subset $L = \bigcup_{n=1}^{N} (\mathbf{x}_n + \mathbb{Z}^s)$ of $\mathbb{R}^s$ is an s-dimensional lattice. Judicious choices of N-point lattice rules yield the desired faster convergence rates. Expository accounts of the theory of lattice rules are given in the books of (Niederreiter 1992, Chap. 5) and Sloan and Joe (1994) as well as in the survey articles of Cools and Nuyens (2008) and Hickernell (1998b).

In recent work, Dick (2007, 2008) has shown that one can also employ (t, m, s)-nets with additional uniformity properties in order to achieve faster convergence rates for smooth integrands.

26.5 Effective Dimension

26.5.1 Definition

In view of (26.6), the QMC method for numerical integration performs asymptotically better than the Monte Carlo method for any dimension s. However, in practical terms, the number N of integration nodes cannot be taken too large, and then already for moderate values of s the size of the factor $(\log N)^s$ may wipe out the advantage over the Monte Carlo method. On the other hand, numerical experiments with many types of integrands have shown that even for large dimensions s the QMC method will often lead to a convergence rate $O(N^{-1})$ rather than $O(N^{-1}(\log N)^s)$ as predicted by the theory, thus beating the Monte Carlo method by a wide margin. One reason may be that the Koksma-Hlawka inequality is in general overly pessimistic. Another explanation can sometimes be given by means of the nature of the integrand f. Even though f is a function of s variables, the influence of these variables could differ greatly. For numerical purposes, f may behave like a function of much fewer variables, so that the numerical integration problem is in essence a low-dimensional one with a faster convergence rate. This idea is captured by the notion of effective dimension.

We start with the ANOVA decomposition of a random variable $f(\mathbf{u}) = f(u_1, \ldots, u_s)$ on I^s of finite variance. This decomposition amounts to writing f in the form

$$f(\mathbf{u}) = \sum_{K \subseteq \{1,\ldots,s\}} f_K(\mathbf{u}),$$

where $f_\varnothing$ is the expected value of f and each $f_K(\mathbf{u})$ with $K \neq \varnothing$ depends only on the variables u_i with $i \in K$ and has expected value 0. Furthermore, f_{K_1} and f_{K_2} are orthogonal whenever $K_1 \neq K_2$. Then the variance $\sigma^2(f)$ of f decomposes as

$$\sigma^2(f) = \sum_{K \subseteq \{1,\ldots,s\}} \sigma^2(f_K).$$

The following definition relates to this decomposition.

Definition 6. Let d be an integer with $1 \leq d \leq s$ and r a real number with $0 < r < 1$. Then the function f has *effective dimension d at the rate r in the superposition sense* if

$$\sum_{|K| \leq d} \sigma^2(f_K) \geq r\sigma^2(f).$$

The function f has *effective dimension d at the rate r in the truncation sense* if

$$\sum_{K \subseteq \{1,\ldots,d\}} \sigma^2(f_K) \geq r\sigma^2(f).$$

Values of r of practical interest are $r = 0.95$ and $r = 0.99$, for instance. The formalization of the idea of effective dimension goes back to the papers of Caflisch et al. (1997) and Hickernell (1998b). There are many problems of high-dimensional numerical integration arising in computational finance for which the integrands have a relatively small effective dimension, one possible reason being discount factors which render variables corresponding to distant time horizons essentially negligible. The classical example here is that of the valuation of mortgage-backed securities (see Caflisch et al. 1997; Paskov 1997). For further interesting work on the ANOVA decomposition and effective dimension, with applications to the pricing of Asian and barrier options, we refer to Imai and Tan (2004), Lemieux and Owen (2002), Liu and Owen (2006) and Wang and Sloan (2005).

26.5.2 Weighted QMC Methods

A natural way of capturing the relative importance of variables is to attach weights to them. More generally, one may attach weights to any collection of variables, thus measuring the relative importance of all projections – and not just of the one-dimensional projections – of the given integrand. This leads then to a weighted version of the theory of QMC methods, an approach that was pioneered by Sloan and Woźniakowski (1998).

Given a dimension s, we consider the set $\{1,\dots,s\}$ of coordinate indices. To any nonempty subset K of $\{1,\dots,s\}$ we attach a weight $\gamma_K \geq 0$. To avoid a trivial case, we assume that not all weights are 0. Let γ denote the collection of all these weights γ_K. Then we introduce the *weighted star discrepancy* $D^*_{N,\gamma}$ which generalizes Definition 1. For $\mathbf{u} = (u_1,\dots,u_s) \in I^s$, we abbreviate the interval $\prod_{i=1}^s [0,u_i)$ by $[\mathbf{0},\mathbf{u})$. For any nonempty $K \subseteq \{1,\dots,s\}$, we let $\mathbf{u}_K$ denote the point in I^s with all coordinates whose indices are not in K replaced by 1. Now for a point set P consisting of N points from I^s, we define

$$D^*_{N,\gamma} = \sup_{\mathbf{u}\in I^s} \max_K \gamma_K |R([\mathbf{0},\mathbf{u}_K);P)|.$$

We recover the classical star discrepancy if we choose $\gamma_{\{1,\dots,s\}} = 1$ and $\gamma_K = 0$ for all nonempty proper subsets K of $\{1,\dots,s\}$. With this weighted star discrepancy, one can then prove a weighted analog of the Koksma-Hlawka inequality (see Sloan and Woźniakowski 1998).

There are some special kinds of weights that are of great practical interest. In the case of *product weights*, one attaches a weight η_i to each $i \in \{1,\dots,s\}$ and puts

$$\gamma_K = \prod_{i\in K} \eta_i \quad \text{for all } K \subseteq \{1,\dots,s\},\ K \neq \varnothing. \tag{26.9}$$

In the case of *finite-order weights*, one chooses a threshold k and puts $\gamma_K = 0$ for all K of cardinality larger than k.

The theoretical analysis of the performance of weighted QMC methods requires the introduction of weighted function spaces in which the integrands live. These can, for instance, be weighted Sobolev spaces or weighted Korobov spaces. In this context again, the weights reflect the relative importance of variables or collections of variables. The papers Kuo (2003), Sloan (2002), and Sloan and Woźniakowski (1998) are representative for this approach.

26.5.3 Tractability

The analysis of the integration error utilizing weighted function spaces also leads to powerful results on tractability, a concept stemming from the theory of information-based complexity. The emphasis here is on the performance of multidimensional numerical integration schemes as a function not only of the number N of integration nodes, but also of the dimension s as $s \to \infty$. Let $\mathcal{F}_s$ be a Banach space of integrands f on I^s with norm $\| f \|$. Write

$$L_s(f) = \int_{I^s} f(\mathbf{u})\, d\mathbf{u} \qquad \text{for } f \in \mathcal{F}_s.$$

Consider numerical integration schemes of the form

$$\mathcal{A}(f) = \sum_{n=1}^{N} a_n f(\mathbf{x}_n) \tag{26.10}$$

with real numbers $a_1, \ldots, a_N$ and points $\mathbf{x}_1, \ldots, \mathbf{x}_N \in I^s$. The QMC method is of course a special case of such a scheme. For $\mathcal{A}$ as in (26.10), we write $\mathrm{card}(\mathcal{A}) = N$. Furthermore, we put

$$\mathrm{err}(\mathcal{A}, \mathcal{F}_s) = \sup_{\|f\| \leq 1} |L_s(f) - \mathcal{A}(f)|.$$

For any $N \geq 1$ and $s \geq 1$, the Nth minimal error of the s-dimensional numerical integration problem is defined by

$$\mathrm{err}(N, \mathcal{F}_s) = \inf\{\mathrm{err}(\mathcal{A}, \mathcal{F}_s) : \mathcal{A} \text{ with } \mathrm{card}(\mathcal{A}) = N\}.$$

The numerical integration problem is called *tractable* if there exist constants $C \geq 0$, $e_1 \geq 0$, and $e_2 > 0$ such that

$$\mathrm{err}(N, \mathcal{F}_s) \leq C s^{e_1} N^{-e_2} \| L_s \|_{\mathrm{op}} \quad \text{for all } N \geq 1, s \geq 1,$$

where $\| L_s \|_{\mathrm{op}}$ is the operator norm of L_s. If, in addition, the exponent e_1 may be chosen to be 0, then the problem is said to be *strongly tractable*.

Tractability and strong tractability depend very much on the choice of the spaces $\mathcal{F}_s$. Weighted function spaces using product weights have proved particularly effective in this connection. Since the interest is in $s \to \infty$, product weights are set up by choosing a sequence $\eta_1, \eta_2, \ldots$ of positive numbers and then, for fixed $s \geq 1$, defining appropriate weights γ_K by (26.9). If the η_i tend to 0 sufficiently quickly as $i \to \infty$, then in a Hilbert-space setting strong tractability can be achieved by QMC methods based on Halton, Sobol', or Niederreiter sequences (see Hickernell and Wang 2002; Wang 2002). Further results on (strong) tractability as it relates to QMC methods can be found e.g. in Hickernell et al. (2004a), Hickernell et al. (2004b), Sloan et al. (2002), Sloan et al. (2004), Wang (2003) and Woźniakowski (2000).

26.6 Randomized QMC Methods

26.6.1 Scrambling Low-Discrepancy Sequences

Conventional QMC methods are fully deterministic and thus do not allow statistical error estimation as in Monte Carlo methods. However, one may introduce an element of randomness into a QMC method by randomizing (or "scrambling") the deterministic integration nodes used in the method. In this way one can combine the advantages of QMC methods, namely faster convergence rates, and those of Monte Carlo methods, namely the possibility of error estimation.

Historically, the first scrambling scheme is *Cranley-Patterson rotation* which was introduced in Cranley and Patterson (1976). This scheme can be applied to any point set in I^s. Let $\mathbf{x}_1, \ldots, \mathbf{x}_N \in I^s$ be given and put

$$\mathbf{y}_n = \{\mathbf{x}_n + \mathbf{v}\} \quad \text{for } n = 1, \ldots, N,$$

where $\mathbf{v}$ is a random vector uniformly distributed over I^s and $\{\cdot\}$ denotes reduction modulo 1 in each coordinate of a point in $\mathbb{R}^s$. This scheme transforms low-discrepancy point sets into low-discrepancy point sets.

A scrambling scheme that is tailored to Halton sequences (see Sect. 26.4.1) was proposed by Wang and Hickernell (2000). Here the initial point of the sequence is sampled randomly from the uniform distribution on I^s, whereas the construction of the subsequent points imitates the dynamics of the generation of the Halton sequence. Further devices for scrambling Halton sequences were studied by Mascagni and Chi (2004) and Vandewoestyne and Cools (2006), among others.

A sophisticated randomization of (t, m, s)-nets and (t, s)-sequences is provided by *Owen scrambling* (see Owen 1995). This scrambling scheme works with mutually independent random permutations of the digits in the b-adic expansions of the coordinates of all points in a (t, m, s)-net in base b or a (t, s)-sequence in base b. The scheme is set up in such a way that the scrambled version of a (t, m, s)-net, respectively (t, s)-sequence, in base b is a (t, m, s)-net, respectively (t, s)-sequence, in base b with probability 1. Further investigations of this scheme, particularly regarding the resulting mean square discrepancy and variance, were carried out e.g. by Hickernell and Hong (1999), Hickernell and Yue (2001), and Owen (1997a,b, 1998b).

Since Owen scrambling is quite time consuming, various faster special versions have been proposed, such as a method of Matoušek (1998) and the method of *digital shifts* in which the permutations in Owen scrambling are additive shifts modulo b and the shift parameters may depend on the coordinate index $i \in \{1, \ldots, s\}$ and on the position of the digit in the digit expansion of the coordinate. In the binary case $b = 2$, digital shifting amounts to choosing s infinite bit strings $\mathbf{B}_1, \ldots, \mathbf{B}_s$ and then taking each point $\mathbf{x}_n$ of the given (t, m, s)-net or (t, s)-sequence in base 2 and bitwise XORing the binary expansion of the ith coordinate of $\mathbf{x}_n$ with $\mathbf{B}_i$ for $1 \leq i \leq s$. Digital shifts and their applications are discussed e.g. in Dick and Pillichshammer (2005) and L'Ecuyer and Lemieux (2002). The latter paper presents also a general survey of randomized QMC methods and stresses the interpretation of these methods as variance reduction techniques.

Convenient scrambling schemes are also obtained by operating on the generating matrices of (t, s)-sequences constructed by the digital method. The idea is to multiply the generating matrices by suitable random matrices from the left or from the right in such a way that the value of the parameter t is preserved. We refer to Faure and Tezuka (2002) and Owen (2003) for such scrambling schemes. Software implementations of randomized low-discrepancy sequences are described in Friedel and Keller (2002) and Hong and Hickernell (2003) and are integrated into the Java library SSJ available at

http://www.iro.umontreal.ca/~simardr/ssj

which contains also many other simulation tools.

26.6.2 Hybrid Sequences

Another way of combining the advantages of QMC methods and Monte Carlo methods was proposed by Spanier (1995) in the context of high-dimensional problems. The idea here is to sample a relatively small number of dominating variables of the integrand by low-discrepancy sequences and the remaining variables by independent and uniformly distributed random variates. The number of dominating variables could be related to the effective dimension of the integrand (see Sect. 26.5.1), and these dominating variables are efficiently captured by low-discrepancy sequences. The sampling from independent and uniformly distributed random variates is realized in practice by using sequences of pseudorandom numbers. Mixing low-discrepancy sequences and sequences of pseudorandom numbers in this way results in what is called a *hybrid sequence*.

The star discrepancy of hybrid sequences can be analyzed from the probabilistic and from the deterministic point of view. Probabilistic results on the star discrepancy of hybrid sequences were obtained in Ökten (1996) and Ökten et al. (2006), and the latter paper discusses also the relevance of the results for computational finance.

The study of deterministic discrepancy bounds for hybrid sequences was initiated by Niederreiter (2009a). A typical case is the mixing of Halton sequences (see Sect. 26.4.1) with linear congruential sequences, which are classical sequences of pseudorandom numbers (see Knuth 1998, Chap. 3 and Niederreiter 1978). Let $\mathbf{x}_1, \mathbf{x}_2, \ldots$ be an s-dimensional Halton sequence. Furthermore, choose a large prime p and integers $g_1, \ldots, g_m, a_1, \ldots, a_m$ with $\gcd(g_j, p) = \gcd(a_j, p) = 1$ for $1 \leq j \leq m$. Then consider the hybrid sequence

$$(\mathbf{x}_n, \{g_1^n a_1/p\}, \ldots, \{g_m^n a_m/p\}) \in [0,1]^{s+m}, \quad n = 1, 2, \ldots .$$

Under suitable conditions on the parameters, a nontrivial discrepancy bound for this sequence is shown in Niederreiter (2009a). Another interesting case is the mixing of Kronecker sequences (see Sect. 26.4.7) with linear congruential sequences. This case is also treated in Niederreiter (2009a). Many further cases of hybrid sequences are discussed in Niederreiter (2009a) and in the more recent paper (Niederreiter 2009b).

26.6.3 Improving the Efficiency

In the context of Monte Carlo methods, several techniques for variance reduction were developed in order to improve the efficiency of Monte Carlo methods. The affinity between randomized QMC methods and Monte Carlo methods suggests to try analogous techniques in the framework of randomized QMC methods. There are indeed successful schemes for variance reduction in randomized QMC methods, and we mention latin supercube sampling (see Owen 1998a) and control variates (see Hickernell et al. 2005) as typical examples.

One can also attempt to manipulate the integrand in such a way that it gains additional desirable properties without changing the value of the integral over I^s. For instance, one may want to periodize the integrand so that it becomes a periodic function with period interval I^s, the idea being that then stronger theoretical error bounds become available. This is an attractive device in the framework of lattice rules (see Sect. 26.4.7). A summary of periodization techniques can be found in (Sloan and Joe 1994, Sect. 2.12).

Another strategy is to manipulate the integrand so as to reduce the effective dimension of the numerical integration problem. Several sophisticated methods are available for this purpose. The key ideas here are bridge sampling and principal component analysis. Bridge sampling was introduced by Caflisch and Moskowitz (1995) and Moskowitz and Caflisch (1996) in the context of QMC methods and the terminology refers to a stochastic process known as a Brownian bridge. Various refinements of this technique have been proposed over the years. We refer to Lin and Wang (2008) for recent work on this topic. Principal component analysis is a standard method in multivariate statistics for allocating importance to the initial coordinates of multidimensional data. Its use for the reduction of the effective dimension of problems in computational finance was proposed by Acworth et al. (1998).

26.6.4 Applications to Computational Finance

Among the various QMC methods, randomized QMC methods are probably the most widely used in computational finance. It is perhaps instructive to include some historical remarks here.

The application of Monte Carlo methods to challenging problems in computational finance was pioneered by the paper of Boyle (1977) from 1977. Although QMC methods were already known at that time, they were not applied to computational finance because it was thought that they would be inefficient for problems with high dimensions occurring in this area.

A breakthrough came in the early 1990s when Paskov and Traub applied QMC integration to the problem of pricing a 30-year collateralized mortgage obligation provided by Goldman Sachs; see Paskov and Traub (1995) for a report on this work.

This problem required the computation of ten integrals of dimension 360 each, and the results were astounding. For the hardest of the ten integrals, the QMC method achieved accuracy 10^{-2} with just 170 nodes, whereas the Monte Carlo method needed 2,700 nodes for the same accuracy. When higher accuracy is desired, the QMC method can be about 1,000 times faster than the Monte Carlo method. Later on, it was realized that one important reason why QMC methods work so well for the problem of pricing mortgage-backed securities is that this problem has a low effective dimension because the discount factors diminish the influence of the later years in the 30-year span. For further work on the pricing of mortgage-backed securities, we refer to Caflisch et al. (1997), Paskov (1997), and Tezuka (1998).

Applications of QMC methods to option pricing were first considered in the technical report of Birge (1994) and the paper of Joy et al. (1996). These works concentrated on European and Asian options. In the case of path-dependent options, if the security's terminal value depends only on the prices at s intermediate times, then after discretization the expected discounted payoff under the risk-neutral measure can be converted into an integral over the s-dimensional unit cube I^s.

A related problem in which an s-dimensional integral arises is the pricing of a multiasset option with s assets; see the paper Acworth et al. (1998) in which numerical experiments comparing Monte Carlo and QMC methods are reported for dimensions up to $s = 100$. This paper discusses also Brownian bridge constructions for option pricing. Related work on the pricing of multiasset European-style options using QMC and randomized QMC methods was carried out in Lai and Spanier (2000), Ross (1998), Tan and Boyle (2000), and comparative numerical experiments for Asian options can be found in Boyle et al. (1997) and Ökten and Eastman (2004).

Due to its inherent difficulty, it took much longer for Monte Carlo and QMC methods to be applied to the problem of pricing American options. An excellent survey of early work on Monte Carlo methods for pricing American options is presented in Boyle et al. (1997). The first important idea in this context was the *bundling algorithm* in which paths in state space for which the stock prices behave in a similar way are grouped together in the simulation. Initially, the bundling algorithm was applicable only to single-asset American options. Jin et al. (2007) recently extended the bundling algorithm in order to price high-dimensional American-style options, and they also showed that computing representative states by a QMC method improves the performance of the algorithm.

Another approach to pricing American options by simulation is the *stochastic mesh method*. The choice of mesh density functions at each discrete time step is crucial for the success of this method. The standard mesh density functions are mixture densities, and so in a Monte Carlo approach one can use known techniques for generating random samples from mixture densities. In a QMC approach, these random samples are replaced by deterministic points whose empirical distribution function is close to the target distribution function. Work on the latter approach was carried out by Boyle et al. (2001, 2002, 2003) and Broadie et al. (2000). Another application of QMC methods to the pricing of American options occurs in *regression-based methods*, which are typically least-squares Monte Carlo methods.

Here Caflisch and Chaudhary (2004) have shown that QMC versions improve the performance of such methods.

26.7 An Example

As an example which illustrates the efficiency of QMC methods, we discuss the problem of valuing an Asian call option. We are grateful to Gunther Leobacher of the University of Linz for working out this example. Consider a share whose price S follows a geometric Brownian motion such that, under the risk-neutral measure,

$$S_t = S_0 \exp(\sigma W_t + (r - \frac{\sigma^2}{2})t),$$

where W is a standard Brownian motion, r is the (constant) riskless interest rate, and σ is the (constant) volatility.

We want to price an option with payoff at time T given by

$$\max\left(\frac{1}{n}\sum_{k=1}^{n} S_{kT/n} - K, 0\right), \tag{26.11}$$

where the strike price K is a constant. The form of the payoff may be the result of the discretization of the average of S over the time interval $[0, T]$ or the option may in fact depend on the average over finitely many instants.

General valuation theory states that the arbitrage-free price at time 0 of the option with payoff (26.11) is given by the expectation

$$E\left(e^{-rT}\max\left(\frac{1}{n}\sum_{k=1}^{n} S_{kT/n} - K, 0\right)\right),$$

see for example Glasserman (2004). There is no simple closed-form expression for this expectation, and so one has to resort to numerical methods. For the case we consider here there are efficient methods for computing close estimates, see for example Rogers and Shi (1995) and the references therein. However, these methods heavily rely on the simplicity of the model, especially on the fact that there is only one Brownian motion involved. For this reason, our example and generalizations of it, such as Asian basket options (Dahl and Benth 2002) or Asian options in a Lévy market (Leobacher 2006), have retained their popularity as benchmarks for QMC methods.

In our numerical algorithm we used the Box-Muller method (Box and Muller 1958) for the generation of independent normal variables from uniform variables. Then the paths were generated using the Brownian bridge algorithm. The parameters were set to $S_0 = 100$, $K = 100$, $T = 1$, $r = 0.02$, $\sigma = 0.1$, $n = 16$.

Figure 26.1 compares the results for several Monte Carlo and QMC rules. We plot the base-10 logarithm of the standard deviation from the exact value over 20 runs against the base-10 logarithm of the number of paths generated. Thereby the exact value is computed using Sobol' points with $2^{23} \approx 8.4 \cdot 10^6$ paths. For

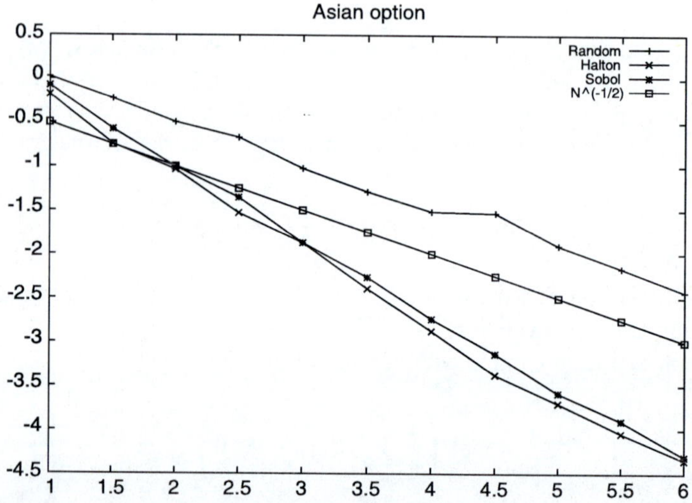

Fig. 26.1 Standard deviation of valuation errors for pseudorandom numbers, Halton and Sobol' sequence, respectively, plotted against number of paths used. Both scales are logarithmic. For better comparison also the sequence $N^{-1/2}$ is depicted

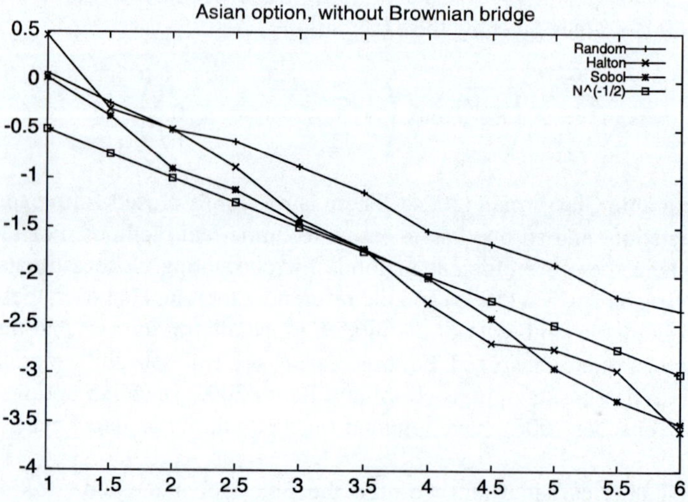

Fig. 26.2 Standard deviation of valuation errors for pseudorandom numbers, Halton and Sobol' sequence, respectively, plotted against number of paths used. Both scales are logarithmic. For better comparison also the sequence $N^{-1/2}$ is depicted

better comparison also the line $N^{-1/2}$ is drawn. As we can see from the graph, the Monte Carlo method shows almost exactly the predicted behavior of convergence order $O(N^{-1/2})$. Both QMC methods show superior convergence of order close to $O(N^{-1})$.

Finally, a word of caution is in order. While QMC methods hardly ever perform worse than Monte Carlo methods, they sometimes provide little advantage when used without care. Figure 26.2 shows the same comparison as Fig. 26.1, but now without the Brownian bridge algorithm. While the QMC methods still outperform plain Monte Carlo, the results are much worse than before. The reason for this, informally, is that the Brownian bridge algorithm reduces the effective dimension of the problem, see Caflisch and Moskowitz (1995) and Moskowitz and Caflisch (1996).

References

Acworth, P., Broadie, M., & Glasserman, P. (1998). A comparison of some Monte Carlo and quasi Monte Carlo techniques for option pricing. In H. Niederreiter et al. (Eds.), *Monte Carlo and quasi-Monte Carlo methods 1996* (pp. 1–18). New York: Springer.

Atanassov, E. I. (2004). On the discrepancy of the Halton sequences. *Mathematica Balkanica, 18*, 15–32.

Birge, J. R. (1994). Quasi-Monte Carlo approaches to option pricing. Technical Report 94–19, Department of Industrial and Operations Engineering, University of Michigan, Ann Arbor, MI.

Box, G. E. P., & Muller, M. E. (1958). A note on the generation of random normal deviates. *Annals of Mathematical Statistics, 29*, 610–611.

Boyle, P. P. (1977). Options: A Monte Carlo approach. *Journal of Financial Economics, 4*, 323–338.

Boyle, P., Broadie, M., & Glasserman, P. (1997). Monte Carlo methods for security pricing. *Journal of Economic Dynamics and Control, 21*, 1267–1321.

Boyle, P. P., Kolkiewicz, A. W., & Tan, K. S. (2001). Valuation of the reset options embedded in some equity-linked insurance products. *North American Actuarial Journal, 5*(3), 1–18.

Boyle, P. P., Kolkiewicz, A. W., & Tan, K. S. (2002). Pricing American derivatives using simulation: A biased low approach. In K. T. Fang, F. J. Hickernell, & H. Niederreiter (Eds.), *Monte Carlo and quasi-Monte Carlo methods 2000* (pp. 181–200). Berlin: Springer.

Boyle, P. P., Kolkiewicz, A. W., & Tan, K. S. (2003). An improved simulation method for pricing high-dimensional American derivatives. *Mathematics and Computers in Simulation, 62*, 315–322.

Bratley, P. & Fox, B. L. (1988). Algorithm 659: Implementing Sobol's quasirandom sequence generator. *ACM Transactions on Mathematical Software, 14*, 88–100.

Bratley, P., Fox, B. L., & Niederreiter, H. (1992). Implementation and tests of low-discrepancy sequences. *ACM Transactions on Modeling and Computer Simulation, 2*, 195–213.

Bratley, P., Fox, B. L., & Niederreiter, H. (1994). Algorithm 738: Programs to generate Niederreiter's low-discrepancy sequences. *ACM Transactions on Mathematical Software, 20*, 494–495.

Broadie, M., Glasserman, P., & Ha, Z. (2000). Pricing American options by simulation using a stochastic mesh with optimized weights. In S. P. Uryasev (Ed.), *Probabilistic constrained optimization: Methodology and applications* (pp. 26–44). Dordrecht: Kluwer.

Caflisch, R. E., & Chaudhary, S. (2004). Monte Carlo simulation for American options. In D. Givoli, M. J. Grote, & G. C. Papanicolaou (Eds.), *A celebration of mathematical modeling* (pp. 1–16). Dordrecht: Kluwer.

Caflisch, R. E., & Moskowitz, B. (1995). Modified Monte Carlo methods using quasi-random sequences. In H. Niederreiter & P. J.-S. Shiue (Eds.), *Monte Carlo and quasi-Monte Carlo methods in scientific computing* (pp. 1–16). New York: Springer.

Caflisch, R. E., Morokoff, M., & Owen, A. (1997). Valuation of mortgage-backed securities using Brownian bridges to reduce effective dimension. *The Journal of Computational Finance, 1*, 27–46.

Cools, R. & Nuyens, D. (2008). A Belgian view on lattice rules. In A. Keller, S. Heinrich, & H. Niederreiter (Eds.), *Monte Carlo and quasi-Monte Carlo methods 2006* (pp. 3–21). Berlin: Springer.

Cranley, R., & Patterson, T. N. L. (1976). Randomization of number theoretic methods for multiple integration. *SIAM Journal on Numerical Analysis, 13*, 904–914.

Dahl, L. O., & Benth, F. E. (2002). Fast evaluation of the Asian basket option by singular value decomposition. In K. T. Fang, F. J. Hickernell, & H. Niederreiter (Eds.), *Monte Carlo and quasi-Monte Carlo methods 2000* (pp. 201–214). Berlin: Springer.

Dick, J. (2007). Explicit constructions of quasi-Monte Carlo rules for the numerical integration of high-dimensional periodic functions. *SIAM Journal on Numerical Analysis, 45*, 2141–2176.

Dick, J. (2008). Walsh spaces containing smooth functions and quasi-Monte Carlo rules of arbitrary high order. *SIAM Journal on Numerical Analysis, 46*, 1519–1553.

Dick, J. & Niederreiter, H. (2008). On the exact t-value of Niederreiter and Sobol' sequences. *Journal of Complexity, 24*, 572–581.

Dick, J. & Pillichshammer, F. (2005). Multivariate integration in weighted Hilbert spaces based on Walsh functions and weighted Sobolev spaces. *Journal of Complexity, 21*, 149–195.

Faure, H. (1982). Discrépance de suites associées à un système de numération (en dimension s). *Acta Arithmetica, 41*, 337–351.

Faure, H. & Tezuka, S. (2002). Another random scrambling of digital (t, s)-sequences. In K. T. Fang, F. J. Hickernell, & H. Niederreiter (Eds.), *Monte Carlo and quasi-Monte Carlo methods 2000* (pp. 242–256). Berlin: Springer.

Fishman, G. S. (1996). *Monte Carlo: Concepts, algorithms, and applications*. New York: Springer.

Fox, B. L. (1986). Algorithm 647: Implementation and relative efficiency of quasirandom sequence generators. *ACM Transactions on Mathematical Software, 12*, 362–376.

Friedel, I. & Keller, A. (2002). Fast generation of randomized low-discrepancy point sets. In K. T. Fang, F. J. Hickernell, & H. Niederreiter (Eds.), *Monte Carlo and quasi-Monte Carlo methods 2000* (pp. 257–273). Berlin: Springer.

Glasserman, P. (2004). *Monte Carlo methods in financial engineering*. New York: Springer.

Halton, J. H. (1960). On the efficiency of certain quasi-random sequences of points in evaluating multi-dimensional integrals. *Numerische Mathematik, 2*, 84–90, 196.

Hickernell, F. J. (1998a). A generalized discrepancy and quadrature error bound. *Mathematics of Computation, 67*, 299–322.

Hickernell, F. J. (1998b). Lattice rules: How well do they measure up? In P. Hellekalek & G. Larcher (Eds.), *Random and quasi-random point sets* (pp. 109–166). New York: Springer.

Hickernell, F. J. & Hong, H. S. (1999). The asymptotic efficiency of randomized nets for quadrature. *Mathematics of Computation, 68*, 767–791.

Hickernell, F. J., & Wang, X. Q. (2002). The error bounds and tractability of quasi-Monte Carlo algorithms in infinite dimension. *Mathematics of Computation, 71*, 1641–1661.

Hickernell, F. J. & Yue, R.-X. (2001). The mean square discrepancy of scrambled (t, s)-sequences. *SIAM Journal on Numerical Analysis, 38*, 1089–1112.

Hickernell, F. J., Sloan, I. H., & Wasilkowski, G. W. (2004a). On tractability of weighted integration for certain Banach spaces of functions. In H. Niederreiter (Ed.), *Monte Carlo and quasi-Monte Carlo methods 2002* (pp. 51–71). Berlin: Springer.

Hickernell, F. J., Sloan, I. H., & Wasilkowski, G. W. (2004b). The strong tractability of multivariate integration using lattice rules. In H. Niederreiter (Ed.), *Monte Carlo and quasi-Monte Carlo methods 2002* (pp. 259–273). Berlin: Springer.

Hickernell, F. J., Lemieux, C., & Owen, A. B. (2005). Control variates for quasi-Monte Carlo. *Statistical Science*, *20*, 1–31.

Hong, H. S. & Hickernell, F. J. (2003). Algorithm 823: Implementing scrambled digital sequences. *ACM Transactions on Mathematical Software*, *29*, 95–109.

Imai, J. & Tan, K. S. (2004). Minimizing effective dimension using linear transformation. In H. Niederreiter (Ed.), *Monte Carlo and quasi-Monte Carlo methods 2002* (pp. 275–292). Berlin: Springer.

Jiang, X. F. (2007). *Quasi-Monte Carlo methods in finance*. Ph.D. dissertation, Northwestern University, Evanston, IL.

Jin, X., Tan, H. H., & Sun, J. H. (2007). A state-space partitioning method for pricing high-dimensional American-style options. *Mathematical Finance*, *17*, 399–426.

Joy, C., Boyle, P. P., & Tan, K. S. (1996). Quasi-Monte Carlo methods in numerical finance. *Management Science*, *42*, 926–938.

Knuth, D. E. (1998). *The art of computer programming*: Vol. 2. *Seminumerical algorithms* (3rd ed.). Reading, MA: Addison-Wesley.

Kritzer, P. (2006). Improved upper bounds on the star discrepancy of (t,m,s)-nets and (t,s)-sequences. *Journal of Complexity*, *22*, 336–347.

Kuipers, L. & Niederreiter, H. (1974). *Uniform distribution of sequences*. New York: Wiley; reprint, Mineola, NY: Dover Publications (2006).

Kuo, F. Y. (2003). Component-by-component constructions achieve the optimal rate of convergence for multivariate integration in weighted Korobov and Sobolev spaces. *Journal of Complexity*, *19*, 301–320.

Lai, Y. Z. & Spanier, J. (2000). Applications of Monte Carlo/quasi-Monte Carlo methods in finance: Option pricing. In H. Niederreiter & J. Spanier (Eds.), *Monte Carlo and quasi-Monte Carlo methods 1998* (pp. 284–295). Berlin: Springer.

Larcher, G. & Niederreiter, H. (1995). Generalized (t,s)-sequences, Kronecker-type sequences, and diophantine approximations of formal Laurent series. *Transactions of the American Mathematical Society*, *347*, 2051–2073.

L'Ecuyer, P. (2004). Polynomial integration lattices. In H. Niederreiter (Ed.), *Monte Carlo and quasi-Monte Carlo methods 2002* (pp. 73–98). Berlin: Springer.

L'Ecuyer, P. & Lemieux, C. (2002). Recent advances in randomized quasi-Monte Carlo methods. In M. Dror, P. L'Ecuyer, & F. Szidarovszky (Eds.), *Modeling uncertainty: An examination of stochastic theory, methods, and applications* (pp. 419–474). Boston: Kluwer.

Lemieux, C. & Owen, A. B. (2002). Quasi-regression and the relative importance of the ANOVA components of a function. In K. T. Fang, F. J. Hickernell, & H. Niederreiter (Eds.), *Monte Carlo and quasi-Monte Carlo methods 2000* (pp. 331–344). Berlin: Springer.

Leobacher, G. (2006). Stratified sampling and quasi-Monte Carlo simulation of Lévy processes. *Monte Carlo Methods and Applications*, *12*, 231–238.

Lin, J. Y. & Wang, X. Q. (2008). New Brownian bridge construction in quasi-Monte Carlo methods for computational finance. *Journal of Complexity*, *24*, 109–133.

Liu, R. X. & Owen, A. B. (2006). Estimating mean dimensionality of analysis of variance decompositions. *Journal of the American Statistical Association*, *101*, 712–721.

Mascagni, M. & Chi, H. M. (2004). On the scrambled Halton sequence. *Monte Carlo Methods and Applications*, *10*, 435–442.

Matoušek, J. (1998). On the L_2-discrepancy for anchored boxes. *Journal of Complexity*, *14*, 527–556.

Mayor, D. J. S. & Niederreiter, H. (2007). A new construction of (t,s)-sequences and some improved bounds on their quality parameter. *Acta Arithmetica*, *128*, 177–191.

Moskowitz, B. & Caflisch, R. E. (1996). Smoothness and dimension reduction in quasi-Monte Carlo methods. *Mathematical and Computer Modelling*, *23*(8–9), 37–54.

Niederreiter, H. (1973). Application of diophantine approximations to numerical integration. In C. F. Osgood (Ed.), *Diophantine approximation and its applications* (pp. 129–199). New York: Academic Press.

Niederreiter, H. (1978). Quasi-Monte Carlo methods and pseudo-random numbers. *Bulletin of the American Mathematical Society, 84*, 957–1041.

Niederreiter, H. (1987). Point sets and sequences with small discrepancy. *Monatshefte für Mathematik, 104*, 273–337.

Niederreiter, H. (1988). Low-discrepancy and low-dispersion sequences. *Journal of Number Theory, 30*, 51–70.

Niederreiter, H. (1992). *Random number generation and quasi-Monte Carlo methods*. Philadelphia, PA: SIAM.

Niederreiter, H. (2003). Error bounds for quasi-Monte Carlo integration with uniform point sets. *Journal of Computational and Applied Mathematics, 150*, 283–292.

Niederreiter, H. (2005). Constructions of (t, m, s)-nets and (t, s)-sequences. *Finite Fields and Their Applications, 11*, 578–600.

Niederreiter, H. (2008). Nets, (t, s)-sequences, and codes. In A. Keller, S. Heinrich, & H. Niederreiter (Eds.), *Monte Carlo and quasi-Monte Carlo methods 2006* (pp. 83–100). Berlin: Springer.

Niederreiter, H. (2009a). On the discrepancy of some hybrid sequences. *Acta Arithmetica, 138*, 373–398.

Niederreiter, H. (2009b). Further discrepancy bounds and an Erdös-Turán-Koksma inequality for hybrid sequences. *Monatshefte für Mathematik, 161*, 193–222.

Niederreiter, H. & Özbudak, F. (2007). Low-discrepancy sequences using duality and global function fields. *Acta Arithmetica, 130*, 79–97.

Niederreiter, H. & Pirsic, G. (2001). Duality for digital nets and its applications. *Acta Arithmetica, 97*, 173–182.

Niederreiter, H. & Xing, C. P. (1996a). Low-discrepancy sequences and global function fields with many rational places. *Finite Fields and Their Applications, 2*, 241–273.

Niederreiter, H. & Xing, C. P. (1996b). Quasirandom points and global function fields. In S. Cohen & H. Niederreiter (Eds.), *Finite fields and applications* (pp. 269–296). Cambridge, UK: Cambridge University Press.

Niederreiter, H. & Xing, C. P. (1998). Nets, (t, s)-sequences, and algebraic geometry. In P. Hellekalek & G. Larcher (Eds.), *Random and quasi-random point sets* (pp. 267–302). New York: Springer.

Niederreiter, H. & Xing, C. P. (2001). *Rational points on curves over finite fields: theory and applications*. Cambridge, UK: Cambridge University Press.

Ökten, G. (1996). A probabilistic result on the discrepancy of a hybrid-Monte Carlo sequence and applications. *Monte Carlo Methods and Applications, 2*, 255–270.

Ökten, G. & Eastman, W. (2004). Randomized quasi-Monte Carlo methods in pricing securities. *Journal of Economic Dynamics and Control, 28*, 2399–2426.

Ökten, G., Tuffin, B., & Burago, V. (2006). A central limit theorem and improved error bounds for a hybrid-Monte Carlo sequence with applications in computational finance. *Journal of Complexity, 22*, 435–458.

Owen, A. B. (1995). Randomly permuted (t, m, s)-nets and (t, s)-sequences. In H. Niederreiter & P. J.-S. Shiue (Eds.), *Monte Carlo and quasi-Monte Carlo methods in scientific computing* (pp. 299–317). New York: Springer.

Owen, A. B. (1997a). Monte Carlo variance of scrambled net quadrature. *SIAM Journal on Numerical Analysis, 34*, 1884–1910.

Owen, A. B. (1997b). Scrambled net variance for integrals of smooth functions. *The Annals of Statistics, 25*, 1541–1562.

Owen, A. B. (1998a). Latin supercube sampling for very high-dimensional simulations. *ACM Transactions on Modeling and Computer Simulation, 8*, 71–102.

Owen, A. B. (1998b). Scrambling Sobol' and Niederreiter-Xing points. *Journal of Complexity, 14*, 466–489.

Owen, A. B. (2003). Variance with alternative scramblings of digital nets. *ACM Transactions on Modeling and Computer Simulation, 13*, 363–378.

Paskov, S. H. (1997). New methodologies for valuing derivatives. In M. A. H. Dempster & S. R. Pliska (Eds.), *Mathematics of derivative securities* (pp. 545–582). Cambridge, UK: Cambridge University Press.

Paskov, S. H. & Traub, J. F. (1995). Faster valuation of financial derivatives. *Journal of Portfolio Management, 22*(1), 113–120.

Pirsic, G. (2002). A software implementation of Niederreiter-Xing sequences. In K. T. Fang, F. J. Hickernell, & H. Niederreiter (Eds.), *Monte Carlo and quasi-Monte Carlo methods 2000* (pp. 434–445). Berlin: Springer.

Rogers, L. C. G. & Shi, Z. (1995). The value of an Asian option. *Journal of Applied Probability, 32*, 1077–1088.

Ross, R. (1998). Good point methods for computing prices and sensitivities of multi-asset European style options. *Applied Mathematical Finance, 5*, 83–106.

Sloan, I. H. (2002). QMC integration – beating intractability by weighting the coordinate directions. In K. T. Fang, F. J. Hickernell, & H. Niederreiter (Eds.), *Monte Carlo and quasi-Monte Carlo methods 2000* (pp. 103–123). Berlin: Springer.

Sloan, I. H. & Joe, S. (1994). *Lattice methods for multiple integration.* Oxford, UK: Oxford University Press.

Sloan, I. H. & Woźniakowski, H. (1998). When are quasi-Monte Carlo algorithms efficient for high dimensional integrals? *Journal of Complexity, 14*, 1–33.

Sloan, I. H., Kuo, F. Y., & Joe, S. (2002). On the step-by-step construction of quasi-Monte Carlo integration rules that achieve strong tractability error bounds in weighted Sobolev spaces. *Mathematics of Computation, 71*, 1609–1640.

Sloan, I. H., Wang, X. Q., & Woźniakowski, H. (2004). Finite-order weights imply tractability of multivariate integration. *Journal of Complexity, 20*, 46–74.

Sobol', I. M. (1967). Distribution of points in a cube and approximate evaluation of integrals. *USSR Computational Mathematics and Mathematical Physics, 7*(4), 86–112.

Spanier, J. (1995). Quasi-Monte Carlo methods for particle transport problems. In H. Niederreiter & P. J.-S. Shiue (Eds.), *Monte Carlo and quasi-Monte Carlo methods in scientific computing* (pp. 121–148). New York: Springer.

Tan, K. S. & Boyle, P. P. (2000). Applications of randomized low discrepancy sequences to the valuation of complex securities. *Journal of Economic Dynamics and Control, 24*, 1747–1782.

Tezuka, S. (1998). Financial applications of Monte Carlo and quasi-Monte Carlo methods. In P. Hellekalek & G. Larcher (Eds.), *Random and quasi-random point sets* (pp. 303–332). New York: Springer.

Vandewoestyne, B. & Cools, R. (2006). Good permutations for deterministic scrambled Halton sequences in terms of L_2-discrepancy. *Journal of Computational and Applied Mathematics, 189*, 341–361.

Vandewoestyne, B. & Cools, R. (2008). On obtaining higher order convergence for smooth periodic functions. *Journal of Complexity, 24*, 328–340.

Wang, X. Q. (2002). A constructive approach to strong tractability using quasi-Monte Carlo algorithms. *Journal of Complexity, 18*, 683–701.

Wang, X. Q. (2003). Strong tractability of multivariate integration using quasi-Monte Carlo algorithms. *Mathematics of Computation, 72*, 823–838.

Wang, X. Q. & Hickernell, F. J. (2000). Randomized Halton sequences. *Mathematical and Computer Modelling, 32*, 887–899.

Wang, X. Q. & Sloan, I. H. (2005). Why are high-dimensional finance problems often of low effective dimension? *SIAM Journal on Scientific Computing, 27*, 159–183.

Woźniakowski, H. (2000). Efficiency of quasi-Monte Carlo algorithms for high dimensional integrals. In H. Niederreiter & J. Spanier (Eds.), *Monte Carlo and quasi-Monte Carlo methods 1998* (pp. 114–136). Berlin: Springer.

Xing, C. P. & Niederreiter, H. (1995). A construction of low-discrepancy sequences using global function fields. *Acta Arithmetica, 73*, 87–102.

Chapter 27
Introduction to Support Vector Machines and Their Applications in Bankruptcy Prognosis

Yuh-Jye Lee, Yi-Ren Yeh, and Hsing-Kuo Pao

Abstract We aim at providing a comprehensive introduction to Support Vector Machines and their applications in computational finance. Based on the advances of the statistical learning theory, one of the first SVM algorithms was proposed in mid 1990s. Since then, they have drawn a lot of research interests both in theoretical and application domains and have became the state-of-the-art techniques in solving classification and regression problems. The reason for the success is not only because of their sound theoretical foundation but also their good generalization performance in many real applications. In this chapter, we address the theoretical, algorithmic and computational issues and try our best to make the article self-contained. Moreover, in the end of this chapter, a case study on default prediction is also presented. We discuss the issues when SVM algorithms are applied to bankruptcy prognosis such as how to deal with the unbalanced dataset, how to tune the parameters to have a better performance and how to deal with large scale dataset.

27.1 Introduction

Finance classification problems occur in credit scoring, company rating, and many fields. One of the most important task is to predict bankruptcy before the disaster. In the era of Basel Committee on Banking Supervision (Basel II), a powerful tool for bankruptcy prognosis can always help banks to reduce their risks. On one

Y.-J. Lee (✉) · H.-K. Pao
Department of Computer Science and Information Engineering, National Taiwan University of Science and Technology, Taipei 10607, Taiwan
e-mail: yuh-jye@mail.ntust.edu.tw; pao@mail.ntust.edu.tw

Y.-R. Yeh
Research Center for Information Technology Innovation, Academia Sinica, Taipei 11529, Taiwan
e-mail: yryeh@citi.sinica.edu.tw

J.-C. Duan et al. (eds.), *Handbook of Computational Finance*, Springer Handbooks of Computational Statistics, DOI 10.1007/978-3-642-17254-0_27,

hand, the tool must be precise with high accuracy, but also easily adaptable to the bank's objectives regarding the relation of false acceptances (Type I error) and false rejections (Type II error). The prognosis has become even more important since the Basel II established borrowers' rating as the crucial criterion for minimum capital requirements of banks. The methods for generating rating figures have developed significantly over the last 10 years Krahnen and Weber (2001).

Parametric statistical models can be used for finance classification. The first introduced model of this type was discriminant analysis (DA) for univariate Beaver (1966) and multivariate models Altman (1968). After DA, the logit and probit approach for predicting default were proposed in Martin (1977) and Ohlson (1980). These approaches rely on the a priori assumed functional dependence between risk of default and predictor. One of the weakest points of DA is that it requires a linear functional, or a preshaped polynomial functional dependence. Such restrictions often fail to meet the reality of observed data. Semi-parametric models as in Hwang et al. (2007) are between conventional linear models and non-parametric approaches. Other than that, nonlinear classification methods such as Support Vector Machines (SVMs) or neural networks Tam and Kiang (1992) and Altman (1994) are even stronger candidates to meet these demands as they go beyond conventional discrimination methods. In this chapter, we concentrate on providing a comprehensive introduction to SVMs and their applications in bankruptcy prognosis.

In the last decade, significant advances have been made in support vector machines (SVMs) both theoretically, by using statistical learning theory; as well as algorithmically, by applying some optimization techniques Burges (1998), Cristianini and Shawe-Taylor (1999), Lee and Mangasarian (2001), Mangasarian (2000), Schölkopf and Smola (2002), Smola and Schölkopf (2004). SVMs have been successfully developed and have become powerful tools for solving data mining problems such as classification, regression and feature selection. In classification problems, an SVM determine an optimal separating hyperplane that classifies data points into different categories. Here, "optimality" refers to the sense that the *separating hyperplane* has the best generalization ability for unseen data points, based on statistical learning theory. With the help of nonlinear kernel functions, SVMs can discriminate between complex data patterns by generating a highly nonlinear separating hyperplane. The nonlinear extension of SVMs makes them applicable to many important real world problems such as character recognition, face detection, analysis of DNA microarrays, breast cancer diagnosis and prognosis Cao and Tay (2003), Min and Lee (2005), and the problem of bankruptcy prognosis as we will see.

The goal of this chapter is to provide a comprehensive introduction to SVMs and their applications in bankruptcy prognosis. The remainder of the chapter is organized as follows. Section 27.2 introduces the basic ideas and the typical formulation of SVM. Some variants of SVMs are discussed in Sect. 27.3 to solve problems of many kinds. Section 27.4 details some implementation issues and techniques. We discuss solving SVMs in primal and dual forms. In Sect. 27.5, to deal with real world problems, we talk about some practical issues of using SVMs.

In Sect. 27.6, we apply SVMs on a problem of bankruptcy prognosis. Then, in Sect. 27.7, we summarize our conclusions.

27.2 Support Vector Machine Formulations

In this section, we first introduce the basic idea of SVM and give the formulation of *linear* support vector machine. Even the linear version looks too simple to be powerful enough for real applications, it has a non-trivial nonlinear extension. The concept of nonlinear extension of SVM is a milestone for dealing with nonlinear problems and it has a great influence on the machine learning community in this couple of decades. All the details of nonlinear extension, including the "kernel trick" and Mercer's theorem, are introduced in this section. In the end, we discuss the actual risk bound to show the insight behind SVM induction.

27.2.1 The Formulation of Conventional Support Vector Machine

In this article, we mainly confine ourselves to binary classification problems, which focus on classifying data into two classes. Given a dataset consisting of m points in the n-dimensional real space $\mathbb{R}^n$, each with a class label y, $+1$ or -1, indicating one of two classes, $\mathbf{A}_+$, $\mathbf{A}_- \subseteq \mathbb{R}^n$ where the point belongs, we want to find the decision boundary between the two classes. For the multi-class case, many strategies have been proposed. They either decompose the problem into a series of binary classification or formulate it as a single optimization problem. We will discuss this issue in Sect. 27.5. In notation, we use capital boldface letters to denote a matrix, lower case boldface letters to denote a column vector, and low case light face letters to denote scalars. The data points are denoted by an $m \times n$ matrix $\mathbf{A}$, where the ith row of the matrix corresponds to the ith data point. We use a column vector $\mathbf{x}_i$ to denote the ith data point. All vectors indicate column vectors unless otherwise specified. The transpose of a matrix $\mathbf{M}$ is denoted by $\mathbf{M}^\top$.

27.2.1.1 Primal Form of Conventional SVM

We start with a strictly linearly separable case, i.e. there exists a hyperplane which can separate the data $\mathbf{A}_+$ and $\mathbf{A}_-$. In this case we can separate the two classes by a pair of parallel *bounding planes*:

$$\begin{aligned} \mathbf{w}^\top\mathbf{x} + b &= +1\,, \\ \mathbf{w}^\top\mathbf{x} + b &= -1\,, \end{aligned} \tag{27.1}$$

where $\mathbf{w}$ is the normal vector to these planes and b determines their location relative to the origin. The first plane of (27.1) bounds the class $\mathbf{A}_+$ and the second plane bounds the class $\mathbf{A}_-$. That is,

$$\begin{aligned} \mathbf{w}^\top \mathbf{x} + b &\geq +1\,, \quad \forall\, \mathbf{x} \in \mathbf{A}_+\,, \\ \mathbf{w}^\top \mathbf{x} + b &\leq -1\,, \quad \forall\, \mathbf{x} \in \mathbf{A}_-\,. \end{aligned} \tag{27.2}$$

According to the statistical learning theory Vapnik (2000), SVM achieves a better prediction ability via maximizing the margin between two bounding planes. Hence, the "hard margin" SVM searches for a separating hyperplane by maximizing $\frac{2}{\|\mathbf{w}\|_2}$. It can be done by means of minimizing $\frac{1}{2}\|\mathbf{w}\|_2^2$ and the formulation leads to a quadratic program as follows:

$$\begin{aligned} &\min_{(\mathbf{w},b)\in\mathbb{R}^{n+1}} \quad \frac{1}{2}\|\mathbf{w}\|_2^2 \\ &\text{s.t.} \quad y_i(\mathbf{w}^\top \mathbf{x}_i + b) \geq 1\,, \quad \text{for } i = 1, 2, \ldots, m\,. \end{aligned} \tag{27.3}$$

The linear separating hyperplane is the plane

$$\mathbf{w}^\top \mathbf{x} + b = 0\,, \tag{27.4}$$

midway between the bounding planes (27.1), as shown in Fig. 27.1a. For the linearly separable case, the *feasible region* of the above minimization problem (27.3) is nonempty and the objective function is a quadratic convex function; therefore, there exists an optimal solution, denoted by $(\mathbf{w}^*,\ b^*)$. The data points on the bounding planes, $\mathbf{w}^{*\top}\mathbf{x} + b^* = \pm 1$, are called *support vectors*. It is not difficult to see that, if we remove any point that is not a support vector, the training result will remain the same. This is a nice property of SVM learning algorithms. For the purpose of data compression, once we have the training result, all we need to keep in our database are the support vectors.

If the classes are not linearly separable, in some cases, two planes may bound the two classes with a "soft margin". That is, given a nonnegative slack vector variable $\boldsymbol{\xi} := (\xi_1, \ldots, \xi_m)$, we would like to have:

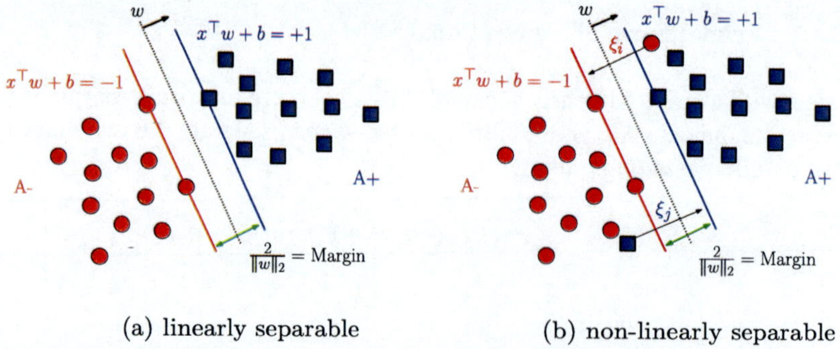

Fig. 27.1 The illustration of linearly separable and non-linearly separable SVMs

$$\begin{aligned} \mathbf{w}^\top \mathbf{x}_i + b + \xi_i &\geq +1\,, \quad \forall\, \mathbf{x}_i \in \mathbf{A}_+ \\ \mathbf{w}^\top \mathbf{x}_i + b - \xi_i &\leq -1\,, \quad \forall\, \mathbf{x}_i \in \mathbf{A}_-\,. \end{aligned} \tag{27.5}$$

The 1-norm of the slack vector variable $\boldsymbol{\xi}$, $\sum_{i=1}^{m} \xi_i$, is called the penalty term. In principle, we are going to determine a separating hyperplane that not only correctly classifies the training data, but also performs well on test data. We depict the geometric property in Fig. 27.1b. With a soft margin, we can extend (27.3) and produce the conventional SVM (Vapnik 2000) as the following formulation:

$$\begin{aligned} \min_{(\mathbf{w},b,\boldsymbol{\xi}) \in \mathbb{R}^{n+1+m}} \quad & \frac{1}{2}\|\mathbf{w}\|_2^2 + C \sum_{i=1}^{m} \xi_i \\ \text{s.t.} \quad & y_i(\mathbf{w}^\top \mathbf{x}_i + b) + \xi_i \geq 1\,, \\ & \xi_i \geq 0, \quad \text{for } i = 1, 2, \ldots, m\,, \end{aligned} \tag{27.6}$$

where $C > 0$ is a positive parameter that balances the weight of the penalty term $\sum_{i=1}^{m} \xi_i$ and the margin maximization term $\frac{1}{2}\|\mathbf{w}\|_2^2$. Alternatively, we can replace the penalty term by the 2-norm measure as follows:

$$\begin{aligned} \min_{(\mathbf{w},b,\boldsymbol{\xi}) \in \mathbb{R}^{n+1+m}} \quad & \frac{1}{2}\|\mathbf{w}\|_2^2 + C \sum_{i=1}^{m} \xi_i^2 \\ \text{s.t.} \quad & y_i(\mathbf{w}^\top \mathbf{x}_i + b) + \xi_i \geq 1\,, \\ & \text{for } i = 1, 2, \ldots, m\,. \end{aligned} \tag{27.7}$$

The 1-norm penalty is considered less sensitive to outliers than the 2-norm penalty, therefore it receives more attention in real applications. However, mathematically the 1-norm is more difficult to manipulate such as when we need to compute the derivatives.

27.2.1.2 Dual Form of Conventional SVM

The conventional support vector machine formulation (27.6) is a standard convex quadratic program Bertsekas (1999), Mangasarian (1994), Nocedal and Wright (2006). The Wolfe dual problem of (27.6) is expressed as follows:

$$\begin{aligned} \max_{\boldsymbol{\alpha} \in \mathbb{R}^m} \quad & \sum_{i=1}^{m} \alpha_i - \frac{1}{2} \sum_{i=1}^{m} \sum_{j=1}^{m} y_i y_j \alpha_i \alpha_j \langle \mathbf{x}_i, \mathbf{x}_j \rangle \\ \text{s.t.} \quad & \sum_{i=1}^{m} y_i \alpha_i = 0\,, \\ & 0 \leq \alpha_i \leq C \quad \text{for } i = 1, 2, \ldots, m\,, \end{aligned} \tag{27.8}$$

where $\langle \mathbf{x}_i, \mathbf{x}_j \rangle$ is the inner product of $\mathbf{x}_i$ and $\mathbf{x}_j$. The primal variable $\mathbf{w}$ is given by:

$$\mathbf{w} = \sum_{\alpha_i > 0} y_i \alpha_i \mathbf{x}_i \,. \tag{27.9}$$

Each dual variable α_i corresponds to a training point $\mathbf{x}_i$. The normal vector $\mathbf{w}$ can be expressed in terms of a linear combination of training data points which have corresponding positive dual variables α_i (namely, the support vectors). By the Karush-Kuhn-Tucker complementarity conditions Bertsekas (1999), Mangasarian (1994):

$$\begin{array}{lll} 0 \leq \quad \alpha_i & \perp & y_i(\mathbf{w}^\top \mathbf{x}_i + b) + \xi_i - 1 \geq 0 \\ 0 \leq C - \alpha_i & \perp & \xi_i \geq 0 \quad , \quad \text{for } i = 1, 2, \ldots, m \,, \end{array} \tag{27.10}$$

we can determine b simply by taking any training point $\mathbf{x}_i$, such that $i \in I := \{k \mid 0 < \alpha_k < C\}$ and obtain:

$$b = y_i - \mathbf{w}^\top \mathbf{x}_i = y_i - \sum_{j=1}^{m} (y_j \alpha_j \langle \mathbf{x}_j, \mathbf{x}_i \rangle) \,. \tag{27.11}$$

In the dual form, SVMs can be expressed by the form of inner product. It implies that we only need the information of the inner product of the data when expressing the formulation and decision function of SVM. This important characteristic carries SVMs to their nonlinear extension in a simple way.

27.2.2 *Nonlinear Extension of SVMs via Kernel Trick*

In many cases, a dataset, as collected in a vector form full of attributes, cannot be well separated by a linear separating hyperplane. However, it is likely that the dataset becomes linearly separable after mapped into a higher dimensional space by a nonlinear map. A nice property of SVM methodology is that we do not even need to know the nonlinear map explicitly; still, we can apply a linear algorithm to the classification problem in the high dimensional space. The property comes from the dual form of SVM which can express the formulation in terms of inner product of data points. By taking the advantage of dual form, the "kernel trick" is used for the nonlinear extension of SVM.

27.2.2.1 Kernel Trick

From the dual SVM formulation (27.8), all we need to know is simply the inner product between training data vectors. Let us map the training data points from the input space $\mathbb{R}^n$ to a higher-dimensional feature space $\mathcal{F}$ by a nonlinear map Φ.

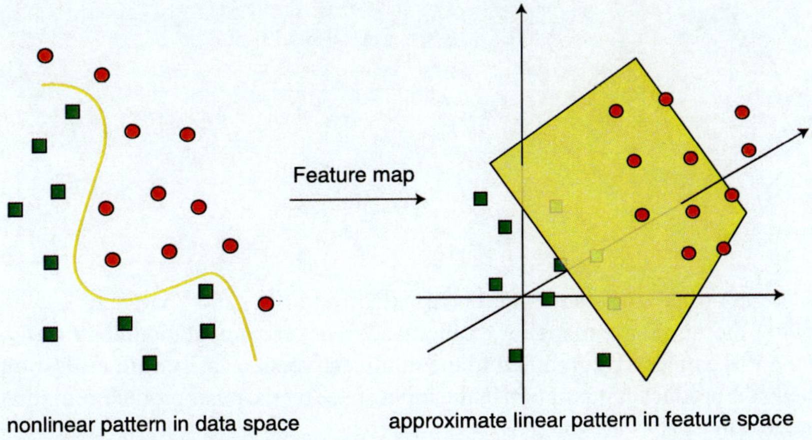

Fig. 27.2 The illustration of nonlinear SVM

The training data $\mathbf{x}$ in $\mathcal{F}$ becomes $\Phi(\mathbf{x}) \in \mathbb{R}^{\ell}$ where ℓ is the dimensionality of the feature space $\mathcal{F}$. Based on the above observation, if we know the inner product $\Phi(\mathbf{x}_i)^{\top}\Phi(\mathbf{x}_j)$ for all $i, j = 1, 2, \ldots, m$, then we can perform the linear SVM algorithm in the feature space $\mathcal{F}$. The separating hyperplane will be linear in the feature space $\mathcal{F}$ but is a nonlinear surface in the input space $\mathbb{R}^n$ (see Fig. 27.2).

Note that we do not need to know the nonlinear map Φ explicitly. It can be achieved by employing a kernel function. Let $k(\mathbf{x}, \mathbf{z}) : \mathbb{R}^n \times \mathbb{R}^n \to \mathbb{R}$ be an inner product kernel function satisfying *Mercer's condition* Burges (1998), Cherkassky and Mulier (1998), Courant and Hilbert (1953), Cristianini and Shawe-Taylor (1999), Vapnik (2000), positive semi-definiteness condition (see Definition 1). We can construct a nonlinear map Φ such that $k(\mathbf{x}_i, \mathbf{x}_j) = \Phi(\mathbf{x}_i)^{\top}\Phi(\mathbf{x}_j)$ where $i, j = 1, 2, \ldots, m$. Hence, the linear SVM formulation can be used on $\Phi(\mathbf{x})$ in the feature space $\mathcal{F}$ by replacing the $\langle \mathbf{x}_i, \mathbf{x}_j \rangle$ in the objective function of (27.8) with a nonlinear kernel function $k(\mathbf{x}_i, \mathbf{x}_j)$. The resulting dual nonlinear SVM formulation becomes:

$$\begin{aligned} \max_{\alpha \in \mathbb{R}^m} \quad & \sum_{i=1}^{m} \alpha_i - \frac{1}{2} \sum_{i=1}^{m} \sum_{j=1}^{m} y_i y_j \alpha_i \alpha_j k(\mathbf{x}_i, \mathbf{x}_j) \\ \text{s.t.} \quad & \sum_{i=1}^{m} y_i \alpha_i = 0 \\ & 0 \le \alpha_i \le C \quad \text{for } i = 1, 2, \ldots, m\,. \end{aligned} \tag{27.12}$$

The nonlinear separating hyperplane is defined by the solution of (27.12) as follows:

$$\sum_{j=1}^{m}(y_j \alpha_j k(\mathbf{x}_j, \mathbf{x}_i)) + b = 0\,, \tag{27.13}$$

where

$$b = y_i - \sum_{j=1}^{m}(y_j \alpha_j k(\mathbf{x}_j, \mathbf{x}_i)), \;\; i \in I := \{k \mid 0 < \alpha_k < C\}\,. \tag{27.14}$$

The "kernel trick" makes the nonlinear extension of linear SVM possible without knowing the nonlinear mapping explicitly. Whatever computation code ready for linear SVM can also be modified to the nonlinear version easily with a substitution of the inner product computation in the input space by the inner product computation in the feature space.

27.2.2.2 Mercer's Theorem

The basic idea of kernel trick is replacing the inner product between data points by the kernel function $k(\mathbf{x}, \mathbf{z})$. However, it is not always possible for a given function $k(\mathbf{x}, \mathbf{z})$ to reconstruct its corresponding nonlinear maps. We can answer the question by the so-called *Mercer's condition* (Vapnik 2000). We conclude this section with *Mercer's condition* and two examples of kernel function.

Definition 1 (Mercer's condition). Let $k(\mathbf{s}, \mathbf{t}) : \mathbb{R}^n \times \mathbb{R}^n \to \mathbb{R}$ be a continuous symmetric function and X be a compact subset of $\mathbb{R}^n$. If

$$\int_{X \times X} k(\mathbf{s}, \mathbf{t}) f(\mathbf{s}) f(\mathbf{t})\, d\mathbf{s}\, d\mathbf{t} \geq 0, \;\; \forall f \in \ell_2(X)\,, \tag{27.15}$$

where the Hilbert space $\ell_2(X)$ is the set of functions f such that

$$\int_X f(\mathbf{t})^2\, d\mathbf{t} < \infty\,. \tag{27.16}$$

then the function k satisfies *Mercer's condition.*

This is equivalent to say that the kernel matrix $\mathbf{K}(\mathbf{A}, \mathbf{A})$ in our application is positive semi-definite (Cristianini and Shawe-Taylor 1999), where $\mathbf{K}(\mathbf{A}, \mathbf{A})_{ij} = k(\mathbf{x}_i, \mathbf{x}_j)$ for $i, j = 1, 2, \ldots, m$. Below are two most popular kernel functions in real applications. The choice of kernel functions may rely on the result of a *cross-validation* or model selection procedure.

Example 1. Polynomial Kernel

$$k(\mathbf{x}, \mathbf{z}) = (\mathbf{x}^\top \mathbf{z} + b)^d\,, \tag{27.17}$$

where d denotes the degree of the exponentiation.

Example 2. Gaussian (Radial Basis) Kernel

$$k(\mathbf{x}, \mathbf{z}) = e^{-\gamma \|\mathbf{x}-\mathbf{z}\|_2^2}, \tag{27.18}$$

where γ is the width parameter of Gaussian kernel.

27.2.3 A Bound on Actual Risk

The main goal of the classification problem is to predict the label of new unseen data points correctly. That is, we seek for a classifier $f(\mathbf{x}, \alpha)$ with output values 1 and −1 that can minimize the following test error:

$$R(\alpha) = \int \frac{1}{2}|y - f(\mathbf{x}, \alpha)|\, dP(\mathbf{x}, y), \tag{27.19}$$

where $\mathbf{x}$ is an instance and y is the class label of $\mathbf{x}$, with $(\mathbf{x}, y)$ drawn from some unknown probability distribution $P(\mathbf{x}, y)$, and α is an adjustable parameter of f. The error, so called the actual risk, in which we are interested can represent the true mean error but it needs to know what $P(\mathbf{x}, y)$ is. However, estimating $P(\mathbf{x}, y)$ is usually not possible so that (27.19) is not very useful in practical usage. The usual way is to approximate the actual risk by using the empirical risk:

$$R_{emp}(\alpha) = \frac{1}{2m}\sum_{i=1}^{m} |y_i - f(\mathbf{x}_i, y)|. \tag{27.20}$$

This empirical risk is obtained by considering only a finite number of training data. Looking for a model that fits the given dataset usually is not a good way to do. There always exists a model that can classify the training data perfectly as long as there is no identical data points that have different labels. However this model might overfit the training data and perform poorly on the unseen data. There are some bounds governing the relation between the capacity of a learning machine and its performance. It can be used for balancing the *model bias* and *model variance*. Vapnik (2000) proposed a upper bound for $R(\alpha)$ with probability $1 - \eta$ as follows:

$$R(\alpha) \le R_{emp}(\alpha) + \sqrt{\frac{h(\log(2m/h) + 1) - \log(\eta/4)}{m}}, \tag{27.21}$$

where η is between 0 and 1, m is the number of instances, h is a non-negative integer called the Vapnik Chervonenkis (VC) dimension. The second term on the right-hand side of (27.21) is called the VC confidence.

The upper bound in (27.21) gives a principle for choosing a learning model for a given task. Thus given several different learning models and a fixed, sufficiently small η, choosing a model that minimizes the right-hand side is equivalent to choosing a model that gives the lowest upper bound on the actual risk. Note that the VC confidence is a monotonic increasing function of h. This means that a complicated learning model may also have a high upper bound on the actual risk. In general, for non zero empirical risk, one wants to choose that learning model which minimizes the right-hand side of (27.21). This idea of balancing the model complexity and empirical risk is considered in SVMs. The objective functions of (27.6) and (27.7) can be interpreted as the upper bound of actual risk in (27.21) Burges (1998), Vapnik (2000). Basically, SVM defines a trade-off between the quality of the separating hyperplane on the training data and the complexity of the separating hyperplane. Higher complexity of the separating hyperplane may cause overfitting and lead to poor generalization. The positive parameter C which can be determined by a *tuning procedure* such as cross-validation, plays the role of balancing this trade-off. We will discuss the issue in more details in Sect. 27.5.

27.3 Variants of Support Vector Machines

Since the typical SVM was proposed for the first time in late 1990s, to deal with various kinds of applications, many variants of SVM have been proposed. The different formulations of SVM have their own approaches in dealing with data. In this section, we will introduce some of them, as well as their properties and applications.

27.3.1 Smooth Support Vector Machine

In contrast to the conventional SVM of (27.6), smooth support vector machine (SSVM) Lee and Mangasarian (2001) minimizes the square of the slack vector $\boldsymbol{\xi}$. In addition, the SSVM prefers a solution with a small value of b (also in 2-norm). That leads to the following minimization problem:

$$\begin{aligned}
\min_{(\mathbf{w},b,\boldsymbol{\xi})\in\mathbb{R}^{n+1+m}} \quad & \frac{1}{2}(\|\mathbf{w}\|_2^2 + b^2) + \frac{C}{2}\sum_{i=1}^{m}\xi_i^2 \\
\text{s.t.} \quad & y_i(\mathbf{w}^\top\mathbf{x}_i + b) + \xi_i \geq 1 \\
& \xi_i \geq 0, \quad \text{for } i = 1, 2, \ldots, m\,.
\end{aligned} \tag{27.22}$$

As a solution of (27.22), $\boldsymbol{\xi}$ is given by $\xi_i = \{1 - y_i(\mathbf{w}^\top\mathbf{x}_i + b)\}_+$ for all i where the plus function x_+ is defined as $x_+ = \max\{0, x\}$. Thus, we can replace ξ_i in (27.22) by $\{1 - y_i(\mathbf{w}^\top\mathbf{x}_i + b)\}_+$. It converts the problem (27.22) into an unconstrained minimization problem as follows:

$$\min_{(\mathbf{w},b)\in\mathbb{R}^{n+1}} \frac{1}{2}(\|\mathbf{w}\|_2^2 + b^2) + \frac{C}{2}\sum_{i=1}^{m}\{1 - y_i(\mathbf{w}^\top \mathbf{x}_i + b)\}_+^2 . \tag{27.23}$$

Compared to (27.22), this formulation reduces the number of variables from $n+1+m$ to $n+1$; however, the objective function to be minimized is no longer twice differentiable. In SSVM, we prefer a twice differentiable form so that a fast Newton method can be applied. We approximate the plus function x_+ by a smooth p-function:

$$p(x,\beta) = x + \frac{1}{\beta}\log(1 + e^{-\beta x}) , \tag{27.24}$$

where $\beta > 0$ is the smooth parameter which controls the "steepness" of the curve or how close it is to the original plus function x_+. By replacing the plus function x_+ with a very accurate approximation p-function gives the SSVM formulation:

$$\min_{(\mathbf{w},b)\in\mathbb{R}^{n+1}} \frac{1}{2}(\|\mathbf{w}\|_2^2 + b^2) + \frac{C}{2}\sum_{i=1}^{m} p(\{1 - y_i(\mathbf{w}^\top \mathbf{x}_i + b)\}, \beta)^2 , \tag{27.25}$$

The objective function in problem (27.25) is strongly convex and infinitely differentiable. Hence, it has a unique solution and can be solved by using a fast *Newton-Armijo* algorithm (discussed in the implementation part, Sect. 27.4). For the nonlinear case, this formulation can be extended to the nonlinear version by utilizing the kernel trick as follows:

$$\min_{(\mathbf{u},b)\in\mathbb{R}^{m+1}} \frac{1}{2}(\|\mathbf{u}\|_2^2 + b^2) + \frac{C}{2}\sum_{i=1}^{m} p([1 - y_i\,\{\sum_{j=1}^{m} u_j k(\mathbf{x}_i, \mathbf{x}_j) + b\}], \beta)^2 , \tag{27.26}$$

where $k(\mathbf{x}_i, \mathbf{x}_j)$ is a kernel function. The nonlinear SSVM classifier $f(\mathbf{x})$ can be expressed as follows:

$$f(\mathbf{x}) = \sum_{u_j \neq 0} u_j k(\mathbf{x}_j, \mathbf{x}) + b . \tag{27.27}$$

27.3.2 Reduced Smooth Support Vector Machine

In these days, very often we have classification or regression problems with large-scale data, such as the data from network traffic, gene expressions, web documents, etc. To solve large-scale problems by SVM, the full kernel matrix will be very large, so it may not be appropriate to use the full matrix when dealing with (27.26). In order to avoid facing such a large full matrix, we brought in the *reduced* kernel technique (Lee and Huang 2007). The key idea of the reduced kernel technique is to randomly select a small portion of data and to generate a thin rectangular kernel matrix, then to use this much smaller rectangular kernel matrix to replace the full kernel matrix. In the process of replacing the full kernel matrix by a reduced kernel, we use the Nyström approximation (Smola and Schölkopf 2000; Williams and Seeger 2001) for the full kernel matrix:

$$\mathbf{K}(\mathbf{A},\mathbf{A}) \approx \mathbf{K}(\mathbf{A},\tilde{\mathbf{A}})\mathbf{K}(\tilde{\mathbf{A}},\tilde{\mathbf{A}})^{-1}\mathbf{K}(\tilde{\mathbf{A}},\mathbf{A})\,, \tag{27.28}$$

where $\tilde{\mathbf{A}}_{\tilde{\mathbf{m}}\times\mathbf{n}}$ is a subset of $\mathbf{A}$ and $\mathbf{K}(\mathbf{A},\tilde{\mathbf{A}}) = \tilde{\mathbf{K}}_{\mathbf{m}\times\tilde{\mathbf{m}}}$ is a reduced kernel. Thus, we have

$$\mathbf{K}(\mathbf{A},\mathbf{A})\mathbf{u} \approx \mathbf{K}(\mathbf{A},\tilde{\mathbf{A}})\mathbf{K}(\tilde{\mathbf{A}},\tilde{\mathbf{A}})^{-1}\mathbf{K}(\tilde{\mathbf{A}},\mathbf{A})\mathbf{u} = \mathbf{K}(\mathbf{A},\tilde{\mathbf{A}})\tilde{\mathbf{u}}\,, \tag{27.29}$$

where $\tilde{\mathbf{u}} \in \mathbb{R}^{\tilde{m}}$ is an approximated solution of $\mathbf{u}$ via the reduced kernel technique. By using the approximation, reduced SVM randomly selects a small subset $\tilde{\mathbf{A}}$ to generate the basis functions $\mathcal{B}$:

$$\mathcal{B} = \{1\} \cup \{k(\cdot,\tilde{x}^i)\}_{i=1}^{\tilde{m}}\,.$$

The formulation of reduced SSVM, hence, is expressed as follows:

$$\min_{\tilde{u},b,\xi} \frac{1}{2}(\|\tilde{u}\|_2^2 + b^2) + \frac{C}{2}\sum_{i=1}^{\tilde{m}} p([1 - y_i\{\sum_{j=1}^{\tilde{m}} \tilde{u}_j k(\mathbf{x}_i,\tilde{\mathbf{x}}_j) + b\}],\beta)^2 \tag{27.30}$$

and its decision function is in the form

$$f(x) = \sum_{i=1}^{\tilde{m}} \tilde{u}_i k(\mathbf{x},\tilde{\mathbf{x}}_i) + b\,. \tag{27.31}$$

The reduced kernel method constructs a compressed model and cuts down the computational cost from $\mathcal{O}(m^3)$ to $\mathcal{O}(\tilde{m}^3)$. It has been shown that the solution of reduced kernel matrix approximates the solution of full kernel matrix well (Lee and Huang 2007).

27.3.3 Least Squares Support Vector Machine

The least squares support vector machine (Suykens and Vandewalle 1999) considers the equality constraints which make the formulation of the classification problem in the sense of least squares as follows:

$$\min_{(\mathbf{w},b,\xi)\in\mathbb{R}^{n+1+m}} \frac{1}{2}\|\mathbf{w}\|_2^2 + C\sum_{i=1}^{m}\xi_i^2 \tag{27.32}$$
$$\text{s.t. } \xi_i = 1 - y_i(\mathbf{w}^\top\mathbf{x}_i + b) \text{ for } i = 1,2,\ldots,m\,.$$

The same idea, called proximal support vector machine, is also proposed simultaneously in Fung and Mangasarian (2001), with adding the square of the bias term b in the objective function. With the least squares form, one can obtain the solution of the classification problem via solving a set of linear equations. Consider the Lagrangian function of (27.32):

$$\mathcal{L}(\mathbf{w}, b, \boldsymbol{\xi}; \boldsymbol{\alpha}) = \frac{1}{2}\|\mathbf{w}\|_2^2 + C\sum_{i=1}^{m} \xi_i^2 - \sum_{i=1}^{m} \alpha_i [y_i(\mathbf{w}^\top \mathbf{x}_i + b) - 1 + \xi_i], \quad (27.33)$$

where $\alpha_i \in \mathbb{R}$ are Lagrange multipliers. Setting the gradient of $\mathcal{L}$ to zeros gives the following Karush-Kuhn-Tucker optimality conditions:

$$\mathbf{w} = \sum_{i=1}^{m} \alpha_i y_i \mathbf{x}_i \quad (27.34)$$

$$\sum_{i=1}^{m} \alpha_i y_i = 0$$

$$\alpha_i = C\xi_i, \ i = 1, \ldots, m$$

$$y_i(\mathbf{w}^\top \mathbf{x}_i + b) - 1 + \xi_i = 0,$$

which are equivalent to the following linear equations:

$$\begin{bmatrix} I & 0 & 0 & -\hat{\mathbf{A}}^\top \\ 0 & 0 & 0 & -\mathbf{y}^\top \\ 0 & 0 & C\mathbf{I} & -\mathbf{I} \\ \hat{\mathbf{A}} & \mathbf{y} & \mathbf{I} & 0 \end{bmatrix} \begin{bmatrix} \mathbf{w} \\ b \\ \boldsymbol{\xi} \\ \boldsymbol{\alpha} \end{bmatrix} = \begin{bmatrix} 0 \\ 0 \\ 0 \\ \mathbf{1} \end{bmatrix}, \quad (27.35)$$

or, equivalently,

$$\begin{bmatrix} 0 & -\mathbf{y}^\top \\ \mathbf{y} & \hat{\mathbf{A}}\hat{\mathbf{A}}^\top + \frac{1}{C}\mathbf{I} \end{bmatrix} \begin{bmatrix} b \\ \boldsymbol{\alpha} \end{bmatrix} = \begin{bmatrix} 0 \\ \mathbf{1} \end{bmatrix}, \quad (27.36)$$

where $\hat{\mathbf{A}} = [\mathbf{x}_1 y_1; \mathbf{x}_2 y_2; \ldots; \mathbf{x}_m y_m]$, $\mathbf{y} = [y_1; y_2; \ldots; y_m]$, and $\mathbf{1} = [1; 1; \ldots; 1]$. From (27.36), the nonlinear least squares SVM also can be extended via the inner product form. That is, the nonlinear least squares SVM solves the following linear equations:

$$\begin{bmatrix} 0 & -\mathbf{y}^\top \\ \mathbf{y} & \mathbf{K}(\mathbf{A}, \mathbf{A}) + \frac{1}{C}\mathbf{I} \end{bmatrix} \begin{bmatrix} b \\ \boldsymbol{\alpha} \end{bmatrix} = \begin{bmatrix} 0 \\ \mathbf{1} \end{bmatrix}, \quad (27.37)$$

where $\mathbf{K}(\mathbf{A}, \mathbf{A})$ is the kernel matrix. Equations (27.36) or (27.37) gives an analytic solution to the classification problem via solving a system of linear equations. This brings a lower computational cost by comparing with solving a conventional SVM while obtaining a least squares SVM classifier.

27.3.4 1-norm Support Vector Machine

The 1-norm support vector machine replaces the regularization term $\|\mathbf{w}\|_2^2$ in (27.6) by a ℓ_1-norm of $\mathbf{w}$. The ℓ_1-norm regularization term is also called the LASSO penalty (Tibshiran 1996). It tends to shrink the coefficients $\mathbf{w}$'s towards zeros in particular for those coefficients corresponding to redundant noise features (Zhu et al. 2004). This nice feature will lead to a way of selecting the important attributes in our prediction model. The formulation of 1-norm SVM is described as follows:

$$\begin{aligned} \min_{(\mathbf{w},b,\boldsymbol{\xi})\in\mathbb{R}^{n+1+m}} \quad & \|\mathbf{w}\|_1 + C\sum_{i=1}^{m}\xi_i \\ \text{s.t.} \quad & y_i(\mathbf{w}^\top\mathbf{x}_i + b) + \xi_i \geq 1 \\ & \xi_i \geq 0, \quad \text{for } i = 1, 2, \ldots, m\,. \end{aligned} \tag{27.38}$$

The objective function of (27.38) is a piecewise linear convex function. We can reformulate it as the following linear programming problem:

$$\begin{aligned} \min_{(\mathbf{w},\mathbf{s},b,\boldsymbol{\xi})\in\mathbb{R}^{n+n+1+m}} \quad & \sum_{j=1}^{n} s_j + C\sum_{i=1}^{m}\xi_i \\ \text{s.t.} \quad & y_i(\mathbf{w}^\top\mathbf{x}_i + b) + \xi_i \geq 1 \\ & -s_j \leq w_j \leq s_j, \quad \text{for } j = 1, 2, \ldots, n\,, \\ & \xi_i \geq 0, \quad \text{for } i = 1, 2, \ldots, m\,, \end{aligned} \tag{27.39}$$

where s_j is the upper bound of the absolute value of w_j. At the optimal solution of (27.39) the sum of s_j is equal to $\|\mathbf{w}\|_1$.

The 1-norm SVM can generate a very sparse solution $\mathbf{w}$ and lead to a parsimonious model. In a linear SVM classifier, solution sparsity means that the separating function $f(\mathbf{x}) = \mathbf{w}^\top\mathbf{x} + b$ depends on very few input attributes. This characteristic can significantly suppress the number of the nonzero coefficients $\mathbf{w}$'s, especially when there are many redundant noise features (Fung and Mangasarian 2004; Zhu et al. 2004). Therefore the 1-norm SVM can be a very promising tool for *variable selection*. In Sect. 27.6, we will use it to choose the important financial indices for our bankruptcy prognosis model.

27.3.5 ε-Support Vector Regression

In regression problems, the response $\mathbf{y}$ belongs to real numbers. We would like to find a linear or nonlinear regression function, $f(\mathbf{x})$, that tolerates a small error in fitting the given dataset. It can be achieved by utilizing the ε-insensitive loss function that sets an ε-insensitive "tube" around the data, within which errors are discarded.

We start with the linear case, that is the regression function $f(\mathbf{x})$ defined as $f(\mathbf{x}) = \mathbf{w}^\top \mathbf{x} + b$. The SVM minimization can be formulated as an unconstrained problem given by:

$$\min_{(\mathbf{w},b,\xi)\in\mathbb{R}^{n+1}} \frac{1}{2}\|\mathbf{w}\|_2^2 + C\sum_{i=1}^{m} |\xi_i|_\varepsilon , \tag{27.40}$$

where $|\xi_i|_\varepsilon = \max\{0, |\mathbf{w}^\top \mathbf{x}_i + b - y_i| - \varepsilon\}$, represents the fitting errors and the positive control parameter C here weights the tradeoff between the fitting errors and the flatness of the linear regression function $f(\mathbf{x})$. Similar to the idea in SVM, the regularization term $\|\mathbf{w}\|_2^2$ in (27.40) is also applied for improving the generalization ability. To deal with the ε-insensitive loss function in the objective function of the above minimization problem, conventionally, it is reformulated as a constrained minimization problem defined as follows:

$$\min_{(\mathbf{w},b,\xi,\xi^*)\in R^{n+1+2m}} \frac{1}{2}\|\mathbf{w}\|_2^2 + C\sum_{i=1}^{m}(\xi_i + \xi_i^*) \tag{27.41}$$

$$\begin{aligned} \text{s.t.}\quad & \mathbf{w}^\top \mathbf{x}_i + b - y_i \le \varepsilon + \xi_i , \\ & -\mathbf{w}^\top \mathbf{x}_i - b + y_i \le \varepsilon + \xi_i^* , \\ & \xi_i, \xi_i^* \ge 0 \text{ for } i = 1, 2, \ldots, m . \end{aligned}$$

This formulation (27.41) is equivalent to the formulation (27.40) and its corresponding dual form is

$$\begin{aligned} \max_{\alpha,\hat{\alpha}\in\mathbb{R}^m} \quad & \sum_{i=1}^{m}(\hat{\alpha}_i - \alpha_i)y_i - \varepsilon\sum_{i=1}^{m}(\hat{\alpha}_i + \alpha_i) \\ & -\sum_{i=1}^{m}\sum_{j=1}^{m}(\hat{\alpha}_i - \alpha_i)(\hat{\alpha}_j - \alpha_j)\langle \mathbf{x}_i, \mathbf{x}_j\rangle , \\ \text{s.t.}\quad & \sum_{i=1}^{m}(\hat{u}_i - u_i) = 0 , \\ & 0 \le \alpha_i, \hat{\alpha}_i \le C , \quad \text{for } i = 1, \ldots, m . \end{aligned} \tag{27.42}$$

From (27.42), one also can apply the kernel trick on this dual form of ε-SVR for the nonlinear extension. That is, $\langle \mathbf{x}_i, \mathbf{x}_j\rangle$ is directly replaced by a kernel function $k(\mathbf{x}_i, \mathbf{x}_j)$ as follows:

$$\begin{aligned} \max_{\alpha,\hat{\alpha}\in\mathbb{R}^m} \quad & \sum_{i=1}^{m}(\hat{\alpha}_i - \alpha_i)y_i - \varepsilon\sum_{i=1}^{m}(\hat{\alpha}_i + \alpha_i) \\ & -\sum_{i=1}^{m}\sum_{j=1}^{m}(\hat{\alpha}_i - \alpha_i)(\hat{\alpha}_j - \alpha_j)k(\mathbf{x}_i, \mathbf{x}_j) , \end{aligned} \tag{27.43}$$

$$\text{s.t.} \sum_{i=1}^{m}(\hat{\alpha}_i - \alpha_i) = 0\,,$$

$$0 \leq \alpha_i, \hat{\alpha}_i \leq C\,, \quad \text{for } i = 1, \ldots, m\,.$$

with the decision function $f(\mathbf{x}) = \sum_{i=1}^{m}(\hat{\alpha}_i - \alpha_i)k(\mathbf{x}_i, \mathbf{x}) + b$.

Similar to the smooth approach in SSVM, the formulation (27.40) can be modified slightly as a smooth unconstrained minimization problem. Before we derive the smooth approximation function, we show some interesting observations:

$$|x|_\varepsilon = (x - \varepsilon)_+ + (-x - \varepsilon)_+ \tag{27.44}$$

and

$$(x - \varepsilon)_+ \cdot (-x - \varepsilon)_+ = 0 \ \text{ for all } x \in \mathbb{R} \text{ and } \varepsilon > 0\,. \tag{27.45}$$

Thus we have

$$|x|_\varepsilon^2 = (x - \varepsilon)_+^2 + (-x - \varepsilon)_+^2\,. \tag{27.46}$$

It is straightforward to replace $|x|_\varepsilon^2$ by a very accurate smooth approximation given by:

$$p_\varepsilon^2(x, \beta) = (p(x - \varepsilon, \beta))^2 + (p(-x - \varepsilon, \beta))^2\,. \tag{27.47}$$

We use this approximation p_ε^2-function with smoothing parameter β to obtain the smooth support vector regression (ε-SSVR) (Lee et al. 2005):

$$\min_{(\mathbf{w},b)\in\mathbb{R}^{n+1}} \frac{1}{2}(\|\mathbf{w}\|_2^2 + b^2) + \frac{C}{2}\sum_{i=1}^{m} p_\varepsilon^2(\mathbf{w}^\top\mathbf{x}_i + b - y_i, \beta)\,, \tag{27.48}$$

where $p_\varepsilon^2(\mathbf{w}^\top\mathbf{x}_i + b - y_i, \beta) \in \mathbb{R}$. For the nonlinear case, this formulation can be extended to the nonlinear ε-SSVR by using the kernel trick as follows:

$$\min_{(\mathbf{u},b)\in\mathbb{R}^{m+1}} \frac{1}{2}(\|\mathbf{u}\|_2^2 + b^2) + \frac{C}{2}\sum_{i=1}^{m} p_\varepsilon^2\Big(\sum_{j=1}^{m} u_j K(\mathbf{x}_j, \mathbf{x}_i) + b - y_i, \beta\Big)\,, \tag{27.49}$$

where $k(\mathbf{x}_i, \mathbf{x}_j)$ is a kernel function. The nonlinear ε-SSVR decision function $f(\mathbf{x})$ can be expressed as follows:

$$f(x) = \sum_{i=1}^{m} u_i k(\mathbf{x}_j, \mathbf{x}) + b\,. \tag{27.50}$$

Note that the reduced kernel technique also can be applied to ε-SSVR while encountering a large scale regression problem.

27.4 Implementation of SVMs

The support vector machine, either in its primal formulation (27.6) or dual formulation (27.8), is simply a standard convex quadratic program (for the nonlinear SVM, the kernel function $k(\mathbf{x}, \mathbf{x})$ used in (27.12) has to satisfy *Mercer's condition* in order to keep the *convexity* of the objective function). The most straightforward way for solving it is to employ a standard quadratic programming solver such as CPLEX (C.O. Inc. 1992), or using an interior point method for quadratic programming (Ferris and Munson 2003). Because of the simple structure of the dual formulation of either linear (27.8) or nonlinear (27.12) SVM, many SVM algorithms are operated in the dual space. However solving SVMs in the primal form can also be efficient (Lee and Mangasarian 2001; Lee et al. 2005; Lee and Huang 2007; Chapelle 2007), such as the approaches to solve SSVM, SSVR, and RSVM which were introduced in the previous section. In the following, we will introduce main methods in solving SVMs in their primal and dual forms.

27.4.1 SVMs Training in the Primal Form

The standard way to solve SVMs in the primal is reformulating (27.6) or (27.7) as an unconstrained minimization problem:

$$\min_{(\mathbf{w},b)\in\mathbb{R}^{n+1}} \tfrac{1}{2}\|\mathbf{w}\|_2^2 + C\textstyle\sum_{i=1}^m L(y_i, \mathbf{w}^\top\mathbf{x} + b)\,, \tag{27.51}$$

with the loss function $L(y, f(\mathbf{x})) = \max(0, 1 - y_i f(\mathbf{x}))^p$. Note that the decision function can be written as a linear combination of data points such as $\mathbf{w} = \sum_i^m u_i\mathbf{x}_i$. Thus, we can rewrite the nonlinear form for SVMs in the primal by utilizing kernel trick as follows

$$\min_{(\mathbf{u},b)\in\mathbb{R}^{m+1}} \tfrac{1}{2}\mathbf{u}^\top\mathbf{K}(\mathbf{A},\mathbf{A})\mathbf{u} + C\textstyle\sum_{i=1}^m L(y_i, \sum_{j=1}^m u_j k(\mathbf{x}_i, \mathbf{x}_j) + b)\,, \tag{27.52}$$

or in another slightly different form based on the generalized SVM Mangasarian (2000):

$$\min_{(\mathbf{u},b)\in\mathbb{R}^{m+1}} \tfrac{1}{2}\mathbf{u}^\top\mathbf{u} + C\textstyle\sum_{i=1}^m L(y_i, \sum_{j=1}^m u_j k(\mathbf{x}_i, \mathbf{x}_j) + b)\,. \tag{27.53}$$

For solving unconstrained minimization problems, Newton-like optimization methods are widely used, so we only focus on solving the minimization problems via Newton method here. The Newton method needs the objective function to be twice differentiable to calculate the Hessian matrix. One way is to replace the loss function by a twice differentiable approximated function. We take SSVM as an example to illustrate the idea. SSVM adopts the quadratic loss and uses p-function to smooth the quadratic loss function as follows:

$$\min_{(\mathbf{w},b)\in\mathbb{R}^{n+1}} \boldsymbol{\Psi}(\mathbf{w},b) := \frac{C}{2}\sum_{i=1}^{m} p(\{1 - y_i(\mathbf{w}^\top \mathbf{x}_i + b)\}, \beta)^2 + \frac{1}{2}(\|\mathbf{w}\|_2^2 + b^2),$$

for the linear case; and for the nonlinear case, the function becomes:

$$\min_{(\mathbf{u},b)\in\mathbb{R}^{m+1}} \boldsymbol{\Psi}(\mathbf{u},b) := \frac{C}{2}\sum_{i=1}^{m} p([1 - y_i\{\sum_{j=1}^{m} u_j k(\mathbf{x}_i,\mathbf{x}_j) + b\}], \beta)^2 + \frac{1}{2}(\|\mathbf{u}\|_2^2 + b^2).$$

Once reformulating SVM as an unconstrained minimization problem with twice differentiable objective function $\boldsymbol{\Psi}(\mathbf{w},b)$ (or $\boldsymbol{\Psi}(\mathbf{u},b)$), Newton-Armijo optimization method is applied to obtain the solution. The Armijo condition

$$\boldsymbol{\Psi}(\mathbf{w}_i, b_i) \leq \boldsymbol{\Psi}(\mathbf{w}_{i-1}, b_{i-1}) - \eta\lambda \mathbf{d}^\top \nabla \boldsymbol{\Psi}(\mathbf{w}, b)$$

is applied here to avoid the divergence and oscillation in Newton method where η is assigned with a small value. For the nonlinear case, one only needs to replace the original data $\mathbf{A}$ by the kernel data $\mathbf{K}(\mathbf{A},\mathbf{A})$ or reduced kernel data $\mathbf{K}(\mathbf{A},\tilde{\mathbf{A}})$ and simply obtains the solution without revising the algorithm.

27.4.2 SVMs Training in the Dual Form

The most popular strategy in solving SVM with dual form is the decomposition method (Osuna et al. 1997; Joachims 1999; Fan et al. 2005; Glasmachers and Igel 2006). The decomposition method is designed to avoid the access of the full kernel matrix while searching for the optimal solution. This method iteratively selects a small subset of training data (the working set) to define a quadratic programming subproblem. The solution of current iteration is updated by solving the quadratic programming subproblem, defined by a selected working set $\mathcal{W}$, such that the objective function value of the original quadratic program strictly decreases at every iteration. The decomposition algorithm only updates a fixed size subset of multipliers α_i, while the others are kept constant. The goal is not to identify all of the active constraints in order to run the optimizer on all of them, but is rather to optimize the global problem by only acting on a small subset of data at a time.

Suppose α' are the coefficients of data belonging to the current working set $\mathcal{W}$. One can reformulate (27.8) to a subproblem and solve it iteratively for updating α as follows:

$$\max_{\alpha'} \sum_{i \in \mathcal{B}} \alpha'_i - \frac{1}{2} \sum_{i,j \in \mathcal{B}} y_i y_j \alpha'_i \alpha'_j k(\mathbf{x}_i, \mathbf{x}_j) \tag{27.54}$$

$$\text{s.t. } 0 \le \alpha'_i \le C \quad \text{for } i \in \mathcal{W},$$

$$\sum_{i \in \mathcal{B}} y_i \alpha'_i + \sum_{i \notin \mathcal{B}} y_i \alpha_i = 0.$$

The critical issue of decomposition methods is selecting an appropriate working set. The sequential minimal optimization (SMO) (Platt 1999) which is an extreme case of the decomposition methods. It only selects a working set with smallest size, two data points, at each iteration. The criterion of selecting these two data points is based on the maximum violating pair scheme. Besides, this smallest working size leads the subproblem to a single variable minimization problem which has a analytic form of solution. Different strategies to select the working set lead to different algorithms. Many methods of selecting a appropriate working set has been proposed (Joachims 1999; Fan et al. 2005; Glasmachers and Igel 2006). Some of them also been well implemented, such as SVM^{light}[1] (Joachims 1999) and LIBSVM[2] (Chang and Lin 2001). The LIBSVM provides an advanced working set selection scheme based on the information of second order. Its efficiency in performance has attracted many people to use in their applications.

In a nutshell, decomposition methods take the advantage of sparsity in SVM to adjust the solution with a small minimization problem iteratively. This strategy makes decomposition methods avoid to access the whole full kernel in seeking the solution. On the other hand, the selection of working set is a key factor for the computational cost. Different working sets and their sizes lead to different rates of convergence. The convergence analysis has been carried out in Chang et al. (2000) and Keerthi and Gilbert (2002).

27.5 Practical Issues and Their Solutions in SVMs

In this section, we discuss some practical issues in SVMs. The topics including dealing with the multi-class classification, dealing with unbalanced data distribution, and the strategy of model selection.

27.5.1 Multi-Class Problems

In the previous sections, we only focus on the binary classification problem in SVM. However, the labels might be drawn from several categories in the real world. There

[1] SVM^{light} is available in http://svmlight.joachims.org/.

[2] LIBSVM is available in http://www.csie.ntu.edu.tw/~cjlin/libsvm.

are many methods have been proposed for dealing with the multi-class problem. These methods can simply be divided into two types. One handles the multi-class problem by dividing it into a series of binary classification problems (Vapnik 2000; Platt et al. 2000; Crammer and Singer 2002; Rifkin and Klautau 2004). The other formulates the multi-class problem as a single optimization problem (Vapnik 2000; Weston and Watkins 1999; Crammer and Singer 2001; Rifkin and Klautau 2004).

In the approach of combining a series of binary classifiers, the popular schemes are one-versus-rest, one-versus-one, directed acyclic graph (DAG) (Platt et al. 2000), and error-correcting coding (Dietterich and Bakiri 1995; Allwein et al. 2001; Crammer and Singer 2002). Now suppose we have k classes in the data. In the one-versus-rest scheme, it creates a series of binary classifiers with one of the labels to the rest so we have k binary classifiers for prediction. The classification of new instances for one-versus-rest is using the winner-take-all strategy. That is, we assign the label by the classifier with the highest output value. On the other hand, one-versus-one scheme generates a series of binary classifiers between every pair of classes. It means we need to construct $\binom{k}{2}$ classifiers in the one-versus-one scheme. The classification of one-versus-one is usually associated with a simple voting strategy. In the voting strategy, every classifier assigns the instance to one of the two classes and then new instances will be classified to a certain class with most votes. The DAG strategy is a variant of one-versus-one scheme. It also constructs $\binom{k}{2}$ classifiers for each pair of classes but uses a different prediction strategy. DAG places the $\binom{k}{2}$ classifiers in a directed acyclic graph and each path from the root to a leaf is an evaluation path. In an evaluation path, a possible labeling is eliminated while passing through a binary classification node. A predicted label is concluded after finishing a evaluation path (see Fig. 27.3).

In the error-correcting coding scheme, output coding for multi-class problems consists of two phases. In the training phase, one need to construct a series of binary

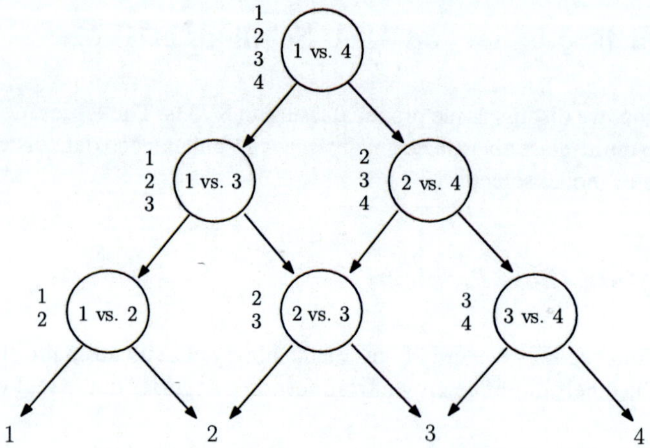

Fig. 27.3 An example of DAG approach in the multi-class problem

classifiers which are based on different partitions of the classes. In the testing phase, the predictions of the binary classifiers are combined to conclude a prediction of a testing instance by using the output coding. Besides, the coding scheme is an issue in the error-correcting coding. There are rich literatures discussing the coding schemes (Dietterich and Bakiri 1995; Allwein et al. 2001; Crammer and Singer 2002). The reader could get more details in these literatures.

The single machine approach for multi-class problem is first introduced in Vapnik (2000) and Weston and Watkins (1999). The idea behind this approach is still using the concept of maximum margin in binary classification. The difference of single machine formulation is that it considers all regularization terms together and pays the penalties for a misclassified instance with a relative quantity evaluated by different models. It means that each instance is associated with $m(k-1)$ slack values if we have m instances and k classes. For understanding the concept more, we display the formulation of single machine approach in Weston and Watkins (1999):

$$\min_{\mathbf{w}_1,\ldots,\mathbf{w}_k\in\mathbb{R}^n,\xi\in\mathbb{R}^{m(k-1)}} \quad \sum_{i=1}^{k}\|\mathbf{w}_i\| + C\sum_{i=1}^{m}\sum_{j\notin y_i}\xi_{ij} \tag{27.55}$$
$$\text{s.t.}\quad \mathbf{w}_{y_i}^{\top}\mathbf{x}_i + b_{y_i} \geq \mathbf{w}_j^{\top}\mathbf{x}_i + b_j + 2 - \xi_{ij}\,,$$
$$\xi_{ij} \geq 0\,.$$

Except for this basic formulation, some further formulations have also been proposed (Vapnik 2000; Crammer and Singer 2001; Rifkin and Klautau 2004). In a nutshell, the single machine approach could give all the classifiers simultaneously in solving a single optimization problem. However, the complicated formulation also brings a higher complexity for solving it.

27.5.2 Unbalanced Problems

In reality, there might be only a small portion of instances belonging to a class compared to the number of instances with the other label. Due to the small share in a sample that reflects reality, using SVMs on this kind of data may tend to classify every instance as the class with the majority of the instances. Such models are useless in practice. In order to deal with this problem, the common ways start off with more balanced training than reality can provide.

One of these methods is a down-sampling strategy (Chen et al. 2006) and work with balanced (50%/50%)-samples. The chosen bootstrap procedure repeatedly randomly selects a fixed number of the majority instances from the training set and adds the same number of the minority instances. One advantage of down-sampling strategy is giving a lower cost in the training phase because it removes lots of data points in the majority class. However, the random choosing of the majority instances might cause a high variance of the model.

In order to avoid this unstable model building, a over-sampling scheme (Härdle et al. 2009) could also be applied to reach a balanced sample. The over-sampling scheme duplicates the number of the minority instances a certain number of times. It considers all the instances in hand and generates a more robust model than the down-sampling scheme. Comparing the computational cost with down-sampling strategy, over-sampling suffers a higher cost in the training phase while increasing the size of training data.

To avoid the extra cost in the over-sampling strategy, one also can apply different weights on the penalty term. In other words, one need to assign a higher weight (higher C) on the minority class. This strategy of assigning different weights gives the equivalent effect with the over-sampling strategy. The benefit of assigning different weights is that it does not increase the size of training data while achieving a balanced training. However, using this strategy needs to revise the algorithm a little bit. In down-sampling and over-sampling strategies, the thing that one needs to do is adjusting the proportions of training data. Hence, down-sampling and over-sampling strategies are easier to be applied for basic users in practical usage.

27.5.3 Model Selection of SVMs

Choosing a good parameter setting for a better generalization performance of SVMs is the so called model selection problem. Model selection is usually done by minimizing an estimate of generalization error. This problem can be treated as finding the maximum (or minimum) of a function which is only vaguely specified and has many local maxima (or minima).

Suppose the Gaussian kernel

$$K(\mathbf{x}, \mathbf{z}) = e^{-\gamma \|\mathbf{x}-\mathbf{z}\|_2^2},$$

is used where γ is the width parameter. The nonlinear SVM needs to be assigned two parameters C and γ. The most common and reliable approach for model selection is exhaustive grid search method. The exhaustive grid search method forms a two dimension uniform grid (say $p \times p$) of points in a pre-specified search range and find a good combination (C, γ). It is obvious that the exhaustive grid search can not effectively perform the task of automatic model selection due to its high computational cost.

Except for the exhaustive grid search method, many improved model selection methods have been proposed to reduce the number of trials in parameter combinations (Keerthi and Lin 2003; Chapelle et al. 2002; Larsen et al. 1998; Bengio 2000; Staelin 2003; Huang et al. 2007). Here we focus on introducing the 2-stage uniform design model selection (Huang et al. 2007) because of its good efficiency. The 2-stage uniform design procedure first sets out a crude search for a highly likely candidate region of global optimum and then confines a finer second-stage search

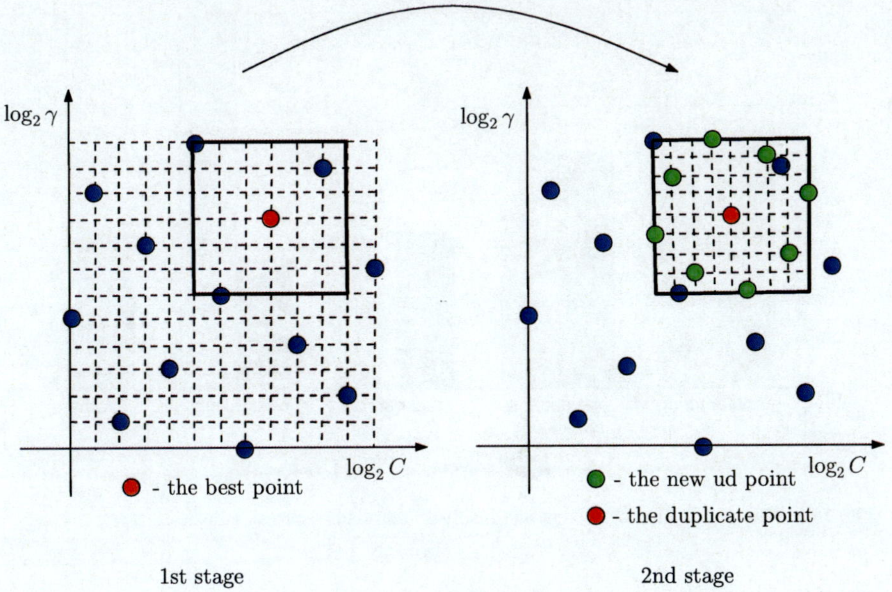

Fig. 27.4 The nested UD model selection with a 13-points UD at the first stage and a 9-points UD at the second stage

therein. At the first stage, we use a 13-runs UD sampling pattern (see Fig. 27.4) in the appropriate search range proposed above. At the second stage, we halve the search range for each parameter coordinate in the log-scale and let the best point from the first stage be the center point of the new search box. Then we use a 9-runs UD sampling pattern in the new range. Moreover, to deal with large sized datasets, we combine a 9-runs and a 5-runs sampling pattern at these two stages. The performance in Huang et al. (2007) shows merits of the nested UD model selection method. Besides, the method of nested UDs is not limited to 2 stages and can be applied in a sequential manner and one may consider a finer net of UDs to start with.

27.6 A Case Study for Bankruptcy Prognosis

To demonstrate the use of SVM, we focus on the problem of bankruptcy prognosis as our case study. The studied data set is `CreditReform` where we are given financial company information and the goal is to predict the possibility of bankruptcy for the companies. The study includes applying the nonlinear SSVM with a reduced kernel, feature selection via 1-norm SVM, conquering the unbalanced problem by over-sampling technique, and model selection by the 2-stage nested uniform design method.

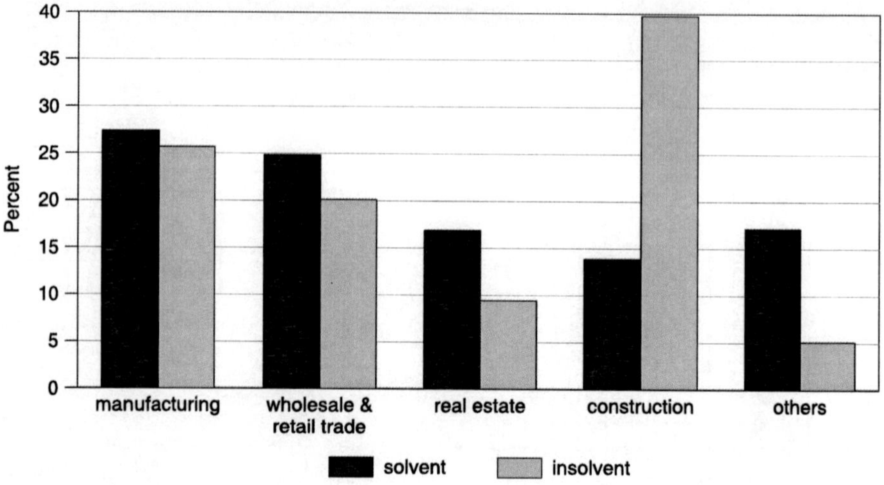

Fig. 27.5 The distribution of solvent and insolvent companies across industries

27.6.1 Data Description

The `CreditReform` database consists of 20,000 financially solvent and 1,000 insolvent German companies observed once in the period from 1997–2002. Although the companies were randomly selected, the accounting data in 2001 and 2002 are the majority. Approximately 50% of the observations come from this period. Figure 27.5 shows the distribution of solvent and insolvent companies across different industries.

A company is described by a set of attributes that includes several balance sheet and income statement items. The attributes include:

- AD (Amortization and Depreciation)
- AP (Accounts Payable)
- AR (Account Receivable)
- CA (Current Assets)
- CASH (Cash and Cash Equivalents)
- CL (Current Liabilities)
- DEBT (Debt)
- EBIT (Earnings before Interest and Tax)
- EQUITY (Equity)
- IDINV (Growth of Inventories)
- IDL (Growth of Liabilities)
- INTE (Interest Expense)
- INV (Inventories)
- ITGA (Intangible Assets)
- LB (Lands and Buildings)

- NI (Net Income)
- OI (Operating Income)
- QA (Quick Assets)
- SALE (Sales)
- TA (Total Assets)
- TL (Total Liabilities)
- WC (Working Capital (=CA-CL))

The companies may appear in the database several times in different years; however, each year of balance sheet information is treated as a single observation. The data of the insolvent companies were collected 2 years prior to their insolvency. The company size is measured by its total assets. We construct 28 ratios to condense the balance sheet information (see Table 27.1). However, before dealing with the data set, some companies whose behavior is very different from others (outliers) are ignored in order to make the dataset more compact. The complete pre-processing procedure is described as follows:

1. We excluded companies whose total assets were not in the range of 10^5–10^7 euros. There are 967 insolvent companies remain and 15,834 solvent companies remain.
2. In order to compute the accounting ratios AP/SALE, OI/TA, TL/TA, CASH/TA, IDINV/INV, INV/SALE, EBIT/TA and NI/SALE, we have removed companies with zero denominators (remaining insolvent: 816; solvent 11,005, after the pre-processing in previous step).
3. We dropped outliers. That is, the insolvent companies with extreme values of financial indices are removed (remaining insolvent: 811; solvent: 10,468).

Table 27.1 The definition of accounting ratios used in the analysis

Variable	Ratio	Indicator for	Variable	Ratio	Indicator for
X1	NI/TA	Profitability	X15	CASH/TA	Liquidity
X2	NI/SALE	Profitability	X16	CASH/CL	Liquidity
X3	OI/TA	Profitability	X17	QA/CL	Liquidity
X4	OI/SALE	Profitability	X18	CA/CL	Liquidity
X5	EBIT/TA	Profitability	X19	WC/TA	Liquidity
X6	(EBIT+AD)/TA	Profitability	X20	CL/TL	Liquidity
	EBIT/SALE		X21	TA/SALE	Activity
X7	EQUITY/TA	Profitability	X22	INV/SALE	Activity
X8	(EQUITY-ITGA)/	Leverage	X23	AR/SALE	Activity
X9	(TA-ITGA-CASH-LB)	Leverage	X24	AP/SALE	Activity
X10	CL/TA	Leverage	X25	Log(TA)	Size
X11	(CL-CASH)/TA	Leverage	X26	IDINV/INV	Growth
X12	TL/TA	Leverage	X27	IDL/TL	Growth
X13	DEBT/TA	Leverage	X28	IDCASH/CASH	Growth
X14	EBIT/INTE	Leverage			

Table 27.2 The prediction scenario of our experiments

Scenario	Observation period of training set	Observation period of testing set
S1	1997	1998
S2	1997–1998	1999
S3	1997–1999	2000
S4	1997–2000	2001
S5	1997–2001	2002

After pre-processing, the dataset consists of 11,279 companies (811 insolvent and 10,468 solvent). In all the following analysis, we focus on the revised dataset.

27.6.2 The Procedure of Bankruptcy Prognosis with SVMs

We conduct the experiments in a scenario in which we train the SSVM bankruptcy prognosis model from the data at hand and then use the trained SSVM to predict the following year's cases. This strategy simulates the real task for analysts who may predict the future outcomes by using the data from past years. The experiment setting is described in Table 27.2. The number of periods used for the training set changes from 1 year (S1) to 5 years (S5) as time goes by. All classifiers we adopt in the experiments are reduced SSVM with Gaussian kernels. We need to determine two parameters, the best combination of C and γ for the kernels. In principle, the 2-D grid search will consume a lot of time. In order to cut down the search time, we adopt the nested uniformed design model selection method Huang et al. (2007), introduced in Sect. 27.5.3 to search for a good pair of parameters for the performance of our classification task.

27.6.2.1 Selection of Accounting Ratios via 1-norm SVM

In principle, many possible combination of accounting ratios could be used as explanatory variables in a bankruptcy prognosis model. Therefore, appropriate performance measures are needed to gear the process of selecting the ratios with the highest separating power. In Chen et al. (2006) Accuracy Ratio (AR) and Conditional Information Entropy Ratio (CIER) determine the selection procedure's outcome. It turned out that the ratio "accounts payable divided by sales", X24 (AP/SALE), has the best performance values for a univariate SVM model. The second selected variable was the one combined with X24 that had the best performance of a bivariate SVM model. This is the analogue of forward selection in linear regression modeling. If one keeps on adding new variables one typically observes a declining change in improvement. This was also the case in that work where the performance indicators started to decrease after the model included eight variables. The described selection procedure is quiet lengthy, since there are at

Table 27.3 Selected variables in V1 and V2 (the symbol "plus" means the common variables in V1 and V2)

Variable	Definition	V1	V2
X2+	NI/SALE	x	x
X3+	OI/TA	x	x
X5+	EBIT/TA	x	x
X6	(EBIT+AD)/TA		x
X8	EQUITY/TA		x
X12	TL/TA	x	
X15+	CASH/TA	x	x
X22	INV/SALE	x	
X23	AR/SALE		x
X24+	AP/SALE	x	x
X26	IDINV/INV	x	

least 216 accounting ratio combinations to be considered. We will not employ the procedure here but use the chosen set of eight variables in Chen et al. (2006) denoted as V1. Table 27.3 presents V1 in the first column.

Except for using V1, we also apply 1-norm SVM which will simplify the selection procedure to select accounting ratios. The 1-norm SVM was applied to the period from 1997 to 1999. We selected the variables according to the size of the absolute values of the coefficients **w** from the solution of the 1-norm SVM. We also select eight variables out of 28. Table 27.3 displays the eight selected variables as V2. Note that five variables, X2, X3, X5, X15 and X24 are also in the benchmark set V1. From Tables 27.4 and 27.5, we can the performances of V1 and V2 are quite similar while we need fewer efforts for extract V1.

27.6.2.2 Applying Over-Sampling to Unbalanced Problems

The cleaned data set consists of around 10% of insolvent companies. Thus, the sample is fairly unbalanced although the share of insolvent companies is higher than in reality. In order to deal with this problem, insolvency prognosis models usually start off with more balanced training and testing samples than reality can provide. Here we use over-sampling and down-sampling Chen et al. (2006) strategies, to balance the size between the solvent and the insolvent companies. In the experiments, the over-sampling scheme shows better results in the Type I error rate but has slightly bigger total error rates (see Tables 27.4 and 27.5). It is also obvious, that in almost all models a longer training period works in favor of accuracy of prediction. Clearly, the over-sampling schemes have much smaller standard deviations in the Type I error rate, the Type II error rate, and the total error rate than the down-sampling one. According to this observation, we conclude that the over-sampling scheme will generate a more robust model than the down-sampling scheme.

Table 27.4 The results in percentage (%) of over-sampling for three variable sets (Reduced SSVM with Gaussian kernel)

Set of accounting ratios	Scenario	Type I error rate Mean	Std	Type II error rate Mean	Std	Total error rate Mean	Std
	S1	33.16	0.55	26.15	0.13	26.75	0.12
	S2	31.58	0.01	29.10	0.07	29.35	0.07
	S3	28.11	0.73	26.73	0.16	26.83	0.16
	S4	30.14	0.62	25.66	0.17	25.93	0.15
V1	S5	24.24	0.56	23.44	0.13	23.48	0.13
	S1	29.28	0.92	27.20	0.24	27.38	0.23
	S2	28.20	0.29	30.18	0.18	29.98	0.16
	S3	27.41	0.61	29.67	0.19	29.50	0.17
	S4	28.12	0.74	28.32	0.19	28.31	0.15
V2	S5	23.91	0.62	24.99	0.10	24.94	0.10

Table 27.5 The results in percentage (%) of down-sampling for three variable sets (Reduced SSVM with Gaussian kernel)

Set of accounting ratios	Scenario	Type I error rate Mean	Std	Type II error rate Mean	Std	Total error rate Mean	Std
	S1	32.20	3.12	28.98	1.70	29.26	1.46
	S2	29.74	2.29	28.77	1.97	28.87	1.57
	S3	30.46	1.88	26.23	1.33	26.54	1.17
	S4	31.55	1.52	23.89	0.97	24.37	0.87
V1	S5	28.81	1.53	23.09	0.73	23.34	0.69
	S1	29.94	2.91	28.07	2.15	28.23	1.79
	S2	28.77	2.58	29.80	1.89	29.70	1.52
	S3	29.88	1.88	27.19	1.32	27.39	1.19
	S4	29.06	1.68	26.26	1.00	26.43	0.86
V2	S5	26.92	1.94	25.30	1.17	25.37	1.06

27.6.2.3 Applying the Reduced Kernel Technique for Fast Computation

Over-sampling duplicates the number of insolvent companies a certain number of times. In the experiments, we have to duplicate in each scenario the number of insolvent companies as many times as necessary to reach a balanced sample. Note that in our over-sampling scheme every solvent and insolvent companys information is utilized. This increases the computational burden due to increasing the number of training instances. We employ the reduced kernel technique in Sect. 27.3 to mediate this problem. Here the key idea for choosing the reduced set $\tilde{A}$ is extracting the same size of insolvent companies from solvent companies. This leads to not only the balance both in the data size and column basis bit but also the lower computational cost.

27.6.2.4 Summary

In analyzing `CreditReform` dataset for bankruptcy prognosis, we presented the usage of SVMs in a real case. The results show the selection of accounting ratios via 1-norm SVM can perform as well as the greedy search. The finance indices selected by 1-norm SVM actually can represent the data well in bankruptcy prognosis. The simple procedure of over-sampling strategy also helps to overcome the unbalanced problem while down-sampling will cause a biased model. In accelerating the training procedure, the reduced kernel technique is performed. It helps to build a SVM model in an efficient way without sacrificing the performance in prediction. Finally, the procedure of tuning parameters in a model is usually a heavy work in analyzing data. A good model selection method can help users to decease the long-winded tuning procedure, such as the 2-stage uniform design method used in this case study. In a nutshell, SVMs have been developed maturely. These practical usages presented here not only show the variability and ability of SVMs but also give the basic ideas for analyzing data with SVMs.

27.7 Conclusion

The clear connection to statistic learning theory, efficient performance, and simple usage of SVMs have attracted many researchers to investigate. Many literatures have shown that SVMs are the state of the art in solving classification and regression problems. This reputation has made SVMs be applied in many fields, such as the quantitative finance field. This chapter presented many topics of SVMs as well as a case study in bankruptcy prognosis to give a guide for the usage of SVMs in quantitative finance filed. The possible applications with SVMs are various and potential in the quantitative finance field. The aim is giving that ones can quickly have solutions in their applications with SVMs while there are fertile materials in the wild.

References

Allwein, E. L., Schapire, R. E., & Singer, Y. (2001). Reducing multiclass to binary: A unifying approach for margin classifiers. *Journal of Machine Learning Research, 1*, 113–141.

Altman, E. (1968). Financial ratios, discriminant analysis and the prediction of corporate bankruptcy. *The Journal of Finance, 23*(4), 589–609.

Altman, E. (1994). Giancarlo marco, and franco varetto. Corporate distress diagnosis: Comparisons using linear discriminant analysis and neural networks (the italian experience). *Journal of Banking and Finance, 18*, 505–529.

Beaver, W. (1966). Financial ratios as predictors of failures. *Journal of Accounting Research, 4*, 71–111.

Bengio, Y. (2000). Gradient-based optimization of hyperparameters. *Neural Computation, 12*(8), 1889–1900.

Bertsekas, D. P. (1999). *Nonlinear programming*. MA: Athena Scientific Belmont.

Burges, C. J. C. (1998). A tutorial on support vector machines for pattern recognition. *Data Mining and Knowledge Discover*, *2*(2), 121–167.

Cao, L.-J. & Tay, F. E. H. (2003). Support vector machine with adaptive parameters in financial time series forecasting. *IEEE Transactions on Neural Networks*, *14*(6), 1506–1518.

Chang, C.-C. & Lin, C.-J. (2001). *LIBSVM: A library for support vector machines*. Software available at http://www.csie.ntu.edu.tw/~cjlin/libsvm.

Chang, C.-C., Hsu, C.-W., & Lin, C.-J. (2000). The analysis of decomposition methods for support vector machines. *IEEE Transactions on Neural Networks*, *11*(4), 1003–1008.

Chapelle, O. (2007). Training a support vector machine in the primal. *Neural Computation*, *19*(5), 1155–1178.

Chapelle, O., Vapnik, V., Bousquet, O., & Mukherjee, S. (2002). Choosing multiple parameters for support vector machines. *Machine Learning*, *46*(1), 131–159.

Chen, S., Härdle, W., & Moro, R. (2006). Estimation of default probabilities with support vector machines. *SFB 649 Discussion Paper 2006-077*.

Cherkassky, V. & Mulier, F. (1998). *Learning from data: Concepts, theory, and methods*. New York: Wiley.

C.O. Inc. (1992). *Using the cplex callable library and cplex mixed integer library*. Incline Village, NV.

Courant, R. & Hilbert, D. (1953). *Methods of mathematical physics*. New York: Interscience Publishers.

Crammer, K. & Singer, Y. (2001). Improved output coding for classification using continuous relaxation. In *Proceeding of Advances in Neural Information Processing Systems 13*, 437–443.

Crammer, K. & Singer, Y. (2002). On the learnability and design of output codes for multiclass problems. *Machine Learning*, *47*(2), 201–233.

Cristianini, N. & Shawe-Taylor, J. (1999). *An introduction to support vector machines and other kernel-based learning methods*. New York: Cambridge University Press.

Dietterich, T. G. & Bakiri, G. (1995). Solving multiclass learning problems via error-correcting output codes. *Journal of Artificial Intelligence Research*, *2*, 263–286.

Fan, R.-E., Chen, P.-H., & Lin, C.-J. (2005). Working set selection using second order information for training support vector machines. *Journal of Machine Learning Research*, *6*, 1889–1918.

Ferris, M. C. & Munson, T. S. (2003). Interior-point methods for massive support vector machines. *SIAM Journal of Optimization*, *13*, 783–804.

Fung, G. & Mangasarian, O. L. (2001). Proximal support vector machine classifiers. In *Proceeding of Seventh ACM SIGKDD International Conference on Knowledge Discovery & Data Mining*, San Francisco.

Fung, G. & Mangasarian, O. L. (2004). A feature selection Newton method for support vector machine classification. *Computational Optimization and Applications*, *28*(2), 185–202.

Glasmachers, T. & Igel, C. (2006). Maximum-gain working set selection for svms. *Journal of Machine Learning Research*, *7*, 1437–1466.

Härdle, W., Lee, Y.-J., Schäfer, D., & Yeh, Y.-R. (2009). Variable selection and oversampling in the use of smooth support vector machines for predicting the default risk of companies. *Journal of Forecasting*, *28*(6), 512–534.

Huang, C.-M., Lee, Y.-J., Lin, D. K. J., & Huang, S.-Y. (2007). Model selection for support vector machines via uniform design. *A special issue on Machine Learning and Robust Data Mining of Computational Statistics and Data Analysis*, *52*, 335–346.

Hwang, R.-C., Cheng, K. F., & Lee, J. C. (2007). A semiparametric method for predicting bankruptcy. *Journal of Forecasting*, *26*(5), 317–342.

Joachims, T. (1999). Making large-scale support vector machine learning practical. *Advances in Kernel Methods: Support Vector Learning*, 169–184.

Keerthi, S. S. & Gilbert, E. G. (2002). Convergence of a generalized smo algorithm for svm. *Machine Learning*, *46*(1), 351–360.

Keerthi, S. S. & Lin, C.-J. (2003). Asymptotic behaviors of support vector machines with gaussian kernel. *Neural Computation*, *15*(7), 1667–1689.

Krahnen, J. P. & Weber, M. (2001). Generally accepted rating principles: A primer. *Journal of Banking and Finance*, *25*(1), 3–23.

Larsen, J., Svarer, C., Andersen, L. N., & Hansen, L. K. (1998). Adaptive regularization in neural network modeling. *Lecture Notes in Computer Science 1524*, 113–132.

Lee, Y.-J., Hsieh, W.-F., & Huang, C.-M. (2005). Ssvr: A smooth support vector machine for-insensitive regression. *IEEE Transactions on Knowledge and Data Engineering*, *17*(5), 678–685.

Lee, Y.-J. & Huang, S.-Y. (2007). Reduced support vector machines: A statistical theory. *IEEE Transactions on Neural Networks*, *18*(1), 1–13.

Lee, Y.-J. & Mangasarian, O. L. (2001). SSVM: A smooth support vector machine for classification. *Computational Optimization and Applications*, *20*(1), 5–22.

Mangasarian, O. L. (1994). *Nonlinear programming*. Philadelphia, PA: Society for Industrial and Applied Mathematics.

Mangasarian, O. L. (2000). Generalized support vector machines. *Advances in Large Margin Classifiers*, 135–146.

Martin, D. (1977). Early warning of bank failure: A logit regression approach. *Journal of Banking and Finance*, *1*, 249–276.

Min, J. H. & Lee, Y.-C. (2005). Bankruptcy prediction using support vector machine with optimal choice of kernel function parameters. *Expert Systems with Applications*, *28*(4), 603–614.

Nocedal, J. & Wright, S. J. (2006). *Numerical optimization*. Berlin: Springer.

Ohlson, J. (1980). Financial ratios and the probabilistic prediction of bankruptcy. *Journal of Accounting Research*, *18*(1), 109–131.

Osuna, E., Freund, R., & Girosi, F. (1997). An improved training algorithm for support vector machines. *Neural networks for signal processing VII*, 276–285.

Platt, J. (1999). Sequential minimal optimization: A fast algorithm for training support vector machines. *Advances in Kernel Methods-Support Vector Learning*, 185–208.

Platt, J., Cristianini, N., & Shawe-Taylor, J. (2000). Large margin dags for multiclass classification. *Advances in Neural Information Processing Systems*, *12*, 547–553.

Rifkin, R. & Klautau, A. (2004). In defense of one-vs-all classification. *Journal of Machine Learning Research*, *5*, 101–141.

Schölkopf, B. & Smola, A. J. (2002). *Learning with kernels: Support vector machines, regularization, optimization, and beyond*. Cambridge: MIT.

Smola, A. J. & Schölkopf, B. (2000). Sparse greedy matrix approximation for machine learning. In *Proceedings of the Seventeenth International Conference on Machine Learning*, 911–918.

Smola, A. J. & Schölkopf, B. (2004). A tutorial on support vector regression. *Statistics and Computing*, *14*, 199–222.

Staelin, C. (2003). Parameter selection for support vector machines. Hewlett-Packard Company, Tech. Rep. HPL-2002-354R1.

Suykens, J. A. K. & Vandewalle, J. (1999). Least squares support vector machine classifiers. *Neural Processing Letters*, *9*(3), 293–300.

Tam, K.-Y. & Kiang, M. Y. (1992). Managerial application of neural networks: The case of bank failure prediction. *Management Science*, *38*(7), 926–947.

Tibshiran, R. (1996). Regression shrinkage and selection via the lasso. *Journal of the Royal Statistical Society*, *58*, 267–288.

Vapnik, V. N. (2000). *The nature of statistical learning theory*. New York: Springer.

Weston, J. & Watkins, C. (1999). Support vector machines for multi-class pattern recognition. In *Proceedings of the Seventh European Symposium On Artificial Neural Networks*.

Williams, C. K. I. & Seeger, M. (2001). Using the Nyström method to speed up kernel machines. In *Proceeding of Advances in Neural Information Processing Systems 13*, 682–688.

Zhu, J., Rosset, S., Hastie, T., & Tibshirani, R. (2004). 1-norm support vector machine. In *Proceeding of Advances in Neural Information Processing Systems 16*.

Part V
Software Tools

Chapter 28
MATLAB® as a Tool in Computational Finance

James E. Gentle and Angel Martinez

Abstract MATLAB is a versatile software package used in many areas of applied mathematics, including computational finance. It is a programming language with a large number of functions for Monte Carlo simulation useful in financial analysis. The design of MATLAB allows for flexible data entry, including easy access of financial data from web resources. The graphical capabilities of MATLAB facilitate exploratory analysis, and the wide range of mathematical and statistical functionality provides the financial data analyst with a powerful tool. This article illustrates some of the basic capabilities of MATLAB, with an emphasis on financial applications.

28.1 Introduction

Serious analysis of financial data requires robust software. The first requirement in data analysis is acquisition and management of the data. The amount of data may be massive, so the software must be able to input and manipulate very large datasets. The software must support exploratory data analysis such as simple graphics that provide multiple views of the data. More advanced analyses include fitting of statistical models that range from stationary distributions to stochastic process with time-varying parameters.

J.E. Gentle (✉)
George Mason University, 4400 University Drive, Fairfax, VA 22030, USA
e-mail: jgentle@gmu.edu

A. Martinez
Strayer University, Fredericksburg, VA 22406-1094, USA

J.-C. Duan et al. (eds.), *Handbook of Computational Finance*, Springer Handbooks of Computational Statistics, DOI 10.1007/978-3-642-17254-0_28,

28.1.1 Types of Software

There is a wide range of software available for modeling and analyzing financial data. Prior to modeling and analysis, it is necessary to get the data into a useful format, and then to do some preliminary processing. There is also a number of software programs for data input and preprocessing of the data.

The type of software varies from simple programs for a specific task to general-purpose packages that can perform many tasks following very simple user directions. At the low end of this range are simple scripts written in Python, Perl, Ruby, or some similar dynamic scripting language, especially for the data input and preprocessing tasks. For more complicated mathematical operations, a simple Fortran or C function may be used. At the other end of the range are comprehensive statistical packages such as SAS, Stata, or SPSS, perhaps together with a macro library for the more common data input and analysis tasks.

There is another important dimension along which the type of software varies. This is the amount of "programming" that is done for a specific research task. A one end of this range, there is essentially no programming; the analyst issues a simple command, possibly by clicking an icon or making a choice on a menu. The other end of the range exists because the sheer number of different analysis tasks that may arise in financial research means that there cannot be an app for each – or if there is an app that fits the task perfectly, finding the app is more difficult than developing it. The analyst/researcher occasionally will write some software for a specific task. This may result in a "throw-away" program that serves one useful purpose and then may not be needed again for a long time, if ever. (Of course, if it turns out that the task becomes a commonplace activity, then whatever software is written to address it, should be packaged into a reusable app.)

There are various software packages that can satisfy the needs of a user at any point within the two spectra described above. Any given package, of course, has its strengths and weaknesses and may be more useful at one point within either of the dimensions than it is at another point. One package that satisfies the needs of users very well at many points along either dimension is MATLAB®.

28.1.2 MATLAB® and Gnu Octave

MATLAB is a technical and engineering computing environment that is developed and sold by The MathWorks, Inc. for algorithm development, modeling and simulation, data visualization, and much more. MATLAB can be thought of as an interactive system and a meta programming language, where the basic data element is an array, which can be a scalar, vector, matrix, or multi-dimensional array. In addition to basic array operations and mathematical functions, it offers programming options that are similar to those of other computing languages, such as user-written functions, control flow, conditional expressions, and so on.

The development of MATLAB (from "matrix laboratory") was begun by Cleve Moler, then a professor of computer science and mathematics at the University of New Mexico, in the late 1970s. The purpose was to give students easier access to the computational power of libraries written in Fortran. In the mid 1980s, a private company, The Mathworks, Inc. was formed to distribute and support MATLAB. The website is
`http://www.mathworks.com/`

The documentation that comes with MATLAB is an excellent resource, and PDF user's guides and documentation can be found at
`http://www.mathworks.com/access/helpdesk/help/techdoc/`

In addition to the documentation provided by The Mathworks, there are a number of primers, tutorials, and advanced user guides for MATLAB. The text by Hanselman and Littlefield (2005) is a comprehensive overview of MATLAB. One of the main strengths of MATLAB is the ability to create graphical user interfaces (GUIs) and to visualize data. The book by Marchand and Holland (2003) is an excellent reference for graphics and GUIs in MATLAB.

MATLAB is available for the common platforms (Microsoft Windows, Linux, Mac OS X, and Unix). It is built on an interactive, interpretive expression language that provides a rich set of program control statements for looping, conditional execution, and so on. MATLAB scripts are typically stored in M-files and the large user community has been active in developing and sharing M-files containing code for a wide variety of applications.

Gnu Octave is a freely available open-source package that provides much of the core functionality of MATLAB in a language with essentially the same syntax. The graphical interfaces for Octave are more primitive than those for MATLAB and do not interact as seamlessly with the operating system. Octave is available for free download from
`http://www.gnu.org/software/octave/download.html`
It is also available for all the common platforms. Eaton et al. (2008) gives an overview of the system. This book is also a standard user's guide for Octave.

There are a number of supplements to the basic MATLAB package, called "toolboxes". Some examples that are relevant to financial applications are the Financial Toolbox, Financial Derivatives Toolbox, Datafeed Toolbox, Fixed-Income Toolbox, Econometrics Toolbox and Statistics Toolbox. There are also some user-written functions and toolboxes on many topics; we will provide a partial list of this in a later section. See
`http://www.mathworks.com/matlabcentral/`
for MATLAB code, tutorials, and more.

There are also a number of books on MATLAB programming in specific areas, such as exploratory statistical data analysis and computational statistics, for example, Martinez et al. (2004), and Martinez and Martinez (2007). Although many of the methods of computational finance are general statistical methods applied to financial modeling, in addition to the books on statistical methods in MATLAB, there are also books addressing specific topics in finance, such as Brandimarte (2006) and Huynh et al. (2008).

28.2 Overview/Tutorial of the MATLAB® Language

We provide a brief overview and tutorial of MATLAB to help the reader better understand how it can be used to analyze financial data. This introduction only scratches the surface of what MATLAB can do, and we refer the reader to the other sources mentioned above.

MATLAB will execute under Windows, Linux, and Macintosh operating systems. This introduction will focus on the Windows version, but most of the information applies to all systems. The main MATLAB software package contains many functions for analyzing data of all types.

28.2.1 Getting Around in MATLAB

When MATLAB is started, a desktop environment is opened. This includes several windows, such as the Command Window, the Workspace Browser, the Command History and more. In addition, there is an Editor/Debugger that can be opened using the File menu. This editor can be used for creating MATLAB M-file scripts and functions. The MATLAB environment has many ways to execute commands, including the typical main menu items along the top of the window, toolbar buttons, context menus (right-click in an area), and specialized GUIs.

The Command Window is the main entry point for interacting with MATLAB. The prompt is indicated by a double-arrow, where the user can type commands, execute functions, and see output. The Command History window allows the user to see the commands that were executed for previous sessions, allowing the user to also re-execute commands using various Windows shortcuts (copy/paste, drag and drop, for example). The Workspace Browser shows information about the variables and objects that are in the current workspace, and it includes the ability to edit the variables in a spreadsheet-like interface. Finally, there is a Help window that provides access to documentation, help files, examples, and demos.

One can also access `help` files from the command line. The `help` files provide information about the function and also gives references for other related functions. From the command line, just type `help` *funcname* to get help on a specific function. The command `help general` provides a list of general purpose commands, and the word `help` used alone returns a list of topics. The command `lookfor` *keyword* will do a search of the first comment line of the M-files (on the MATLAB path) and return functions that contain that word.

The user can enter commands interactively at the command line or save them in an M-file. Thus, it is important to know some commands for file management. The commands shown in Table 28.1 can be used for this purpose.

Variables can be created at the command line or in M-file scripts and functions (covered later). A variable name cannot start with a number, and they are case sensitive. So, `Temp`, `temp`, and `TEMP` are all different objects. The variables that

Table 28.1 Basic commands for file management

Command	Usage
`dir, ls`	Shows the files in the current directory
`delete` *filename*	Deletes *filename*
`cd, pwd`	Shows the current directory
`cd dir, chdir`	Changes the current directory
`which` *filename*	Displays the path to *filename*
`what`	Lists the .m and `.mat` files in the current directory

Table 28.2 Basic commands for working with variables

Command	Usage
`who`	Lists all variables in the workspace
`whos`	Lists all variables and information about the variables
`clear`	Removes all variables from the workspace
`clear x y`	Removes variables x and y from the workspace

Table 28.3 Basic commands for working with external data files

Command	Usage
`load` *filename*	Loads all variables in *filename.mat*
`load` *`filename`* `var1`	Loads only `var1` in *filename.mat*
`load` *`filename.txt`* `-ascii`	Loads ascii *filename.txt* stores in the workspace with the same name

are created in a MATLAB session live in the workspace. We already mentioned the Workspace Browser; there are additional commands for workspace management. The commonly used ones are summarized in Table 28.2.

It is also necessary to get data into and out of MATLAB for analysis. One of the simplest ways to get data into MATLAB is to use the `load` command at the prompt; the `save` command works similarly to export your data in the MATLAB `.mat` format or ascii. Table 28.3 shows some of the common ways for loading data.

You can also use the commands in the File menu to load variables and to save the workspace. There is also the usual window for browsing directories and selecting files for importing.

MATLAB uses certain punctuation and characters in special ways. The percent sign denotes a comment line. Characters following the % on any command line is ignored. Commas have many uses in MATLAB; the most important is in array building to concatenate elements along a row. A semi-colon tells MATLAB *not* to display the results of the preceding command. Leaving the semi-colon off can cause a lot of data to be dumped to the command window, which can be helpful when debugging MATLAB programs but in other cases clutters up the window. Three periods denote the continuation of a statement. Comment statements and variable names, however, cannot be continued with this punctuation. The colon is used to specify a sequence of numbers; for example,

```
1:10
```

produces a sequence of numbers 1 through 10. A colon is also used in array indexing to access all elements in that dimension; for example,
`A(i,:)`
refers to the *i*th row of the array `A`.

28.2.2 Data Types and Arithmetic

MATLAB has two main data types: floating point numbers (type `double`) and strings (type `char`). The elements in the arrays or variables will be of these two data types.

28.2.2.1 Basic Data Constructs

The fundamental data element in MATLAB is an array. Arrays can be one of the following:

- The 0×0 empty array that is created using empty brackets: [ ].
- A 1×1 scalar array.
- A row vector, which is a $1 \times n$ array.
- A column vector, which is an $n \times 1$ array.
- A matrix with two dimensions, say $m \times n$ or $n \times n$.
- A multi-dimensional array, say $m \times \ldots \times n$.

Arrays must always be dimensionally conformal and all elements must be of the same data type. In other words, a 2×3 matrix must have three elements on each of its two rows.

In most cases, the data analyst will need to import data into MATLAB using one of the many functions and tools that are available for this purpose. We will cover these in a later section. Sometimes, we might want to type in simple arrays at the command line prompt for the purposes of testing code or entering parameters, etc. Here, we cover some of the ways to build small arrays. Note that these ideas can also be used to combine separate arrays into one large array.

Commas or spaces concatenate elements (an element can be an array) as columns. Thus, we get a row vector from the following:

$$temp = [1, 4, 5];$$

Recall that the semi-colon at the end of the expression tells MATLAB to not print the value of the variable `temp` in the command window. We can concatenate two column vectors `a` and `b` into one matrix, as follows:

$$temp = [a\ b];$$

Table 28.4 Special arrays

Function	Usage
`zeros, ones`	Build arrays containing all 0s or all 1s respectively
`rand, randn`	Build arrays containing uniform or normal random values
`eye`	Create an identity matrix

Using the semi-colon to separate elements of the array tells MATLAB to concatenate elements `a` and `b` as rows. So, we would get a column vector from this command:

$$temp = [1; 4; 5];$$

When we use arrays as building blocks for larger arrays, then the sizes of each array element must be conformal for the type of operation.

There are some useful functions in MATLAB for building special arrays. These are summarized in Table 28.4; look at the help file to learn how each function is used.

Cell arrays and structures are a special MATLAB data type that allow for more flexibility. Cell arrays are array-type structures, where the *contents* of each cell element can vary in size and type (numeric, character, or cell). The cell array has an overall structure that is similar to the basic data arrays we have already discussed. For example, the cells are arranged in rows and columns. If we have a 2×3 cell array, then each of its two rows has to have three cells.

Structures are similar to cell arrays in that they allow one to combine collections of dissimilar data into a single variable. Individual structure elements are addressed by fields. We use the dot notation to access the fields. Each element of a structure is called a record.

As an example, suppose we had a structure called `data` that had the following fields: `name`, `dob`, and `test`. Then we could obtain the information in the tenth record using

```
data(10).name
data(10).dob
data(10).text
```

28.2.2.2 Array Addressing

In Table 28.5, we show some of the common ways to access elements of arrays.

Suppose we have a cell array called `A`. The last line of Table 28.5 shows how to access the contents of the *ij*th cell in `A`. Curly braces are used to get to the contents, and parentheses point to the cells. The two notations can be combined to access part of the contents of a cell. For example, `A{1,1}(1:2)` extracts the first two elements of the vector that is contained in cell `A(1,1)`.

Table 28.5 Addressing: Arrays or cell arrays

Notation	Usage
`a(i)`	Denotes the *i*th element
`A(:,i)`	Addresses the *i*th column. Here, the colon operator tells MATLAB to access all rows
`A(i,:)`	Addresses the *i*th row. Here, the colon tells MATLAB to gather all of the columns
`A(1,3,4)`	Addresses the element indexed at three levels
`A{i,j}`	Addresses the *contents* of the *ij*th cell

Table 28.6 Element-wise arithmetic

Operator	Usage
`.*`	Multiply two arrays element-by-element
`./`	Divide two arrays element-by-element
`.^`	Raise each element of an array to some power

28.2.2.3 Arithmetic Operations

MATLAB has the usual mathematical operators found in programming languages, such as addition (+), subtraction (-), multiplication(*), division(/), and exponentiation (^). These follow the same order of operations that is found in algebra and can be changed using parentheses. MATLAB also follows the conventions found in linear algebra. In other words, the arrays must have the same dimensions when adding or subtracting vectors or matrices, and the operation is carried out element-by-element. If we are multiplying two matrices, `A` and `B`, they must be dimensionally correct; e.g., the number of columns of `A` must be equal to the number of rows of `B`.

In some cases, we might want to multiply two arrays element-by-element. In this case, we would put a period in front of the multiplication operator. We can do the same thing to divide two arrays element-by-element, as well exponentiation. We list these in Table 28.6.

28.3 Writing and Using Functions in MATLAB

MATLAB has many built-in functions that execute commonly used tasks in linear algebra, data analysis, and engineering. Some of these standard functions include trigonometric functions (`sin`, `cos`, `tan`, and so on), `log`, `exp`, specialized functions (Bessel, gamma, beta, and so on), and many more. In this section, we provide an introduction on writing your own functions and programs in MATLAB.

28.3.1 Script Files and Functions

MATLAB programs are saved in M-files. These are text files that contain MATLAB commands and expressions, and they are saved with the `.m` extension. Any text editor can be used to create them, but the one that comes with MATLAB is recommended. It has special color coding and other helpful features to write MATLAB programs that execute correctly.

When script M-files are executed, the commands are implemented just as if they were typed at the prompt. The commands have access to the workspace and any variables created by the script file are in the workspace when the script finishes executing. To execute a script file, simply type the name of the file at the command line or prompt.

Script files and functions have the same `.m` file extension. However, a function has a special syntax for the first line. In the general case, this syntax is

```
function [out1,..., outM] = func_name(in1, ..., inN)
```

A function does not have to be written with input or output arguments. Also, a function can be called with fewer input and/or output arguments, but not more. The function corresponding to the above declaration would be saved in a file called `func_name.m`, and it is invoked using `func_name`.

It is important to understand the scope of MATLAB workspaces. Each function has its own workspace, which is separate from the main MATLAB workspace. Communicating information about variables and their values is accomplished by way of the input and output variables. This concept is very important when writing and debugging functions.

It is always a good idea to put several comment lines at the beginning of a function. This is the information that is returned by the command `help func_name`. At a minimum, this should include a description about what the function does and what types of variables are used for inputs and outputs.

Most computer languages provide features that allow one to control the flow of execution that depends on conditions. MATLAB has similar constructs, and these are listed here:

- For loops
- While loops
- If-else statements
- Switch statement

These should be used sparingly to make the code run fast. In most cases, it is more efficient in MATLAB to operate on an entire array rather than looping through it. It is important to note that most of the functions in MATLAB operate on entire arrays, alleviating the need to loop through the arrays. A brief description of these programming constructs is given below.

The basic syntax for a for loop is

```
for i = array
```

```
        commands
    end
```

The looping variable is given by *i*, and the loop runs one time for each element in the variable *array*. The variable *i* takes on the next value in *array* and the *commands* between the `for` and `end` are executed each time the loop executes. The colon notation is often used to generate a sequence of numbers for the variable array. For example, using `for i = 1:10` means that *i* would take on the values 1 through 10. Several `for` loops can be nested, with each one terminated by an `end` statement.

Unlike a `for` loop, a `while` loop executes an indefinite number of times. The general syntax is

```
    while expression
        commands
    end
```

The *commands* between the `while` and the `end` are executed as along as *expression* is true. (Note that in MATLAB a scalar that is nonzero evaluates to true.) Usually, a scalar entry is used in *expression*, but an array can be used also. In the case of arrays, all elements of the resulting array must be true for the commands to execute.

Sometimes commands must be executed based on a relational test. The if-else statement is suitable in these situations. The basic syntax is

```
    if expression
        commands
    elseif expression
        commands
    else
        commands
    end
```

Only one `end` is required at the end of the sequence of `if`, `elseif` and `else` statements. Commands are executed only if the corresponding *expression* is true. Note that in the simplest case, one might only need to use the following

```
    if expression
        commands
    end
```

In other words, you do not necessarily need to have the `else` constructs.

The `switch` statement is useful if one has a lot of `if`, `elseif` statements in the program. This construct is very similar to that in the C language. The basic syntax is

```
    switch expression
    case value1
        commands    % executes if expression = value1
    case value2
        commands    % executes if expression = value2
    ...
    otherwise
        commands    % executes if nothing else does
    end
```

The *expression* must be either a scalar or a character string.

28.4 Visualization

Here we briefly describe some of the basic plotting capabilities in the main MATLAB package. For more information on the extensive graphics that are available with this software, see the MATLAB documentation and the book by Marchand and Holland (2003).

28.4.1 2-D Plots

The main function used for 2-D plots is `plot`. When the `plot` function is called, it opens a `Figure` window (if one is not already there), scales the axes to fit the data, and plots the points. The default is to plot the points and connect them using straight lines. For example, `plot(x, y)` plots the values in the vector `x` along the horizontal axis and the values in the vector `y` on the vertical axis (the values correspond to `(x, y)` points), connected by straight lines. Thus, the vectors must have the same length. Using just one vector argument to `plot` will show the values of `y` against the index number.

The default line style is a solid line, but one can also use the following: dotted line, dash-dot line, and dashed line. Additionally, other plotting symbols can be used instead of points. There are many choices (e.g., `*`, `x`, `o`); please use `help plot` to get a list of them. MATLAB also provides some pre-defined colors; these can also be found in the documentation on `plot`.

Any number of pairs can be used as arguments to plot. For instance, `plot(x, y1, x, y2)` will create two curves on the same plot. If only one argument is supplied to `plot`, then MATLAB plots the vector versus the index of its values. To create a plot with different markers and line styles, just enclose your choices between single quotes as shown here:

```
plot(x, y, 'b*-.')
```

This command tells MATLAB to plot the points in `x` and `y` as an asterisk and connect them with a blue dash-dot line. As another example, leaving out the line style would produce a scatter plot in the default color: `plot(x, y, '*')`.

The following example shows how to plot the adjusted closing price for Google stock for weeks in 2008. In this case, we do not need to have an explicit `x` vector because the `x` axis is the week number, which is the same as the index.

```
plot(goog08(:,6), '-*')
title('Adjusted Closing Price -- Google')
xlabel('Week Number -- 2008')
ylabel('Adjusted Closing Price')
```

The results are shown in Fig. 28.1.

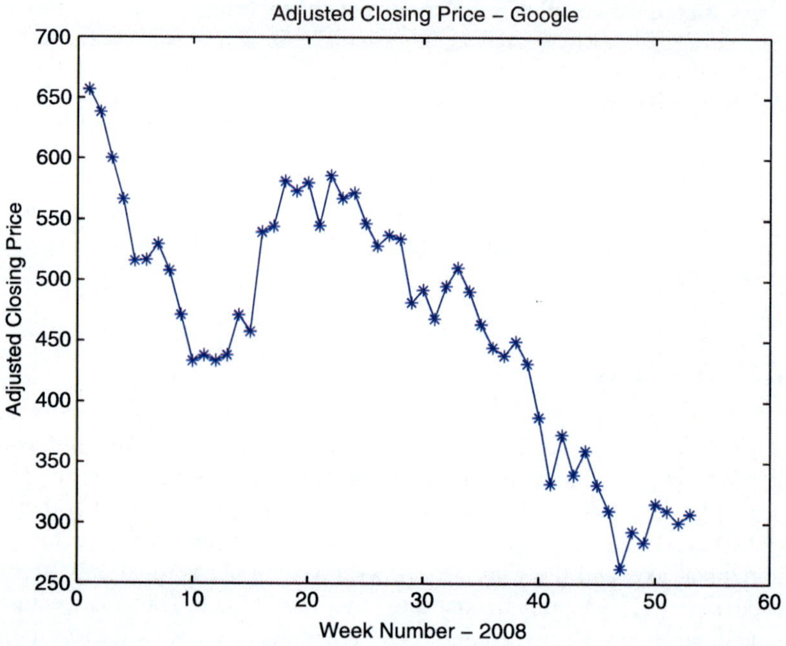

Fig. 28.1 A simple 2-D plot

When a 2-D plot is created, then the `Basic Fitting` tool can be activated from the `Tools` menu in the `Figure` window. This invokes a GUI that provides many options for fitting curves and interpolation using the `x`, `y` data in the plot. Finally, there are many specialized 2-D plots, such as `scatter`, `polar`, and `plotyy`. Also, see `help graph2d` for more functions that can be used on for 2-D plots. Figure 2 illustrates multiple 2-D plots in the same window, and a Fig. 3 shows a scatterplot matrix with histograms along the diagonal cells.

28.4.2 3-D Plots

To plot ordered triples of points, one can use the `plot3` function. It works the same as the `plot` function, except that it requires three vectors for plotting: `plot3(x, y, z)`. All of the concepts, line styles, colors, and plotting symbols apply to `plot3`.

Another form of 3-D graphics that would be of use in computational finance problems is to plot surfaces. In this case, we have a function $z = f(x, y)$, and we want to show the value of z as the surface height. The two main functions for doing this are `surf` and `mesh`. The function `mesh` draws a wireframe of the surface with

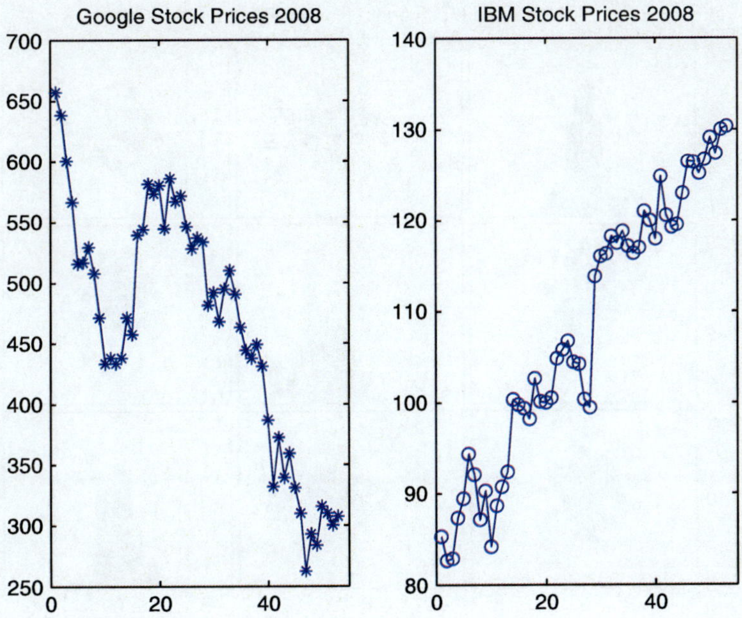

Fig. 28.2 A display with two subplots

the color determined by the value of z (as the default). The function `surf` creates a shaded and faceted surface plot.

There are many options that one can apply to these surface plots, such as changing the color maps, applying shading and lighting, changing the view point, applying transparency, and more. See `help graph3d` for more capabilities and 3-D graphing functions in MATLAB (Fig. 28.2).

28.4.3 Other Useful Plotting Capabilities

What we have described so far is the ability to put one plot or set of axes in a `Figure` window. In some applications, it would be useful to have a way to put several axes or plots in one window. We can do this through the use of the `subplot` function. This creates an $m \times n$ matrix of plots (or axes) in the current `Figure` window. The example provided below shows how to create two plots side-by-side.

```
% Create the left plot
subplot(1, 2, 1)
plot(x, y)
% Create the right plot
subplot(1, 2, 2)
plot(x, z)
```

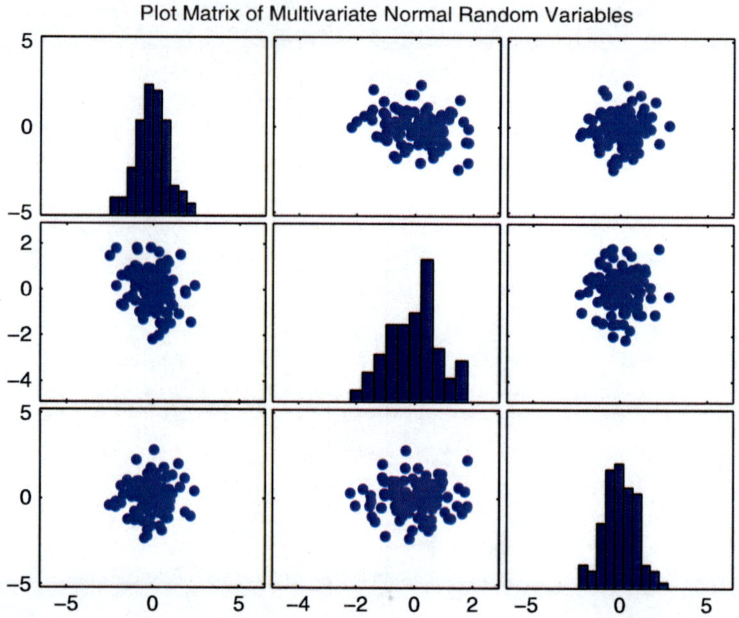

Fig. 28.3 A scatterplot matrix with three variables

The first two arguments to the function `subplot` tell MATLAB about the layout of the plots within the `Figure` window. In the example above, we have a layout with one row and two columns of plots. The third argument tells MATLAB which plot to work with. The plots are numbered from top to bottom and left to right. The most recent plot that was created or worked on is the one affected by any subsequent plotting commands. You can think of the `subplot` function as a pointer that tells MATLAB what set of axes to work with (Fig. 28.3).

One of the most useful plots in data analysis is the scatter plot where (x, y) pairs are displayed as points. We can create a plot matrix of scatter plots when we have more than two variables. The main MATLAB package has a function called `plotmatrix` that will produce this type of plot. The basic syntax for this function is `plotmatrix(X, Y)`, where `X` and `Y` are matrices, which scatter plots the columns of `X` against the columns of `Y`. If `plotmatrix` is called with just one matrix argument, then it produces all pairwise scatter plots of the columns of `X`, with a histogram of the columns along the diagonal (Fig. 28.4).

28.5 Getting Financial Data into MATLAB

Often financial data is available in a spreadsheet or comma-separated-value (csv) format. A useful website for historica stock prices is
`www.finance.yahoo.com`

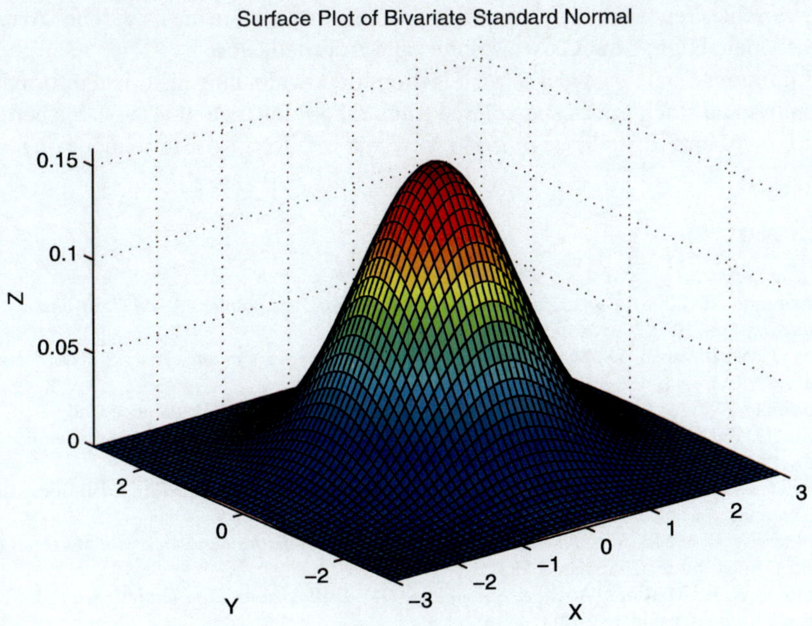

Fig. 28.4 A simple 3-D plot

For example, to get historical information on the Dow Jones Industrial Average, go to
`http://finance.yahoo.com/q/hp?s=%5EDJI`
You can download data from 1928 to the present. There is a link at the bottom of the page that allows you to save the data to a spreadsheet (csv format). You can also get the same information for other entities and stocks by entering the symbol in the text box at the right of the page.

Once it is stored as a csv file, the MATLAB function `csvread` can be used to bring it into the MATLAB workspace. This function, however, expects only numeric data. We can use optional arguments to tell MATLAB where to start (row and column) loading the data. The function `csvread` uses 0 as the staring point for the rows and columns in the file, so to start in the second field in the second row (to skip the header), we specify these starting points as 1 and 1. The following code shows how to import the data:

```
% The file called DowJones1928toNow.csv was
% downloaded from Yahoo Finance.
% The first column contains header information.
% The second column contains dates in text format.
% Now read in the data.
X = csvread('DowJones1928toNow.csv', 1,1);
```

The variables represented by the columns of the matrix `X` are the Dow Jones Average at the Open, High, Low, Close, Volume, and Adjusted Close.

Luminous Logic provides a free function for downloading historical information on individual stock prices and volume from Yahoo! Finance. It is available here `http://luminouslogic.com/matlab-stock-market-scripts`

References

Brandimarte, P. (2006). *Numerical methods in finance and economics: A MATLAB-based introduction* (2nd Ed.). New York: Wiley.

Eaton, J. W., Bateman, D., & Hauberg, S. (2008). *GNU octave manual version 3*. UK: Network Theory Ltd.

Hanselman, D. C. & Littlefield, B. L. (2005). *Mastering MATLAB 7*. NJ: Prentice Hall.

Huynh, H., Lai, V. S., & Soumaré, I. (2008). *Stochastic simulation and applications in finance with MATLAB programs*. New York: Wiley.

Marchand, P. & Holland, O. T. (2003). *Graphics and GUIs with MATLAB* (3rd Ed.). Boca Raton: Chapman & Hall.

Martinez, W. L. & Martinez, A. R. (2007). *Computational statistics handbook with MATLAB* (2nd Ed.). Boca Raton: Chapman & Hall.

Martinez, W. L., Martinez, A. R., & Solka, J. (2004). *Exploaratory data analysis with MATLAB*. Boca Raton: Chapman & Hall.

Chapter 29
R as a Tool in Computational Finance

John P. Nolan

29.1 Introduction

R is a powerful, free program for statistical analysis and visualization. R has superb graphics capabilities and built-in functions to evaluate all common probability distributions, perform statistical analysis, and to do simulations. It also has a flexible programming language that allows one to quickly develop custom analyses and evaluate them. R includes standard numerical libraries: LAPACK for fast and accurate matrix multiplication, QUADPACK for numerical integration, and (univariate and multivariate) optimization routines. For compute intensive procedures, advanced users can call optimized code written in C or Fortran in a straightforward way, without having to write special interface code.

The R program is supported by a large international team of volunteers who maintain versions of R for multiple platforms. In addition to the base R program, there are thousands of packages written for R. In particular, there are dozens of packages for solving problems in finance. Information on implementations on obtaining the R program and documentation are given in Appendix 1.

A New York Times article by Vance (2009a) discussed the quick growth of R and reports that an increasing number of large companies are using R for analysis. Among those companies are Bank of America, Pfizer, Merck, InterContinental Hotels Group, Shell, Google, Novartis, Yale Cancer Center, Motorola, Hess. It is estimated in Vance (2009b) that over a quarter of a million people now use R.

The following three simple examples show how to get free financial data and how to begin to analyze it. Note that R uses the back arrow `<-` for assignment and that the `>` symbol is used as the prompt R uses for input. The following six lines

J.P. Nolan (✉)
Department of Mathematics and Statistics, American University, 4400 Massachusetts Ave, North West, Washington, DC 20016-8050, USA
e-mail: jpnolan@american.edu

J.-C. Duan et al. (eds.), *Handbook of Computational Finance*, Springer Handbooks of Computational Statistics, DOI 10.1007/978-3-642-17254-0_29,

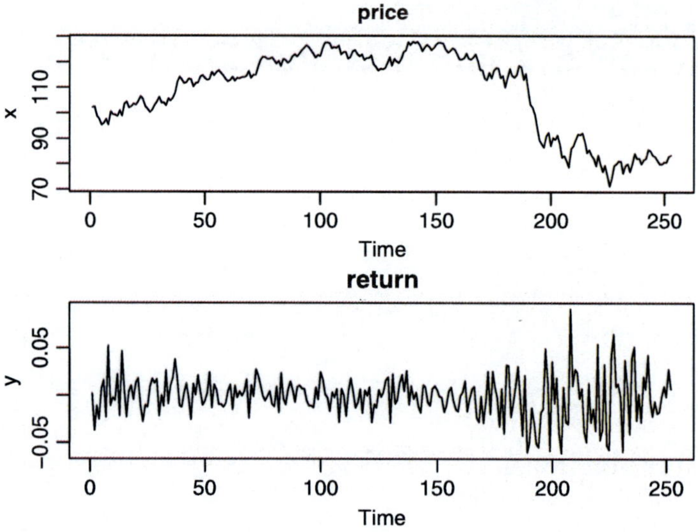

Fig. 29.1 Closing price and return for IBM stock in 2008

of R code retrieve the adjusted closing price of IBM stock for 2008 from the web, compute the (logarithmic) return, plot both time series as shown in Fig. 29.1, give a six number summary of the return data, and then finds the upper quantiles of the returns.

```
> x <- get.stock.price("IBM")
IBM  has 253 values from 2008-01-02 to 2008-12-31
> y <- diff(log(x))
> ts.plot(x,main="price")
> ts.plot(y,main="return")
> summary(y)
     Min.    1st Qu.    Median      Mean   3rd Qu.      Max.
-0.060990 -0.012170 -0.000336 -0.000797  0.010620  0.091390
> quantile(y, c(.9,.95,.99) )
       90%        95%        99%
0.02474494 0.03437781 0.05343545
```

The source code for the function `get.stock.price` and other functions used below are given in Appendix 2. The next example shows more information for 3 months of Google stock prices, using the function `get.stock.data` that retrieves stock information that includes closing/low/high prices as well as volume (Fig. 29.2).

```
> get.stock.data("GOOG",start.date=c(10,1,2008),
  stop.date=c(12,31,2008))
  GOOG  has 64 values from
  2008-10-01 to 2008-12-31
```

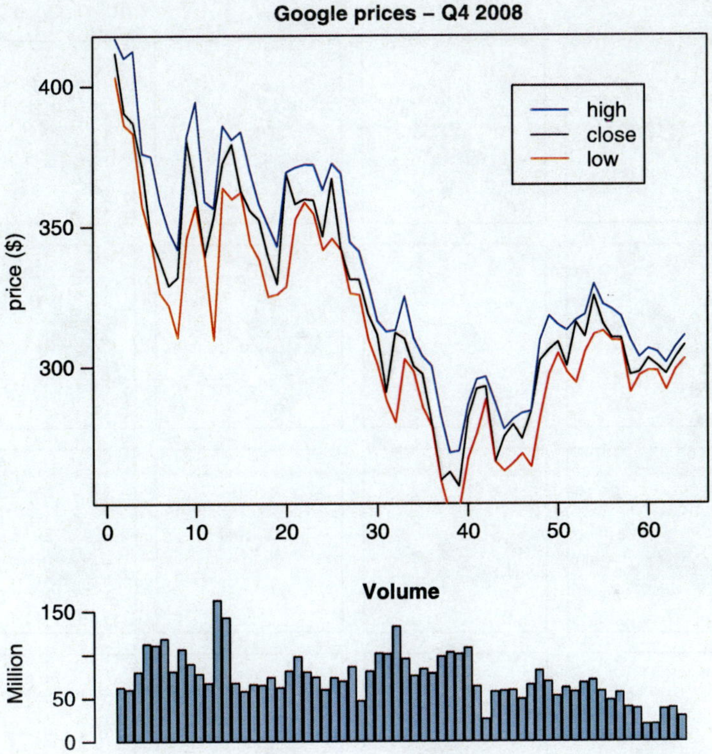

Fig. 29.2 Google stock prices and volume in the fourth quarter of 2008

```
> par(mar=c(1,4,2,2))  # graphing option
> num.fig <- layout(matrix(c(1,2)),heights=c(5,2))
  # setup a multiplot
> ts.plot(x$Close,ylab="price (in $)", main="Google
  prices - 4th quarter 2008")
> lines(x$Low,col="red")
> lines(x$High,col="blue")
> legend(45,400,c("high","close","low"),lty=1,
  col=c("blue","black","red"))
> barplot(x$Volume/100000,ylab="Million",
  col="lightblue",main="\nVolume")
```

Another function `get.portfolio.returns` will retrieve multiple stocks in a portfolio. Dates are aligned and a matrix of returns is the results. The following code retrieves the returns from IBM, General Electric, Ford and Microsoft and produces scatter plots of the each pair of stocks. The last two commands show the mean return and covariance of the returns (Fig. 29.3).

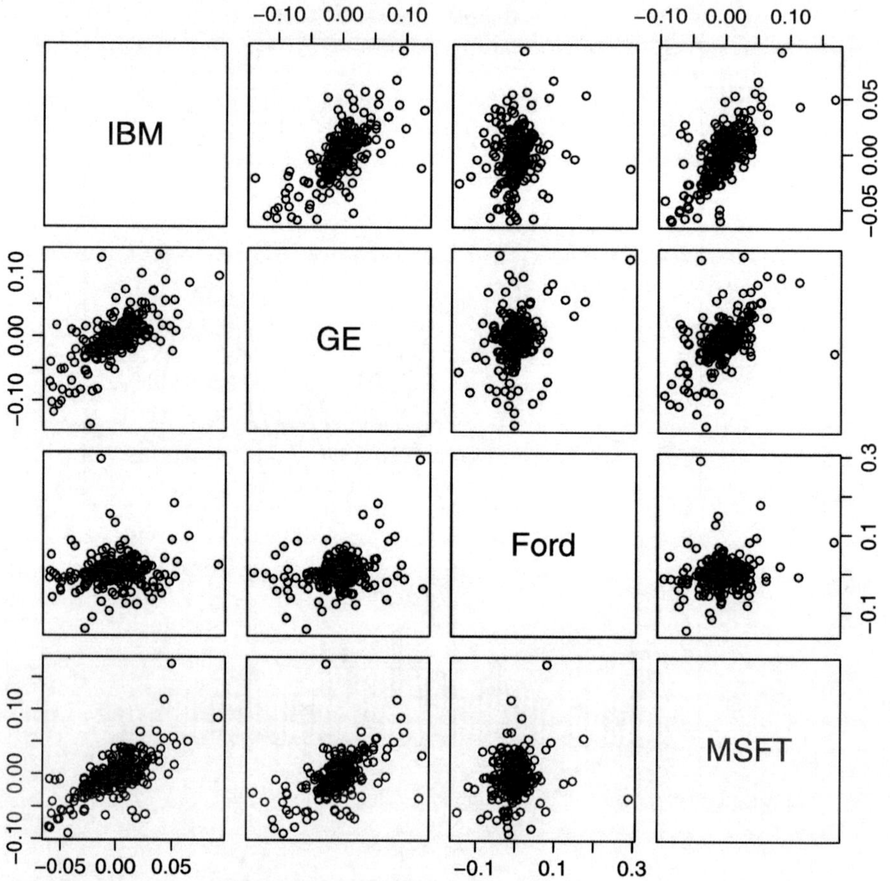

Fig. 29.3 Pairwise scatter plots of returns for four stocks

```
> x <- get.portfolio.returns( c("IBM","GE","Ford","MSFT") )
IBM  has 253 values from 2008-01-02 to 2008-12-31
GE  has 253 values from 2008-01-02 to 2008-12-31
Ford  has 253 values from 2008-01-02 to 2008-12-31
MSFT  has 253 values from 2008-01-02 to 2008-12-31
     253 dates with values for all stocks, 252 returns
     calculated
> pairs(x)
> str(x)
'data.frame':    252 obs. of  4 variables:
 $ IBM : num  0.00205 -0.03665 -0.01068 -0.02495 0.00742 ...
 $ GE  : num  0.00118 -0.02085 0.00391 -0.02183 0.01128 ...
 $ Ford: num  -0.01702 -0.00862 -0.00434 -0.02643 0.00889 ...
 $ MSFT: num  0.00407 -0.02823 0.00654 -0.03405 0.0293 ...
```

```
> mean(x)
          IBM            GE          Ford          MSFT
-0.0007974758 -0.0030421414 -0.0002416205 -0.0022856306
> var(x)
              IBM           GE         Ford         MSFT
IBM  0.0005138460 0.0005457266 0.0001258669 0.0004767922
GE   0.0005457266 0.0012353023 0.0003877436 0.0005865461
Ford 0.0001258669 0.0003877436 0.0016194549 0.0001845064
MSFT 0.0004767922 0.0005865461 0.0001845064 0.0009183715
```

The rest of this paper is organized as follows. Section 29.2 gives a brief introduction to the R language, Sect. 29.3 gives several examples of using R in finance, and Sect. 29.4 discusses the advantages and disadvantages of open source vs. commercial software. Finally, the two appendices give information on obtaining the R program and the R code used to obtain publicly available data on stocks.

29.2 Overview/Tutorial of the R Language

This section is a brief introduction to R. It is assumed that the reader has some basic programming skills; this is not intended to teach programming from scratch. You can find basic help within the R program by using the question mark before a command: `?plot` (alternatively `help("plot")`) will give a description of the `plot` command, with some examples at the bottom of the help page. Appendix 1 gives information on more documentation.

One powerful feature of R is that operations and functions are vectorized. This means one can perform calculations on a set of values without having to program loops. For example, `3*sin(x)+y` will return a single number if x and y are single numbers, but a vector if x and y are vectors. (There are rules for what to do if x and y have different lengths, see below.)

A back arrow `<-`, made from a less than sign and a minus sign, is used for assignment. The equal sign is used for other purposes, e.g. specifying a title in the plots above. Variable and function names are case sensitive, so `X` and `x` refer to different variables. Such identifiers can also contain periods, e.g. `get.stock.price`. Comments can be included in your R code by using a # symbol; everything on the line after the # is ignored. Statements can be separated by a semicolon within a line, or placed on separate lines without a seperator.

29.2.1 *Data Types and Arithmetic*

Variables are defined at run time, not by a formal declaration. The type of a variable is determined by the type of the expression that defines it, and can change from line to line. There are many data types in R. The one we will work with most is the numeric type `double` (double precision floating point numbers). The simplest numeric type is a single value, e.g. `x <- 3`. Most of the time we will be working with vectors, for example, `x <- 1:10` gives the sequence from 1 to 10. The statement `x <- seq(-3,3,0.1)` generates an evenly spaced sequence from −3 to 3 in steps of size 0.1. If you have an arbitrary list of numbers, use the combine command, abbreviated `c(...)`, e.g. `x<- c(1.5, 8.7, 3.5, 2.1, -8)` defines a vector with five elements.

You access the elements of a vector by using subscripts enclosed in square brackets: `x[1]`, `x[i]`, etc. If `i` is a vector, `x[i]` will return a vector of values. For example, `x[3:5]` will return the vector `c(x[3],x[4],x[5])`.

The normal arithmetic operations are defined: $+, -, *, /$. The power function x^p is `x^p`. A very useful feature of R is that almost all operations and functions work on vectors elementwise: `x+y` will add the components of `x` and `y`, `x*y` will multiply the components of `x` and `y`, `x^2` will square each element of `x`, etc. If two vectors are of different lengths in vector operations, the shorter one is repeated to match the length of the longer. This makes good sense in some cases, e.g. `x+3` will add three to each element of `x`, but can be confusing in other cases, e.g. 1:10 + c(1,0) will result in the vector `c(2, 2,4,4,6,6,8,8,10,10)`.

Matrices can be defined with the `matrix` command: `a <- matrix( c(1,5, 4,3,-2,5), nrow=2, ncol=3)` defines a 2×3 matrix, filled with the values specified in the first argument (by default, values are filled in one column at a time; this can be changed by using the `byrow=TRUE` option in the `matrix` command). Here is a summary of basic matrix commands:

`a + b` adds entries element-wise (a[i,j]+b[i,j]),
`a * b` is element by element (not matrix) multiplication (a[i,j]*b[i,j]),
`a %*% b` is matrix multiplication,
`solve(a)` inverts a,
`solve(a,b)` solves the matrix equation a x = b,
`t(a)` transposes the matrix a,
`dim(a)` gives dimensions (size) of a,
`pairs(a)` shows a matrix of scatter plots for all pairs of columns of a,
`a[i,]` selects row i of matrix a,
`a[,j]` selects column j of matrix a,
`a[1:3,1:5]` selects the upper left 3×5 submatrix of a.

Strings can be either a single value, e.g. `a <- "This is one string"`, or vectors, e.g. `a <- c("This", "is", "a", "vector", "of", "strings")`.

Another common data type in R is a data frame. This is like a matrix, but can have different types of data in each column. For example, read.table and read.csv return data frames. Here is an example where a data frame is defined manually, using the `cbind` command, which “column binds” vectors together to make a rectangular array.

```
name <- c("Peter","Erin","Skip","Julia")
age <- c(25,22,20,24)
weight <- c(180,120,160,130)
info <- data.frame(cbind(name,age,weight))
```

A more flexible data type is a list. A list can have multiple parts, and each part can be a different type and length. Here is a simple example:

```
x <- list(customer="Jane Smith",
     purchases=c(93.45,18.52,73.15),
     other=matrix(1:12,3,4))
```

You access a field in a list by using `$`, e.g. `x$customer` or `x$purchases[2]`, etc.

R is object oriented with the ability to define classes and methods, but we will not go into these topics here. You can see all defined objects (variables, functions, etc.) by typing `objects( )`. If you type the name of an object, R will show you it's value. If the data is long, e.g. a vector or a list, use the structure command `str` to see a summary of what the object is.

R has standard control statements. A `for` loop lets you loop through a body of code a fixed number of times, `while` loops let you loop until a condition is true, `if` statements let you execute different statements depending on some logical condition. Here are some basic examples. Brackets are used to enclose blocks of statements, which can be multiline.

```
sum <- 0
for (i in 1:10) {sum <- sum + x[i] }

while (b > 0) { b <- b - 1 }

if (a < b) { print("b is bigger") }
else { print("a is bigger") }
```

29.2.2 General Functions

Functions generally apply some procedure to a set of input values and return a value (which may be any object). The standard math functions are built in: `log, exp, sqrt, sin, cos, tan,` etc. and we will not discuss them specifically. One very handy feature of R functions is the ability to have optional arguments and to specify default values for those optional arguments. A simple example of an optional argument is the log function. The default operation of the statement `log(2)` is to compute the natural logarithm of two. However, by adding an optional second argument, you can compute a logarithm to any base, e.g. `log(2,base=10)` will compute the base 10 logarithm of 2.

There are hundreds of functions in R, here are some common functions:

Function name	Description
`seq(a,b,c)`	Defines a sequence from a to b in steps of size c
`sum(x)`	Sums the terms of a vector
`length(x)`	Length of a vector
`mean(x)`	Computes the mean
`var(x)`	Computes the variance
`sd(x)`	Computes the standard deviation of x
`summary(x)`	Computes the 6 number summary of x (min, quartiles, mean, max)
`diff(x)`	Computes successive differences $x_i - x_{i-1}$
`c(x,y,z)`	Combine into a vector
`cbind(x,y,...)`	"Bind" x, y, ... into the columns of a matrix
`rbind(x,y,...)`	"Bind" x, y, ... into the rows of a matrix
`list(a=1,b="red",...)`	Define a list with components a, b, ...
`plot(x,y)`	Plots the pairs of points in x and y (scatterplot)
`points(x,y)`	Adds points to existing plot
`lines(x,y)`	Adds lines/curves to existing plot
`ts.plot(x)`	Plots the values of x as a times series
`title("abc")`	Adds a title to an existing plot
`par(...)`	Sets parameters for graphing, e.g. par(mfrow=c(2,2)) creates a 2 by 2 matrix of plots
`layout(...)`	Define a multiplot
`scan(file)`	Read a vector from an ascii file
`read.table(file)`	Read a table from an ascii file
`read.csv(file)`	Read a table from an Excel formated file
`objects()`	Lists all objects
`str(x)`	Shows the structure of an object
`print(x)`	Prints the single object x
`cat(x,...)`	Prints multiple objects, allows simple stream formatting
`sprintf(format,...)`	C style formatting of output

29.2.3 Probability Distributions

The standard probability distributions are built into R. Here are the abbreviations used for common distributions in R:

Name	Distribution
binom	Binomial
geom	Geometric
nbinom	Negative binomial
hyper	Hypergeometric
norm	Normal/Gaussian
chisq	χ^2
t	Student t
f	F
cauchy	Cauchy distribution

For each probability distribution, you can compute the probability density function (pdf), the cumulative distribution function (cdf), the quantiles (percentiles = inverse cdf) and simulate. The function names are given by adding a prefix to the distribution name.

Prefix	Computes	Example
d	Density (pdf)	dnorm(x,mean = 0,sd = 1)
p	Probability (cdf)	pnorm(x,mean = 0,sd = 1)
q	Quantiles (percentiles)	qnorm(0.95,mean = 0,sd = 1)
r	Simulate values	rnorm(1,000,mean = 0,sd = 1)

The arguments to any functions can be found from the `arg` command, e.g. `arg(dnorm)`; more explanation can by found using the built-in help system, e.g. `?dnorm`. Many have default value for arguments, e.g. the mean and standard deviation default to 0 and 1 for a normal distribution. A few simple examples of using these functions follow.

```
x <- seq(-5,5,.1)
y <- dnorm(x, mean=1,sd=0.5)
plot(x,y,type='l') # plot a N(1,0.25) density

qnorm(0.975)  # z_{0.25} = 1.96

pf( 2, 5, 3) # P(F_{5,3} < 2) = 0.6984526

x <- runif(10000) # generate 10000 uniform(0,1) values
```

29.2.4 Two Dimensional Graphics

The basic plot is an xy-plot of points. You can connect the points to get a line with type='l'. The second part of this example is shown in Fig. 29.4.

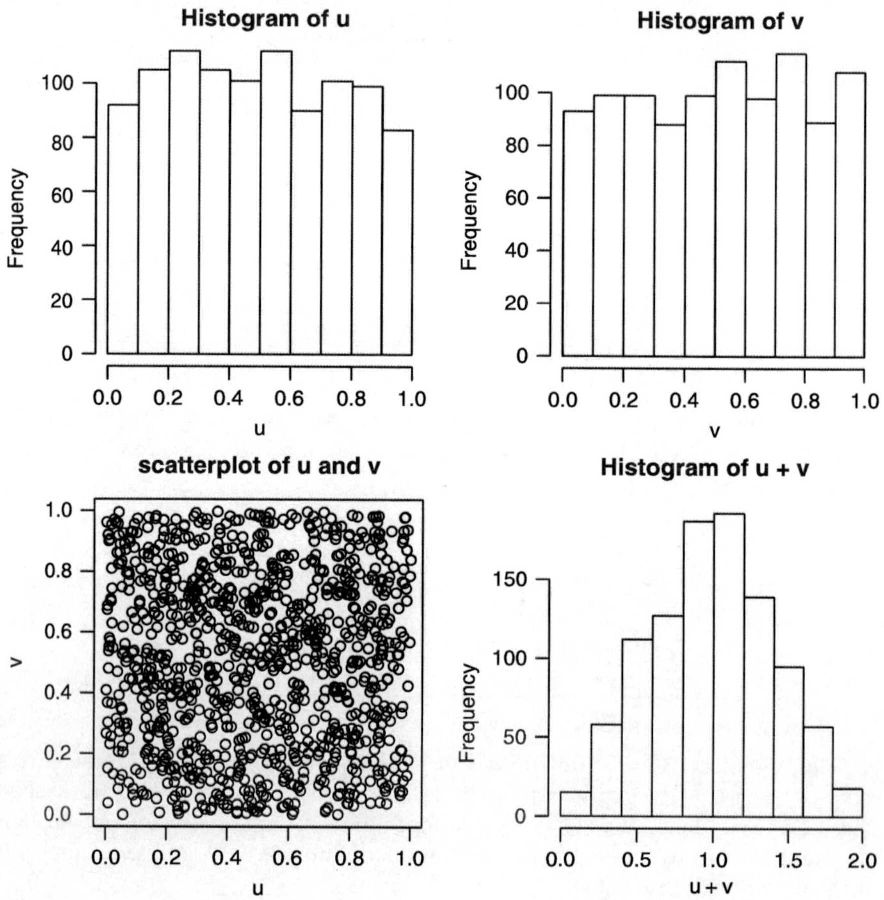

Fig. 29.4 Multiplot showing histograms of u, v, $u + v$, and a scatter plot of (u, v)

```
x <- seq(-10,10,.25)
y <- sin(x)
plot(x,y,type='l')
lines(x,0.5*y,col='red') # add another curve and color
title("Plot of the sin function")

u <- runif(1000)
v <- runif(1000)
par(mfrow=c(2,2)) # make a 2 by 2 multiplot
hist(u)
hist(v)
plot(u,v)
title("scatterplot of u and v")
hist(u+v)
```

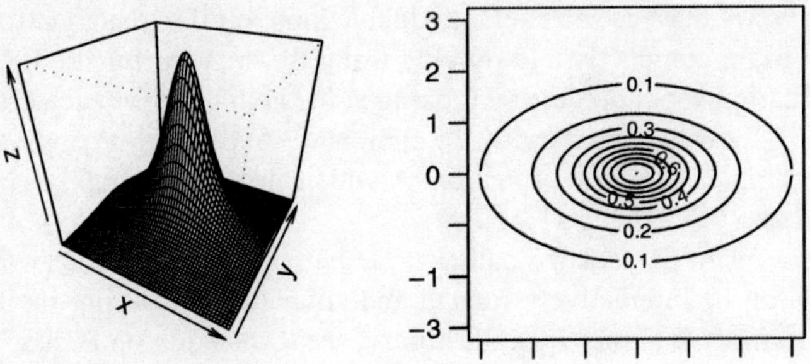

Fig. 29.5 A surface and contour plot

There are dozens of options for graphs, including different plot symbols, legends, variable layout of multiple plots, annotations with mathematical symbols, trellis/lattice graphics, etc. See `?plot` and `?par` for a start.

You can export graphs to a file in multiple formats using "File", "Save as", and select type (jpg, pdf, postscript, png, etc.)

29.2.5 Three Dimensional Graphics

You can generate basic 3D graphs in standard R using the commands `persp`, `contour` and `image`. The first gives a "perspective" plot of a surface, the second gives a standard contour plot and the third gives a color coded contour map. The examples below show simple cases; there are many more options. For static graphs, there are three functions: All three use a vector of x values, a vector of y values, and and matrix z of heights, e.g. `z[i,j]<-f(x[i],y[j])`. Here is one example where such a matrix is defined using the function $f(x, y) = 1/(1 + x^2 + 3y^2)$, and then the surface is plotted (Fig. 29.5).

```
x <- seq(-3,3,.1) # a vector of length 61
y <- x
# allocate a 61 x 61 matrix and fill with f(x,y) values
z <- matrix(0,nrow=61,ncol=61)
for (i in 1:61) {
  for (j in 1:61) {
     z[i,j] <- 1/(1+x[i]^2 + 3*y[j]^2)
  }
}
par(mfrow=c(2,2),pty='s')   # set graphics parameters
persp(x,y,z,theta=30,phi=30) # plot the surface
contour(x,y,z)
image(x,y,z)
```

For clarity, we have used a standard double loop to fill in the z matrix above, one could do it more compactly and quickly using the `outer` function. You can find more information about options by looking at the help page for each command, e.g. `?persp` will show help on the `persp` command. At the bottom of most help pages are some examples using that function. A wide selection of graphics can be found by typing `demo(graphics)`.

There is a recent R package called `rgl` that can be used to draw dynamic 3D graphs that can be interactively rotated and zoomed in/out using the mouse. This package interfaces R to the OpenGL library; see the section on Packages below for how to install and load `rgl`. Once that is done, you can plot the same surface as above with

```
rgl.surface(x,y,z,col="blue")
```

This will pop up a new window with the surface. Rotate by using the left mouse button: hold it down and move the surface, release to freeze in that position. Holding the right mouse button down allows you to zoom in and out. You can print or save an rgl graphic to a file using the `rgl.snapshot` function (use `?rgl.snapshot` for help).

29.2.6 Obtaining Financial Data

If you have data in a file in ascii form, you can read it with one of the R read commands:

- `scan("test1.dat")` will read a vector of data from the specified file in free format.
- `read.table("test2.dat")` will read a matrix of data, assuming one row per line.
- `read.csv("test3.csv")` will read a comma separate value file (Excel format).

The examples in the first section and those below use R functions developed for a math finance class taught at American University to retrieve stock data from the Yahoo finance website. Appendix 2 lists the source code that implements the following three functions:

- `get.stock.data`: Get a table of information for the specified stock during the given time period. A data frame is returned, with Date, Open, High, Low, Close,Volume, and Adj.Close fields.
- `get.stock.price`: Get just the adjusted closing price for a stock.
- `get.portfolio.returns`: Retrieve stock price data for each stock in a portfolio (a vector of stock symbols). Data is merged into a data frame by date, with a date kept only if all the stocks in the portfolio have price information on that date.

All three functions require the stock ticker symbol for the company, e.g."IBM" for IBM, "GOOG" for Google, etc. Symbols can be looked up online at `www.finance.yahoo.com/lookup`. Note that the function defaults to data for 2008, but you can select a different time period by specifying start and stop date, e.g. `get.stock.price("GOOG",c(6,1,2005),c(5,31,2008))` will give closing prices for Google from June 1, 2005 to May 31, 2008.

If you have access to the commercial Bloomberg data service, there is an R package named RBloomberg that will allow you to access that data within R.

29.2.7 Script Windows and Writing Your Own Functions

If you are going to do some calculations more than once, it makes sense to define a function in R. You can then call that function to perform that task any time you want. You can define a function by just typing it in at the command prompt, and then call it. But for all but the simplest functions, you will find to more convenient to enter the commands into a file using an editor. The default file extension is .R. To run those commands, you can either use the `source` command, e.g. `source("mycommands.R")`, or use the top level menu: "File", then "Source R code", then select the file name from the pop-up window.

There is a built in editor within R that is convenient to use. To enter your commands, click on "File" in the top level menu, then "New script". Type in your commands, using simple editing. To execute a block of commands, highlight them with the cursor, and then click on the "run line or selection" icon on the main menu (it looks like two parallel sheets of paper). You can save scripts (click on the diskette icon or use CTRL-S), and open them (from the "File" menu or with the folder icon). If you want to change the commands and functions in an existing script, use "File", then "Open script".

Here is a simple example that fits the (logarithmic) returns of price data in `S` with a normal distribution, and uses that to compute Value at Risk (VaR) from that fit.

```
compute.VaR <- function( S, alpha, V, T ){
# compute a VaR for the price data S at level alpha, value V
# and time horizon T (which may be a vector)

ret <- diff(log(S)) # return = log(S[i]/S[i-1])
mu <- mean(ret)
sigma <- sd(ret)
cat("mu=",mu," sigma=",sigma," V=",V,"\n")
for (n in T) {
  VaR <- -V * ( exp(qnorm( alpha, mean=n*mu,
                sd=sqrt(n)*sigma)) - 1 )
  cat("T=",n, " VaR=",VaR, "\n")}
}
```

Applying this to Google's stock price for 2008, we see the mean and standard deviation of the returns. With an investment of value $V = \$1{,}000$, and 95% confidence level, the projected VaRs for 30, 60, and 90 days are:

```
> price <- get.stock.price( "GOOG" )
GOOG  has 253 values from 2008-01-02 to 2008-12-31
> compute.VaR( price, 0.05, 1000, c(30,60,90) )
mu= -0.003177513  sigma= 0.03444149  V= 1000
T= 30  VaR= 333.4345
T= 60  VaR= 467.1254
T= 90  VaR= 561.0706
```

In words, there is a 5% chance that we will lose more than \$333.43 on our \$1,000 investment in the next 30 days. Banks use these kinds of estimates to keep reserves to cover loses.

29.3 Examples of R Code for Finance

The first section of this paper gave some basic examples on getting financial data, computing returns, plotting and basic analysis. In this section we briefly illustrate a few more examples of using R to analyze financial data.

29.3.1 Option Pricing

For simplicity, we consider the Black-Scholes option pricing formula. The following code computes the value V of a call option for an asset with current price S, strike price K, risk free interest rate r, time to maturity T and volatility σ. It also computes the "Greek" delta: $\Delta = dV/dS$, returning the value and delta in a named list.

```
call.optionBS <- function(S,K,r,T,sigma) {
d1 <- (log(S/K)+(r+sigma^2/2)*T )/ (sigma*sqrt(T))
d2 <- (log(S/K)+(r-sigma^2/2)*T )/ (sigma*sqrt(T))
return(list(value=S*pnorm(d1)-K*exp(-r*T)*pnorm(d2),
            delta=pnorm(d1)))}

> call.optionBS( 100, 105, .02, .25, .1 )
$value
[1] 0.536332
$delta
[1] 0.1974393
a <- call.optionBS( 100,105,.02,seq(0,.25,length=101),.1)
plot(time,a$value,type='l',main="Black-Scholes call option",
    xlab="time to maturity",ylab="value")
```

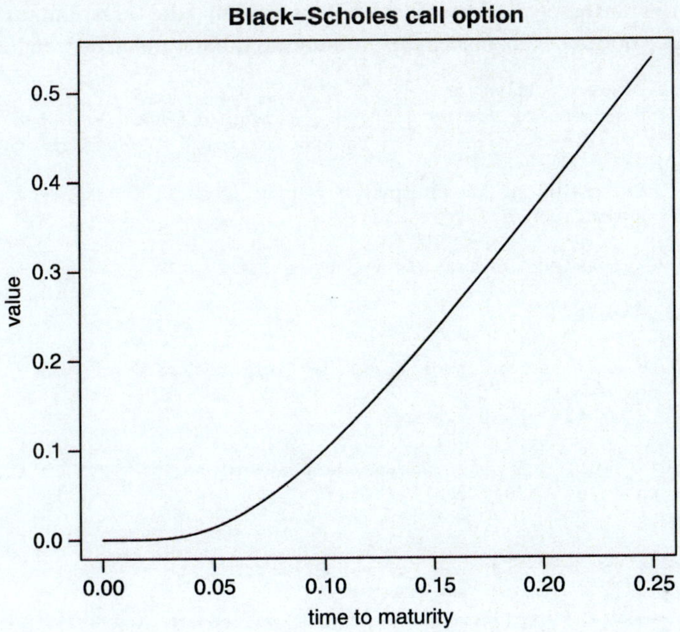

Fig. 29.6 Value of a call option with current price $S = 100$, time to maturity varying from 0 to 1/4 year, risk free rate $r = 0.02$, and volatility $\sigma = 0.1$

When the current price is $S = 100$, the strike price is $K = 105$, interest rate $r = 0.02$, $T = 0.25$ years to maturity and volatility $\sigma = 0.1$, the Black-Scholes price for a call option is \$0.53. Also, the delta is 0.1974, meaning that if the price S increases by \$1, then the price of the option will increase by about \$0.19. The delta values are used in hedging. The last three lines of the code above compute the price of a call option for varying days until maturity, starting at \$0.00 for $T = 0$ and increasing to \$0.53 for 1/4 year until maturity. See Fig. 29.6. Note that this last example works without changing the code for function `call.optionBS( )` because R uses vectorization.

29.3.2 Value-at-Risk for a Portfolio

Above we looked at a function to compute Value-at-Risk (VaR) for a single asset. We can generalize this to a static portfolio of N assets, with proportion w_i of the wealth in asset i. Large financial institutions are required by the Basel II Accords (see Bank for International Settlements 2009) to regularly compute VaR and to hold capital reserves to cover losses determined by these numbers. (In practice, one may adjust the weights dynamically, to maximize return or minimize risk based on performance of the individual assets.) As above, we will assume that the (logarithmic) returns are multivariate normal. This makes the problem easy, but unrealistic (see below).

```
portfolio.VaR <- function( x, w, V, T=1, alpha=0.05) {
# compute portfolio VaR by fitting multivariate normal to returns
# x is a matrix of returns for the portfolio, w are the allocation
  weights
# V = total value of investment, T = time horizon (possibly a vector)
# alpha = confidence level

# fit multivariate normal distribution
mu <- mean(x)
covar <- cov(x)

# compute mean and variance for 1-day weighted returns
mu1 <- sum(w*mu)
var1 <- t(w) %*% covar %*% w

cat("mu1=",mu1,"   var1=",var1,"   alpha=",alpha,"   V=",V, "\nweights:")
for (i in 1:length(symbols)) {cat("   ",symbols[i],":",w[i]) }
cat("\n")

# compute VaR for different time horizons
for (t in T) {
  VaR <- -V * ( exp(qnorm(alpha,mean=t*mu1,sd=sqrt(t*var1))) - 1.0)
  cat("T=",t,"   VaR=",VaR,"\n") }
}
```

Applying this to a portfolio of equal investments in Google, Microsoft, GE and IBM for 2008 data, and an investment of \$100,000, we find the 95% VaR values for 1 day, 5 days and 30 days with the following.

```
> x <- get.portfolio.returns( c("GOOG","MSFT","GE","IBM") )
GOOG  has 253 values from 2008-01-02 to 2008-12-31
MSFT  has 253 values from 2008-01-02 to 2008-12-31
GE  has 253 values from 2008-01-02 to 2008-12-31
IBM  has 253 values from 2008-01-02 to 2008-12-31
     253 dates with values for all stocks, 252 returns
     calculated
> portfolio.VaR( x, c(.25,.25,.25,.25), 100000, c(1,5,30) )
mu1= -0.002325875    var1= 0.0006557245    alpha= 0.05    V= 1e+05
weights:    GOOG : 0.25    MSFT : 0.25    GE : 0.25    IBM : 0.25
T= 1    VaR= 4347.259
T= 5    VaR= 10040.67
T= 30    VaR= 25953.49
```

29.3.3 Are Equity Prices Log-Normal?

It is traditional to do financial analysis under the assumption that the returns are independent, identically distributed normal random variables. This makes the analysis easy, and is a reasonable first approximation. But is it a realistic assumption? In this section we first test the assumption of normalility of returns, then do some graphical diagnostics to suggest other models for the returns. (We will not examine time dependence or non-stationarity, just the normality assumption.)

There are several statistical tests for normality. The R package `nortest`, Gross (2008), implements five omnibus tests for normality: Anderson-Darling, Cramer-von Mises, Lilliefors (Kolmogorov-Smirnov), Pearson chi-square, and Shapiro-Francia. This package must first be installed using the Packages menu as discussed below. Here is a fragment of a R session that applies these tests to the returns of Google stock over a 1 year period. Note that the text has been edited for conciseness.

```
> library("nortest")
> price <- get.stock.price("GOOG")
GOOG  has 253 values from 2008-01-02 to 2008-12-31
> x <- diff(log(price))
> ad.test(x)
    Anderson-Darling test  A = 2.8651, p-value = 3.188e-07
> cvm.test(x)
    Cramer-von Mises test  W = 0.4762, p-value = 4.528e-06
> lillie.test(x)
    Lilliefors test  D = 0.0745, p-value = 0.001761
> pearson.test(x)
    Pearson chi-square test  P = 31.1905, p-value = 0.01272
> sf.test(x)
    Shapiro-Francia test  W = 0.9327, p-value = 2.645e-08
```

All five tests reject the null hypothesis that the returns from Google stock are normal. These kinds of results are common for many assets. Since most traditional methods of computational finance assume a normal distribution for the returns, it is of practical interest to develop other distributional models for asset returns. In the next few paragraphs, we will use R graphical techniques to look at the departure from normality and suggest other alternative distributions.

One of the first things you should do with any data set is plot it. The following R commands compute and plot a smoothed density, superimpose a normal fit, and do a normal QQ-plot. The result is shown in Fig. 29.7. The density plot shows that while the data is roughly mound shaped, it is leptokurtotic: there is a higher peak and heavier tails than the normal distribution with the same mean and standard deviation. The heavier tails are more evident in the QQ-plot, where both tails of the data are noticeably more spread out than the normal model says they should be. (The added line shows perfect linear correlation between the data and normal fit.)

```
> price <- get.stock.price("GOOG")
GOOG  has 253 values from 2008-01-02 to 2008-12-31
> x <- diff(log(price))
> par(mfrow=c(1,2))
> plot(density(x),main="density of Google returns")
> z <- seq(min(x),max(x),length=201)
> y <- dnorm(z,mean=mean(x),sd=sd(x))
```

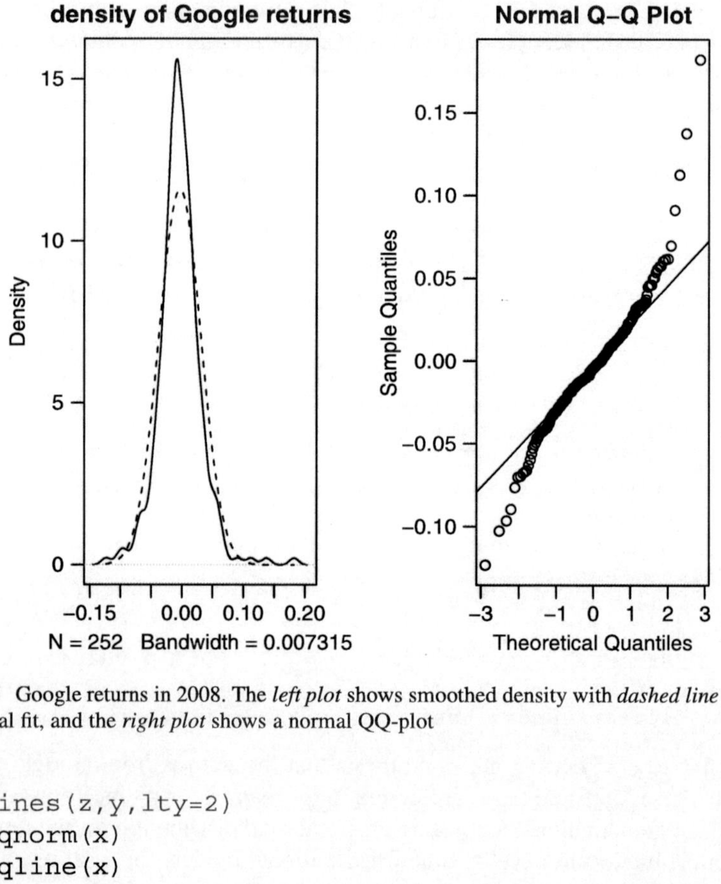

Fig. 29.7 Google returns in 2008. The *left plot* shows smoothed density with *dashed line* showing the normal fit, and the *right plot* shows a normal QQ-plot

```
> lines(z,y,lty=2)
> qqnorm(x)
> qqline(x)
```

So, one question is what kind of distribution better fits the data? The data suggests a model with fatter tails. One popular model is a t-distribution with a few degrees of freedom. The following code fragment defines a function `qqt` to plot QQ-plots for data vs. a t distribution. The results of this for 3, 4, 5 and 6 degrees of freedom are shown in Fig. 29.8. The plots show different behavior on lower and upper tail: 3 d.f. seems to best describe the upper tails, but 4 or 5 d.f. best describes the lower tail.

```
qqt <- function( data, df ){
#  QQ-plot of data vs. a t-distribution with df degrees of freedom
n <- length(data)
t.quantiles <- qt( (1:n - 0.5)/n, df=df )
qqplot(t.quantiles,data,main=paste("t(",df,") Q-Q Plot",sep=""),
   xlab="Theoretical Quantiles",ylab="Sample Quantiles")
qqline(data) }

# diagnostic plots for data with t distribution with 3,4,5,6 d.f.
par(mfrow=c(2,2))
for (df in 3:6) {
  qqt(x,df)
}
```

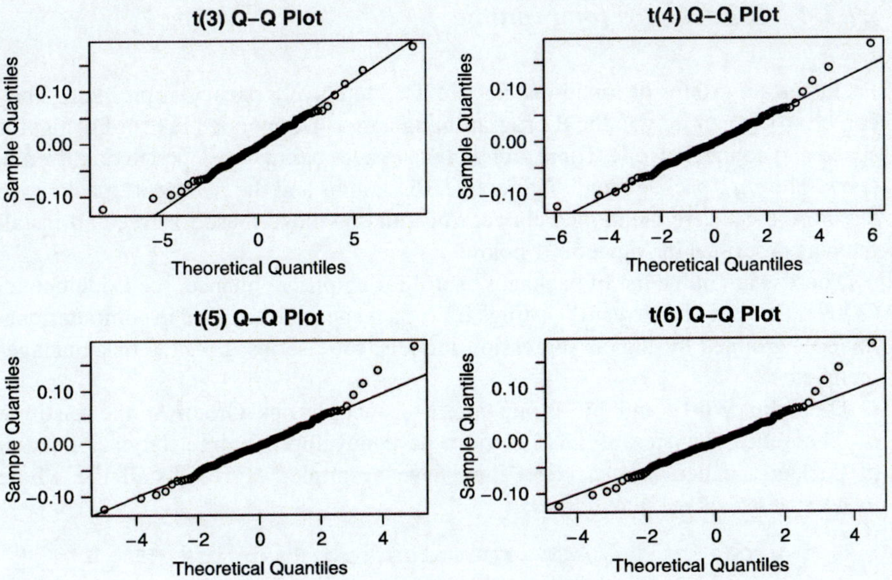

Fig. 29.8 QQ-plots of Google returns in 2008 for t distributions with 3, 4, 5 and 6 degrees of freedom

There are many other models proposed for fitting returns, most of them have heavier tails than the normal and some allow skewness. One reference for these models is Rachev (2003). If the tails are really heavy, then the family of stable distributions has many attractive features, including closure under convolution (sums of stable laws are stable) and the Generalized Central Limit Theorem (normalized sums converge to a stable law).

A particularly difficult problem is how to model multivariate dependence. Once you step outside the normal model, it generally takes more than a covariance matrix to describe dependence. In practice, a large portfolio with many assets of different type can have very different behavior for different assets. Some returns may be normal, some t with different degrees of freedom, some a stable law, etc. Copulas are one method of dealing with multivariate distributions, though the limited classes of copulas used in practice seems to have misled people into thinking they had correctly modeled dependence. In addition to modeling complete joint dependence, there is research on modeling tail dependence. This is a less ambitious goal, but could be especially useful in modeling extreme movements by multiple assets – an event that could cause a catastrophic result.

Realistically modeling large portfolios is an important open problem. The recent recession may have been prevented if practitioners and regulators had better models for returns, and ways to effectively model dependence.

29.3.4 R Packages for Finance

Packages are groups of functions that are used to solve a particular problem. They can be written entirely in the R programming language, or coded in C or Fortran for speed and connected to R. There are many packages being developed to do different tasks. The `rgl` package to do interactive 3-D graphs and the `nortest` package to test normality were mentioned above. You can download these for free and install them as described in Appendix 1 below.

There is an online list of packages useful for empirical finance, see Eddelbuettel (2009). This page has over 100 listings for R packages that are used in computational finance, grouped by topics: regression models, time series, finance, risk management, etc.

Diethelm Würtz and his group at the Econophysics Group at the Institute of Theoretical Physics of ETH Zurich, have developed a free, large collection of packages called `Rmetrics`. They have a simple way to install the whole `Rmetrics` package in two lines:

```
> source("http://www.rmetrics.org/Rmetrics.R")
> install.Rmetrics()
```

29.4 Open Source R Versus Commercial Packages

We end with a brief comparison of the advantages and disadvantages of open source R vs. commercial packages (matlab, Mathematica, SAS, etc.) While we focus on R, the comments are generally applicable to other open source programs.

Cost: Open software is free, with no cost for obtaining the software or running it on any number of machines. Anyone with a computer can use R – whether you work for a large company with a cumbersome purchasing process, are an amateur investor, or are a student in a major research university or in an inner city school, whether you live in Albania or in Zambia. Commercial packages generally cost in the one to two thousand dollar range, making them beyond the reach of many.

Ease of installing and upgrading: The R Project has done an excellent job of making it easy to install the core system. It is also easy to quickly download and install packages. When a new version comes out, users can either immediately install the new version, or continue to run an existing version without fear of a license expiring. Upgrades are simple and free.

Verifiability: Another advantage of open software is the ability to examine the source code. While most users will not dig through the source code of individual routines, anyone can and someone eventually will. This means that algorithms can be verified by anyone with the interest. Code can be fixed or extended by those who want to add capabilities. (If you've ever hit a brick wall with a commercial package that does not work correctly, you will appreciate this feature. Years ago, the author was using a multivariate minimization routine with box constraints from a well known commercial package. After many hours debugging, it was discovered that the problem was in the minimization routine: it would sometimes search outside the specified bounds, where the objective function was undefined. After days of trying to get through to the people who supported this code, and presenting evidence of the problem, they eventually confirmed that it was an issue, but were unwilling to fix the problem or give any work-around.)

Documentation and support: No one is paid to develop user friendly documentation for R, so built-in documentation tends to be terse, making sense to the cognesceti, but opaque to the novice. There is now a large amount of documentation online and books on R, though the problem may still be finding the specific information you want. There are multiple active mailing lists, but with a very heterogeneous group of participants. There are novices struggling with basic features and R developers discussing details of the internals of R. If a bug is found, it will get fixed, though the statement that "The next version of R will fix this problem" may not help much in the short run. Of course, commercial software support is generally less responsive.

Growth: There is a vibrant community of contributors to R. With literally thousands of people developing packages, R is a dynamic, growing program. If you don't like the way a package works, you can write your own, either from scratch or by adapting an existing package from the available source code. A drawback of this distributed development model is that R packages are of unequal quality. You may have to try various packages and select those that provide useful tools.

Stability: Software evolves over time, whether open source or commercial. In a robust open source project like R, the evolution can be brisk, with new versions appearing every few months. While most of R is stable, there are occasionally small changes that have unexpected consequences. Packages that used to work, can stop working when a new version comes out. This can be a problem with commercial programs also: a few years ago matlab changed the way mex programs were built and named. Toolboxes that users developed or purchased, sometimes at a significant cost, would no longer work.

Certification: Some applications, e.g. medical use and perhaps financial compliance work, may require that the software be certified to work correctly and reproducibly. The distributed development and rapid growth of R has made it hard to do this. There is an effort among the biomedical users of R to find a solution to this issue.

Institutional resistance: In some institutions, IT staff may resist putting freeware on a network, for fear that it may be harmful. Also, they are wary of being held responsible for installing, maintaining, and updating software that is not owned/licensed by a standard company.

In the long run, it seems likely that R and other open source packages will survive and prosper. Because of their higher growth rate, they will eventually provide almost all of the features of commercial products. When that point will be reached is unknown. In the classroom, where the focus is on learning and adaptability, the free R program is rapidly displacing other alternatives.

There is a new development in computing that is a blend of free, open source software and commercial support. REvolution Computing (2008) offers versions of R that are optimized and validated, and have developed custom extensions, e.g. parallel processing. This allows a user to purchase a purportedly more stable, supported version of R. It will be interesting to watch where this path leads; it may be a way to address the institutional resistance mentioned above. Another company, Mango Solutions (2009) provides training in R, with specific courses R for Financial Data Analysis. A third company, Inference for R (2009), has an integrated development environment that does syntax highlighting, R debugging, allows one to run R code from Microsoft Office applications (Excel, Word and PowerPoint), and other features. Finally, we mention the SAGE Project. Sage (2008) is an open source mathematics system that includes R. In addition to the features of R, it includes symbolic capabilities to handle algebra, calculus, number theory, cryptography, and much more. Basically, it is a Python program that interfaces with over 60 packages: R, Maxima, the Gnu Scientific Library, etc.

29.5 Appendix 1: Obtaining and Installing R: R Project and Comprehensive R Archive Network

The R Project's website is `www.r-project.org`, where you can obtain the R program, packages, and even the source code for R. The Comprehensive R Archive Network (CRAN) is a coordinated group of over 60 organizations that maintain servers around the world with copies of the R program (similar to the CTAN system for TEX). To download the R program, go to the R Project website and on the left side of the page, click on "CRAN", select a server near you, and download the version of R for your computer type. (Be warned: this is a large file, over 30 mb.) On Windows, the program name is something like R-2.10.0-win32.exe, which is version 2.10.0 of R for 32-bit Windows; newer versions occur every few months and will have higher numbers. After the file is on your computer, execute the program. This will go through the standard installation procedure. For a Mac, the download is a universal binary file (.dmg) for either a PowerPC or an Intel based processor. For linux, there are versions for debian, redhat, suse or ubuntu. The R Project provides free manuals that explain different parts of R. Start on the R homepage `www.r-project.org` and click on Manuals on the left side of the page. A standard starting point is An Introduction to R, which is a PDF file of about 100 pages. There are dozens of other manuals, some of which are translated to 13 different languages.

To download a package, it is easiest to use the GUI menu system within R. Select "Packages" from the main top menu, then select "Install package(s)". You will be prompted to select a CRAN server from the first pop-up menu (pick one near you for speed), then select the package you want to install from the second pop-up menu. The system will go to the server, download a compressed form of the package, and install it on your computer. This part only needs to be done once. Anytime you want to use that package, you have to load it into your session. This is easy to do from the Packages menu: "Load package...", and then select the name of an installed package. You can also use the `library( )` command, as in the example with the `nortest` package above.

If you want to see the source code for R, once you are on the CRAN pages, click on the section for source code.

29.6 Appendix 2: R Functions for Retrieving Finance Data

Disclaimer: these functions are not guaranteed for accuracy, nor can we guarantee the accuracy of the Yahoo data. They are very useful in a classroom setting, but should not be relied on as a basis for investing.

```
# R programs for Math Finance class
# John Nolan, American University   jpnolan@american.edu
#######################################################################
get.stock.data <- function( symbol, start.date=c(1,1,2008),
                         stop.date=c(12,31,2008), print.info=TRUE ) {
# get stock data from yahoo.com for specified symbol in the
# specified time period.  The result is a data.frame with columns for:
#                Date, Open, High, Low, Close,Volume, Adj.Close

url <- paste("http://ichart.finance.yahoo.com/table.csv?a=",
          start.date[1]-1,"&b=",start.date[2],"&c=",start.date[3],
          "&d=",stop.date[1]-1,"&e=",stop.date[2],"&f=",stop.date[3],"&s=",
          symbol,sep="")
x <- read.csv(url)

# data has most recent days first, going back to start date
n <- length(x$Date); date <- as.character(x$Date[c(1,n)])
if (print.info) cat(symbol,"has", n,"values from",date[2],"to",date[1],"\n")

# data is in reverse order from the read.csv command
x$Date <- rev(x$Date)
x$Open <- rev(x$Open)
x$High <- rev(x$High)
x$Low  <- rev(x$Low)
x$Close <- rev(x$Close)
x$Volume <- rev(x$Volume)
x$Adj.Close <- rev(x$Adj.Close)

return(x) }
#######################################################################
get.stock.price <- function( symbol, start.date=c(1,1,2008),
                          stop.date=c(12,31,2008), print.info=TRUE ) {
# gets adjusted closing price data from yahoo.com for specified symbol

x <- get.stock.data(symbol,start.date,stop.date,print.info)

return(x$Adj.Close) }
#######################################################################
```

```
get.portfolio.returns = function( symbols, start.date=c(1,1,2008),
                                  stop.date = c(12,31,2008) ){
# get a table of returns for the specified stocks in the stated time period

n = length(symbols)
for (i in 1:n) {
  t1 = get.stock.data( symbols[i], start.date=start.date, stop.date=stop.date)
  #  need to merge columns, possibly with mismatching dates
  a = data.frame(t1$Date,t1$Adj.Close)
  names(a) = c("Date",symbols[i])
  if (i == 1) {b=a}
  else {b = merge(b,a,sort=FALSE)}
  }
#  leave off the date column
nn = dim(b)[1]
cat("    ",nn,"dates with values for all stocks,",nn-1,"returns calculated\n")
b = b[,2:ncol(b)]
bb = data.frame(apply(b,2,"log.ratio"))
names(bb) = symbols
return(bb) }
```

References

Bank for International Settlements (2009). Basel II: Revised international capital framework. www.bis.org/publ/bcbsca.htm.

Eddelbuettel, D. (2009). Cran task view: Empirical finance. cran.r-project.org/web/views/Finance.html.

Gross, J. (2008). nortest: Tests for Normality. cran.r-project.org/web/packages/nortest/index.html.

Inference for R (2009). Online. inferenceforr.com.

Mango Solutions (2009). Online. www.mango-solutions.com.

Rachev, S. T. (2003). *Handbook of heavy tailed distributions in finance*. Amsterdam: Elsevier.

REvolution Computing (2008). Online. www.revolution-computing.com.

Sage (2008). Open source mathematics system. www.sagemath.org.

Vance, A. (2009a). Data Analysts Captivated by Rs Power. NY Times, page B6. Online at www.nytimes.com/2009/01/07/technology/business-computing/07program.html.

Vance, A. (2009b). R You Ready for R? NY Times website. Online at bits.blogs.nytimes.com/2009/01/08/r-you-ready-for-r/.